SOIL TESTING AND PLANT ANALYSIS

Third Edition

Soil Testing and Plant Analysis

Third Edition

Editor: R. L. Westerman

Editorial Committee: R. L. Westerman
J. V. Baird
N. W. Christensen
P. E. Fixen
D. A. Whitney

Managing Editor: S. H. Mickelson

Editor-in-Chief SSSA: David E. Kissel

Number 3 in the Soil Science Society of America Book Series

Published by: Soil Science Society of America, Inc.
Madison, Wisconsin, USA

1990

The views expressed in this publication represent those of the individual Editors and Authors. These views do not necessarily reflect endorsement by the Publisher(s). In addition, trade names are sometimes mentioned in this publication. No endorsement of these products by the Publisher is intended, nor is any criticism implied of similar products not mentioned.

Reprinted in 1995, 2005, 2007

Soil Science Society of America, Inc.
677 South Segoe Road, Madison, WI 53711-1086 USA

Library of Congress Cataloging-in-Publication Data

Soil testing and plant analysis.—3rd ed./editor, R.L. Westerman.
p. cm.—(Soil Science Society of America book series; no. 3)
Includes bibliographical references and index.
ISBN-10: 0-89118-844-4
ISBN-13: 978-0-89118-844-5
1. Soils—Analysis. 2. Crops—Analysis. 3. Crops—Nutrition. 4. Deficiency diseases in plants—Diagnosis. 5. Soil fertility. 6. Fertilizers. I. Westerman, Robert Lee, 1938- . II. Soil Science Society of America. III. Series.
S593.S743 1990
631.4´1—dc20 90-49801
CIP

Printed in the United States of America.

CONTENTS

FOREWORD

Soil testing and plant analyses have proven to be invaluable tools in the diagnosing of nutritional deficiencies and problems related to plant growth. Each advance in our basic understanding of plant physiology and soil chemistry, and each advance in instrumentation leads to improvements in methodology and interpretation. The Soil Science Society of America has always felt a keen responsibility to ensure that the best and most accurate methods were generally available and accessible to those who need and use them. This volume represents a significant advance from previous editions and reflects the updated analytical methods available, a better understanding of plant nutrition, and a need to use fertilizers and soil amendments even more efficiently and safely.

W. R. Gardner, *president*
Soil Science Society of America

PREFACE

Soil Testing and Plant Analysis was first published in 1967 in a two-part series as a result of a collection of papers presented at several symposia in 1964 and 1965 that dealt with soil testing, plant analysis, and tissue testing. Principles and practices used in soil testing and plant analysis and problems associated with sampling, analysis, interpretation, and correlation of results were discussed. The first edition set the standards for our understanding of the significance of plant analysis as a diagnostic tool and soil testing as a guide for fertilizer use.

In 1973, the second edition was combined into one publication that coincided with a rapid increase in knowledge regarding nutritional status of soils, dynamic cropping systems, equipment, and methodology. This edition further enhanced our overall understanding of soil testing and plant analysis.

More than 17 yr have passed since the last edition was published. During this period, there has been substantial progress in the development of methodology for analysis of soils and plants and interpretation of analytical results for best management practices that promote efficient fertilizer use and optimum yield, yet minimizes the potential of groundwater, lake, and stream pollution. Because of these advances, revision and updating of the subject matter was deemed necessary.

Many of the former authors were asked to participate in the revision of this book. The information included in this book represents a summary of current knowledge and experiences on the use of soil testing and plant analysis as a diagnostic tool for assessing nutritional requirements of crops, efficient fertilizer use, saline-sodic conditions, and toxicity of metals. Discussions also include analytical instrumentation and equipment used in soil testing, plant analysis, and data processing.

This publication is useful in the USA as well as abroad for individuals interested in the operation of public and commercial laboratories, industrial and extension agronomists and scientists involved in research related to soils, crops, forestry, horticulture, ecology, botany, and environmental issues as well as for college student and teacher use.

Appreciation is expressed to Dr. J.D. Beaton, Phosphate and Potash Institute of Canada, for his early efforts in organization and initiation of the revision process. I extend my sincere thanks and appreciation to the many authors who drew from their areas of expertise to provide current discussions of the subject matter and the anonymous reviewers who provided their time in the review process. I also express my appreciation to the members of the editorial committee who devoted their time and talents to complete this revision. Editorial committee members are: R.L. Westerman, Editor, Oklahoma State University, Stillwater, OK; J.V. Baird, North Carolina State University, Raleigh, NC; N.W. Christensen, Oregon State University, Corvallis, OR; P.E. Fixen, South Dakota State University, currently Phosphate and Potash Institute, Brookings, SD; and D.A. Whitney, Kansas State University, Manhattan, KS.

I also express my appreciation to the ASA Headquarters staff, namely Mr. D.M. Kral, Ms. Sherri Mickelson and others, for their assistance in editing manuscripts and making arrangements for publication of this book.

R.L. Westerman, *editor*
Regents Professor
Oklahoma State University
Stillwater, OK

CONTRIBUTORS

Dale E. Baker Professor of Soil Chemistry, Pennsylvania State University, University Park, PA 16802

John E. Bowen Professor, University of Hawaii, Hilo, HI 96720

D. M. Brandon Director and Agronomist, Rice Experiment Station, California Cooperative Rice Research Foundation, Biggs, CA 95917

Gaylon S. Campbell Professor of Soils, Department of Agronomy and Soils, Washington State University, Pullman, WA 99164-6420

Vernon W. Case Coordinator, Soil Testing Laboratory, University of Kentucky, Lexington, KY 40546-0275

Richard B. Corey Professor, Department of Soil Science, University of Wisconsin-Madison, Madison, WI 53706

George A. Cummings Professor of Soil Fertility, Department of Soil Science, North Carolina State University, Raleigh, NC 27695

W. C. Dahnke Professor of Soils, Soil Science Department, North Dakota State University, Fargo, ND 58105

S. J. Donohue Professor, Department of Crop and Environmental Sciences, Virginia Polytechnic Institute and State University, Blacksburg, VA 24061

Harold V. Eck Soil Scientist, USDA-ARS, Conservation and Production Research Laboratory, Bushland, TX 79012

P. E. Fixen Associate Professor, South Dakota State University, Brookings, SD 57006. Currently Regional Director, Potash and Phosphate Institute

C. M. Geraldson Professor (Soils Chemist), Gulf Coast Research and Education Center, University of Florida, Bradenton, FL 34203

S. W. Gettier Supervisor, Soil Testing and Plant Analysis Laboratory, Department of Crop and Soil Environmental Sciences, Virginia Polytechnic Institute and State University, Blacksburg, VA 24061

J. H. Grove Associate Professor, Agronomy Department, University of Kentucky, Lexington, KY 40546-0091

V. A. Haby Professor, Texas Agricultural Experiment Station, Texas A&M University Agricultural Research and Extension Center, Overton, TX 75684

F. Jackson Hills Extension Agronomist, Department of Agronomy and Range Science, University of California, Davis, CA 95616

Robert A. Isaac Professor, Agricultural Services Laboratories, Soil Test Laboratory, University of Georgia, Athens, GA 30605

D. W. James Professor, Department of Soil Science and Meteorology, Utah State University, Logan, UT 84322-4840

Gordon V. Johnson Professor of Agronomy, Department of Agronomy, Oklahoma State University, Stillwater, OK 74078

J. Benton Jones, Jr. Professor, Department of Horticulture, University of Georgia, Athens, GA 30602. Currently President, Benton Laboratories, Inc., Athens, GA 30604

K. A. Kelling Professor of Soil Science, Department of Soil Science, University of Wisconsin-Madison, Madison, WI 53706

W. L. Lindsay University Distinguished Professor, Department of Agronomy, Colorado State University, Fort Collins, CO 80523

D. C. Martens Professor of Soil Chemistry, Crop and Soil Environmental Sciences Department, Virginia Polytechnic Institute and State University, Blacksburg, VA 24061

J. E. Matocha Professor. Texas A&M University Research Center, Corpus Christi, TX 78410

Gordon S. Miner Professor, Soil Science Department, North Carolina State University, Raleigh, NC 27695-7619

S. Miyamoto Professor, Texas A&M University Agricultural Research and Extension Center, El Paso, TX 79927

Robert D. Munson Adjunct Professor, Soil Science Department, University of Minnesota, St. Paul, MN 55108

Werner L. Nelson Retired from Potash and Phosphate Institute, West Lafayette, IN 47906

R. A. Olson (Deceased) Professor, Department of Agronomy, University of Nebraska, Lincoln, NE 68583

T. R. Peck Professor of Soil Chemistry, Agronomy Department, University of Illinois, Urbana, IL 61801

J. D. Rhoades Director, USDA-ARS-PWA, U.S. Salinity Laboratory, Riverside, CA 92501

Timothy L. Righetti Associate Professor, Department of Horticulture, Oregon State University, Corvallis, OR 97331

Jeffrey A. Risser Assistant Professor of Plant and Soil Chemistry, University of Maine, Orono, ME. Currently Research Associate, USEPA/AERL, Athens, GA 30613-7799

M. P. Russelle Soil Scientist, Plant Science Research Unit, USDA-ARS, U.S. Dairy Forage Resesarch Center (Minnesota Cluster) and Department of Soil Science, University of Minnesota, St. Paul, MN 55108

Wayne E. Sabbe Professor of Agronomy, Agronomy Department, University of Arkansas, Fayetteville, AR 72701

E. L. Skidmore Research Hydraulic Engineer, USDA-ARS, Grassland, Soil and Water Research Laboratory, Temple, TX 76502

Earl O. Skogley Professor of Soil Science, Plant and Soil Science Department, Montana State University, Bozeman, MT 59717

P. N. Soltanpour Professor of Soils, Agronomy Department, Colorado State University, Fort Collins, CO 80523

M. Ray Tucker Agronomist, Soil Testing, Agronomic Division NCDA, Blue Ridge Road Center, Raleigh, NC 27611

K. B. Tyler Extension Vegetable Specialist, Department of Vegetable Crops, Kearney Agricultural Research Center, University of California, Parlier, CA 93648

Albert Ulrich Plant Physiologist Emeritus, Department of Soil Science, University of California, Berkeley, CA 94720

W. van Lierop	Soil Scientist, Cookshire, Quebec, Canada J4B 1M0
Regis Voss	Professor of Agronomy, Agronomy Department, Iowa State University, Ames, IA 50011
Darryl D. Warncke	Professor of Soil Fertility and Organic Soil Management, Department of Crop and Soil Sciences, Michigan State University, East Lansing, MI 48824
M. E. Watson	Associate Professor, Research-Extension Analytical Laboratory, Ohio Agricultural Research and Development Center, Ohio State University, Wooster, OH 44691
G. F. Weetman	Professor, Faculty of Forestry, University of British Columbia, Vancouver, British Columbia, Canada V6T 1W5
C. G. Wells	Soil Scientist (retired), USDA-FS, Southeastern Forest Experiment Station, Research Triangle Park, NC 27709
K. L. Wells	Extension Professor (Soils), Department of Agronomy, University of Kentucky, Lexington, KY 40546-0091
D. G. Westfall	Professor, Department of Agronomy, Colorado State University, Fort Collins, CO 80523
D. A. Whitney	Professor, Agronomy Department, Kansas State University, Manhattan, KS 66506
Kris L. Wilder	Research Analyst, Department of Horticulture, Oregon State University, Corvallis, OR 97331. Currently Stable Isotope Analyst, Department of Soil Science, Oregon State University, Corvallis, OR 97331
J. R. Williams	Research Hydraulic Engineer, USDA-ARS, Grassland, Soil and Water Research Laboratory, Temple, TX 76502
Lowell J. Zelinski	Farm Advisor, University of California, Cooperative Extension, Fresno, CA 93702. Currently Consulting Agronomist, 1038 E. Yale Ave., Fresno, CA 93704

Conversion Factors for SI and non-SI Units

To convert Column 1 into Column 2, multiply by	Column 1 SI Unit	Column 2 non-SI Unit	To convert Column 2 into Column 1, multiply by
		Length	
0.621	kilometer, km (10^3 m)	mile, mi	1.609
1.094	meter, m	yard, yd	0.914
3.28	meter, m	foot, ft	0.304
1.0	micrometer, μm (10^{-6} m)	micron, μ	1.0
3.94×10^{-2}	millimeter, mm (10^{-3} m)	inch, in	25.4
10	nanometer, nm (10^{-9} m)	Angstrom, Å	0.1
		Area	
2.47	hectare, ha	acre	0.405
247	square kilometer, km^2 $(10^3\ m)^2$	acre	4.05×10^{-3}
0.386	square kilometer, km^2 $(10^3\ m)^2$	square mile, mi^2	2.590
2.47×10^{-4}	square meter, m^2	acre	4.05×10^3
10.76	square meter, m^2	square foot, ft^2	9.29×10^{-2}
1.55×10^{-3}	square millimeter, mm^2 $(10^{-3}\ m)^2$	square inch, in^2	645
		Volume	
9.73×10^{-3}	cubic meter, m^3	acre-inch	102.8
35.3	cubic meter, m^3	cubic foot, ft^3	2.83×10^{-2}
6.10×10^4	cubic meter, m^3	cubic inch, in^3	1.64×10^{-5}
2.84×10^{-2}	liter, L (10^{-3} m^3)	bushel, bu	35.24
1.057	liter, L (10^{-3} m^3)	quart (liquid), qt	0.946
3.53×10^{-2}	liter, L (10^{-3} m^3)	cubic foot, ft^3	28.3
0.265	liter, L (10^{-3} m^3)	gallon	3.78
33.78	liter, L (10^{-3} m^3)	ounce (fluid), oz	2.96×10^{-2}
2.11	liter, L (10^{-3} m^3)	pint (fluid), pt	0.473

	Mass		
2.20×10^{-3}	gram, g (10^{-3} kg)	pound, lb	454
3.52×10^{-2}	gram, g (10^{-3} kg)	ounce (avdp), oz	28.4
2.205	kilogram, kg	pound, lb	0.454
0.01	kilogram, kg	quintal (metric), q	100
1.10×10^{-3}	kilogram, kg	ton (2000 lb), ton	907
1.102	megagram, Mg (tonne)	ton (U.S.), ton	0.907
1.102	tonne, t	ton (U.S.), ton	0.907
	Yield and Rate		
0.893	kilogram per hectare, kg ha^{-1}	pound per acre, lb $acre^{-1}$	1.12
7.77×10^{-2}	kilogram per cubic meter, kg m^{-3}	pound per bushel, bu^{-1}	12.87
1.49×10^{-2}	kilogram per hectare, kg ha^{-1}	bushel per acre, 60 lb	67.19
1.59×10^{-2}	kilogram per hectare, kg ha^{-1}	bushel per acre, 56 lb	62.71
1.86×10^{-2}	kilogram per hectare, kg ha^{-1}	bushel per acre, 48 lb	53.75
0.107	liter per hectare, L ha^{-1}	gallon per acre	9.35
893	tonnes per hectare, t ha^{-1}	pound per acre, lb $acre^{-1}$	1.12×10^{-3}
893	megagram per hectare, Mg ha^{-1}	pound per acre, lb $acre^{-1}$	1.12×10^{-3}
0.446	megagram per hectare, Mg ha^{-1}	ton (2000 lb) per acre, ton $acre^{-1}$	2.24
2.24	meter per second, m s^{-1}	mile per hour	0.447
	Specific Surface		
10	square meter per kilogram, m^2 kg^{-1}	square centimeter per gram, cm^2 g^{-1}	0.1
1000	square meter per kilogram, m^2 kg^{-1}	square millimeter per gram, mm^2 g^{-1}	0.001
	Pressure		
9.90	megapascal, MPa (10^6 Pa)	atmosphere	0.101
10	megapascal, MPa (10^6 Pa)	bar	0.1
1.00	megagram per cubic meter, Mg m^{-3}	gram per cubic centimeter, g cm^{-3}	1.00
2.09×10^{-2}	pascal Pa	pound per square foot, lb ft^{-2}	47.9
1.45×10^{-4}	pascal Pa	pound per square inch, lb in^{-2}	6.90×10^3

(continued on next page)

Conversion Factors for SI and non-SI Units

To convert Column 1 into Column 2, multiply by	Column 1 SI Unit	Column 2 non-SI Unit	To convert Column 2 into Column 1, multiply by
		Temperature	
1.00 (K − 273)	Kelvin, K	Celsius, °C	1.00 (°C + 273)
(9/5 °C) + 32	Celsius, °C	Fahrenheit, °F	5/9 (°F − 32)
		Energy, Work, Quantity of Heat	
9.52×10^{-4}	joule, J	British thermal unit, Btu	1.05×10^{3}
0.239	joule, J	calorie, cal	4.19
10^{7}	joule, J	erg	10^{-7}
0.735	joule, J	foot-pound	1.36
2.387×10^{-5}	joule per square meter, J m^{-2}	calorie per square centimeter (langley)	4.19×10^{4}
10^{5}	newton, N	dyne	10^{-5}
1.43×10^{-3}	watt per square meter, W m^{-2}	calorie per square centimeter minute (irradiance), cal cm^{-2} min^{-1}	698
		Transpiration and Photosynthesis	
3.60×10^{-2}	milligram per square meter second, mg m^{-2} s^{-1}	gram per square decimeter hour, g dm^{-2} h^{-1}	27.8
5.56×10^{-3}	milligram (H_2O) per square meter second, mg m^{-2} s^{-1}	micromole (H_2O) per square centimeter second, μmol cm^{-2} s^{-1}	180
10^{-4}	milligram per square meter second, mg m^{-2} s^{-1}	milligram per square centimeter second, mg cm^{-2} s^{-1}	10^{4}
35.97	milligram per square meter second, mg m^{-2} s^{-1}	milligram per square decimeter hour, mg dm^{-2} h^{-1}	2.78×10^{-2}
		Plane Angle	
57.3	radian, rad	degrees (angle), °	1.75×10^{-2}

Electrical Conductivity, Electricity, and Magnetism			
10	siemen per meter, S m^{-1}	millimho per centimeter, mmho cm^{-1}	0.1
10^4	tesla, T	gauss, G	10^{-4}
Water Measurement			
9.73×10^{-3}	cubic meter, m^3	acre-inches, acre-in	102.8
9.81×10^{-3}	cubic meter per hour, $m^3 h^{-1}$	cubic feet per second, $ft^3 s^{-1}$	101.9
4.40	cubic meter per hour, $m^3 h^{-1}$	U.S. gallons per minute, gal min^{-1}	0.227
8.11	hectare-meters, ha-m	acre-feet, acre-ft	0.123
97.28	hectare-meters, ha-m	acre-inches, acre-in	1.03×10^{-2}
8.1×10^{-2}	hectare-centimeters, ha-cm	acre-feet, acre-ft	12.33
Concentrations			
1	centimole per kilogram, cmol kg^{-1} (ion exchange capacity)	milliequivalents per 100 grams, meq 100 g^{-1}	1
0.1	gram per kilogram, g kg^{-1}	percent, %	10
1	milligram per kilogram, mg kg^{-1}	parts per million, ppm	1
Radioactivity			
2.7×10^{-11}	becquerel, Bq	curie, Ci	3.7×10^{10}
2.7×10^{-2}	becquerel per kilogram, Bq kg^{-1}	picocurie per gram, pCi g^{-1}	37
100	gray, Gy (absorbed dose)	rad, rd	0.01
100	sievert, Sv (equivalent dose)	rem (roentgen equivalent man)	0.01
Plant Nutrient Conversion			
	Elemental	***Oxide***	
2.29	P	P_2O_5	0.437
1.20	K	K_2O	0.830
1.39	Ca	CaO	0.715
1.66	Mg	MgO	0.602

Chapter 1

The Principles of Soil Testing

T. R. PECK, *University of Illinois, Urbana*

P. N. SOLTANPOUR, *Colorado State University, Fort Collins*

Soil testing is practiced, with some degree of success, in nearly all parts of the world. Success or failure of soil testing varies depending upon the amount and quality of research data available for the calibration and interpretation of the tests. In some areas, agronomists are not entirely satisfied with the reliability or usefulness of soil tests. Quite often this lack of confidence may be traced to expectations that exceed the capability limits of the program. Many misconceptions seem to prevail as to just what soil tests can or cannot do. This chapter includes the realities and principles that form a general philosophy of soil testing.

Soil testing may be generally defined as any chemical or physical measurement that is made on a soil. But through common usage, the term *soil testing* has been given both a more restricted and broader meaning. The term is restricted in the sense that it has come to mean rapid chemical analyses to assess the plant-available nutrient status, salinity and elemental toxicity of a soil. It has broadened to represent a program that includes interpretations, evaluations, fertilizer, and amendment recommendations based on results of chemical analyses and on several other considerations. It is necessary, therefore, to distinguish between those factors that represent technical data and those that are interpretive judgments.

Sixteen elements are essential for crop growth. Three of these elements—N, P, and K—are commonly deficient in soil. Soil pH and salinity also commonly limit plant growth. Secondary and micronutrient deficiencies are found in some soils, with S, Zn, Fe, and B being the most common, but these are usually restricted to special soil areas. Therefore, soil testing predominantly involves N, P, K, and pH with salinity, but secondary and micronutrient analyses vary widely on a regional basis. Farmer acceptance of soil testing strongly depends upon the extent and severity of the limits for crop production, whether the problems are cost effectively treatable and the accuracy with which the tests can be used to predict crop responses with fertilizer and reclamation needs.

The nutrient status of a soil may be evaluated in several ways. Among these methods are (i) field-plot fertilizer trials, (ii) greenhouse pot experiments, (iii) crop symptoms, (iv) plant analysis, (v) rapid tissue or sap analyses, (vi) biologic tests (i.e., growing microorganisms on the soil to detect a nutrient deficiency or toxicity), and (vii) rapid chemical analyses of the soil. Each of these methods has certain limits when applied to individual farm fields or farms on a general service basis. Field-plot fertilizer trials are indispensible and serve as the basis from which the other diagnostic techniques derive a cause-and-effect relationship. However, fertilizer field trials, cannot, in a practical sense, be conducted on every farmer's field and the research results may not be directly transferred from one field to another, except for N in areas of high rainfall (>75 cm/yr). A common transfer carrier is needed. Chemical analyses of the experimental soil and similar analyses of the soil to which the field-plot data are to be applied has proven useful and this is what "soil testing" is. Greenhouse fertilizer trials, besides being costly or time consuming, often give results that cannot be quantitatively extrapolated to the field. Deficiency and toxicity symptoms indicate only the very severe deficiency or toxicity condition. These symptoms usually appear too late in a growing season for a farmer to initiate remedial actions. Total plant analysis and plant sap analysis (rapid tissue tests) are postmortem types of analyses that help explain what was wrong that year on that soil but do not quantitatively predict fertilizer needs. Biologic tests can determine the nutrient status of a soil when properly conducted but are costly and time consuming. In summary, chemical analyses, or soil testing, can be rapid, inexpensive and accurate, and can serve as the transfer agent of field, greenhouse, and laboratory research into predictions of lime and fertilizer needs of a soil before crops are seeded.

Soil testing probably started, in some form or another, as soon as people became interested in how plants grow. The search for the "principle of vegetation" started long before, and actually led to, the development of the science of chemistry. One could classify Liebig (1840) as having been an early worker in the soil-testing field. Like many areas of agriculture, soil testing has had its followers and detractors. From Liebig's time in 1850, until the early 1920s, little progress was made. Dyer (1894), Hilgard (1911), and Burd (1918) made significant contributions to soil chemistry. During the late 1920s and early 1930s, significant contributions to soil testing were made by Bray (1929), Hester (1934), Morgan (1932), Spurway (1932), and Truog (1930). These workers understood the importance of measuring labile instead of total soil nutrient content. Since the late 1940s, soil testing has been widely accepted as an essential tool to formulate a sound lime and fertilizer program. Traditionally, soil testing has involved the evaluation of nutrient deficiencies in soils. Today, however, with the increased emphasis on environmental quality, soil tests are a logical tool to determine areas where adequate or excess fertilization has occurred. It is as important to determine where fertilizers should not be used as it is to determine fertilizer need. Currently, soil testing is used to evaluate salinity and elemental toxicity.

We cannot discuss the principles of soil testing without considering the objectives. Soil testing is an accurate and indispensable tool essential for the assessment of the fertility and productivity status of soils. To us the objectives of soil testing are to: (i) accurately determine the available nutrient status of soils, (ii) clearly indicate to the farmer the seriousness of any deficiency or excess that may exist in terms of the various crops, (iii) form the basis on which fertilizer needs are determined, and (iv) express the results in such a way that they permit an economic evaluation of the suggested fertilizer recommendations. The objectives recognize that soil test results per se are a part of the factual data and must, therefore, be reported in finite terms, while interpretations will vary with different crops and the resulting fertilizer recommendations will include interpretative judgments. Another objective of soil testing may be to identify soils (usually garden soils) with toxic levels of heavy metals leading to the production of unsafe vegetation for human and animal consumption or unsafe environments for work or play. In this chapter, we emphasize the former objective in agronomic production.

A sound soil-testing program requires an enormous amount of background research. This background research should determine, among other things, the significant chemical forms of the available nutrients in the soils of the area, the extractants most suitable for extracting all or part of the available nutrient forms, the relative productive capacity of the soils for the various crops, the differential response of the various rates and methods of fertilizer appliation for the different crops, field-sampling techniques, test procedures, and methodologies. The soundness of the required interpretive judgments will depend almost entirely on the thoroughness and quality of these background studies. Too often, especially in developing countries, soil-testing programs are started without an adequate research background. The chemical methods can be transferred easily from state to state, or country to country, but the research database required for valid interpretations and sound judgments usually cannot. As a result, soil-testing programs often flounder. We are not exaggerating when we say that the success of a soil-testing program is directly proportional to its research backing.

In most states, soil testing is a program that may be divided into four phases: collecting the soil samples, extracting and determining the index of soil fertility, interpreting the analytical results (indexes), and making the fertilizer recommendations. One should, therefore, speak of a soil-testing program and recognize that its success depends as much on good individual judgments as on accurate chemical analyses. Unfortunately, when errors are made the chemical analyses are the popular scapegoat.

In discussing in more detail the principles of soil testing, it is necessary, and more convenient, to do so on the basis of the various phases that comprise the program.

I. FIELD SAMPLING

The soil-testing program starts with the collection of a soil sample, or samples, from a field. The analytical results are expected to represent the entire field. The first basic principle of the soil-testing program is that a field can be sampled in such a way that chemical analyses of collected samples will accurately reflect the field's true nutrient status. This does not mean that all of the samples must, or will, yield the same test results, but rather that the results reflect the true variations within the field. Differences of opinion exist as how best to sample a field to obtain an adequate evaluation of its nutrient level. This is because not enough research has been conducted in the area of soil sampling. Reuss et al. (1977) found that in eastern Colorado a systematic sampling plan was superior to a random one for NO_3-N. Also they discovered a large sampling error in the fields and the need for more intensive sampling. Depth of sampling is also important and should reflect the crop rooting system and nutrient distribution in soils. For mobile elements, deep samples should be taken. For example, Soltanpour et al.(1982) found that shallow (0–30 cm) soil samples were useless for predicting Se content of the dryland wheat (*Triticum aestivum* L.) grain from a soil test. But Se in deep (0–90 cm) samples correlated well with Se in grain. Although Se is not an essential plant nutrient, nevertheless the above example illustrates the importance of deep sampling under certain conditions. What must be recognized is that the entire soil-testing program can never be any more accurate than the accuracy of the soil sample or samples in characterizing the field. A single sample can seldom accurately characterize the nutrient status of a fertilized field.

In the past, before fertilizers were commonly used, it was relatively uncommon to find big differences in nutrient levels in different parts of a given field, except where extreme heterogeneity of soil types existed. Today, especially where fertilizers have been band applied, large differences are often found in the nutrient levels of samples taken from different parts of the same field. These differences usually are not sampling or testing errors but actual variations in fertility patterns. Wide test variations within a field pose the problem of determining or judging whether a single fertilizer recommendation can be prescribed for the entire field, or whether the variations are so great that different recommendations are required for different parts of the field. Sampling systems that were adequate for untreated soils will probably be inadequate for heavily fertilized fields. Many of our present sampling procedures will need to be reexamined, for the realities are, that fertilizer usage will continue, that variability in the fertility patterns within a given field will continue to grow, and that the problem of assessing the true, or average, fertility level for a given field will become more difficult.

II. EXTRACTION AND CHEMICAL ANALYSIS

Once the soil sample has been collected and prepared, the status of plant-available nutrients must be determined. By plant-available nutrient, one usually means the chemical form or forms of an essential plant nutrient in the soil whose variation in amount is reflected in variations in plant growth and yield. It is a basic principle of soil testing that simple rapid chemical analytical procedures can be designed to accurately measure, or be a measure of, the level of plant-available soil nutrients. The two parts for the laboratory analyses are extraction and measurement. Extraction involves the use of a chemical reagent solution to separate from the soil all or a fraction of the plant-available nutrient(s). Strict adherence to laboratory conditions, such as degree of soil pulverization, soil/solution ratio, and rate and length of shaking time are important because of the lack of an equilibrium between extracting solution and soil. Since the test value only has meaning in terms of plant response, should two soil test laboratories, testing the same soil sample and using the same extractant find different results, conditions of soil drying, pulverization, and extraction must be examined to reconcile the differences. Measurement is the determination of the amount of plant-available nutrient extracted. Any proven chemical analysis technique is acceptable. Here-to-fore extraction and measurement procedures have been dedicated to a single nutrient with each handling of a soil sample. In such systems, labor is the dominating cost of soil testing. With the advent of analytical instruments designed for multielement measurement, there is increasing research to develop multinutrient extractants. Several laboratories are already using such a system (Soltanpour, 1985). Many chemical methods have been suggested, and are being used, for the measurement of essential available plant nutrients. Actually, the chemical method used is important only to the extent that it must separate deficient from nondeficient soils and combined with other soil and plant variables predict fertilizer requirements of plants.

In working up and designing a chemical soil test, it is necessary to work with soil samples from fields of known crop response and which give a range in crop response. The method must show a mathematically definable increase in the measured plant-available nutrient as the indicated field crop response decreases. It is impossible to adequately correlate chemical soil test results with crop responses simply by testing large numbers of soils of unknown response. In the absence of soil samples from fields of known response, the chemical methods are correlated with greenhouse work using an index such as an "A-value" (Fried & Dean, 1952) or simply with response to a nutrient under greenhouse conditions. Alternatively, one may use a soil test proven useful in a different region with similar soils as a first approximation for finding a range of responding soils. While greenhouse and laboratory studies are useful in determining whether a method correlates with measuring plant-available form, it is important a method be calibrated with field response.

III. INTERPRETING THE ANALYTICAL RESULTS

Analytical results obtained from chemical analyses of soils must be interpreted meaningfully. This is usually accomplished through some type of a previously determined correlation between test results and known field crop responses. Methods that Cate and Nelson (1965), Nelson and Anderson(1977), and Kiesling and Mullinex (1979) developed can be used to develop the correlations. Therefore, valid correlation studies must precede intelligent interpretations of soil test values. A basic principle of soil testing is that a soil test value can, under most circumstances, be treated and related as an independent variable to the percent yield and response obtained for a specific crop. Percent yield (yield of unfertilized crop as a percentage of fertilized crop) measures response to fertilizer under identical conditions of season, land, soil type, diseases, and weeds. Therefore, these latter variables are not required to be considered in interpreting the soil test. But, these variables are required to be considered for making a fertilizer recommendation specially for mobile nutrients.

At Illinois, the work of Bray (1944, 1945) has shown that the Mitscherlich equation, with minor modifications, can be used successfully to describe the yield associated with different levels of immobile available nutrients in the soil and different amounts of applied fertilizer.

IV. MAKING FERTILIZER RECOMMENDATIONS

In making a valid fertilizer recommendation, eight criteria must be known: (i) the soil test level of the soil, (ii) the crop to be grown, (iii) yield potential and percent sufficiency, (iv) the increase in yield with increasing rates of applied fertilizers, (v) the method of fertilizer application (i.e., banded vs. broadcast), (vi) leguminous crops preceding the crop to be grown in case of N recommendations, (vii) whether animal manure has recently been applied to the field and its potential contribution to nutrient status of the soil, and (viii) the degree of mineralization of native organic matter during plant growth, especially for N and S. Of these, the soil test simply indicates an index of the nutrient level in the soil. The soil test values say nothing about the yield potential of the soil, the season, the management practices, or the amount of fertilizer needed. The accuracy with which a test value can be interpreted will depend on the kind and quality of the field research work on which the correlations are based. The purpose of the soil test value is simply to indicate the starting point—the present level of available nutrients in the soil—on which the subsequent required judgments are based.

To interpret a soil test value, proper correlations of test values with known field responses for the various crops are required; that is, one must be able to determine and express the seriousness of the nutrient deficiency for a particular crop from the soil test value. In addition, crop response to different soil nutrient levels and to increased increments of both broadcast and band-applied fertilizer must be known. This factor requires considerable research, since it must involve combinations of different levels of available soil nutrients, different rates of applied fertilizers, different methods of applying the fertilizer, and different crops.

Economic considerations are important to determine the upper level of fertilization. This is especially true for the value of the expected crop increase as related to the cost of the fertilizer. To make these economic evaluations, an estimate is needed of the forthcoming yield of the particular crop and its probable value in absolute terms. This estimate is extremely difficult and about the best one can do for any given year is to assume average yields and prices. The yield estimate will depend largely on the predetermined productive capacity of the soil, and some estimate of what constitutes an average yield in an average season under specified management systems. Therefore, all the factors used to determine the upper level of fertilization are based essentially on personal judgments. Once the upper level of fertilization for a crop has been decided, the actual amount of fertilizer needed to achieve that goal is the difference between the amount of available nutrients present (estimated from soil test value) and the amounts needed to achieve that upper limit. The amount of actual fertilizer needed is further modified by the method of applying the fertilizer (broadcast and banded), the efficiency with which increasing fertilizer increments have been shown to increase the yield and the kind of crop to be grown. Therefore, the final fertilizer recommendations depend on accurate soil test analyses and on interpretations of the test results based on sound research and judgment. An ideal method would be to recommend fertilizers for different yield levels and prices and allow farmers to choose the yield level and price combination they desire.

V. GENERAL SUMMARY

In a soil-testing program, as in any program, the results depend on the quality of the component parts. The primary objectives of the program are sound fertilizer recommendations for the various crops, yet many of our fertilizer recommendations are subject to criticism. Where then are the most probable sources of error?

The accuracy of most field-sampling techniques cannot be evaluated because of lack of extensive research data. Much greater variability in the fertility patterns of a field, as a result of band-applied fertilizers, may be expected. Additionally, sampling techniques, as well as interpretations of the results, must recognize irregular fertility patterns. Variations of test results within a field are not always because of faulty testing or sampling. They may present a true picture of the field's fertility patterns, which experience, past fertilizer history, and good judgment must reconcile in making the fertilizer recommendation.

Usually, the chemical analyses are the most accurate part of the soil-testing program in the sense that they are chemically reproducible on any given soil sample. Not all chemical methods measure the available form or forms of a nutrient in the soil. Errors are, therefore, more likely because of failure of the method to extract available-nutrient forms and to poor correlations than to the chemical reproducibility of the test.

Fertilizer recommendations depend on good fertilizer research directed towards determining the efficiency of different crops and soils of varying nutrient levels. If the results are expressed in terms of percent sufficiencies, the fertilizer recommendation needs only one personal judgment, that of picking the upper limit of fertilization. However, the selection of the upper limit may involve several judgments and usually depends on economic considerations. The selected upper limit of fertilization may represent that point at which maximum returns per acre are expected: that point at which the last small increment of fertilizer used will give a yield increase just equal to the cost of the last fertilizer increment, or some lower point at which the value of the crop increases may be several times that of the fertilizer costs. Errors in these judgments include those of estimating the economics of expected yield increases in relation to the fertilizer costs, estimating the forthcoming season and the productivity capacity of the soil, evaluating the farmer managerial capacity, and many more. What users of soil testing must recognize is that every fertilizer recommendation is based on some individual's judgment as to what are reasonable expectations for a given situation, and that the expectation will probably not occur in any given year on any given field.

In all soil-testing programs, tables are prepared, or computer programs written, showing soil test results and suggested fertilizer use for various crops. However, judgments on which such tables are based are not always clearly indicated. What is often omitted in reporting soil test results is an indication of how serious the deficiencies are and the degree of response the farmer can expect if he carries out the recommendation. In fact, soil test results and accompanying fertilizer recommendations are almost always reported in such a manner that the farmer cannot evaluate them in terms of his own managerial and economic capabilities. This makes it extremely difficult for either farmers or field people to modify the average judgments used in such tables, or computerized programs, to accommodate their evaluation of the particular local condition.

A good soil-testing program is essential to sound fertilizer use. The soil test value is the starting point. It can indicate how serious the deficiency is and provide an anchor on which many of the subsequent required judgments depend. If a soil test value can be treated or expressed as an independent variable, fewer personal judgments are required and more accurate fertilizer recommendations will be made.

Currently, few soils being farmed do not require efficient fertilizer usage and good management for profitable crop production. On many soils, excess fertilizers are being applied and environmental concerns are being voiced. A sound soil-testing program is today's best and perhaps only way of determining what constitutes adequate, but not excessive, fertilizer use for high and efficient crop production.

REFERENCES

Bray, R.H. 1929. A field test for available phosphorus in soils. Illinois Agric. Exp. Stn. Bull. 337.

Bray, R.H. 1944. Soil-plant relations: I. The quantitative relation of exchangeable potassium to crop yields and to crop response to potash additions. Soil Sci. 58:305–324.

Bray, R.H. 1945. Soil-plant relations: II. Balanced fertilizer use through soil tests for potassium and phosphorus. Soil Sci. 60:463–473.

Burd, J.S. 1918. Water extractions of soils as criteria of their crop-producing power. J. Agric. Res. 12:297–310.

Cate, R.B., Jr., and L.A. Nelson. 1965. A rapid method for correlation of soil test analyses with plant response data. North Carolina Agric. Exp. Stn., Int. Soil Testing Ser., Tech. Bull. 1.

Dyer, B.1894. On the analytical determination of probably available mineral plant food in soil. J. Chem. Soc. 65:115.

Fried, M., and L. Dean. 1952. A concept concerning the measurement of available soil nutrients. Soil Sci. 73:263–271.

Hester, J.B. 1934. Micro-chemical soil tests in connection with vegetable crop production. Virginia Truck Exp. Stn. Bull. 82.

Hilgard, E.W. 1911. Soils. The Macmillan Co., New York.

Kiesling, T.C., and B. Mullinix. 1979. Statistical considerations for evaluating micronutrient tests. Soil Sci. Soc. Am. J. 43:1181–1184.

Liebig, J. von. 1840. Organic chemistry in its application to agriculture and physiology. Playfair, london.

Morgan, M.F. 1932. Microchemical soil tests. Connecticut Agric. Exp. Stn. Bull. 333.

Nelson, L.A., and R.L. Anderson. 1977. Partitioning of soil test-crop response probability. p. 19–38. *In* T.R. Peck et al. (ed.) Soil testing: Correlating and interpreting the analytical results. ASA Spec. Publ. 29. ASA, CSSA, and SSSA, Madison, WI.

Reuss, J.O., P.N. Soltanpour, and A.E. Ludwick. 1977. Sampling distribution of nitrates in irrigated fields. Agron. J. 69:588–592.

Soltanpour, P.N. 1985. Use of ammonium bicarbonate DTPA soil test to evaluate elemental availability and toxicity. Commun. Soil Sci. Plant Anal. 16:323–338.

Soltanpour, P.N., S.R. Olsen, and R.J. Goos. 1982. Effect of nitrogen fertilization of dryland wheat on grain selenium concentration. Soil Sci. Soc. Am. J. 46:430–433.

Spurway, C.H. 1933. Soil testing: A practical system of soil diagnosis. Michigan Agric. Exp. Stn. Bull. 132.

Truog, E. 1930. The determination of the readily available phosphorus in soils. J. Am. Soc. Agron. 22:874–882.

Chapter 2

Physical-Chemical Aspects of Nutrient Availability[1]

RICHARD B. COREY, *University of Wisconsin, Madison*

The instantaneous rate of nutrient uptake by a plant root is generally considered to be a function of the concentration of the dissolved absorbable form(s) of the nutrient at the root surface. Nutrient uptake models have used linear (Nye & Tinker, 1977) or Michaelis-Menten (Barber, 1984) equations to describe this function. Maintaining optimum plant growth over a growing season requires that a certain minimum concentration of the absorbable nutrient species be maintained at root surfaces to maintain required uptake rates. This requires transport from the bulk soil to the root surface to replace nutrients as they are absorbed.

The nutrient resupply system depends on a complex interaction of many chemical, physical, and biological reactions that can be visualized with the help of a comprehensive transport equation. If, for the sake of simplicity, we assume a one-dimensional flow system and select an infinitesimal volume of soil at a distance, x, from an actively absorbing plant root, the total flux, F, of dissolved nutrient toward the root is equal to the sum of convective and diffusive fluxes as shown in Eq. [1]:

$$F = vC_l - (D_l\ \theta f/b)\ dC_L/dx \qquad [1]$$

where:

v = velocity of water flow to the root, m^3(soln) m^{-2}(soil) s^{-1};
C_l = nutrient concentration in solution, mol m^{-3}(soln);
C_L = labile nutrient concentration in soil, mol m^{-3}(soil);
D_l = diffusion coefficient of nutrient in solution, m^2(soln) s^{-1};
θ = volumetric water fraction, m^3(soln) m^{-3}(soil);
f = conductivity factor related to path tortuosity, m^2(soil) m^{-2}(soln); and
b = buffer power = dC_L/dC_l, m^3(soln) m^{-3}(soil).

[1] Contribution from the College of Agricultural and Life Sciences, Univ. of Wisconsin-Madison, Madison, WI 53706.

The negative sign in the diffusive flux term results from the fact that dC_L/dx is negative when flux is positive (flow is from high to low concentration). The term *labile* as used here refers to the amount of dissolved nutrient per unit volume of soil plus the amount of surface-bound nutrient per unit volume of soil that is in reasonably rapid equilibrium with the dissolved nutrient. The labile pool does not include nutrients present as precipitates except for those at the surface of the precipitate that react rapidly with the solution, nor does it include organic forms mineralized or immobilized by microorganisms. These forms are treated as separate source-sink terms in Eq. [2]. A source is defined here as any reaction that increases the concentration of a nutrient in the soil solution at a given point, and a sink is defined as any reaction that decreases the nutrient concentration at that point. Actually, the equilibrium between the labile-adsorbed and dissolved forms could also be treated as a source-sink reaction, but because these reactions are reasonably fast, they will be included in the diffusive transport term as a component of the buffer power.

A solute-conservation equation can be derived from Eq. [1] and the source-sink terms. If $D_l\ \theta f/b$ and temperature are assumed constant, the equation can be written as:

$$(\partial C_l/\partial t)_x = v\,(\partial C_l/\partial x)_t - (D_l\ \theta f/b)\,(\partial^2 C_L/\partial x^2)_t + (\partial C_p/\partial t)_x + (\partial C_O/\partial t)_x + (\partial C_V/\partial t)_x \qquad [2]$$

where:

C_p = nutrient in precipitates, mol m^{-3} (soil);
C_O = nutrient in organic form, mol m^{-3} (soil);
C_V = dissolved nutrient in potentially volatile form, mol m^{-3} (soil);
x = distance from the root surface, m; and
t = time.

When integrated using appropriate initial and boundary conditions, Eq. [2] can be used to calculate the concentration gradient away from the root as a function of time (Nye & Tinker, 1977). Because of the complexity of this differential equation, numerical methods are needed to solve it. However, this equation in its present form contains the various physical, chemical, and biological variables that affect the concentration and transport of specific nutrients and, therefore, will be used to focus the discussion that follows.

I. REACTIONS AFFECTING NUTRIENT AVAILABILITY TO ROOTS

A schematic representation of some important physical, chemical, and biological processes occurring in soils is shown in Fig. 2–1. The descriptions of these processes that follow are general in nature because the complexity of the soil system makes a rigorous theoretical approach to nutrient availability impractical given our present inability to define accurately the compo-

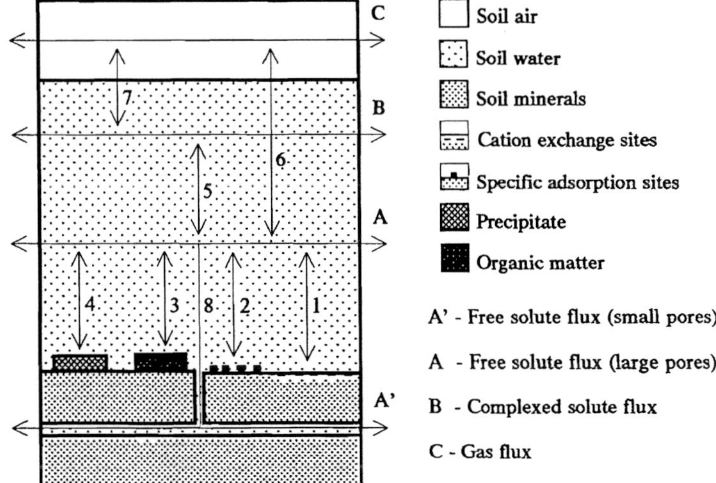

Fig. 2-1. Schematic representation of transport processes and source-sink reactions controlling solute transport in soil systems. Numbered reactions are as follows: (1) cation exchange, (2) specific adsorption, (3) precipitation/dissolution, (4) biological mobilization/immobilization, (5) complexation, (6) volatilization/solution of free solute, (7) volatilization/solution of complexed solute, (8) equilibration of solute between big and small pores. Oxidation/reduction reactions are not included in the figure, but they are important.

nents and properties of the system. Therefore, simple mathematical models of soil processes will be used in preference to more complex mechanistic models on the assumption that the coefficients of any model will have to be derived by curve fitting in a real-world soil system and that there is little improvement associated with the use of the more complex models. References giving more rigorous, theoretical treatments of the processes will be cited where appropriate. The two books of Sposito (1981, 1984) are particularly recommended for their comprehensive, rigorous coverage of the thermodynamics of soil solutions and surface reactions.

A. Complexation

Some plant nutrients, particularly the trace metals, form complexes (coordination compounds) with organic ligands (reaction 5 in Fig. 2-1). Chelates are complexes in which a metal ion acts as an acceptor of two or more electron-pairs donated by a single organic-ligand molecule. A single- or multiple-ring structure is formed, each ring consisting of the metal ion and usually four to six atoms from the ligand molecule. Some of these chelates are extremely stable, and the concentrations of strongly chelated metals, particularly Fe^{3+} and Cu^{2+}, can be many orders of magnitude greater than concentrations of free metal ions in soil solutions (Fujii et al., 1983). Chelating ligands usually carry a negative charge, but after reaction with the metal ion the charge may be negative, neutral, or positive depending on the initial negative charge of the ligand, the positive charge of the metal ion, and the number of H^+ ions displaced by the metal ion from electron-pair donor sites on the ligand. Positively charged chelates are not common in soil solutions because they are strongly adsorbed on soil cation-exchange sites. Chelate stability at constant metal-ion activity decreases with decreasing pH because of the tendency of H^+ to form electron-pair bonds with the ligand thereby displacing metal ions. However, if the activity of the metal ion is controlled by the solubility of a metal hydroxide, as with $Fe(OH)_3$, the increase in metal-ion activity with decreasing pH may more than offset the metal-displacing effect of the increasing H^+ activity.

Intact metal chelates do not appear to be absorbed by plant roots. Instead, the uptake rate seems to depend on the activity of the free metal ion at the root surface (Checkai et al., 1987). However, if equilibrium between free metal ion and chelate is approached at the root surface, the effective convective transport (first term in Eq. [1]) to the root will be a function of the total dissolved metal concentration rather than the free-ion concentration. This will also be true for the buffer power, *b,* in the diffusive transport term (second term in Eq. [1]). In both cases, the presence of the chelating ligand results in a greater transport rate than would be found with the existing free-ion concentration but no chelate. Chelation is particularly important in transporting nutrient metals of low solubility such as Fe. In fact, Fe-stressed plants have been reported to release Fe-complexing ligands into the soil solution to promote transport to the root (Romheld & Marschner, 1986). Other chelating agents such as citrate are common in rhizosphere soil (Marschner, 1986).

Addition of a chelating agent does not always result in increased availability of the chelated metals. In fact, adding chelating ligands can significantly lower free-metal-ion activities in the soil solution if those activities are not buffered sufficiently by labile solid-phase metals. Free-metal-ion activities are especially sensitive to addition of chelating agents in solution cultures where no solid-phase buffering is present. The ultimate effect of chelates on availability in soils will depend on whether the increased rate of transport to the root offsets the decrease in free-metal-ion activity in the bulk soil solution.

Dissolved ion pairs and small-molecular-weight complexes (single electron-pair bonds) should also contribute to transport in the soil solution, but their effects on absorption across cell membranes have not been well defined.

B. Adsorption on Particle Surfaces

Adsorption reactions on particle surfaces take the form of cation exchange on negatively charged surface sites (or, more rarely, anion exchange on positively charged sites) and specific adsorption of cations on electron-pair donor sites or anions on electron-pair acceptor sites. Most specific-adsorption sites also have some electrostatic contribution to the bond, and some involve mainly multiple ionic bonds where steric factors determine specificity [e.g., K^+ "fixation" in collapsed vermiculties and in mica edge sites

(McLean & Watson, 1985)]. Cation-exchange and specific-adsorption reactions are shown as reactions 1 and 2, respectively, in Fig. 2–1.

Cation-exchange sites may be associated with silicate minerals, oxide/hydroxide minerals or organic matter. In the commonly occurring expanding 2:1 layer silicate minerals most of the charge arises from isomorphous substitution of cations of lower valence for cations of higher valence in the octahedral or tetrahedral layers within the mineral. At the broken edges of silicate minerals or the surfaces of oxide/hydroxide minerals, fractional charges may be associated with electron-pair sites on surface oxygen ions. On organic matter, negative charge results when protons dissociate from surface weak-acid functional groups, primarily carboxyls and phenols.

Charges arising from isomorphous substitution are insulated from the soil solution and are not affected by solution properties including pH. They are referred to as permanent charges. Charges arising at surfaces are affected by solution properties, particularly pH and specifically adsorbed ions. These are referred to as pH-dependent charges. At high pH, pH-dependent sites on mineral surfaces have fractional negative charges. As the pH decreases or the concentration of specifically adsorbed cations increases, H^+ or other specifically adsorbed cations will occupy some of the sites, and those sites will assume a partial positive charge. Therefore, as the pH decreases, the ratio of negative to positive charge sites decreases until a point of zero net charge is reached. Further decrease in pH will then produce a net positive charge on the surface. Alternatively, increasing pH or concentration of specifically adsorbed anions increases the ratio of negative to positive sites.

Association of H^+ with ionized weak-acid functional groups on organic matter changes the charge from -1 to 0. Since there are no fractional charges on the organic matter sites and since additional protonation to a $+1$ charge does not occur in soil pH ranges, organic matter sites that are negatively charged at high pH do not become positively charged at low pH. However, organic matter does have N-containing functional groups such as amines in which the N has a free electron pair available for bonding H^+ (or other electron-pair acceptor). The protonated site would have a $+1$ charge. Based on pK values of about 9.2 for dissociation of the proton from NH_4^+ and 7.2 and 10.0 for dissociation of the two protons of protonated ethylenediamine (Sillen & Martell, 1964), one would expect that the amine groups on the organic matter should be protonated and have positive charges in the normal soil pH range. These positive charges are usually not noticed in soils, possibly because the positively charged amine groups are attracted to negative mineral sites and neutralized or are effectively neutralized by adjacent negative sites on the organic matter itself.

The attraction of cation-exchange sites for cations in solution increases with charge density and with nearness of the charge source to the particle surface. The relative strength of bonding of the individual cations depends on the force of attraction of the negative surface, the charge on the cation, the ionic strength of the solution, and the distance between the negative charge plane and the cation nucleus that depends on cation radius, hydration energy, and polarizability of the dehydrated cation. Generally, selectivity for a par-

Table 2–1. Relative strength of bonding of important exchangeable cations to cation exchange sites on planar surfaces of montmorillonite. (Order may differ somewhat on different exchangers). Cations normally present in significant amounts are underlined.

Slightly acid to alkaline soils:

$$Ba^{2+} \geq Sr^{2} \geq \underline{Ca^{2+}} > \underline{Mg^{2+}} > Rb^{+} > \underline{K^{+}} = NH_4^{+} > \underline{Na^{+}}$$

Very acid soils:

$$\underline{Al^{3+}} \gg Ba^{2+} \geq Sr^{2+} \geq \underline{Mn^{2+}} \geq \underline{Ca^{2+}} > \underline{Mg^{2+}} > Rb^{+} > \underline{K^{+}}$$

$$= NH_4^{+} > H^{+} \geq \underline{Na^{+}}$$

Neutral anaerobic soils:

$$Ba^{2+} \geq \underline{Fe^{2+}} \geq \underline{Mn^{2+}} \geq Sr^{2+} \geq \underline{Ca^{2+}} > \underline{Mg^{2+}} > Rb^{+} > \underline{K^{+}} = \underline{NH_4^{+}} > \underline{Na^{+}}$$

ticular cation on cation-exchange sites is greater the higher the charge on the cation and, for cations of similar charge, the lower the hydration energy and the greater the polarizability. Because large cations bind water molecules less strongly than do small cations, there is a greater probability that a large cation can lose an associated water molecule and get closer to the particle surface than a hydrated smaller cation. However, if the attractive force is strong enough, the smaller cations may also be dehydrated and thus be preferentially adsorbed. Concentrating the soil solution (e.g., by evaporation) increases adsorption of ions of lower valence relative to those of higher valence in accord with the well-known ratio law (Schofield, 1947). Table 2–1 gives the usual strength-of-bonding order of cations normally found (or added with land-applied wastes) in soils that vary in pH and redox status. The proportions of the different cations found on the exchange will depend on relative solution activities as well as strength of bonding.

Several equations have been derived to describe the relationship between cation ratios in solution and on the exchange sites (Sposito, 1981, 1984). Most of them are similar for homovalent exchange but differ for heterovalent exchange. The Gapon equation is one of the simplest and will be used here to demonstrate the factors that affect the buffer power term in Eq. [2]. For more theoretically sound approaches, see Sposito (1981, 1984). The Gapon mass action equation for heterovalent cation exchange using the Ca^{2+}-K^{+} couple for illustration is given in Eq. [3] where X is a single negatively charged surface site:

$$Ca_{1/2}X + K^{+} = KX + 1/2\ Ca^{2+} \qquad [3]$$

The Gapon selectivity coefficient, K^{G}, is derived from this equation:

$$K^{G} = [KX]\ (Ca^{2+})^{1/2}/[Ca_{1/2}X]\ (K^{+}) \qquad [4]$$

where [ ] denote concentrations and ( ) activities.

In soils, each separate component that contributes to the cation-exchange capacity will likely have a different K^{G}. For the K^{+}-Ca^{2+} system, the relative affinity of the different sites for K probably increase in the order: organic < edge charge < smectites < vermiculite < mica edges. The K^G determined for a soil sample will be a weighted average of the K^Gs of the individual components and will probably be valid only over a fairly narrow range of $(K^+)/(Ca^{2+})^{1/2}$.

The effects of the variables in Eq. [4] on K activity are important because the K-uptake rate is a function of (K^+) at the root surface. These effects can be seen more clearly by solving Eq. [4] for (K^+):

$$(K^+) = \{1/K^G\}\{[KX]/[Ca_{1/2}X]\}(Ca^{2+})^{1/2}. \quad [5]$$

This equation shows that the K activity in solution varies directly with the ratio of K to other cations on the exchange sites and the activities of competing ions (represented here by Ca^{2+}) in the soil solution, and inversely with the selectivity coefficient.

Because the total labile K per unit volume of soil is $[KX] + \theta[K^+]$, the K buffer power, b_K, can be calculated by substituting $\gamma_+[K]$ for (K^+) in Eq. [4] where γ_+ is the activity coefficient for K^+, solving Eq. [4] for $[KX]$, adding $\theta[K^+]$ to both sides of the equation to convert the left side to labile K, and differentiating the labile K with respect to $[K^+]$.

$$b_K = d\{[KX] + \theta[K^+]\}/d[K^+]$$
$$= K^G\ \gamma_+[Ca_{1/2}X]/(Ca^{2+})^{1/2} + \theta. \quad [6]$$

Thus, the K buffer power varies directly with the selectivity coefficient, the cation-exchange capacity not satisfied by K^+, and the volumetric water content, and inversely with the activities of competing cations. If there is no adsorbed phase of a nutrient (as with NO_3^-) the buffer power is equal to the volumetric water fraction, θ.

Buffer powers for individual cations in complex mixtures normally found in soils can be calculated from solution concentrations and the exchangeable concentration of one of the cations if all of the competitive Gapon selectivity coefficients are known (Weitholter, 1983). However, any buffer-power values used in plant-uptake models should be derived using Gapon constants that are valid for the adsorption site mix and the solution composition of the soil in question. Effects of variation in solution composition over the growing season should also be considered.

In Tables 2-2 and 2-3, cations and anions, respectively, are grouped broadly according to relative selectivities on specific-adsorption sites. If a competitive Langmuir model (Veith & Sposito, 1977) is used, the mathematical derivation of buffer powers for nutrients that are specifically adsorbed is similar to that for the cation exchange system. The major difference is the much greater range in selectivity exhibited by the specifically adsorbed nutrients. For example, results of Fujii et al. (1983) suggest that the selectivity coefficient for the Cd^{2+}-Ca^{2+} exchange on the most selective sites in soils is about 10^5 in favor of Cd^{2+}. However, as the Cd^{2+}/Ca^{2+} ratio in

Table 2–2. Relative strength of bonding of cations on complexing sites and on specific adsorption sites of minerals.

No bond	Weak	Strong
Na^+	Mg^{2+}	H^+, Ag^+, Al^{3+}
K^+	Ca^{2+}	Cd^{2+}, Co^{2+}, Cr^{3+}
NH_4^+	Mn^{2+}	Cu^+, Cu^{2+}, Fe^{2+}
Rb^+	Sr^{2+}	Fe^{3+}, Hg^+, Hg^{2+}
Cs^+	Ba^{2+}	Ni^{2+}, Zn^{2+}, Pb^{2+}

Table 2–3. Relative strength of bonding of anions on adsorption sites.

No bond	None to weak	Weak to moderate	Strong
NO_3^-	SO_4^{2-}	$H_2BO_3^-$	OH^-
Cl^-	SeO_4^{2-}	MoO_4^{2-}	$H_2PO_4^-$
	CrO_4^{2-}		$H_2AsO_4^-$
			AsO_2^-
			$HSeO_3^-$
			F^-

creases, the selective sites become saturated with Cd, and eventually Cd competes with Ca on cation-exchange sites where the selectivities are about equal (Hendrickson & Corey, 1981). The effect of pH is also more pronounced with specific adsorption reactions because H^+ competes directly and strongly with specifically adsorbed cations for the sites and OH^- competes similarly with specifically adsorbed anions such as $H_2PO_4^-$.

One potentially important factor affecting buffer power calculations for adsorption systems that is rarely mentioned in the literature is pore size. If a uniform surface-adsorption-site density is assumed, the ratio of surface-adsorption sites to volume in a cylindrical pore will be inversely proportional to the pore radius. Therefore, at a constant solution composition, which would maintain a constant fraction of the sites saturated with a given solute, the amount of adsorbed solute per unit volume of pore would also be inversely proportional to the pore radius. If the dissolved solute is insignificant compared with the amount adsorbed, as is frequently the case with strongly adsorbed solutes such as $H_2PO_4^-$, the local buffer power will also be inversely proportional to the pore radius. This has a dramatic effect on the distance that a solute will diffuse in a given time because the mean-square diffusion distance at a given time is inversely proportional to the square root of the buffer power (Jost, 1960). This slow diffusive transport through small pores is shown as transport process A′ in Fig. 2–1, and the equilibration of solute between large and small pores is shown as reaction 8.

C. Precipitation

Solubility-product criteria can control activities of plant nutrients that form precipitates in soils (reaction 3 in Fig. 2–1). If a plant nutrient tends to form a precipitate under soil conditions, the following sequence usually occurs in the precipitation process as the concentration of the nutrient in

Table 2-4. Common precipitates in soil systems (underlined) and other substances that can form similar precipitates or coprecipitates with the common precipitates.

Hydrous oxides	Carbonates	Phosphates	Sulfates	Sulfides
$\underline{Fe^{3+}}$	Ba^{2+}	$\underline{Fe^{3+}}$	$\underline{Ca^{2+}}$	$\underline{Fe^{2+}}$
$\underline{Al^{3+}}$	$\underline{Ca^{2+}}$	$\underline{Al^{3+}}$	Ba^{2+}	Ag^{+}
Ca^{2+}	Cd^{2+}	$\underline{Ca^{2+}}$	Hg_2^{2+}	As^{3+}
Co^{2+}	Co^{2+}	$\underline{Fe^{2+}}$	Pb^{2+}	Cd^{2+}
Cr^{3+}	Cu^{2+}	Ag^{+}	Sr^{2+}	Co^{2+}
Cu^{2+}	Fe^{2+}	Ba^{2+}	SeO_4^{2-}	Cr^{3+}
Fe^{2+}	Hg_2^{2+}	Cd^{2+}		Cu^{+}
Hg^{2+}	Mg^{2+}	Co^{2+}		Cu^{2+}
Hg_2^{2+}	Mn^{2+}	Cr^{3+}		Hg_2^{2+}
Mn^{2+}	Ni^{2+}	Cu^{2+}		Hg^{2+}
Ni^{2+}	Pb^{2+}	Fe^{2+}		Mn^{2+}
Pb^{2+}	Sr^{2+}	Hg_2^{2+}		Ni^{2+}
VO^{2+}	Zn^{2+}	Mn^{2+}		Pb^{2+}
Zn^{2+}		Ni^{2+}		Sn^{2+}
		Pb^{2+}		Zn^{2+}
		Sr^{2+}		
		Zn^{2+}		
		AsO_4^{3-}		
		F^{-}		

the soil solution increases (Corey, 1981): (i) adsorption on particle surfaces, (ii) supersaturation with respect to the reactants in the precipitation reaction, (iii) nucleation of the precipitate on the particle surfaces, and (iv) crystal growth. It is only when the precipitate is present (stages 3 and 4) that the solubility product of the precipitation reaction controls the activities of the reacting species. Even then, significant undersaturation or supersaturation of reactants in the soil solution may occur if addition or loss of reactants occurs faster than the reattainment of equilibrium with the precipitate.

The major classes of precipitates found in soils are silicates, oxides, hydroxides, carbonates, phosphates, sulfates, and sulfides. Specific precipitates in these classes (except silicates) are listed in Table 2-4. The underlined elements in Table 2-4 form the common precipitates and the others are usually found as coprecipitated minor constituents within the major precipitates (Corey, 1981). Solubility products for most of the pure precipitates listed can be found in Lindsay (1979).

The coprecipitation process can best be illustrated with an example. Both Ca and Mn form carbonate precipitates. If Ca is present at relatively high concentration in the soil solution and Mn at relatively low concentration, and the CO_3^{2-} concentration is raised to the point that $CaCO_3$ starts to precipitate, Mn will be coprecipitated along with the Ca even though the solubility product of $MnCO_3$ is not exceeded. The ratio of Ca to Mn in the precipitate will depend on the ratio in solution and their relative affinities for surface adsorption during the crystal-growth process. If the ratio of

Ca^{2+} to Mn^{2+} remained constant and the ions were freely diffusible within the precipitate, the relationship between the ratio of ions in solution to the ratio in the precipitate could be described by a separation factor, D (Sposito, 1981):

$$D = [(CaCO_3)_s/(MnCO_3)_s]/[(Mg^{2+})/(Ca^{2+})] \quad [7]$$

where the parentheses around the two solid phases indicate activities of those phases which can be approximated by their mole fractions. However, if the Mn^{2+} concentration is not well buffered and the ions are not freely diffusible within the precipitate, the relative rate of Mn depletion from solution will differ from that of Ca so that the Ca/Mn ratio in solution will change continuously during precipitation, and the Ca/Mn ratio in the central part of the precipitate will differ from that at the surface. If equilibrium between the particle surface and the soil solution is ultimately attained, the solution activity of Ca, the major constituent, can be predicted quite accurately from solubility-product calculations, but this is not the case with Mn. It is generally true that solution activities of solutes that are minor constituents in solid solutions cannot be predicted accurately from precipitate composition and solubility product criteria (Garrels & Christ, 1965).

The rate at which nutrient ions are released from precipitates depends on the surface exposed and the rate-limiting step for dissolution (Sparks, 1987). The rate-limiting step may be the rate of detachment from the particle surface (surface control) or the rate of diffusion from the saturated solution at the particle surface to the bulk solution (diffusion control). For many precipitates, particularly those that form discrete crystals, the surface exposed is small relative to the concentration of potential adsorption sites for the precipitate constituents on soil-particle surfaces. For example, crystals of variscite, $AlPO_4 \cdot 2H_2O$, might form in an acid soil. That soil would also contain potential P-adsorbing sites on edges of silicate minerals and on sesquioxide surfaces. The number of adsorbing sites per unit volume of soil would probably be orders of magnitude greater than the number of sites exposed on the surface of the variscite particles. If equilibrium were attained, the activity of P in solution would be controlled by the variscite solubility product, but the major source of buffering against rapid changes in P activity (the buffer power) would be the surface adsorption sites. Therefore, in Eq. [2], precipitation reactions are considered to be source-sink reactions within the infinitesimal soil volume, and the total labile forms, which include the precipitate surface, are assumed to be the labile solid-phase component of the buffer power.

An exception to the model of precipitates providing long-term solubility control of adsorbable/precipitatable solutes and adsorbed phases providing the major short-term buffering can be found with the iron oxides. Iron oxide precipitates form coatings on particle surfaces and, indeed, are major adsorbing surfaces for many specifically adsorbed nutrients. According to

Lindsay (1979), this "$Fe(OH)_3$(soil)" has a solubility product intermediate between that of amorphous $Fe(OH)_3$ and γFe_2O_3, and, because of its large surface, it probably controls the Fe^{3+} activity in soils in the short term as well as the long term.

D. Mineralization and Immobilization by Organisms

For some plant nutrients such as N, S, B, and P, a major input to the labile pool is mineralization of organic matter containing significant concentrations of those nutrients. Conversely, some labile nutrients can be immobilized during the decomposition process by incorporation into microbial cells if the nutrient concentration in the organic material is not sufficient to meet the needs of the microorganism doing the decomposing. The details of these reactions are covered in numerous texts (e.g., Tate, 1987; Paul & Clark, 1989) and will not be discussed here. In Eq. [2], these reactions are considered to be source-sink reactions that do not contribute to buffer power because no equilibrium exists between solution and solid phases. Mineralization/immobilization rates are determined by the factors that affect the activities of the organisms involved.

E. Liquid-Gas Interactions

Some of the plant nutrients such as C, H, O, N, and S may exist in gaseous form in the soil. Under aerobic conditions, CO_2 H_2O, O_2, and N_2 are common constituents of the soil air. Under anaerobic conditions, CH_4, H_2, NO, N_2O, and H_2S may form and free O_2 may disappear. At high pH, volatilization of NH_3 may become significant.

The interchange between gaseous molecules in the soil air and those dissolved in the soil solution is governed, at low concentrations, by Henry's law. According to this law, the solubility of a gas in the soil solution is proportional to the partial pressure of the gas in the soil air. Volatile constituents can be transported relatively rapidly to the atmosphere if there is sufficient air permeability because diffusion coefficients in air are about 10^4 times those in water. Therefore, the rate of gas diffusion through soils depends on the number and continuity of air-filled pores which, in turn, are complex functions of the water content. The ability of gas diffusivity models to predict gaseous diffusion in soils is limited by the uncertainty of this relationship between gas permeability and water content (Collin & Rasmuson, 1988). Gaseous transport is represented by process C in Fig. 2–1, and the equilibrium between solution and gas phases is shown as reactions 6 and 7.

Reactions involving liquid-gas interchange are treated as source-sink reactions in Eq. [2] because many of them are microbially mediated and are not reversible.

F. Oxidation/Reduction (Redox)

Redox reactions involve the transfer of electrons. The redox potential is a measure of the tendency of an atom, ion, radical, or molecule to donate or accept electrons. A high redox (reduction) potential indicates a strong tendency to accept electrons from an electron donor with lower redox potential. If the electron transfer occurs, the electron acceptor is converted from an oxidized to a reduced state and the electron donor is converted from a reduced to an oxidized state. Many substances remain in metastable redox states because the activation energy required to make the reaction proceed is not available and this prevents them from achieving thermodynamic equilibrium. In many of these cases, microbial enzymes can act as catalysts to make the reactions proceed to the equilibrium state.

In soil systems, the source of electrons that drives most redox reactions is the microbial respiration process. In respiration, electrons are transferred from C to an electron acceptor with the release of energy. If O_2 is available at the respiration site, it is the preferred electron acceptor because the greatest energy release is then obtained. If O_2 is not available, the organisms usually select the electron acceptor that gives the next greatest energy release, and this electron acceptor is reduced in the process (Harris, 1982). Based on the redox potentials as given by Lindsay (1979), the usual energy-release order of potential electron acceptors in soil systems at pH 7 is

$$\underline{O_2} > \underline{NO_3^-} > Cu^{2+} > \underline{MnO_2} > \underline{SO_4^{2-}} > \underline{Fe_2O_3} > \underline{CO_2} > MoO_4^{2-} > \underline{H^+},$$

with the underlined elements being the most important electron sinks. In aerobic soils, autotrophic bacteria may oxidize reduced forms of some of these nutrients to obtain energy using O_2 as the electron acceptor.

Redox reactions do not enter in directly to Eq. [2] or Fig. 2–1, but they do affect many of the reactions shown there. Any time the rate of O_2 used in respiration exceeds the transport of O_2 to the respiration site, reduction of alternative electron acceptors proceeds. This results in a change in the chemical species of individual nutrients. In his Tables 13.3 and 13.4, Sposito (1989) distinguishes between plant-nutrient elements that are directly affected by changes in redox potential and those that are indirectly affected. Generally, those directly affected undergo changes in oxidation state in response to changes in the redox status, whereas those indirectly affected undergo changes in concentration or chemical species without a change in oxidation state. Nutrients indirectly affected include those that can be adsorbed or precipitated on Fe(III) or Mn(IV) hydroxy solids (released to the solution if the solid phase dissolves), those that form inorganic sulfides and those that compete with Fe^{2+} or Mn^{2+} for cation-exchange or specific-adsorption sites in reduced systems.

G. Plant Root Effects

Reactions at the root-soil interface affect the environment of the rhizosphere. The plant root is the ultimate sink for both nutrients and water in nutrient-uptake models. In addition to this function, the root changes the rhizosphere environment through exudation of CO_2, H^+, HCO_3^- and organic compounds. The CO_2 is a product of root respiration, whereas H^+ and HCO_3^- are excreted to maintain electrical neutrality when the root absorbs more equivalents of cations than anions (H^+ release) or more equivalents of anions than cations (HCO_3^- release). The change in rhizosphere pH as a result of these excretions can be as much as 2 pH units (Marschner & Romheld, 1983). Rhizosphere pH changes of this magnitude have a marked effect on solubilities of many nutrients. A major factor determining whether H^+ or HCO_3^- will be excreted is whether N is taken up as NO_3^- (HCO_3^- release) or NH_4^+ (H^+ release) or fixed from atmospheric N_2 (H^+ release).

Organic compounds released by roots fall into three categories: low-molecular-weight compounds (free exudates), high-molecular-weight gelatinous compounds (mucigel), and sloughed-off tissues and their lysates (Marshner, 1986, p. 454). Some of the low-molecular-weight compounds are complexing agents, and some of the compounds of higher molecular weight exhibit cation-exchange and specific-adsorption sites. All of the compounds serve as C sources for microbial respiration and cause an increased O_2 demand in the rhizosphere compared with the bulk soil.

II. SUMMARY

Optimum nutrient availability to plants over a growing season requires an adequate sustained supply of absorbable forms of the essential nutrients at the root surface. This involves mobilization of nutrients from labile forms adsorbed on particle surfaces (exchangeable and specifically adsorbed forms), from dissolution of precipitates, or from mineralization of organic forms with subsequent transport to the roots via convection or diffusion. A solute-conservation equation was used to illustrate the reactions that would occur within an infinitesimal volume near an actively absorbing root. These include the transport processes, convection and diffusion, and the source-sink reactions of adsorption-desorption on cation-exchange and specific-adsorption sites, precipitation-dissolution, mineralization-immobilization of organic forms, and volatilization-solution across the air-water interface. Soluble complexed forms enhance transport even though they are not absorbed as such. Redox reactions change the kinds and amounts of dissolved and precipitated nutrient species in addition to affecting root respiration. Exudates from plant roots affect pH, concentration of complexing ligands, and oxygen demand in the rhizosphere. An understanding of how all of these reactions interact is needed in developing a nutrient-uptake model that can be applied over a variety of soil conditions. It is also helpful in designing soil-testing methods and interpreting their results.

REFERENCES

Barber, S.A. 1984. Soil nutrient bioavailability. John Wiley and Sons, New York.

Checkai, R.T., R.B. Corey, and P.A. Helmke. 1987. Effects of ionic and complexed metal concentrations on plant uptake of cadmium and micronutrient cations from solution. Plant Soil 99:335–345.

Collin, M., and A. Rasmuson. 1988. A comparison of gas diffusivity models for unsaturated porous media. Soil Sci. Soc. Am. J. 52:1559–1565.

Corey, R.B. 1981. Adsorption vs. precipitation. p. 161–182. *In* M.A. Anderson and A.J. Rubin (ed.) Adsorption of inorganics at solid-liquid interfaces. Ann Arbor Sci. Publ., Ann Arbor, MI.

Fujii, R., L.L. Hendrickson, and R.B. Corey. 1983. Ionic activities of trace metals in sludge-amended soils. p. 179–190. *In* E.A. Jenne and R.E. Wildung (ed.) Biological availability of trace metals. Chemical estimation, ecological and health implications. Proc. Hanford Life Sci. Symp. 21st, Richland, WA. 1981. Sci. Total Environ. Elsevier Publ. Co., Amsterdam.

Garrels, R.M., and C.L. Christ. 1965. Minerals, solutions, and equilibrium. Harper and Row, New York.

Harris, R. 1982. Energetics of nitrogen transformations. p. 833–890. *In* F.J. Stevenson et al. (ed.) Nitrogen in agricultural soils. Agronomy Monogr. 22. ASA, CSSA, and SSSA, Madison, WI.

Hendrickson, L.L., and R.B. Corey. 1981. Effect of equilibrium metal concentrations on apparent selectivity coefficients of soil complexes. Soil Sci. 131:163–171.

Jost, W. 1960. Diffusion in solids, liquids and gases. Academic Press, London.

Lindsay, W.L. 1979. Chemical equilibria in soils. Wiley-Interscience, New York.

Marschner, H. 1986. Mineral nutrition in higher plants. Academic Press, London.

Marschner, H., and V. Romheld. 1983. In-vivo measurement of root-induced pH changes at the soil-root interface. Z. Pflanzenphysiologie 111:241–251.

McLean, E.O., and M.E. Watson. 1985. Soil measurements of plant-available potassium. p. 277–308. *In* R.D. Munson (ed.) Potassium in agriculture. ASA, CSSA, and SSSA, Madison, WI.

Nye, P.H., and P.B. Tinker. 1977. Solute movement in the soil-root system. Univ. of California Press, Los Angeles.

Paul, E.A., and F.E. Clark. 1989. Soil microbiology and biochemistry. Academic Press, San Diego.

Romheld, V., and H. Marschner. 1986. Evidence for a specific uptake system for iron phytosiderophores in roots of grasses. Plant Physiol. 80:175–180.

Schofield, R.K. 1947. A ratio law governing the equilibrium of cations in the soil solution. Proc. 11th Int. Congr. Pure Appl. Chem. 3:257–361.

Sillen, L.G., and A.E. Martell. 1964. Stability constants of metal-ion complexes. Spec. Publ. 17. The Chemical Soc., London Burlington House, W.I.

Sparks, D.L. 1987. Kinetics of soil chemical processes: Past progress and future needs. p. 61–74. *In* L.L. Boersma et al. (ed.) Future developments in soil science research. SSSA, Madison, WI.

Sposito, G. 1981. The thermodynamics of soil solutions. Oxford Univ. Press, New York.

Sposito, G. 1984. The surface chemistry of soils. Oxford Univ. Press, New York.

Sposito, G. 1989. The chemistry of soils. Oxford Univ. Press, New York.

Tate, R.L. III. 1987. Soil organic matter. John Wiley and Sons, New York.

Veith, J.A., and G. Sposito. 1977. On the use of the Langmuir equation in the interpretation of adsorption phenomena. Soil Sci. Soc. Am. J. 41:697–702.

Wietholter, S. 1983. Predicting potassium uptake by corn in the field using the strontium nitrate soil testing method and a diffusion-controlled uptake model. Ph.D. diss., Univ. of Wisconsin-Madison (Diss. Abstr. 83:21780).

Chapter 3

Soil Sample Collection and Handling: Technique Based on Source and Degree of Field Variability

D. W. JAMES, *Utah State University, Logan*

K. L. WELLS, *University of Kentucky, Lexington*

The reproducibility, or analytical precision, of laboratory tests done on a composite sample from a field may be very high. However, the precision of tests on different samples from the same field is often very low. This is related to both vertical and horizontal variability in the field.

Field researchers have long been aware of soil heterogeneity (Waynick, 1918). Early researchers used an effective but time-consuming method to evaluate soil heterogeneity. This was known as the uniformity trial (LeClerg et al., 1962). The method was simply to plant a crop and manage it uniformly throughout the growing season. The field would be divided into small segments and crop yield measured on each segment. Crop yield variability among segments was taken as the measure of soil fertility heterogeneity in the field. The crop yield data obtained from a uniformity trial were plotted on a map, and field segments having similar yields were connected by smooth lines. This represented a soil fertility isoline or contour map. LeClerg et al. (1962) drew two general conclusions from the early uniformity trials:

1. Soil fertility variations are not distributed randomly, but they are, to some degree, systematic; that is, contiguous field segments are more likely to be alike than are segments separated by some distance.
2. Soil fertility is seldom distributed so systematically that it can be described by a mathematical formula.

The conclusions are equally valid today in describing the fundamental problems associated with soil sampling. Nowadays, soil fertility is most often estimated from chemical analyses of soil samples. Accordingly, soil sampling methodology must be critically evaluated to obtain an understanding of true field characteristics with respect to variability.

 Soil Testing and Plant Analysis, 3rd ed.—SSSA Book Series, no. 3.

Some principles of soil sampling were outlined by Cline (1945) and reviewed by Petersen and Calvin (1965). These authors described the following soil sampling schemes: (i) simple random, (ii) stratified random, and (iii) systematic. They also showed methods of calculating the population means, variances, and confidence limits for the various sampling schemes.

The mean $\bar{x}$, variance S^2, standard deviation (SD), and coefficient of variation (CV) are defined as follows:

$$\bar{x} = \Sigma x/n \qquad \text{CV} = \text{SD}/x$$

$$S^2 = \{\Sigma x^2 - [(\Sigma x)^2/n]\}/(n - 1)$$

$$\text{SD} = \sqrt{S^2}$$

where n = number of individuals sampled.

Rigney and Reed (1945), Reed and Rigney (1947), and Jacob and Klute (1956) used components of S^2 among samples within an area, and S^2 among determinations on the same sample. In general, they concluded that differences between determinations on the same sample were small when compared to differences among samples. Another general conclusion was that different soil properties (e.g., pH, P, K, and lime requirement) had different patterns of variation. These workers also showed how to determine the best or most economical sampling scheme for any given field using estimates of S^2 obtained from previous work on the same field.

Beckett and Webster (1971) reviewed data on the lateral variability of soil properties with particular emphasis on soil series and soil mapping. They showed that S^2 and CV increased with the size of the area sampled. They reported, in addition, that up to one-half of the S^2 in a whole field may be present within any square meter area of the field.

The foregoing discussion illustrates the historical usage of classical statistics in describing field variability. Unfortunately, statistical tools commonly applied have severe limitations for some purposes. First, use of the SD assumes that the samples represent a random population. In the field, this is rarely, if ever, true because soil variability is patterned, as stated above (LeClerg et al., 1962). Second, the statistical mean should be accompanied by information on the range of values, low to high, and also the *location* in the field of the low and high values of the soil property in question.

A systematic, i.e., nonrandom, sample is required to determine the field locations and distributions of low and high values for any soil parameter. This approach is obviously more time consuming and requires analysis of more soil samples than the simple random (composite) sample or some variation of the stratified sample. James and Dow (1972), Dow and James (1973), and Dow et al. (1973) demonstrated the utility of intensive (e.g., square grid) sampling in fields that have extensive areas of subsoil exposure. The simple composite sample, in these cases, had practically no value in managing the whole field for uniformly high crop production.

To overcome wide variations in soil fertility levels resulting either from land-forming processes or expansion of field boundaries to include areas that were previously managed differently, a sampling and statistical procedure known as Regionalized Variable Theory (RVT) has recently been evaluated as a technique for making better fertilizer recommendations. Use of RVT requires some type of intensive field sampling procedure. With this technique, it is possible to predict S^2 at any *unsampled* point *in that field* without further sampling. The limitation to this technique is that results are specific to the field on which RVT was used and cannot be readily extrapolated to other fields. In other words, RVT results are site specific. However, use of this technique holds promise for better estimating fertilizer needs for situations mentioned above.

Since RVT is a new technique, a brief explanation of the process is in order. Use of RVT includes two processes, semivariograms and kriging, for segregating total field variability into three or four categories, namely nugget, sill, drift, and residual or truly random S^2. These properties are illustrated in Fig. 3-1 and 3-2. An example of the application of RVT to soils work is that of Webster and Burgess (1980) and Assmus et al. (1985). Up to this point, RVT has been used more frequently in assaying soil physical parameters (Folorunso & Rolston, 1985; Kachanoski et al., 1985; Vieira et al., 1981) but it is finding its way into soil fertility and plant nutrition studies as well (Knighton, 1983; Knighton & James, 1985; Tabor et al., 1985). Sabbe and Marx (1987) present an excellent detailed explanation on use of RVT for sampling fields for use in evaluating soil fertility needs.

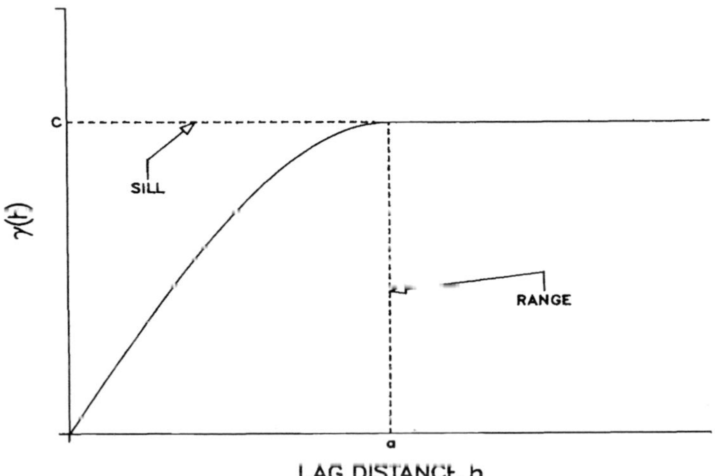

Fig. 3-1. Shape of the semi-variogram using the spherical model. Semi-variance is dependent on distance (h) between lag distance zero and *a*. At lag distance *a* and beyond semi-variance is constant, this is also referred to as the sill. Lag distance is distance between any selected reference point and any other point in the field. (Adapted from Clark, 1979.)

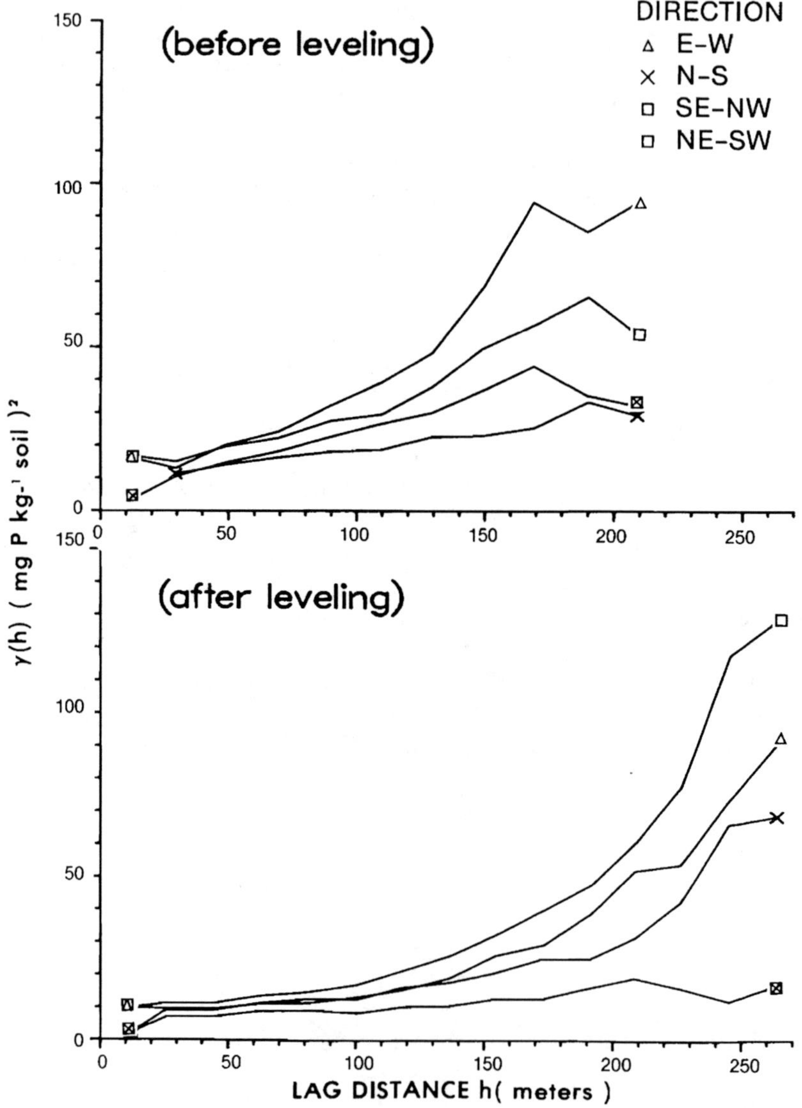

Fig. 3-2. Semi-variograms of soil test P for four directions before and after land leveling. The nugget effect is demonstrated here where the semi-variance is not zero at zero distance. Drift is demonstrated by increasing semi-variance with distance. In these exmaples drift varied with direction. (Adapted from Knighton & James, 1985.)

I. SOURCE AND DEGREE OF FIELD VARIABILITY

Soil variability may be classified as to vertical and horizontal direction; short, medium, or long distance; and whether the variability is natural or induced by human operations on the land.

A. Natural Variation

Soil-forming processes may cause sharply contrasting differences in the soil profile, particularly the A and B horizons. These differences relate to organic matter, pH, texture, cation-exchange capacity, and plant nutrient availability. In humid-subhumid soils, the pH usually stays the same or decreases with depth. In arid-semiarid soils, however, calcium carbonate tends to accumulate in the subsurface layers. Interbedded rock parent materials can also result in sharply contrasting differences among soils in the same field.

Ordinarily, soil samples represent the surface or plow layer, but if the subsoil is exposed by erosion, land leveling, or smoothing, sharp differences in soil properties may occur in very short horizontal distances. This may severely compound the problem of obtaining a "representative" sample of the field. An awareness that the potential exists for such variability is obviously a prerequisite to designing a proper sampling procedure.

Natural variation across somewhat longer distances is associated with slope and aspect. Soil tends to be shallower on the crest of knolls and deeper on the upland flats and lower slopes. Aspect may affect absorption and reflection of solar radiation, which in turn, affects soil temperature, water relations, and associated plant growth. Soil sampling across gradients in slope and aspect will result in an average that may mask important soil differences.

B. Human Variation

Tilling of the land and managing crops can have profound effects on field variability. The practice of land leveling, frequently done to facilitate surface irrigation in drier regions, or terracing in humid regions can result in induced variation. Tillage practices, too, will affect soil erodability and have the tendency to expose subsoils.

Different cropping systems and associated fertilizer management practices may lead to differences in organic matter amount and type as well as different residual fertilizer patterns.

Peculiar kinds of variability can be induced by fertilizer application methods. Jensen and Pesek (1962a, b) illustrated problems in crop productivity associated with non-uniform fertilizer application. It is not usually difficult to obtain uniform fertilizer distribution if the proper equipment is used. Even though fertilizer may be uniformly applied, there may be a significant degree of induced small-scale heterogeneity. The following examples illustrate these points.

Consider P fertilizer applied uniformly as a topdressing without any mixing or incorporation. The result is a large difference, depending on the rate of fertilization, between a thin P-enriched soil 2- to 3-cm layer at the surface and the soil immediately below the surface. The topdressed P may be incorporated by discing or plowing. The effect will be some redistribution with depth depending on the type of tillage involved. But tillage may or may not lead to uniform vertical distribution of fertilizer to the depth at which it is incorporated (Hurlburt & Menzel, 1953).

Another sharp contrast in soil variability may be observed when fertilizer is injected, banded, and sidedressed. This leads to high variability along lines normal to the direction of fertilizer application. This cyclic variability results in high concentrations of the fertilizer element at the point of injection, separated by somewhat uniform concentrations of the element between the lines of injection. This kind of variability will be diminished by subsequent tillage operations (e.g., plowing) that tend to stir the plow layer.

The nature of the applied fertilizer itself has an effect on soil heterogeneity. For example, nitrogen, despite the specific form applied, tends to nitrify in soil over time to nitrate, a completely soluble form. Nitrates diffuse in the soil solution with infiltrating water, leading to more soil homogeneity in comparisons with the high heterogeneity induced at the time of fertilizer application. It is apparent that soil nitrate may vary greatly over time at different points in the soil profile following fertilizer applications.

Other fertilizer elements contrast sharply with N in terms of soil mobility. For example, potassium reacts readily with the exchange complex and, except in soils with very low cation exchange capacity, will essentially be immobilized. Phosphorus fertilizer reacts rapidly upon soil application with other chemical constituents (e.g., Ca, Al, and Fe) to form fairly insoluble and, therefore, immobile compounds. Generally, researchers agree that P will diffuse in the soil away from the fertilizer pellet and enrich a soil volume no more than about two or three diameters of the original pellet. The result is very high heterogeneity on a microscale. Again, tillage operations completed after P fertilizer equilibration will mix the P-enriched soil.

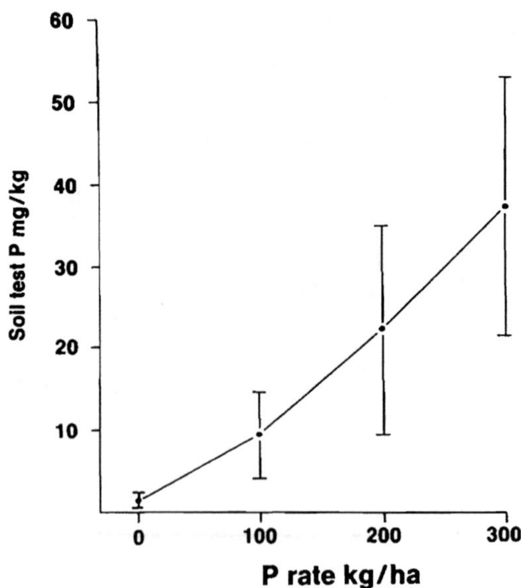

Fig. 3-3. Micro-variation: variability of soil test P in 30 soil cores of 2.5-cm diam. as related to P fertilizer rate. The fertilizer was applied broadcast-plowdown in spring and the soil was sampled in autumn. Vertical line segments represent the 95% confidence limits. (Adapted from James & Dow, 1972.)

C. Point-to-Point Variability

Soil differences between points on the landscape present the basic challenge in designing an effective soil-sampling procedure. Soil variability is classified into three distance categories—micro-, meso-, and macro-variability to describe appropriate soil-sampling technique.

Micro-variation refers to soil variation between points separated from 0 to 0.05 m. Figure 3–3 illustrates this type of variability. The data of Fig. 3–3 were obtained by analyzing individual soil cores, that were about 2.5 cm in diameter and 25-cm deep, for available P ($NaHCO_3$ method). Figure 3–3 shows that the average effect of fertilization was a smooth curvilinear increase in soil test P (STP). But STP varied widely (i.e., micro-variation) among soil cores on all plots except the ones receiving no fertilizer.

Meso-variation is soil variation between points separated by 0.05 to 2 m. This kind of variability is demonstrated in Fig. 3–4 which shows that distribution of STP along a 1.78-m sampling transect that covered two potato (*Solanum tuberosum* L.) rows. Extreme soil variability on the mesoscale may be induced by fertilizer placement.

Macro-variation is soil variation between points separated by distance >2 m. Macro-variation is associated with natural soil processes, but human's soil management practices also can have a significant influence on this type of variation. Figure 3–5 illustrates management-induced macro-variability. The mean whole-field STP was 8.40 mg kg^{-1} with SD = 4.84. A moderate rate of fertilizer P would be recommended for this level of STP. However, 19.2% of the field was very high in available P and would yield no immediate benefit to a fertilizer application. On the other hand, 34.8% of the field was low in STP needing a moderately high rate of fertilizer P. In addition, 18.4% of the field was very low in P and needed a high rate of fertilizer

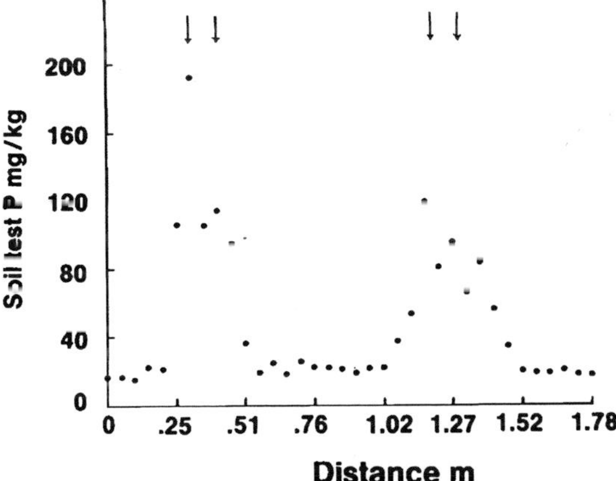

Fig. 3–4. Meso-variation: scatter diagram of soil test P in individual cores (2.5-cm diam.) collected on a transect that spanned two potato rows. Arrows indicate approximate positions of sidedressed P fertilizer at planting time. Soil samples were collected in autumn after harvest. Dispersion of soil test P values was related to local mixing by the digging equipment at harvest time. (Adapted from James & Dow, 1972.)

to completely eliminate soil fertility as a limiting factor to crop growth. In other words, if the field mean STP was the only criterion for fertilization, only 27.6% of the whole field would receive the appropriate amount of fertilizer. Thus, sampling methodology clearly has significant practical implications, especially where the field is very heterogeneous.

Another problem associated with low-P soil in Fig. 3–5 is the soil-texture differential between surface and subsoil. The surface soil at this site is clas-

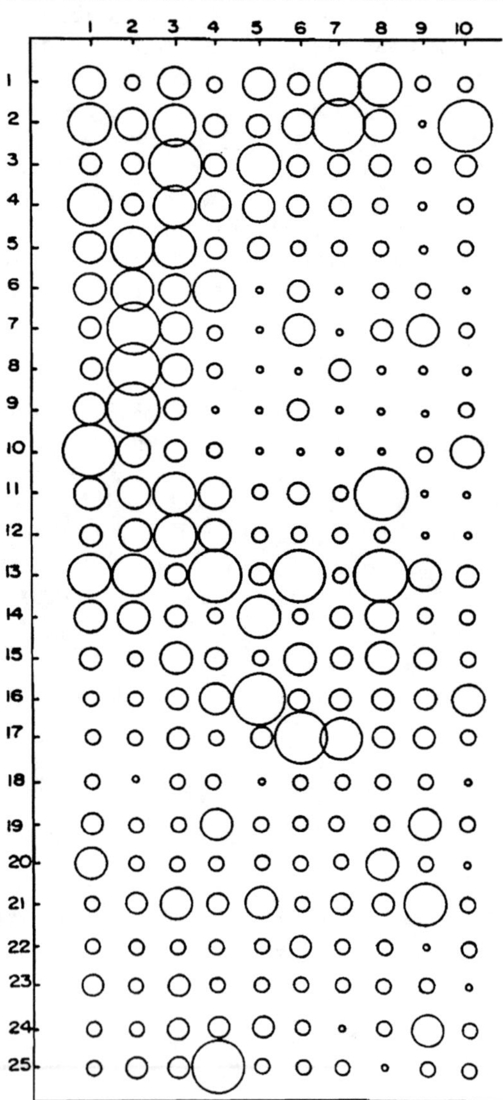

Fig. 3–5. Macro-variation: soil test P variability determined from point soil samples collected on a 15.2-m square grid after land leveling. Sampling column and row numbers are indicated on margins. For data shown, no. of samples = 230, $\overline{X}$ = 12.6 mg of P kg^{-1}, SD = 62.43, range = 2 to 46. The circles, small to large, represent the STP categories <5, 6 to 10, 11 to 15, 16 to 20, and >20 mg of P kg^{-1}. (D.W. James, 1983, unpublished data.)

sified as silty clay loam. The subosil is clay. Because of the variable texture in the smoothed field, surface irrigation water infiltration rate and soil water-holding capacity would be as variable as the STP. Limitations in P availability may be overcome through judicious applications of fertilizer. However, no practical treatment is known that will decrease the spatial variability of soil water intake rate and holding capacity, at least in the short term.

II. SAMPLING METHODS

Soil-sampling procedure should be adapted to the degree of variability on all scales, micro-, meso- and macro. In this sense, it is apparent that some fore-knowledge of field conditions is required to determine what is appropriate. This fore-knowledge would include such things as exposure of subsoil from land leveling or erosion; kind and amount of fertilizer applied in previous seasons, and whether the fertilizer was applied broadcast, sidedress, band; and the degree of soil mixing that may have occurred subsequent to fertilization by plowing, disking, tillage to control weeds, and rowhilling. Sampling techniques are broadly defined in the following discussion of uniform and non-uniform field conditions.

A. Uniform Fields

Uniform fields signify field areas that are similar in regard to slope, aspect, management history, cropping and fertilization, and if possible, crop appearance during the growing season. *Uniform* requires that macro- and meso-variations be nonsignificant. Sampling procedures that fully satisfy these conditions involve the collection of randomly selected soil cores that are mixed together into one sample. This is the customary random composite soil sample.

To be truly random, the soil cores should be taken from field segments within the uniform area that are randomly selected. It is usually not practical, however, to divide a field area into segments and, after numbering them, sample a randomly selected set of segments. In practice, it is common to collect soil cores by following a zigzag path where a conscious effort is made to force the path into corners and along edges as well as the central parts of the area being sampled.

Thus, realistically, "random composite soil samples" are subject to judgment in terms of both field segregation into uniform areas and the path followed to collect the sample. When the field history is well known, this type of sampling scheme can be completely adequate in terms of developing an appropriate fertilizer management program.

Micro-variation will be adequately controlled if the number of soil cores collected is sufficient, e.g., 25 to 30 cores per sample and if the composited cores are thoroughly crushed and mixed.

B. Non-uniform Fields

Nonrandom sampling: Where macro-variation is large, a nonrandom soil sampling procedure is recommended. A major objective in nonrandom sampling is to understand not only the average field condition but also the extreme high and low values, and what is more important, the specific locations of the field extremes. By its nature, nonrandom soil sampling requires numerous point soil samples. To do this, a field grid is developed by placing marks at regular intervals in two directions and collecting soil samples at the grid line intersections. Spacing between grids will vary with the degree of detail needed to satisfy the sampling objectives. Typical grid spacings will lie between 15 and 30 m. Within this range, the number of samples per hectare would lie between 45 and 12 depending on where the beginning grid point is with respect to field boundaries.

A point soil sample is collected from each grid point. This consists of 8 to 10 cores from a circle of about 1-m diam. centered over the point. The number of soil cores should provide a total soil volume of practical size to perform the needed chemical or physical analyses. Eight to 10 cores are usually sufficient to overcome micro-variation.

Each point sample is analyzed and the results are plotted on a field map in relation to their grid location. Soil test isolines (contours) are then drawn which stratify the field into selected soil test ranges. After establishing the field location of soil test categories, fertilizer rates can be adjusted for each field stratum. As an example, soil test P strata can be visualized in the field represented by Fig. 3–5. If, in this example, rates of P fertilizer are applied appropriate to the STP conditions, then the STP will become more homogeneous and the intensive grid-sampling procedure would not need to be repeated in subsequent years.

Random sampling: High meso-variation, the type of soil variability peculiar to injection or band-applied fertilizer, has always been a special challenge in regard to obtaining representative soil samples. With minimum tillage and no-tillage practices becoming more widespread meso-variation must be studied more intently if soil sampling is to be adequate for diagnostic soil testing.

Band-applied fertilizer generates two interlacing populations consisting of (i) the bulk soil not affected by fertilization, and (ii) compact bands spaced at regular intervals that are highly enriched by fertilization. The bulk soil would be characterized by one set of $\bar{X}$ and SD values and the enriched band soil would be characterized by another set of $\bar{X}$ and SD values. As a practical matter, these two populations need to be treated as one so as to obtain an adjusted average soil fertility test value that truly reflects the overall conditions encountered by plant roots.

Soil sampling of a field with significant meso-variation requires many soil cores to obtain the proper averaging of the two soil populations. The soil sample is analogous to the above-described random composite sample except that a much greater number of soil cores are needed. In addition, the primary sample would be much larger, requiring special handling to thoroughly crush and mix the collected soil cores followed by sample reduction through a multiple splitting process.

The specific number ofsoil cores needed to satisfy the conditions of significant meso-variation requires further investigation but it is evident that it will depend on the following parameters:

1. Distance between fertilizer bands.
2. Width of the fertilizer band.
3. Diameter of the sample coring tube.
4. $\bar{X}$ and SD within the band.
5. $\bar{X}$ and SD of the bulk soil between bands.

It is apparent that the number of individual soil cores in the intensive composite sample will need to be four- to fivefold larger than the number needed for the customary random composite sample.

III. DEPTH OF SAMPLING

Soil-sampling depth is most often the plow layer, e.g., 0 to 17 cm (0–6⅔ in.) or 0 to 20 cm (0–8 in.). Traditionally, the plow layer is generally estimated to weigh 357 Mg ha^{-1} (2 000 000 lb/acre). Many studies have been reported in the literature that show incorporation of lime and the immobile fertilizer nutrients to plow depth on acid and infertile soils to be effective in increasing crop yields as compared to not liming or fertilizing. In a review of research reported on depth of lime incorporation, Wells (1980) concluded that, with few exceptions, nearly all yield increase because of liming was obtained by incorporation within the plow layer. Deeper incorporation of lime rarely improved yields. As pointed out by Shoemaker (1964), this depth (volume) factor needs to be considered when sampling shallower or deeper than the plow layer to adjust rates of lime and fertilizer to the proper soil volume basis. The concern about adjustment of recommended rates on a soil volume basis is because of widespread use of tillage implements which till deeper than 17 to 20 cm. There is as much or more concern about proper manipulation of rates and volume with advent in recent years of the various conservation tillage practices that do not mix lime and fertilizer throughout the traditional plow layer.

A. Conventional Tillage

In conventional tillage, soil is somewhat mixed to the depth to which the primary tillage implement penetrates and additionally mixed in the surface 7 to 10 cm (3–4 in.) by secondary tillage as with a disk harrow. By this process, lime and the immobile fertilizer nutrients become somewhat uniformly mixed throughout the plow layer over a few years. Figures 3–6 and 3–7, made from data reported by Randall (1980), show the effect of the tillage method on long-term soil test values for P and K. As shown, concentra-

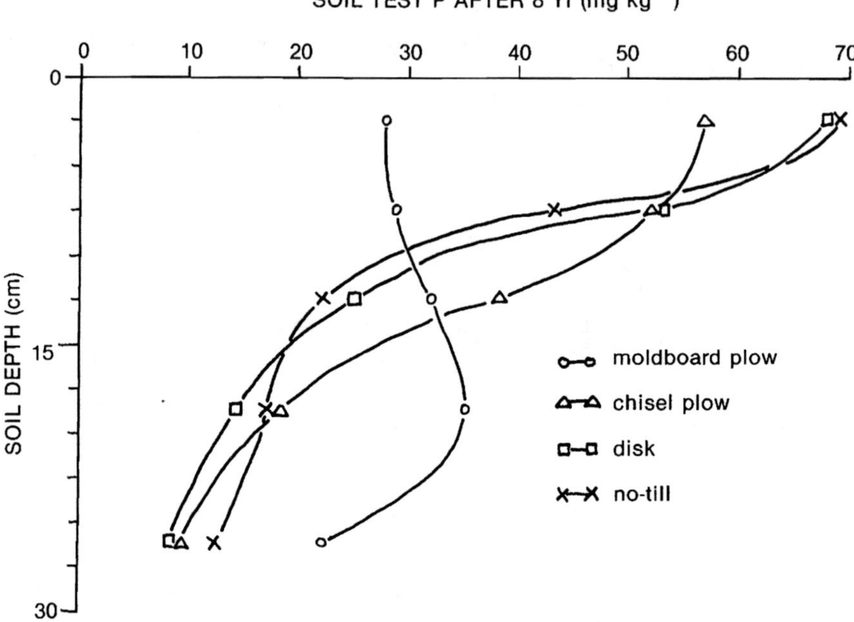

Fig. 3-6. Effect of tillage system on distribution of soil test P levels. (Adapted from Randall, 1980.)

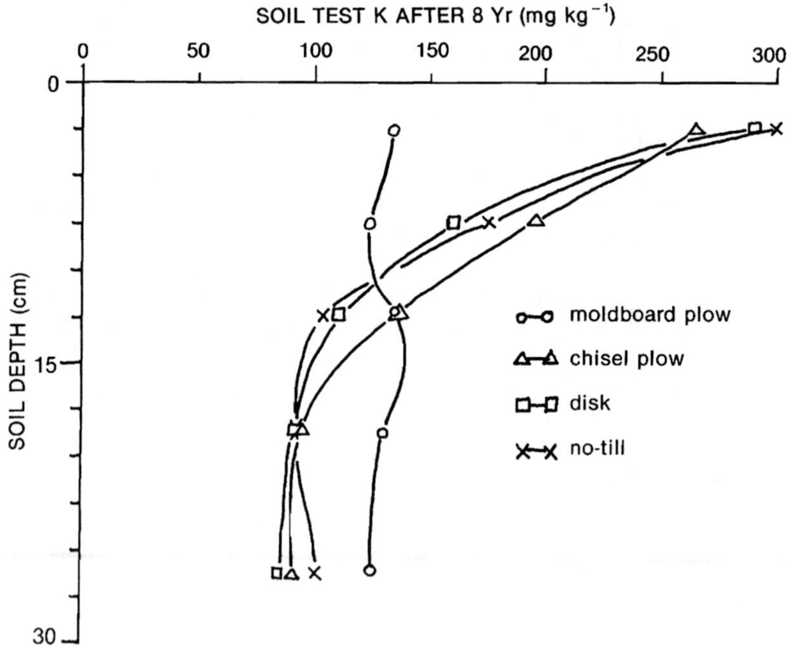

Fig. 3-7. Effect of tillage system on distribution of soil test K levels. (Adapted from Randall, 1980.)

tion of available P and K is much more uniform throughout the plow layer with conventional plow-disk tillage as compared to the conservation tillage methods of chisel plow-disk, disk only, or no till. For routine sampling of conventionally tilled soils, the plow layer sample depth is considered satisfactory for determining level of soil acidity and immobile nutrients.

In the semiarid and some humid areas, however, where sampling for N content is recommended, deeper samples are necessary. As Meisinger (1984) summarized, recommended depths for soil nitrate sampling range from 30 to 180 cm (12–72 in.), but most often from 60 to 120 cm (24–48 in.). These depths correspond to those suggested by Nelson et al. (1967), James (1971, 1978), and James et al. (1977) for nitrate testing in irrigated soils. Variation in depth recommended for deep sampling for nitrates was because of site-specific factors such as type of crop, type of soil, and prevalent soil conditions.

B. Conservation Tillage

Concern over soil and water conservation has resulted in development of several systems which involve much less mechanical tillage than the traditional plow-disk systems. This results in mixing of a much smaller volume of soil and a concentration of lime and immobile nutrients at shallower depths than from conventional tillage systems. Although there are several conservation tillage systems being used, chisel plowing followed by spring disking, disking only, and no-till are the most common. Ridge-till is becoming important in some areas. The effect which continuous use of these systems has on depth to which P and K are incorporated is shown in Fig. 3–6 and 3–7. These results are similar to other published reports about influence of conservation tillage on depth to which lime and immobile nutrients are incorporated. Both disk only and no-till result in a highly concentrated layer of soil test extractable P and K in the surface 0 to 7.5 cm (0–3 in.) with much lower concentrations below 7.5 cm (3 in.) as compared to conventional moldboard plow tillage. Chisel plowing incorporates P and K a little deeper than no-till or disking, but not as deep as moldboard plowing.

This raises the question of how to sample conservation tillage fields to best determine lime and fertilizer needs. Whitney (1982) reported on a survey from several states, that most had modified their recommended sampling depth for conservation tillage to shallower depths than the traditional plow layer depth. He suggested sampling such fields to a depth of 5 to 10 cm (2–4 in.). Such shallow sampling, particularly for no-till planted crops is currently recognized as more effectively evaluating soil fertility status than deeper sampling. Wells (1985) has reported good corn (*Zea mays* L.) yields from surface application of P and K on low testing no-till fields even though test levels below 7.5 cm (3 in.) remain low. Mengel (1982) showed increased root distribution in the surface layer of no-till corn fields which presumably accounts for this.

In addition, this shallower sampling procedure more effectively defines surface acidity levels that have been shown by Kells et al. (1980) to influence activity of some herbicides, particularly the *s*-triazines. Deeper sampling dilutes the extremely acid conditions that develop in the soil immediately below the surface mulch residues which accumulate in continuous no-till systems. In some cases, even shallower sampling at the 0 to 5 cm (0–2 in.) depth is used to more effectively describe the surface acidity of no-till fields. Mengel (1982) recommended two sampling procedures for conservation tillage corn: (i) sample the 0 to 10 cm (0–4 in.) and the 10 to 20 cm (4–8 in.) depths for all reduced tillage fields and on all no-till fields where N is injected below the soil surface, and (ii) sample no-till fields where N is surface applied at the 0 to 5 cm (0–2 in.) and 5 to 20 cm (2–8 in.) depths.

Another conservation tillage technique, ridge-till, is a rapidly growing practice in some areas and presents a unique problem in soil sampling. As described by Randall (1982), surface-applied nutrients are swept from row middles onto the ridge during the ridging process, enriching the ridge fertility at the expense of the row middles. This management practice leads to a type of meso-variability, and intensive sampling would be needed to obtain a representative sample. Moncrief et al. (1984) recommend a 0 to 15 cm (0–6 in.) sampling depth for ridge-till. They recommend sampling after planting but before ridging to lower variability. If sampling cannot be done before ridging, they suggest taking samples half way up the ridge.

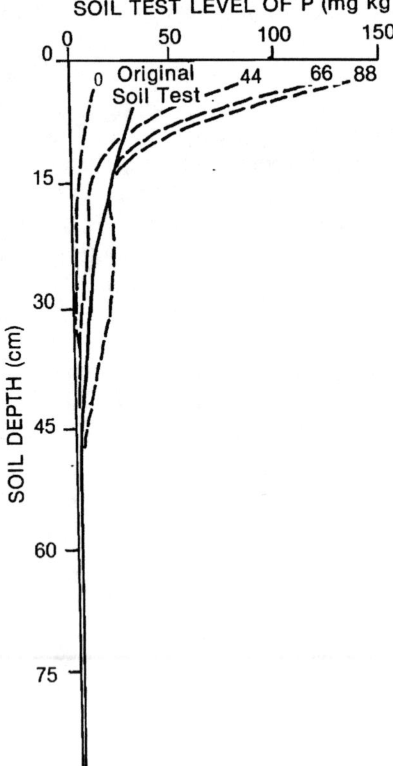

Fig. 3–8. Soil test P levels in an alfalfa sod following 6 yr annual topdressing of P at 44, 66, and 88 kg ha^{-1}. (Adapted from Vaught et al., 1977.)

Currently, the prevailing thought in sampling conservation tilled fields is to sample at a shallower depth than for conventional tillage to monitor surface acidity and buildup of immobile nutrients. Concurrent sampling below the shallow sample to a depth of 15 to 20 cm (6–8 in.) enables monitoring of nutrient movement deeper into the soil and the degree of stratification.

C. Permanent Sod

Vast acreages of sodland exist throughout the USA that are fertilized and managed for hay or pasture production. Typically, sampling of the traditional plow layer has been recommended when fields are conventionally plowed to establish the various sod-forming crops. Lime and fertilizer are often recommended for topdress application onto sods between intervals of establishment or re-establishment. Annual topdressings of sodland, as with conservation tillage, result in a buildup of immobile nutrients in the surface 0 to 5 cm (0–2 in.). Wells and Parks (1961) and Vaught et al. (1977) showed that annual topdressings of P and K to alfalfa on silt loam surface-textured soils resulted in little movement of P and K below 7.5 cm (3 in.) even at high annual rates over several years. Figures 3–8 and 3–9 from Vaught et al. (1977)

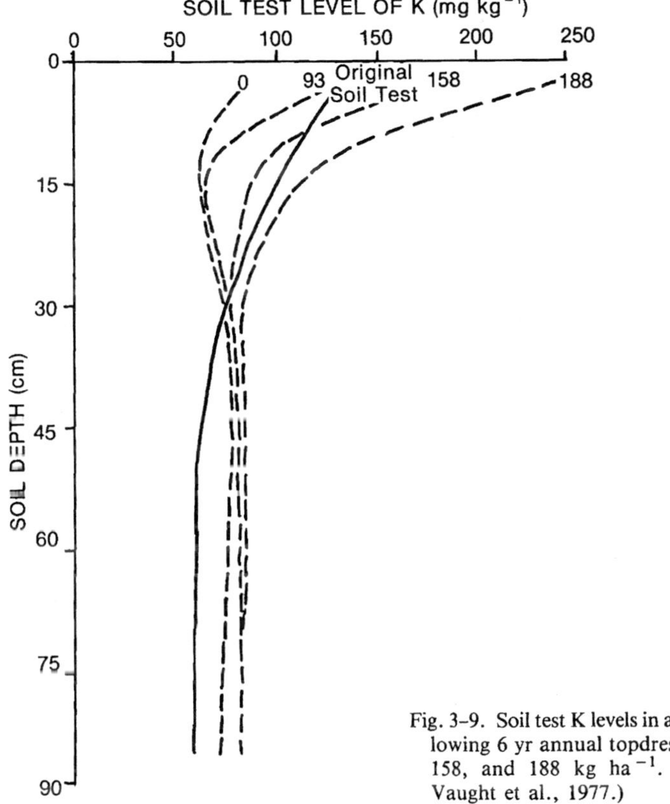

Fig. 3–9. Soil test K levels in an alfalfa sod following 6 yr annual topdressing of K at 93, 158, and 188 kg ha^{-1}. (Adapted from Vaught et al., 1977.)

illustrate soil-profile distribution of soil test P and K levels after 6 yr of annual topdressings. High yields of 13.4 t alfalfa dry wt. ha^{-1} (6 ton/acre) were maintained at the lowest level of fertilizer P and K applied annually.

With the advent of no-till drills for seeding forage crops in recent years and with shallow tillage to 7.5 to 10 cm (3–4 in.) often used to renovate pastures, many sod fields rarely have lime and fertilizer mixed into the traditional plow layer by primary tillage tools. Consequently, Thom et al.(1982) in Kentucky recommend sampling of such fields only to the 10 cm (4 in.) depth.

IV. EFFECT OF TIME OF YEAR ON SAMPLE COLLECTION

As discussed by Peck and Melstead (1973), seasonal variation in soil test values should be expected because of variations in factors that influence mineral accumulation by plants and recharge of the soil solution content of minerals as plants remove them from solution. They point out that it is difficult to quantify how much this can change soil test values at different times of the year. Their review of published literature indicated that soil acidity increased in acid soils during the growing season (May–September) but did not consistently vary in alkaline soils. Data from tobacco experiments in Kentucky by Reneau et al. (1968) on an acid soil showed similar results. Soil samples of the 0 to 15 cm (0–6 in.) depth of non-irrigated unfertilized soil taken at weekly intervals from April through August showed that pH values dropped from an initial value of 6.0 in April to 5.3 by mid-May and varied between 5.3 and 5.6 throughout the remainder of the summer. When fertilizer was applied, the resultant salt effect dropped pH to 5.1 to 5.2 by mid-June where it remained throughout July and August. On soil that was irrigated, pH dropped to the 4.8 to 5.2 range. Additional studies on tobacco by Sims and Atkinson (1974) showed similar results, with soil pH dropping 0.2 to 0.6 units during the summer, the magnitude of drop being influenced by fertilization. It would seem reasonable to expect a seasonal drop in pH values from spring to autumn because of the soluble salt effect resulting from mineralization of organic residues, weathering of minerals, or application of soluble fertilizers. In areas where there is enough rainfall during the winter to leach soluble salts deeper into the soil, pH values would generally be expected to return to near the level from which they dropped.

Although there is little published data on seasonal effect on soil test readings for P and K, the general consensus is that they would likely decline somewhat from spring to autumn. Childs and Jencks (1967) reported from studies on three soils in West Virginia that P and K soil test values declined progressively from annual high values in November to December to annual low values in July. The differences in readings were great enough to influence the amount of P and K fertilizer recommendations for crop production. Lime requirement and pH tests in their study showed the same seasonal effect as previously described.

In general, soil samples taken in late summer or early fall will probably test lower in pH and plant-available P and K than if taken any other time of year. Highest readings would likely be measured on samples taken during the winter or early spring months. Recommended rates of lime and fertilizer based on late summer or early fall samples would likely be higher than those based on winter or early spring samples. Particularly for medium or lower testing soils in P and K and on very acid soils, the late summer or early fall samples may more accurately reflect the need for lime, P, and K than samples taken during the winter or early spring.

V. SOIL SAMPLING TOOLS

A wide array of sampling equipment is available which will perform well in pulling soil samples, ranging from hand tools to vehicle-mounted hydraulic driven power probes or augers. The main consideration in selecting a tool for sampling is that it can easily be cleaned between composite samples taken in a field and that it accurately samples the intended depth. Hand probes (tubes) are probably the most widely used tool. Preferably, they should be of stainless steel construction to minimize oxide contamination of samples, which is particularly important in sampling for micronutrients such as Fe and Zn. Auger-type sampling tools (screw or bucket augers) are more effective than probes or tubes in sampling gravelly or rocky soils.

The container into which composite soil samples are placed is an extremely important component of sampling. A clean, plastic bucket is probably the best container to use since it is light weight, easily cleaned, will not contaminate the sample with oxides, and can easily be used to mix up the composite samples taken in each field. Peck and Melsted (1973) described good sampling equipment as that which should:

1. Take a small enough equal volume of soil from each subsampling site so that the composite sample will be of an appropriate size to process for analysis.
2. Be easy to clean.
3. Be adaptable to dry sandy soil as well as moist sticky soil.
4. Be rust resistant and durably constructed to resist bending or breakage.
5. Be relatively easy to use and thus provide for fairly rigid sampling of a field.

As they point out, the most important feature of a sampling tool is that it will provide uniform cores or slices of equal volume at all spots within the composite sampling area.

VI. HANDLING SAMPLES

A. Prevention of Contamination

Great care should be taken to prevent sample contamination during the process of collection and sampling handling. Common sources of contamination are dirty sampling tools, dirty containers, cigarette or pipe ashes, drying samples on dirty paper or in dusty places. Galvanized metal containers should never be used for samples to be tested for Zn. Empty coffee cans can be a major source of Zn contamination. Likewise, samples to be tested for B should not be placed in or on Kraft paper bags since such paper can be a B contaminant. The best preventative for contamination is use of clean tools, clean plastic buckets, clean plastic bags, and use of containers supplied by testing laboratories for packaging the sample to be sent to the lab.

B. Mixing

The subsamples or cores of soil taken from a field should be thoroughly mixed. Probably the most difficult subsamples to mix are those which were taken either too wet or too dry for crumbling by hand. Samples too dry to break apart by hand must be crushed in some manner before the subsamples can be mixed well, while samples taken too wet must be dried to the point they can either be hand or mechanically crushed. If soil is at the proper moisture content to permit hand crushing, the final mixing of the subsamples can be made easier by crushing each sample as it is taken from the sampling tool and placed into the carrying container. The basic principle to keep in mind is that in reducing the volume of subsamples taken from 1 to 4 L (approx. 1–4 qt) to about one-half liter (approximately 1 pint) for packaging and sending to the laboratory, the amount packaged must represent an average composition of all the subsamples taken. Unless the subsamples are uniformly crushed to a fairly small particle size (2–5 mm) (1/8 to 1/4 in.), it will be difficult to thoroughly mix the subsamples into one homogeneous composite sample.

C. Drying Samples

If samples taken from a field are too wet to crush for mixing, they should be dried sufficiently for crushing to a friable, uniform particle size to prepare the composite sample. It is not necessary to take the sample to complete dryness for this purpose. Wet samples should be dried at temperatures no greater than 35 to 50 °C (approximately 100–120 °F). Higher drying temperatures can alter the nutrient solubility of the organic and mineral fractions of the soil. Great care should be taken to prevent contamination during the drying process.

D. Splitting the Composite for Laboratory Analysis

As previously mentioned, the composite sample taken from a field will be too large for sending to the soil testing lab, and must be split into a smaller volume. This requires that the composite sample be uniformly mixed prior to splitting. One unbiased technique to use for sample splitting would be to split the mixed composite into halves, split one of the halves in half and if still too large for the sample container, split the sample in half again to obtain a subsample of the whole composite for sending to the laboratory. Several variations of such a splitting technique are possible.

REFERENCES

Assmus, R.A., P.E. Fixen, and P.D. Evenson. 1985. Detection of soil phosphorus spatial variability through the use of semivariograms and strip sampling. J. Fert. Issues 2(4):136–143.

Beckett, P.H.T., and R. Webster. 1971. Soil variability: a review. Soils Fertilizers 34:1–15.

Childs, F.D., and E.M. Jencks. 1967. Effect of time and depth of sampling upon soil test results. Agron. J. 59:537–540.

Clark, I. 1979. Practical geostatistics. Applied Sci. Publ., London.

Cline, M.G. 1945. Principles of soil sampling. Soil Sci. 58:275.

Dow, A.I., and D.W. James. 1973. Intensive soil sampling: A principle of soil fertility management in intensive irrigation agriculture. Washington Agric. Exp. Stn. Bull. 781.

Dow, A.I., D.W. James, and T.S. Russell. 1973. Soil variability in central Washington and sampling for soil fertility tests. Washington Agric. Exp. Stn. Bull. 788.

Folorunso, O.A., and D.E. Rolston. 1985. Spatial and spectral relationships between field-measured denitrification gas fluxes and soil properties. Soil Sci. Soc. Am. J. 49:1087–1093.

Hammond, L.C., W.L. Pritchett, and V. Chew. 1958. Soil sampling in relation to soil heterogeneity. Soil Sci. Soc. Am. Proc. 22:548–552.

Hurlburt, W.C., and R.C. Menzel. 1953. Soil mixing characteristics of tillage implements. Agric. Eng. 34:702–708.

Jacob, W.C., and A. Klute. 1956. Sampling soils for physical and chemical properties. Soil Sci. Soc. Am. Proc. 20:170–172.

James, D.W. 1971. Soil fertility relationships of sugarbeets in central Washington. Washington Agric. Exp. Stn. Tech. Bull. 68.

James, D.W. 1978. Diagnostic soil testing for nitrogen availability: The effects of nitrogen fertilizer rate, time, method of application and cropping pattern on residual soil nitrogen. Utah Agric. Exp. Stn. Bull. 497.

James, D.W., and A.I. Dow. 1972. Source and degree of soil variation in the field: the problem of sampling for soil tests and estimating soil fertility status. Washington Agric. Exp. Stn. Bull. 749.

James, D.W., F.J. Francom, R.F. Wells, and D.C. Sisson. 1977. Control of soil fertility for high sugarbeet yield and quality. Utah Agric. Exp. Stn. Bull. 496.

Jensen, D., and J. Pesek. 1962a. Inefficiency of fertilizer use resulting from non-uniform spatial distribution: I. Theory. Soil Sci. Soc. Am. Proc. 26:170–173.

Jensen, D., and J. Pesek. 1962b. Inefficiency of fertilizer use resulting from non-uniform spatial distribution: II. Yield losses under selected distribution patterns. Soil Sci. Soc. Am. Proc. 26:174–178.

Kachanoski, R.G., D.E. Rolston, and E. de Jong. 1985. Spatial variability of a cultivated soil as affected by past and present microtopography. Soil Sci. Soc. Am. J. 49:1082–1087.

Kells, J.J., C.E. Rieck, R.L. Blevins, and W.M. Muir. 1980. Atrazine dissipation as affected by surface pH and tillage. Weed Sci. 28:101–104.

Knighton, R.E., and D.W. James. 1985. Soil test phosphorus as a regionalized variable in leveled land. Soil Sci. Soc. Am. J. 49:675–679.

Knighton, R.E. 1983. Changes in variability of soil test phosphorus with land leveling. M.S. thesis. Utah State Univ., Logan.

LeClerg, E.L., W.H. Leonard, and A.G. Clark. 1962. Field plot technique. 2nd ed. Burgess Publ. Co., Minneapolis.

Leo, M.W.J. 1963. Heterogeneity of soil sampling. J. Agric. Food Chem. 11:432–434.

Meisinger, J.J. 1984. Evaluating plant available nitrogen in soil-crop systems. p. 391–416. *In* Nitrogen in crop production. ASA, Madison, WI.

Mengel, D.B. 1982. Developing fertilizer programs for conservation tillage. *In* Proc. Indiana Plant Food and Agric. Chem. Conf., Purdue Univ., West Lafayette, IN. 14–15 Dec. Purdue Univ., West Lafayette, IN.

Moncrief, J.F., W.E. Fenster, and G.W. Rehm. 1984. Effect of tillage on fertilizer management. p. 45–56. *In* Conservation tillage for Minnesota. Univ. of Minnesota Agric. Ext. Serv. Publ. AG-BU-2402.

Nelson, C.E., M.A. Mortensen, and R.E. Early. 1967. NH_4 vs. NO_3 fertilization of field corn and related studies. Washington Agric. Exp. Stn. Bull. 685.

Peck, T.R., and S.W. Melsted. 1973. Field sampling for soil testing. p. 67–75. *In* Soil testing and plant analysis. SSSA, Madison, WI.

Petersen, R.G., and L.D. Calvin. 1965. Sampling. p. 54–72. *In* C.A. Black et al. (ed.) Methods of soil analysis. Part 1. Agronomy Monogr. 9. ASA, Madison, WI.

Randall, G.W. 1980. Fertilization practices for conservation tillage. *In* Proc. 32nd Annu. Fert. and Agric. Chem. Dealers Conf., Des Moines, IA. 8–9 Jan. Iowa State Univ., Ames.

Randall, G.W. 1982. Strip tillage systems-fertilizer mangement. *In* Farm Agric. Resources Management Conf. on Conservation Tillage. Iowa Stae Univ. Ext. Publ. CE-1755.

Reed, J.F., and J.A. Rigney. 1947. Soil sampling from fields of uniform and non-uniform appearance and soil types. J. Am. Soc. Agron. 39:26–40.

Reneau, R.B., J.L. Ragland, and W.O. Atkinson. 1968. Effects of ammonium nitrate and the growth of burley tobacco plants on soil pH. Tobacco 166 (12):34–37.

Rigney, J.A., and J.F. Reed. 1945. Some factors affecting the accuracy of soil sampling. Soil Sci. Soc. Am. Proc. 10:257–259.

Sabbe, W.E., and D.B. Marx. 1987. Soil sampling: Spatial and temporal variability. p. 1–14. *In* Soil testing: Sampling, correlation, calibration, and interpretation. SSSA Spec. Publ. 21. SSSA, Madison, WI.

Shoemaker, H. 1964. Fit lime to plowing depth. Better Crops Plant Food 48(1):24–27.

Sims, J.L., and W.O. Atkinson. 1974. Soil and plant factors influencing accumulation of dry matter in burley tobacco growing in soil made acid by fertilizer. Agron. J. 66:775–778.

Tabor, J.A., A.W. Warrick, D.E. Myers, and D.A. Pennington. 1985. Spatial variability of nitrate in irrigated cotton: II. Soil nitrate and correlated variables. Soil Sci. Soc. Am. J. 49:390–394.

Thom, W.O., K.L. Wells, and L. Murdock. 1982. Taking soil test samples. Kentucky Coop. Ext. Serv., Lexington, Publ. AGR-16.

Vaught, H.C., K.L. Wells, and K.L. Driskill. 1977. Alfalfa response to varying rate of phosphorus and potassium fertilization on deep, red, limestone-derived soils of the Pennyroyal area in Kentucky. p. 1–7. *In* Agronomy notes. Vol. 10(6). Dep. of Agron., Univ. of Kentucky, Lexington.

Vieira, S.R., D.R. Nielson, and J.W. Biggar. 1981. Spatial variability of field-measured infiltration rate. Soil Sci. Soc. Am. J. 45:1040–1048.

Waynick, D.D. 1918. Variability in soils and its significance to past and future soil investigations: I. A statistical study of nitrification in soils. Agric. Sci. 3:243–270.

Webster, R., and T.M. Burgess. 1980. Optimal interpolation and isorithmic mapping of soil properties: I. The semi-variogram and punctual kriging. Soil Sci. 31:315–331.

Wells, K.L. 1980. Influence of tillage systems and placement on effectiveness of liming materials. p. 62–67. *In* Proc. of Natl. Conf. on Agric. Limestone, Nashville, TN. 16–18 Oct. Bull. Y-166. TVA Natl. Fert. Develop. Ctr., Muscle Shoals, AL.

Wells, K.L. 1985. Soil tests and conservation-till: Are they compatible? Solutions 29(7):34–45.

Wells, K.L., and W.L. Parks. 1961. Vertical distribution of soil phosphorus and potassium on several established alfalfa stands that received various rates of annual fertilization. Soil Sci. Soc. Am. Proc. 25:117–120.

Whitney, D.A. 1982. Soil sampling techniques under reduced tillage systems. p. 279. *In* Agronomy abstract. ASA, Madison, WI.

Chapter 4

Soil Test Correlation, Calibration, and Recommendation[1]

W. C. DAHNKE, *North Dakota State University, Fargo*

R. A. OLSON, *University of Nebraska, Lincoln*

The basic aim of soil testing is to assess nutrient status, thereby identifying current and potential need for fertilization, monitor the effects of cropping practices on soil fertility, and assist in developing fertilizer recommendations.

There is no doubt about the existence of a relationship between soil nutrient status and crop growth. The relationship, however, is often obscured by the influence of many other growth factors. Knowledge of the soil's ability to supply nutrients, the amount of nutrients required for crop growth, and the influence applied nutrients have on crop growth is all needed to improve fertilizer recommendations. It must be understood, however, that the influence of fertilizer on crop growth during the growing season will largely depend on weather conditions during the growing season and on crop management intensity. Variable response to applied nutrients from unpredictable growing season conditions is probably the main reason why so many processes, systems, and models have been proposed to convert basic soil-fertility data into soil-testing programs.

In addition to using soil tests to help identify crop production constraints such as nutrient deficiencies, they also are used to identify toxic levels of particular nutrients and other elements or of soluble salts in general. Also, they are used to determine soil pH, lime requirement, and organic matter.

In this chapter, soil test development will be divided into the following steps: (i) correlation, (ii) calibration, and (iii) interpretation of data to develop recommendations. These terms are not clearly defined in the literature and are sometimes used interchangeably.

In this text, *correlation* is defined as the process used to determine if a soil nutrient, as extracted by a soil test, and crop response to added nutrient are so related that one directly implies the other. In other words, correlation

[1]Contribution from the Dep. of Soil Science, North Dakota State Univ., Fargo, ND 58105 and the Dep. of Agronomy, Univ. of Nebraska-Lincoln, NE 68583.

is the process of selecting the best soil test for the soils of the area. Soil test *calibration* is the process of ascertaining the degree of limitation to crop growth or the probability of getting a growth response to applied nutrient at any soil test level. The amount of extractable nutrient is usually expressed as low, medium, high, or as a range of critical concentration. The final step is to develop fertilizer *recommendations*. This step is often spoken of as soil test interpretation. Individual interpretation and experience are evident in this step and are the cause of much controversy and confusion in soil testing. Examples will be given demonstrating several ways to interpret the same data.

In spite of these problems, site specific estimates of nutrient status, as provided by soil tests, can be the most efficient method for the rational use of fertilizers in producing crops.

I. SOIL TEST CORRELATION

Soil test correlation is the process of determining whether there is a relationship between plant uptake of a nutrient or yield and the amount of nutrient extracted by a particular soil test (Corey, 1987). This relationship can be determined either mathematically or graphically.

Correlation research is usually conducted in two steps: an exploratory fertilization trial in the greenhouse with many soils followed by trials with fewer, carefully selected soils in the field. The advantage of growth chamber studies is better control of environmental factors and soil homogeneity. Greenhouse results, however, may not be directly transferable to field conditions (Corey, 1987).

A. Soil Extraction

Most soil tests involve the extraction of nutrient from a soil sample. One of the first soil tests was that of Daubeny (1845). It involved extracting the soil with carbonated water. This test was never put to practical use because of analytical difficulties. Many other extracting solutions have been suggested and used to extract nutrients from soil samples. The reason some extractants fail today is not because of analytical problems but because the amount of nutrient extracted does not correlate with either plant nutrient uptake or yield.

In addition to the extractant itself, the other important aspect of a soil test is how the soil is extracted. Some of the important factors of a soil test procedure, such as soil/solution ratio, extraction time, shaking speed, shape of extraction flask, are discussed by Grava (1980) and Munter (1988). A soil test not only refers to the type of solution used in extracting a nutrient but also how the extraction is accomplished.

The following sections discuss methods used to correlate the amount of nutrient extracted from the soil with plant nutrient uptake or yield of greenhouse and field trials.

B. Greenhouse Trials

Greenhouse trials are an excellent tool to determine the relationship between nutrient availability as measured by soil test and nutrient uptake by plants. Greenhouse trials are better than field trials for this purpose because the influence of uncontrolled variables such as subsoil, climate, and soil variability can largely be eliminated. It is important that many soils be used to increase confidence that the results will be accurate and not because of chance. It is also important to select soils with a wide range in nutrient availability and that the nutrient range approaches a normal distribution. This is especially true if results are analyzed mathematically because the usual method of calculating correlation coefficients assumes a normal distribution of the variables (Snedecor & Cochran, 1971).

In the case of nutrients such as nitrate-N and chlorine that can readily move in the soil or are mineralized from organic matter, field correlation studies are necessary.

1. Mathematical Correlation

Correlation coefficients for various soil tests can be compared directly. A correlation coefficient of $r = 1.0$ indicates a perfect correlation and a coefficient of zero indicates no relationship. In addition to the determination of correlation coefficients, a regression analysis is often done to mathematically express the change in nutrient uptake as soil test changes (Corey, 1987). This enables one to compare the relationship between nutrient uptake and soil test. When there are numerous variables that may influence nutrient uptake or yield, various multiple regression techniques have been used in an attempt to determine the relationship. These techniques are often used to consider interactions.

2. Cate-Nelson Method of Soil Test Correlation

In contrast to mathematically determining correlation, the simple graphic method of Cate-Nelson (Cate & Nelson, 1965) has practical advantages in addition to showing whether good correlation exists, it splits the data into two populations (soil test calibration). Thus, it acknowledges the basic capability of soil tests, which is to separate soils that are likely to respond to added nutrient from those that are unlikely to respond. In addition, outlying data points can be identified readily in a scatter diagram and isolated for further study. When used to analyze greenhouse yield data, it can give a rough approximation of the critical soil test level, which will be of aid in selecting field sites.

The Cate-Nelson Method plots percentage yield against soil test (Fig. 4-1) to give a visual indication of the reliability of the soil test. If the majority of points on the graph have a low percentage yield when the soil test is low and a high percentage yield when the soil test is high, the soil test being studied accurately predicts the need or lack of need for a nutrient. In other words, the soil test level is well correlated with response to added nutrients. Many points in the negative quadrants (Fig. 4–3) indicate the soil test is not well suited to the soils of the area or that there is no correlation between soil test values and plant response to the added nutrient. Steps in the Cate-Nelson Graphical Method are:

1. Determine percentage yield values for each fertilizer rate trial.

$$\text{Percentage yield} = \frac{\text{Yield at zero level of nutrient studied}}{\text{Yield when all nutrients are adequate}} \times 100$$

2. Determine soil test values for nutrient being studied.
3. Plot percentage yield (Y-axis) vs. soil test value (X-axis) on arithmetic graph paper (Fig. 4–1).
4. Draw two intersecting perpendicular lines on a clear piece of plastic. Label the top right and bottom left quadrant with a + and the bottom right and top left with a – (Fig. 4–2).
5. Place the overlay over the graph moving it horizontally or vertically until the maximum number of points are in the two positive quadrants (Fig. 4–3).
6. If nearly all of the points are in the positive quadrants, the test is accurately predicting response. Soils with a low soil test should have a low percentage yield (large response to added nutrient) and soils with a high soil test should have a high percentage yield (little or no response to added nutrient). If many of the points cannot be placed in the positive quadrants, look for another test or try to discover why these soils are outliers. The horizontal line in Fig. 4–3 will usually fall between a percentage yield of 80 to 90. Percentage yields above this range are usually not statistically significant increases in yield. Occasionally, percentage yields of more than 100 are obtained.

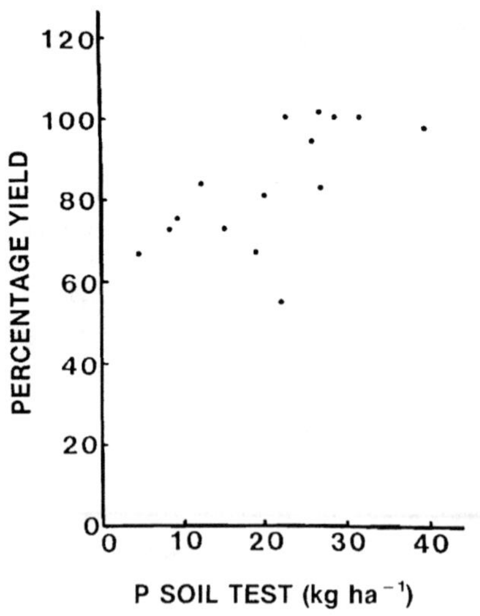

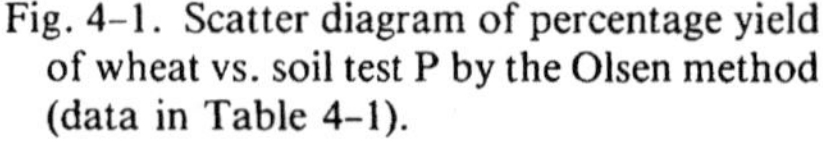
Fig. 4–1. Scatter diagram of percentage yield of wheat vs. soil test P by the Olsen method (data in Table 4–1).

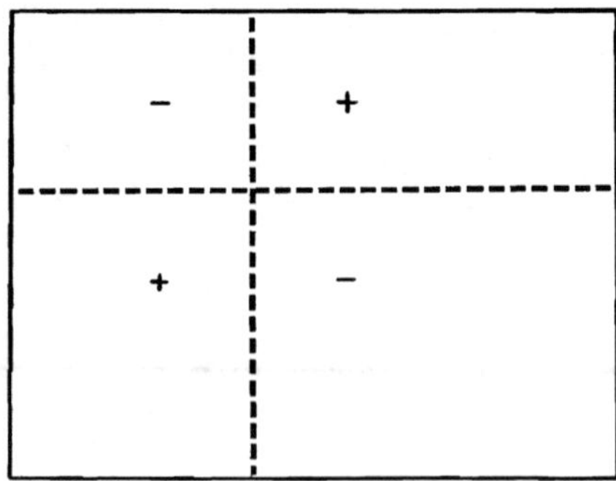

Fig. 4–2. Format for clear plastic overlay used in the Cate-Nelson graphical method of soil test correlation.

In addition to simplicity (Waugh et al., 1973; Ayodele & Agboola, 1985), this method:

1. Indicates whether or not there is a good correlation between soil test and yield response. If nearly all of the points fall in the positive quadrants, there is a good correlation. If points are equally distributed in all quadrants, the correlation is poor.

2. Indicates the soil test critical level, the point above which response to added nutrient is unlikely (Fig. 4–3).

3. Gives results that are not biased by a few outliers. Outliers can be identified readily for further study. Outliers may be caused by complete unavailability of added nutrients, a high level of the nutrient in the subsoil in the case of field trials, or some points may be in the upper left quadrant because another factor was more limiting. It must be remembered that the purpose of greenhouse trials is only to determine if there is a relationship between a soil test and response to added nutrients. Field trails must be conducted to determine how well the test performs in the field and to obtain final calibration and interpretation data.

C. Field Trials

Preliminary greenhouse data are useful to compare different extraction methods but field trials are necessary for final selection. Plant growth and yield are functions of many variables that can be grouped into soil, crop, climate, and management categories (Fitts, 1955). Poorer correlations are obtained with field trials because the uncontrolled variables are less uniform in the field than in the greenhouse. Correlations using field data can be improved, however, by plotting relative yield rather than absolute yield, yield

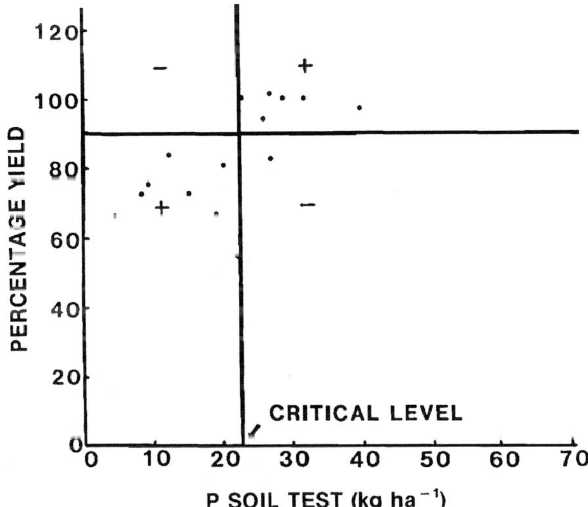

Fig. 4–3. Scatter diagram of percentage yield vs. soil test showing the soil test critical level determined by the Cate-Nelson procedure.

response, or nutrient uptake as a function of soil test values. Relative yield, also known as percentage yield, is the yield of the treatment with adequate but not excessive amounts of all nutrients other than the one being correlated, divided by the maximum yield from the nutrient studied with optimum amounts of other nutrients times 100. The reason relative yield results in a better correlation is that the influence of some of the uncontrolled variables is eliminated (Bartholomew, 1972). Relative yield cannot be used to determine economic rates of nutrient application because it does not take into consideration absolute yield level.

Field trials should be multiple rate so that the information obtained can be used not only for correlation and calibration but also for developing recommendations from the soil test that is finally selected. As with the greenhouse study, field sites chosen should range from low to high in the nutrient under investigation. Because most soils are inherently variable, a separate soil sample should be taken from each plot if at all possible. It will help explain yield variability within a site.

Use five or more rates of the nutrient being studied. The lowest rate should be zero and the highest rate should be more than enough to obtain maximum yield. It is important that the interval between rates be relatively small and well spaced within the response range to determine the point or area where no further response occurs. The number and size of plots per site should be selected to minimize effects of soil variability and yet be practical.

More reliable information can be obtained by controlling interactions when it is possible. Although emphasis has been placed on nutrient interactions, they are of little importance when other growth influencing factors are maintained at favorable levels. Much of the interaction occurs when comparing the O rate of a single nutrient with adequate rates of another nutrient (Anderson & Nelson, 1975).

All treatments should be replicated three or more times, depending on the crop. High plant population crops such as wheat (*Triticum aestivum* L.) and barley (*Hordeum vulgare* L.) require fewer replicates than low plant population crops such as potato (*Solanum tuberosum* L.). In production areas where crop response varies widely with year and location, it is of greater value to have more sites and fewer replicates than a lot of replicates and few sites. A more complete discussion of the establishment of field trials is given by Hunter and Fitts (1969).

Field data as well as greenhouse data can be analyzed by plotting relative yield vs. soil test to determine how well each soil extractant being investigated predicts added nutrient response. The best extractant will have most of the points in the lower left (Cate-Nelson graphic technique) and in the upper right quadrants. The vertical dividing line between quadrants is the critical level, the soil test below which response to added nutrient is very likely. The horizontal line approximately separates those trials with a statistically significant response from those with a nonsignificant response.

II. SOIL TEST CALIBRATION

Soil test calibration is the process of ascertaining the meaning of the soil test measurement in terms of crop response (Bray, 1936, 1937; Olson et al., 1958; Corey, 1987; Rouse, 1967). The purpose of soil test calibration is to describe the soil test results in easily understood terminology and to simplify the process of making fertilizer recommendations by placing soils in response categories. The terminology often used to describe categories is very low, low, medium, high, and very high concentration ranges.

A. Continuous Curves

A common procedure used in establishing soil test categories is to plot relative yield or yields vs. soil test and fit a continuous curve to the points. The curve is then divided into several categories such as low, medium, and high or it is divided into fertility indices (Cope & Rouse, 1973). The basis for the division into classes is subjective and arbitrary since continuous regression models have no inflection point as a justification for making a division.

B. Probability Approach

The Cate-Nelson correlation procedure, described in a previous section, has a fundamental advantage. In addition to showing whether there is a good correlation, it also separates soil response data into two populations, those likely to respond and those unlikely to respond to specific nutrient additions. The soil test value where this split occurs is known as the soil test critical level (Fig. 4-3). Soils testing below this level usually respond to added nutrient and those testing above this level usually do not respond to added nutrient. This fundamental separation recognizes the basic fact that a soil test cannot predict yield or the absolute amount of response. A soil test can only be used to determine the probability that a response will occur (Fitts, 1955; Fitts & Nelson, 1956). Cate and Nelson (1971) also developed a simple statistical procedure for partitioning soil test data into two classes.

The next step is to divide the soil test data into more than two categories (Fitts, 1955, Olson et al., 1954, 1958) if there are sufficient field data to permit more separations. If three categories are used, one could be a *low* category where the probability of a response is great and the amount of fertilizer recommended is large, a *medium* category where there is a 50% probability of getting a response, and a *high* category where there is a small probability of a response. None or only a small amount of starter fertilizer is recommended on soils in the high category.

When using the Cate-Nelson procedure, for example, the medium category could bracket the critical level. Soil test levels below this could be called low (response to applied fertilizer likely) and soil test levels above this could be called high (response to applied fertilizer unlikely). An alternative would be the Cate-Nelson Analysis of Variance Method (Nelson & Anderson, 1977). This statistical procedure can be used to establish three or more classes.

As more data are obtained, more soil test categories can be established. Many soil testing laboratories use five categories or more. Also, as more data are obtained categories can be established for peculiar soils and each crop. The main purpose of soil test categories is to convert the soil test numbers or indices into terms that will give a grower some indication of the nutrient status of his soil.

Another reason for establishing soil test categories is to simplify the fertilizer recommendation process when laboratory staff use tables in making fertilizer recommendations for growers. However, at present most recommendations are made from equations in computers. As a result, simplification for staff efficiency and accuracy is no longer a major concern.

III. SOIL TEST RECOMMENDATIONS

Given the background of correlation and calibration described and having selected procedural methods that will best serve the soil and other environmental conditions for the area involved, the final step becomes that of interpreting and making recommendations from the specific soil test value. It perhaps goes without saying that the individual making the interpretation should be thoroughly acquainted with the crops and soils of the region for fine tuning the recommendation. As an example, he will know and give recognition to the fact that subsoil of certain soils may be well endowed with nutrient reserves and fail to respond even with a rather low-surface soil test. He will recognize the impact that fallowing will have on certain nutrients and how this will modify the recommendation compared with a continuous cropping situation. He will understand and adjust for the varied responsiveness of different crops to a specified nutrient, give allowance for known allelochemical effects of crop rotation, and potential nutrient sources not measured in the soil test value. In short, the person making recommendations must be well versed in agronomic principles associated with crop production in addition to being a qualified soil chemist.

Inadequate attention to the various local production factors gives rise to some of the observed differences in recommendations coming from different laboratories that have tested the same soil sample. Significant differences in interpretation also come from varied philosophies on what soil test values mean and what fertilization should accomplish.

A. Crop Production Models and Response Curves

A discussion of the processes involved in the development of a soil test would not be complete without examining crop production models and the shape of the curves drawn when plotting yield vs. applied nutrients.

Some have tried to define crop yield by theoretical models such as (1) $Y = F(X_1, X_2, X_3, \ldots X_n)$ (NAS-NRC, 1961) where Y is crop yield and X_i are the factors affecting plant growth. If all of these factors could be measured, predicted, or controlled, the yields of crops could be predicted accurately by this model. Many factors, however, cannot be measured, predicted, or controlled. Therefore, models using factors that cannot be measured or predicted must be considered as approximations with little or no predictive ability (Passioura, 1973; Perrin, 1976).

The fact a relationship is often found between nutrient level, one of many factors influencing growth, and crop growth indicates soil fertility has an important influence on productivity. It is this relationship that allows soil nutrient extractions to be used as a guide in the fertilization of crops.

The way nutrients influence plant growth is often discussed in conjunction with the shape of a response curve. The discussion usually centers around whether the response is linear or curvilinear.

The linear model was first proposed by Liebig and later broadened by Blackman (1905). It is commonly known as the "Law of the Minimum" and states that yield increases until the minimum factor becomes the factor limiting growth. According to this concept, yield is directly proportional to the amount of the limiting nutrient in the soil.

A curvilinear model was first proposed by Wollny (Gauch, 1972) but later developed by Mitscherlich (1928) and Willcox (1944). This became known as Mitscherlich's Law of Diminishing Returns or simply as the Law of Diminishing Returns. It states that a much larger increase in yield is obtained per unit of applied nutrient when the nutrient is deficient as compared to when that nutrient is nearly adequate. Curvilinear models have been accepted by many, especially for slightly soluble nutrients (Bray, 1963). One point that needs to be recognized about response curves is that when yield data from different sites or years are averaged it always results in a curvilinear response. This would be true even if a linear response were obtained at each individual field trial (Fig. 4–4). In other words, one should be careful when combining data from different experiments to obtain an overall response curve.

While curvilinear crop responses to a nutrient do occur, the way the data are analyzed will strongly influence the conclusions reached. For instance,

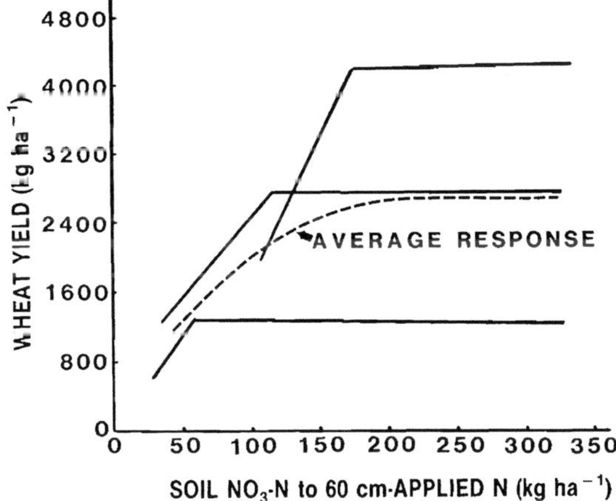

Fig. 4–4. Linear-plateau responses (solid lines) become curvilinear (dashed line) when averaged.

a curvilinear response can be obtained when the data from the replicates of a non-uniform site are averaged even though the response was linear on individual replicates. Another case where response sometimes appears curvilinear is when too few levels of nutrient are applied to adequately define the response (Sparrow, 1979a, b).

Willcox (1949) felt that the more than 27 000 field fertilizer response trials conducted in Germany in the 1930s and reported by Mitscherlich and Gericke gave definite proof that crops respond asymptotically. As indicated above, this is not proof since averaged data from many sites or years always result in an exponential response curve. This again is not to say that curvilinear responses do not occur (Fig. 4–5) but that the occurrence may be less frequent than commonly believed.

Willcox (1949) believed that curvilinear responses (the Mitscherlich equation) only occur when all growing conditions are ideal. The opposite may be more likely. Boyd et al. (1976) and Boyd (1970) presented evidence that curvilinear responses occur because of disease, pests, excess amounts of other nutrients, or other non-ideal situations. Bray (1963) believed crop response to N is linear and response to P and other slightly soluble nutrients is curvilinear. While it is generally agreed that response to N is linear, there are also many examples in the literature of linear response to slightly soluble nutrients, such as P (Boyd et al., 1976; Boyd, 1970; Sparrow, 1979a, b; Waggoner & Norvell, 1979).

Crop response curves can be visualized as a series of curves each with a different maximum or plateau depending on site and year. The potential (maximum) yield is not known and cannot be predicted for any site or year. The plateau or maximum yield on a particular site will be determined largely by soil, crop, weather, and management. Weather is not predictable and many management factors cannot be quantified, therefore the grower is probably most qualified to estimate the plateau or maximum yield (commonly called

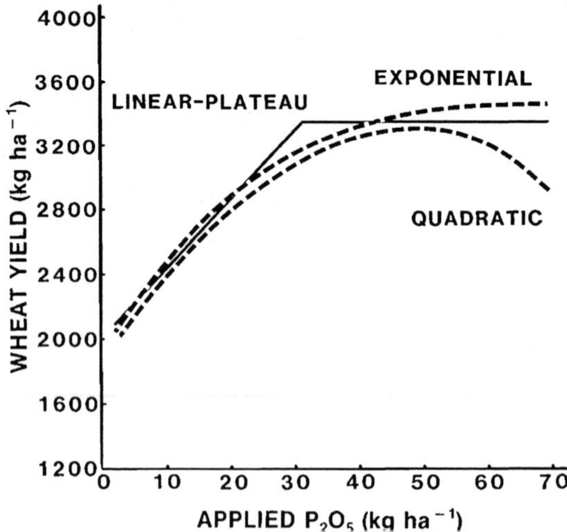

Fig. 4–5. Shapes of plant yield response curves to applied nutrients.

yield goal) for a particular field. The role of the soil testing laboratory then becomes one of determining the amount of nutrient needed in addition to the amount already in the soil so that nutrients are not a limiting factor in reaching the yield goal (Dahnke et al., 1984). Economics alone cannot be used to determine the best fertilizer rate because yield, prices, and the shape of the response curve are not predictable for the coming growing season.

If the response curve cannot be predicted for a given growing season, it follows that the Law of Diminishing Returns or any other curve cannot be used to choose the most economical rate of fertilization. Fertilization rate should be based on soil factors, crop to be grown, and the probability of reaching a certain yield goal. The probability of favorable economics of fertilization is dependent on a well-correlated soil test that has been correctly calibrated and interpreted. In the case of wheat in North Dakota, it takes about 3.7 kg ha^{-1} of N and 0.25 kg ha^{-1} of P to grow an extra 100 kg of wheat (Dahnke et al., 1985). At 1987 prices for hard red spring wheat of $0.09 per kilogram and N fertilizer of $0.25 per kilogram of N, a $5 to $7 return can be expected for each dollar invested in fertilizer on responsive soils.

B. Examples of Data Interpretation for Making Fertilizer Recommendations

At the present time, there are no established rules to determine what response model is most appropriate to use in the interpretation of fertilizer response data from field trials. Much of the literature on fertilizer response discusses some version of the Law of Diminishing Returns. In this section, several models will be used to interpret the field data from sites 1 to 9 in Table 4–1.

1. Mitscherlich-Bray Function

The observation by some that when equal increments of a nutrient are applied to a crop, the yield response becomes smaller from each increment than from the preceding increment led to various attempts to describe this curvilinear yield response by a mathematical formula. The most famous was that of Mitscherlich (1928) who developed his theory in 1906. Mitcherlich's equation was modified by Bray (1936) to the following form:

$$\log (A - y) = \log A - c_1 b_1 \qquad [1]$$

where A is the yield possible when all nutrients are present in adequate quantities; y is the yield when the soil test for nutrient b_1 is less than adequate and c_1 is the proportionality constant.

This equation is sometimes referred to as the *percent yield* equation (Melsted & Peck, 1977) because percentage yield as well as actual yields can be used in the equation (Fig. 4–6 and 4–7). Following is an expanded version of the equation:

Table 4-1. Hard red spring wheat response to applied P fertilizer at several sites in North Dakota.

Site	P test	P_2O_5 applied	Yield
	kg ha^{-1}		
1	15	0	1277
		11	1546
		22	1680
		34	1680
		45	1680
		67	1814
2	4	0	1344
		11	1613
		22	1949
		34	1983
		45	2016
		67	2016
3	20	0	1814
		11	2016
		22	2083
		34	2083
		45	2083
		67	2285
4	9	0	2016
		11	2218
		22	2419
		34	2520
		45	2621
		67	2688
5	27	0	2083
		11	2285
		22	2554
		34	2554
		45	2486
6	12	0	2150
		11	2285
		22	2486
		34	2554
		45	2554
7	8	0	2150
		11	2418
		22	2688
		34	2957
		45	2957
8	19	0	2218
		11	2490
		22	2755
		34	3040
		45	3293
9	24	0	3369
		11	3537
		22	3477
		45	3779
		72	3813
10	19	0	2083
		11	2419
		22	2890
		34	3091
		45	3091
11	31	0	1298
		22	1231
		45	1284
		72	1304
12	29	0	3510
		22	3618
		36	3450
13	28	0	1520
		22	1594
		45	1594
14	23	0	2609
		11	2623
		22	2589
		45	2589
		72	2576
15	40	0	3201
		18	3201
		30	3416
		42	3248
		54	3302
16	32	0	4970
		120	4593
		240	5084

$$\log (A - y) = \log A - c_1 b_1 - cx \text{ (Melsted \& Peck, 1977)} \qquad [2]$$

where c is the efficiency factor for applied fertilizer and x is the quantity of fertilizer needed. This equation is meant to be used to determine the amount of fertilizer needed to raise the percent yield from one point to another. Figure 4-7 indicates that it will take approximately 18 kg ha^{-1} of P_2O_5 to increase the yield from 95% of maximum to 98%. Using the wheat yield data in Table

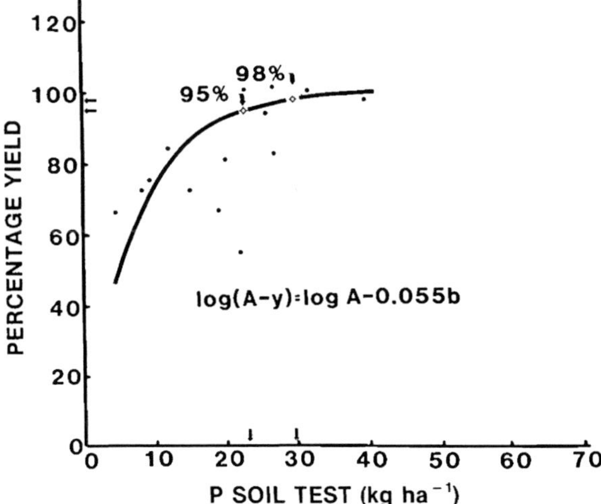

Fig. 4–6. The Mitscherlich-Bray function relating P soil test level to percentage yield of wheat.

4–1, this would be equal to increasing the yield by 114 kg ha^{-1} of wheat. This points out one of the frequent criticisms of the Mitscherlich—Bray function and other curvilinear models. They recommend too much fertilizer in relation to the amount of yield increase obtained at or near maximum yield.

2. Other Curvilinear Models

a. Yield Response. Another method of interpreting field data for fertilizer recommendations is to plot yield response vs. amount of applied nutrient (Hauser, 1973). The data are first grouped according to the soil test categories established by soil test calibration. In this case, the data from Table

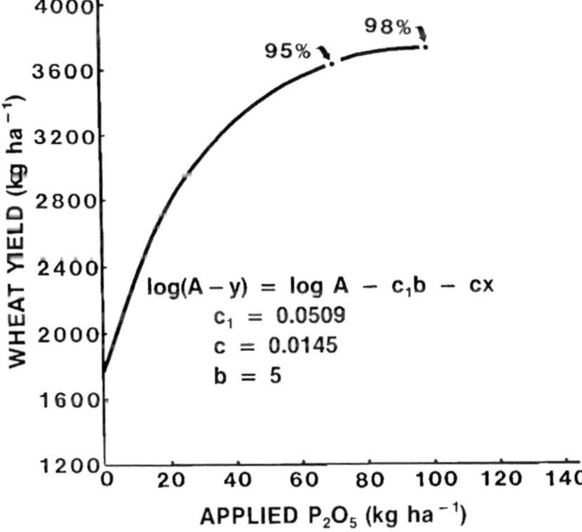

Fig. 4–7. The Mitscherlich-Bray function relating applies P_2O_5 to wheat yield at a soil test of 5 kg ha^{-1} by the Olsen test.

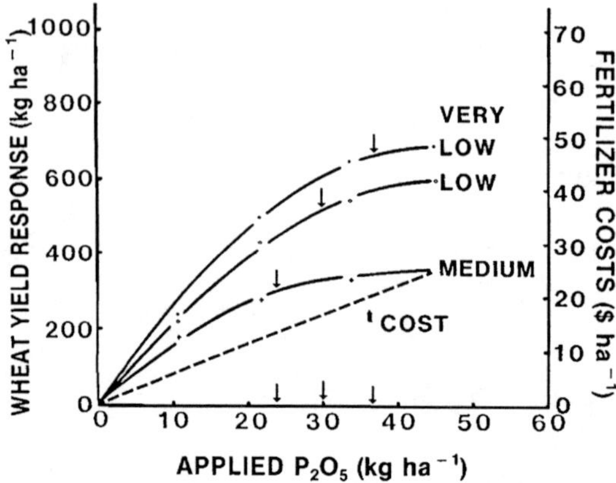

Fig. 4–8. Interpretation graph showing the relationship between wheat response to applied P for three soil test levels.

4–1 are grouped into very low (0–10 kg of P ha^{-1}), low (11–21 kg of P ha^{-1}) and medium (22–34 kg of P ha^{-1}) categories. The average yield response for trials in each of these categories is then plotted against the amount of applied nutrient (Fig. 4–8). The most economical rate of fertilizer application is determined by plotting cost of applied fertilizer on the same graph (Fig. 4–8). The optimum fertilizer rate is the point on the curve where marginal revenue equals marginal cost. The major advantage of this method is simplicity. The fact that the resulting recommendations have no relation to total yield is a disadvantage.

b. Yield. Another procedure for soil test interpretation is to use curvilinear regression to express yield as a function of soil test and applied fertilizer. This method is usually unsatisfactory because yield is determined by many factors other than the nutrient variable (Colwell, 1967). As with the Mitscherlich equation, which is similar to this method, the fertilizer recom-

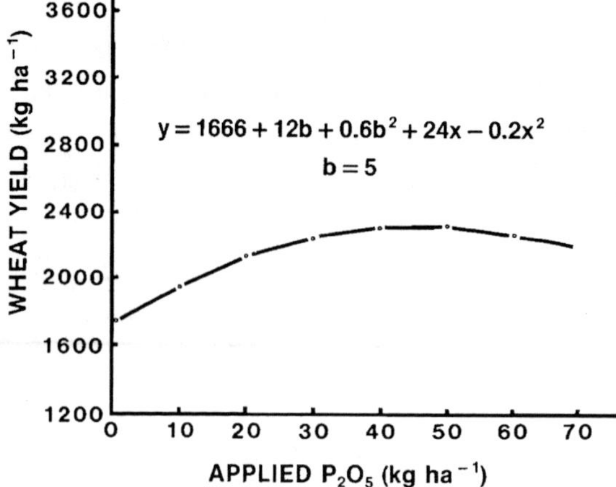

Fig. 4–9. Combining yield data from all sites using curvilinear regression.

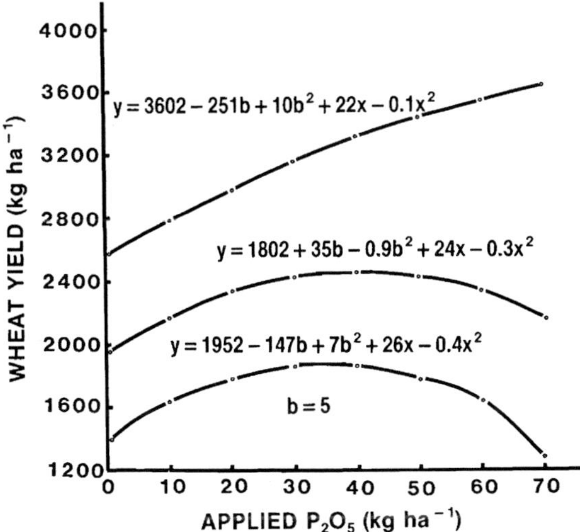

Fig. 4–10. Using curvilinear regression after the data was separated into three different yield levels.

mendation increases rapidly in relation to yield increase near the point of maximum yield (Fig. 4–9). The other point evident with this method is that a very small portion of the yield range of sites 1 to 9 (Table 4–1) is represented by the resulting curve (Fig. 4–9). A greater portion of the yield range can be taken into consideration by placing the data into high, medium, and low yield groups (Fig. 4–10). Note that the amount of applied P_2O_5 needed to reach maximum yield increases from the lowest yielding group to the highest yielding group (Fig. 4–10). The next obvious step would be to look at the yield response curves for individual sites (Fig. 4–11), which leads to the following method of soil test interpretation.

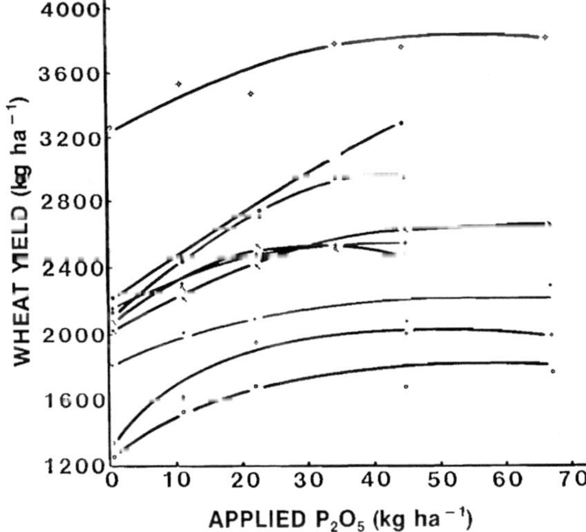

Fig. 4–11. Wheat yield response to P data from each site represented by smooth hand fitted curves.

Table 4–2. Hard red spring wheat response to applied N fertilizer at several sites in North Dakota.

Site	Soil NO_3-N to 60 cm + applied N	Yield
	kg ha^{-1}	
1	84	2218
	106	2554
	129	2554
	151	2621
2	30	1142
	58	1814
	86	2150
	114	2688
	142	2621
3	99	1814
	155	3562
	211	3965
	267	4368
	323	4032
4	110	2554
	166	3427
	222	3763
	278	3763
	334	3629
5	34	806
	56	1277
	78	1277
	101	1277
6	60	1277
	83	1747
	105	1882
	128	1882

3. Plateau Yield Points

In a previous section it was mentioned that although many prefer to represent response data with continuous curves, in many cases a response can be as well represented by a linear-plateau curve. This is true for both-soluble and slightly soluble nutrients in the soil. The phosphate response data from the first nine sites in Table 4–1 are plotted and the points fitted using continuous curves in Fig. 4–11 and linear-plateau curves in Fig. 4–12. The same comparison is made using N response data (Table 4–2) in Fig. 4–13 and Fig. 4–14. In both cases, linear-plateau curves fit the data as well or better than a continuous curve. The linear-plateau curve has the advantage of more readily indicating the point at which maximum yield is reached.

Since yields often increase linearly to the point of near maximum yield, it is logical from an economic point of view to fertilize to this point. A line drawn through the plateau-yield points (Table 4–3, Fig. 4–12) can, therefore, represent the fertilizer recommendations that should be made for various yield goals on, in this casc, P-dcficicnt soils. The following equation represents the line through the plateau-yield points in Fig. 4–12:

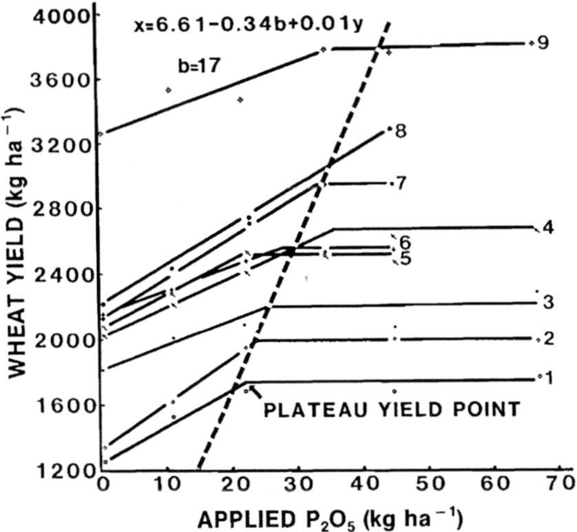

Fig. 4-12. Wheat yield response to P data from each site represented by hand fitted linear-plateau curves with a line drawn through plateau yield points.

$$x = 6.61 - 0.34b + 0.01\,y \tag{3}$$

where x is kg ha^{-1} of P_2O_5 recommended, b is P soil test, and y is yield goal.

Recognizing that the relationship between soil test level and yield is tenuous because of the fact that soil fertility is only one of many factors influencing yield, the line through the plateau yield points is assigned a soil test value near or slightly below the critical level (Fig. 4-3). The area to the right of this line represents more responsive sites (a lower soil test) whereas the area to the left represents less responsive sites (Fig. 4-15). The final step

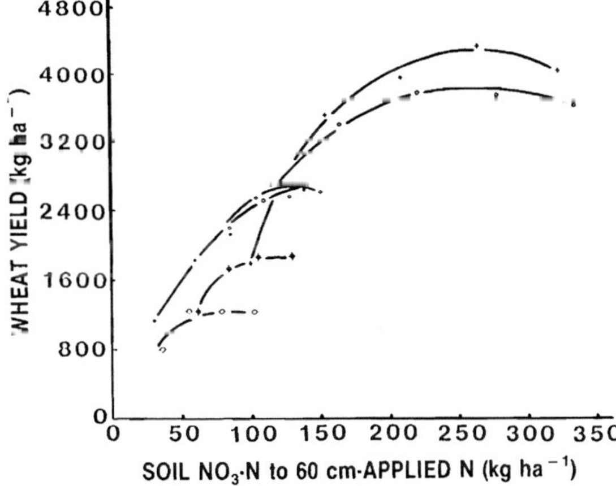

Fig. 4-13. Wheat yield response to N data from each site represented by smooth hand fitted curves.

Table 4-3. Soil tests and plateau yields for nine hard red spring wheat field trials in North Dakota.

Site	P soil test	Plateau yield	Applied P_2O_5 for plateau yield
	kg ha^{-1}		
1	15	1750	22
2	4	2000	24
3	20	2200	26
4	9	2680	37
5	27	2550	23
6	12	2580	28
7	8	2960	35
8	19	3300	45
9	24	3780	35

is to write an equation for the nutrient recommendation using soil test and the grower's yield goal. In this example, the recommendation equation for the very low, low, and medium testing fields in Fig. 4-15 is:

$$x = y\,(0.0163 - 0.0003b) \qquad [4]$$

where x is the recommendation in kg ha^{-1} of P_2O_5, y is the yield goal of the grower in kg ha^{-1} and b is the soil test.

4. Comparison of Interpretation Methods

A comparison of the recommendations for some of the methods discussed above is shown in Fig. 4-16. These recommendations were all developed using the same data and demonstrate how much fertilizer recommendations vary for near maximum yield depending on the methods used to interpret the data. Figure 4-16 clearly shows that for near-maximum yield

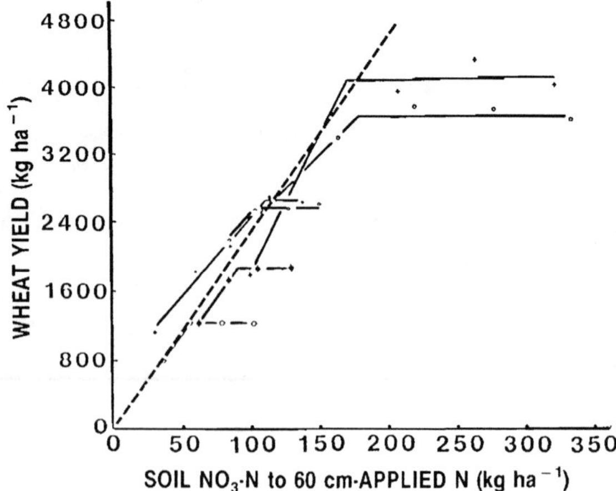

Fig. 4-14. Wheat yield response to N data from each site represented by hand fitted linear-plateau curves with a line drawn through plateau yield points.

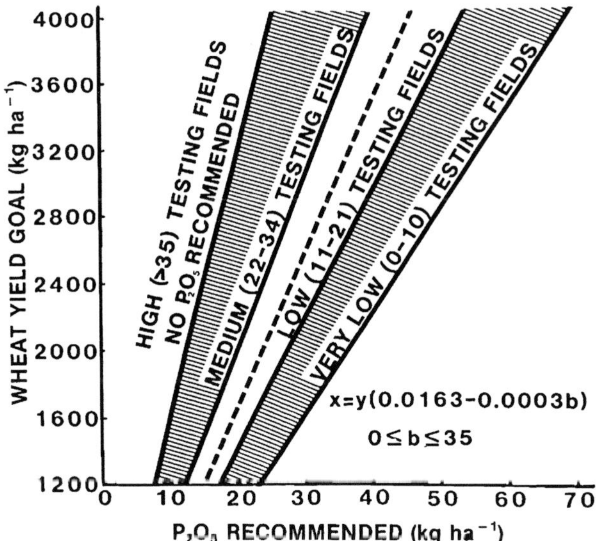

Fig. 4-15. Very low, low, medium, and high soil test recommendations made using the Plateau Yield method. The dashed line through the middle of the low testing category represents the plateau yield points of Fig. 4-12.

the Mitcherlich method makes the highest recommendations. It also shows the Yield Response method makes recommendations that do not change much as soil test changes. The Plateau Yield method recommends about half as much P at a very low test level as the Mitscherlich method.

The method of drawing a line through the Plateau Yield or maximum yield points has also been used with N response data for wheat (Table 4-3 and Fig. 4-14) and potato (Table 4-4 and Fig. 4-17) in North Dakota.

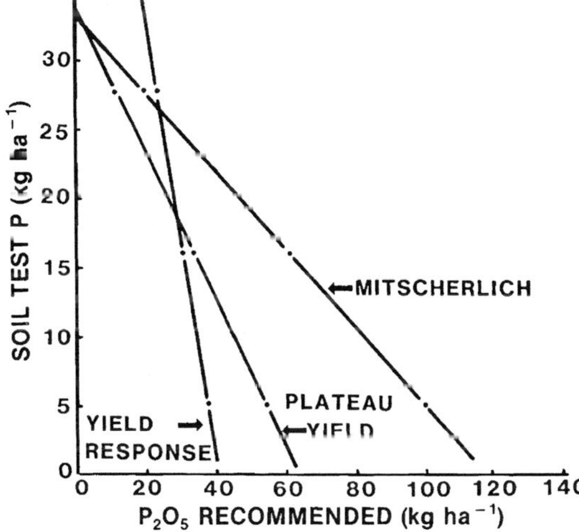

Fig. 4-16. Comparison of P recommendations from three different methods of soil test data interpretation.

Table 4–4. Potato response to applied N fertilizer at several sites in North Dakota.

Site	Soil NO_3-N to 60 cm + applied N	Yield
	kg ha^{-1}	t ha^{-1}
1	56	20.8
	84	24.6
	112	24.4
	168	22.0
2	56	24.6
	84	27.1
	112	26.8
	168	26.2
3	32	16.7
	59	18.3
	87	22.2
	143	22.4
4	56	28.2
	84	28.8
	112	30.4
	168	29.0
5	56	31.4
	84	32.1
	112	35.2
	168	33.4
6	26	18.7
	82	27.8
	138	31.1
	194	30.8
	250	32.6

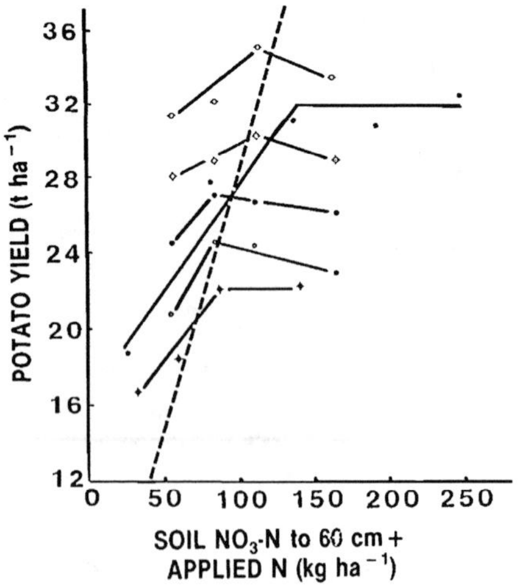

Fig. 4–17. Line representing plateau yield points of potato response to applied N.

The Plateau Yield data interpretation method is different from other methods in that no attempt is made to make recommendations from an average response curve. Recommendations are made from a line that represents the average plateau yield points of responsive sites. In essence, whether the yield response curve is linear or curvilinear is of little importance because only plateau yield points are used in developing the recommendations.

C. Cation Saturation Ratio

A series of reports during the 1940s and 1950s proposed ideal proportions of the major exchangeable cations in soil (Bear et al., 1945; Bear & Toth, 1948; Graham, 1959). The proposed ranges were 65 to 75% Ca^{2+}, about 10% Mg^{2+}, 2.5 to 5% K^+, and 10 to 2% H^+, or approximate ratios of 7:1 for Ca/Mg, 15:1 for Ca/K, and 3:1 for Mg/K. In contrast to methods just discussed, this method does not use crop response data other than the original work that resulted in the idea.

The above suggests base saturation should be in the range of 80 to 90%. Many other early studies had shown a relation between the percentage saturation of an individual cation and its availability to plants (Allaway, 1945; Chu & Turk, 1949), and that the uptake of a given cation could affect the uptake of other cations (Van Ittalie, 1938; York et al., 1954). There was, accordingly, a theoretical basis for the cation saturation ratio concept and it afforded a convenient arithmetic manipulation of analytical data with considerable appeal. As a result, the concept has been in common use by many organizations providing soil testing services and fertilizer recommendations.

It is surprising that the cation saturation concept has received the credibility accorded to it in consideration of other early and recent literature accounts on the issue. Hunter and associates in New Jersey (Hunter, 1949) could find no ideal Ca/Mg or Ca/K ratios for alfalfa (*Medicago sativa* L.), nor did Foy and Barber (1958) find yield response of corn (*Zea mays* L.) to varying K/Mg ratio in Indiana. Similar Ohio studies of alfalfa and German millet (*Setaria italica* L. Beauv.) revealed no yield effects from varied Ca/Mg ratios (McLean & Carbonell, 1972), nor were any Mg or K responses obtained with corn on four Nebraska soils having widely varying ratios of Ca/K, Ca/Mg, and Mg/K (Olson et al., 1982). Neither was the concept of any value in predicting K response of crops grown on coastal plain soils of Delaware (Liebhardt, 1981). Perhaps the most comprehensive study refuting the concept is that of McLean and associates (Eckert & McLean, 1981; McLean et al., 1983) where varied levels of Ca, Mg, and K relative to each other were employed in a complete factorial design with crops in both growth chamber and field experiments. They concluded the ratio had essentially no impact on yields except at extremely wide ratios where a deficiency of one element was caused by excesses of others. They emphasized the need for assuring sufficient levels of each cation rather than attempting adjustment to an ideal cation saturation ratio that does not exist.

D. Implications of Interpretation

1. Surface vs. Profile Soil Testing

Sampling only surface soils has been traditional in soil testing for the obvious reasons that a major portion of crop root systems exist there and surface samples are easy to collect. There is mounting evidence, however, of the need for collecting and analyzing samples from as much of the crop rooting profile as possible (Murdock & Engelbert, 1958; Herron et al., 1971; Reuss & Rao, 1971; Hill et al., 1978; Leggett, 1959; Soper & Huang, 1963; Young et al., 1967). The need applies especially for very soluble nutrients like NO_3-N and SO_4-S and most particularly for the drier cropping regions where drainage through the rooting profile rarely occurs. Residual mineral N in the rooting profile has, however, proved to be a dominant factor in irrigated agriculture controlling rate of fertilizer N required for optimum production of corn (Olson et al., 1976; Herron et al., 1971) and sugar-beet (*Beta vulgaris* L.) (Reuss & Rao, 1971). This suggests that profile mineral N is a factor as well for crop production in humid climatic regions with deep, medium to fine-textured and well-drained soils.

Deep rooted long growing season crops like corn and sugarbeets are capable of using residual N to depths of at least 180 cm so long as deep moisture is available and the upper profile does not contain abundant available N (Gass et al., 1971; Peterson et al., 1979). Taking into consideration the use of deep residual N by the crop to be grown will reduce the amount that ultimately could become a groundwater contaminant.

Subsoil reserves of slightly soluble nutrients like P and K can also control response to fertilizers containing those nutrients whatever the surface soil test value may be (Murdock & Engelbert, 1958; Olson et al., 1985). Some laboratories adjust fertilizer recommendations in accordance with known subsoil nutrient status of specific soil series without actually analyzing subsoil samples.

2. Rapid Build-up and Maintenance

The rapid build-up and maintenance concept of fertilization promotes the application of sufficiently high rates of deficient slightly soluble nutrients, like P or K, to raise the soil test level in 1 or 2 yr. This is followed by an annual application equivalent to the amount likely to be removed by the crop to be grown. The intention is that the farmer will be protected against any possible yield loss because of nutrient deficiency. This is a common approach to soil test interpretation and crop fertilization. When combined with the cation balance concept or the Mitscherlich equation, fertilizer rates can become quite liberal.

Trade journals are filled with reports justifying higher fertilizer rates for building and maintaining soil fertility while achieving maximum economic yields, even to the extent of advocating that the soil test never be allowed to drop to the point where crop response to applied fertilizer occurs. It is difficult to find any documentation in the literature of long-term, appropriately designed experiments that economically justify maintenance above the critical level. The rapid build-up and maintenance concept discounts the inherent nutrient delivery capacity of a soil's native mineral reserves, which with most soils other than sands is large for most nutrients. It is these native reserves that allow the non-eroded soil of the Morrow Plots to produce corn yields of more than 3000 kg ha^{-1} after 100 yr and more than 5000 kg ha^{-1} when rotated with soybean [*Glycine max* (L.) Merr.] in both cases with no fertilizer applied (Welch, 1976). Further, if the maintenance aspect of this concept of fertilization were to follow its precepts, there would need to be a recommendation for replacement of all 13 soil-derived nutrients likely to be taken up in the projected crop yield and not just N, P, and K. Complete adherence to this system of interpretation would essentially eliminate the need for further soil testing from the nutrient standpoint once the initial build-up had been accomplished. It has its most logical application for soil regions depleted in surface soil nutrients and essentially devoid of inherent subsoil nutrient reserves, such as the most highly weathered soils of warm and humid regions with high nutrient-fixing capacities (Yost et al., 1979).

3. Sufficiency Level

The sufficiency approach to soil test interpretation is similar to the rapid build-up and maintenance with the following exceptions:

1. The rate of build-up of the soil fertility level is much slower.
2. The soil fertility level is built to a much lower level.

Test results are classified low, medium, and high with associated probabilities of yield response to applied fertilizer (McLean, 1977; Olsen et al., 1954; Bray & Kurtz, 1945; Olson et al., 1954; Cope & Rouse, 1973; Dahnke, 1985; Cate & Nelson, 1965). The high or very high category, being above the level of yield response, receives no fertilizer recommendation. The medium category receives a small recommendation because responses are infrequent in this category. Fields in the low category receive more fertilizer because the soil's supply is not adequate. The actual amount recommended on medium testing fields is usually more or less equal to maintenance recommendation. The amount of fertilizer recommended on very low testing fields often is more than maintenance, resulting in a slow build-up in the soil. Most of the research establishing the efficacy of soil testing has focused on these calibration ranges, and the majority of university laboratories retain this system.

The sufficiency level approach is the most conservative among the three systems under consideration. Whereas the cation ratio and the rapid build-up and maintenance level concepts center on fertilizing the soil, the sufficiency level concept emphasizes different cut off levels for different crops.

There is no basis for a belief that sufficiency level interpretation procedures result in soil mining. Fertilizing most soils at nominal rates in accordance with likelihood of crop response still provides a gradual buildup of soil test values to a sufficiency level (Olson et al., 1982; Havlin et al., 1984; Cope, 1981; McLean, 1977; Mattingly, 1973). Yost et al. (1979) found on an oxisol total yield and P uptake were similar after four crops of corn for periodic broadcast and annual band treatments in which the same total amount of P had been applied. This indicates that even on oxisols, which are capable of precipitating large amounts of P, the sufficiency approach may be a better approach than the build-up and maintenance approach.

4. Economic and Environmental Implications

The interpretations made from soil tests have an immediate economic impact on the farmer as well as long-term agronomic and environmental impacts. In the first case, it goes without saying that the farmer's economic interests are best served by achieving the highest yield possible under his cropping conditions with the least amount of fertilizer. These criteria are not being met by many soil testing programs at the present time, probably because of the rapid build-up and maintenance concept of interpretation used by those laboratories. Rapid build-up and maintenance programs simply do not fit with economics for the majority of temperate region soils for reasons already elaborated. Its use should probably be restricted to very highly weathered soils high in sesquioxide and 1:1 type clays.

Long-term agronomic damage can result from excessive applications of certain nutrients related to plant physiological disturbances at the plant root surface and elsewhere within plant structures. One of the more commonly observed interactions of this kind is P-induced Zn deficiency. It can be readily corrected by applying Zn fertilizer, but possibly at unnecessary cost to the farmer. More serious is the induction of Fe deficiency, particularly on high pH soils, with overuse of P and Zn. Another type of excess is the toxicities to crops that can result from overuse, especially from continued applications of Cu and B. Crop removal of excess Cu can be a long process.

Finally, soil test interpretations that result in overly liberal fertilizer applications can have environmental impacts for which agricultural producers will be condemned by the rest of society. Contamination of streams, lakes, and groundwaters by nutrients and pesticides is occurring with increasing frequency causing public concern. No one appreciates the eutrophication of surface waters heavily committed to recreational use, nor is the presence of any foreign material in drinking water acceptable to the average citizen.

There is no question that soil erosion and nutrient runoff has contributed to eutrophication of surface waters or that fertilizer N is responsible for some groundwater nitrate. It seems logical that a major role of soil testing from this point onward should be one of assisting with the control of nutrient pollution in the environment (Ellis & Olson, 1986).

REFERENCES

Allaway, W.H. 1945. Availability of replaceable calcium from different types of colloids as affected by degree of calcium saturation. Soil Sci. 59:207–217.

Anderson, R.L., and L.A. Nelson. 1975. A family of models involving intersecting straight lines and concomitant experimental designs useful in evaluating response to fertilizer nutrients. Biometrics 31:303–318.

Ayodele, O., and A. Agboola. 1985. Calibration of available P in soils from derived savannah zone of western Nigeria. Fert. Res. 6:121–129.

Bartholomew, W.V. 1972. Soil nitrogen-supply processes and crop requirements. North Carolina State Univ. Int. Soil Testing Ser. Tech. Bull. 6.

Bear, F.E., A.L. Price, and J.L. Malcolm. 1945. Potassium needs of New Jersey soils. New Jersey Agric. Exp. Stn. Bull. 721.

Bear, F.E., and S.J. Toth. 1948. Influence of Ca on availability of other soil cations. Soil Sci. 65:69–74.

Blackman, F.F. 1905. Optima and limiting factors. Ann. Bot. 19:281–295.

Boyd, D.A. 1972. Some recent ideas on fertilizer response curves. p. 461–473. *In* Role of fertilization in the intensification of agricultural production, Antibes, France. 15–18 Sept. 1970. Proc. 9th Int. Congr. Potash Inst., Berne.

Boyd, D.A., L.T.K. Yuen, and P. Needham. 1976. Nitrogen requirement of cereals. 1. Response curves. J. Agric. Sci. 87:149–162.

Bray, R.H. 1936. Calibrating soil tests for available potassium. Soil Sci. Soc. Am. Proc. 1:225–231.

Bray, R.H. 1937. New concepts in the chemistry of soil fertility. Soil Sci. Soc. Am. Proc. 2:175–179.

Bray, R.H. 1963. Confirmation of the nutrient mobility concept of soil plant relationships. Soil Sci. 95:124–130.

Bray, R.H., and L.T. Kurtz. 1945. Determination of total organic and available forms of phosphorus in soil. Soil Sci. 59:39–46.

Cate, R.B., Jr., and L.A. Nelson. 1965. A rapid method for correlation of soil test analyses with plant response data. North Carolina State Univ. Int. Soil Testing Series Tech. Bull. 1.

Cate, R.B., Jr., and L.A. Nelson. 1971. A simple statistical procedure for partitioning soil test correlation data into two classes. Soil Sci. Soc. Am. Proc. 35:658–660.

Chu, T.S., and L.M. Turk. 1949. Growth and nutrition of plants as affected by degree of base saturation of different types of clay minerals. Michigan Agric. Exp. Stn. Tech. Bull. 214.

Colwell, J.D. 1967. The calibration of soil tests. J. Austr. Inst. Agric. Sci. (Dec.) 321–330.

Cope, J.T., Jr. 1981. Effects of 50 years of fertilization with phosphorus and potassium on soil test levels and yields of six locations. Soil Sci. Soc. Am. J. 45:342–347.

Cope, J.T., Jr., and R.D. Rouse. 1973. Interpretation of soil test results. p. 35–54. *In* L.M. Walsh and J.D. Beaton (ed.) Soil testing and plant analysis. Revised ed. SSSA, Madison, WI.

Corey, R.B. 1987. Soil test procedures: Correlation. p. 15–22. *In* J.R. Brown (ed.) Soil testing: Sampling, correlation, calibration, and interpretation. SSSA Spec. Publ. 21. SSSA, Madison, WI.

Dahnke, W.C. 1985. Soil test correlation, calibration and interpretation. North Dakota State Univ. Agric. Exp. Stn. Bull. 517.

Dahnke, W.C., L.J. Swenson, R.J. Goos, and A.G. Leholm. 1984. Choosing a crop yield goal. North Dakota State Univ. Coop. Ext. Serv. Circ. S-F-822.

Dahnke, W.C., E.H. Vasey, and C.D. Fanning. 1985. North Dakota fertilization tables and equations based on soil test levels and yield goals. North Dakota State Univ. Coop. Ext. Ser. Circ. SF-882.

Daubeny, C. 1845. VII Memoir on the the rotation of crops and on the quantity of inorganic matters abstracted from the soil by various plant under different circumstances. R. Soc. London. Philos. Trans. 135:179–253.

Eckert, D.J., and E.O. McLean. 1981. Basic cation saturation ratios as a basis for fertilizing and liming agronomic crops: I. Growth chamber studies. Agron. J. 73:795–799.

Ellis, B.G., and R.A. Olson. 1986. Economic, agronomic and environmental implications of fertilizer recommendations. Michigan State Univ., North Central Agric. Exp. Stn. Reg. Res. Pub. 310.

Fitts, J.W. 1955. Using soil tests to predict a probable response from fertilizer application. Better Crops Plant Food 39(3):17–20.

Fitts, J.W., and W.L. Nelson. 1956. The determination of lime and fertilizer requirements of soils through chemical tests. p. 241–282. *In* A.G. Norman (ed.) Adv. in agronomy. Vol. 8. Academic Press, New York.

Foy, C.D., and S.A. Barber. 1958. Magnesium deficiency and corn yield on two acid Indiana soils. Soil Sci. Soc. Am. Proc. 22:145–148.

Gass, W.B., G.A. Peterson, R.D. Hauck, and R.A. Olson. 1971. Recovery of residual nitrogen by corn (*Zea mays* L.) from various soil depths as measured by 15N tracer techniques. Soil Sci. Soc. Am. Proc. 35:290–294.

Gauch, H.G. 1972. Inorganic plant nutrition. Dowden, Hutchinson and Ross, Stroudsburg, PA.

Graham, E.R. 1959. An explanation of theory and methods of soil testing. Missouri Agric. Exp. Stn. Bull. 734.

Grava, J. 1980. Importance of soil extraction techniques. p. 9–11. *In* W.C. Dahnke (ed.) Recommended chemical soil test procedures for the North Central Region. (Revised), North Dakota Agric. Exp. Stn., Bull. 499.

Hauser, G.F. 1973. Guide to the calibration of soil tests for fertilizer recommendations. Soils Bull. 18. FAO, Rome.

Havlin, J.L., D.G. Westfall, and H.M. Golus. 1984. Six years of phosphorus and potassium fertilization of irrigated alfalfa on calcareous soils. Soil Sci. Soc. Am. J. 48:331–336.

Herron, G.M., A.F. Dreier, A.D. Flowerday, W.L. Colville, and R.A. Olson. 1971. Residual mineral N accumulation in soil and its utilization by irrigated corn (*Zea mays* L.) Agron. J. 63:322–327.

Hill, F.J., F.E. Broadbent, and M. Fried. 1978. Timing and rate of fertilizer nitrogen for sugarbeets related to nitrogen uptake and pollution potential. J. Environ. Qual. 71:368–372.

Hunter, A.H., and J.W. Fitts. 1969. Soil test interpretation studies: Field trials. North Carolina State Univ. Int. Soil Testing Ser. Tech. Bull. 5.

Hunter, A.S. 1949. Yield and composition of alfalfa as affected by variations in the Ca/Mg ratio in the soil. Soil Sci. 67:53–62.

Leggett, G.E. 1959. Relationships between wheat yield, available moisture, and available nitrogen in eastern Washington dryland areas. Washington Agric. Exp. Stn. Bull. 609.

Liebhardt, W.C. 1981. The basic cation saturation concept and lime and potassium recommendations on Delaware's coastal plain soils. Soil Sci. Soc. Am. J. 45:544–549.

Mattingly, C.G.E. 1973. Residual and cumulative value of super phosphate in a three-course rotation. Rothamsted Report for 1973, Part I: 56–58.

McLean, E.O. 1977. Contrasting concepts in soil test interpretation: Sufficiency levels of available nutrients versus basic cation saturation ratios. p. 39–54. *In* T.R. Peck et al. (ed.) Soil testing: Correlating and interpreting the analytical results. ASA Spec. Publ. 29. ASA, CSSA, and SSSA, Madison, WI.

McLean, E.O., and M.D. Carbonell. 1972. Calcium, magnesium, and potassium ratios in two soils and their effects upon yields and nutrient content of German millet and alfalfa. Soil Sci. Soc. Am. Proc. 36:927–930.

McLean, E.O., R.C. Hartwig, D.J. Eckert, and G.B. Triplett. 1983. Basic cation saturation ratios as a basis for fertilizing and liming agronomic crops. II. Field studies. Agron. J. 75:635–639.

Melsted, S.W., and T.R. Peck. 1977. The Mitscherlich-Bray growth function. p. 1–18. *In* T.R. Peck et al. (ed.) Soil testing: Correlating and interpreting the analytical results. ASA Spec. Publ. 29. ASA, CSSA, and SSSA, Madison, WI.

Mitscherlich, E.A. 1928. The second approximation of the effect of growth factors. Z. F. Pflanzenernaehr. 12:281.

Munter, R.C. 1988. Laboratory factors affecting the extractability of nutrients. *In* W.C. Dahnke (ed.) Recommended chemical soil test procedures for the North Central Region. North Dakota Agric. Exp. Stn. Bull. 499 (Revised).

Murdock, J.T., and L.E. Engelbert. 1958. The importance of subsoil phosphorus to corn. Soil Sci. Soc. Am. Proc. 22:53–57.

National Academy of Science, National Research Council. 1961. Status and methods of research in economic and agronomic aspects of fertilizer response and use. Publ. 918. U.S. Gov. Print. Office, Washington, DC.

Nelson, L.A., and R.L. Anderson. 1977. Partitioning of soil test-crop response probability. p. 19–38. *In* T.R. Peck et al. (ed.) Soil testing: Correlating and interpreting the analytical results. ASA Spec. Publ. 29. ASA, CSSA, and SSSA, Madison, WI.

Olsen, S.R., C.V. Cole, F.S. Watanabe, and L.A. Dean. 1954. Estimation of available P in soils by extraction with sodium bicarbonate. USDA Circ. 939. U.S. Gov. Print. Office, Washington, DC.

Olson, R.A., A.F. Dreir, and R.C. Sorenson. 1958. The significance of subsoil and soil series in Nebraska soil testing. Agron. J. 50:185–188.

Olson, R.A., K.D. Frank, E.J. Deibert, A.F. Dreier, D.H. Sander, and V.A. Johnson. 1976. Impact of residual mineral N in soil on grain protein yields of winter wheat and corn. Agron. J. 68:769–772.

Olson, R.A., K.D. Frank, P.H. Grabouski, and G.W. Rehm. 1982. Economic and agronomic impacts of varied philosophies of soil testing. Agron. J. 74:492–499.

Olson, R.A., M.B. Rhodes, and A.F. Dreier. 1954. Available phosphorus status of Nebraska soils in relation to series clasification, time of sampling and method of measurement. Agron. J. 46:175-180.

Olson, R.A., W.R. Raun, Yang Shou Chun, and J. Skopp. 1985. N managemet and interseeding effects on N use efficiency of irrigated corn and grain sorghum. Agron. J. 78:856-862.

Passioura, J.G. 1973. Sense and nonsense in crop simulation. J. Aust. Inst., Agric. Sci. 39:181-183.

Perrin, R.K. 1976. The value of information and the value of theoretical models in crop response research. Am. J. Agric. Econ. 58:54-61.

Peterson, G.A., F.N. Anderson, G.E. Varvel, and R.A. Olson. 1979. Uptake of 15N labelled nitrate by sugar beets (*Beta vulgares*) from depths greater than 180 cm. Agron. J. 71:371-372.

Reuss, J.O., and P.S.C. Rao. 1971. Soil nitrate nitrogen levels as an index of nitrogen fertilizer needs of sugarbeets. J. Am. Soc. Sugar Beet Technol. 16:461-470.

Rouse, R.D. 1967. Organizing data for soil test interpretation. p. 115-123. *In* G.W. Hardy et al. (ed.) Soil testing and plant analysis. Part 1. SSSA. Spec. Publ. 2. SSSA, Madison, WI.

Soper, R.J., and P.M. Huang. 1963. The effect of nitrate nitrogen in the soil profile on the response of barley to fertilizer nitrogen. Can. J. Soil Sci. 43:350-358.

Sparrow, P.E. 1979a. Nitrogen response curves of spring barley. J. Agric. Sci. 92:307-317.

Sparrow, P.E. 1979b. The comparison of five response curves for representing the relationship between the annual dry-matter yield of grass herbage and fertilizer nitrogen. J. Agric. Sci. 93:513-520.

Snedecor, G.W., and W.G. Cochran. 1971. Statistical methods. 6th Ed. The Iowa State Univ. Press, Ames.

Van Ittalie, T.B. 1938. Cation equilibria in plants in relation to the soil. Soil Sci. 46:175-186.

Waggoner, P.E., and W.A. Norvell. 1979. Fitting the Law of the Minimum to fertilizer applications and crop yields. Agron. J. 71:352-354.

Waugh, D.L., R.B. Cate, Jr. and L.A. Nelson. 1973. Discontinuous models for rapid correlation, interpretation and utilization of soil analysis and fertilizer response data. Int. Soil Fertil. Evaluation and Improve. Ser. North Carolina State Univ. Tech. Bull. 7.

Welch, L.F. 1976. The Morrow Plots—hundred years of research. Ann. Agron. 27(5-6)881-890.

Willcox, O.W. 1944. Yield depression effect of fertilizers and its measurement by the universal yield diagram. J. Am. Soc. Agron. 36:20-31.

Willcox, O.W. 1949. Verification of the Mitscherlich effect law. Agron. J. 41:225-229.

York, E.T., R. Bradfield, and M. Peech. 1954. Influence of lime and potassium on yields and cation composition of plants. Soil Sci. 77:53-63.

Yost, R.S., E.J. Kamprath, E. Lobato, and G. Naderman. 1979. Phosphorus response of corn on an Oxisol as influenced by rates and placement. Soil Sci. Soc. Am. J. 43:338-343.

Young, R.A., J.L. Ozbun, A. Bauer, and E.H. Vasey. 1967. Yield response of spring wheat and barley to nitrogen fertilizer in relation to soil and climatic factors. Soil Sci. Soc. Am. Proc. 31:407-410.

Chapter 5

Soil pH and Lime Requirement Determination

W. VAN LIEROP, *British Columbia Ministry of Agriculture and Fisheries, Kelowna*

A potentiometrically determined soil pH is essentially an index of hydrogen ion (H^+) activity in solution at equilibrium with soil particles. However, H^+ activity is not the only factor contributing to the potential developed by a glass/calomel electrode combination in the presence of soil particles. Soil pH is, nonetheless, a measure of the intensity of acidity or alkalinity and is used in deciding whether a soil needs liming or acidification. While soil pH may not be a thermodynamically well-defined test, it is useful for the following reasons: soil pH determination is relatively rapid, precise, and inexpensive; values are easily interpreted and relatively well understood; and soil pH is broadly related to the availability of some elements that are required by, or are toxic to plants.

Mineral soils that have pH[1] values below 4 contain free acids generally arising from the oxidation of S or S-containing compounds. Furthermore, soils that have values below about pH 5.5 likely contain exchangeable Al that may be present at sufficiently high levels to be toxic to plants (Thomas, 1967; Hoyt & Nyborg, 1971a). These soils may also contain toxic levels of Mn (Adams & Wear, 1957; Morris, 1948; Ouellette & Dessureaux, 1958; Foy, 1964). Although it is possible to find soils with a pH below 7 that contain unreacted limestone (free carbonates), it is difficult to predict from soil pH values when soils start accumulating large quantities. Past experience suggests that at pH 7.2 soils will occasionally contain sufficient carbonates to neutralize the acid in the Bray-1 P-extracting solution (25 $cmol_c$ L^{-1} soil; Bray & Kurtz, 1945). In any event, soils with a pH situated between 7.0 and 8.5 contain free carbonates. Higher pH values, up to 9.9, are occasionally encountered in soils from semiarid and arid regions.These contain, in addition to free limestone, varying amounts of carbonates and bicarbonates of alkali metals, commonly Na.

Soil pH is an intensity measurement of soil acidity or alkalinity, and as such, does not indicate the amount H or hydroxyl ions present. An analogy for the intensity parameter is air pressure in a container. Although, pres-

[1] Unless otherwise indicated soil pH refers to measurements made in water.

 Soil Testing and Plant Analysis, 3rd ed.—SSSA Book Series, no. 3.

sure is proportionally related to the amount of air in containers of uniform size, pressure alone does not indicate the quantity of air present in containers of different volumes. Similarly, pH can indicate the amount of acidity present in soils with identical cation-exchange properties. However, pH does not indicate the amount of acidity or free-carbonates present in soils with differing cation-exchange properties. For example, a coarse-textured soil containing little organic matter (OM) (low cation-exchange capacity, CEC) with a pH of 4.0 may have a lime requirement of only 1 t ha^{-1} for attaining pH 6.5, while a fine-textured soil having the same pH and containing a high OM level may require 25 t to achieve the same pH. Similarly, alkaline soils may contain greatly different amounts of free carbonates at identical pH levels.

It is interesting that the term *neutral* soil is frequently used, but has no precise meaning. Strictly speaking, a soil with a pH of 7.0 has an equivalent activity of H and hydroxyl ions in solution at equilibrium with soil particles as found in freshly distilled water. However, distilled water at equilibrium with the 0.03% carbon dioxide (CO_2) in the atmosphere will have a pH value of about 5.7. Soil air contains from 10 to 100 times more CO_2 (Bradfield, 1941). Accordingly, water at equilibrium with these higher CO_2 levels can have pH values situated between about 5.7 and 4.5. As soil neutrality is undefined, why do we wish to raise the pH of acid soils to between 6.5 and 7.0? Precise reasons may not exist, and justifications are probably rooted in early concepts of soil chemistry. The relatively low cost of liming to a pH between 6.5 to 7.0 in many places may have influenced the common supposition that liming to that level is desirable for efficient crop production.

Generally, soils with a pH around 7.0 have a better supply of bases and possibly other plant nutrients than those that have been leached to a lower pH level. However, it may be erroneous to conclude, as Adams and Pearson (1967) pointed out, that soil-supplied plant nutrients become more available to plants by liming acid (leached) soils to a pH near 6.5 or 7.0. Similarly, Woodruff (1967) stated that pH values recommended by most agencies are far above those required by plants to produce optimum yields. Current agronomic thinking appears to favor the concept that acid soils be limed to a sufficiently high pH to suppress Al and Mn toxicity and, thereby, allow crops to achieve optimum yields. However, a large proportion of soils will continue to be limed to around pH 6.5 where liming costs for achieving or maintaining that standard are relatively low, because it has gained widespread acceptance.

Soil pH is a measure of the intensity of soil acidity (or alkalinity) but does not reveal the amount that will react with limestone when raising pH to a desired level. Buffer-pH lime requirement (LR) tests have been developed for estimating the amount of limestone required by a plow layer of soil (a weight or volume contained in a defined area) to achieve a desired pH. In fact, the LR of a soil is defined as the rate of liming material required by a plow layer to raise its pH to a specified value (Soil Science Society of America, 1987). A LR is a capacity index of soil acidity as it assesses the amount to neutralize for achieving a desired pH. Soil tests are also available for determining whether soils contain plant-toxic levels of Al or Mn and so decide whether liming is needed (Hoyt & Nyborg, 1971a, b, 1987). However, if levels are found to be sufficiently high to affect plant growth, a buffer-pH method is then used to determine the LR to raise pH sufficiently to neutralize toxic elements.

As indicated by Schollenberger and Simon (1945), the amount of H displaced from soil colloids depends on the final pH of a soil at equilibrium. Effectively, as pH is raised, successive amounts of covalently held H^+ are displaced from exchange sites for reaction with dissolved limestone. Accordingly, LR increases with increasing target pH; though it is difficult to accurately measure a LR to achieve a pH much above 6.5, presumably because free carbonates start accumulating in some soils around this pH.

Hydrogen ions are not released directly from permanent exchange sites located on mineral soil colloids for reaction with dissolved limestone (Schofield, 1949; Coleman et al., 1959; Pratt & Bair, 1962). According to this definition, exchangeable ions are held by pH independent electrostatic linkages. Hydrogen is often said, therefore, to be nonexchangeable. Nonetheless, H^+ is released by soil colloids when increasing pH by liming; this occurs through two pathways mainly. The first is hydrolysis of Al displaced from permanent exchange sites by increasing soil pH with liming (also of Fe at pH levels below 4 as indicated by McLean, 1982). The Al and Fe so displaced will form metal hydroxides and liberate H^+ for further reaction and neutralization. This source of H^+ is important on very acid mineral soils. The second source of H^+ originates from the dissociation of covalently held H from mineral and organic soil colloids that are progressively released as soil pH increases. Acidity so liberated can be named pH-dependent acidity (from pH-dependent exchange sites). The pH-dependent sites are located on soil OM, possibly allophane, and layer silicate-sesquioxide complexes (Coleman & Thomas, 1964; Schwertmann & Jackson, 1963; Volk & Jackson, 1964). Organic matter is probably the most important source of pH-dependent acidity as suggested by the study of Helling et al. (1964). They found the average contribution of clay and OM to CEC to be about equal at pH 2.5, and that their contribution increased linearly with increasing pH. The contribution of OM, however, increased at a significantly faster rate, being about six times higher than clay at pH 8.0. The increase in CEC is caused by a release of pH-dependent acidity. The latter will react with liming material when increasing soil pH. Of course, neutralization of acidity by liming acid Histosols will, in most instances, largely displace pH-dependent acidity.

Permanent negative exchange sites on mineral soil colloids are countered by metal cations, including Al, which are held by electrostatic forces. The sites are created by isomorphous substitution of cations within the structure of layer silicate minerals by cations of lower charge, thus providing soil colloids with a residual charge (Schofield, 1949; Coleman et al., 1959; Coleman & Thomas, 1967). These exchange sites are called *permanent* because they are present over a relatively wide range of pH values. Permanent sites are considered saturated with bases when positively charged monomeric- and hydroxy-Al have been displaced by other bases, principally Ca or Mg.

As Clark (1965, 1966b) pointed out, there is no consistent relationship between pH and the concentration of soluble Al in soils. Nonetheless, liming acid mineral soils to a pH around 5.5 or above will neutralize Al forms that might contribute to phytotoxicity. This conclusion is suggested by the work of Pratt and Bair (1962), the corrected lime potential data gathered for a wide range of soils by Turner and Clark (1965) and Clark and Nichol (1966), and the work of Coleman and Thomas (1967), Kamprath (1970), and Hoyt and Nyborg (1971a, 1972, 1987). Liming increases soil pH, which in turn causes displacement of Al ions from permanent exchange sites. Displaced Al ions subsequently hydrolyze, polymerize, and eventually precipitate as complex Al-compounds. Crops can tolerate low concentrations of Al as indicated by Adams and Pearson (1967), Kamprath (1970), Evans and Kamprath (1970), and Reeve and Sumner (1970a, b). Presumably, through negating the effect of Al toxicity by liming to a pH of around 5.5, or higher, and by eliminating possible Mn toxicity that might also occur in acid soils (van Lierop et al., 1982), crop yields are no longer limited by the effects of soil acidity.

I. pH DETERMINATION

A. Soil pH Concepts

Sorensen (1909) defined pH as being the negative logarithm of H^+ concentration. This notation has been maintained; though, H^+ potential has replaced concentration to adapt it to electromotive cell thermodynamics. A voltage change of about 59 mV is produced with each 10-fold change in H^+ activity (pH unit). The voltage is defined as zero at pH 7, positive at <7.0, and negative above that value. The linear relationship between voltage and pH enables a pH meter to interpolate sample values automatically after calibration.

The potential produced by an ideal H or glass electrode follows the relationship predicted by the Nernst equation. The potential increases as H^+ activity increases. Ion activity is defined as the proportion of total ions contributing to a measured potential. However, the activity of a cation cannot be determined independently from its accompanying anion. Activity coefficients, therefore, corresponds to average activity of species present. This difficulty has led the National Bureau of Standards (NBS) in the USA to calibrate and certify buffer standards for an operational pH scale (Bates, 1973; Durst, 1975).

The NBS uses a H and silver silver-chloride electrode combination for defining standards. This combination avoids liquid-junction potential errors. However, the H electrode is cumbersome and unreliable for soil pH determination. Accordingly, soil pH is determined routinely with a glass combined with a calomel reference electrode. It consists of measuring soil-solution emf and comparing it to defined buffer standard values. Accuracy depends on the differences in liquid junction-potential error between standards and sample.

B. Glass Electrode

The glass electrode is sensitive and reversible to H^+, and is free of oxidation-reduction interferences. However, unlike the H electrode, it has a very high internal resistance. It consists of a thin pH responsive glass bulb fused to a stem of nonresponsive glass. This construction ensures that readings are independent of immersion depth as long as the responsive area is covered with sample. The responsive glass consists of a thin conditioned membrane that allows the potential developed in its gel layer to be reflected to the internal electrode. A silver-chloride electrode immersed in HCl or a buffered chloride solution is commonly used as an internal electrode, though a calomel electrode can also be used.

During the conditioning process, constituent Na^+ and other ions dissolve and leach from the glass forming a matrix of interstices or gel layer. The vacated ion spaces can be filled by H^+ making the bulb selectively responsive to its activity. Entry of H^+ into the gel layer causes these to reflect a corresponding potential to the internal electrode. Porbably not unlike a voltage being impressed across the dielectric of a capacitor. The responsiveness of common glass electrodes is relatively unimpaired by Na^+ and other cations between pH 1 and 9.5 (Bates, 1973). As the pH of most soils fall within this range, it is able to provide accurate pH readings.

C. Glass Electrode Care and Rejuvenation

Accurate soil pH measurement with a glass electrode requires quick exchange of H^+ between hydrated gel layer and sample. As gel-layer water content is related to responsiveness, a glass electrode should not be placed in dehydrating agents. Furthermore, new electrodes or those that have been stored dry should be conditioned before usage by soaking the bulb in water overnight preferably.

Sluggish or impaired electrode performance can be verified simultaneously by noting uncorrected span and response time at calibration. To check span, immerse the electrodes in a first buffer and set the pH meter to its specified value, for instance pH 6.86; rinse the electrodes with distilled water and measure the pH of a second calibration buffer (pH 4.01). The pH meter should be able to provide the second pH without significant slope (percent efficiency) correction. A span correction of around 4% or more suggests that rejuvenation may improve response and accuracy. Response time is examined by determining whether a stable reading is obtained within about 30 s ± 0.05 pH. If a much longer time is required, the electrode should be examined for cracks; if it is, it should be replaced, it not, it may be rejuvenated.

Glass electrodes are rejuvenated by cleaning, or dissolving and removing a very thin glass layer from the pH-responsive area. Electrode impairment may be caused by a layer of spongy responsive glass having collapsed, or some contaminant sealing the gel layer. Rejuvenation should be started by immersing the tip in about 0.1 *M* HCl for 15 to 20 s, rinsing with distilled water, and repeating the treatment using a similar concentration of KOH.

The HCl-KOH treatments should be repeated several times. If electrode response is not restored, it may require a more drastic $NH_4 \cdot HF$ treatment. However, it dissolves a small amount of surface glass and shortens electrode life. It is accomplished by immersing the pH-sensitive bulb for 30 to 45 s in 20% $NH_4 \cdot HF$ contained in a small plastic vessel (it is a hazardous compound). Rapidly rinse the electrode tip in distilled water and dip in about 6 *N* HCl, followed by another rinsing. Subsequently, set the tip in water, overnight preferably.

D. Reference Electrodes

Hydrogen ion activity is determined with a glass electrode combined with a calomel or silver silver-chloride reference electrode (Bates, 1973). A calomel electrode consists of a platinum element in contact with mercury mercurous-chloride (calomel) surrounded by a KCl solution. The KCl is saturated with calomel and an excess of calomel ensures that it remains saturated with changes in temperature. On the other hand, the silver silver-chloride electrode consists of a silver element coated with a layer of silver-chloride. It is most commonly, like the calomel electrode, surrounded by a KCl solution. However, the surrounding KCl solution in this case is saturated with AgCl. Electrical contact between the calomel (or silver silver-chloride) electrode and test solution is maintained through a salt bridge formed by a small orifice permitting minute KCl flow (the liquid junction). However, the silver silver-chloride electrode can also be used without salt bridge to determine soil pH as shown by Clark (1964, 1966a). This type of electrode is employed by the NBS to define pH standards (Durst, 1975). Its main advantage is that test values are unaffected by liquid-junction potential or suspension effect errors.

A potassium chloride (KCl) concentration can vary a great deal in calomel electrodes without affecting accuracy. Saturated KCl appear to be the preferred concentration in North America; however, 3.5 *M* KCl seems to be more extensively used in Europe (Bates, 1973). Values measured with either solutions are practically identical because of the comparative nature of pH determination.

According to Willard et al. (1974), the 0.1 and 1.0 *M* KCl calomel electrodes are more accurate because they reach their equilibrium potential more rapidly and are less affected by changes in temperature than the more concentrated or saturated types. Considering the better properties of the 0.1 *M* KCl concentration, it is interesting that saturated KCl has almost replaced it (Bates, 1973). This replacement suggests a misunderstanding of the importance of KCl concentration for reliable pH measurements. Presumably, because of the ease of ensuring a constant KCl concentration, has the saturated KCl calomel electrode found preference in North America.

A junction potential error is always present at the liquid junction. Its magnitude cannot be easily established; however, it is generally accepted as being small in solutions. The error consists of a potential produced by an uneven diffusion between K^+ and Cl^-. A more serious source of error in soils, and colloidal suspensions, is the suspension effect that is additive to the junction potential. A suspension effect is manifested by differences between supernatant, and sediment or suspension pH. Generally, measured pH values decrease from supernatant to suspension.

Although the magnitude of a suspension effect may be more important than junction potential, it is not possible to distinguish between these errors. Whether either or both are caused by a Donnan-type potential produced by a modified mobility of anions relative to cations has not been resolved unequivocally. Nonetheless, these difficulties are associated with soil pH, and have received considerable attention (Loosjes, 1950; Jenny et al., 1960; Coleman et al., 1951; Peech & McDevit, 1951; Peech et al., 1953; Overbeek, 1953; Raupach, 1954, 1957; Bower, 1961; Clark, 1966a; Thomas & Hargrove, 1984) and others.

Peech et al. (1953) and Peech (1965a) suggest that the calomel electrode be placed in the supernatant when routinely measuring soil pH. Although this step may minimize errors, many soils need a 1:2 soil/water ratio instead of the 1:1 Peech (1965a) recommended to obtain sufficient supernatant depth. Even then, supernatants contain suspended colloidal particles. Junction potential and suspension effect errors can only be controlled effectively in two ways: 1, measuring pH in 0.01 *M* $CaCl_2$ or more concentrated salt solutions (Coleman et al., 1951; Clark, 1966a); or 2, using an Ag-AgCl electrode reference electrode without liquid junction (Clark, 1964, 1966a).

E. Reference Electrode Selection, Care, and Rejuvenation

Although liquid junctions designs vary, they can be classed roughly into four groups according to flow rates. At 50 to 150 $\mu L\ h^{-1}$ the ground glass sleeve junction has the fastest KCl flow rate of commercially available electrodes. This junction type is preferred for determining soil pH and buffer pH. However, it can be awkward because the sleeve slips occasionally. Porous ceramic or asbestos fiber junctions may, therefore, be preferred. These have modest flow junctions (5–10 $\mu L\ h^{-1}$) and are also recommended for measuring soil pH. Unfortunately, LR buffers have a propensity to diffuse into lower flow junctions. Contamination of electrode solution affects operation and accuracy. Electrodes with very low flow junctions (0.5–3 $\mu L\ h^{-1}$; i.e., controlled crack bead junction) and gel-filled electrodes (without KCl flow) are not recommended for soils. These electrodes were designed for applications where higher flow rates are detrimental to the sample.

Electrolyte level should be maintained as high as possible to prevent backflow and electrode solution contamination. Entrapped air bubbles possibly causing air-locks should be avoided when filling a dry electrode with KCl. The internal solution should be allowed to come to equilibrium after filling before use. Correct sample contact and an operational liquid junction are required for accurate measurements.

A blocked liquid junction is probably the most common calomel electrode problem. Symptoms are sluggish and creeping readings; though these are also present when measuring the pH of poorly buffered samples, particularly those varying by several pH units and displaying a memory effect. A correctly flowing junction requires regular replenishment of KCl. Operation can be verified with an ohmmeter (note: many digital voltmeters cannot make this test). Fill the electrode with KCl to above the filling hole. Immerse the liquid junction into a small beaker containing sufficient KCl solution of similar concentration. Place one ohmmeter lead tip into the internal solution through the filling hole and the other into the immersing solution. Typically, a resistance of a properly functioning liquid junction varies between 10 to 14 Kohm. An electrode having twice this reading, or more, needs service.

Correct junction operation should be checked when filling, if not daily. A ground-glass and reverse ground-glass sleeve junction is reestablished easily and rapidly by moving the sleeve, allowing some KCl to flow out, and firmly reseating the sleeve. The advantage of the reverse sleeve is that it does not fall off the electrode (break) as easily. On the other hand, porous ceramic and asbestos fiber-type junctions can be examined using air pressure applied by squeezing a bulb or plastic KCl filling bottle that seals against the filling hole. This pressure is often successful in demonstrating or re-establishing the liquid junction. A minute KCl flow, or a mixture of KCl and fine air bubbles will appear at the tip of a properly functioning junction. The junction of a dry electrode should be moistened by setting it in water overnight. Low flow rate junctions often suffer from blockage by soil particles. Removing the clogged portion of a junction tip may be accomplished by rubbing with waterproof fine emery cloth. This drastic treatment can sometimes repair a reluctant electrode that cannot be rejuvenated otherwise.

II. FACTORS THAT INFLUENCE MEASURED pH

A soil pH is produced by the activity of H^+ in solution to which are added the influence of several other factors. These are discussed under the following subheadings: A. soil/solution ratios, B. soluble salts, C. suspension effect, D. carbonic acid, E. drying, and F. seasonal pH fluctuation. Although discussed separately, many factors are closely interrelated. The objective of this section is to promote standard pH determination procedures, and minimize variations by factors other than changes in H^+ activity. Some verifications were made and regression data included.

A. Soil/Solution Ratios and Sample Size

Hissink (1930), as cited by Peech (1965a), reported adoption of a 1:2.5 ratio which suggest that the importance of a consistent procedure was recognized for achieving comparable results. Early work with potentiometric soil pH determination seemed concerned with moisture contents approaching field conditions. Huberty and Haas (1940) and Chapman et al. (1941) found that soil pH varied from about 0.5 to 1.5 pH with changing moisture contents. Higher moisture contents than necessary for producing a sticky point were required to ensure stable and reproducible results (Chapman et al., 1941).

Similarly, Turner and Nichol (1958) reported that a thinner ratio than 2:1 was necessary to generate reproducible values. More concentrated ratios than the 1:1 recommended by Peech et al.(1947) and Peech (1965a) appear less commonly used; though Jackson (1958) mentions a water saturation percentage and Hesse (1971) expresses preference for a saturated paste moisture content.

A consistent soil/water ratio is important for obtaining reproducible and comparable pH values. However, utilization of low moisture contents which rely on sample appearance (crumbly, sticky point, flow point, and saturated paste) are not advised for the following reasons: 1, moisture content varies with sample texture and OM content; 2, sample preparation is laborious; 3, moisture content is often subjective; 4, low moisture contents aggravate junction potential errors; 5, low moisture contents provide unreliable electrode-solution contact; and 6, electrode malfunction and breakage risk are higher when inserting into a paste.

Probably the most commonly recommended soil/water ratio for determining soil pH is 1:1 w/v (Peech et al., 1947; Peech, 1965a; McLean, 1973, 1982). Using a 20-g soil sample is recommended by Peech et al. (1947), Jackson (1958), and Peech (1965a). Soil pH, measured at a given soil/solution ratio, is not affected by sample size but may be by electrode position (supernatant, suspension, or sediment). A comparison of the pH of 60 soils determined in 1:1 w/v (Y; weighed samples) and 1:1 v/v (X; scooped) soil/water ratios ranged from 3.59 to 8.81 (Y), and 3.56 to 8.79 (X) with means of 6.28 and 6.25, respectively ($Y = 1.01X + 0.05$; $r = 0.999^{**}$, significant at $P = 0.01$; $s_{y.x} = 0.06$), revealed that weighing was unnecessary for precise pH measurement, as similar regression parameters were obtained between procedures and replicates of a procedure.

Using a 1:1 ratio usually provides sufficient supernatant to measure pH when using 20-g samples. However, slurries or pastes are often produced at this ratio when soils contain high clay or OM levels. A 1:2 ratio is probably more practical for routine pH determination when soils have a wide range in properties. Average pH generally increases somewhat by diluting soils from 1:1 to 1:2 ratio. This was confirmed by also determining the pH of 60 soils mentioned earlier at a 1:2 ratio. Values ranged from pH 3.56 to 8.79 with a mean of 6.25, and 3.64 to 9.11 with a mean of 6.39 for 1:1 and 1:2 (v/v) soil/water ratios, respectively. Although average pH increased slightly, differences were small enough to justify not using different interpretative norms: ($Y = 1.00X + 0.05$; $r = 0.999^{**}$; $s_{y.x} = 0.07$). Data was gathered with a ground-glass sleeve junction reference and a glass electrode combination.

Soil pH measurement in 0.01 *M* $CaCl_2$ relies on a 1:2 soil/solution ratio (Jackson, 1958; Peech, 1965a; Hesse, 1971) originally recommended by Schofield and Taylor (1955). Values are not sensitive to fairly wide changes in soil/solution ratio (Schofield & Taylor, 1955; Turner & Nichol, 1968; Clark, 1964; Ryti, 1965; White, 1969). It is unnecessary, therefore, to weigh soil samples for pH determination when using this solution. Puri and Asghar (1938) reported using soil/1 *N* KCl ratios ranging from 1:2.5 to 1:25 and found little effect on the pH of acid, but a significant effect on that of calcareous soils. Jackson (1958) and Hesse (1971) recommend using a 1:2.5 w/v soil/1 *N* KCl ratio. Collins et al. (1970) used a 1:2 soil/1 *N* KCl ratio. Little, if any, dilution effect was reported between pH measurements made at 1:1 and 1:2 ratios with 0.01 *M* $CaCl_2$ and 1 *N* KCl for mineral and Histosols by van Lierop and Tran (1979) and van Lierop (1981a), respectively.

1. Recommendation

A 1:2 or 1:2.5 v/v soil/solution ratio is suggested for determining pH of soils with a wide range of properties whether using water, 0.01 *M* $CaCl_2$, or 1 *N* KCl as the suspending solution. Changing a soil/solution ratio over a narrow range does not affect pH values for interpretative purposes. Accordingly, weighing soils is not generally justifiable for routine pH measurement. The ratio and suspending medium used should be specified when reporting results.

B. Soluble Salts and Lime Potential

The increase in pH produced by diluting soils from a 1:1 to a 1:2 soil/solution ratio is not directly related to acid dilution, but is caused by a decrease in H^+ dissociation with lower solution ionic strength. This conclusion is based on lime-potential findings which demonstrate that by adjusting pH for changes in activity of Ca and Mg in solution (the predominant cations in most acid soils) a remarkably constant value results (Schofield & Taylor, 1955). It was coined the lime potential and defined as being: pH − ½p(Ca + Mg). Effectively, in its nonlogarithmic form, the lime potential bears a readily recognizable kinship to the ratio law of Schofield (1947) $[H^+/(Ca^{2+} + Mg^{2+})^{1/2}]$. Soil pH and ½p(Ca + Mg) increase in value with dilution but their difference, the lime potential, remains constant over a relatively wide range of ratios and electrolyte concentrations. The constancy of the lime potential has been confirmed when Ca and Mg activity are determined (Turner & Nichol, 1958, 1962; Clark, 1964). The lime potential may also fluctuate less seasonally as it corrects for changes in solution salt content. In any event, a relatively dilution-independent pH value is obtained by measuring in 0.01 *M* $CaCl_2$, though there is little point in routinely subtracting 1.14 (½pCa in 0.01 *M* $CaCl_2$) from measured values as Thomas (1967) pointed out. The constancy of the lime potential appears desirable, however, determining ½p(Ca + Mg) is onerous and difficult to justify for a routine interpretative index of soil acidity.

The difference in pH between water and 0.01 *M* $CaCl_2$ measurements is often assumed to be about 0.5 pH unit. Although that may be a good average value, actual disparity varies from zero (similar values) to more than a pH unit, depending on soil salt content. Generally, pH differences decrease as salt content increases. The effect of salt level on pH divergence is suggested by lime-potential findings. However, the effect of salt concentration on the disparity between water and 0.01 *M* $CaCl_2$ pH values was also clearly demonstrated by Ryti (1965).

The effect of solution ion level on the size of discrepancy between water and 0.01 *M* $CaCl_2$ pH was verified by contrasting values of 30 relatively low-salt soils with conductivities $\leq$0.1 dS m^{-1} (1:2 v/v soil-water extracts) against those of a second group of soils with higher conductivities situated between 0.1 and 8 dS m^{-1}. Regression parameters derived between pH values from low-salt soils measured in water (Y; 1:2 v/v ratio; range = 4.49–8.52; y = 6.34) and 0.01 *M* $CaCl_2$ (X; 1:2 v/v ratio; range = 3.63–7.69; x = 5.54) produced the following: $Y = 0.99X + 0.83$; $r = 0.99^{**}$; $s_{y.x} = 0.21$. On the other hand, regression parameters from high-salt soils pH were different: (Y; 1:2 v/v soil/water ratio; range = 4.13–9.48; y = 6.77) and 0.01 *M* $CaCl_2$ (X; 1:2 v/v ratio; range = 4.18–8.34; x = 6.30): $Y = 1.086X - 0.07$; $r = 0.97^{**}$; $s_{y.x} = 0.34$. The difference between these regression parameters emphasizes the role of solution ionic level on the disparity between water and 0.01 *M* $CaCl_2$ pH values.

Correcting for salt content by entering extract conductivity as a second variable in fitting a multiple linear-regression equation decreased the standard error of estimate ($s_{y.x}$) from 0.34 to 0.23. Addition of solution conductivity as a contributing variable significantly improved the precision of the regression equation for predicting water from 0.1 *M* $CaCl_2$ pH values.

Lower pH generally results from using 1 *N* KCl instead of water. Puri and Asghar (1938) found that values were about 1.5 pH unit lower. However, results reported by Collins et al. (1970), van Lierop and Tran (1979), and van Lierop (1981a) suggest that the disparity is closer to a pH unit with acid low-salt soils. Values in 1 *N* KCl are generally lower than in 0.01 *M* $CaCl_2$; however, it is not uncommon to see little difference or the reverse with calcareous soils. The reason for a smaller difference or reversal with calcareous soils was thought to be related to a change in predominant cations on dissolution of soil carbonates.

Differences between water and 1 *N* KCl pH values are affected by salt levels as well as pH. The latter was suggested by the slope of a regression equation fitted between pH data of 40 soils measured in water (Y; 1:2 v/v ratio; range = 3.90–8.88; y = 6.85) and 1 *N* KCl (X; 1:2 v/v ratio; range = 3.40–8.05; x = 6.02): $Y = 0.86X + 1.70$; $r = 0.97^{**}$; $s_{y.x} = 0.28$. The slope suggests that pH disparity decreases with increasing soil pH.

1. Recommendations

The principal advantage in determining soil pH in 0.01 *M* $CaCl_2$ or 1 *N* KCl is that values are independent of electrode position (no suspension effect). As with measurements made in water, samples need not be weighed, as narrow changes in soil/solution ratios have little effect on pH. Values measured in 0.01 *M* $CaCl_2$, or 1 *N* KCl should rely on another interpretative scale as suggested by Davies (1971). Water, 0.01 *M* $CaCl_2$, and 1 *N* KCl values are closely correlated with reported coefficients situated between $r \approx 0.97^{**}$ and 0.99^{**} (Ryti, 1965; Collins et al., 1970; Davies, 1971; van Lierop, 1981a).

C. Junction Potential

It is assumed that a suspension effect is an aggravated junction potential caused by impeded mobility of K^+ relative to Cl^+ by large electrostatically charged soil colloids. Synonymous usage of these errors is made in this section. Refer to section I.D for an introduction to its mechanism.

The theory that attributes junction potential to a Donnan emf generated by an impeded mobility of K^+ relative to Cl^- was developed by Overbeek (1953). Experimental results supporting it were obtained by Coleman et al. (1951), Bloksma (1957), and Bower (1961). The suspension effect is illustrated by van Olphen (1963) who relates that by keeping the salt bridge of a first calomel electrode in the supernatant and that of another in the sediment a Donnan potential is observed. Marshall (1964), as cited by Thomas and Hargrove (1984), attributed this potential difference to a Donnan potential caused by a higher H^+ activity in the solution in proximity of the sediment relative to the supernatant. If such were the case, substituting two glass electrodes for the calomel units should result in a potential difference. Such a potential has not been reported.

Generally, suspension pH is lower than supernatant pH. Peech et al. (1953) indicate that junction potentials rarely exceed 0.25 pH unit. They and Peech (1965a) suggest that the liquid junction be located in the supernatant after precipitation of soil particles when measuring pH. Unfortunately, obtaining a clear supernatant usually takes considerably longer than allocated for measuring pH. Its absence means that the size of the suspension effect can be as high as reported in the literature. Maximum values of 0.9 and 0.5 pH unit were reported for mineral soils by Coleman et al. (1951) and Ryti (1965), respectively. Similarly, high values of 1.1 pH unit were observed with acid Histosols by van Lierop and MacKenzie (1977). The magnitude of the suspension effect can vary from being negligible to over a pH unit, and appears to be largely infuenced by soil salt content. Coleman et al. (1951) reported that soil pH in 1 *N* KCl remained unaffected by electrode position in the sample. Clark (1966a) reported finding no suspension effect when salt content was higher than 0.005 *M* $CaCl_2$.

Strangely, a negative junction potential was reported by Peech et al. (1953) as being caused by a change in the nature of the predominant cations. Ryti (1965) also reported higher suspension than supernatant pH for a soil with a pH above 7.0. Similarly, Raupach (1954) reports suspension pH values that are as much as 1.4 pH unit higher than supernatant values for soils with pH $\geq$ 7.0.

The pH of many calcareous soils increases when stirring a supernatant into a suspension of soil particles. This anomalous suspension effect, however, may be attributed to rash acceptance of an apparently steady value. Readings of reluctant calcareous soils may require up to 30 min to settle to near-equilibrium value. Slow stabilization is probably not just related to low H^+ activity, but appears attributable to low sample buffering. The reverse suspension effect observed with the verification in this study was thought to be produced by an absence of equilibrium pH, because similar reverse suspension effects were observed with soils suspended in 0.01 *M* $CaCl_2$, 1 *N* KCl, and 2 *N* NaCl. It is generally accepted that suspension effects are absent when measuring pH in salt solutions. This suggested that weak sample buffering was responsible for a slow response time, hence anomalous suspension effect. Clark (1964) has indicated that to determine pH of calcareous soils accurately may be difficult.

The time needed for measuring pH of calcareous soils can be reduced to that required for acid soils by swirling soil/solution mixtures with a magnetic stirrer. Considerable hysteresis can be experienced when measuring pH of a calcareous soil without stirring, especially after measuring an acid soil. Initially, pH readings creep relatively rapidly when changing to a weakly buffered calcareous soil; however, the rate of pH change slows after the first minutes, and a rash reading is easily accepted because it seems stable compared to acid soils. Yet, it can be in error by as much as 1.5 pH unit.

1. Recommendations

Soil pH measurements in 0.01 *M* $CaCl_2$ or 1 *N* KCl are free of suspension effect errors. Values are thus independent of electrode position. Calcareous soil/solution mixtures should be stirred during pH determination.

D. Effect of Carbon Dioxide on Soil pH

The pH of distilled water at equilibrium with carbonic acid at partial pressure of 0.03% CO_2 (pCO_2) in the atmosphere is about 5.72 (Bradfield, 1941). However, a soil atmosphere contains much higher pCO_2 pressures than the air above it. This increase is largely a result of soil biological activity and restricted soil ventilation. Bradfield (1941) suggests that soils contain from 10 to 100 times more CO_2 than the atmosphere above it, and values as high as 12% CO_2 have been proposed (Simmons, 1939). Higher pCO_2 pressures impose higher soil-solution carbonic and bicarbonic acid contents. In turn, these higher contents lower soil pH and increase the concentration of Ca and Mg in solution (Simmons, 1939; Whitney & Gardner, 1943; Turner & Clark, 1956).

The effect of pCO_2 pressures on soil pH have been studied mostly with calcareous soils. Whitney and Gardner (1943) report changes between 1 and 2.5 pH units by increasing the pCO_2 a 100-fold. Although, the change in pH can be predicted from pCO_2 pressures and the solubility product constant of calcareous material, these vary in different soils, and free carbonates are often more soluble than calcite (Cole, 1957; Doner & Pratt, 1969; Olsen & Watanabe, 1959; Clark, 1964). Increasing pCO_2 pressures also affect pH of noncalcareous near-neutral soils (Nichol & Turner, 1957); increasing pCO_2 pressure by a 100-fold, decreased soil pH by about one unit. The effect of CO_2 variation on acid soil pH has not been as extensively studied; however, it can be assumed from lime-potential findings that increasing carbonic and bicarbonic acid concentrations increases Ca and Mg concentrations in solution. These increases, decrease soil pH.

1. Recommendation

Although pCO_2 changes can affect soil pH significantly, air- or oven-drying samples reduces pCO_2 pressures to that in the atmosphere. As the pCO_2 pressure in the atmosphere can be considered constant, drying samples before analysis should eliminate the effect of variable pCO_2 on soil pH.

E. Effect of Drying on Soil pH

Soils are usually dried, crushed, and sieved before analysis. Baver (1927) reported a decrease between 0 and 0.6 pH by air drying. Bailey (1932) concluded that pH was generally lowered somewhat by air drying. Similarly, Huberty and Haas (1940) and Collins et al. (1970) found that oven drying decreased pH further. Bowser and Leat (1958) also observed that pH decreased an average of about 0.4 pH unit with drying, but noted that it increased with a calcareous soil. Average decreases of 0.15 and 0.5 pH unit were reported for drying acid Histosols by Davis and Lawton (1947) and van Lierop and MacKenzie (1977), respectively.

Interestingly, the decrease in pH with drying appears reversible. Comparing pH values of 100 acid and calcareous soils measured 1 d after adding water and 7 d later indicated that average pH increased by about 0.25 unit. The increase was assumed to be related to reducing conditions produced by saturating soils. Although, little precise information is available concerning the causes of acidity fluctuation with drying or wetting soils, its reversible nature suggests that oxidation and reduction of S compounds have an important role. In any event, drying-wetting cycles promote OM mineralization, and eventually produce a salt effect on soil pH. However, drying soils before analysis provides a constant soil/solution ratio when measuring pH, hence less dilution of soil salts, and a lower pH with higher salt.

The decrease in pH with drying can be quite important in soils containing large amounts of sulfides. Hesse (1971) mentions changes of 2 pH units between the wet and dry season in an East Pakistan soil. Greater changes can be observed on acid sulfate clays (often called *cat-clays* or mud-clays) from around pH 7 when flooded to pH 2 during the dry season (Moormann, 1963; Dost, 1973; Bohn et al., 1985). The oxidation of S compounds are largely responsible for such wide variations in pH, though the oxidation of Fe compounds and hydrolyses of Al and Fe have important roles.

1. Recommendation

It is recommended that soils be air or oven dried before determining pH. Drying affects soil pH, but changes are not generally agronomically significant unless soils contain large amounts of sulfides. Wide changes in pH, however, can result from drying soils containing sufficient oxidizable S compounds. Whether pH of these soils is determined on field-moist or dried samples, or both, depends on the objectives of the tests.

F. Seasonal pH Fluctuation

In view of the many factors that affect soil pH during the growing season, it is not surprising that it fluctuates during the year and from year to year. Baver (1927) and Huberty and Haas (1940) noted that pH vaired about a unit during the growing season and that variations seemed related to prevailing moisture regime. Bowser and Leat (1958) found that pH varied by as much as 2 units during the growing season on a calcareous soil, and that moisture and pH fluctuations appeared interrelated. Generally, pH gradually increases and decreases during periods of high and low rainfall, respectively (van der Paauw, 1962). Although, fluctuations of field-moist soils may be partially attributable to changes in soil atmosphere pCO_2 pressure during periods of high biological activity, low pH tends to occur during summer months when moisture levels are often lower and presumably soil aeration is better. In any event, drying soils, before pH determination, reduces the effect of pCO_2 to a homogeneous value, yet pH fluctuations are observed on air- or oven-dried soils. The pH of acid Histosols varies like that of mineral soils (van Lierop & MacKenzie, 1977). They reported variations up to 1 pH unit on limed but only 0.2 units on unlimed treatments with field-moist samples. Their findings support the assumption that variations in carbonic acid concentrations are implicated in pH fluctuations of field-moist samples. Be that as it may, the principal factors contributing to pH fluctuations have been discussed under drying and salt effects. The importance of soil-solution salt content on pH fluctuation was confirmed by Collins et al. (1970).

1. Recommendations

Seasonal fluctuation indicates that pH is a changing soil property. Many factors affect its value. Fluctuation is an integrated response to all components contributing to pH value during a period. If anything, it underlines the desirability of using a uniform and well-documented procedure so that factors other than changes in H^+ activity have a minimal impact. A desirable procedure is precise, rapid, and objective; however, its properties and relationship to other procedures should be known.

III. pH MEASUREMENT

A. Buffer Standards

Certified buffer pH standards can be obtained from the Office of Standard Reference Materials, National Bureau of Standards, Washington, DC, USA 20234. Standards can also be purchased as solutions, tablets, or packets for dissolution from chemical supply houses. They may also be prepared advantageously from high purity chemicals. Adding a preservative consisting of 1 mL of chloroform or toluene, or about 1 g of thymol L^{-1} is recommended to discourage mold growth.

1. KHP Buffer, pH 4.01

Dry potassium hydrogen phthalate (1-$KOCOC_6H_4$-2-COOH; FW = 204.23) at 110 °C for 2 h. Dissolve 10.21 g in distilled water and dilute to 1 L. The pH of this standard is 3.999, 4.002, 4.008, 4.015, and 4.024 at 15, 20, 25, 30, and 35 °C, respectively.

2. Phosphate Buffer, pH 6.86

Dry KH_2PO_4 and Na_2HPO_4 at 110 °C for 2 h. Dissolve 3.44 g of KH_2PO_4 and 3.55 g of Na_2HPO_4 in distilled water and dilute to 1 L. The pH of this buffer is 6.900, 6.881, 6.865, 6.853, and 6.844 at 15, 20, 25, 30, and 35 °C, respectively.

3. Borax Buffer, pH 9.18

Dry $Na_2B_4O_7 \cdot 10H_2O$ at 110 °C for 2 h. Dissolve 3.81 g in distilled water and dilute to 1 L. The pH of this standard is 9.276, 9.225, 9.180, 9.139, and 9.102 at 15, 20, 25, 30, and 35 °C, respectively.

B. pH-Meter Calibration

Use pH 6.86 and 4.01 buffer standards for calibration and verification if few calcareous soils are present. However, a pH 9.18 buffer should be added to check operational span when a significant proportion of calcareous soils are tested.

As many different types of pH meters are in use, only general calibration principles can be described. Exact steps are provided in the operation manual supplied with the pH meter. Generally, calibration should emulate the following approach. Set average operating temperature (usually 25 °C). Insert electrodes in the pH 6.86 buffer standard, gently swirl buffer, and set value when reading has stabilized using the calibration control. Subsequently, rinse electrodes with distilled water and insert into pH 4.01 buffer while swirling gently: a stable reading can be adjusted using the slope (percent efficiency) correction. If a slope correction of about 4% or greater is required see sections I.C and I.E on electrode care and rejuvenation. A calibrated pH meter will read exact standard values without further adjustment. Accuracy should be verified regularly during operation. Electrodes are stored in distilled water when the pH meter is not in use.

C. Measuring pH in Water

1. Procedure

Scoop 10-mL soil samples into appropriate beakers, add 20-mL aliquots of distilled water, and stir thoroughly. Allow the mixture to stand between 15 to 30 min and measure pH. Sluggish calcareous samples should be swirled with a magnetic stirrer to enhance rapid and accurate pH measurement.

D. Measuring pH in 0.01 *M* $CaCl_2$

1. Reagents

a. Prepare a 1 *M* $CaCl_2$ stock solution ($CaCl_2 \cdot 2H_2O$; FW = 147.02) by dissolving 147.02 g in 1 L of distilled water.

b. A 0.01 *M* $CaCl_2$ solution is prepared by diluting 10 mL of 1 *M* $CaCl_2$ stock solution to 1 L. One liter suffices for 50 samples.

2. Procedure

Scoop 10-mL soil samples into appropriate beakers, add 20-mL aliquots of 0.01 *M* $CaCl_2$, and stir thoroughly. Allow the mixture to stand between 15 to 30 min and measure pH. Sluggish calcareous samples should be swirled with a magnetic stirrer to enhance rapid and accurate measurement.

E. Colorimetric Soil pH

Soil pH was determined colorimetrically before pH meters became widely available. The more sensitive indicators and proper technique provide pH values within about 0.3 pH unit of potentiometric values (Davis & Lawton, 1947; Collins et al., 1970; Hesse, 1971). This type of precision is adequate for many applications. Field people often rely on the convenience of these procedures. Information about preparation and use of indictors can be obtained from Jackson (1958), Bates (1973), Woodruff (1961), Peech (1965a), and Hesse (1971).

IV. LIME REQUIREMENT DETERMINATION

A. Introduction

A LR is the amount of limestone ($CaCO_3$) needed by a plow layer of acid soil to increase its pH to a desired level. The contribution of pH-dependent acidity to LR gradually expands as soil pH is increased. A LR increases, therefore, as a target pH rises. At a pH above about 5.5, the permanent CEC of mineral soil colloids is saturated with bases and no longer contributes to LR.

Rapid buffer-pH LR methods measure a proportion of the acidity neutralized by $CaCO_3$. The more accurate methods rely on exhaustive displacement (leaching) of acidity to measure LR (Peech, 1965b). Disparity of LR values between various buffer-pH procedures have two principal causes. First, calibration accuracy affects recommendations. Calibration should adjust for incomplete measurement of acidity by a buffer. Second, the discrepancy between soil-buffer and target pH affects the amount and proportion of acidity measured. The buffer's initial pH and buffering capacity also influence equilibrium pH and, consequently, the amount and proportion of acidity included in a measurement.

Accurate LR recommendations depend on three interrelated postulates. First, a buffer-pH procedure should be accurately[2] calibrated. Accuracy is particularly important for determining the LR of soils with low-buffering capacities as these may be limed (overlimed) with a single application. Second, accurate liming requires uniform application. It also requires that the rate be incorporated in the correct volume or weight of soil. Third, the use of a general liming factor affects accuracy. A factor is often included in recommendations to attempt correction for possible lower reactivity of commercial limestone relative to finely ground $CaCO_3$ used to calibrate or verify a method. An arbitrary general factor may not be justified as limestone solubility is related to its nature and grinding fineness.

Space restrictions allow only a partial presentation of the many methods in use; an overview of operational assumptions and relative accuracy for the principal procedures has, however, been included. Three buffer-pH methods with updated calibrations are advanced, though others are referenced and can be used effectively. The three methods were selected because they are well known, relatively simple to use, and versatile.

B. Lime Requirement: Volume or Weight Basis?

Buffer-pH LR determination methods are calibrated or verified with liming rates expressed in meq $CaCO_3$ 100 g^{-1} or mL^{-1} of soil. More recently, LR rates have been reported in $cmol_c$ kg^{-1}, which are equivalent SI units. The LR calibrations for the proposed procedures are in metric tonnes of $CaCO_3$ 2 million L^{-1} of soil (a furrow-layer of 1 ha to a depth of 20 cm). These are equivalent in magnitude to $cmol_c$ L^{-1} soil. Corresponding rates in tons $acre^{-1}$ to 20-cm depth are derived by multiplying recommendations by 0.45. The merits and drawbacks of various approaches of calculating LR are discussed by van Lierop (1989).

C. Shoemaker, McLean, and Pratt Procedure

1. Principles

The Shoemaker, McLean, and Pratt (SMP) (Shoemaker et al., 1961) single-buffer procedure has been widely adopted and found particularly accurate for soils needing more than 8 to 10 t limestone ha^{-1} (McLean et al., 1966, 1978; Tran & van Lierop, 1981a, 1982). It has often been described as more sensitive to acidic Al than the Woodruff buffer (McLean et al., 1960, 1964, 1966; Shoemaker et al., 1961; Pionke et al., 1968). A greater sensitivity was assumed to be the reason for the better accuracy of the SMP buffer at higher LR values. However, it should then also be more precise[2] as high LR soils generally contain larger amounts of acidic Al. The SMP buffer is consistently as closely correlated with reference LR values as the Woodruff, which suggests that it responds to the same forms of acidity (McLean et al.,

[2] A procedure is accurate when measured and true values agree. It is precise when measured values are reproducible and proportionately related to true values.

1960, 1966, 1978; Pionke et al., 1968; Webber et al., 1977; Fox, 1980; Tran & van Lierop, 1981a, 1982; Loynachan, 1981; Brown & Cisco, 1984; Alabi et al., 1986).

Similar correlation coefficients between reference LR and soil-buffer pH values suggest that the SMP and Woodruff buffers are equally precise. This fact and the comparison of corresponding accuracy zones (i.e., the Woodruff is relatively accurate at low LR but recommends too little for high LR soils while the SMP buffer recommends too high LR when soils require low levels but is relatively accurate for high LR) suggest two particularly important conclusions. First, the dissimilar accuracy between zones is probably not related to inherent differences between buffers, but rather to corresponding calibration procedures. Although, Shoemaker et al. (1961) did not describe the procedure used for calibrating the SMP buffer, they apparently employed a regression method for relating soil-buffer pH to reference LR values. This procedure is considered largely responsible for its greater accuracy at higher LRs. The second conclusion suggested by the relative accuracy zones is that the relationship between soil-buffer pH and LR, unlike between buffer pH and additions of strong acid, is not linear but curvilinear (Fig. 5-1).

Curvilinearity between soil-buffer pH and LR values increases with increasing difference between initial buffer and target pH (desired soil pH; Tran & van Lierop, 1981a, 1982). The principal reason for curvilinearity is that buffer-pH procedures measure a greater proportion of soil acidity from low than high LR soils. Conversely, linearity is optimized by using similar initial buffer and target pH. Superfluous pH-dependent acidity is measured when soil-buffer pH is higher than target pH.

Curvilinearity between incubation and measured LR values was indicated by the data of Loynachan (1981) for the SMP and Woodruff buffers. It was also suggested by the data of Webber et al. (1977), and the discussion of Alabi et al.(1986). Furthermore, it was confirmed by Nômmik (1983) and

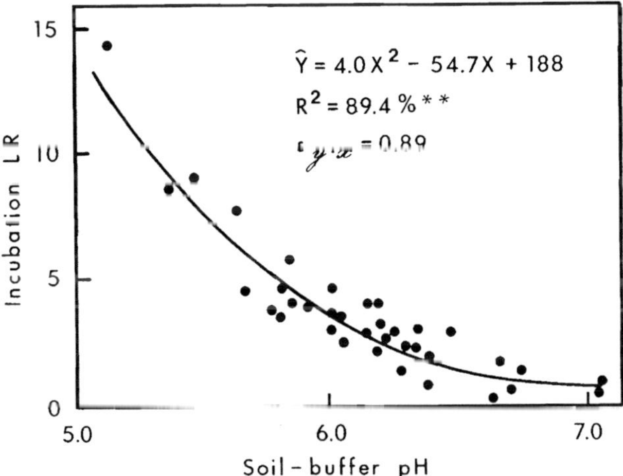

Fig. 5-1. Curvilinear relationship between soil-buffer pH and LR values to attain pH 5.5 for the SMP buffer. (From Tran & van Lierop, 1982.)

Soon and Bates (1986). A double-buffer procedure tends largely to correct for curvilinearity by inter- or extrapolating a desired change in soil pH against a calibrated change in soil-buffer pH. In any event, the accuracy of the SMP single-buffer procedure can be improved by using a curvilinear instead of a linear calibration. Curvilinearity is relatively small when target pH is 6.5, but it gradually increases as target pH decreases to 5.5.

McLean was aware that the SMP buffer measured a decreasing proportion of soil acidity as soil-buffer pH values decreased (McLean et al., 1960, 1964, 1966, 1977; McLean, 1973, 1978, 1982). A SMP double-buffer procedure was proposed by McLean et al. (1977, 1978) for improving the accuracy of LR determination of low-exchange capacity soils. However, a single-buffer procedure is more rapid, versatile, and lower in cost. Accordingly, the SMP-buffer calibrations proposed by Shoemaker et al. (1961) and McLean (1973, 1982) were adapted for curvilinearity by combining calibrations with those obtained by Tran and van Lierop (1981a, 1982) to improve accuracy at low LR values and for achieving soil pH values of 5.5, 6.0, 6.5, and 7.0. Calibrations are in metric tonnes of $CaCO_3$ 2-million L^{-1} of soil (1 ha to a depth of 20 cm): see section IV.B for modifying rates to fit other measurement units.

2. Equipment and Reagents

1. pH meter equipped with a glass and ground-glass junction reference electrode.
2. Standard buffers, pH 6.86 and 4.01.
3. Beakers.
4. Mechanical shaker.
5. Lime requirement buffer.
6. Automatic pipette.
7. A 10-mL scoop soil measure.

3. Shoemaker, McLean, and Pratt Buffer Preparation

1. Ten litters of SMP buffer (500 samples) are prepared by dissolving the following chemicals in about 5 L distilled water while stirring vigorously:
   a. 18.0 g of *p*-nitrophenol ($NO_2C_6H_4OH$; FW = 139.11).
   b. 30.0 g of potassium chromate (K_2CrO_4; FW = 194.2).
   c. 531 g of calcium chloride dihydrate ($CaCl_2 \cdot 2H_2O$; FW = 147.02).
2. Dissolve 20.0 g of calcium acetate [$(CH_3COO)_2Ca \cdot H_2O$; FW = 176.19] in about 2 L distilled water.
3. Mix solutions 1 and 2; continue stirring vigorously for about 2 h.
4. Add 100 mL of TEA[3] stock solution.
5. Dilute to 10 L and continue stirring for 6 to 8 h.

[3]Triethanolamine (TEA) stock solution is used because TEA [$N(CH_2 \cdot CH_2OH)_3$; FW = 149.19] is very viscous and difficult to pipette accurately. It is prepared by diluting 280.5 g of TEA (250 mL·1.122 g mL^{-1} = 280.5 g) to 1 L.

Table 5-1. Relationships between soil-SMP-buffer pH and lime requirement (LR) values to achieve pH 5.5, 6.0, 6.5, and 7.0 of mineral soils.

Soil-buffer pH	LR†			
	pH 5.5	pH 6.0	pH 6.5	pH 7.0
6.9	0.5	0.6	0.7	0.9
6.8	0.6	1.0	1.2	1.5
6.7	0.7	1.4	1.8	2.2
6.6	0.9	1.8	2.5	2.8
6.5	1.2	2.3	3.3	3.6
6.4	1.6	2.9	4.0	4.4
6.3	2.0	3.5	4.9	5.2
6.2	2.5	4.2	5.7	6.0
6.1	3.1	4.9	6.6	7.0
6.0	3.8	5.6	7.5	8.0
5.9	4.5	6.5	8.5	9.0
5.8	5.3	7.3	9.5	10.0
5.7	6.1	8.2	10.5	11.2
5.6	7.0	9.2	11.6	12.4
5.5	8.0	10.2	12.7	13.6
5.4	9.1	11.3	14.0	14.9
5.3	10.2	12.4	15.0	16.2
5.2	11.4	13.6	16.2	17.6
5.1	12.7	14.8	17.5	19.0
5.0	14.0	16.1	18.8	20.4
4.9	15.5	17.4	20.1	22.0

† Lime requirement in metric tonnes $CaCO_3$ ha^{-1} for a furrow layer of 20-cm depth (2 million L) soil.

6. Adjust pH to 7.50 ± 0.02 by with either 4 *N* NaOH or HCl as required.
7. Filter buffer through a fiberglass membrane if necessary.

4. Verification

Buffering capacity is checked by titrating 20 mL of SMP buffer from pH 7.50 to 5.00 with 0.1 *N* HCl. It should be 0.28 ± 0.005 $cmol_c$ HCl/pH. Correctness affects accuracy.

5. Procedure

Measure 10 mL of dry soil samples into beakers (see footnote 4 for a description of volumetric soil measurement). Add 20 mL of SMP buffer, stir with a glass rod to wet samples thoroughly. Shake soil-buffer samples for 15 min at about 200 cycles min^{-1}. Let samples stand an additional 15 min and measure soil-buffer pH. Obtain LR values for desired target pH from Table 5-1 or calculate with equation(s) from Table 5-2.

[4] A volume of soil is measured with a stainless steel cylindrical ladle filled to overflowing by scooping, tapped twice or thrice with a stainless steel rod to eliminate large voids, levelled, and transferred to a beaker. Avoid filling scoop against a side of the container.

Table 5-2. Regression equations relating soil-buffer pH (BpH) to LR values for achieving pH 5.5, 6.0, 6.5, and 7.0 of mineral soils.

Equations†	$\%R^2$	$s_{y.x}$
LR pH 5.5		
LR(SMP) = $3.498(BpH)^2 - 48.77(BpH) + 170.46$	99.6	0.33
LR(WDF)‡ = $5.524(BpH)^2 - 78.265(BpH) + 277.1$	99.7	0.29
LR(MEO) = $1.6(BpH)^2 - 25.2(BpH) + 96.3$	99.9	0.12
LR(MEH) = $55.1 - 8.86(BpH)$	99.9	0.02
LR pH 6.0		
LR(SMP) = $2.573(BpH)^2 - 38.76(BpH) + 145.58$	99.8	0.25
LR(WDF) = $1.178(BpH)^2 - 26.234(BpH) + 124.9$	99.8	0.25
LR(MEH) = $75.4 - 11.75(BpH)$	99.9	0.02
LR pH 6.5		
LR(SMP) = $1.867(BpH)^2 - 31.82(BpH) + 131.23$	99.9	0.06
LR(WDF) = $-1.063(BpH)^2 - 1.156(BpH) + 58.7$	99.9	0.09
LR(MEH) = $3(BpH)^2 - 47.5(BpH) + 180.2$	99.9	0.15
LR pH 7.0		
LR(SMP) = $2.455(BpH)^2 - 39.50(BpH) + 156.58$	99.9	0.11

† Equations were derived from tabulated data. See corresponding tables for application limits.

‡ WDF, MEO, and MEH represent Woodruff, Mehlich original, and Mehlich-buffer calibrations, respectively.

6. Comments

All buffer-pH procedures extract acidity when initial buffer pH is higher than soil pH. An accurate buffer-pH procedure can estimate a LR within about 0.25 unit of target pH. This tolerance probably approaches the accuracy limitation of the better procedures when used with soils having a wide range of properties. Calculated LR precision should only be expressed to one place after the decimal. Rare instances of soil-buffer pH values above 6.9 are observed on coarse-textured soils that need liming, in such cases recommend a minimum rate. Also see section IV.F.1 for applicable comments.

D. Woodruff Single Buffer Method

1. Principles

Woodruff (1947, 1948) probably acquired his buffer-pH LR test concepts from Brown (1943). His is the first widely used test and has been evaluated many times. The procedure allows rapid and convenient LR determinations. Woodruff (1948) stated that though the quantity indicated by the test was sufficient to bring soil pH to 7.0, heterogeneity of mixing limestone in the field and the presence of slowly reacting larger particles provided the needed insurance that soils would not be completely neutralized. He indicated that the concept of liming soils to a pH around 6.0 to 6.5 was fairly well established and that this buffer would achieve that end.

According to Brown and Cisco (1984), Woodruff produced a new LR determination buffer in the mid-1960s. Though its development is not well documented, it is discussed nonetheless under section IV.F. It is well to remember that buffer accuracy is imparted by calibration and use, and not particularly composition. Woodruff (1948) stated the following concept concerning measurement of exchange acidity with a buffer: "If the depression in pH is restricted to small values, it is an absolute measure of the exchangeable hydrogen." The effect of buffering capacity on acid-extraction efficiency is supported by buffer-pH experience.

McLean et al. (1960) evaluated the Woodruff buffer and suggested that its accuracy could be improved using regression procedures. McLean et al. (1966) found that it recommended too little for soils with high LR but that it seemed accurate at lower rates. A similar differentiation was encountered between low and high LR soil groups by Tran and van Lierop (1981a). From this differentiation they suggested that the relationship between soil-buffer pH and LR was more accurately described by a curvilinear line.

Webber et al. (1977) evaluated the Woodruff buffer for determining the LR to 5.5 and 6.0 of Canadian soils and found it was as precise as the SMP (Shoemaker et al., 1961) buffer. Loynachan (1981) compared it and the SMP buffer procedure and found that LR values were closely correlated (r = 0.99**). However, plotting his Woodruff (or SMP) against incubation LR values indicated a strong curvilinear relationship. A curvilinear relationship was observed for the Woodruff buffer by Tran and van Lierop (1981a, 1982): calibration curvilinearity has been discussed for the SMP-SB procedure and comments are applicable to the Woodruff procedure. Fox (1980) also evaluated the Woodruff procedure and found, like others, that it was as precise as any, that it was quite accurate at low values, but that it underestimated high LR. The Woodruff buffer was also studied by Brown and Cisco (1984) and Alibi et al. (1986) and their findings confirm those of others. The Woodruff buffer is generally evaluated for determining LR to achieve pH 6.5, but Tran and van Lierop (1982) also verified it for determining LR to pH 6.0 and 5.5. They found it as precise as any for pH 6.0, but that it can be a little less precise than some for determining LR to pH 5.5. In conclusion, the original Woodruff buffer calibration is not accurate, improved calibrations derived by Tran and van Lierop (1981a, 1982) are, therefore, provided for determining the LR to pH 5.5, 6.0, and 6.5 for mineral soils. These are in metric tonnes of $CaCO_3$ 2 million L^{-1} of soil (1 ha to a depth of 20 cm): see section IV.B for modifying rates to fit other measurement units.

2. Equipment and Reagents: See section IV.C.2.

3. Preparation of Woodruff (1948) Buffer

1. Ten liters of Woodruff buffer (500 samples) are prepared by dissolving the following chemicals in about 9 L of distilled water while stirring vigorously and diluting to 10 L:
   a. 80 g of p-nitrophenol ($NO_2C_6H_4OH$; FW = 139.11).
   b. 6.2 g of magnesium oxide (light MgO; FW = 40.3).
   c. 400 g of calcium acetate [$(CH_3COO)_2Ca \cdot H_2O$; FW = 176.19].
2. Adjust pH to 7.0 by titrating with 4 N NaOH or HCl as required.

Table 5–3. Relationship between soil-buffer pH and lime requirement (LR) values to achieve pH 5.5, 6.0, and 6.5 of mineral soils with the Woodruff buffer.

Soil-buffer pH	LR†		
	pH 5.5	pH 6.0	pH 6.5
6.8	0.3	1.0	1.7
6.7	0.7	2.0	3.3
6.6	1.2	3.1	4.8
6.5	1.8	4.1	6.3
6.4	2.5	5.2	7.8
6.3	3.3	6.4	9.2
6.2	4.2	7.5	10.7
6.1	5.2	8.7	12.1
6.0	6.4	9.9	13.5
5.9	7.6	11.1	14.8
5.8	9.0	12.4	16.2
5.7	10.5	13.6	17.6
5.6	12.0	14.9	18.9
5.5	13.7	16.2	20.2
5.4	15.5	17.6	21.5

† The LR recommendations to achieve pH 5.5, 6.0, and 6.5 were obtained from buffer-pH calibration verifications by Tran and van Lierop (1981a, 1982).

4. Verification

Buffering capacity should be checked by titrating 20 mL of Woodruff buffer from pH 7.0 to 6.0 with 0.1 *N* HCl. It should be 0.70 ± 0.02 $cmol_c$ HCl/pH.

5. Procedure

Measure 10 mL of dry soil into beakers (see footnote 4 for a description of volumetric soil measurement). Add 20 mL of Woodruff buffer using an automatic pipette. Stir with a glass rod to wet sample thoroughly and allow it to stand for 15 to 30 min. Stir the soil-buffer mixture anew and read pH. Obtain LR values for desired target pH from Table 5–3 or calculate using equation(s) from Table 5–2.

6. Comments: See applicable comments in section IV.C.6.

E. Mehlich Single-Buffer Method

1. Principles

Mehlich (1976) calibrated his buffer for assessing the LR to neutralize permanent (neutral-salt) exchangeable acidity (EA). This is the acidity implied in restricting crop growth on acid mineral soils (Kamprath, 1970; Evans & Kamprath, 1970; Reeve & Sumner,1970a, b; Mehlich, 1976; Mehlich et al., 1976). A curvilinear equation was found necessary to accurately predict

Table 5-4. Relationships between soil-buffer pH and Mehlich lime requirements (LR), and LR values to pH 5.5, 6.0, and 6.5 for mineral soils with the Mehlich buffer.

Soil-buffer pH	Mehlich	LR† pH 5.5	LR† pH 6.0	LR† pH 6.5
6.5	0.4	−2.4	−0.9	−0.6
6.4	0.9	−1.5	0.2	0.3
6.3	0.9	−0.7	1.4	1.2
6.2	1.9	0.2	2.6	2.2
6.1	2.4	1.1	3.8	3.2
6.0	3.0	2.0	4.9	4.3
5.9	3.6	2.9	6.1	5.4
5.8	4.2	3.8	7.3	6.6
5.7	4.9	4.6	8.5	7.9
5.6	5.6	5.5	9.6	9.2
5.5	6.3	6.4	10.8	10.6
5.4	7.1	7.3	12.0	12.0
5.3	7.9	8.2	13.2	13.6
5.2	8.7	9.1	14.4	15.1
5.1	9.6	10.0	15.5	16.7
5.0	10.5	10.9	16.7	18.4
4.9	11.4	11.7	17.9	20.2
4.8	12.4	12.6	19.0	22.0
4.7	13.4	13.5	20.2	23.9
4.6	14.4	14.4	21.4	25.8
4.5	15.5	15.3	22.6	27.8
4.4	16.5	16.2	23.7	29.8
4.3	17.7	17.0	24.9	32.0
4.2	18.8	18.0	26.1	34.1
4.1	20.0	18.8	27.3	36.4
4.0	21.2	19.7	28.4	38.7
3.9	22.5	20.6	29.6	41.0

† Lime requirement calibrations for Mehlich LR values were obtained from Mehlich (1976), to pH 5.5 and 6.0 from Tran and van Lierop (1982), and pH 6.5 by Ssali and Nuwamanya (1981).

LR from buffer-pH values. Calibration of this buffer differs from others in that LR recommendations are meant to produce optimum yields rather than achieve a pH. However, when comparing LR values obtained with its original calibration against incubation values to achieve pH 5.5, Tran and van Lierop (1982) found that it was the most accurate among procedures tested. The probable reason is that neutral-salt EA tends to predominate in acid soils at pH levels lower than 5.5. A remarkable similarity between LR values recommend by Mehlich (1976) and derived by Tran and van Lierop (1982) for attaining pH 5.5 was found, confirming the validity of its original calibration (Table 5-4).

Although, the Mehlich buffer method is precise, its calibration may not be accurate to attain pH 6.0 or 6.5 (Ssali & Nuwamanya, 1981; McLean et al., 1978; Tran & van Lierop, 1981a, 1982). This inaccuracy is expected from using EA values for calibration. This buffer-pH procedure recommends about 50, 59, and 60% of reference values to obtain pH 6.5 (McLean et al., 1978; Tran & van Lierop, 1981a; Ssali & Nuwamanya, 1981, respectively). Nonetheless, it is particularly well suited for determining the LR for neutralizing acidity harmful to crop productivity and will generally recommend sufficient limestone to achieve a pH slightly above 5.5. This pH is sufficient to eliminate possible Al toxicity. As LR values to achieve other pH levels may be desired in some instances, calibrations for achieving pH 5.5, 6.0, and 6.5 of mineral soils are provided. Calibrations are in metric tonnes of $CaCO_3$ 2 million L^{-1} of soil (1 ha to a depth of 20 cm): see section IV.B for modifying rates to fit other measurement units.

2. Equipment and Reagents: See section IV.C.2.

3. Preparation of Mehlich Buffer

1. Ten liters of Mehlich buffer (500 samples) are prepared by dissolving the following chemicals in about 7 L of distilled water while stirring vigorously:
   a. 12.5 mL of glacial acetic acid (CH_3COOH; FW = 60.05).
   b. 90 mL of TEA stock solution.[3]
   c. 215 g of ammonium chloride (NH_4Cl; FW = 53.49).
   d. 100 g of barium chloride ($BaCl_2 \cdot 2H_2O$; FW = 244.28).
2. Dissolve 90 g of sodium glycerophosphate, also named sodium salt of beta-glycerophosphoric acid [$(HOCH_2)_2CHOPO_3Na_2 \cdot 5H_2O$; FW = 306.12] in about 2 L of distilled water.
3. Mix solutions 1 and 2 while swirling vigorously. Allow it to cool to room temperature, then dilute to 10 L and mix thoroughly.
4. Adjust pH of buffer to 6.60 ± 0.04 with glacial acetic acid or TEA stock solution as required.

4. Verification

Buffering capacity is checked by mixing 20 mL of Mehlich buffer with 10 mL 0.1 *N* HCl-$AlCl_3$ solution. The pH of the resulting mixture should be 4.1 ± 0.05. The 0.1 *N* HCl-$AlCl_3$ solution is made by dissolving 4.024 g $AlCl_3 \cdot 6H_2O$ in 100 mL 0.05 *N* HCl.

5. Procedure

Measure 10 mL of dry soil into beakers (see footnote 4 for a description of volumetric soil measurement). Add 20 mL of buffer, stir with a glass rod to wet sample thoroughly, allow to stand for 1 h, and measure pH. Obtain LR values for desired target pH from Table 5–4 or calculate using equation(s) from Table 5–2.

6. Comments

Buffer concentration is one-half that advanced by Mehlich (1976) as LR is determined only on selected samples. Negative LR values in Table 5–4 in dicate that liming is not required. A soil-buffer pH close to 6.6 is occasionally observed with coarse-textured soils needing liming, in such case recommend a minimum rate. See section IV.C.6 for applicable comments.

F. New Woodruff Single-Buffer Method

1. Principles

Mention of the new Woodruff (NW) buffer was made by McLean (1973). It appears to have been developed in the mid-1960s and is described by Brown et al. (1977). According to regression parameters provided by Brown and Cisco (1984) and Alabi et al. (1986), NW buffer is as precise as the original but recommends about 1.6 times higher LR. Given that it provides higher recommendations with the same effective calibration range as the original, suggests that it displaces soil acidity more efficiently, and consequently meets its maximum range more frequently.

According to Brown and Cisco (1984), the NW buffer is more accurate than the original for determining the LR to pH 7.0. They obtained reference LR values, however, by measuring soil pH in salt solutions varying in concentrations between 0.01 to 0.2 M $CaCl_2$ with a brief $Ca(OH)_2$ titration to pH 7.0. It is difficult to ascertain whether their conclusion is correct as these procedures affect reference LR values. Regression parameters from Alabi et al. (1986) suggest that the NW-buffer method recommends higher LR values than required to achieve pH 6.5. Recondite as facts may be when comparing results of studies using different methods, an interesting observation is that both studies found that the SMP single-buffer procedure recommended higher values than the NW procedure. The SMP procedure has been shown to overestimate LR of soils with low LR according to McLean et al. (1966, 1978) and Tran and van Lierop (1981a, 1982), inasmuch as low LR soils were preponderantly studied by Brown and Cisco (1984) and Alabi et al. (1986), suggests that NW-buffer LR values are reasonable for low LR soils. Some uncertainty exists, however, as it recommends considerably higher values than the original, which is considered about right for determining low LR soils.

Regression parameters and data reported by Alabi et al. (1986) suggest that the original and the NW (as other procedures) measure a LR when liming is not necessary. This peculiarity is inherent in buffer-pH procedures and emphasizes the need for using soil pH for deciding whether a LR test is required. Buffer-pH procedures occasionally will not indicate a LR on acid soils with low-buffering capacities, and conversely, procedures will occasionally measure a LR for soils with high-buffering capacities when it has a sufficiently high pH (hence does not need liming). This apparent peculiarity is sometimes ascribed to using soils that are different in nature than those originally used to calibrate a procedure. However, buffer-pH procedures rely on measuring a calibrated portion of soil acidity that can react with limestone to achieve a target pH. Such a target pH should preferably be situated between soil and initial buffer pH. It is hoped that the NW-buffer procedure will be studied further, though, it will give satisfactory results with its calibration.

G. Adams and Evans Single-Buffer Method

1. Principles

The Adams and Evans (A-E) method was developed for measuring the LR of Red-Yellow Podzolic soils (Ultisols) that have low LRs and which may be affected by crop yield reduction with overliming (Adams & Evans, 1962). The method was developed because other buffers were not satisfactory for determining the LR of these low-exchange capacity soils. Calibration of the A-E buffer is realized by calculating LR values from a general relationship between soil pH and base *unsaturation*, and the ability of the buffer to indicate the amount of exchange acidity to be neutralized, and hence LR. Inasmuch as the relationship between pH and percent base saturation is known to vary widely with soils (Shaw, 1952), Adam and Evans stated correctly that any general application of such a relationship to soil LR should be made with caution. The difficulty is that percent base saturation and unsaturation are relative intensity terms, while LR is a capacity index of soil acidity. Apparently, calibration values are meant to increase base saturation to 75% and decrease base unsaturation to 25%. These values were shown to correspond to the acidity displaced by ammonium acetate (pH 7.0). Validity of this approach was verified by incubating some soils with increments of $Ca(OH)_2$ and comparing incubation with estimated LR values. It is interesting to note that Peech (1965b) also reported moderately accurate LR values derived from a generalized pH-base saturation relationship. To such a relationship, Adams and Evans incorporated measured buffer acidity to determine LR to pH 6.5 (Adams & Evans, 1962; Hajek et al., 1972).

According to McLean (1982), the A-E buffer is very sensitive and particularly useful for soils with low LRs. It appears to be widely used in the southern USA (Adams, 1984). Furthermore, its accuracy has been verified quite extensively. Fox (1980) evaluated the A-E method and concluded that it tended to overestimate LR, though these were well correlated with incubation values. Similarly, Tran and van Lierop (1981a) found that it was suitable essentially for low LR soils, but that it was not as precise as some for determining higher LR. They also found that the A-E method overestimated LR, and suggested that high initial buffer pH (pH 8.0) could be responsible. Using this initial buffer pH enables the buffer to include pH-dependent acidity between pH 6.5 and 8.0 that need not be neutralized. More recently, Alabi et al. (1986) confirmed that the A-E method overestimated LR of coarse-textured soils. Although it appears to overestimate LRs, the A-E calibration includes a liming factor of 1.5. Since most verifications have been carried out either with $CaCO_3$ or $Ca(OH)_2$-incubations without commenting on the liming factor, it may well be responsible for recommending rates above those required to achieve pH 6.5.

H. Nômmik Single-Buffer Method

1. Principles

The Nômmik single-buffer method is the most recently introduced buffer-pH LR procedure (Nômmik, 1983). Relatively little is known about its use or verification. Its buffering components consist of imidazole, maleic acid, and acetate, with K and Na as acidity exchangers. The buffer was developed to enable exchangeable Ca, Mg, and Mn determination in the extract by EDTA titration. Characterizing of soil CEC, in addition to determining LR, may have been a primary goal in developing this procedure as reference LR values to pH 7.0 were used for calibration. These were derived, however, from a single $Ca(OH)_2$-incubation treatment (10 wk) by calculating LR values to pH 7.0 by inter- or extrapolation. This procedure assumes a linear relationship between added base and change in pH.

The Nômmik buffer is probably as precise as any; however, its accuracy is unknown, as perplexing results were reported for a comparison between incubation and values measured with the SMP-SB (McLean, 1973) and Yuan double-buffer methods (Yuan, 1974). Nômmik measured somewhat higher LR values with the Yuan procedure than indicated by incubation. However, the Yuan procedure does not fully measure incubation LR (McLean et al., 1978; Tran & van Lierop, 1981a, 1982). A difference between the Nômmik and Yuan procedure is quite possible; however, an incertitude arises because lower LR values were measured with the SMP than by incubation. It is generally recognized that the SMP-SB procedure recommends higher than reference values for soils having low to medium LRs (McLean et al., 1966, 1978; Tran & van Lierop, 1981a; Brown & Cisco, 1984; Alabi et al., 1986).

Nômmik also found that single-buffer procedures react with proportionately more acidity from low than high LR soils. Curvilinearity was reported for the relationship between soil-buffer pH and incubation LR values by Tran and van Lierop (1981a, 1982), Soon and Bates (1986), the data of Loynachan (1981), and Webber et al. (1977). As indicated for the NW buffer, it is hoped that this procedure will be studied further, though it will give satisfactory results with its current calibration.

I. Double-Buffer Lime Requirement Determination Procedures

1. Principles

Double-buffer LR determination was introduced by Yuan (1974, 1976) and subsequently adapted by McLean et al. (1977, 1978) for use with the SMP buffer. It is said double-buffer procedures differ from single-buffer procedures in that the former weighs the characteristic buffering capacity of a soil to be limed. Be that as it may, it is generally accepted that single-buffer procedures also measure sample-buffering capacity to determine LR. As some confusion about the working mechanism of these procedures arises from these statements, operational principles are developed more explicitly to study differences. It is apparent from diagrams provided that double-buffer techniques rely on triangulation. However, exact differences between assumptions used for double- and single-buffer methods are not clear from these.

Single- and double-buffer procedures measure the quantity of acidity neutralized by liming to attain a desired pH, which is defined as LR. However, buffer-pH procedures do not measure all the acidity involved (McLean et al., 1977, 1978; Tran & van Lierop, 1981a, 1982). In the case of single-buffer methods, the amount of acidity, and by corollary, the LR to a target pH is determined from the relationship between soil-buffer pH and incubation values preferably established by regression techniques. On the other hand, double-buffer procedures, as originally developed by Yuan (1974, 1976), rely on three fundamental assumptions (operational principles). The first is that changes in soil pH with additions of base or buffer are linear. Second, the change in soil pH produced by adding buffers is extrapolated to the neutralization of soil acidity by $CaCO_3$. Third, the buffers completely displace and assess the acidity that is neutralized by limestone. Although none of these fundamental assumptions is entirely correct, their adoption has resulted in the development of precise and relatively accurate LR-determination procedures.

The Yuan and SMP-double-buffer (Yuan, 1974, 1976; McLean et al., 1977, 1978) use the first principle. Its adoption theoretically allows double-buffer procedures to determine LR values to any selected target pH situated between current soil pH and about 6.5 to 7.0. The flexibility of target pH selection appears to be the principal advantage in favor of double-buffer procedures. Single-buffer methods, however, can be calibrated in LR steps of about 0.5 pH unit; these steps satisfy practical needs and may be close to the accuracy resolution of LR-determination procedures in any event. Therefore, the flexibility advantage in favor of double-buffer procedures may be largely apparent. The Yuan- and SMP-double-buffer procedures use the second principle which is usually described as measuring the "individual" buffering capacity of a soil. This measurement relies on inter- or extrapolating the amount of acidity displaced by the buffers, as indicated by their change in pH, into a LR. This principle is usually represented by triangulation diagrams. The third principle is not used by the SMP-double-buffer procedure. McLean et al. (1977, 1978) recognized that buffer procedures displace a proportion instead of the total amount of acidity. They corrected, therefore, for incomplete soil-acidity displacement by multiplying measured values by a constant derived from incubation values using regression techniques.

The main advantage claimed in favor of double-buffer procedures is their greater accuracy at low LR values. This characteristic is especially valuable for avoiding overliming of soils with low-buffering capacities. The SMP-double buffer is more accurate than the Yuan-double buffer procedure for determining higher LR values mainly because it uses a correction factor to adjust for incomplete acidity displacement. It so avoids the large inaccuracies produced by the erroneous assumption associated with the third operational principle. As the only advantage in favor of double-buffer procedures appears to be their greater accuracy at low values, either the Yuan- or SMP-double-buffer procedure can be equally effective.

J. Yuan Double-Buffer Method

1. Principles

The amount of acidity neutralized by $CaCO_3$ when increasing soil pH is measured in three steps. First, two soil-buffer pH values are determined on different samples with buffers having initial pH values each of 7.0 and 6.0, respectively. Second, the soil-buffering index alpha (α) is derived by calculating the ratio of the amount of acidity displaced relative to the change in pH produced by the buffers. The acidity displaced is calculated from the assumption that 1 $cmol_c$ acidity changes buffer pH by one unit. The change in soil pH produced by the buffers is calculated by subtracting the soil-buffer pH of the pH 6.0 buffer from that obtained with the pH 7.0 buffer. Therefore, α is the slope of a relationship between acidity measured per unit change in pH produced by the buffers. Third, LR is calculated from the slope (α), by extrapolating a desired change in soil pH from that measured with the buffers.

Yuan apparently accepted the operational principles because differences between measured and $Ca(OH)_2$-titration LR values were small (Yuan, 1974, 1976). His data suggests that the double-buffer procedure measured an average of about 90% of reference values. McLean et al. (1978) verified the Yuan-double-buffer procedure and found measured LR values too low; this finding was confirmed by Tran and van Lierop (1981a, 1982). The low-measured LR values signify that the third operational principle for double-buffer procedures is incorrect and buffers do not displace all acidity that reacts with $CaCO_3$ when increasing soil pH to the target value. Tran and van Lierop (1981a, 1982) suggested that the accuracy of the Yuan-double-buffer procedure could be improved substantially by incorporating a correction factor to adjust for the incomplete measurement of soil acidity. They found that the Yuan-double-buffer procedure was as precise as any for determining the LR to pH 6.5 and 6.0, but slightly less precise for 5.5. Fox (1980) evaluated the Yuan- and SMP-double-buffer procedures and also found them about equally precise.

2. Comments

The Yuan-double-buffer method is as precise as any and appears to be quite accurate for determining the LR of poorly buffered soils. Double-buffer LR determination procedures are about twice as costly and time consuming as single-buffer methods. Although these procedures do not require a table or regression equation to relate soil-buffer pH values to LR, calculations are extensive.

K. SMP Double-Buffer Method

1. Principles

The SMP buffer (section IV.C) was adapted by McLean et al. (1977, 1978) to a double-buffer methodology similar to that proposed by Yuan

(1974). This approach was selected for improving the accuracy of LR determination for low-buffering capacity soils. However, the SMP-single buffer does not follow the operating principles used by the Yuan-double buffer procedure exactly. McLean et al. (1977, 1978) tried several procedural variations: they concluded that double buffer, as single-buffer, procedures do not measure all the acidity neutralized by $CaCO_3$ either. They, therefore, included a proportionality factor into the SMP-double-buffer calibration similar to that needed for single-buffer calibrations. This factor corrects for partial acidity displacement. The proportionality factor is derived from incubation data using regression techniques. Therefore, the SMP-double buffer generally recommends higher LR rates for soils with moderate and high LR levels than the Yuan-double-buffer procedure.

McLean (1982) preferred a double-buffer variation called the single-buffer two-pH technique. It provides essentially the same results as the two-buffer two-pH technique proposed by Yuan (1974) but allows saving in materials and possibly time. McLean (1982) proposed its use for routine LR determination of soils with low-buffering capacities. Indeed this variation lends itself to combine using the SMP-double-buffer modification with the SMP-single-buffer procedure for low LR samples that might benefit from additional accuracy. The one-buffer two-pH technique relies on determining soil-buffer pH for only one sample; rapidly and accurately adding sufficient HCl to the soil-buffer mixture to depress buffer pH to 6.0, and subsequently, after an appropriate reaction period, obtaining the second soil-buffer pH value.

The SMP-double-buffer procedure has been evaluated by several workers over the years (Fox, 1980; Ssali & Nuwamanya, 1981; Tran & van Lierop, 1981a, 1982; Alabi et al., 1986; Soon & Bates, 1986). Generally, it has been found to be as closely correlated with reference LR values as the SMP-single-buffer procedure, suggesting that the procedure is as precise as any. Better accuracy relative the SMP-single-buffer procedure with its original calibration was confirmed for determining LR values to pH 6.5 for low-buffering capacity soils by Tran and van Lierop (1981a). They also found that it was about as accurate as the SMP-single-buffer procedure for determining the LR to achieve pH 6.5 of soils with higher requirements. However, Tran and van Lierop (1982) subsequently discovered that the correction factor proposed by McLean et al. (1977, 1978) did not hold particularly well for determining LR to attain pH 5.5 or 6.0. They found similar inaccuracies associated with the Yuan-double-buffer procedure. Although, confirmation of this finding would be useful, it nonetheless suggests that double-buffer procedures may not be more accurate than single-buffer procedures for measuring the LR to achieve different soil pH levels. Still, the SMP-double-buffer adaptation, like the Yuan-double-buffer procedure, can offer improved accuracy for determining the LR of soils with low-buffering capacities.

2. Comments

See sections IV.C.6, IV.F.1, and IV.I.1 for applicable comments. The SMP-double-buffer procedure has the same disadvantages as the Yuan-double buffer as it is more onerous than a single buffer procedure. However, the single-buffer two-pH variation can be combined with single buffer methodology for selectively determining LR of soils with low-buffering capacities, which could benefit from double-buffer methodology.

V. INDIRECT LIME REQUIREMENT DETERMINATION METHODS

A. Principles

Indirect LR-determination procedures rely on estimating a LR from soil properties without directly measuring acidity. These methods identify soil components that contribute most importantly to LR. Indirect procedures can be precise but are generally more onerous than buffer-pH procedures. Nonetheless, they are occasionally convenient for determining a LR value from available soil test data. Joret et al. (1934) proposed the following equation relating soil OM and clay content to LR: 0.11 [% clay + (5 × % OM)]. The relationship suggests that OM contributes about five times more acidity than clay. However, soil clay often exceeds OM content by a factor of 5, and then clay would contribute as much or more to LR as OM. Helling et al. (1964) indicated that soil OM contributes about 2.2 times as much to CEC per gram of material as clay at pH 6.0.

Keeney and Corey (1963) found that clay content or exchangeable Al had relatively little influence on LR. They formulated the following equation relating a desired change in pH and soil OM content to LR: (pH 6.5 − soil pH) × (%OM). Organic matter content appears as the principal component in LR for a given change in pH. Keeney and Corey (1963) concluded that their soils contained too little exchangeable Al to influence LR. Presumably because of this, they also found that clay content had little effect. Subsequently, Pionke and Corey (1967) studied the interrelation between acidic Al, clay, and OM contents. They concluded that exchangeable-Al concentration was primarily related to soil pH. They demonstrated that exchangeable-Al concentrations decreased exponentially as soil pH increased and that little or no Al remained at soil pH of about 4.5 (1 *N* KCl) or above—equivalent to pH of about 5.5 (water). A similar relationship between soil pH and exchangeable-Al concentrations was observed by Thomas (1967), van Lierop et al. (1982), Nômmik (1983), and others. Aluminum will not generally contribute to LR when soil pH is about 5.5 (water) or above. Accordingly, the equation suggested by Keeney and Corey may be useful for determining LR when soils are between about pH 5.5 and 6.3 (water).

Rémy and Marin-Laflèche (1974) modified the equation proposed by Joret et al. (1934) by adding a change-in-pH factor similar to Keeney and Corey (1963). However, the Rémy and Marin-Lafleche equation retained OM and clay content and emphasizes the role of clay in LR. Pionke et al. (1968) also studied soil properties contributing to LR for a wider selection of soils and produced an expression that includes clay, exchangeable Al, OM, and change in pH. Gathering data to calculate a LR with such an expression unfortunately does not rate it as a quick-test procedure, and may only be of interest for identifying soil components participating in LR. A LR can generally be more easily obtained using a buffer-pH procedure.

Indirect LR-determination studies suggest that the most important contributing factors to LR are exchangeable Al, OM, and clay contents and the desired change in soil pH. These factors contribute collectively to a LR and cannot be neutralized independently by liming. Accordingly, if liming soils to neutralize potentially toxic levels of Al is envisaged, part of the applied limestone will react with Al but the remainder will neutralize acidity released by other soil components, this neutralization scheme will proceed until soil pH is sufficiently high to complete Al precipitation.

More accurate and complex equations are available for predicting LR for more acid soils than proposed by Keeney and Corey (1963), according to Tran and van Lierop (1981b). As indirect procedures do not offer advantages over buffer-pH procedures, the equation suggested by Keeney and Corey is offered for use when an alternate procedure needing minimal data would be useful. These authors included a typical liming factor (1.6) in their equation. However, its general use may not be justified as suggested in section IV.A. Furthermore, verification by Tran and van Lierop (1981b) suggest the Keeney and Corey equation recommends on the average 130% more limestone than indicated by incubation. The equation is reproduced with its original liming factor.

$$\text{LR (6.5)} = 1.6\,(6.5 - \text{soil pH}) \times (\%\text{OM}). \quad [1]$$

2. Comments

The indirect LR estimation procedure proposed by Keeney and Corey (1963) is advanced for use when buffer-pH values are not available and a LR recommendation is required. The procedure is fairly accurate and relies on common soil tests.

VI. LIME REQUIREMENT DETERMINATION OF ACID HISTOSOLS

A. Principles

Acid Histosols are usually limed to a pH around 5.2 to 5.4 (water), which is agronomically comparable to 6.5 for mineral soils (McLean, 1971, 1973, 1982; van Lierop, 1983). A pH of 5.4 (water) corresponds to approximately 5.0 in 0.01 *M* $CaCl_2$ (van Lierop, 1981a). Although this pH appears relatively low, crop growth is not affected by excessive acidity. Generally, Histosols contain little extractable Al, and, therefore, probably produce optimum yields at lower pH levels. However, different calibrations are required for determining LR with buffer-pH procedures. Not just because of a different target value, but more importantly, because the nature of their exchange complex is different (McLean, 1973, 1982; Mehlich, 1976; van Lierop, 1983; Nômmik, 1983).

The Mehlich and Nômmik buffers were originally calibrated for determining the LR of Histosols (Mehlich, 1976; Nômmik, 1983). However, calibration of the Mehlich buffer was not achieved with incubation reference values, and Nômmik probably chose pH 7.0 for CEC characterization rather than determining a LR to achieve optimum yields. Unfortunately, liming Histosols to pH 7.0 is too high for optimum crop production and minimizing subsidence (van Lierop, 1981a, 1983).

Although McLean (1973, 1982) published a relationship between the SMP-single buffer soil-buffer pH and LR values to achieve pH 5.2 for organic soils, details of the calibration appear not to have been published. Few buffer-pH verifications have been carried out for Histosols. However, verifications of the single-buffer calibrations were studied by van Lierop (1983). He concluded that the Mehlich-buffer calibration was accurate for determining the LR of acid Histosols. It provides LR values for achieving pH to 5.4 (water). On the other hand, the SMP-single-buffer calibration proposed by McLean (1973, 1982) was found to recommend too high $CaCO_3$ rates for achieving pH 5.2 (water) for moderate and low requirement soils. Accordingly, improved calibrations and procedural limitations are discussed for the Mehlich, SMP, and Woodruff buffer-pH procedures by van Lierop (1983).

Errors in LR determination for acid Histosols probably have three principal causes: (i) erroneous-assumed bulk density (BD) values, (ii) inaccurate buffer calibration, and (iii) irreversible sample drying. Histosols ought not be dried and weighed for determining LR because the value for a desired plow-layer depth can only be calculated if the bulk density is known. A BD (volume weight) value for Histosols can vary from about 0.1 to 0.7 g mL^{-1} (Kaila, 1956; van Lierop, 1981b). Furthermore, drying increases BD by an average of about 200%; however, field-moist BDs are usually not closely correlated with dried soil values (van Lierop, 1981b; Boelter, 1964). A furrow layer cannot be accurately defined for Histosols without adjusting for differences in BD. Measuring a LR of scooped dried samples will probably not improve accuracy of recommendations, as BD increases with drying, and its modified value is not precisely related to what it was in the original field-moist state. To compound difficulties, dried Histosols are often difficult to rehydrate completely, and CEC properties are likely altered as some drying seems irreversible.

Obviously, LR determination of acid Histosols ought to be made on volumetrically measured field-moist soil samples having a representative field density to improve recommendation accuracy, as inaccuracies are largely related to changes of BD between moist field and dried laboratory samples. The importance of sample BD on LR determination accuracy was recognized by van Lierop (1981b), who proposed a reconstituted BD procedure to simulate original field-moist density in the laboratory. The procedure was used to verify and calibrate the Mehlich (1976), SMP (Shoemaker et al., 1961), and Woodruff (1948) buffers for determining the LR of acid Histosols. Interestingly, a LR for achieving any other target pH between current soil pH and about 6.0 (water) can be derived from pH 5.4 values as the change in pH is linearly related to amounts of $CaCO_3$ added.

B. Comments

The SMP, original Woodruff, and Mehlich buffers have similar precision and accuracies for determining the LR of acid Histosols (van Lierop, 1983). Accordingly, it is suggested that if one of these buffers is being used for determining the LR of mineral soils, that it be used with its improved calibration and procedure for organic and Histosols. Buffer-pH LR procedures are about as accurate for acid Histosols as mineral soils. Calibrations are in metric tonnes of $CaCO_3$ 2 million L^{-1} of soil (1 ha to a depth of 20 cm): see section IV.B for information about modifying rates to fit other measurement units.

VII. INDIRECT LIME REQUIREMENT DETERMINATION OF ACID HISTOSOLS

A. Principles

The LR of Histosols containing about 50% or more OM can be determined accurately, and sometimes more conveniently by using the following indirect procedure. It was derived from BD, pH, and incubation-LR data of the Histosols used for calibrating and verifying buffer-pH procedures by van Lierop (1983). The relationship between LR and BD was suggested by the linear changes in pH with increasing additions of $CaCO_3$. Slopes of incubation graphs derived by plotting soil pH against additions of $CaCO_3$ changed with BD. Accordingly, a regression equation was fitted to the relationship between changes in soil pH adjusted to a constant soil BD (δpH $\times$ BD = β) and incubation LR values (Fig. 5–2). The relationship implies that a given amount of liming produces the same change in pH with different Histosols when these are adjusted for differences between BD.

$$\mathrm{LR} = 35\beta \quad [2]$$

$$\beta = \delta\mathrm{pH} \times \mathrm{BD} \quad [3]$$

$$r = 0.971^{**};\ s_{y.x} = 1.72. \quad [4]$$

The relationship between β and LR values is not very different from that suggested by Keeney and Corey (1963) for determining the LR of mineral soils as they assumed that LR was a function of a desired change in pH (δpH) and soil OM content. The equation estimates the amount of acidity, hence the LR, of acid Histosols. It allows computation of LR for achieving any

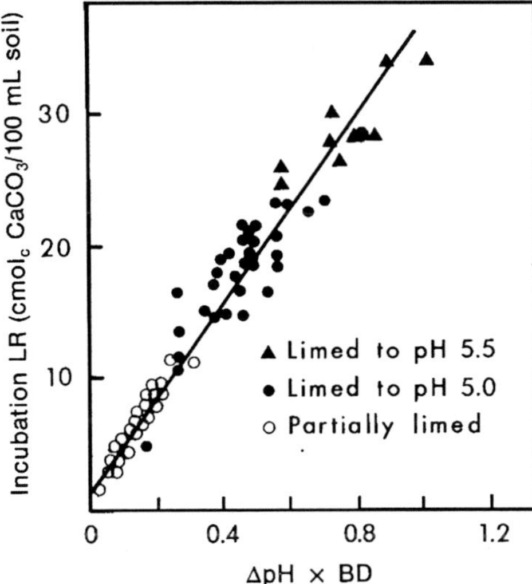

Fig. 5-2. Relationship between LR and β-values (δpH × BD) for Histosols in metric tonnes $CaCO_3$ ha^{-1} (2 million L of soil).

selected target pH between current value up to about 6.0 (water) for soils containing more than about 50% OM. Furthermore, the equation appears to be able to estimate LR when soil pH is determined in 0.01 *M* $CaCl_2$, or when a soil has been partially limed. Calibrations are in metric tonnes of $CaCO_3$ 2 million L^{-1} of soil (1 ha to a depth of 20 cm): see section IV.B for information about modifying rates to fit other measurement units.

B. Procedure

Although no buffer is required to determine LR with this procedure, a soil BD is necessary for calculating the β-value. As field BD is rarely available, the reconstituted BD procedure proposed by van Lierop (1981b) is suggested. Its value is determined by compressing soils into a sturdy (about 17 mL; ½ oz) medicine cup with slightly inclined walls. Actual cup size is not critical but ought to have about the same shape. Soils are compacted into the cup by hand until either no more soil can be pressed in, or water comes to the surface. Thereafter, excess soil is trimmed and the compressed soil transferred to a drying vessel, dried at 105 °C, weighed, and the BD is calculated. Subsequently, the β-value is obtained by multiplying BD by the desired change in soil pH (δpH). Estimated LR is obtained by multiplying β-value by 35. Lime requirement values should not be expressed with greater precision than one place after the decimal.

C. Comments

This procedure may be convenient for determining LR of acid Histosols when relatively few samples are involved. Although no buffer is necessary, drying time and space, as well as a balance are required. The procedure has an accuracy similar to a buffer-pH method. Field-moist soil samples, or actual field BD values are required to obtain accurate estimates.

VIII. ALUMINUM AS CRITERION FOR LIMING ACID SOILS

A. Introduction

Aluminum was implicated in the chemistry of acid mineral soils by early research (Jenny, 1961; McLean, 1965). Although Al is abundant in soils, only exchangeable and solution Al occurring at acid pH values have a bearing on LR. Ionic forms in the soil solution are of interest because they are responsible for Al toxicity and poor crop growth on acid soils when present at sufficiently high levels (Ragland & Coleman, 1959; Kamprath, 1970; Evans & Kamprath, 1970; Reeve & Sumner, 1970a, b; Hutchinson & Hunter, 1970; Martini et al., 1974; Hoyt & Nyborg, 1971a; Hoyt & Webber, 1974; Penney et al., 1977; Webber et al., 1982). The amount of Al on the permanent soil-exchange sites is largely influenced by soil pH (Jenny, 1961; McLean, 1976). Aluminum may occupy a large proportion of the permanent exchange sites of highly weathered soils such as Oxisols and Ultisols (Evans & Kamprath, 1970; Reeve & Sumner, 1970a, b). However, the permanent exchange sites are fully saturated with bases other than Al when soil pH is near 5.5 (water) or higher as indicated by extensive lime and corrected lime-potential studies (Clark, 1965, 1966b; Clark & Hill, 1964; Clark & Nichol, 1966; Turner & Nichol, 1962; Turner & Clark, 1965; Singh, 1972). Liming to pH 5.5 ensures elimination of possible Al toxicity, though it will result in liming some soils unnecessarily as they may not contain toxic concentrations of Al (Reeve & Sumner, 1970b). The concentration of exchangeable Al in soils appears to decrease exponentially as soil pH increases (Pionke & Corey, 1967; MacLeod & Jackson, 1967; Thomas, 1967; van Lierop et al., 1982; Nômmik, 1983).

Soil OM has some influence on the concentration of Al affecting plants as it complexes Al and thus mitigates its toxicity (Schnitzer & Skinner, 1963a, b, 1964; Clark & Nichol, 1966; Evans & Kamprath, 1970). Generally, as soil OM increases, less Al is found in the soil solution at a given pH (Evans & Kamprath, 1970). However, a further mitigating effect of complexed-Al ions on its toxicity to crops was suggested by the work of Foy and Brown (1964) who demonstrated that complexing agents increase the solubility but decrease Al toxicity to Al-sensitive plants. Unfortunately, Al levels that may not affect growth of a crop may be toxic to another, as crops and cultivars have different tolerance to Al toxicity (Foy, 1964; Foy et al., 1965, 1969).

Aluminum is commonly determined using either "flame" (AA or ICAP), colorimetric, or titration procedures (Yuan, 1959; Lin & Coleman, 1960; McLean, 1965; Rich, 1970; Dewan & Rich, 1970; Barnhisel & Bertsch, 1982). However, it should be pointed out that flame procedures include, in addition to extracted ionic-Al, polymerized-Al, and Al-OM complexes since these-forms are destroyed, ionized, and detected by flame. Clark (1965) emphasized the importance of using certain colorimetric means, such as 8-hydroxyquinoline, for determining Al in soil extracts as he found these procedures to reflect ionic-Al concentrations more accurately. Titration and colorimetric procedures seem to have been used more frequently in the cited literature, possibly because exchangeable Al is generally extracted with a concentrated neutral salt that often causes burner salt accumulation with many flame units. Such buildup gradually affects flame intensity and causes signal drift. Consequently, a flame procedure may provide unreliable readings. A buildup requires more frequent equipment cleaning and adjustment. Furthermore, some AA units are less sensitive and precise than titration or colorimetric procedures for determining Al.

Crop yield responses to liming are closely related to exchangeable-Al reductions, though, sensitive plants appear to have some tolerance to low concentrations. Consequently, exchangeable Al does not have to be completely precipitated by liming to enable maximum yield production (Ragland & Coleman, 1959; Hutchinson & Hunter, 1970; Kamprath, 1970; Martini et al., 1974; Reeve & Sumner, 1970b; Hoyt & Webber, 1974; Penney et al., 1977; Webber et al., 1982). A tolerance by crops suggests that convenient determination procedures, like titration or flame, may be employed to determine potentially harmful levels, as a low background—caused by innocuous complexed, polymerized, or low level of exchangeable or ionic forms—would not affect crop productivity.

The main advantage that favors using Al as a liming criterion is that smaller amounts of liming material are required to precipitate plant-toxic Al levels by liming to pH 5.5 than to higher soil pH values. Lower rates have considerable practical benefit where relative costs of liming are high. Many places do not have inexpensive sources for liming agricultural soils, and occasionally, under extensive agricultural production, liming costs may be higher than land costs. Under such circumstances, using Al as liming criterion may result in significant yield improvement and economic advantage. It is often said that liming to a higher pH has potentially beneficial effects on nutrient availability and microbiological activity (McLean, 1970, 1971), and liming to a higher pH does occasionally result in increased yields. However, liming a very acid soil at high levels can also decrease yields by inducing plant nutrient deficiencies. As it is not the objective of this section to clarify these schools of thought, suffice perhaps to say that if liming costs are low, slowly liming an acid soil to a pH around 6.0 will not adversely affect yields. However, where liming costs are high using Al, or Al and pH levels to decide whether liming is required has advantages. Generally, maximum yields are attained when plant-toxic levels of Al are eliminated and plant nutrients are in adequate supply. Liming at these low levels, however, may require more frequent additions to maintain soil pH.

B. Approaches and Procedures

1. Neutralization of Possible Aluminum by Liming to pH 5.5

a. Principles. Probably the simplest approach when using Al as liming criterion is to assume that soils that have a pH < 5.5 contain potentially toxic levels. Liming these soils to a pH ≥ 5.5 neutralizes it and enables obtaining maximum crop yields when plant nutrient supplies are adequate. This approach relies on measuring LR to pH 5.5 with a buffer-pH procedure, and subsequently liming the soil. The original calibration of the Mehlich buffer-pH procedure (Mehlich, 1976) was developed to neutralize acidity harmful to crop growth which includes potentially toxic levels of Al. Consequently, it or another proposed buffer-pH procedure can be used to ensure optimal yields at relatively low-liming costs. However, some mineral soils with a pH between 4.9 and 5.5 (water) do not contain crop-toxic levels of Al, these would be limed without producing increased yields. Liming these soils, however, may increase pH, improve cropping opportunities, and delay further acidification.

C. Liming to pH 5.5 when Exchangeable Aluminum is High

1. Principles

This approach is more onerous as an additional test is required to determine whether soils with pH values between 4.9 and 5.5 contain suffciently high levels of Al to be plant-toxic and hinder growth. It averts liming soils that might not respond to liming with increased yields. Soils with pH ≤ 5.5 and containing higher-exchangeable-Al levels than a selected norm are tested for LR to pH 5.5 using a buffer-pH procedure, and subsequently limed.

Moschler et al. (1960) obtained maximum alfalfa yields when exchangeable-Al levels were below 0.2 $cmol_c$ Al kg^{-1} of soil. Ragland and Coleman (1959) reported good sorghum [*Sorghum bicolor* (L.) Moench] growth when limed soils contained ≤0.1 $cmol_c$ exchangeable Al kg^{-1}. However, oat, alsike clover, and grass yields were not increased by liming soils that initially contained about 1 $cmol_c$ Al kg^{-1}. Reeve and Sumner (1970a, b) proposed using an exchangeable-Al index (EAI) and suggested that sorghum yields were not increased by liming, if soils contained <0.2 $cmol_c$ Al kg^{-1} soil. Hoyt and Nyborg (1971a, 1987) found the concentration of exchangeable Al in soils to be a good indicator for predicting response to liming. Furthermore, data of Hoyt and Webber (1974) and Webber et al. (1982) indicates that maximum barley yields were obtained when exchangeable-Al concentrations were 0.2 $cmol_c$ kg^{-1} or lower.

Survey of the above studies suggests that maximum yield of relatively Al-sensitive crops like alfalfa, soybean, and barley is realized when the exchangeable-Al level is lower than 0.1 $cmol_c$ Al kg^{-1} of soil. Accordingly, this level was selected as the testing norm. Crops more tolerant to Al can be grown successfully at this or higher Al concentrations. Soils that have a pH ≤ 5.5 and contain more 1 *N* KCl or 0.2 *N* NH_4Cl extracted Al than 0.1 $cmol_c$ kg^{-1} of soil (about 10 μg of Al g^{-1} or mL^{-1} of soil) are limed. Lower Al levels are found in soils with a pH between 4.9 and 5.5 (Clark, 1965, 1966b; Clark & Hill, 1964; Clark & Nichol, 1966; Turner & Nichol, 1962; Turner & Clark, 1965; Singh, 1972; Hoyt & Webber, 1974; Penny et al., 1977; Webber et al., 1982). However, using this approach, soils containing lower Al levels need not be limed or tested for LR.

2. Equipment

1. Buchner funnels and appropriate Whatman no. 42 or equivalent filters.
2. Mechanical shaker.
3. Automatic pipette.
4. 2.5-mL scoop soil measure.

3. Reagents

Either 1 *N* KCl: dissolve 74.56 g of KCl (FW = 74.56) per liter.
Or 0.2 *N* NH_4Cl: dissolve 10.5 g of NH_4Cl (FW = 53.49) per liter.

4. Procedures

1 *N* KCl exchangeable Al: Measure 2.5 mL[4] dried and sieved (≤2 mm) soils with pH between 4.9 and 5.5 into Erlenmeyer flasks. Add 25 mL of 1 *N* KCl and shake 30 min at about 180 cycles per minute. Filter with gentle suction. Determine Al by one of the procedures described by McLean (1965) or Barnhisel and Bertsch (1982).

0.2 *N* NH_4Cl Extracted Al: This procedure relies on the EAI proposed by Reeve and Sumner (1970a). Measure 2.5 mL[4] dried and sieved (≤2 mm) soils with pH between 4.9 and 5.5 into Erlenmeyer flasks. Add 25 mL of 0.2 *N* NH_4Cl and shake for 2 min at about 180 cycles min^{-1}. Filter with gentle suction. Determine Al by one of the procedures described by McLean (1965) or Barnhisel and Bertsch (1982).

5. Comments

Soils with pH values between about 4.9 and 5.5 and containing ≳0.1 $cmol_c$ Al L^{-1} extracted with 1 *N* KCl or 0.2 *N* NH_4Cl are tested for LR with a buffer-pH procedure and limed to pH 5.5. Soils with pH < 4.9 are tested for LR and limed to pH 5.5.

D. Liming to pH 5.5 when Soil-Solution Aluminum is High

1. Principles

This approach is similar to section VIII.C except that *soil solution* is substituted for *exchangeable* Al as additional test criterion. Soils with pH values between 4.9 and 5.5 are tested for soil-solution (soluble) Al to select those that contain plant-toxic levels. The additional test allows choosing soils that might respond to liming with increased yields. Soluble Al is extracted with a dilute neutral salt solution as 0.01 *M* $CaCl_2$. Soils containing higher concentrations of soluble Al than the selected norm are tested for LR using a buffer-pH procedure, and subsequently limed to achieve pH 5.5.

Although it is difficult to determine from the literature whether improved liming response recognition is obtained with *soluble* instead of *exchangeable* Al as the screening norm, as researched soils were probably not randomly selected. Perusal of published data suggest, nonetheless, that a proportion of soils limed by using the 0.1 $cmol_c$ exchangeable-Al criterion, would not be, had a 1 μg (0.01 *M* $CaCl_2$) soluble Al norm been used instead. A further advantage favoring 0.01 *M* $CaCl_2$, is that it does not cause burner salt-buildup problems with flame techniques.

Extracting 0.01 *M* $CaCl_2$ soluble Al is realized at the same soil/solution ratio and concentration as for determining soil pH; accordingly both tests can be carried out on the same sample by filtering pH samples and determining Al in the extracts. Particularly, since Al concentrations extracted by 0.01 *M* $CaCl_2$ are not extraction-time sensitive (Hoyt & Webber, 1974). However, instead of using soil pH values ranging between 4.9 to 5.5 (water) as selection parameters for determining soluble Al, corresponding 0.01 *M* $CaCl_2$ pH situated between about 4.4 and 5.0 should be used instead.

Interest in soluble Al as screening norm evolved because exchangeable Al is not always a good indicator for determining crop response to liming (Ragland & Coleman, 1959; Adams & Lund, 1966; Kamprath, 1970; Evans & Kamprath, 1970; Martini et al., 1974; Webber et al., 1982). There are several possible reasons for this. First, plants have some tolerance to Al; therefore, levels need not be brought to zero. Second, high levels of exchangeable Al do not necessarily translate into high soil-solution levels that can affect plant growth (Evans & Kamprath, 1970). Milder extraction solutions such as 0.01 *M* $CaCl_2$ were proposed to measure potentially toxic concentrations of soluble Al (Hoyt & Nyborg, 1971a; Webber et al., 1982). Thirdly, Al impairs Ca and Mg absorption, and high levels of the latter seem to counteract its toxic effect. This was the reason that percentage Al saturation was suggested as an index for predicting yield response to liming (Adams & Pearson, 1967; Kamprath, 1970; Evans & Kamprath, 1970; Martini et al., 1974; Webber et al., 1982). However, obtaining a percentage Al or base saturation is more laborious, proposed values vary widely, and serve only to emphasize that exchangeable Al does not generally need to be completely neutralized to achieve maximum crop yields (Adams & Pearson, 1967; Kamprath, 1970; Evans & Kamprath, 1970; Martini et al., 1974; Penney et al., 1977; Webber et al., 1982).

Soil-solution Al concentration was determined in 0.01 *M* $CaCl_2$ for corrected lime-potential studies (Clark, 1965, 1966b; Clark & Hill, 1964; Clark & Nichol, 1966; Turner & Nichol, 1962; Turner & Clark, 1965; Singh, 1972). Soluble Al is the principal toxic substance in acid soils (Ragland & Coleman, 1961; Hourigan et al., 1961; Kamprath, 1970; Reeve & Sumner, 1970a, b), and soil-solution Al is an index of its potenital toxicity. Accordingly, Hoyt and Nyborg (1971a, b) proposed using 0.01 *M* $CaCl_2$ at a 1:2 soil/solution ratio as a diagnostic test for determining potentially toxic concentrations of Al and Mn in acid soils. Subsequently, the procedure was refined by demonstrating that Al values obtained with Clark's 5-d procedure (Clark, 1965) and 16- and 1-h extraction times variations provided equally good correlations between Al and crop yield data (Hoyt & Nyborg, 1971a, 1972). The extraction time was shortened further from 1 h to 5 min as the procedure is not time sensitive (Hoyt & Webber, 1974). Doubling extracting solution concentration from 0.01 to 0.02 M $CaCl_2$ gave equally good correlations between Al and crop data (Hoyt & Nyborg, 1972; Hoyt & Webber, 1974). The more concentrated 0.02 M $CaCl_2$ solution is suggested because it removes about twice as much Al and may provide better analytical precision (Hoyt & Nyborg, 1972; Hoyt & Webber, 1974; Webber et al., 1977; Hoyt & Nyborg, 1987). However, a 0.01 M $CaCl_2$ solution has been used as effectively over the years (Hoyt & Nyborg, 1971a, 1972; Hoyt & Webber, 1974; Webber et al., 1977, 1982), and has the advantage of being more versatile as it can be used to measure soil pH and extract soluble Al from the same sample. It was selected, therefore, for diagnosing soluble Al levels in acid soils.

Unfortunately, neither proponents of exchangeable or soil-solution Al have proposed a test norm vociferously for deciding whether Al concentrations are sufficiently high to affect yields. However, a standard was used (1 μg of Al mL^{-1} soil extracted with 0.02 M $CaCl_2$) when verifying LR-determination procedures by Webber et al. (1977). Study of published data suggest that yields are not affected by excessive acidity when soluble Al levels in 0.01 M $CaCl_2$ are lower than about 1 μg of Al mL^{-1} soil (Hoyt & Nyborg, 1971a, 1972; Hoyt & Webber, 1974; Webber et al., 1982). Utilization of this norm does not imply that maximum yields cannot be attained at higher Al levels in some soils. The norm is meant for determining whether a soil-solution Al level is low enough not to affect yield of sensitive crops. Validity of the 1 μg of Al mL^{-1} soil (0.01 M $CaCl_2$) test norm for accomplishing this objective was confirmed by P.B. Hoyt (1988, personal communication). Furthermore, comparison of corresponding soluble or 0.1 $cmol_c$ exchangeable-Al norms to yield data revealed that the 1 μg of soluble Al mL^{-1} soil test criterion is applicable to a wide range of soils.

2. Equipment: See section VIII.C.2.

3. Reagents: See section III.D.1.

4. Procedure

Measure 10 mL[5] of dried and sieved ($\leq$2 mm) soils with pH between about 4.4 and 5.0 (0.01 M $CaCl_2$) or between 4.9 and 5.5 (water) into Erlenmeyer flasks. Add 20 mL 0.01 M $CaCl_2$ and shake 5 min at about 180 cycles min^{-1}. Filter using gentle suction. Determine Al by atomic absorption (Webber, 1974; Webber et al., 1982), or by a procedure described by McLean (1965) or Barnhisel and Bertsch (1982).

5. Comments

Soils with pH values between about 4.5 and 5.0 (0.01 M $CaCl_2$) or about 4.9 and 5.5 (water) containing more than about 1 μg of soluble Al mL^{-1} soil are tested for LR with a buffer-pH procedure and limed to pH 5.5. Soils with lower pH values are tested for LR and limed to pH 5.5.

E. Calculating Lime Requirement Rates from Exchangeable-Aluminum Levels

1. Principles

The effect of exchangeable Al on crop yield was introduced in section VIII.C.1. It was proposed then that exchangeable Al be used as additional test criterion to determine whether levels were high to warrant liming. If they were high then a buffer-pH test would be used to determine LR, and the soil limed to pH 5.5. In this section, an alternative approach that relies on calculating LR levels from the concentration of exchangeable Al is proposed. Using this approach, exchangeable Al is determined on soils with pH $\leq$ 5.5.

Exchangeable-Al levels as high as 30 and 26 $cmol_c$ Al kg^{-1} soil have been reported by Ragland and Coleman (1959) and Mengel and Kamprath (1978), respectively. However, most values are situated between 0 and 4 $cmol_c$ kg^{-1} soil. The advantage in using exchangeable Al for deriving LR is that liming rates expand as Al levels and toxicity increase. Accordingly, soils containing high or no exchangeable Al will be limed at corresponding rates. Nonetheless, the fundamental reason for interest in this approach is that rates so derived are usually much lower than required for reducing soluble Al to nontoxic levels or increasing soil pH to 5.5. For example, a soil containing 3 $cmol_c$ exchangeable Al would require liming at an equivalent 3 t $CaCO_3$ ha^{-1} to grow relatively Al-tolerant crops. While this example conveys the principle of the exchangeable-Al approach, a liming factor is often included in LR rates to increase these and thereby enable growing more Al-sensitive crops. Data supplied by Webber et al. (1977) suggests that an average of about one-fifth the limestone required to achieve pH 5.5 would be applied using exchangeable Al. Their data also shows that an average of one-third the limestone required to reduce soluble-Al to nontoxic levels would be applied using rates based on exchangeable Al without liming factor.

At pH values below 5.5 (water) most of the soil-buffering capacity is related to exchangeable Al (Jackson, 1963; Kamprath, 1970; Reeve & Sumner, 1970a, b), and others. Consequently, liming at rates equivalent to the concentration of exchangeable Al react primarily with Al as shown by Kamprath (1970). He proposed using exchangeable-Al levels for calculating liming rates by multiplying values by either 1.5 or 2.0 for very Al-sensitive crops. Kamprath's work, Ragland and Coleman (1950), and Evans and Kamprath (1970) support the premise that such rates are ample for obtaining maximum yields. Although this liming scheme appears valid for determining rates for heavily leached soils, such as those used in these studies, it is not clear how widely applicable this approach is for liming different soils. Section V.A discusses the contribution of various soil properties in regulating LR. In any event, data from Webber et al. (1977) suggest that rates based on exchangeable Al, as proposed by Kamprath (1970), can be high enough to achieve pH of 5.5 or higher and therefore maximum yields. However, using this liming scheme will result in liming insufficiently for neutralizing Al in some soils. Still, Al need not be completely precipitated for achieving maximum yields for most crops. This liming scheme appears to offer the possibility of achieving maximum yields for many crops with the lowest liming rates. Of course, if liming costs are relatively low, using this approach offers little or no advantage. However, for extensive agricultural production where liming is a relatively important production cost, as it is in many parts of the world, using exchangeable Al for determining rates offers a reasonable alternative approach for maintaining or improving production and soil quality.

2. **Equipment:** Same as described in section VIII.C.2.

3. **Reagent:** Prepare 1 *N* KCl as described in section VIII.C.3.

4. **Procedure**

Extract 1 *N* KCl exchangeable Al as described in section VIII.C.4. Liming rates are calculated using the following equations suggested by Kamprath (1970) and discussed by McLean (1982).

$$LR(I) = Al \quad [5]$$

$$LR(II) = 1.5Al \quad [6]$$

$$LR(III) = 2Al \quad [7]$$

where LR(I), LR(II), and LR(III) represent the LR in metric tonnes of $CaCO_3$ ha^{-1} (2 million L) for crops having most, moderate, and least Al tolerance, respectively. The Al concentrations in these equations are expressed in $cmol_c$ Al L^{-1} of soil. Aluminum is considered trivalent and fully ionized. These assumptions generally result in liming at higher rates than if true ionic Al concentration were considered.

5. **Comments**

This LR-determination approach has particular applicability where large tracts of land suitable for extensive agricultural production are being under utilized because of excessive acidity and high-liming costs. Liming rates based on Al as a criterion will enhance productivity at the lowest possible cost and slow the rate of soil acidification. This liming approach will provide optimal yields in many situations and is a step in the right direction for others.

IX. LIMING RECOMMENDATIONS AND LIMING

A LR is the amount of $CaCO_3$ required by a hectare furrow-layer of soil to neutralize acidity and increase pH to a selected target value or neutralize exchangeable Al. To effect neutralization efficiently and accurately, the liming rate has to be incorporated uniformly into a furrow-layer, which is often assumed to have a depth of 20 cm (2 million L ha^{-1}; van Lierop,

1989). To achieve that end, the liming material must first be spread precisely and uniformly over the soil surface, and subsequently mixed thoroughly into the furrow layer. Generally, it is difficult to satisfy these requirements with a single application or incorporation. However, certain management practices favor attaining the desired end as efficiently as possible. One of these is to adjust the liming rate to fit tillage depth. For example, if applied limestone is only worked into 10 cm (4 in.) soil, the recommended rate should be reduced by one-half because only one-half the assumed furrow layer is being limed. On the other hand, if limestone is left on the soil surface to gradually find its way down, the rate can best be reduced to a minimum application, followed by others as it moves down the profile. Limestone so applied will gradually neutralize acidity at depth by moving through the profile at an approximate rate of 1 to 2 cm y^{-1} where moisture and drainage are adequate (Brown et al., 1956). Not adjusting liming rates for differences between assumed and actual furrow-layer depths may cause localized overliming, particularly if rates and target values are high. The only advantage to liming to a higher pH than required for achieving maximum yields is that a longer period may lapse before liming is required again for maintaining soil pH.

The principal liming inaccuracies are caused by uneven spreading and incorporation. Generally, spreading uniformity is improved by applying a LR in two or more increments. Similarly, incorporation uniformity is improved by increasing the intensity and number of tillings. For example, a more accurate liming will result if a LR of 15 t is applied in three separate additions of 5 t of limestone ha^{-1}. Whether a LR is applied in several increments or as a single application depends of course on the size of the initial LR rate. Perhaps a single application should $\not>$5 to 6 t ha^{-1}. Incremental liming need not cost more, but may require more than one growing season to complete. However, using fewer amendment additions is justified when soils are tilled infrequently at depth: taking advantage of such an opportunity may be advantageous. It is well to remember when partially liming a furrow layer, that the greatest impact on yields results if liming is restricted to the upper portion of the soil (Hourigan et al., 1961; Lathwell & Peech, 1965). The applied liming material will gradually move down and neutralize acidity at depth while the upper soil layer is relimed to maintain pH (Brown et al., 1956).

Portioning a LR into several increments will enhance its distribution through the profile and assist in maintaining surface pH (Brown et al., 1956). Principally, because large LRs ought to be applied in multiple increments, the calibration accuracy of buffer-pH procedure is less important for soils with high requirements. Presumably, those soils will be tested again before a last corrective LR increment is broadcast to monitor pH progress and decide on future management. On the other hand, highly accurate buffer-pH calibrations are required to determine the LR of soils with low-buffering capacities as they will usually be limed with a single application. Accuracy of determination and application is, therefore, particularly important at low rates.

Limestone can neutralize soil acidity quite rapidly. In fact, soil pH can be increased by several units in a matter of minutes when liming a very acid soil, though a stable pH will not result for some time. Generally, most of the change in pH occurs shortly after liming. Subsequently, it continues to increase more gradually and eventually plateaus. Thereafter, it remains there for a period, to finally slowly decline (fluctuating during the course of this generalized cycle). Field studies suggest that soil pH usually peaks within 8 to 12 wk after liming. However, it appears more difficult to increase the pH of a soil with a relatively high than low value. For example, increasing the pH from pH 6.5 to 7.0 is more difficult than from pH 5 to 6.5. Limestone reactivity probably decreases because liming has lowered the intensity of acidity present, as indicated by an increased pH. Occasionally, a soil is limed with dolomitic limestone to correct a low-available Mg level, as this may be the least-expensive source for that nutrient. Liming with dolomitic limestone offers some advantages when soil Mg levels are low, and can be as effective as calcitic limestone for correcting pH. Whether it is used is largely a matter of relative cost.

Measured LRs are often adjusted by multiplying $CaCO_3$-LR values by a constant, between 1.4 and 1.5, referred to as a liming factor (Shoemaker et al., 1961; Woodruff, 1948; Adams & Evans, 1962; McLean, 1973, 1982). It is used partially to correct for the inability to achieve a homogeneous dispersion of liming material through a furrow layer, and partially because larger particles in commercial limestone dissolve more slowly. At the risk of breaking with established practice, it is difficult to justify using a typical liming factor blindly, as commercial limestones often have agricultural values (AV) (Murphy & Follett, 1978) that are not significantly different from 100%. The AV of available limestones probably ought to be known so that the better grades can be selected. In any event, correcting LR recommendations by using an arbitrary AV does not seem to offer advantages, particularly if liming is carried out in multiple applications as recommended.

Several procedures have been suggested to evaluate the efficacy of limestone for neutralizing soil acidity (Barber, 1984). Efficacy of a liming material is affected by its solubility. However, more importantly its solubility is largely related to surface area exposed to chemical reaction. This area increases in inverse proportion to particle fineness (i.e., halving the particle size doubles the surface area). A practical approach for evaluating agricultural limestones is achieved by rating sieve fractions and summing their effect. Such a rating procedure compensates for differences in overall fineness. Although, sieves and ratings vary somewhat from one procedure to another, the Ohio and Canadian methods described by Tisdale and Nelson (1966) and Tisdale et al. (1985), or the procedure proposed by Murphy and Follett (1978) are fairly typical. Admittedly, the accuracy of LR recommendations may be improved by adjusting the rate for the agricultural value of the particular liming produce used, as proposed by Murphy and Follett (1978). Probably such adjustments would only improve the accuracy of liming if a limestone had an AV that is significantly lower than about 90%. However, values are rarely known for a specific limestone, even less for a specific lot. Accordingly, using a customary liming factor does little to improve liming accuracy. Its only material contribution to accuracy is to increase recommended rates proportionately (about 50%).

Hydrated lime is occasionally substituted for limestone; however, a recommended LR should then be reduced by 26% (multiply LR by 0.74). Hydrated lime costs more, and is occasionally used because it is believed to neutralize soil acidity and increase pH more rapidly than limestone. If the effect is to be more rapid, it should be incorporated into the furrow layer as soon after application as possible. If left on the soil surface for any length of time, it will quickly absorb CO_2 from the air to become $CaCO_3$. Although, hydrated lime can neutralize soil acidity more rapidly under some circumstances, generally it is not more efficient than a limestone having the equivalent particle-size distribution. However, hydrated lime is often more finely pulverized than limestone and may, therefore, act more quickly.

ACKNOWLEDGMENT

Preparation of chapter 5 was requested and started while the author was employed by the British Columbia Ministry of Agriculture.

REFERENCES

Adams, F. 1984. Crop responses to liming in the southern United States. p. 211–266. *In* F. Adams (ed.) Soil acidity and liming. 2nd ed. Agronomy Monogr. 12. ASA, Madison, WI.

Adams, F., and C.E. Evans. 1962. A rapid method for measuring the lime requirement of Red-Yellow Podzolic soils. Soil Sci. Am. Proc. 26:355–357.

Adams, F., and Z.F. Lund. 1966. Effect of chemical activity of soil solution aluminum on cotton root penetration of acid subsoils. Soil Sci. 101:193–198.

Adams, F., and R.W. Pearson. 1967. Crop response to lime in the southern United States and Puerto Rico. p. 161–206. *In* R.W. Pearson and F. Adams (ed.) Soil acidity and liming. Agronomy Monogr. 12. ASA, Madison, WI.

Adams, F., and J.I. Wear. 1957. Manganese toxicity and soil acidity in relation to crinkle leaf in cotton. Soil Sci. Soc. Am. Proc. 21:305–308.

Alabi, K.E., R.C. Sorensen, D. Knudsen, and G.W. Rehm. 1986. Comparison of several lime requirement methods on coarse textured soils of Northeastern Nebraska. Soil Sci. Soc. Am. J. 50:937–941.

Bailey, E.H. 1932. The effect of air drying on the hydrogen ion concentration in the soils of the United States and Canada. USDA Tech. Bull. 291. U.S. Gov. Print. Office, Washington, DC.

Barber, S.A. 1984. Liming materials and practices. p. 171–209. *In* F. Adams (ed.) Soil acidity and liming. 2nd ed. Agronomy Monogr. 12. ASA, CSSA, and SSSA, Madison, WI.

Barnhisel, R., and P.M. Bertsch. 1982. Aluminum. p. 275–300. *In* A.L. Page et al. (ed.) Methods of soil analysis. Part 2. 2nd ed. Agronomy Monogr. 9. ASA and SSSA, Madison, WI.

Bates, R.G. 1973. Determination of pH: Theory and practice. 2nd ed. John Wiley and Sons, New York.

Baver, L.D. 1927. Factors affecting the hydrogen ion concentration in soils. Soil Sci. 23:399–414.

Bloksma, A.H. 1957. An experimental test of Overbeek's treatment of the suspension effect. J. Colloid Sci. 12:135–143.

Boelter, D.H. 1964. Water storage characteristics of several peats in situ. Soil Sci. Soc. Am. Proc. 28:433–435.

Bohn, H.L., B.L. McNeal, and G.A. O'Conner. 1985. Soil chemistry. 2nd ed. John Wiley and Sons, New York.

Bower, C.A. 1961. Studies on the suspension effect with the sodium electrode. Soil Sci. Soc. Am. Proc. 25:18–21.

Bowser, W.E., and J.N. Leat. 1958. Seasonal pH fluctuations in a Gray Wooded soil. Can. J. Soil Sci. 38:128–133.

Bradfield, R. 1941. Calcium in the soil: 1. Physico-chemical relations. Soil Sci. Soc. Am. Proc. 6:8–15.

Bray, R.H., and L.T. Kurtz. 1945. Determination of total, organic, and available forms of phosphorus in soils. Soil Sci. 59:39–45.

Brown, B.A., R.I. Munsell, R.F. Holt, and A.V. King. 1956. Soil reactions at various depths as influenced by time since application and amounts of limestone. Soil Sci. Soc. Am. Proc. 20:518–522.

Brown, I.C. 1943. A rapid method of determining exchangeable hydrogen and total exchangeable bases in soils. Soil Sci. 56:353–357.

Brown, J.R., and J.R. Cisco. 1984. An improved Woodruff buffer for estimation of lime requirements. Soil Sci. Soc. Am. J. 48:587–592.

Brown, J.R., J. Garett, and T.R. Fisher. 1977. Soil testing in Missouri. Univ. of Missouri-Columbia Ext. Div. Ext. Circ. 923.

Chapman, H.D., J.H. Axley, and D.S. Curtis. 1941. The determination of pH at soil moisture levels approximating field conditions. Soil Sci. Soc. Am. Proc. (1940) 5:191–200.

Clark, J.S. 1964. An examination of the pH of calcareous soils. Soil Sci. 98:145–151.

Clark, J.S. 1965. The extraction of exchangeable cations from soils. Can. J. Soil Sci. 45:311–322.

Clark, J.S. 1966a. The pH values of soils suspended in dilute salt solutions. Soil Sci. Soc. Am. Proc. 30:11–14.

Clark, J.S. 1966b. The lime potential and percent base saturation of some representative podzolic and brunisolic soils in Canada. Soil Sci. Soc. Am. Proc. 30:93–97.

Clark, J.S., and R.G. Hill. 1964. The pH-percent base saturation relationships of soils. Soil Sci. Soc. Am. Proc. 28:490–492.

Clark, J.S., and W.E. Nichol. 1966. Lime potential—percent base saturation relations of acid surface horizons of mineral and organic soils. Can. J. Soil Sci. 46:281–285.

Cole, C.V. 1957. Hydrogen and calcium relationships of calcareous soils. Soil Sci. 83:141–150.

Coleman, N.T., and G.W. Thomas. 1964. Buffer curves of acid clays as affected by the presence of ferric iron and aluminum. Soil Sci. Soc. Am. Proc. 28:187–190.

Coleman, N.T., and G.W. Thomas. 1967. The basic chemistry of soil acidity. p. 1–41. *In* R.W. Pearson and F. Adams (ed.) Soil acidity and liming. Agronomy Monogr. 12. ASA, Madison, WI.

Coleman, N.T., S.B. Weed, and R.J. McCracken. 1959. Cation-exchange capacity and exchangeable cations in Piedmont soils of North Carolina. Soil Sci. Soc. Am. Proc. 23:146–149.

Coleman, N.T., D.E. Williams, T.R. Nielsen, and H. Jenny. 1951. On the validity of interpretations of potentiometrically measured soil pH. Soil Sci. Soc. Am. Proc. (1950) 15:106–110.

Collins, J.B., E.P. Whiteside, and C.E. Cress. 1970. Seasonal variability of pH and lime requirements in several southern Michigan soils when measured in different ways. Soil Sci. Soc. Am. Proc. 34:56–61.

Davies, B.E. 1971. A statistical comparison of pH values of some English soils after measurement in both water and 0.01 *M* calcium chloride. Soil Sci. Soc. Am. Proc. 35:35–36.

Davis, J.F., and K. Lawton. 1947. A comparison of the glass electrode and indicator methods for determining the pH of organic soils and the effect of time, soil water ratio, and air drying on glass electrode results. J. Am. Soc. Agron. 39:719–723.

Dewan, H.C., and C.I. Rich. 1970. Titration of acid soils. Soil Sci. Soc. Am. Proc. 34:38–44.

Doner, H.E., and P.F. Pratt. 1969. Solubility of calcium carbonate precipitated in aqueous solutions of magnesium and sulfate salts. Soil Sci. Soc. Am. Proc. 33:690–693.

Dost, H. (ed.). 1973. Acid sulphate soils. Vols. 1 and 2. Publ. 18. Int. Inst. for Land Reclamation and Improvement, Wageningen, Netherlands.

Durst, R.A. 1975. Standardization of pH measurement. Natl. Bureau of Standards Spec. Publ. 260-53. U.S. Gov. Print. Office, Washington, DC.

Evans, C.E., and E.J. Kamprath. 1970. Lime response as related to percent Al saturation, solution Al, and organic matter content. Soil Sci. Soc. Am. Proc. 34:893–896.

Foy, C.D.1964. Toxic factors in acid soils of the southeastern United States as related to the response of alfalfa to lime. USDA-ARS Production Res. Rep. 80. U.S. Gov. Print. Office, Washington, DC.

Foy, C.D., W.H. Armiger, L.W. Briggle, and D.A. Reid. 1965. Differential aluminum tolerance of wheat and barley varieties in acid soils. Agron. J. 57:413–417.

Foy, C.D., and J.C. Brown. 1964. Toxic factors in acid soils: II. Differential aluminum tolerance of plant species. Soil Sci. Soc. Am. Proc. 28:27–32.

Foy, C.D., A.L. Flemming, and W.H. Armiger. 1969. Aluminum tolerance of soybean varieties in relation to calcium nutrition. Agron. J. 61:505–511.

Fox, R.H. 1980. Comparison of several lime requirement methods for agricultural soils in Pennsylvania. Commun. Soil Sci. Plant Anal. 11:57–69.

Hajek, B.F., F. Adams, and J.T. Cope, Jr. 1972. Rapid determination of exchangeable bases, acidity, and base saturation for soil characterization. Soil Sci. Soc. Am. Proc. 36:436–438.

Helling, C.S., G. Chesters, and R.B. Corey. 1964. Contribution of organic matter and clay to soil cation exchange capacity as affected by the pH of the saturation solution. Soil Sci. Soc. Proc. Am. 28:517–520.

Hesse, P.R. 1971. A textbook of soil chemical analysis. Chemical Publ. Co., New York.

Hissink, D.J. 1930. Report of the Committee on Soil Reaction Measurements of the International Society of Soil Science. Soil Res. 2:141–144.

Hourigan, W.R., R.E. Franklin, Jr., E.O. McLean, and D.R. Bhumbla. 1961. Growth and Ca uptake by plants as affected by rate and depth of liming. Soil Sci. Soc. Am. Proc. 25:491–494.

Hoyt, P.B., and M. Nyborg. 1971a. Toxic metals in acid soils: 1. Estimation of plant-available aluminum. Soil Sci. Soc. Am. Proc. 35:236–240.

Hoyt, P.B., and M. Nyborg. 1971b. Toxic metals in acid soil: II. Estimates of plant-available manganese. Soil Sci. Soc. Am. Proc. 35:241–244.

Hoyt, P.B., and M. Nyborg. 1972. Use of dilute calcium chloride for the extraction of plant-available aluminum and manganese from acid soil. Can. J. Soil Sci. 52:163–167.

Hoyt, P.B., and M. Nyborg. 1987. Field calibration of liming responses of four crops using pH, Al, and Mn. Plant Soil 102:21–25.

Hoyt, P.B., and M.D. Webber.1974. Rapid measurement of plant-available aluminum and manganese in acid Canadian soils. Can. J. Soil Sci. 54:53–61.

Huberty, M.R., and A.R.C. Haas. 1940. The pH of soil as affected by soil moisture and other factors. Soil Sci. 49:455–478.

Hutchinson, F.E., and A.S. Hunter. 1970. Exchangeable aluminum in two soils as related to lime treatment and growth of six crop species. Agron. J. 62:702–704.

Jackson, M.L. 1958. Soil chemical analysis. Prentice-Hall, Englewood Cliffs, NJ.

Jackson, M.L. 1963. Aluminum bonding in soils: A unifying principle in soil science. Soil Sci. Soc. Am. Proc. 27:1–10.

Jenny, H. 1961. Reflections on the soil acidity merry-go-round. Soil Sci. Soc. Am. Proc. 25:428–432.

Jenny, H., T.R. Nielsen, N.T. Coleman, and D.E. Williams. 1950. Concerning the measurement of pH, ion activities, and membrane potentials in colloidal systems. Science 112:164–167.

Joret, G., H. Malterre, and M. Cabazan. 1934. L'appréciation des besoins en chaux des sols de limon d'après leur état de saturation en bases échangeables. Ann. Agron. 22:453–479.

Kaila, A. 1956. Determination of the degree of humification of peat samples. J. Sci. Agric. Soc. Finl. 28:18–25.

Kamprath, E.J. 1970. Exchangeable aluminum as a criterion for liming leached mineral soils. Soil Sci. Soc. Am. Proc. 34:252–254.

Keeney, D.R., and R.B. Corey. 1963. Factors affecting the lime requirement of Wisconsin soils. Soil Sci. Soc. Am. Proc. 27:277–280.

Lathwell, D.J., and M. Peech. 1965. Interpretation of chemical soil tests. New York (Cornell Univ.) Agric. Exp. Stn. Bull. 995.

Lin, C., and N.T. Coleman. 1960. The measurement of exchangeable aluminum in soils and clays. Soil Sci. Soc. Am. Proc. 24:444–446.

Loosjes, R. 1950. pH-meting in suspensies. Chem. Weekbl. 46:902–906.

Loynachan, T.E. 1981. Lime requirement methods for cold regions. Soil Sci. Soc. Am. J. 45:77–80.

MacLeod, L.B., and L.P. Jackson. 1967. Water-soluble and exchangeable aluminum in acid soils as affected by liming and fertilization. Can. J. Soil Sci. 47:203–210.

Marshall, C.E. 1964. The physical chemistry and mineralogy of soils. Vol. 1. John Wiley and Sons, New York.

Martini, J.A., R.A. Kochhann, O.J. Siqueira, and C.M. Borkert. 1974. Response of soybean to liming as related to soil acidity, Al, and Mn toxicities, and P in some oxisols of Brazil. Soil Sci. Soc. Am. Proc. 38:616–620.

McLean, E.O. 1965. Aluminum. p. 978–998. *In* C.A. Black et al. (ed.) Methods of soil analysis. Part 2. Agronomy Monogr. 9. ASA, Madison, WI.

McLean, E.O. 1970. Lime requirement of soils—inactive toxic substances or favorable pH range. Soil Sci. Soc. Am. Proc. 34:363–364.

McLean, E.O. 1971. Potentially beneficial effects from liming: Chemical and physical. Soil Crop Sci. Soc. Fla. Proc. 31:189–196.

McLean, E.O. 1973. Testing soils for pH and lime requirement. *In* L.M. Walsh and J.D. Beaton (ed.) Soil testing and plant analysis. SSSA, Madison, WI.

McLean, E.O. 1976. Chemistry of soil aluminum. Commun. Soil Sci. Plant Anal. 7:619–636.

McLean, E.O. 1978. Principles underlying the practice of determining lime requirements of acid soils by use of buffer methods. Commun. Soil Sci. Plant Anal. 9:699–715.

McLean, E.O. 1982. Soil pH and lime requirement. p. 199–224. *In* A.L. Page et al. (ed.) Methods of soil analysis. Part 2. 2nd ed. Agronomy Monogr. 9. ASA and SSSA, Madison, WI.

McLean, E.O., S.W. Dumford, and F. Coronel. 1966. A comparison of several methods of determining lime requirements of soils. Soil Sci. Soc. Am. Proc. 30:26–30.

McLean, E.O., D.J. Eckert, G.Y. Reddy, and J.F. Trierweiler. 1978. An improved SMP soil lime requirement method incorporating double-buffer and quick-test features. Soil Sci. Soc. Am. J. 42:311–316.

McLean, E.O., W.R. Hourigan, H.E. Shoemaker, and D.R. Bhumbla. 1964. Aluminum in soils: V. Form of aluminum as a cause of soil acidity and a compliction in its measurement. Soil Sci. 97:119–126.

McLean, E.O., J.F. Trierweiler, and D.J. Eckert. 1977. Improved SMP buffer method for determining lime requirement of acid soils. Commun. Soil Sci. Plant Anal. 8:667–675.

Mehlich, A. 1976. New buffer pH method for rapid estimation of exchangeable acidity and lime requirements of soils. Commun. Soil Sci. Plant Anal. 7:253–263.

Mehlich, A., S.S. Bowling, and A.L. Hatfield. 1976. Buffer pH acidity in relation to nature of soil acidity and expression of lime requirement. Commun. Soil Sci. Plant Anal. 7:253–263.

Mengel, D.B., and E.J. Kamprath. 1978. Effect of soil pH and liming on growth and nodulation of soybeans in Histosols. Agron. J. 70:959–963.

Moormann, F.R. 1963. Acid sulfate soils (cat-clays) of the tropics. Soil Sci. 95:271–275.

Morris, H.D. 1948. The soluble manganese content of acid soils and its relation to the growth and manganese content of sweet clover and lespedeza. Soil Sci. Soc. Am. Proc. 13:362–371.

Moschler, W.W., G.D. Jones, and G.W. Thomas. 1960. Lime and soil acidity effects on alfalfa growth in a Red-Yellow podzolic soil. Soil Sci. Soc. Am. Proc. 24:507–509.

Murphy, L.S., and H. Follett. 1978. Liming. Taking another look at the basics. Agrichemistry 22:22–26.

Nichol, W.E., and R.C. Turner. 1957. The pH of non-calcareous near-neutral soils. Can. J. Soil Sci. 37:96–101.

Nômmik, H. 1983. A modified procedure for rapid determination of titrable acidity and lime requirement in soils. Acta Agric. Scand. 33:337–348.

Olsen, S.R., and F.S. Watanabe. 1959. Solubility of calcium carbonate in calcareous soils. Soil Sci. 88:123–129.

Ouellette, G.J., and L. Dessureaux. 1958. Chemical composition of alfalfa as related to degree of tolerance to manganese or aluminum. Can. J. Plant Sci. 38:206–214.

Overbeek. J.Th.G. 1953. Donnan-E.M.F. and suspension effect. J. Colloid Sci. 8:593–605.

Peech, M. 1965a. Hydrogen-ion activity. p. 914–926. *In* C.A. Black et al. (ed.) Methods of soil analysis. Part 2. Agronomy Monogr. 9. ASA, Madison, WI.

Peech, M. 1965b. Lime requirement. p. 927–932. *In* C.A. Black et al. (ed.) Methods of soil analysis. Part 2. Agronomy Monogr. 9. ASA, Madison, WI.

Peech, M., and W.F. McDevit. 1951. Discussion of the paper on the validity of interpretations of potentiometrically soil pH. Soil Sci. Soc. Am. Proc. (1950) 15:112–114.

Peech, M., R.A. Olsen, and G.H. Bolt. 1953. The significance of potentiometric measurements involving liquid junction in clay and soil. Soil Sci. Soc. Am. Proc. 17:214–218.

Penney, D.C., M. Nyborg, P.B. Hoyt, W.A. Rice, B. Siemens, and D.H. Laverty. 1977. An assessment of the soil acidity problem in Alberta and Northeastern British Columbia. Can. J. Soil Sci. 57:157–164.

Pionke, H.B., and R.B. Corey. 1967. Relations between acidic aluminum and soil pH, clay and organic matter. Soil Sci. Soc. Am. Proc. 31:749–752.

Pionke, H.B., R.B. Corey, and E.E. Schulte. 1968. Contributions of soil factors to lime requirement and lime requirement tests. Soil Sci. Soc. Am. Proc. 32:113–117.

Pratt, P.F., and F.L. Bair. 1962. Cation-exchange properties of some acid soils of California. Hilgardia 33:689–706.

Puri, A.N., and A.G. Asghar. 1938. Influence of salts and soil water ratio on pH value of soils. Soil Sci. 46:249–257.

Ragland, J.L., and N.T. Coleman. 1959. The effect of aluminum and calcium on root growth. Soil Sci. Soc. Am. Proc. 23:355–357.

Raupach, M. 1954. The errors involved in pH determination in soils. Aust. J. Agric.Res. 5:716–729.

Raupach, M. 1957. Investigations into the nature of soil pH. Soil Publ. 9. CSIRO, Melbourne, Australia.

Reeve, N.G., and M.E. Sumner. 1970a. Effects of aluminum toxicity and phosphorus fixation on crop growth on Oxisols in Natal. Soil Sci. Soc. Am. Proc. 34:263–267.

Reeve, N.G., and M.E. Sumner. 1970b. Lime requirements of Natal Oxisols based on exchangeable aluminum. Soil Sci. Soc. Am. Proc. 34:595–598.

Rémy, J.C., and A. Marin-Laflèche. 1974. L'analyse de terre: réalisation d'un programme d'interprétation automatique. Ann. Agron. 25:607–632.

Rich, C.I. 1970. Conductometric and potentiometric titration of exchangeable Al. Soil Sci. Soc. Am. Proc. 34:31–38.

Ryti, R. 1965. On the determination of soil pH. Maataloustiet. Aikak. 37:51–60.

Schnitzer, M., and S.I.M. Skinner. 1963a. Organo-metallic interactions in soils: 1. Reactions between a number of metal ions and the organic matter of a Podzol Bh horizon. Soil Sci. 96:86–93.

Schnitzer, M., and S.I.M. Skinner. 1963b. Organo-metallic interactions in soils: 2. Reactions between different forms of iron and aluminum and the organic matter of a Podzol Bh horizon. Soil Sci. 96:181–190.

Schnitzer, M., and S.I.M. Skinner. 1964. Organo-metallic interactions: 3. Properties of iron- and aluminum-organic-matter complexes, prepared in the laboratory and extracted from a soil. Soil Sci. 98:197–203.

Schofield, R.K. 1947. A ratio law governing the equilibrium of cations in the soil solution. Proc. 11th Int. Congr. Pure Appl. Chem. 3:257–261.

Schofield, R.K. 1949. Effect of pH on electric charges carried by clay particles. J. Soil Sci. 1:1–8.

Schofield, R.K., and A.W. Taylor. 1955. The measurement of soil pH. Soil Sci. Soc. Am. Proc. 19:164–167.

Schollenberger, C.J., and R.H. Simon. 1945. Determination of exchange capacity and exchangeable bases in soil—ammonium acetate method. Soil Sci. 59:13–24.

Schwertmann, U., and M.L. Jackson. 1963. Hydrogen-aluminum clays: A third buffer range appearing in potentiometric titrations. Science (Washington, DC) 139:1052–1054.

Shaw, W.M. 1952. Report on exchangeable hydrogen in soils. Interrelationships between calcium sorption, exchangeable hydrogen, and pH values of certain soils and subsoils. J. Assoc. Agric. Chem. 35:597–621.

Shoemaker, H.E., E.O. McLean, and P.F. Pratt. 1961. Buffer methods for determining the lime requirement of soils with appreciable amounts of extractable aluminum. Soil Sci. Soc. Am. Proc. 25:274–277.

Simmons, C.F. 1939. The effect of carbon dioxide pressure on the equilibrium of the system hydrogen colloidal clay-H_2O-$CaCO_3$. J. Am. Soc. Agron. 31:638–648.

Singh, S.S. 1972. The effect of temperature on the iron activity product $(AL)(OH)_3$ and its relation to lime potential and degree of base saturation. Soil Sci. Soc. Am. Proc. 36:47–50.

Soil Science Society of America. 1987. Glossary of soil science terms. SSSA, Madison, WI.

Soon, Y.K., and T.E. Bates. 1986. Determination of the lime requirement for acid soils in Ontario using the SMP buffer methods. Can. J. Soil Sci. 66:373–376.

Sorensen, S.P.L. 1909. Enzyme studies: II. The measurement and importance of the hydrogen ion concentration in enzyme reaction. C.R. Trav. Lab. Carlsberg 8:1.

Ssali, H., and J.K. Nuwamanya. 1981. Buffer pH methods for estimation of lime requirement of tropical acid soils. Commun. Soil Sci. Plant Anal. 12:643–659.

Thomas, G.W. 1967. Problems encountered in soil testing methods. p. 37–54. *In* Soil testing and plant analysis. Part 1. SSSA Spec. Publ. 2. SSSA, Madison, WI.

Thomas, G.W., and W.H. Hargrove. 1984. The chemistry of soil acidity. p. 3–56. *In* F. Adams (ed.) Soil acidity and liming. 2nd ed. Agronomy Monogr. 12. ASA, CSSA, and SSSA, Madison, WI.

Tisdale, S.L., and W.L. Nelson. 1966. Soil fertility and fertilizers. 2nd ed. Macmillan Publ. Co., New York.

Tisdale, S.L., W.L. Nelson, and J.D. Beaton. 1985. Soil fertility and fertilizers. 4th ed. Macmillan Publ. Co., New York.

Tran, T.S., and W. van Lierop. 1981a. Evaluation and improvement of buffer-pH lime requirement methods. Soil Sci. 131:178–188.

Tran, T.S., and W. van Lierop. 1981b. Evaluation des méthodes de détermination du besoin en chaux en relation avec les propriétes physiques et chimiques des sols acides. Sci. Sol 12:253–267.

Tran, T.S., and W. van Lierop. 1982. Lime requirement determination for attaining pH 5.5 and 6.0 of coarse textured soils using buffer-pH methods. Soil Sci. Soc. Am. J. 46:1008–1014.

Turner, R.C., and J.S. Clark. 1956. The pH of calcareous soils. Soil Sci. 82:337–341.

Turner, R.C., and J.S. Clark. 1965. Lime potential and degree of base saturation of soils. Soil Sci. 99:194–199.

Turner, R.C., and W.E. Nichol. 1958. The pH of strongly acid soils. Can. J. Soil Sci. 38:63–68.

Turner, R.C., and W.E. Nichol. 1962. A study of the lime potential: 1. Conditions for the lime potential to be independent of salt concentration in aqueous suspensions of negatively charged clays. Soil Sci. 93:374–382.

van der Paauw, E. 1962. Periodic fluctuations of soil fertility, crop yields, and responses to fertilization as affected by alternating periods of low and high rainfall. Plant Soil 17:155–182.

van Lierop, W. 1981a. Conversion of organic soil pH values measured in water, 0.01 *M* $CaCl_2$, or 1 *N* KCl. Can. J. Soil Sci. 61:577–579.

van Lierop, W. 1981b. Laboratory determination of field bulk density for improving fertilizer recommendations of organic soils. Can. J. Soil Sci. 61:475–482.

van Lierop, W. 1983. Lime requirement determination of acid organic soils using buffer-pH methods. Can. J. Soil Sci. 63:411–423.

van Lierop, W. 1989. Effect of assumptions on accuracy of analytical results and liming recommendations when testing a volume or weight of soil. Commun. Soil Sci. Plant Anal. 20:121–137.

van Lierop, W., and A.F. MacKenzie. 1977. Soil pH measurement and its application to organic soils. Can. J. Soil Sci. 57:55–64.

van Lierop, W., and T.S. Tran. 1979. L'acidité et le besoin en chaux: 1. Mesure du pH du sol. Agriculture 36:9–12.

van Lierop, W., T.S. Tran, G. Banville, and S. Morissette. 1982. Effect of liming on potato yields as related to pH, Al, Mn, and Ca. Agron. J. 74:1050–1055.

van Olphen, H. 1963. An introduction to clay colloid chemistry. Interscience Publ. John Wiley and Sons, New York.

Volk, V.V., and M.L. Jackson. 1964. Inorganic pH dependent cation exchange charge of soils. Clays Clay Miner. 12:281–285.

Webber, M.D. 1974. Atomic absorption measurements of Al in plant digests and neutral salt extracts of soils. Can. J. Soil Sci. 54:81–87.

Webber, M.D., P.B. Hoyt, and D. Corneau. 1982. Soluble Al, exchangeable Al, base saturation and pH in relation to barley yield on Canadian acid soils. Can. J. Soil Sci. 62:397–405.

Webber, M.D., P.B. Hoyt, M. Nyborg, and D. Corneau. 1977. A comparison of lime requirement methods for acid Canadian soils. Can. J. Soil Sci. 57:361–370.

White, R.E. 1969. On the measurement of soil pH. J. Austr. Inst. Agric. Sci. 35:3–14.

Whitney, R.S., and R. Gardner. 1943. The effect of carbon dioxide on soil reaction. Soil Sci. 55:127–141.

Willard, H.H., L.L. Merritt, Jr., and J.A. Dean. 1974. Instrumental methods of analysis. D. Van Nostrand Co., New York.

Woodruff, C.M. 1947. Determination of exchangeable hydrogen and lime requirement of the soil by means of the glass electrode and a buffered solution. Soil Sci. Soc. Am. Proc. 12:141–142.

Woodruff, C.M. 1948. Testing soils for lime requirement by means of a buffered solution and the glass electrode. Soil Sci. 66:53–63.

Woodruff, C.M. 1961. Brom cresol purple as an indicator of soil pH. Soil Sci. 91:272.

Woodruff, C.M. 1967. Crop response to lime in the midwestern United States and Puerto Rico. p. 207–231. *In* R.W. Pearson and F. Adams (ed.) Soil acidity and liming. Agronomy Monogr. 12. ASA, Madison, WI.

Yuan, T.L. 1959. Determination of exchangeable hydrogen in soils by titration method. Soil Sci. 88:164–167.

Yuan, T.L. 1974. A double buffer method for the determination of lime requirement of acid soils. Soil Sci. Soc. Am. J. 38:437–440.

Yuan, T.L. 1976. Anomaly and modification of pH-acidity relationships in the double buffer method for lime requirement determination. Soil Sci. Soc. Am. J. 40:800–802.

Chapter 6

Testing Soils for Available Nitrogen[1]

W. C. DAHNKE, *North Dakota State University, Fargo*

GORDON V. JOHNSON, *Oklahoma State University, Stillwater*

Crops are probably more often deficient in N than in any other element and yet there are no widely accepted methods of testing soils for this nutrient other than testing for residual nitrate-nitrogen (NO_3-N) and ammonium-nitrogen (NH_4-N). This is largely because 97 to 99% of the N in the soil is present in complex organic compounds and slowly becomes available to plants through microbial decomposition. The problems of developing a test for available N are that: (i) The rate at which microorganisms decompose soil organic matter (OM) is dependent on temperature, moisture, aeration, type of organic matter, pH and other factors and (ii) the inorganic forms of N produced are subject to leaching, fixation, denitrification, and other losses. Thus, it becomes difficult to predict either when N will become available, how much will become available or what will happen to it.

The broad spectrum of conditions affecting availability of N to growing crops may be identified at one extreme by cold-humid environments. Under these conditions, soil OM content is high and mineralization of organic N supplies a significant portion of the total requirement for the short-season crops typical of this environment. In these situations, soil OM could be expected to provide a suitable N availability index. Hot, dry environments, typical of the southwestern USA, represent the other extreme. Under these conditions, cultivated crops are almost always irrigated and high yielding. Soil OM levels are very low (<1.0%) and mineralizable N provides only a small portion of the total N required for high-yielding crops. In these situations, a test of soil NO_3-N is a good direct measure of available N. Between these extremes there is a myriad of conditions under which available soil N may be identified by combinations of indirect and direct measures. In some cases, even the best application of existing tests may provide only a poor measure of available N.

[1] Contribution from the Dep. of Soil Science, North Dakota State Univ., Fargo, ND 58105 and Coop. Ext. Serv., Div. of Agric., Oklahoma State Univ., Stillwater, OK.

Because of the importance of N in crop production, many agronomists and soil scientists have worked on soil test methods for N. Research on this subject has been reviewed by Harmsen and Van Schreven (1955), Bremner (1965a), Harmsen and Kolenbrander (1965), Dahnke and Vasey (1973), Keeney (1982), and Stanford (1982). Most of the studies discussed in these reviews were concerned with developing procedures to predict the amount of N that will become available during the growing season by mineralization of OM.

The following will briefly discuss that topic but will also discuss some of the testing methods for residual inorganic N and how the test results are used to aid growers in N fertilization.

I. NITROGEN-AVAILABILITY INDEXES

Nitrogen-availability tests can be either biological or chemical. Biological tests are considered by many to be the most reliable because living organisms are used, but they have the disadvantage of usually taking much more time to conduct than chemical tests.

Nitrogen-availability indexes are a measure of the potential of a soil to supply N to plants when conditions are ideal for mineralization. They do not include existing inorganic N present when the soil was sampled. Surface soils usually contain between 0.08 and 0.4% total N, mostly in the organic form (Bremner, 1965b; Stevenson, 1982). The amount that mineralizes each year is commonly in the range of 1 to 3% (Broadbent, 1984) but depends on the amount and type of recent plant residue and environmental conditions. One reason it has been difficult to correlate these tests with plant growth is that the rate at which mineralization of N takes place under field conditions is controlled by several unpredictable environmental factors such as temperature, moisture, and aeration. In spite of this, there are many instances in which favorable results were obtained using N-availability tests (Allison & Sterling, 1949; Fitts et al., 1953; Hanway & Dumenil, 1955; Munson & Stanford, 1955; Cook et al., 1957; Kresge & Merkle, 1957; Saunder et al., 1957; Eagle & Matthews, 1958; Olson et al., 1960; Eagle, 1961; Clement & Williams, 1962; Gasser & Williams, 1963; Eagle, 1963; MacLean, 1964; Keeney & Bremner, 1966a; Robinson, 1968; Stanford & Legg, 1968; Stanford & Smith, 1976; Shumway & Atkinson, 1978; Fox & Piekielek, 1978; Powers, 1980; Saito & Ishii, 1987).

The data indicate that long-term biological indexes (Stanford & Smith, 1972) will give more useful results than short-term biological indexes. This is because short-term incubation tests are influenced to a greater extent by sample pretreatment (Stanford et al., 1974). Preincubation conditions that influence the results of incubation tests were reviewed by Bremner (1965a). He pointed out that incubation results for mineralizable-N in soils are influenced by method of sampling, drying, grinding, sieving, storing, and incubating. The biological N index recommended by Keeney (1982) is that of Waring and Bremner (1964) which determines the amount of NH_4-N produced under waterlogged conditions. The soil sample is incubated at 40 °C for 7 d, and the amount of NH_4-N formed is determined. The amount of NH_4-N in the soil before incubation is also determined. Mineralizable N is calculated as the difference between the results of these analyses.

Most of the N being mineralized in soil obviously comes from a fraction of the soil organic matter that is easily decomposed. Therefore, it is likely that a mild acid or alkaline extractant could make a suitable chemical extractant (Henkinson, 1968; MacLean, 1964; Keeney & Bremner, 1966b; Stanford & DeMar, 1970). A chemical availability index could have the advantage of being simple or rapid. Keeney (1982) recommends the procedure of Stanford and DeMar (1969) as modified by Smith and Stanford (1971) and Stanford and Smith (1976). This procedure involves the determination of NH_4-N formed when a soil sample is autoclaved for 16 h at 121 °C.

A procedure that is currently being used in Europe (Nemeth, 1979) is electro-ultrafiltration (EUF). This procedure is an adaptation of the electrodialysis technique described by Mattson (1926). The basic difference is that EUF uses a polycarbonate micropore filter as a membrane and suction to remove the secondary products of electrodialysis. The removal of these secondary products from the soil solution minimizes the effect of electrodialysis on soil pH. This procedure measures NO_3-N as well as other forms of N.

Trials on 4000 farms in Austria showed a close correlation between EUF-N values of soil samples taken in June and July and sugarbeet (*Beta vulgaris* L.) yield and quality the following year (Nemeth, 1979). It was found that 1 mg of EUF extractable N per 100 g of soil extracted at 20 °C and 200 V was equal to 30 kg of fertilizer N ha^{-1} (Wiklicky, 1982).

II. RESIDUAL INORGANIC NITROGEN

The potential of a soil to mineralize N compounds as measured by N availability indexes, should be fairly constant from year to year unless the type of crop or amount of crop residue changed. The amount of residual NO_3-N will be influenced by these and many other factors. Therefore, it is necessary to test for residual NO_3-N each year. Residual inorganic N soil tests are most appropriate shortly before planting to early in the growing season (Stanford, 1982).

For many years, the amount of NO_3-N in the soil was ignored or dismissed in the scientific literature as not being important because it is variable or low in concentration (20 kg ha^{-1} or less). For these reasons, Harmsen and van Schrevan (1955) and Bremner (1965) thought that a measure of initial N was of little or no value. In recent years, the inorganic N content of the rooting zone (instead of only the 0–15 cm depth) has been reexamined following the example of early workers such as King and Whitson (1901, 1902). In the following sections, the use of the initial inorganic N tests in predicting crop response to N will be discussed.

A. Nitrate-Nitrogen

While the use of the residual NO_3-N test has become well established in western Canada (Soper & Huang, 1963) and the Great Plains of the USA (Leggett, 1959; Young et al., 1967; Carson, 1975; Geist et al., 1970; Herron et al., 1977) in the last 20 yr, a test for NO_3-N was first proposed in the early 1900s (King & Whitson, 1901, 1902; King, 1905; Buckman, 1910; Call, 1914).

The study of NO_3-N is complicated by the fact that: (i) this form of N can move up or down in the soil profile in response to drying and wetting conditions; (ii) NO_3-N can be rapidly immobilized by soil microorganisms if a suitable source of energy is present, only to reappear after a short time as the microorganisms reduce the C/N ratio of the energy source (Harmsen & Kolenbrander, 1965); and (iii) up to 70% of applied fertilizer N can be lost by denitrification (Firestone, 1982). In spite of these difficulties, numerous studies since 1950 (Dahnke & Vasey, 1973; Carter et al., 1974; Carson, 1975; Magdoff et al., 1984; Gelderman et al., 1988) have shown that tests for residual NO_3-N are helpful in determining N fertilizer needs of crops. A residual NO_3-N test is especially useful for short-season, fast growing crops that do not allow much time for mineralization to take place. It is also especially useful when the amount of residual NO_3-N in the soil is high at planting time. In this situation, it would not be beneficial to add fertilizer N. However, when the soil is low in residual NO_3-N, the amount to apply cannot be accurately predicted. Thus, there is a need for an N-availability index to improve N fertilization.

B. Ammonium Nitrogen

Many environmental factors, such as aeration, temperature, moisture content, pH and soil nutrient content affect the production of ammonium nitrogen (NH_4-N) from OM. Many investigators agree that reduced aeration, as in high moisture soils, high temperatures, and low pH all favor ammonification over nitrification (Harmsen & Kolenbrander, 1965). Thus, when conditions are less favorable for plant growth it is more common to find an accumulation of NH_4-N in the soil. Under favorable growing conditions, NH_4-N levels are usually quite low except for variable periods after the application of an NH_4 fertilizer. Not many soil testing programs include NH_4-N in testing for residual inorganic N.

C. Soil Sampling for Residual Inorganic Nitrogen

The number of soil samples that need to be taken per field, depth of profile samples, time of sampling, and how the samples are handled are all important factors in a successful soil testing program.

The number of subsamples that should be taken to represent a field for residual NO_3-N testing depends on the accuracy and precision desired. A study by Swenson et al. (1984) indicated it is necessary to take approximately 20 subsamples per field in North Dakota to obtain an accuracy of $\pm 15\%$

for NO_3-N at a precision level of 80%. The number of subsamples needed for that accuracy and precision varied little as field size increased from 10 to 40 ha. This reflects the fact that the level of NO_3-N varies as much over relatively short distances as it does over long distances in a particular field. To accurately map soil fertility would probably require sampling fields on a grid of <30 m.

The profile depth to sample for residual NO_3-N depends on climate, soil, and crop to be grown. Samples are usually taken to a depth of 60 cm (Smith, 1977) but many studies have found that the correlation between residual NO_3-N and crop yield increases as depth of sampling increases within the rooting depth. More humid climates may require deeper sampling. The soil type will influence depth of rooting and water movement.

The time to sample depends on climate. In relatively cold and dry areas, such as the Northern Great Plains, soils can be sampled for residual NO_3-N from early fall to planting the following spring (Dahnke & Vasey, 1973). In more humid areas, the best sampling time may be after the crop has started to grow (Magdoff et al., 1984).

Shortly after a soil sample is taken, it should be treated to stop mineralization. It is especially important to do this when the soil is sampled in late fall or winter in temperate climates. If cold (< 10 °C) soil is brought into a warm building for any amount of time, the sample will no longer represent the status in the field. Mineralization will take place in the sample but will not be taking place in the field. The most common and practical way to stop mineralization is to air dry the sample by spreading it out in a thin layer. Other methods that have been used are freezing or the addition of a biological inhibitor, such as toluene (Bremner, 1965; Storrier, 1966). These latter methods are inconvenient or ineffective.

III. METHODS OF DETERMINING RESIDUAL NITROGEN

There are many methods for the determination of NH_4, nitrite (NO_2) and NO_3-N in soils. The advantages and disadvantages of some of these methods were discussed by Keeney and Nelson (1982).

Details of the procedures listed in Table 6–1 are given in the paper by Keeney and Nelson (1982).

An additional procedure not discussed in their paper is ion chromatography (Dick & Tabatabai, 1979; Nieto & Frankenberger, 1985). This method is rapid, sensitive, and precise. It can be used to simultaneously detect chloride, NO_2, NO_3, and sulfate (SO_4). The time and sensitivity will be largely determined by the type and length of the ion exchange column.

IV. CROP NITROGEN REQUIREMENT

Knowledge of the amount of N removed by a crop is necessary for the interpretation of a N soil test (Johnson, 1982). Crop N removal is obtained by multiplying the concentration of N in the crop by the crop yield. Crop

Table 6-1. Methods for the determination of inorganic forms of N in soil extracts. (After Keeney & Nelson, 1982.)

Method of analysis	Concentration range in extract	Interferences	Comments
	mg L^{-1}		
		NH_4-N	
Indophenol blue	0.005–20	Calcium and Mg	Simple and rapid
Ion electrode	0.1–1400	Volatile amides	Sensitive, rapid
Steam distillation with MgO	0.1–1000	Volatile amides	Rapid, widely used
Microdiffusion with MgO	0.1–100	None	Simple
		NO_2-N	
Griess-Ilosvay	0.1–12	None	Sensitive
Steam distillation with Devarda alloy after NH_2SO_3H		Alkaline soils or extracts	Insensitive
		NO_3-N	
Reduction to NO_2 by Cd; Griess-Ilosvay method	0.01–20	None	Very sensitive, problems with reduction step
Ion electrode	2–1400	Chloride, bromide, nitrite, iodide, sulfide, ionic strength	Less accurate than many methods
Reduction to NH_3	1–1000	Labile amides, phosphate, nitrite	Measure NH_3 by titration or colorimetric methods
Microdiffusion with Devarda alloy	0.1–10	Nitrite	Insensitive, slow
Ion chromatography†	0.1–50	None	Sensitive, rapid

† Added to above table by authors.

yield in this case is the plant material removed from the field during harvest. Nitrogen contained in the residue remaining in the field may be assumed to be recycled and, for the most part, sufficient to meet the N requirement of nonharvested portions of succeeding crops. This assumption is most valid in continuous production of the same crop.

The degree to which calculated crop N removal will be satisfied by a predetermined level of available soil N will be influenced by many factors. These factors include: (i) competition for N (immobilization) by soil microflora and weeds; (ii) N leaching below the effective root zone; (iii) failure of roots to grow into or absorb N from the same depth the soil was tested; (iv) cropping and N fertilization history; (v) mineralization or organic soil N; (vi) volatilization of N; (vii) crop type; (viii) crop vigor; and (ix) other growing conditions.

Hence, because the soil-crop system is dynamic, the efficiency with which crops use applied N is variable and dependent on local conditions. Hauck (1973) has reported values of 50 to 75% for crop recovery of applied N as representative of agricultural crops based on ^{15}N studies. The upper value may be appropriate when losses by volatilization and leaching are minimal and immobilization and mineralization are in balance. Lower values are more

Table 6-2. Available soil N requirement in relation to wheat grain yield.

Yield	Crop N† removal	Available N requirement‡
	kg ha^{-1}	
1345	30	44
2015	46	66
2690	62	88
3360	76	109
4030	92	131
4705	108	153
5375	122	175
6050	138	197
6720	153	220

† Assumes 13% crude protein. ‡ Assumes 70% N-use efficiency.

appropriate when losses or transformation to unavailable forms are expected to be high. Whatever value is used, it becomes the link between crop N removal and soil test N and allows one to calibrate the soil test. As an example, the values in Table 6-2 have been calculated for wheat (*Triticum aestivum* L.) grain at 13% crude protein (N × 5.7) using an N use efficiency of 70%. If one assumes the N measured by a soil test of NO_3-N or predicted from an index value will be available to meet the crop N requirement, then Table 6-2 may be used directly to interpret the soil test.

It is obvious from Table 6-2 that to use an appropriate value for the available soil N requirement, one must first identify the production level or yield goal.

V. YIELD GOAL

The yield goal is best identified by the grower. Growers best understand the degree that they can manage or control those factors other than available N which will ultimately determine crop yield (Dahnke, 1973; Dahnke et al., 1988). Growers, however, frequently do not distinguish between yield goal and yield average. This can be a costly error in lost production when weather, especially rainfall in dryland farming, varies greatly from one season to another.

Under conditions typical of the Great Plains region of the USA, the upper yield limit of nonlegumes is usually determined by the amount and frequency of rainfall during the growing season. For example, the 10-yr yield average of wheat on a particular field may be 2015 kg ha^{-1} (30 bu acre^{-1}), but range from 1000 kg in a poor rainfall year to 3000 kg in a good rainfall year. It would seem logical to apply fertilizer N for an average yield of 2015 kg ha^{-1} each year. In this way, N not used in poor years would be available as carry-over or residual to meet the crop needs in good years. The fallacy of this approach is that it assumes the good and poor years are normally distributed and the poor years will precede the good years. Using this approach, if several good years occur without intervening poor years, yield may be limited by lack of available N.

A more reasonable approach, and one which reduces the risk of limiting yield from lack of N, is to set the yield goal at the highest observed yield of the last 5 to 10 yr. In this instance, the common error will be that N is applied in excess of the crop requirement in poor years. Under dryland conditions, a poor year is usually a dry year, and the excess N will not be lost by leaching and should be identified by the N soil test for the next crop. An adequate, but not excessive, fertilizer rate should result in 10 to 20 kg of NO_3-N ha^{-1} available at the end of the growing season in most years (Johnson, 1981).

When the crop and season allow split applications of N, the total N requirement (identified by a yield goal) may be modified based on changing conditions during the growing season. Especially in dryland farming, N use efficiency may be greatly improved by adjusting the N supply in relation to unusual soil moisture conditions that develop during the growing season.

When crops are grown under irrigation or in areas where rainfall is not limiting, the yield average may serve well as the yield goal. This assumes, of course, that N deficiency has not been a common yield limiting factor of the past.

VI. NITROGEN SOIL TEST INTERPRETATIONS

The reactions of N in a soil-plant-water system are complex involving many reactions and interactions. One approach that has been used to try to integrate the many factors involved in making a N recommendation is mathematical modeling (Tanji, 1982). The problem with this approach in a soil testing program is that to be useful the complexities have to be simplified and "at times, the simplifications are so gross that the results are not very meaningful" (Tanji, 1982). Therefore, the approaches actually used in soil testing programs are simplified and often empirical.

A. Available Nitrogen Indexes

The need for a N availability index has already been indicated by the extensive list of researchers publishing on this approach to estimating crop N needs. Since these indexes, whether biological or chemical, are indirect measures or estimates of available N they must be calibrated. The calibration usually involves field research measuring yield, N uptake or both. Most often the calibration allows one to arrive at a numerical value for available N without having actually measured it. In some instances, the estimation of available N may be circumvented by a calibration that relates the index to a N fertilizer need at a specified crop yield level.

The most common availability index is soil OM. An example of how this index is used to estimate available N adjustments is shown in Table 6–3 as used by the Univ. of Missouri (Buchholz et al., 1981). This table estimates the amount of N that will become available based on soil organic matter content, texture, cation exchange capacity, and whether a cool- or warm-

Table 6-3. Nitrogen rate adjustments used by the Univ. of Missouri based upon soil texture, organic matter (OM), and time of major crop growth (Buchholz et al., 1981).

Soil texture	Cation exchange capacity	OM	Cool-season crops	Warm-season crops
	$cmol_c\ kg^{-1}$	%	—— kg ha^{-1} (lb $acre^{-1}$) ——	
Sands to sandy-loams	10	0.5	11(10)	22(20)
		1.0	22(20)	45(40)
		1.5	34(30)	67(60)
Silt loams to loams	10–18	2.0	22(20)	45(40)
		3.0	34(30)	67(60)
		4.0	45(40)	90(80)
Clay loams to clays	18	2.0	11(10)	22(20)
		3.0	17(15)	34(30)
		4.0	22(20)	45(40)
		5.0	28(25)	56(50)

season crop is being produced. Interpretation of the availability index (%OM) simply consists of subtracting the amount of available N estimated from the amount of N required for a specified production level. As an example, an N requirement of 88 kg would be indicated from Table 6–2 for a wheat yield goal of 2690 kg ha^{-1}. A loam soil with 3% OM would be expected to supply 34 kg ha^{-1} for wheat (cool-season crop). The difference between N required (88) and estimated available N (34) is the fertilizer N requirement, in this case 54 kg of N ha^{-1}.

The advantage of using an index such as percentage organic matter is that it is relatively stable from year to year. However, this is also a major disadvantage as it fails to indicate N carryover from excessive fertilization or crop failure.

B. Residual Inorganic Nitrogen

Direct measure of residual available soil N (NO_3-N) has had its greatest acceptance in arid and semiarid agricultural production. The limited rainfall of these regions lessens the change of NO_3-N loss by leaching or denitrification. Soil samples are usually taken soon after harvest (northern regions) or just before planting (southern regions) to avoid ambiguities caused by immobilization or mineralization. Thoroughness and consistency in sampling are critical to success in using the soil test. When 20 or more cores are composited for separate samples of the surface and subsoil of relatively uniform fields in monoculture, then tests results, with time, can become a useful guide to determining N fertilizer rates.

Interpretation of the NO_3-N soil test can be straightforward once the information for crops (Table 6–2) is available. The amount of available soil N indicated by the soil test is simply subtracted from the available N requirement, as determined from the yield goal. The difference is the amount of N which must be made available to the crop before or during the growing season:

$$\text{N requirement} - \text{soil test N} = \text{fertilizer N}. \qquad [1]$$

The above calculation represents the approach which has been used in North Dakota (Dahnke, 1973) and Oklahoma (Johnson & Tucker, 1982) for interpretation of NO_3-N soil test. A similar approach has been used in Nebraska (R.A. Wiese & G.W. Hergert, 1986, personal communication) for interpretation of this soil test. The Nebraska interpretation employs an equation to calculate the N requirement as a function of the yield goal, crop N content, and a constant 56 kg ha^{-1} (50 lb acre^{-1}) fertilizer N addition. This calculated N level is adjusted upward for decreasing N-use efficiency as yield goal increases. Similarily, the N requirement tables used in Oklahoma reflect a decrease in N-use efficiency at high yield levels.

C. Interpretation Adjustments

Production systems involving crop rotation and varied types and times of tillage do not lend themselves to straightforward interpretation of available-N tests. The principal problem centers around the seasonally changing conditions for mineralization and immobilization of N. Reducing the number of tillage operations and delaying incorporation of crop residues at a given N regime will promote transformation of residual N from mineral forms to organic forms. In these situations, such as changing from conventional tillage to minimum tillage, N-use efficiency will be lower until equilibrium is reached with a higher soil OM level. Available-N soil test levels will be low and responsive fertilizer N rates higher than for conventional tillage.

The concept of a small labile organic N pool (active) in equilibrium with a large stable organic N pool (passive) on the one hand and a small mineral N pool on the other hand, first proposed by Janssen (1958), helps explain the apparent shortcomings of the NO_3-N soil test. Whether the active and passive pools are distinct entities or a continuum as suggested by Broadbent (1984), soil OM is a buffer against maintenance of a significant change in the level of mineral N in the soil. Consequently, excessively high rates of N fertilizer do not immediately result in high residual NO_3-N soil test levels. If the level of N in the OM pool is low or the size of the OM pool is small relative to what the environment will support, then it may take several years of seemingly excessive N fertilization before significant residual N will be detected by the soil test. Jacobsen and Westerman (1988) has shown the N requirement for maximum grain yield of wheat to be 1.2 times higher for no tillage compared to conventional tillage. This also happens when the yield goal for a field, and associated N requirement, is suddenly changed from much below the potential yield to a yield which the climate and modern cultural practices can support.

Contrary to the above scenario, when the level of N in the organic pool is high and the pool is large relative to what the environment (or cultural practices) will support, then high yields and measureable residual NO_3-N may occur without use of fertilizer N. Examples of this are easily demonstrated when a crop such as wheat is grown after plowing down alfalfa that has been in production 5 yr or more.

Interpretation of the N soil test is not always straightforward. However, the test can be a useful tool for identifying or monitoring adequate N fertilizer rates even in cropping systems with many variables. The keys to interpretation are understanding when during the season residual NO_3-N is most apt to be present for sampling and determining the crop N requirement from a realistic yield goal. When attention is given to these factors the test has been found to have promise even outside the Great Plains region (Magdoff, 1984).

REFERENCES

Allison, F.E., and L.D. Sterling. 1949. Nitrate formation from soil organic matter in relation to total nitrogen and cropping practices. Soil Sci. 67:239–252.

Bremner, J.M. 1965a. Nitrogen availability indexes. p. 1324–1345. *In* C.A. Black et al. (ed.) Methods of soil analysis. Part 2. Agronomy Monogr. 9. ASA, Madison, WI.

Bremner, J.M. 1965b. Organic nitrogen in soils. p. 93–149. *In* M.W. Bartholomew and F.E. Clark (ed.) Soil nitrogen. Agronomy Monogr. 10. ASA, Madison, WI.

Broadbent, F.E. 1984. Plant use of soil nitrogen. p. 171–182. *In* R.D. Hauck (ed.) Nitrogen in crop production. ASA, CSSA, and SSSA, Madison, WI.

Buchholz, D.D., J.R. Brown, R.G. Hanson, H.N. Wheaton, and J.D. Garrett. 1981. Soil test interpretations and recommendations handbook. Univ. of Missouri, Columbia.

Buckman, H.O. 1910. Moisture and nitrate relations in dry-land agriculture. J. Am. Soc. Agron. 2:121–138.

Call, L.E. 1914. The effect of different methods of preparing a seed bed for winter wheat upon yield, soil moisture, and nitrates. J. Am. Soc. Agron. 6:249–259.

Carson, P.L. 1975. Recommended nitrate-nitrogen tests. *In* W.C. Dahnke (ed.) Recommended chemical test procedures for the north-central region. North Dakota Agric. Exp. Stn. Bull. 499.

Carter, J.N., M.E. Jensen, and S.M. Bosma. 1974. Determining nitrogen fertilizer needs for sugar beets from residual nitrate and mineralizable nitrogen. Agron. J. 66:319–323.

Clement, C.R., and T.E. Williams. 1962. An incubation technique for assessing the nitrogen status of soils newly ploughed from keys. J. Soil Sci. 13:82–91.

Cook, F.D., F.G. Warder, and J.L. Doughty. 1957. Relationship of nitrate accumulation to yield response of wheat in some Saskatchewan soils. Can. J. Soil Sci. 37:84–88.

Dahnke, W.C. 1973. Yield goal approach to fertilizer recommendations. Annual Agricultural Conference and Short Course. North Dakota Agric. Assoc., Fargo.

Dahnke, W.C., L.J. Swenson, R.J. Goos, and A.G. Leholm. 1988. Choosing a crop yield goal. circular SF-822 (Revised). North Dakota State Univ., Ext. Serv., Fargo.

Dahnke, W.C., and E.H. Vasey. 1973. Testing soils for nitrogen. p. 97–114. *In* L.M. Walsh and J.D. Beaton (ed.) Soil testing and plant analysis. SSSA, Madison, WI.

Dick, W.A., and M.A. Tabatabai. 1979. Ion chromatographic determination of sulfate and nitrate in soils. Soil Sci. Soc. Am. J. 43:899–901.

Eagle, D.J. 1961. Determination of the nitrogen status of soils in the West Midlands. J. Sci. Food Agric. 12:712–717.

Eagle, D.J. 1963. Response of winter wheat to nitrogen and soil nitrogen status. J. Sci. Food Agric. 14:391–394.

Eagle, D.J., and B.C. Matthews. 1958. Measurement of nitrate-supplying power of soils by an incubation method and correlation with crop yield response. Can. J. Soil Sci. 38:161–170.

Firestone, M.K. 1982. Biological denitrification. p. 289–326. *In* F.J. Stevenson (ed.) Nitrogen in agricultural soils. Agronomy Monogr. ASA, CSSA, and SSSA, Madison, WI.

Fitts, J.W., W.V. Bartholomew, and H. Heidel. 1953. Correlation between nitrifiable nitrogen and yield response of corn to nitrogen fertilization on Iowa soils. Soil Sci. Soc. Am. Proc. 17:119–122.

Fox, R.H., and W.P. Piekielek. 1978. Field testing of several nitrogen availability indexes. Soil Sci. Soc. Am. J. 42:747–750.

Gasser, J.K.R., and R.J.B. Williams. 1963. Soil nitrogen: VII. Correlations between measurements of nitrogen status of soil and nitrogen % and nitrogen content of crops. J. Sci. Food Agric. 14:269–277.

Geist, J.M., J.O. Reuss, and D.D. Johnson. 1970. Prediction of nitrogen fertilizer requirements of field crops. II. Application of theoretical models to malting barley. Agron. J. 62:385–389.

Gelderman, R.H., W.C. Dahnke, and L. Swenson. 1988. Correlation of several soil N indices for wheat. Commun. Soil Sci. Plant Anal. 19(6):755–772.

Hanway, J., and L. Dumenil. 1955. Predicting nitrogen fertilizer needs of Iowa soils: III. Use of nitrate production together with other information as a basis for making nitrogen fertilizer recommendations for corn in Iowa. Soil Sci. Soc. Am. Proc. 19:77–80.

Harmsen, G.W., and G.J. Kolenbrander. 1965. Soil inorganic nitrogen. p. 43–92. *In* W.V. Bartholomew and F.E. Clark (ed.) Soil nitrogen. Agronomy Monogr. 10. ASA, Madison, WI.

Harmsen, G.W., and D.A. Van Schreven. 1955. Mineralization of organic nitrogen in soil. Adv. Agron. 7:299–398.

Hauck, R.D. 1973. Nitrogen tracers in nitrogen cycle studies—past use and future needs. J. Environ. Qual. 2:317–327.

Herron, G.M., A.F. Dreier, A.D. Flowerday, W.L. Colville, and R.A. Olson. 1971. Residual mineral N accumulation in soil and its utilization by irrigated corn (*Zea mays* L.) Agron. J. 63:322–327.

Jacobsen, J.S., and R.L. Westerman. 1988. Nitrogen fertilization in winter wheat tillage systems. J. Prod. Agric. 1:235–239.

Janssen, S.L. 1958. Tracer studies on nitrogen transformations in soil with special attention to mineralization-immobilization relationships. Ann. R. Agric. Coll. Swed. 24:101–361.

Jenkinson, D.S. 1968. Chemical tests for potentially available nitrogen in soil. J. Sci. Food Agric. 19:160–168.

Johnson, G.V. 1981. Soil fertility: An analogy. Crops Soils 34(3):8–11.

Johnson, G.V. 1982. Soil test interpretations—available nitrogen and small grain production. Oklahoma State Univ. Ext. Facts No. 2232.

Johnson, G.V., and B.B. Tucker. 1982. Oklahoma State University soil test calibration. Ext. Facts No. 2225.

Keeney, D.R. 1982. Nitrogen-availability indices. p. 711–733. *In* A.L. Page et al. (ed.) Methods of soil analysis. Part 2. 2nd ed. Agronomy Monogr. 9. ASA and SSSA, Madison, WI.

Keeney, D.R., and J.M. Bremner. 1966a. Comparison of laboratory methods of obtaining an index of soil nitrogen availability. Agron. J. 58:498–503.

Keeney, D.R., and J.M. Bremner. 1966b. A chemical index of soil nitrogen availability. Nature (London) 211:892–893.

Keeney, D.R., and D.W. Nelson. 1982. Nitrogen—inorganic forms. p. 643–698. *In* A.L. Page et al. (ed.) Methods of soil analysis. Part 2. 2nd ed. Agronomy Monogr. ASA and SSSA, Madison, WI.

King, F.H. 1905. Investigations in soil management. USDA Bur. of Soils, Bull. 26. U.S. Gov. Print. Office, Washington, DC.

King, F.H., and A.R. Whitson. 1901. Development and distribution of nitrates and other soluble salts in cultivated soils. Univ. of Wisconsin Agric. Exp. Stn. Bull. 85.

King, F.H., and A.R. Whitson. 1902. Development and distribution of nitrates in cultivated soils. (Second paper). Univ. of Wisconsin Agric. Exp. Stn. Bull. 93.

Kresge, C.B., and F.G. Merkle. 1957. A study of the validity of laboratory techniques in appraising the available nitrogen producing capacity of soils. Soil Sci. Soc. Am. Proc. 21:516–521.

Leggett, G.E. 1959. Relationship between wheat yield, available moisture, and available nitrogen in eastern Washington dry land areas. Washington Agric. Exp. Stn. Bull. 609.

MacLean, A.A. 1964. Measurement of nitrogen supplying-power of soils by extraction with sodium bicarbonate. Nature (London) 203:1307–1308.

Magdoff, F.R., D. Ross, and J. Amadon. 1984. A soil test for nitrogen availability to corn. Soil Sci. Soc. Am. J. 48:1301–1304.

Mattson, S. 1926. Electrodialysis of the colloidal soil material and the exchangeable bases. J. Agric. Res. 33:553–567.

Munson, R.D., and G. Stanford. 1955. Predicting nitrogen fertilizer needs of Iowa soils: IV. Evaluation of nitrate production as a criterion of nitrogen availability. Soil Sci. Soc. Am. Proc. 19:464–468.

Nemeth, K. 1979. The availability of nutrients in the soil as determined by electro-ultrafiltration (EUF). Adv. Agron. 31:155–188.

Nieto, K.F., and W.T. Frankenberger, Jr. 1985. Single column ion chromatography: I. Analysis of inorganic anions in soils. Soil Sci. Soc. Am. 49:587–592.

Olson, R.A., M.W. Meyer, W.C. Lamke, A.D. Woltemath, and R.E. Weiss. 1960. Nitrate production rate as a soil test for estimating fertilizer nitrogen requirement of cereal crops. Int. Congr. Soil Sci. Trans. 7th 2:463–470.

Power, R.F. 1980. Mineralizable soil nitrogen as an index of nitrogen availability to forest trees. Soil Sci. Soc. Am. J. 44:1314–1320.

Robinson, J.B.D. 1968. A simple available soil nitrogen index: I. Laboratory and greenhouse studies. J. Soil Sci. 19:269–279.

Saito, M., and K. Ishii. 1987. Estimation of soil nitrogen mineralization in corn-grown fields based on mineralization parameters. Soil Sci. Plant Nutr. 33(4):555–566.

Saunder, D.H., B.S. Ellis, and A. Hall. 1957. Estimation of available nitrogen for advisory purposes in southern Rhodesia. J. Soil Sci. 8:301–312.

Shumway, J., and W.A. Atkinson. 1978. Predicting nitrogen fertilizer response in unthinned stands of Douglas-fir. Commun. Soil Sci. Plant Anal. 9:529–539.

Smith, C.M. 1977. Interpreting inorganic nitrogen soil tests: Sample depth, soil water, climate, and crops. p. 85–98. *In* T.R. Peck et al. (ed.) Soil testing: Correlating and interpreting the analytical results. ASA Spec. Publ. 29. ASA, Madison, WI.

Smith, S.J., and G. Stanford. 1971. Evaluation of a chemical index of soil nitrogen availability. Soil Sci. 111:228–232.

Soper, R.J., and P.M. Huang. 1963. The effect of nitrate nitrogen in the soil profile on the response of barley to fertilizer nitrogen. Can. J. Soil Sci. 43:350–358.

Stanford, G. 1982. Assessment of soil nitrogen availability. p. 651–688. *In* F.J. Stevenson (ed.) Nitrogen in agricultural soils. Agronomy Monogr. 22. ASA, CSSA, and SSSA, Madison, WI.

Stanford, G., J.N. Carter, and S.J. Smith. 1974. Estimates of potentially mineralizable soil nitrogen base on short-term incubations. Soil Sci. Soc. Am. Proc. 38:99–102.

Stanford, G., and W.H. DeMar. 1969. Extraction of soil organic nitrogen by autoclaving in water. I. The NaOH-distillable fraction as an index of nitrogen availability in soils. Soil Sci. 107:203–205.

Stanford, G., and W.H. DeMar. 1970. Extraction of soil organic nitrogen by autoclaving in water. II. Diffusible ammonia, an index of soil nitrogen availability. Soil Sci. 109:190–196.

Stanford, G., and J.O. Legg. 1968. Correlation of soil nitrogen availability indexes with N uptake by plants. Soil Sci. 105:320–326.

Stanford, G., and S.J. Smith. 1972. Nitrogen mineralization potentials of soils. Soil Sci. Soc. Am. Proc. 36:465–472.

Stanford, G., and S.J. Smith. 1976. Estimating potentially mineralizable soil nitrogen from a chemical index of soil nitrogen availability. Soil Sci. 122:71–76.

Stevenson, F.J. 1982. Origin and distribution of nitrogen in soil. p. 1–42. *In* F.J. Stevenson (ed.) Nitrogen in agricultural soils. Agronomy Monogr. 22. ASA, CSSA, and SSSA, Madison, WI.

Storrier, R.R. 1966. The pre-treatment and storage of soil samples for nitrogen analyses. J. Aust. Inst. Agric. Sci. 32:106–113.

Storrier, R.R. 1967. The estimation of available soil nitrogen. J. Aust. Inst. Agric. Sci. 33:278–283.

Swenson, L.J., W.C. Dahnke, and D.D. Patterson. 1984. Sampling for soil testing. North Dakota State Univ. Dep. of Soil Sci. Res. Rep. 8.

Tanji, K.K. 1982. Modeling of the soil nitrogen cycle. p. 721–772. *In* F.J. Stevenson (ed.) Nitrogen in agricultural soils. Agronomy Monogr. ASA, CSSA, and SSSA, Madison, WI.

Waring, S.A., and J.M. Bremner. 1964. Ammonium production in soil under waterlogged conditions as an index of nitrogen availability. Nature (London) 201:951–952.

Wiklicky, L. 1982. Application of the EUF procedure in sugar beet cultivation. Plant Soil 64:115–127.

Young, R.A., J.L. Ozbun, A. Bauer, and E.H. Vasey. 1967. Yield response of spring wheat and barley to nitrogen fertilizer in relation to soil and climate factors. Soil Sci. Soc. Am. Proc. 31:407–410.

Chapter 7

Testing Soils for Phosphorus[1]

P. E. FIXEN, *Potash and Phosphate Institute, Brookings, South Dakota*

J. H. GROVE, *University of Kentucky, Lexington*

The major purpose of testing soils for P is to determine the quantity of supplemental P required to prevent economic loss of crop value because of P deficiency. A soil test provides an index of the plant-available P in a soil which is in turn used to predict the amount of supplemental P needed.

A second purpose for testing soils for P is to monitor the quantity of available P present over time. This information is useful for evaluation of fertilization practices and in making decisions about waste disposal.

Phosphorus exists in soils in a multitude of chemical forms. These all contribute to varying extents to the plant-available pool. The quantity of plant-available P is not a distinct value for a given soil. It varies with several plant root characteristics and with environmental conditions that influence both soil and plant parameters. Therefore, predicting the quantity of plant-available P in a soil is no small task. However, several excellent P extraction procedures have been developed that correlate well with P uptake in controlled environments. It is the intent of the following discussion to review these facets of testing soils for P including interpretation of the test results. In the process, an attempt will be made to indicate what existing soil test methodology can and cannot do.

I. SOIL PHOSPHORUS

The transformations of applied P occur within a framework of soil P forms and reactions that can be described in a general way as the P cycle (Smeck, 1985). An example is provided in Fig. 7-1. The generally low level of P in the solution phase has led to much work describing the relationships of solution P, the pool from which the root draws the plant's P nutrition, to the other P pools interacting with solution phase P (Olsen et al., 1977;

[1] Contribution of the Plant Sci. Dep., South Dakota State Univ., Brookings, SD 57007 and the Agronomy Dep., Univ. of Kentucky, Lexington, KY 40546.

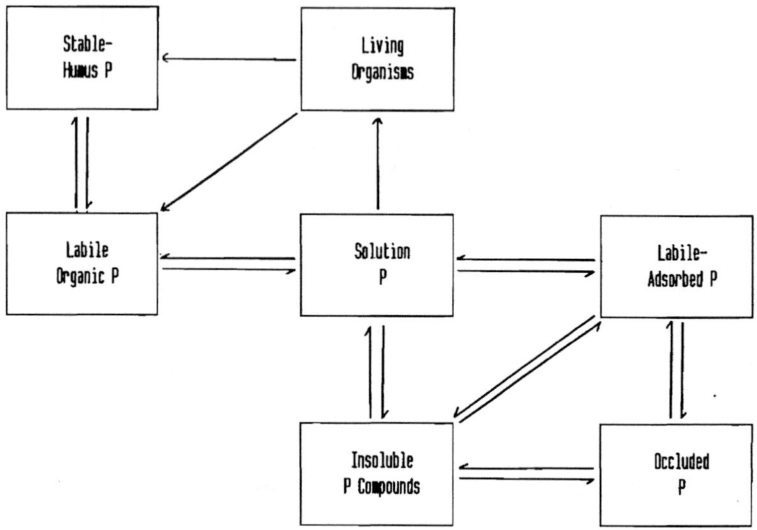

Fig. 7-1. Possible P transformations. (After White, 1980; Barrow, 1983; and Smeck, 1985.)

Sample et al., 1980). While much of the soil science literature has concentrated on inorganic P transformations, there would appear to be a growing interest in organic P transformations as well (Dalal, 1977; Anderson, 1980). This seems to reflect a desire to understand soil P availability as a part of the agro-ecosystem, a marriage of agricultural and ecological research emphases (Smeck, 1985).

This discussion will touch upon all the pools in Fig. 7-1, but with an emphasis on those aspects important to soil testing. Further, the divisions among the inorganic solid phases is more than a little arbitrary, given the heterogeneity of soil components. The removal of P from soil solution subsequent to soluble fertilizer P addition is more likely a continuum of processes, from adsorption to precipitation, that results in a continuum of chemical states (i.e., adsorbed, chemisorbed, amorphous-crystalline-occluded, precipitated), for the transformed P (Sample et al., 1980; Barber, 1984).

When describing the status of P in soils, it becomes apparent that the definitions of the various forms (pools) of P are dependent on the investigatory techniques chosen. Therefore, this discussion will use investigatory techniques as a framework of presentation.

A. Adsorbed Forms of Phosphorus

Schofield (1955) argued that continued research emphasis on the action of P extracting solutions on soils would not lead to a better understanding of the fundamental chemistry of available soil P. Since Schofield's paper, there have been many studies regarding the mechanism(s) of P retention by soil carbonates, silicates and hydrous oxide clay minerals, and soil organic matter. As the literature on this topic is voluminous, only representative reports will be cited here.

Adsorbed P is often equated with the labile soil phosphate most available to plant roots; i.e., most readily in equilibrium with the soil solution. Phosphorus adsorption is delineated from P precipitation in that the suspension's initial solution phase composition does not exceed the solubility product of any *known* P compound. One approach to measuring this soil P fraction is the use of ^{32}P in isotopic exchange studies (Olsen & Khasawneh, 1980; Wolf et al., 1986). Though this technique is somewhat more sophisticated, the extension of the kinetic information obtained to an understanding of actual P-bonding mechanisms has proved illusive as results are highly dependent on method and the mathematical formula used (White, 1976; Olsen & Khasawneh, 1980). This problem led Wolf et al. (1986) to conclude that isotopic exchange should not be used on soils with high amorphous Fe contents and consequently higher P fixation tendencies.

Adsorption curves (rather than isotherms, see White, 1980) and the equations to describe them are derived from reacting minerals and soils with solutions at various P concentrations and have been a popular means of describing P adsorption and proposing P bonding mechanisms (Barrow, 1980b; Olsen & Khasawneh, 1980; Sample et al., 1980). Because the technique is a simple one, adsorption curves have been used on the largest number of soils and minerals.

Uehara and Gillman (1981) proposed that differences in P adsorption could be accounted for by differences in: (i) specific surface; (ii) colloid capacity to occlude P; (iii) concentration or type of species, usually anions, that compete with P by adsorbing to, or dissolving, the adsorption site; and (iv) colloid surface reactivity.

Differences in P adsorption because of surface reactivity are largely the result of the reactions of P with Al, Ca, and Fe cations that are themselves part of, or strongly sorbed to, colloid surfaces (Wild, 1950; Thomas & Peaslee, 1973). Such differences separate soils into broad classes on the basis of the reactions of their colloid suite with soluble P. Geographically, as rainfall, temperature, and weathering increase, the role of Ca in P adsorption reactions diminishes and that of Al and Fe rises.

On the most highly weathered soils, P adsorption is thought to be largely related to the presence of hydrous Al and Fe oxides. In studies with pure oxides, P adsorption curves often resolve into at least two and sometimes three regions of adsorbing surface where distinctly different bonding mechanisms are thought operative (Bache, 1964; Ryden et al., 1977). White (1980) has been critical of such an approach to mechanism determination, arguing that two (or more) surface models are empirical and no better than models that assume a continuous spectrum of adsorption strength. This latter feature would be especially true of heterogeneous mixtures (soils).

Data from traditional P adsorption curves have been combined with additional information on changes in surface charge, solution pH, and concentrations of both counter ions and background electrolytes to better describe P bonding mechanisms on hydrous oxides of Al and Fe as well as other minerals (White, 1980). The ligand exchange of phosphate for surface aquo and hydroxyl groups bonded to Al and Fe has been shown to result in monodentate, bidentate, or binuclear forms of adsorbed P (Fig. 7-2a). The bidentate and binuclear types of bonding should be less reversible than monodentate. Hingston et al. (1974) have proposed that P sorbed to Al is more labile than that adsorbed to Fe. Ainsworth et al. (1985) demonstrated that ^{32}P-isotopic exchange rates on goethite increased with Al substitution for Fe in the hydrous oxide lattice.

As suspension pH rises, phosphate adsorption is less favored by the greater negative charge at the oxide surface (Fig. 7-2b) and the reduced polarization of the metal-oxygen bond. Martin and Smart (1987) used x-ray photoelectron spectroscopy (XPS) to confirm both the pH dependence and binuclear nature of P adsorption to goethite. Goldberg and Sposito (1985) reviewed the evidence for multiple attachment of phosphate to hydroxylated surfaces and found it wanting, primarily because the analytical techniques used to confirm the existence of binuclear complexation (XPS and infrared spectroscopy among them) required dry, severely evacuated samples that are unlike those same colloid surfaces in soils where water is present.

Adsorption of P in lime amended, acid, oxidic soils has not always behaved as predicted from pure oxide models; P sorption increasing, rather than decreasing, with increasing pH (Amarasiri & Olsen, 1973; Mokwunye, 1975; Friesen et al., 1980). Haynes (1983), using the E horizon of a Spodosol, was able to reverse increased P sorption with greater pH by air drying the soil samples after lime addition. Sims and Ellis (1983a) found that as the lime incubation time progressed the energy of subsequent adsorbed P bonding was reduced even though the adsorption maxima increased, regardless of whether the limed soil samples were subjected to wet and dry cycles during incubation.

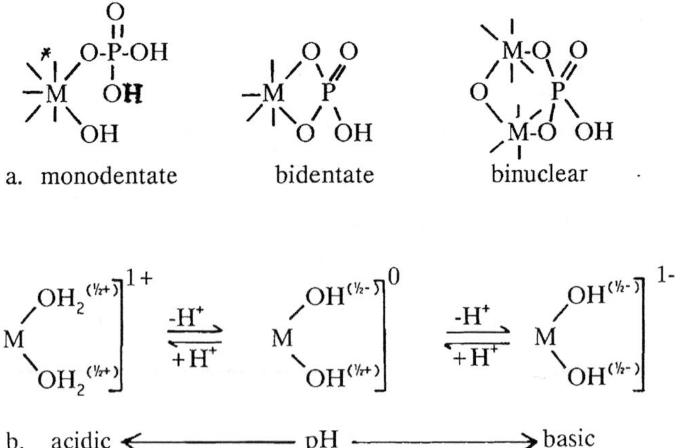

Fig. 7-2. Types of P bonding (a) and pH-dependent surface charge characteristics (b) associated with hydrous oxides. (After Sample et al., 1980; Uehara & Gillman, 1981; and White, 1980.) *M = Al, Fe.

Closer examination of this phenomena on model systems using Al-resins (Robarge & Corey, 1979), Al-gels (Sims & Ellis, 1983b), Al-peats (Bloom, 1981; White & Thomas, 1981), and acid montmorillonites (Coleman et al., 1960; Traina et al., 1986a, b) indicates that the existence of sufficient exchangeable Al^{3+} prior to liming, the formation of amorphous hydroxy-Al polymers after liming, and the reaction of these polymers with P to form sorbed/precipitated complexes of an OH/Al/P composition near 2:1:1 is the likely chain of events. Haynes (1984) summarized the different P sorption response patterns observed in terms of the loss of surface area because of further crystallization of the hydroxy-Al with both aging and drying after liming. Haynes (1984) also notes that further work is needed as both inorganic and organic anions have been reported to slow hydroxy-Al crystallization and this could result in increased P adsorption despite drying.

Exchangeable bases and silicate clay minerals strongly influence P adsorption in soils of weak to moderate acidity (minimal exchangeable Al). It is widely known that P adsorption increases as both counter-ion valence and ionic strength increase in silicate clay systems (Sample et al., 1980; White, 1980; Velayutham, 1980). Some P is likely adsorbed at the broken edges of the silicate clay lattice by ligand exchange at Al (White, 1980).

Though Ca has long been known to enhance P adsorption, the clay-Ca-P linkage model was thought inappropriate (Velayutham, 1980). Further observations on Ca enhanced P adsorption in both silicate and oxide clay systems (Helyar et al., 1976; Haynes, 1983; Smillie et al., 1987) have led to the proposal that Ca and P sorption are complementary (Smillie et al., 1987), interacting as a Ca-P surface complex (Helyar et al., 1976). This mechanism is distinct from that proposed to explain the role of Ca on P fixation in acid soils by the displacement and hydrolysis of reactive Al (Robarge & Corey, 1979; Traina et al., 1986b).

In calcareous soils P adsorption occurs on soil carbonates, whose impurities result in greater specific surface than that of pure calcite (White, 1980). Though adsorption is thought to dominate the initial reaction of soluble P with calcareous soils at low P concentrations, precipitation of P as a Ca-P compound is the ultimate fate (Barrow, 1980a). The adsorbed P sites serve as nucleation points for the formation of a basic Ca-P compound (White, 1980). There is still some question regarding the initial Ca-P compound formed subsequent to P adsorption (Sample et al., 1980; Freeman & Rowell, 1981).

B. Phosphorous Desorption

No discussion of P adsorption is complete without some reference to P desorption. Phosphorus desorption curves often do not coincide with their adsorption counterparts (Fig. 7-3) resulting in what has been termed *P sorption hysteresis* (Uehara & Gillman, 1981). This irreversibility or fixation of adsorbed P has been ascribed to several processes, including precipitation (Veith & Sposito, 1977), occlusion (Uehara & Gillman, 1981), solid state diffusion (Barrow, 1983) and bidentate or binuclear bonding with the colloid surface (Hingston et al., 1974; Taylor & Ellis, 1978).

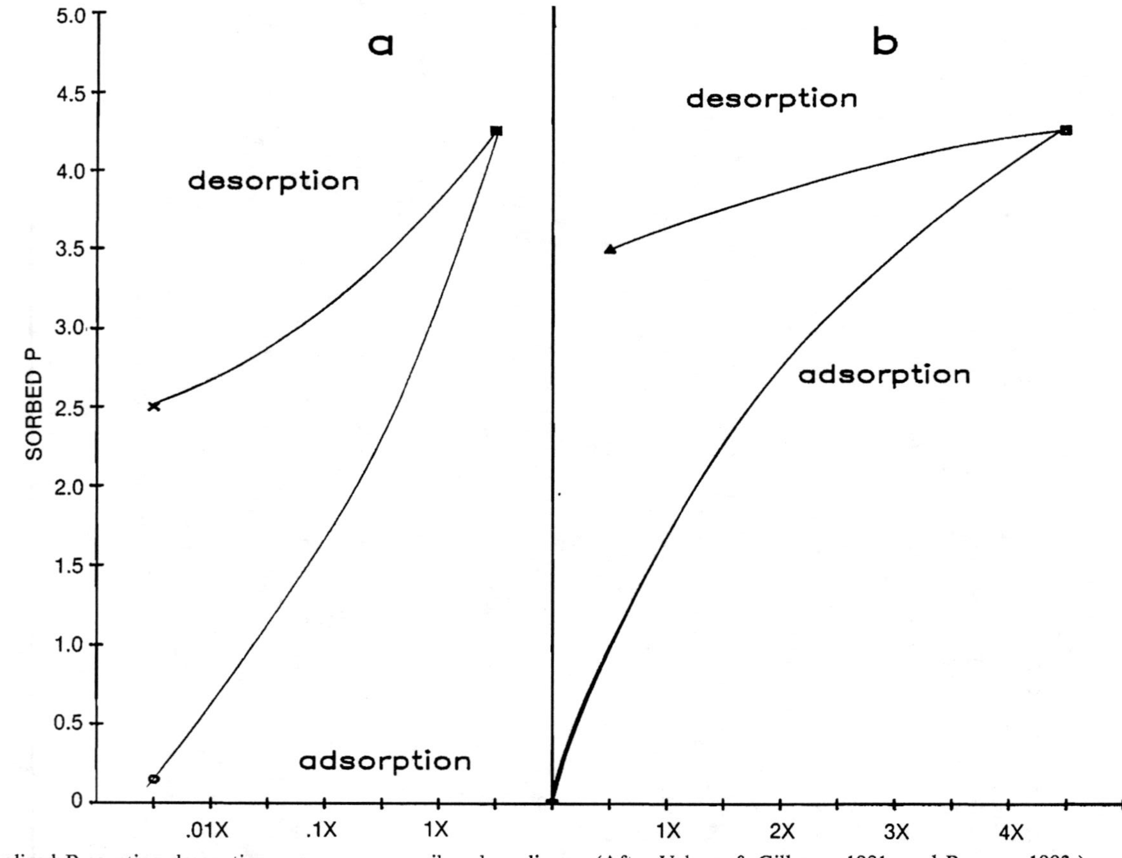

Fig. 7–3. Idealized P sorption-desorption curves: a = semilog, b = linear. (After Uehara & Gillman, 1981; and Barrow, 1983.)

It is clear that P sorption hysteresis is a function of time, initial P application rate, colloid type, and investigative approach (Barrow, 1980b). Anion exchange resins have been used to study P desorption (Bache & Ireland, 1980; Olsen & Khasawneh, 1980; Wolf et al., 1986), as have adsorption-desorption curves (White, 1980; Barrow, 1983).

Desorption curves in laboratory solutions often do not match those determined on the basis of crop removal (White, 1980). This would indicate that our knowledge of plant root effects on the abiotic pools and processes outlined in Fig. 7-1 is still incomplete. As more knowledge on P bonding mechanisms is acquired, our understanding of soil P release should improve as well.

C. Precipitated Forms of Phosphorus

When P fertilizer is applied to soil as highly water soluble granules or as droplets of a suspension or solution the nearby soil particles are awash in a solution of very high P concentration. The solubility product of one or more P compounds is likely to be exceeded and precipitation ensues (Sample et al., 1980). Many different types of reaction products are possible, though most involve Al, Ca, Fe, Mg, NH_4 or K as the counter ion to HPO_4 or PO_4 (Sample et al., 1980).

Initial reaction products are likely to undergo transformation to less soluble forms with time. Thomas and Peaslee (1973), in their summary on soil P testing, came to three general conclusions no less valid today:

1. The quantity of inorganic P in any given soil is related to the soil's parent material.
2. The form of P present, whether bonded to Al, Ca, or Fe, is related to the degree of weathering.
3. Added soluble P will eventually be found in precipitated forms not unlike those of native P (Fig. 7-4).

In calcareous soils, soluble P should ultimately be transformed to an apatite, though the presence of free iron oxides in these soils can complicate the outcome (Ryan et al., 1985). In most acid soils, iron phosphates are the least soluble, most stable form of soil P.

Two approaches to the understanding of solid phase soil P are in general use, solubility product-chemical potential evaluations and chemical fractionations of soil P. Evaluation of solid phase P compounds in soil using solubility or ion activity products is thoroughly discussed by Lindsay and Vlek (1977). In general, the composition of the solution in equilibrium with the soil (mineral) is determined and that point plotted on a phase diagram (Fig. 7-5). If the point lies above the nearest line, then the solution is supersaturated relative to that mineral and precipitation can occur. A point below the line indicates undersaturation of the solution and dissolution is plausible.

This approach has been used to characterize several soil and fertilizer P transformations. Olsen et al. (1977, 1983) reported that cropping 23 calcareous and alkaline soils of eastern Colorado resulted in the dissolution and

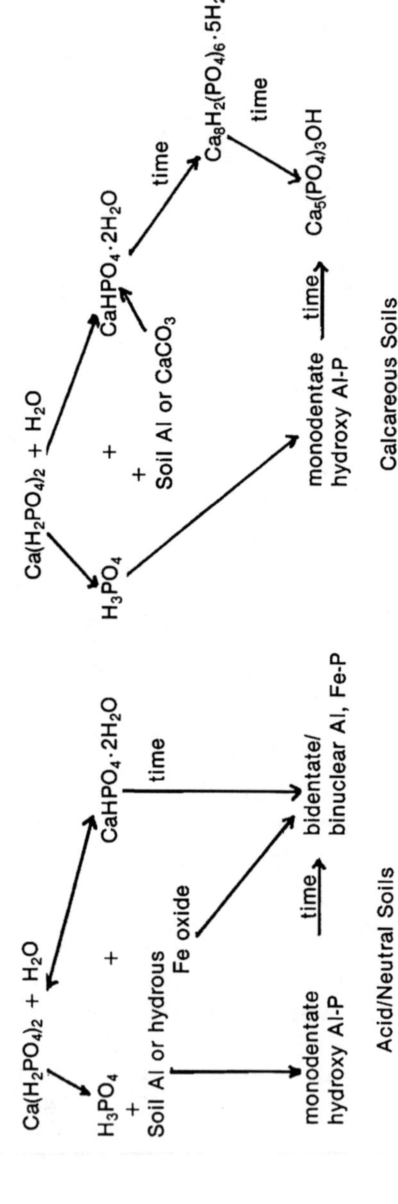

Fig. 7–4. Transformations of monocalcium phosphate in soils. (Modified from Thomas & Peaslee, 1973).

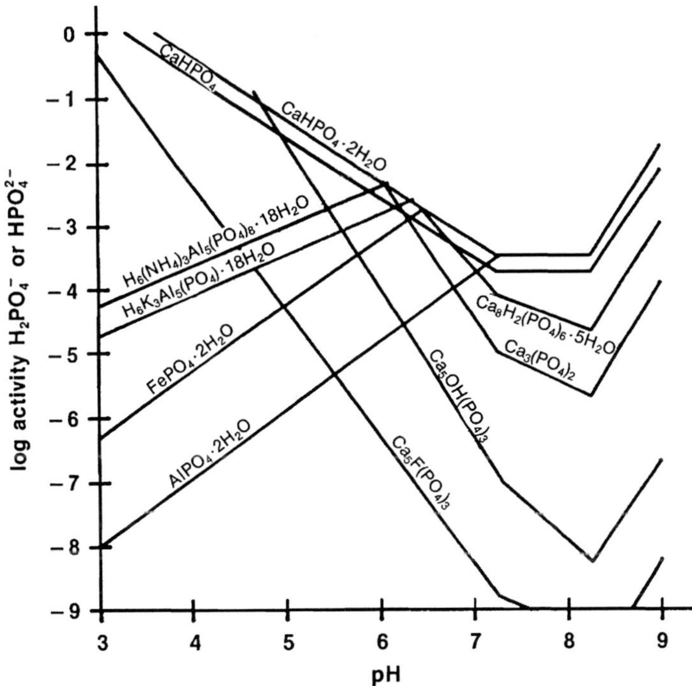

Fig. 7–5. A "unified" phase diagram for several phosphate compounds. (Taken from Lindsay & Vlek, 1977.)

plant utilization of octacalcium phosphate $[Ca_8H_2(PO_4)_6 \cdot 5H_2O]$ and suggested that soils exhibiting equilibrium with this P compound would have a high P status. A similar observation was reported by Adepoju et al. (1986) on 12 southern California soils. Fixen et al. (1983) and Havlin and Westfall (1984) concluded that P solubility was being controlled by β-tricalcium phosphate $[\beta\text{-}Ca_3(PO_4)_2]$ in two calcareous soils with reduced P availability.

Phase diagrams are usually used in soil P systems at or near equilibrium, but not always. Hanson and Westfall (1985) used such diagrams to follow the reaction(s) of injected ammonium polyphosphate over a 90-d sampling period. They concluded that superior plant P nutrition from dual injection of NH_3 and ammonium polyphosphate resulted from the high initial pH, which caused the formation of readily plant-available calcium phosphate reaction products in their moderately acid soils.

Using phase diagrams to determine the dominant stable soil P compound has generally been more successful in calcareous soils. Harrison and Adams (1987) could not determine whether any P mineral was controlling solution phase P concentrations in an Ultisol limed to several pH levels and P fertilized at several rates. This may be because adsorbed P is more important over a comparable time frame in noncalcareous soils (Barrow, 1980a). Blanchar and Stearman (1984) have demonstrated that the combination of ion activity from the solution phase with an estimate of the activity of P sorbed to soil Al compounds (the solid phase P activity) could help predict P solubility subsequent to P addition on eight noncalcareous Missouri soils.

Ion activity product-phase diagrams assume that the P minerals indicated on the phase diagram are indeed those of importance in a particular soil-fertilizer P system. Soil P compounds may contain impurities or be sufficiently amorphous so as to significantly change the solubility product from that posted in the literature for a pure, crystalline P compound. In addition, although there may be several soil P compounds in a sample, only the most soluble will be determined. Organic P compounds are not determined. Most importantly, the technique assumes the system is at or near equilibrium, even if only metastable, between soil and extract. Solid phase transformations may occur very slowly and the extract can be far from equilibrium (Olsen & Khasawneh, 1980; Sample et al., 1980).

Soil P fractionation makes use of chemical extraction/digestion techniques, applied in sequence to a single sample, to strip away different classes of P compounds one group at a time. The most popular approach is that of Chang and Jackson (1957), which first separated soluble P using NH_4Cl, then Al-P compounds with NH_4F (pH 7.0). The sample is then subjected to NaOH extraction for Fe-P forms and H_2SO_4 dissolution for Ca-P minerals. The final step is reduction of Fe in insoluble compounds with $Na_2S_2O_4$ in the presence of citrate to release occluded, "reductant soluble" soil P. Modifications to the basic scheme are detailed by Olsen and Khasawneh (1980) and Olsen and Sommers (1982). Other soil P fractionation techniques involve fewer steps (Tiessen & Stewart, 1983; Sharpley, 1986; Weil et al., 1988) or reflect a greater emphasis on the organic P fraction (Hedley et al., 1982).

Fractionation has been put to several purposes. Ryan et al. (1985) found that P sorption in several calcareous soils from Lebanon was related more to the presence of free iron oxides than to soil carbonates. Enwezor (1977) related plant growth to each of several soil P fractions, as well as other indices of plant-available P, and found all fractions accounted, to some extent, for plant performance. The direct relationship between many of the chemical constituents used in fractionation and those used as plant-available P extractants has led to many such comparisons.

McCallister et al. (1987) used a fractionation scheme to follow transformation(s) of applied P on four Nebraska Mollisols after 12 yr of cropping at several P-fertilization rates. Annual P application resulted in greater fixed P than biennial applications at equivalent (over time) fertilization rates. This observation was not buttressed by greater crop response to biennial P applications. Hooker et al. (1980) used fractionation in a similar way on five calcareous soils, finding that the added P initially sorbed by Fe and Al bearing surfaces was then lost to less soluble forms more quickly when the soil carbonate level was >18 g kg^{-1} (1.8%).

Fractionation-extraction techniques are still the approach of choice in many evaluations of soil organic P transformations, whether because of cultivation (Hedley et al., 1982; Dick, 1983; Tiessen & Stewart, 1983; Sharpley & Smith, 1985; Weil et al., 1988), crop rotation (O'Halloran et al., 1987), or soil genesis/weathering (Smeck, 1985; Sharpley et al., 1987). These studies are facilitated by newly proposed fractionation sequences that purport to better describe organic P fractions (Bowman & Cole, 1978; Hedley et al., 1982).

Soil P fractionation has been sharply criticized by Olsen and Khasawneh (1980), who argue that such extractions are rarely selective, incapable of distinguishing among metastable intermediate reaction products, and overly dependent on the simplistic classification of soil P as Ca-P, Fe-P, and Al-P. Nevertheless, much P research continues to rely on such procedures.

D. Organic Phosphorus

There have been several recent reviews regarding organic P. The reader may refer to Dalal (1977) and Anderson (1980) for more detailed information on this subject. Organic P is an important reservoir for soil P, ranging-from 20 to 80% of total soil P (Dalal, 1977). Organic P compounds are largely phosphate esters. Major classes of these, in order of decreasing resistance to mineralization (decreasing quantity found in soils) are: inositol phosphates, phospholipids, and the nucleic acids (Dalal, 1977; Anderson, 1980). About one-half of the organic P fraction consists of unidentified compounds. In soils with substantial levels of organic matter, soluble organic P may constitute one-half of the P found in the soil solution (Barber, 1984). Despite such observations, the role of organic P in crop P nutrition has often been viewed as a small one.

New information on nutrient cycling (McGill & Cole, 1981; Smeck, 1985) has placed greater emphasis on organic P and related transformations. Of particular interest to soil testing are agro-ecosystems where organic P does indeed play a critical role in plant (and animal) P nutrition. Some of these include pastures-grasslands (Rixon, 1966; Cole et al., 1977; Sharpley, 1985) and agricultural soils in parts of Africa (Adepetu & Corey, 1976; Enwezor, 1977; Anderson, 1980). In several of these studies, organic P has been somewhat related to extractable, soil test P (Adepetu & Corey, 1976; Enwezor, 1977; Sharpley, 1985; Sharpley et al., 1987). This relationship may warrant special consideration in more highly weathered soils (Sharpley et al., 1987).

II. FACTORS AFFECTING PLANT AVAILABILITY OF SOIL PHOSPHORUS

The quantity of P available to plants at any instance is determined by both soil and plant properties. These properties are in turn influenced by several environmental conditions.

A. Soil Properties

Classically, the factors defining soil P availability have been intensity (soil solution P concentration); quantity (amount of solid phase P that is capable of entering the soil solution); capacity ($\partial Q/\partial I$); dissolution/desorption rate; and diffusion (Dalal & Hallsworth, 1976). These factors are interdependent and influenced by the relative saturation of the P adsorption maximum (Gunary & Sutton, 1967). This relationship had led to several studies where phosphate sorption or relative saturation were successfully used to predict fertilizer P requirements (Woodruff & Kamprath, 1965; Peaslee & Fox, 1978). Due to the time and labor involved in such determinations, however, these methods are generally not acceptable for routine diagnostic laboratories.

More recently, the classical factors have been recombined as part of a mechanistic model for nutrient uptake that involves both soil and plant parameters (Claassen & Barber, 1976). In this approach, the soil P factors are C_{li}, the initial concentration in the soil solution; b, the buffer power; and De, the effective diffusion coefficient. The relationship between these factors for a Raub silt loam (fine-silty, mixed, mesic Aquic Argiudolls) is shown in Fig. 7-6. The model has accurately predicted P uptake in both growth chambers and in the field (Schenk & Barber, 1980; MacKay & Barber, 1985b).

Soil pH exerts indirect effects on soil P availability through the factors mentioned previously. Larsen et al. (1965) showed that the half-life of labile P from fertilizer additions decreased significantly as pH increased for noncalcareous mineral soils with initial pH levels from 5.8 to 8.0. Formation of apatitic minerals was cited as a possible cause. An equilibration study of a Lucedale sandy loam soil (fine-loamy, siliceous, thermic Rhodic Paleudults) from Alabama with an initial pH of 5.2, showed that soil solution P levels increased with liming up to a pH of 5.7, then decreased at 6.2 and decreased further at 6.6 provided the lime was added before the P (Soltanpour et al., 1974). Adams (1984) later pointed out that the increase in P solubility from liming at the lower pH levels was likely due to precipitation of exchangeable Al^{3+} as suggested by Coleman et al. (1960).

Such factors as Fe content, carbonate content, carbonate specific surface area, clay content, and organic P content, have been shown to also indirectly affect the availability of soil P (Soper & El Bagouri, 1964; Fixen & Ludwick, 1982; Holford, 1977; Cole et al., 1977). These effects have been discussed previously.

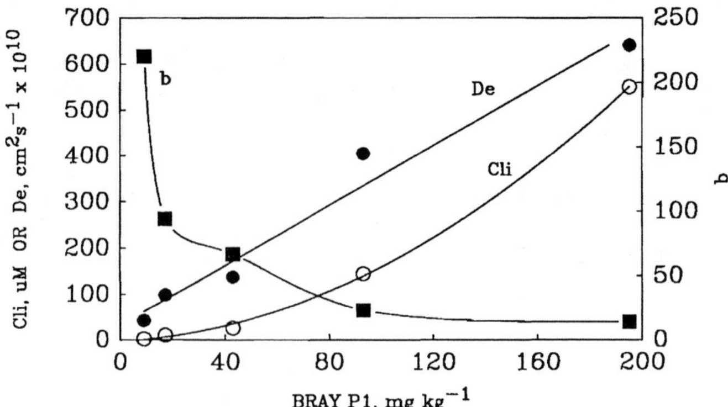

Fig. 7-6. Soil P supply parameters for a Raub silt loam. (Modified from Yao & Barber, 1986.)

B. Plant Characteristics

Generally <1% of the soil volume is occupied by plant roots (Barber, 1984). Therefore, any factor influencing root system size or morphology will likely effect the quantity of soil P that is available to a plant. Simulated sensitivity analysis has demonstrated the large impact rate of root elongation has on P uptake (Silberbush & Barber, 1983).

Characteristics of individual roots, such as root hair length and density (Itoh & Barber, 1983); mycorrhiza infection (Tinker, 1984); and uptake kinetic parameters (Anghinoni et al., 1981) also influence a plant's ability to remove P from soils. Such differences in root system characteristics and individual root properties, together with differences in root/shoot ratios and other characteristics, are responsible for the differing P management requirements of species and, in some cases, cultivars (Baligar & Barber, 1979).

C. Environmental Conditions

Many of the soil and plant parameters discussed above are influenced by environmental conditions such as temperature and moisture. Temperature directly influences soil P in several ways. In a review article, Sutton (1969) described studies demonstrating increases in isotopically dilutable P, labile P, and resin exchangeable P with increased temperature. Regression studies have indicated that the importance of organic P in predicting P availability increases with temperature (Eid et al., 1951). A Nebraska study on the effect of time of sampling on several soil test P measurements revealed that soil test levels were approximately 25% higher in the early spring than they were in the preceding fall. The authors concluded that this change was due to environmental conditions unfavorable for P mineralization and replacement of P used by plants prior to the fall sampling (Olson et al., 1954). Temperature also influences several of the plant characteristics mentioned earlier. Root elongation rates and root uptake kinetic parameters are sensitive to temperature changes.

Many studies have been conducted on the effect of soil moisture on P availability and response to applied P. Power et al. (1961) reported that 53% of the variation in fertilizer response of spring wheat (*Triticum aestivum* L.) on medium P soils was because of variation in available soil moisture. Grain yield response to seed placed P increased 29 kg ha^{-1} for every centimeter increase in available soil water at seeding (1.09 bu in.$^{-1}$). If precipitation between tillering and heading growth stages was also considered, 81% of the response variation was explained.

A greenhouse experiment on millet demonstrated that P uptake increased as soil moisture increased at a soil temperature of 27 °C, but not at 16 °C (Mack & Barber, 1960). This type of interaction between environmental factors has frequently made it difficult to understand P response under field conditions.

More recent investigations on the effects of soil moisture on root growth and P uptake have aided our understanding of how moisture content and plant P uptake are related. MacKay and Barber (1985a, b) showed that P uptake by corn (*Zea mays* L.) decreased at soil moisture contents above and below field capacity because of effects on both P diffusion and root growth rates. Uptake of P dropped more than did total plant weight when soil moisture was decreased below field capacity indicating that the decrease in uptake was not likely due only to decreased demand. They also demonstrated that the negative effects of low soil moisture contents may be partially offset by increases in root hair growth. A sensitivity analysis by these authors using the Claassen-Barber (1976) simulation model indicated that root growth rate was the uptake parameter most influenced by soil moisture content (Barber & MacKay, 1985).

D. The Role of Labile Phosphorus

Clearly, the quantity of plant-available P in a soil is a dynamic entity that can only be crudely estimated by any soil extractant. Our greatest reasonable expectation of a soil P extractant is that it be highly correlated with labile P. The definition of labile P is discussed at length by Olsen and Khasawneh (1980). Conceptually, labile P is soil P in rapid equilibrium with soil solution P and is usually assumed to be equivalent to the fraction of soil P which is isotopically exchangeable with ^{32}P within a given period.

Even if a soil extractant is perfectly correlated with labile P, it would likely explain only a fraction of the P response variability in the field when diverse soils, climatic conditions, and cultural practices are included. The quantity of labile P in a soil is just one of several factors that determine the plant availability of soil P.

III. DETERMINING PLANT-AVAILABLE SOIL PHOSPHORUS

Based on the discussion in the previous section, it is obvious that labile and plant-available P are not directly superimposable concepts, not precisely defined (nor observed) as discrete portions of soil P, and depend on time and technique (Olsen & Khasawneh, 1980). While it is evident that techniques such as resin desorption, isotopic exchange, and quantity/intensity curve formation offer alternative approaches for plant-available P assessment, none have proven uncomplicated enough in practice, theory, or interpretation to make serious inroads into common soil test usage. The overwhelmingly largest fraction of soil samples are tested for available P by extraction with dilute solutions.

A. Selection Criterion for Available Soil Phosphorus Extractions

Kamprath and Watson (1980) have outlined the objectives of P soil tests. First, the soil test result should place the sample (field) in a predetermined group or category to make a fertilizer P recommendation. The terms low, medium, and high equate to one such classification system. Second, the test result should indicate the likelihood of an economic return to fertilizer use. This goes hand-in-hand with the classification system; a low test result indicating a high probability of response and vice-versa. Additional economic information on fertilizer and commodity prices is required to successfully meet this goal. Third, the soil test result should be directly related to the soil's P supply capability. This direct relationship, in addition to indicating whether fertilizer is needed, can be used to calculate the rate of fertilizer P to be added. This latter calculation assumes a knowledge of how added soluble P changes P test values for soils of the region(s) served by the testing laboratory.

Criterion for a suitable extractant have been detailed in several reports. Thomas and Peaslee (1973) felt that the extractant should (i) remove a reproducible and consistent proportion of the soil's plant-available P and (ii) reflect the extent and nature of reactions between soils and added P. Barrow (1980b) also felt that a soil test was an integration of several factors affecting soil P status into one numerical value. He included soil P buffering capacity with P quantity and soil P reaction time in his list of factors. Holford (1980) demonstrated that soil test P removal should be inversely related to buffer capacity, though this depended on whether plant-available P was being made available via desorption or dissolution. Kamprath and Watson (1980) noted that the extractant should (i) extract a proportionate fraction of plant-available P from soils differing in other properties; (ii) the procedure should work with reasonable accuracy and speed; and (iii) the soil test P value determined should be correlated with crop growth response to P or P uptake.

B. Extractant Mode of Action

Though Thomas and Peaslee (1973) state that several soil test extractants were developed empirically prior to the release of the P fractionation method (Chang & Jackson, 1957), the fractionation technique is often the basis of explaining extractant mode of action. A knowledge of extractant-soil interaction chemistry is most useful when understanding divergent extractant behavior because of (i) differences in available P sources on soils of otherwise similar properties (Smyth & Sanchez, 1982; Admont et al., 1986) or (ii) differences among soils fertilized with a common P source (Varvel et al., 1981).

There has been a divergence between soil P chemistry and soil P testing. More is known about soil P bonding than is used in soil testing. Little additional information exists regarding the bonding of P removed by common chemical extractants. In terms of new definitions of soil P, are we extracting monodentate, binuclear, amorphous occluded or crystalline P compounds? There is a need for detailed studies on the nucleophilic/electrophilic behavior of extractant constituents on P sorbed to model surface materials. Progress in this area is not furthered by the highly empirical approach seemingly being employed in the development of multiple element extraction techniques.

Kamprath and Watson (1980) proposed four general reactions contributing to P release from soil by extractants. These were (i) acid dissolution, (ii) anion exchange, (iii) cation complexation, and (iv) cation hydrolysis.

Both strong (HCl, HNO_3, and H_2SO_4) and weak (acetic, citric, and lactic) acids have been used to extract soil P. Acid concentrations used are usually low, resulting in extracting solution pH values generally below 5 for weak acid extractants, and near pH 2 for strong acid extractants. All three inorganic P forms are dissolved to some extent by acids, where Ca-P > Al-P > Fe-P describes the decreasing order of solubility (Mehlich, 1978a). If the duration of soil-extractant contact is kept short, Fe-P dissolution is minimized (Thomas & Peaslee, 1973).

Other anions can replace adsorbed P if their size, binding strength, or concentrations are appropriate to the quantity and type of colloid surface present. They also serve to minimize readsorption of P if the extraction time is kept short (5–10 min). Examples are acetate, bicarbonate, citrate, lactate, and sulfate (Mehlich, 1978a; Kamprath & Watson, 1980).

Fluoride and certain organic anions (citrate and lactate) complex Al. Extractants containing these anions release P from Al-P compounds. Bicarbonate precipitates soluble Ca as $CaCO_3$, causing release of Ca-P. Fluoride also results in $CaHPO_4$ dissolution with concomitant precipitation of CaF_2. Thomas and Peaslee (1973) argue that bicarbonate (HCO_3) and F remove similar soil P forms, though F is more competitive/aggressive.

The distinction between anion competition at the colloid surface and ligand competition in the solution is probably artificial as the same cations (Al, Ca, and Fe) and anions (acetate and sulfate) are often involved. Cation hydrolysis occurs at high pH values where the hydroxyl anion (OH) dissolves a portion of the Al-P and Fe-P by hydrolysis of Al and Fe. This is a specific form of the anion/ligand competition reactions described earlier. Extractions using buffered HCO_3 solutions can remove some Al-P and Fe-P from soils because of their higher OH concentrations (Maida, 1978).

Conflicting claims regarding the selectivity of the fractionation technique (Thomas & Peaslee, 1973; Olsen & Khasawneh, 1980) can be extended to discussions on the specificity of a particular extractant for particular forms of soil P. Even the simplest P extracting solutions offer more than one mode of action in P removal. Chang and Juo (1963) observed that the distribution of soil P amongst the several fractions affected the correlation of extractable P for a particular method with a particular soil P fraction. As the soil P becomes distributed more evenly amongst more fractions the relationship between any P extraction and a P fraction for a given soil falters. This also applies to situations where soil diversity is so great as to require "grouping" before meaningful relationships between soil P extractions and soil P fractions can be undertaken (Sen Tran & Giroux, 1985; Sharpley et al., 1987).

The release of P from soil organic compounds can be a significant source of plant-available P (Adepetu & Corey, 1976; Enwezor, 1977; Sharpley, 1985). The extraction of available inorganic P from such soils has been shown to be related to the organic P fraction. Again, the exact mechanism is unknown, but the most successful extractant contains both fluoride and dilute acid.

The implication is that either acid hydrolysis of organic P esters or breakup of organo-metallic (Al and Fe) bound P is occurring. Sharpley (1985) reported that the organic P-extractable P relationship for a given soil was strongly related to concurrently extractable inorganic P concentrations.

The analytical outcome (amount of P removed) of any extraction is determined by technique factors in addition to the chemical composition of the extracting solution. Increasing the solution to soil ratio can increase P extractability (Randall & Grava, 1971) as can lengthening the extraction time (Breland & Sierra, 1962). Raising or lowering the concentration(s) of one or more solution constituents can markedly influence soil P removal (Thomas & Peaslee, 1973). These kinds of differences in soil testing technique as well as other, more subtle, factors such as shaking speed, extraction container size and shape, and sample size contribute to such variation (Grava, 1975).

C. Extractants Used to Determine Available Soil Phosphorus

Extractants commonly used in soil test laboratories for both past and present P evaluation are listed in Table 7-1. Earlier surveys of procedures in use in the USA indicate that the Bray P_1, Mehlich I, and Olsen extractions were often used (Bingham, 1962; Jones, 1973). A more recent survey suggests that the Mehlich III procedure is gaining wider acceptance (SRSTIEG, 1984).

Though it is generally accepted that acid extractants are more appropriate for acid soils and bicarbonate containing reagents more suitable for calcareous soils, there is evidence that the Olsen ($NaHCO_3$) extraction performs reasonably well on acid soils (Farina & Channon, 1979; Smyth & Sanchez, 1982).This may be related to release of P from Fe surfaces to both plants

Table 7-1. Soil test extractants commonly used to determine available soil P.

Extractant name(s)	Extractant composition	Reference
AB-DTPA	1 M NH_4HCO_3 + 0.005 M DTPA – pH 7.5	Soltanpour & Schwab, 1977
Bray P_1	0.03 M NH_4F + 0.025 M HCl	Bray & Kurtz, 1945
Bray P_2	0.03 M NH_4F + 0.1 M HCl	Bray & Kurtz, 1945
Citric acid	1% citric acid	Dyer, 1894
Egnor	0.01 M Ca lactate + 0.02 M HCl	Egner et al., 1960
ISFEIP (Hunter)	0.25 M $NaHCO_3$ + 0.01 M NH_4F + 0.01 M EDTA – pH 8.5	ISFEIP, 1972
Mehlich I (Double acid, North Carolina)	0.05 M HCl + 0.0125 M H_2SO_4	Mehlich, 1953
Mehlich II	0.015 M NH_4F + 0.2 M CH_3COOH 0.2 M NH_4Cl + 0.012 M HCl	Mehlich, 1978b
Mehlich III	0.015 M NH_4F + 0.2 M CH_3COOH 0.25 M NH_4NO_3 + 0.013 M HNO_3	Mehlich, 1984
Morgan	0.54 M CH_3COOH + 0.7 M $NaC_2H_3O_2$ – pH 4.8	Morgan, 1941
Olsen	0.5 M $NaHCO_3$ – pH 8.5	Olsen et al., 1954
Truog	0.001 M H_2SO_4 + $(NH_4)_2SO_4$ – pH 3	Truog, 1930

and HCO_3 solutions in these soils (Maida, 1978). Maida (1978) also reported that the Bray P_1 procedure did not seem to recover Fe-P, but rather Al-P and Ca-P, on the Malawi soils studied.

Michaelson and Ping (1986) reported that the amount of P recovered by weak acid-fluoride extractants (Mehlich II, III, and Bray P_1) was strongly related to oxalate extractable Al + Fe in 10 soils. They also found Mehlich III to be superior to the other extractants when volcanic ash and loess-derived soils were pooled together. No calcareous soils were included in their report.

As extraction procedures have changed over time, the early solutions used to determine the availability of more than one nutrient element (Dyer, 1894; Morgan, 1941) were replaced by those more optimally tuned to evaluate soil P availability alone (Bray & Kurtz, 1945; Olsen et al., 1954). These changes were accompanied by appropriate research in correlation and calibration. More recently the wheel has turned again, and soil test laboratories attempting to be more cost effective have pressed for procedures to determine multiple nutrient element availability (Mehlich, 1978b; Soltanpour & Schwab, 1977; Eik & Hanway, 1986). This latest shift has not always been accompanied by the requisite correlation and calibration research; rather the P recovered in the new extractant was correlated with that of the old procedure used by the lab using several samples often taken from the general pool of samples arriving at the soil testing laboratory each day. The regression equation so developed is then used to determine new criterion for fertilizer P recommendation. This approach, which reduces or eliminates the difficult correlation/calibration step in fertilizer P recommendation development, has its faults (see section IV, "Correlation of Phosphorus Soil Tests.")

D. Other Chemical Approaches to Available Soil Phosphorus Evaluation

The chemical extractants discussed previously are better related to the soil P quantity and buffer capacity factors than to soil P intensity; the P found in the soil solution. Solution phase P is determined using water or dilute salt solutions (Kamprath & Watson, 1980; Luscombe et al., 1979). Solution P concentrations are often low and dedication to analytical technique is required. Still, soil solution P is well related to plant P nutritional status. Other factors do influence the solution P level associated with P nutritional adequacy. Most notable are soil texture; sandy soils will require a higher solution P concentration for adequacy (Olsen & Watanabe, 1963); mineralogy of the clay colloid fraction (Uehara & Gillman, 1981) and the plant species being evaluated (Fox et al., 1974).

Kovar and Barber (1988) advocated several additions of P to moistened portions of each sample to predict the effect of added P on both solution P (P intensity) and the P subsequently desorbed using an anion exchange resin (P quantity). This results in a site-specific Q/I curve from each sample. McLean et al. (1982) proposed a similar procedure where only extractable P (P quantity) was determined. As the proposed equilibration period was short (2 h), a correction for the longer fertilizer-soil reaction time found in the field was used (McLean, 1985). Lee and Bartlett (1977) used a similar technique to arrive at the Phosphorus Fertilizer Index on each sample.

Menon et al. (1989) impregnated paper strips with iron hydroxide and used those strips to desorb P from soil samples in suspension for 16 h. Phosphorus sorbed to the strips was removed by H_2SO_4 extraction. Available P test values were better related to greenhouse culture corn dry matter yield and P uptake than those of Bray P_1, Bray P_2, Mehlich I, Olsen, water, and resin extraction. Van der Zee et al. (1987) have described the kinetics of the strip extraction.

E. Modeling the Plant Availability of Soil Phosphorus

Another approach to available P assessment is to model the release of soluble P from other forms of residual soil P. Barrow (1980b) outlined several mechanisms behind this phenomena and summarized the information in terms of a descriptive model that simulates the initially rapid, but ultimately slow, gradual decline in available P after fertilizer addition while taking into account P removal by the harvested crop. This model, and a seemingly more mechanistic model proposed by Wolf et al. (1987), do not require a measurement of plant available P by soil test methods. They rely more upon crop response and a crop P budget developed on both fertilized and unfertilized soil, as well as previously determined empirical constants, to describe the soil transformation(s) of applied soluble P. Other models rely on soil test P information (Cox et al., 1981; Jones et al., 1984a; Sharpley et al., 1984) for initialization or evaluation.

Models that rely on crop response to residual P suffer for lack of sufficient long-term field trials with information on both P uptake and soil P transformations (Janssen et al., 1987). Models that use extractable P can be tested using successive cropping in the greenhouse to exhaust extractable P as an alternative approach to field evaluation. This has been criticized by several researchers because plants often remove more P than the decline in extractable soil P would indicate in such greenhouse trials (Novais & Kamprath, 1978; Cox et al., 1981; Adepoju et al., 1982; Fixen & Ludwick, 1982; Aquino & Hanson, 1984). This has also been observed on occasion in the field environment when the unfertilized crop yield and, therefore, plant-available P, continued to decline, but the extractable P value had levelled off (Hooker et al., 1983).

Barrow (1980b) indicated that there was a need to determine the relevant soil properties that cause differences among soils in the release of residual P. Greenhouse exhaustion trials can be used effectively to this end (Novais & Kamprath, 1978; Adepoju et al., 1982; Fixen & Ludwick, 1982; Holford, 1982; Aquino & Hanson, 1984). Properties such as clay content, pH, carbonate content, hydrous oxide content, and soluble Ca have been proposed to be related to P removal in one or more of these studies. The P cycling model of Jones et al. (1984a) makes use of several soil physical and chemical properties in predicting the residuality of fertilizer P (Sharpley et al., 1984). Cox and Lins (1984) have proposed that clay content be included in fertilizer rate recommendation equations, based on the model of Cox et al. (1981), and strengthened this proposal with additional evidence (Lins et al., 1985).

Models that seek to predict changes in extractable soil P over time (Cox et al., 1981; Jones et al., 1984a, b) to make appropriate fertilizer recommendations are subject to some of the same vagaries as those soil test procedures. Initializing values are subject to differences in choice of extraction technique and in technique execution. The model may "model" the extractant's performance, but not plant-available P. For example, Barrow (1980b) suggests that bicarbonate extractable P would decline less rapidly than actual P availability on "weakly buffered" soils. Conversely, on soils with greater buffer capacity the decline in soil test P might exceed that observed by the crop. This was reported by Hooker et al. (1983) at one location using Bray P_1 to evaluate P availability to corn.

Because of a preponderance of short-term research, information is lacking on P residuality. White (1980) pointed out that P deficiency is relatively rare in most developed countries and that continued P applications to P rich agricultural soils are both uneconomical and environmentally unsound. There is a need for more "cropping down" studies so that future P fertilizer needs for such soils could be predicted with better confidence. If models are to become more important in agronomic/environmental management (reducing reliance on conventional soil testing?) more validation work must be done. Research is needed on "between-model" comparisons as well.

IV. CORRELATION OF PHOSPHORUS SOIL TESTS

The general process of soil test correlation was discussed previously (see, chapter 4 in this book, by Dahnke and Olson). Numerous correlation studies of P soil tests have been conducted with the majority of them being greenhouse experiments.

A. Greenhouse and Field Studies

As discussed earlier, greenhouse studies are sometimes preferred because the effects of uncontrolled variables can be nearly eliminated. A few precautions, however, must be taken when P tests are being correlated in the greenhouse. Plant populations or duration of the growth period should be selected to minimize competition between adjacent roots. Excessive root competition, beyond that normally encountered under field conditions, tends to overestimate the importance of the buffer capacity factor and underestimate the importance of the intensity factor (Nye & Tinker, 1977). Thus, high root densities would unduly favor tests that correlate well with the buffer capacity. Also, care must be taken to keep root zone temperatures comparable to field conditions. Unusually high or low temperatures could differentially alter reaction kinetics of both organic and inorganic P forms, influence root activity, and bias the resulting correlations.

Field experiments can also be used in P correlation studies, however, correlation coefficients will normally be lower. A multitude of climatic interactions, cultural practice differences, plant differences, and soil factors influence P uptake and crop yield under field conditions. Field correlation studies, however, have the advantage of being conducted under the same diverse conditions under which the selected test will eventually be used.

B. Comparison of Methods

A summary of the results of correlation studies conducted on 10 soil tests is reported in Table 7-2. Numerous availability parameters and their transformations were used in these experiments. Included are crop P uptake, P concentration, "A" value, resin-P, crop yield, and relative yield. In several cases, the relationship between soil test value and the availability parameter was nonlinear, resulting in log transformations or use of polynomial functions to describe the correlation. In other studies, the authors use nonfunctional correlation techniques such as the Cate-Nelson approach (see chapter 4 in this book, by Dahnke and Olson).

A wide range in correlation coefficients was found for all the soil tests evaluated (Table 7-3). Nearly every soil test varied from no significant correlation with availability to explaining more than 90% of the variability in P availability (r^2). Resin methods appear to be the exception because the lowest r value reported was 0.69. This may indicate that the resin procedures are suitable over a broader range of soil properties than the other methods. A similar conclusion was drawn by Sibbesen (1983) after reviewing and summarizing 29 published papers where the anion exchange resin method was among the P tests used. The range in r values in Table 7-3 illustrates the importance of selecting the appropriate soil test for a given set of soils. Nearly any soil test will fail if used on inappropriate soils.

C. Relationships between Soils and Soil Tests

Several soil properties influence how well specific soil tests correlate with P availability. Some of these properties are quite predictable once the basic chemistry of the extractant soil mixture is understood. Other properties are more subtle relative to their effects on soil test success.

The presence of carbonates in soils has often been viewed as a problem for strong acid extractants such as the Bray and Kurtz methods or the Mehlich I procedure. Several studies included in Table 7-2 show a reduction in r for the Bray P_1 test compared to the $NaHCO_3$ test for calcareous soils (Blanchar & Caldwell, 1964, Nesse & Grava, 1986). This reduction has generally been attributed to neutralization of the acid by $CaCO_3$ followed by precipitation of the fluoride by the released Ca. Thus, the extractants' potential for removing P is reduced. However, in other correlation studies the Bray P_1 and $NaHCO_3$ tests have performed similarly on calcareous soils in Colorado and Nebraska (Olsen et al., 1954). This suggests that properties other than total $CaCO_3$ content are involved in determining if the acid tests

Table 7-2. Correlation coefficients of soil test extractable P with various P-availability parameters.

Region or category	No. of observations	Availability parameter	Extractant† Bray P 1	Bray P 2	Olsen	AB-DTPA	Mehlich I	Mehlich II	Water or dilute salts	Resin	Reference
			r								
			Greenhouse studies								
Nigeria											
Alfisols	30	P uptake by rice	0.38	0.35	0.33		0.26		0.21‡		Oko & Agboola, 1974
Oxisols	30	P uptake by rice	0.32	0.38	0.31		0.32		0.20‡		Oko & Agboola, 1974
Poorly drained	30	P uptake by rice	0.43	0.43	0.29		0.25		0.03‡		Oko & Agboola, 1974
Combined	90	P uptake by rice	0.37	0.22	0.32		0.27		0.03‡		Oko & Agboola, 1974
Ohio											
Native pH <5.5	10	P uptake by alfalfa	0.98§	0.99§	0.96§						Thompson & Pratt, 1954
Native pH ≥5.5	8	P uptake by alfalfa	0.93§	0.81	0.81						Thompson & Pratt, 1954
Combined	18	P uptake by alfalfa	0.92	0.63	0.87						Thompson & Pratt, 1954
Pennsylvania	44	P uptake, corn	0.66						0.68¶		Baker & Hall, 1967
Colorado											
Calcareous	30#	A value, oat	0.90		0.94						Olsen et al., 1954
0-9.6% $CaCO_3$	38	A value, millet	0.80		0.87				0.97‡		Olsen et al., 1954
New Jersey	10	P uptake, tomato	0.70		0.88		0.73		0.72‡		van Diest, 1963
USA & Saskatchewan	74	A value, millet	0.75		0.94				0.99‡		Olsen et al., 1954
Oregon	30	A value, sudan			0.87				0.75‡		Olsen et al., 1954
Australia	18	P uptake, wheat	0.52						0.45¶		Dalal & Hallsworth, 1976
New South Wales											
7 soil orders	30	P uptake, clover	0.77		0.92		0.46				Holford, 1980
Colorado											
0.05-8% $CaCO_3$	23	P uptake from multiple crops	0.88							0.92	Bowman et al., 1978
Colorado	11	Relative yield††			0.71	0.72			0.77¶		Labhsetwar & Soltanpour 1985
Minnesota											
Noncalcareous	7	P uptake, oat	0.86		0.85				0.96‡	0.73	Blanchar & Caldwell, 1964
Calcareous	7	P uptake, oat	0.23		0.73				0.95‡	0.69	Blanchar & Cladwell, 1964
Connecticut	14‡‡	Alfalfa yield	0.90				0.91				Griffin & Lorton, 1970
New York	21	P uptake, multiple			0.65					0.87	Lathwell et al., 1958
New Zealand	5	Perennial ryegrass yield			0.72				0.92§		Luscombe et al., 1979

Syria										
All soils	270	Relative wheat yield increase§			−0.03					Matar & Samman, 1975
Basaltic	33	Relative wheat yield increase§			−0.62					Matar & Samman, 1975
Euphrates	16	Relative wheat yield increase§			−0.89					Matar & Samman, 1975
Khabaur	13	Relative wheat yield increase§			−0.81					Matar & Samman, 1975
Limestone	39				−0.64					Matar & Samman, 1975
Nebraska										
All soils	37	A value, oat	0.96		0.95					Olson et al., 1954
Calcareous	11	A value, oat	0.72		0.74					Olson et al., 1954
Wisconsin	90	P uptake, corn	0.13					0.43§§		Wendt & Corey, 1981
North Dakota	29	Sudangrass yield increase	−0.79		−0.81					Zubriski, 1971
CO, MT, NE	30	A value, oat	0.94		0.96			0.96‡		Olsen et al., 1954
North Carolina	72	P uptake, millet	0.83		0.70	0.88	0.84			Mehlich, 1978b
Scotland	40	P uptake, oat			0.75			0.61¶¶		Williams, 1966
Brazil	8	P uptake, rice				0.41			0.98	van Raij et al., 1986
						Field studies				
British Columbia										
pH 5.4–6.6	66	Alfalfa P conc.##	0.37	0.31	0.50					John et al., 1967
pH 6.7–7.0	65	Alfalfa P conc.##	0.46	0.11	0.48					John et al., 1967
pH 7.1–8.2	61	Alfalfa P conc.##	0.79	0.34	0.76					John et al., 1967
Combined	192	Alfalfa P conc.##	0.63	0.29	0.66					John et al., 1967
Australia	8	Wheat grain yield			0.80			0.61¶		Dalal & Hallsworth, 1976
Texas	23	Sorghum yield increase†††			0.69					Onken et al., 1980
South Dakota										
pH < 6.6	15	Log of % yield increase‡‡‡	−0.77		−0.74					Fixen & Carson, 1978
pH 6.6–7.0	30	Log of % yield increase‡‡‡	−0.17		−0.21					Fixen & Carson, 1978
pH > 7.1	20	Log of % yield increase‡‡‡	−0.59		−0.61					Fixen & Carson, 1978
Nebraska-all	40	Rel. wheat yield increase	−0.63		−0.58					Olson et al., 1954
Calcareous	13	Rel. wheat/oat increase	−0.63		−0.58					Olson et al., 1954
Brazil	28	Rel. cotton yield							0.85	van Raij et al., 1986

† See previous section for soil test procedure description.
‡ Water.
§ Quadratic.
¶ 0.01 M $CaCl_2$.
# Three soil series, 10 treatments per series.
†† Two alfalfa cuttings; r is calculated from r^2 of Cate-Nelson two-class model.
‡‡ Two soil series, seven treatments per series; log of P extracted.
§§ 0.001 M $SrCl_2$.
¶¶ Log water extractable.
Log of soil test value.
††† r is calculated from Cate-Nelson, three class model.
‡‡‡ 45 spring wheat, 13 barley, 8 oat, 8 winter wheat.

Table 7-3. Range in correlation coefficients reported across studies for the common P soil tests.

Soil test	Correlation coefficient†	
	Minimum	Maximum
Bray P_1	0.13	0.98
Olsen	0.03	0.96
Mehlich I	0.25	0.91
Water or salts	0.03	0.99
Resin	0.69	0.98

† Selected from Table 7-2.

fail. Because these other factors have not been completely identified, use of the acid tests is usually avoided on calcareous soils. Widening the soil to solution ratio to 1:50 or 1:100 has improved correlations on neutral and calcareous soils and apparently lessens the problem (Blanchar & Caldwell, 1964; Randall & Grava, 1971; Fixen & Carson, 1978).

The pH of noncalcareous soils has been identified as another factor influencing soil test performance, although not consistently. A field study in British Columbia showed that correlations were higher for alkaline soils than acid soils for both Bray P_1 and $NaHCO_3$ tests (John et al., 1967). A South Dakota study showed that correlations were lowest in the pH range of 6.6 to 7.0 for both tests (Fixen & Carson, 1978). An Ohio study demonstrated only slight reductions in correlations when the pH exceeded 5.5 for Bray P_1 and $NaHCO_3$ tests (Thompson & Pratt, 1954).

Holford (1980) demonstrated that P buffer capacity had a marked effect on how well Bray P_1 and $NaHCO_3$ soil tests correlated with P uptake. Both tests were highly correlated with uptake from weakly and moderately buffered soils (r = 0.84–0.95) but were not correlated with P uptake from strongly buffered soils. The author indicated that this lack of correlation may mean that a mechanism other than adsorption was dominant in buffering some of these soils.

The combined effects of carbonate content, pH, buffer capacity, and other factors on correlation coefficients can be readily observed by comparing correlations between soil series. In a Wisconsin study, the correlation coefficient for both $SrCl_2$ extractable P and Bray P_1 varied considerably across soil series and when the series were combined correlations dropped markedly (Table 7-4). Grouping the 90 soils by previous crop significantly increased correlation coefficients for the $SrCl_2$ test. Also, the soil with the lowest correlation coefficient, the Withee, was the only series with a significant mix of previous crops. Thus, cropping history may be another factor influencing correlations and perhaps P uptake.

D. Correlation between Methods

An approach commonly used when considering a new soil test procedure is to correlate the results of the standard test with the new or alternate test. Results of such comparisons are summarized for Bray P_1 and $NaHCO_3$

Table 7–4. Simple correlations coefficients for soil test vs. P uptake correlations (Wendt & Corey, 1981).

Soil†	Observations	Correlation coefficient	
		$SrCl_2$-P	Bray P_1
Plainfield S	19	0.93**	0.86**
Plano Sil	15	0.98**	0.87**
Fayette Sil	15	0.85**	0.91**
Withee Sil	18	0.41	0.49*
Kewaunee Sicl and Hibbing Sicl	18	0.86**	0.71**
Total soils	90	0.43**	0.13
Previous crop:			
Row crop	50	0.81**	--
Alfalfa	40	0.64**	--

*,** $P = 0.05$ and 0.01, respectively.
† S, sand; Sil, silt loam; Sicl, silty clay loam.

procedures in Table 7–5. The correlation coefficients are generally high, but variation does occur in slopes and intercepts between the same procedures.

Correlation of the results from two extractants is a useful first-step technique in evaluating a new procedure. These studies can be done quickly on many soils and are relatively inexpensive. However, the results need to be interpreted cautiously. Table 7–6 shows the possible correlation coefficients between Y (yield) and Z (new soil test) given correlation coefficients between X (standard soil test) and Y and between X and Z. This table shows that as the X-Z correlation coefficients decrease, the possible range in Y-Z correlation coefficients increases rapidly and the correlations quickly become totally unreliable.

Table 7–7 demonstrates an additional concern. Examination of correlation slopes for individual soils as opposed to that for the group of individual samples indicates that the Mehlich III procedure extracts less P than Bray P_1 on the Maury soil, nearly the same on the Trappist, more on the Newark and the group of lab samples, and a great deal more than Bray P_1 on the Belknap. Without further correlation and calibration research, equivalent fertilizer P recommendations based upon the equation for the large group of lab samples would result in overferilization of Maury (fine, mixed, mesic Typic Paleudalfs) and Trappist (clayey, mixed, mesic Typic Hapludults) soils and underfertilization of the Belknap series (coarse-silty, mixed, acid, mesic Acric Fluvaquents). Producers owning such soils are not well served by such changes—and do not even know it. Further, this makes optimization of P fertilizer inputs to achieve economic/environmental goals difficult and reduces the value of soil testing as a management practice.

Correlations between soil tests can be useful in screening a new procedure. If the r value is low between the new test and the standard test (assuming the standard test is very well correlated with P uptake), it is doubtful that the new procedure will be acceptable and further evaluation is not likely justifiable. However, if the r value is high between the new and standard procedures, considering the correlation coefficients in Table 7–2, one can-

Table 7-5. Correlation of Bray P_1 and Olsen P soil tests with other methods.

Method (Y)	Region	No. of observations	pH range	*r*	Intercept	Slope	Reference
				Bray P_1 (X)			
Olsen	Florida	7	4.1-7.5	0.95	6.1 mg kg^{-1}	0.13	Breland & Sierra, 1962
Olsen	Nebraska	15‡	5.5-8.2	0.97	1.1 mg kg^{-1}	0.43	Olsen et al., 1954
Olsen	Washington	46	--	0.87	12.2 mg kg^{-1}	0.35	Kuo & Jellum, 1985
Olsen	South Dakota	165	6.0-8.0	0.95	3.9 mg kg^{-1}	0.68	Malo & Gelderman, 1984
Olsen	North Cent. USA	91	4.3-6.8§	0.85	2.9 mg kg^{-1}	0.30	Wolf & Baker, 1985
Olsen	MN, CCE < 7%	14	7.7-8.4	0.97	−2.7 mg kg^{-1}	0.72	Nesse & Grava, 1986
Mehlich I	Florida	7	4.1-7.5	0.84	9.1 mg kg^{-1}	0.44	Breland & Sierra, 1962
Mehlich I†	Missouri	5	6.2-6.9	0.33	--	--	Aquino & Hanson, 1984
Mehlich II	Missouri	5	6.2-6.9	0.91	--	--	Aquino & Hanson, 1984
Mehlich II	Illinois	191	--	0.97	--	--	T.R. Peck, 1984, personal commun.
Mehlich II	International	122	<7.2	0.97	--	--	Mehlich, 1978b
Mehlich III	North Cent. USA	91	4.3-6.8§	0.98	4.2 mg kg^{-1}	0.87	Wolf & Baker, 1985
Mehlich III	SE USA	105	3.8-7.5	0.96			Mehlich, 1984
Mehlich III	Oklahoma	310	5.0-7.6	0.97	−16 kg ha^{-1}	1.12	Hanlon & Johnson, 1984
AB-DTPA	Oklahoma	310	5.0-7.6	0.94	−4 kg ha^{-1}	0.21	Hanlon & Johnson, 1984
Bray P_2†	Missouri	5	6.2-6.9	0.37	--	--	Aquino & Hanson, 1984
$SrCl_2$†	Missouri	5	6.2-6.9	0.91	--	--	Aquino & Hanson, 1984
Water	Florida	7	4.1-7.5	0.91	0.61 mg kg^{-1}	0.024	Breland & Sierra, 1962
				Olsen			
Mehlich I	Florida	7	4.1-7.5	0.82	−4.1 mg kg^{-1}	3.0	Breland & Sierra, 1962
Mehlich I	North Cent. USA	91	4.3-6.8§	0.93	0.62 mg kg^{-1}	1.52	Wolf & Baker, 1985
Mehlich II	International	122	<7.2	0.89	--	--	Mehlich, 1978b
Mehlich III	North Cent. USA	91	4.3-6.8§	0.88	−6.91 mg kg^{-1}	3.08	Wolf & Baker, 1985
AB-DTPA	Colorado	481	7.0-9.2	0.92	−0.63 mg kg^{-1}	0.51	Soltanpour & Schwab, 1977
AB-DTPA	Colorado	11	6.4-7.8	0.98	−2.4 mg kg^{-1}	0.54	Labhsetwar & Soltanpour, 1985
EDTA	Colorado	11	6.4-7.8	0.91	−0.75 mg kg^{-1}	3.0	Labhsetwar & Soltanpour, 1985
EDTA	India	20	5.5-8.7	0.90	--	--	Sahrawat, 1977
$CaCl_2$	Colorado	11	6.4-7.8	0.91	0.020 mg kg^{-1}	0.027	Labhsetwar & Soltanpour, 1985
log $CaCl_2$	Scotland	40	5.0-6.4	0.68	--	--	Williams, 1966
Water	Florida	7	4.1-7.5	0.85	−0.0080 mg kg^{-1}	0.16	Breland & Sierra, 1962
Log water	Scotland	40	5.0-6.4	0.66	--	--	Williams, 1966

† Slope of regression lines of extractable P vs. P removal in greenhouse study. § pH measured in 0.01 *M* $CaCl_2$.
‡ Series means used to calculate *r*; No. of locations = 80.

Table 7-6. Range in possible correlation coefficients between y and z calculated from given correlation coefficients between x and z (Schulte & Hodgson, 1987).

Simple correlation		Range in correlation
(x) Test 1*(z) yield	(x)Test 1*(y) Test 2	between test 2(y) & yield (z)
	0.98	0.87–0.99
	0.95	0.81–0.99
0.95	0.90	0.72–0.99
	0.85	0.64–0.97
	0.80	0.57–0.95
	0.98	0.80–0.97
	0.95	0.72–0.99
0.90	0.90	0.62–0.99
	0.85	0.54–0.99
	0.80	0.46–0.98
	0.98	0.73–0.94
	0.95	0.64–0.97
0.85	0.90	0.54–0.99
	0.85	0.44–1.00
	0.80	0.36–1.00
	0.98	0.66–0.90
	0.95	0.57–0.95
0.80	0.90	0.46–0.98
	0.85	0.36–1.00
	0.80	0.28–1.00

Table 7-7. Correlation between Mehlich III (y) and Bray P_1 (x) extractable P for individual experimental field research sites (Thom, 1985; J.H. Grove, 1987, unpublished data) and many individual soil samples received by the Univ. of Kentucky Soil Test Laboratory.

Sample group	No. of observations	M(III) vs. Bray P_1	*r*
	Individual sites		
Maury silt loam	36	$y = 0.90x - 4.0$	0.996
Trappist silt loam	56	$y = 0.94x + 1.0$	0.979
Newark silt loam	18	$y = 1.29x + 9.3$	0.990
Belknap silt loam	160	$y = 1.43x + 3.9$	0.978
	Individual samples		
Soil test lab	510	$y = 1.20x - 1.1$	0.959

not predict how well correlated the new procedure will be with P uptake. Correlations between soil tests can, in some cases, be used to eliminate alternate P procedures but rarely should be used to accept new procedures. It is easy to appreciate the importance of maintaining soil sample collections for which plant P availability parameters are known.

V. CALIBRATION AND INTERPRETATION

General aspects of calibration and interpretation are covered in chapter 4 of this book by Dahnke and Olson. Factors more specific to soil test P calibration and interpretation will be discussed here. Calibration begins subsequent to correlation of the desired soil test procedure with plant P uptake, dry matter production (yield), crop quality, or other factors of interest.

A. Soil Test Phosphorus Calibration

Calibration of a soil test P procedure involves giving meaning to a numerical extractable P value in terms of soil P availability and the likelihood of crop response to fertilizer P additions. The extractable soil P level is only one of several factors influencing plant response to P fertilizer. If all other factors are held constant, an excellent relationship between soil test P and relative yield will be obtained. Easily done in the greenhouse; this often means one location of one soil series for 1 yr in field research.

Factors complicating calibration may or may not be under control of the manager. Tillage systems can influence the soil test P level thought critical for crop response. Fixen et al. (1987) reported adequate P nutrition to corn at lower soil test P levels for no-tillage than moldboard plowed systems. It is not yet clear whether the surface stratification of P associated with no-tillage or shallow surface tillage sufficiently reduces soil P contact to influence P availability. The presence/distribution of moisture conserving residues probably plays a key role in such effects. The current literature is not consistent, but some have concluded that shallow surface soil samples are adequate to characterize P availability in such systems (Touchton et al., 1982). Surface accumulation of P was less dramatic and less likely to result in greater uptake than was that for K in another report (Blevins et al., 1986) comparing no-tillage and moldboard plowing.

Mineralization of organic P may complicate soil test P calibration. Whether driven by differences in amount and quality of organic matter, tillage system (Vivekanandan & Fixen, 1988), or liming (Lathwell, 1979), organic P release may be involved in cases where little or no crop response to fertilizer P at low soil test P values has been found (Havlin et al., 1984).

Climate, especially moisture availability, can introduce variability in crop response to available P at a single location when several years of information are combined (Randall et al., 1986; Thom, 1985) for calibration purposes. Phosphorus acquisition is strongly related to soil moisture, temperature, and texture because the nutrient is largely immobile; moving to the root surface via diffusion in water films on particle surfaces.

The largest factor complicating the calibration process is differences in soils themselves. In practice, correlation work often results in a compromise. The soil test procedure chosen works reasonably well on most soils in the area for which the test will be used by the testing laboratory. Diverse soil physiographic regions are often present. The procedure does not always work on all soils (Varvel et al., 1981) and calibration often begins by first dividing soils into different groups (response categories) *prior* to determining the level(s) of extractable P critical to P availability. Combining as few as two soils often means that the calibration loses precision to accommodate modest differences in the crop-soil P response relationship (Randall et al., 1986).

Soil properties used in prerecommendation grouping are those related to either soil P quantity or soil P buffer capacity (Holford, 1980, 1982), especially texture or the quantity of active Fe and Al (Lins et al., 1985; Bahl & Singh, 1986). Soil clay content also influences soil P availability in calcareous soils, where the solubility of Ca-P compounds is thought important to P release (Olsen et al., 1983; Fixen & Ludwick, 1982).

Correlation methodology may confuse the calibration step. Should the single large initial P application needed on some soils be included in the calibration? Rather, should calibration be based on subsequent P availability behavior that is generally similar to that of other well-fertilized soils (Lathwell, 1979; Engelstad & Terman, 1980; Uehara & Gillman, 1981)? Some investigators include but 1 yr of correlation information (Peaslee, 1978), which precludes an evaluation of the hysteresis in soil test P as cropping continues and P fertilization ceases. Soil test P has been found to fall more rapidly per unit of P removed in soybean [*Glycine max* (L.) Merr.] grain than it rose per unit of fertilizer P applied (Thom, 1985).

Ultimately, calibration results in soil test values or ranges that predict crop response to P. Phosphorus soil tests are best at predicting the probability of P response, poorer at predicting the magnitude of any response, and weakest at determining the exact rate of fertilizer P needed for optimum economic performance in any given year or field. This outcome has resulted in many approaches to the conversion of extractable soil P values into P fertilizer rate recommendations. There is a degree of uncertainty and subjectiveness to all. In chapter 4, Dahnke and Olson discuss two calibration procedures, one involving continuous functions (Mitscherlich, quadratic, and logarithmic), the other discontinuous functions (Cate-Nelson, linear plateau). Both approaches result in only a single unambiguous P critical level. Arbitrary ranges corresponding to low, medium, and high P availability can and often are superimposed on these models, however. Sanchez (1976) also thoroughly reviews the topic.

B. Soil Tests as Relative Values

The use of class intervals like low, medium, and high is usually guided by crop response, fertilizer use economics, a practical sense of farm fertilizer use rates, and the vagaries of acquiring a representative sample. Some have advocated "index" systems; a series of discontinuous functions whose discontinuities occur at intervals corresponding to low, medium, etc. (Cope, 1973), or a continuous function akin to the Mitscherlich function in shape (Hatfield, 1972; Fisher, 1974). Use of the Mitscherlich function to describe growth response to immobile nutrients has been discussed by Melsted and Peck (1973).

Another development, the "boundary line approach" has recently been proposed (Evanylo & Sumner, 1987). A scatter diagram relating yield to a soil test nutrient level is developed and lines are drawn at the boundary of the scatter. This diagram tends to illuminate peak yield regions where nutrient availability is apparently optimal. Inspection of several of the diagrams contained in the Evanylo and Sumner paper (1987) suggests that a statistical approach defining the boundaries (and excluding outliers) would be useful and more suggestive of an actual response surface.

Further refinement of the yield × soil test nutrient populations to take "nutrient balance" into account was attempted by Evanylo et al. (1987). This was an extension of Diagnosis and Recommendation Integrated System (DRIS) methodology from plant tissue nutrient composition to available soil nutrient loading. The norms (developed for soybean on highly weathered soils) were not totally successful and the authors concluded that soil nutrient balance was not critical to soybean yield determination on these soils.

Evanylo and Sumner (1987) decry the general lack of information regarding nutrient optimization from factorial-type experiments, and suggest that the optimal level derived from single factor experiments is a characteristic of the basal availability of other nutrients. Factorial experimentation is useful but logistically cumbersome. Silva (1981) has compiled several alternative designs that provide a good deal more information on nutrient interactions than single factor trials without the resource demands of full-factorial designs.

C. Soil Test Phosphorus Interpretation/Recommendation

The final step, development of a P fertilizer rate recommendation, considers several factors in addition to the previously developed calibration relationship. First and foremost, a philosophical choice must be made. Is the soil or the crop to be fertilized? The former is often described as buildup and maintenance, correction and maintenance, or simple insurance of adequacy of soil P nutrition. The latter (fertilizing the crop) is termed *sufficiency level*, sufficient level of available nutrients (SLAN), or merely the sufficiency approach.

The correction and maintenance approach presumes that high levels of P nutrition should be maintained to maximize potential gain of all crops to be grown in the field. Once such high levels are reached, soil P availability is maintained by correcting for P removal as well as P fixation. This approach conserves soil P reserves for future generations, albeit at a high level of P availability. In practice, this approach is most economical on soils with little tendency to irretrievably fix applied P and are, therefore, corrected and maintained with modest rates of fertilizer P. If regular soil sampling is done, and nutrient removal monitored, this becomes part of a nutrient "log" for a given field. This approach does tend to minimize any potential adverse effects from the occasional unrepresentative sample as well as the uncertainty of P soil test interpretation in the gray area of medium P availability.

The sufficiency approach is based upon the need of the current crop, with some common sense consideration of other components of the crop rotation under use in the field. This approach emphasizes enhanced soil P solubility at the time of greatest crop demand rather than season-long P availability. It is generally used in conjunction with P fertilizer placement to further reduce soil P interaction and fixation. There is often a slow adjustment of the soil test P level to the optimum because the sufficiency concept recognizes the soil's competitive sinks for applied P at low soil test P levels. That optimum level is often in the medium range because of the interval between soil samples and the generally conservative nature of the approach.

Maintaining crop yield levels using sufficiency criterion and low to medium soil test P levels are well documented. Hooker et al. (1983) found that corn yields were maintained near the maximum by an annual application of 20 kg of P/ha and an average Bray P_1 value of 7 and 12 mg of P/kg on two soils for 20 and 12 yr, respectively. Touchton et al. (1982) observed that 32 kg of P/ha per yr was sufficient for wheat and double-cropped soybean each year for 3 yr. Soil test P levels were low to medium (Mehlich I extraction) at this rate. Whitney et al. (1985) examined the relationship between corn or soybean yield response and soil P (Mehlich I extraction) in Alabama and concluded little response was to be found at medium test levels. Cope (1981), reporting on 50 yr of response of five crops at six locations, determined that low to medium P availability was adequate for all these crops. In Nebraska, Olson et al. (1982) conducted a direct comparison of fertilizer recommendations from several laboratories differing in philosophy. Though surface soil P increased sharply at several locations receiving correction and maintenance P applications (as well as those of other nutrients), corn yields were not improved over those garnered with the more conservative sufficiency approach where P availability was maintained at medium levels.

The sufficiency approach is most appropriate when cash flow is limited and land tenure questionable. Environmental concerns must also be considered. Maintaining excessive soil P availability may contribute to surface water quality degradation should runoff or erosion occur.

The sufficiency approach requires more refinement and fertilizer recommendations based on this approach must consider additional factors. Some of these include differences in P source solubility, crop species, and differences in subsoil P availability among soils. Englestad and Terman (1980) ascribe differences in crop response because of fertilizer P source solubility primarily to differences in early growth of the crop in question. When the growing season is limiting such early growth, differences may be more important. Smyth and Sanchez (1982) found that rock phosphate was not able to provide adequate P for soybean on an Oxisol. Annual banded superphosphate applications were required as well.

Crop species are well known to vary in their fertilizer P response on the same soil. This observation takes on more significance when the sufficiency approach to fertilizer P recommendations is being used. Ozanc (1980) has summarized several mechanisms behind such observations. In general, most research suggests that crop P requirement increases in the order: soybean [*Glycine max* (L.) merr.] < grain sorghum [*Sorghum vulgare* (L.) Pers.] = corn [*Zea mays* (L.)] < alfalfa [*Medicago sativa* (L.)] < wheat [*Triticum aestivum* (L.)]. Oat [*Avena sativa* (L.)] and clover (*Triticum* sp.) vary quite widely in this series, depending upon the observer (deMooy et al., 1973; Peaslee, 1978; Hanway & Olson, 1980). Lathwell (1979) and Uehara and Gillman (1981) have summarized similar observations on more highly weathered soils.

Subsoil P status can influence crop P fertilization requirement. Plant uptake of subsoil P has been demonstrated (Murdock & Englebert, 1958; Pothulari et al., 1986). Differences in subsoil P availability have been observed, even mapped, in some states (Hanway & Olson, 1980; Schulte, 1986).

The P fertilizer rate recommendation for a field must be based on local correlation and calibration information for soils, crops, and cultural practices appropriate to the region. Specific interpretation and recommendation information is too voluminous for a general treatise of this nature. Such material can be obtained from soils research and extension personnel at most State agricultural experiment stations and extension services.

VI. OUTLOOK

Large sample numbers in some laboratories and the development of analytical techniques (e.g., flow injection or continuous flow analysis, ion chromatography, and plasma emission spectrometry) continue to create great interest in universal extractants as discussed earlier. Caution must be exercised to avoid sacrificing the predictive ability of a procedure to expedite laboratory operations. Likewise, we must be careful not to reject improved procedures that have greater predictive ability because they are awkward in a lab's existing operation.

Stewart and Sharpley (1987) recently wrote "The limitations of present soil test methods for P and S are that they only measure inorganic forms in the soil and do not properly account for the contribution from mineralization of organic forms and past management histories. . .The challenge to soil science is to develop a complete understanding of P and S cycles in the soil for the development of more firmly based predictive relationships." As cultural systems such as tillage and rotations become more diverse, this limitation will become more critical and must be addressed.

Since 1960, limited progress has been made in how we interpret P soil tests. However, during this period we have learned much about the soil P cycle, ion transport to roots, and mechanisms of plant uptake. The challenge for soil testing in the future will be to integrate this new knowledge into our interpretation programs. The demands placed on soil P testing and interpretations will be much greater in the future than they have been in the past. This demand will include precise supplemental P requirement predictions for economic and environmental reasons as well as increased demands for P testing for waste disposal guidelines in highly populated areas. Research must continue on improved procedures and more sophisticated interpretation if these demands are to be met.

REFERENCES

Adams, F. 1984. Crop response to lime in the southern United States. p. 211–265. *In* F. Adams (ed.) Soil acidity and liming. 2nd ed. Agronomy Monogr. 12. ASA, CSSA, and SSSA, Madison, WI.

Adepetu, J.A., and R.B. Corey. 1976. Organic phosphorus as a predictor of plant available P in soils of southern Nigeria. Soil Sci. 122:159–164.

Adepoju, A.Y., P.F. Pratt, and S.V. Mattigod. 1982. Availability and extractability of phosphorus from soils having high residual phosphorus. Soil Sci. Soc. Am. J. 46:583–588.

Adepoju, A.Y., P.F. Pratt, and S.V. Mattigod. 1986. Relationship between probable dominant phosphate compound in soil and phosphorus availability to plants. Plant Soil 92:47–54.

Admont, P.H., R. Boniface, J.C. Fardeau, M. Jahiel, and C. Morel. 1986. Currently used methods for measuring available soil P: Their use in assessing the fertilizer value of rock phosphates. Fert. Agric. 92:39–50.

Ainsworth, C.C., M.E. Sumner, and V.J. Hurst. 1985. Effect of aluminum substitution in goethite on phosphorus adsorption: I. Adsorption and isotopic exchange. Soil Sci. Soc. Am. J. 49:1142–1149.

Amarasiri, S.L., and S.R. Olsen. 1973. Liming as related to solubility of P and plant growth in an acid, tropical soil. Soil Sci. Soc. Am. Proc. 37:716–720.

Anderson, G. 1980. Assessing organic phosphorus in soils. p. 411–431. *In* F.E. Khasawneh et al. (ed.) The role of phosphorus in agriculture. ASA, CSSA, and SSSA, Madison, WI.

Anghinoni, I., V.C. Baligar, and S.A. Barber. 1981. Growth and uptake rates of P, K, Ca and Mg in wheat. J. Plant Nutr. 3:923–933.

Aquino, B.F., and R.G. Hanson. 1984. Soil phosphorus supplying capacity evaluated by plant removal and available phosphorus extraction. Soil Sci. Soc. Am. J. 48:1091–1096.

Bache, B.W. 1964. Aluminum and iron phosphate studies relating to soils. II. Reactions between phosphate and hydrous oxides. J. Soil Sci. 15:110–116.

Bache, B.W., and C. Ireland. 1980. Desorption of phosphate from soils using anion exchange resins. J. Soil Sci. 31:297–306.

Bahl, G.S., and N.T. Singh. 1986. Phosphorus diffusion in soils in relation to some edaphic factors and its influence on P uptake by maize and wheat. J. Agric. Sci. 107:335–341.

Baker, D.E., and J.K. Hall. 1967. Measurements of phosphorus availability in acid soils of Pennsylvania. Soil Sci. Soc. Am. J. 31:662–667.

Baligar, V.C., and S.A. Barber. 1979. Genotypic differences of corn for ion uptake. Agron. J. 71:870–873.

Barber, S.A. 1984. Soil nutrient bioavailability. John Wiley and Sons, New York.

Barber, S.A., and A.D. MacKay. 1985. Sensitivity analysis of the parameters of a mechanistic mathematical model affected by changing moisture. Agron. J. 77:528–931.

Barrow, N.J. 1980a. Differences among some North American soils in the rate of reaction with phosphate. J. Environ. Qual. 9:644–648.

Barrow, N.J. 1980b. Evaluation and utilization of residual phosphorus in soils. *In* F.E. Khasawneh et al. (ed.) The role of phosphorus in agriculture. ASA, CSSA, and SSSA, Madison, WI.

Barrow, N.J. 1983. On the reversibility of phosphate sorption by soils. J. Soil Sci. 34:751–758.

Bingham, F.T. 1962. Chemical soil tests for available phosphorus. Soil Sci. 94:87–95.

Blanchar, R.W., and A.C. Caldwell. 1964. Phosphorus uptake by plants and readily extractable phosphorus in soils. Agron. J. 56:218–221.

Blanchar, R.W., and G.K. Stearman. 1984. Ion products and solid-phase activity to describe phosphate sorption by soils. Soil Sci. Soc. Am. J. 48:1253–1258.

Blevins, R.L., J.H. Grove, and B.K. Kitur. 1986. Nutrient uptake of corn grown using moldboard plow or no-tillage soil management. Commun. Soil Sci. Plant Anal. 17:401–417.

Bloom, P.R. 1981. Phosphorus adsorption by an aluminum peat complex. Soil Sci. Soc. Am. J. 45:267–272.

Bowman, R.A., and C.V. Cole. 1978. An exploratory method for fractionation of organic phosphorus from grassland soils. Soil Sci. 125:95–101.

Bowman, R.A., S.R. Olsen, and F.S. Watanabe. 1978. Greenhouse evaluation of residual phosphate by four phosphorus methods in neutral and calcareous soils. Soil Sci. Soc. Am. J. 42:451–454.

Bray, R.H., and L.T. Kurtz. 1945. Determination of total, organic and available forms of phosphorus in soils. Soil Sci. 59:39–45.

Breland, H.L., and F.A. Sierra. 1962. A comparison of the amounts of phosphorus removed from different soils by various extractants. Soil Sci. Soc. Am. Proc. 26:348–350.

Chang, S.C., and M.L. Jackson. 1957. Fractionation of soil phosphorus. Soil Sci. 84:133–144.

Chang, S.C., and A.S.R. Juo. 1963. Available phosphorus in relation to forms of phosphorus in the soils. Soil Sci. 95:91–96.

Claassen, N., and S.A. Barber. 1976. Simulation model for nutrient uptake from soil by a growing plant root system. Agron. J. 68:961–964.

Cole, C.V., G.S. Innis, and J.W.B. Stewart. 1977. Simulation of phosphorus cycling in semi-arid grasslands. Ecology 58:1–15.

Coleman, N.T., J.T. Thorup, and W.A. Jackson. 1960. Phosphate sorption reactions that involve exchangeable Al. Soil Sci. 90:1–7.

Cope, J.T., Jr. 1973. Use of a fertility index in soil test interpretation. Commun. Soil Sci. Plant Anal. 4:137–146.

Cope, J.T., Jr. 1981. Effects of 50 years of fertilization with phosphorus and potassium on soil test levels and yields at six locations. Soil Sci. Soc. Am. J. 45:342–347.

Cox, F.R., E.J. Kamprath, and R.E. McCollum. 1981. A descriptive model of soil test nutrient levels following fertilization. Soil Sci. Soc. Am. J. 45:529–532.

Cox, F.R., and I.D.G. Lins. 1984. A phosphorus soil test interpretation for corn grown on acid soils varying in crystalline clay content. Commun. Soil Sci. Plant Anal. 15:1481–1491.

Dalal, R.C. 1977. Soil organic phosphorus. Adv. Agron. 29:83–117.

Dalal, R.C., and E.G. Hallsworth. 1976. Evaluation of the parameters of soil phosphorus availability factors in predicting yield response and phosphorus uptake. Soil Sci. Soc. Am. J. 40:541–545.

deMooy, C.J., J.L. Young, and J.D. Kaap. 1973. Comparative response of soybeans and corn to phosphorus and potassium. Agron. J. 65:851–855.

Dick, W.A. 1983. Organic carbon, nitrogen, and phosphorus concentrations and pH in soil profiles as affected by tillage intensity. Soil Sci. Soc. Am. J. 47:102–107.

Dyer, B. 1894. On the analytical determination of probable available mineral plant food in soils. Trans. Chem. Soc. 65:115–167.

Egner, H., H. Riehm, and W.R. Domingo. 1960. Untersuchungen uber die chemishe bodenanalyse als grundlage fur die beurteilung des nahrstoffzustandes der boden. II. Chemische extraktions-methoden zur phosphor—und kalimbestimmung kungl. Lantbrukshoegsk. Ann. 26:204–209.

Eid, M.T., C.A. Black, and O. Kempthorne. 1951. Importance of soil organic and inorganic phosphorus to plant growth at low and high soil temperatures. Soil Sci. 71:361–370.

Eik, K., and J.J. Hanway. 1986. Simultaneous extraction of P and K from Iowa soils with Bray 1 solution containing NH_4Cl. Commun. Soil Sci. Plant Anal. 17:1203–1225.

Engelstad, D.P., and G.L. Terman. 1980. Agronomic effectiveness of phosphate fertilizers. p. 311–332. *In* F.E. Khasawneh et al. (ed.) The role of phosphorus in agriculture. ASA, CSSA, and SSSA, Madison, WI.

Enwezor, W.O. 1977. Soil testing for phosphorus in some Nigerian soils 3. Forms of phosphorus in soils of southeastern Nigeria and their relationship to plant available phosphorus. Soil Sci. 124:27–33.

Evanylo, G.K., and M.E. Sumner. 1987. Utilization of the boundary line approach in the development of soil nutrient norms for soybean production. Commun. Soil Sci. Plant Anal. 18:1379–1401.

Evanylo, G.K., M.E. Sumner, and W.S. Letzsch. 1987. Preliminary development and testing of DRIS soil norms for soybean production. Commun. Soil Sci. Plant Anal. 18:1355–1377.

Farina, M.P.W., and P. Channon. 1979. A comparison of several P availability indexes. Gewasproduksie 8:165–169.

Fisher, T.R. 1974. Some considerations for interpretation of soil tests for phosphorus and potassium. Univ. Missouri Agric. Exp. Stn. Res. Bull. 1007.

Fixen, P.E., and P.L. Carson. 1978. Relationship between soil test and small grain response to P fertilization in field experiments. Agron. J. 70:838–844.

Fixen, P.E., B.G. Faber, and M. Vivekanandan. 1987. Phosphorus management as influenced by tillage systems in eastern South Dakota. p. 201–212. *In* J.T. Batchelor (ed.) Proc. Symp. Fluid Fert. Foundation, Clearwater Beach, FL. 16–18 Mar. Natl. Fert. Solution Assoc., St. Louis.

Fixen, P.E., and A.E. Ludwick. 1982. Residual available phosphorus in near-neutral and alkaline soils: II. Persistence and quantitative estimation. Soil Sci. Soc. Am. J. 46:335–338.

Fixen, P.E., A.E. Ludwick, and S.R. Olsen. 1983. Phosphorus and potassium fertilization of irrigated alfalfa on calcareous soils: II. Soil phosphorus solubility relationships. Soil Sci. Soc. Am. J. 47:112–117.

Fox, R.L., R.K. Nishimoto. J.R. Thompson, and R.S. de la Pena. 1974. Comparative external phosphorus requirements of plants growing in tropical soils. Int. Congr. Soil Sci. Trans. 10th, 1974 4:232–239.

Freeman, J.S., and D.L. Rowell. 1981. The adsorption and precipitation of phosphate onto calcite. J. Soil Sci. 32:75-84.

Friesen, D.K., A.S.R. Juo, and M.H. Miller. 1980. Liming and lime-phosphorus-zinc interactions in two Nigerian Ultisols: I. Interactions in the soil. Soil Sci. Soc. Am. J. 44:1221-1226.

Goldberg, S., and G. Sposito. 1985. On the mechanism of specific phosphate adsorption by hydroxylated mineral surfaces: A review. Commun. Soil Sci. Plant Anal. 16:801-821.

Grava, J. 1975. Causes for variation in phosphorus soil tests. Commun. Soil Sci. Plant Anal. 6:129-138.

Griffin. G.F., and R.E. Lorton. 1970. Phosphorus availability on two soils as determined by several methods. Agron. J. 62:336-341.

Gunary, D., and C.D. Sutton. 1967. Soil factors affecting plant uptake of phosphate. J. Soil Sci. 18:167-173.

Hanlon, E.A., and G.V. Johnson. 1984. Bray/Kurtz, Mehlich III, AB/D and ammonium acetate extractions of P, K and Mg in four Oklahoma soils. Commun. Soil Sci. Plant Anal. 15:277-294.

Hanson, R.L., and D.G. Westfall. 1985. Orthophosphate solubility transformations and availability from dual applied nitrogen and phosphorus. Soil Sci. Soc. Am. J. 49:1283-1289.

Hanway, J.J., and R.A. Olson. 1980. Phosphate nutrition of corn, sorghum, soybeans, and small grain. p. 681-692. *In* F.E. Khasawneh et al. (ed.) The role of phosphorus in agriculture. ASA, CSSA, and SSSA, Madison, WI.

Harrison, R.B., and F. Adams. 1987. Solubility characteristics of residual phosphate in a fertilized and limed Ultisol. Soil Sci. Soc. Am. J. 51:963-969.

Hatfield, A.L. 1972. Soil test reporting: A nutrient index system. Commun. Soil Sci. Plant Anal. 3:425-436.

Havlin, J.L., and D.G. Westfall. 1984. Soil test phosphorus and solubility relationships in calcareous soils. Soil Sci. Soc. Am. J. 48:327-330.

Havlin, J.L., D.G. Westfall, and H.M. Golus. 1984. Six years of phosphorus and potassium fertilization of irrigated alfalfa on calcareous soils. Soil Sci. Soc. Am. J. 48:331-336.

Haynes, R.J. 1983. Effect of lime and phosphate applications on the adsorption of phosphate, sulfate, and molybdate by a Spodosol. Soil Sci. 135:221-227.

Haynes, R.J. 1984. Lime and phosphate in the soil-plant system. Adv. Agron. 37:249-315.

Hedley, M.J., J.W.B. Stewart, and B.S. Chauhan. 1982. Changes in inorganic soil phosphorus fractions induced by cultivation practices and by laboratory incubations. Soil Sci. Soc. Am. J. 46:970-976.

Helyar, K.R., D.A. Munns, and R.G. Barau. 1976. Adsorption of phosphate by gibbsite. II. Formation of a surface complex involving divalent cations. J. Soil Sci. 27:315-323.

Hingston, F.J., A.M. Posner, and J.P. Quirk. 1974. Anion adsorption by goethite and gibbsite. II. Desorption of anions from hydrous oxide surfaces. J. Soil Sci. 25:16-26.

Holford, I.C.R. 1977. Soil properties related to phosphate buffering in calcareous soils. Commun. Soil Sci. Plant Anal. 8:125-137.

Holford, I.C.R. 1980. Greenhouse evaluation of four phosphorus soil tests in relation to phosphate buffering and labile phosphate in soils. Soil Sci. Soc. Am. J. 44:555-559.

Holford, I.C.R. 1982. Effects of phosphate sorptivity on long-term plant recovery and effectiveness of fertilizer phosphate in soils. Plant Soil 64:225-236.

Hooker, M.L., R.E. Gwin, G.M. Herron, and P. Gallagher. 1983. Effects of long-term, annual applications of N and P on corn grain yields and soil chemical properties. Agron. J. 75:94-99.

Hooker, M.L., G.A. Peterson, D.H. Sander, and L.A. Daigger. 1980. Phosphate fractions in calcareous soils as altered by time and amounts of added phosphate. Soil Sci. Soc. Am. J. 44:269-277.

International Soil Fertility Evaluation and Improvement Program. 1972. Annual report. Soil Sci. Dep. North Carolina State Univ., Raleigh, NC.

Itoh, S., and S.A. Barber. 1983. Phosphorus uptake by six plant species as related to root hairs Agron. J. 75:457-461.

Janssen, B.H., D.J. Lathwell, and J. Wolf. 1987. Modeling long term crop response to fertilizer phosphorus. II. Comparison with field results. Agron. J. 79:452-458.

John, M.K., A.L. vanRyswyk, and J.L. Mason. 1967. Effect of soil order, pH, texture and organic matter on the correlation between phosphorus in alfalfa and soil test values. Can. J. Soil Sci. 47:157-161.

Jones, C.A., C.V. Cole, A.N. Sharpley, and J.R. Williams. 1984a. A simplified soil and plant phosphorus model: I. Documentation. Soil Sci. Soc. Am. J. 48:800-805.

Jones, C.A., A.N. Sharpley, and J.R. Williams. 1984b. A simplified soil and plant phosphorus model: III. Testing. Soil Sci. Soc. Am. J. 48:810–813.

Jones, J.B., 1973. Soil testing in the United States. Commun. Soil Sci. Plant Anal. 4:307–322.

Kamprath, E.J., and M.E. Watson. 1980. Conventional soil and tissue tests for assessing the phosphorus status of soils. p. 433–469. *In* F.E. Khasawneh et al. (ed.) The role of phosphorus in agriculture. ASA, CSSA, and SSSA, Madison, WI.

Kovar, J.L., and S.A. Barber. 1988. Phosphorus supply characteristics of 33 soils as influenced by seven rates of phosphorus addition. Soil Sci. Soc. Am. J. 52:160–165.

Kuo, S., and E.J. Jellum. 1985. Evalution of four phosphorus soil tests and their relationship to corn yields for some western Washington soils. Washington State Univ. Res. Bull. 0965.

Labhsetwar, V.K., and P.N. Soltanpour. 1985. A comparison of NH_4HCO_3-DTPA, $NaHCO_3$, $CaCl_2$, and Na_2-EDTA soil tests for phosphorus. Soil Sci. Soc. Am. J. 49:1437–1440.

Larsen, S., D. Gunary, and C.D. Sutton. 1965. The rate of immobilization of applied phosphate in relation to soil properties. J. Soil Sci. 16:142–148.

Lathwell, D.J. (ed.). 1979. Phosphorus response on Oxisols and Ultisols. Cornell Univ. Int. Agric. Bull. 33.

Lathwell, D.J., N. Sanchez, D.J. Lisk, and P. Peech. 1958. Availability of soil phosphorus as determined by several chemical methods. Agron. J. 50:366–369.

Lee, Y.S., and R.J. Bartlett. 1977. Assessing phosphorus fertilizer needs based on intensity-capacity relationships. Soil Sci. Soc. Am. J. 41:710–712.

Lindsay, W.L., and P.L.G. Vlek. 1977. Phosphate minerals. p. 639–692. *In* J.B. Dixon and S.B. Weed (ed.) Minerals in soil environments. SSSA, Madison, WI.

Lins, I.D.G., F.R. Cox, and J.J. Nicholaides, III. 1985. Optimizing phosphorus fertilization rates for soybeans grown on Oxisols and associated Entisols. Soil Sci. Soc. Am. J. 49:1457–1460.

Luscombe, P.C., J.K. Syers, and P.E.H. Gregg. 1979. Water extraction as a soil testing procedure for phosphate. Commun. Soil Sci. Plant Anal. 10:1361–1369.

Mack, A.R., and S.A. Barber. 1960. Influence of temperature and moisture on soil phosphorus: II. Effect prior to and during cropping on soil phosphorus availability for millet. Soil Sci. Soc. Am. J. 24:482–484.

MacKay, A.D., and S.A. Barber. 1985a. Effect of soil moisture and phosphate level on root hair growth of corn roots. Plant Soil 86:321–331.

MacKay, A.D., and S.A. Barber. 1985b. Soil moisture effects on root growth and phosphorus uptake by corn. Agron. J. 77:519–523.

Maida, J.H.A. 1978. Phosphate availability indices related to phosphate fractions in selected Malawi soils. J. Sci. Food Agric. 29:423–428.

Malo, D.D., and R.H. Gelderman. 1984. Portable soil test laboratory results compared to standard soil test values. Commun. Soil Sci. Plant Anal. 15:909–927.

Martin, R.R., and R.St.C. Smart. 1987. X-ray photoelectron studies of anion adsorption on goethite. Soil Sci. Soc. Am. J. 51:54–56.

Matar, A.E., and M. Samman. 1975. Correlations between $NaHCO_3$-extractable P and response to P fertilization in pot tests. Agron. J. 67:616–618.

McCallister, D.L., C.A. Shapiro, W.R. Raun, F.N. Anderson, G.W. Rehm. O.P. Engelstad, M.P. Russelle, and R.A. Olson. 1987. Rate of phosphorus and potassium buildup/decline with fertilization for corn and wheat on Nebraska Mollisols. Soil Sci. Soc. Am. J. 51:1646–1652.

McGill, W.B., and C.V. Cole. 1981. Comparative aspects of cycling of organic C, N, S and P through soil organic matter. Geoderma 26:267–286.

McLean, E.O. 1985. Improved lime, equilibration soil test methods for assessing buffering or fixation tendencies of individual soils. Commun. Soil Sci. Plant Anal. 16:1229–1257.

McLean, E.O., T.O. Oloya, and S. Mostaghimi. 1982. Improved corrective fertilizer recommendations based on a two-step alternative usage of soil tests: I. Recovery of soil-equilibrated phosphorus. Soil Sci. Soc. Am. J. 46:1193–1197.

Mehlich, A. 1953. Determination of P, Ca, Mg, K, Na and NH_4. North Carolina Soil Testing Div. Mimeo, Raleigh.

Mehlich, A. 1978a. Influence of fluoride, sulfate and acidity on extractable phosphorus, calcium, magnesium and potassium. Commun. Soil Sci. Plant Anal. 9:455–476.

Mehlich, A. 1978b. New extractant for soil test evaluation of phosphorus, potassium, magnesium, calcium, sodium, manganese and zinc. Commun. Soil Sci. Plant Anal. 9:477–492.

Mehlich, A. 1984. Mehlich 3 soil test extractant: A modification of Mehlich 2 extractant. Commun. Soil Sci. Plant Anal. 15:1409–1416.

Melsted, S.W., and T.R. Peck. 1977. The Mitscherlich-Bray growth function. p. 1-18. *In* T.R. Peck et al. (ed.) Soil testing: Correlating and interpreting the analytical result. ASA Spec. Publ. 29. ASA, CSSA, and SSSA, Madison, WI.

Menon, R.G., L.L. Hammond, and H.A. Sissingh. 1989. Determination of plant-available phosphorus by the P_i soil test. Soil Sci. Soc. Am. J. 53:110-115.

Michaelson, G.J., and C.L. Ping. 1986. Extraction of phosphorus from the major agricultural soils of Alaska. Commun. Soil Sci. Plant Anal. 17:275-297.

Mokwunye, U. 1975. The influence of pH on the adsorption of phosphate by soils from the Guinea and Sudan savannah zones of Nigeria. Soil Sci. Soc. Am. Proc. 39:1100-1102.

Morgan, M.F. 1941. Chemical soil diagnosis by the universal testing system. Connecticut Agric. Exp. Stn. Bull. 450.

Murdock, J.T., and L.E. Engelbert. 1958. The importance of subsoil phosphorus to corn. Soil Sci. Soc. Am. Proc. 22:53-57.

Nesse, P., and J. Grava. 1986. Correlation of several tests for phosphorus with resin extractable phosphorus on alkaline soils. p. 218-220. *In* A report on field research in soils. Univ. of Minnesota Agric. Exp. Stn. Misc. Publ. 2 (revised).

Novais, R., and E.J. Kamprath. 1978. Phosphorus supplying capacities of previously heavily fertilized soils. Soil Sci. Soc. Am. J. 42:931-935.

Nye, P.H., and P.B. Tinker. 1977. Solute movement in the soil-root system. Univ. of California Press, Berkeley.

O'Halloran, I.P., J.W.B. Stewart, and R.G. Kachanoski. 1987. Influence of texture and management practices on the forms and distribution of soil phosphorus. Can. J. Soil Sci. 57:147-163.

Oko, B.F.D., and A.A. Agboola. 1974. Comparison of different phosphorus extractants in soils of the Western State of Nigeria. Agron. J. 66:39-642.

Olsen, S.R., R.A. Bowman, and F.S. Watanabe. 1977. Behavior of phosphorus in the soil and interactions with other nutrients. Phosphorus Agric. 70:31-46.

Olsen, S.R., C.V. Cole, F.S. Watanabe, and L.A. Dean. 1954. Estimation of available phosphorus in soils by extraction with sodium bicarbonate. USDA Circ. 939. U.S. Gov. Print. Office, Washington, DC.

Olsen, S.R., and F.E. Khasawneh. 1980. Use and limitations of physical-chemical criteria for assessing the status of phosphorus in soils. p. 361-410. *In* F.E. Khasawneh et al. (ed.) The role of phosphorus in agriculture. ASA, CSSA, and SSSA, Madison, WI.

Olsen, S.R., and L.E. Sommers. 1982. Phosphorus. p. 403-430. *In* A.L. Page et al. (ed.) Methods of soil analysis. Part 2. 2nd ed. Agronomy Monogr. 9. ASA and SSSA, Madison, WI.

Olsen, S.R., and F.S. Watanabe. 1963. Diffusion of phosphorus as related to soil texture and plant uptake. Soil Sci. Soc. Am. Proc. 27:648-653.

Olsen, S.R., F.S. Watanabe, and R.A. Bowman. 1983. Evaluation of fertilizer phosphorus residues by plant uptake and extractable phosphorus. Soil Sci. Soc. Am. J. 47:952-958.

Olsen, S.R., F.S. Watanabe, H.R. Cosper, W.E. Larson, and L.B. Nelson. 1954. Residual phosphorus availability in long-time rotations on calcareous soils. Soil Sci. 78:141-151.

Olson, R.A., K.D. Frank, P.H. Grabouski, and G.W. Rehm. 1982. Economic and agronomic impacts of varied philosophies of soil testing. Agron. J. 74:492-499.

Olson, R.A., M.B. Rhodes, and A.F. Drier. 1954. Available phosphorus status of Nebraska soils in relation to series classification, time of sampling and method of measurement. Agron. J. 46:175-180.

Onken, A.B., R. Matheson, and E.J. Williams. 1980. Evaluation of EDTA-extractable phosphorus as a soil test procedure. Soil Sci. Soc. Am. J. 44:783-786.

Ozanne, P.G. 1980. Phosphate nutrition of plants—a general treatise. p. 559-589. *In* F.E. Khasawneh et al. (ed.) The role of phosphorus in agriculture. ASA, CSSA, and SSSA, Madison, WI.

Peaslee, D.E. 1978. Relationships between relative crop yields, soil test phosphorus levels, and fertilizer requirements for phosphorus. Commun. Soil Sci. Plant Anal. 9:429-442.

Peaslee, D.E., and R.L. Fox 1978. Phosphorus fertilizer requirements as estimated by phosphate sorption. Commun. Soil Sci. Plant Anal. 9:975-993.

Pothulari, J.V., D.E. Kissel, D.A. Whitney, and S.J. Thien. 1986. Phosphorus uptake from soil layers having different soil test phosphorus levels. Agron. J. 78:991-994.

Power, J.F., P.L. Brown, T.J. Army, and M.G. Klages. 1961. Phosphorus responses by dryland spring wheat as influenced by moisture supplies. Agron. J. 53:106-108.

Randall, G.W., S.D. Evans, and W.W. Nelson. 1986. High phosphorus and potassium rates in a corn-soybean rotation. p. 153-164. *In* A report on field research in soils. Univ. of Minnesota Agric. Exp. Stn. Misc. Publ. 2.

Randall, G.W., and J. Grava. 1971. Effect of soil: Bray no. 1 ratios on the amount of phosphorus extracted from calcareous Minnesota soils. Soil Sci. Soc. Am. Proc. 35:112–114.

Rixon, A.J. 1966. Soil fertility changes in a red-brown earth under irrigated pastures II. Changes in phosphorus. Aust. J. Agric. Res. 17:317–325.

Robarge, W.P., and R.B. Corey. 1979. Adsorption of phosphate by hydroxy-aluminum species on a cation exchange resin. Soil Sci. Soc. Am. J. 43:481–487.

Ryan, J., H.M. Hasan, M. Baasiri, and H.S. Tabbara. 1985. Availability and transformation of applied phosphorus in calcareous Lebanese soils. Soil Sci. Soc. Am. J. 49:1215–1220.

Ryden, J.C., J.R. McLaughlin, and J.K. Syers. 1977. Mechanisms of phosphate sorption by soils and hydrous ferric oxide gel. J. Soil Sci. 28:72–92.

Sahrawat, K.L. 1977. EDTA extractable phosphorus in soils as related to available and inorganic phosphorus forms. Commun. Soil Sci. Plant Anal. 8:281–287.

Sample, E.C., R.J. Soper, and G.J. Racz. 1980. Reactions of phosphate fertilizers in soils. p. 263–310. *In* F.E. Khasawneh et al. (ed.) The role of phosphorus in agriculture. ASA, CSSA, and SSSA, Madison, WI.

Sanchez, P.A. 1976. Properties and management of soils in the Tropics. John Wiley and Sons, New York.

Schenk, M.K., and S.A. Barber. 1980. Potassium and phosphorus uptake by corn genotypes grown in the field as influenced by root characteristics. Plant Soil 54:65–76.

Schofield, R.K. 1955. Can a precise meaning be given to 'available' soil phosphorus? Soils Fert. 18:373–375.

Schulte, E.E. 1986. How can dealers react to very high soil tests? p. 147–150. *In* K.A. Kelling (ed.) Proc. 1986 Fertilizer, Ag Lime Pest Manage. Conf., Madison, WI. 21–23 Jan. Univ. Wisconsin Coop. Ext. Serv., Madison, WI.

Schulte, E.E., and P.R. Hodgson. 1987. Suitability of the Mehlich-3 extractant for multielement analysis in soils of the North-Central states. *In* T.R. Peck (ed.) Proc. 11th Soil-Plant Analyst's Workshop, St. Louis. 27–28 Oct. Univ. of Illinois, Urbana.

Sen Tran, T., and Giroux, M. 1985. Comparison de differentes methodes d'extraction du P assimilable en relation avec les proprietes chimiques et physiques des sols du Quebec. Can. J. Soil Sci. 65:35–46.

Sharpley, A.N. 1985. Phosphorus cycling in unfertilized and fertilized agricultural soils. Soil Sci. Soc. Am. J. 49:905–911.

Sharpley, A.N. 1986. Disposition of fertilizer phosphorus applied to winter wheat. Soil Sci. Soc. Am. J. 50:953–958.

Sharpley, A.N., C.A. Jones, C. Gray, and C.V. Cole. 1984. A simplified soil and plant phosphorus model: II. Prediction of labile, organic, and sorbed phosphorus. Soil Sci. Soc. Am. J. 48:805–809.

Sharpley, A.N., and S.J. Smith. 1985. Fractionation of inorganic and organic phosphorus in virgin and cultivated soils. Soil Sci. Soc. Am. J. 49:127–130.

Sharpley, A.N., H. Tiessen, and C.V. Cole. 1987. Soil phosphorus forms extracted by soil tests as a function of pedogenesis. Soil Sci. Soc. Am. J. 51:362–365.

Sibbesen, E. 1983. Phosphate soil tests and their suitability to assess the phosphate status of soil. J. Sci. Food Agric. 34:1368–1374.

Silberbush, M., and S.A. Barber. 1983. Sensitivity of simulated phosphorus uptake to parameters used by a mechanistic-mathematical model. Plant Soil 74:93–100.

Silva, J.A. (ed.). 1981. Experimental designs for predicting crop productivity with environmental and economic inputs for agrotechnology transfer. Dep. Paper 49. Hawaii Inst. Trop. Agric. Human Resources, Univ. of Hawaii, Honolulu.

Sims, J.T., and B.G. Ellis. 1983a. Adsorption and availability of phosphorous following the application of limestone to an acid, aluminous soil. Soil Sci. Soc. Am. J. 47:888–893.

Sims, J.T., and B.G. Ellis. 1983b. Changes in phosphorus adsorption associated with aging of aluminum hydroxide suspensions. Soil Sci. Soc. Am. J. 47:912–916.

Smeck, N.E. 1985. Phosphorus dynamics in soils and landscapes. Geoderma 36:185–199.

Smillie, G.W., D. Curtin, and J.K. Syers. 1987. Influence of exchangeable calcium on phosphate retention by weakly acid soils. Soil Sci. Soc. Am. J. 51:1169–1172.

Smyth, T.J., and P.A. Sanchez. 1980. Effects of lime, silicate, and phosphorus applications to an Oxisol on phosphorus sorption and ion retention. Soil Sci. Soc. Am. J. 44:500–505.

Smyth, T.J., and P.A. Sanchez. 1982. Phosphate rock and superphosphate combinations for soybeans in a Cerrado Oxisol. Agron. J. 74:730–735.

Soltanpour, P.N., F. Adams, and A.C. Bennett. 1974. Soil phosphorus availability as measured by displaced soil solutions, calcium-chloride extracts, dilute-acid extracts, and labile phosphorus. Soil Sci. Soc. Am. J. 38:225–228.

Soltanpour, P.N., and A.P. Schwab. 1977. A new soil test for simultaneous extraction of macro- and micro-nutrients in alkaline soils. Commun. Soil Sci. Plant Anal. 8:195–207.

Soper, R.J., and I.H.M. El Bagouri. 1964. The effect of soil carbonate level on the availability of added and native phosphorus in some calcareous soils. Can. J. Soil Sci. 44:337–344.

Southern Regional Soil Testing Information Exchange Group. 1984. Procedures used by state soil testing laboratories in the southern region of the United States. Oklahoma State Univ. Agric. Exp. Stn. South. Coop. Ser. Bull. 190.

Stewart, J.W.B., and A.N. Sharpley. 1987. Controls on dynamics of soil and fertilizer phosphorus and sulfur. p. 101–121. *In* Soil fertility and organic matter as critical components of production systems. SSSA Spec. Publ. 19. ASA and SSSA, Madison, WI.

Sutton, C.D. 1969. Effect of low soil temperature on phosphate nutrition of plants—A review. J. Sci. Food Agric. 20:1–3.

Taylor, R.W., and B.G. Ellis. 1978. A mechanism of phosphate adsorption on soil and anion exchange resin surfaces. Soil Sci. Soc. Am. J. 42:432–436.

Thom, W.O. 1985. Soil test interpretation with corn and soybeans on a Belknap silt loam. Univ. Kentucky Agric. Exp. Stn. Bull. 720.

Thomas, G.W., and D.E. Peaslee. 1973. Testing soils for phosphorus. p. 115–132. *In* L.M. Walsh and J.D. Beaton (ed.) Soil testing and plant analysis. Revised ed. SSSA, Madison, WI.

Thompson, L.F., and P.F. Pratt. 1954. Solubility of phosphorus in chemical extractants as indexes to available phosphorus in Ohio soils. Soil Sci. Soc. Proc. 18:467–470.

Tiessen, H., and J.W.B. Stewart. 1983. Particle-size fractions and their use in studies of soil organic matter: II. Cultivation effects on organic matter composition in size. Soil Sci. Soc. Am. J. 47:509–514.

Tinker, P.B. 1984. The role of microorganisms in mediating and facilitating the uptake of plant nutrients from soil. Plant Soil 76:77–91.

Touchton, J.T., W.L. Hargrove, R.R. Sharpe, and F.C. Boswell. 1982. Time, rate, and method of phosphorus application for continuously double-cropped wheat and soybeans. Soil Sci. Soc. Am. J. 46:861–864.

Traina, S.J., G. Sposito, D. Hesterberg, and U. Kafkafi. 1986a. Effects of ionic strength, calcium, and citrate on orthophosphate solubility in an acidic, montmorillonitic soil. Soil Sci. Soc. Am. J. 50:623–627.

Traina, S.J., G. Sposito, D. Hesterberg, and U. Kafkafi. 1986b. Effects of pH and organic acids on organophosphate solubility in an acidic, montmorillonitic soil. Soil Sci. Soc. Am. J. 50:45–52.

Truog, E. 1930. Determination of the readily available phosphorus of soils. J. Am. Soc. Agron. 22:874–882.

Uehara, G., and G.P. Gillman. 1981. The mineralogy, chemistry and physics of tropical soils with variable charge. Westview Press, Boulder, CO.

van der Zee, S.E.A.T.M., L.G.J. Fokkink, and W.H. van Riemsdijk. 1987. A new technique for assessment of reversibly adsorbed phosphate. Soil Sci. Soc. Am. J. 51:599–604.

van Diest, A., 1963. Soil test correlation studies on New Jersey soils: 1. Comparison of seven methods for measuring labile inorganic soil phosphorus. Soil Sci. 96:261–266.

van Raij, B., J.A. Quaggio, and N.M. da Silva. 1986. Extraction of phosphorus, potassium, calcium, and magnesium from soils by an ion exchange resin procedure. Commun. Soil Sci. Plant Anal. 17:547–566.

Varvel, G.E., F.N. Anderson, and G.A. Peterson. 1981. Soil test correction problems with two phosphorus methods on similar soils. Agron. J. 73:516–520.

Veith, J.A., and G. Sposito. 1977. Reactions of aluminosilicates, aluminum hydrous oxides, and aluminum oxide with o-phosphate: the formation of X-ray amorphous analogs of variscite and montebrasite. Soil Sci. Soc. Am. J. 41:870–876.

Velayutham, M. 1980. The problem of phosphate fixation by minerals and colloids. Phosphorus Agric. 77:1–8.

Vivekanandan, M., and P.E. Fixen. 1988. Cropping system effects on phosphorus response of corn. Proc. 18th North Central Ext.-Ind. Soil Fert. Workshop. 4:95–103.

Weil, R.R., P.W. Benedetto, L.J. Sikora, and V.A. Bandel. 1988. Influence of tillage practices on phosphorus distribution and forms in three Ultisols. Agron. J. 80:503–509.

Wendt, R.C., and R.B. Corey. 1981. Available P determination by equilibration with dilute $SrCl_2$. Commun. Soil Sci. Plant Anal. 12:557–568.

White, R.E. 1976. Concepts and methods in the measurement of isotopically exchangeable phosphate in soil. Phosphorus Agric. 67:9–16.

White, R.E. 1980. Retention and release of phosphate by soil and soil constituents. p. 71–114. *In* P.B. Tinker (ed.) Soils and agriculture. John Wiley and Sons, New York.

White, R.E., and G.W. Thomas. 1981. Hydrolysis of aluminum on weakly acidic organic exchanges: Implications for phosphate adsorption. Fert. Res. 2:159–167.

Whitney, D.A., J.T. Cope, and L.F. Welch. 1985. Prescribing soil and crop nutrient needs. p. 25–52. *In* O.P. Engelstad (ed.) Fertilizer technology and use. 3rd ed. SSSA, Madison, WI.

Wild, A., 1950. The retention of phosphate by soil. A review. J. Soil Sci. 1:221–238.

Williams, E.G. 1966. The intensity and quantity aspects of phosphate status and laboratory extraction values. An. Edafol. Agrobiol. 26:525–546.

Wolf, A.M., and D.E. Baker. 1985. Comparisons of soil test phosphorus by Olsen, Bray PI, Mehlich I and Mehlich III methods. Commun. Soil Sci. Plant Anal. 16:467–484.

Wolf, A.M., D.E. Baker, and H.B. Pionke. 1986. The measurement of labile phosphorus by the isotopic dilution and anion resin methods. Soil Sci. 141:60–70.

Wolf, J., C.T. deWit, B.H. Janssen, and D.J. Lathwell. 1987. Modeling long-term crop response to fertilizer phosphorus. I. The model. Agron. J. 79:445–451.

Woodruff, J.R., and E.J. Kamprath. 1965. Phosphorus adsorption maximum as measured by the Langmuir isotherm and its relationship to phosphorus availability. Soil Sci. Soc. Am. J. 29:148–150.

Zubriski, J.C. 1971. Relationships between forms of soil phosphorus, some indexes of phosphorus availability and growth of sudangrass in greenhouse trials. Agron. J. 63:421–425.

Yao, J., and S.A. Barber. 1986. Effect of one phosphorus rate placed in different soil volumes on P uptake and growth of wheat. Commun. Soil Sci. Plant Anal. 16:467–484.

Chapter 8

Testing Soils for Potassium, Calcium, and Magnesium[1]

V. A. HABY, *Texas Agricultural Experimental Station, Texas A&M University System, Overton*

M. P. RUSSELLE, *USDA-ARS, University of Minnesota, St. Paul*

EARL O. SKOGLEY, *Montana State University, Bozeman*

High-yielding crops may contain quantities of K in excess of the amount of N. Potassium content in the aboveground portion of most forage, grain, oil, fruit, vegetable, and specialty crops ranges from a low of 40 to 50 kg of K ha^{-1} in flax (*Linum usitatissimum*) to more than 500 kg of K ha^{-1} in alfalfa (*Medicago sativa* L.), napiergrass (*Pennisetum purpureum* Schumach.), table beet (*Beta vulgaris* L.), and pineapple [*Ananas comosus* (L.) Merr.]. Banana crops (*Musa paradisiacac* L. var. *sapightum*) may contain 1400 kg of K ha^{-1} in aboveground plant parts. Most commercial crops, at near-maximum economic yield, contain 100 to 300 kg of K ha^{-1} in aerial parts. Generally, crop roots have proportionately equivalent requirements. When total crop demand is considered, soils must supply large quantities of K for plant uptake during rapid growth. The rate of K accumulation normally exceeds that of dry matter accumulation. This fact further stresses the soil's capacity to supply adequate K during certain stages of plant growth.

Crop plant Ca content will normally be less than one half as much as K. Deficiencies of Ca as a nutrient are uncommon. Neutral and alkaline soils normally contain adequate Ca, while acid soils are usually limed to provide a favorable pH for most crops. Crops with unusually high requirements for Ca (e.g., alfalfa and peanut, *Arachis hypogaea* L.), may require more Ca or specialized Ca management to produce top yields. Magnesium content may be only one-half or less as much as Ca. Magnesium deficiencies have been reported with increasing frequency in recent years. Magnesium deficiency in livestock (hypomagnesaemia, commonly known as grass tetany)

[1] Joint contribution of the USDA-ARS and the Texas, Minnesota, and Montana Agric. Exp. Stn. TAES Journal Article No. TA 24654.

can occur if forages are deficient in Mg or contain an imbalance of Mg with other cation nutrients.

Generalized areas of K-, Ca-, and Mg-responsive soils in the USA and Canada are mapped in Fig. 8–1 and 8–2. These maps, based on responses to a survey of public laboratories and on the literature (Whitney, 1975; Dumanski et al., 1982; Pearen, 1984; Henry, 1985; Mahler et al., 1985; Penney, 1985), reflect both the experience and the philosophy of soil-testing laboratory directors. They have been drawn to correspond in general with major soil boundaries and agricultural areas. However, divisions between areas of responsive and nonresponsive soils sometimes are arbitrary; that is, they may follow political boundaries, rather than recognized changes in soil characteristics.

There are instances of crop response to fertilizer K, Ca, and Mg that cannot be predicted from soil characteristics alone. For example, small grain responses to fertilizer K are highly dependent on seasonal weather conditions in Montana. This is due partly to soil water and temperature influences on

Fig. 8–1. Generalized areas of soil in the USA and Canada on which crops frequently respond to K applications (from a survey of public soil-testing laboratories conducted by Russelle). See text for further details.

the rate of K diffusion to plant roots (Schaff & Skogley, 1982a; Skogley & Schaff, 1985). Potassium deficiency in cotton (*Gossypium hirsutum* L.) has been induced by disease (Ashworth et al., 1982). It is difficult to separate the effect of Ca from the liming effect. Calcium is usually supplied as calcitic or dolomitic limestone to acid, potentially Ca-deficient soils. In the arid West, Ca responses are often found on soils of high Na content (e.g., in Alberta and California). Magnesium deficiencies can be due to high K/Mg ratios in soils (e.g., in Nevada), rather than to a lack of Mg per se.

Crops grown on sandy soils in many parts of North America frequently will respond to applications of K, Ca, and Mg, particularly under irrigated conditions. Crops also have significantly different requirements for these elements. A soil may have insufficient K supply for alfalfa, but maximum corn (*Zea mays* L.) grain yields may be achieved without additional K. Under these conditions, consider Fig. 8-1 and 8-2 *highly* generalized. They should not be used to predict crop responses in specific situations.

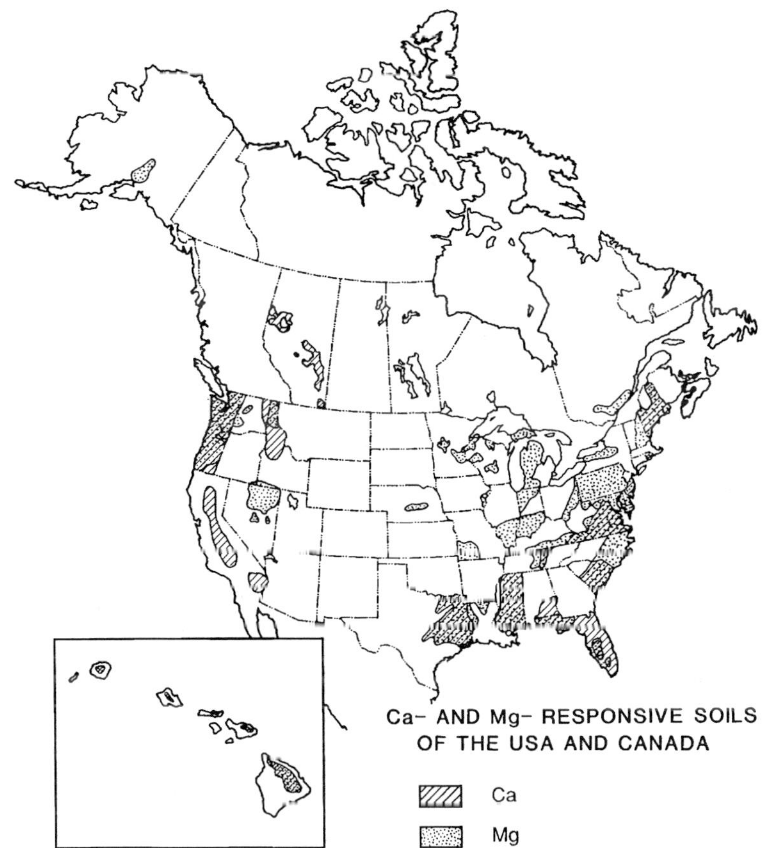

Fig. 8-2. Generalized areas of soils in the USA and Canada on which crops frequently respond to applications of Ca and Mg (from a survey of public soil-testing laboratories conducted by Russelle). See text for further details.

Analytical procedures for plant-available soil K, Ca, and Mg must provide a good relationship between test results and actual nutrient availability during the growing season for a specific crop. "Mass flow" and "diffusion" are the two processes through which nearly all of the K, Ca, and Mg ions are moved from their original position in the soil solution to the root surface where they can be taken up by the plant (see chapter 3 in this book, by James and Wells). Other equilibria and dynamic aspects of "exchangeable" or "slowly available" ions also influence true nutrient availability to crops. Nutrient extraction procedures that do not directly account for these soil processes and for the influence of soil, site, and weather conditions on each process will, at best, be estimates or indexes of nutrient availability. However, it is necessary that analytical procedures be simple and rapid. They are useful tests if the extracted nutrients relate closely to the amounts of K, Ca, and Mg that actually become available to plants during the growing season. The extreme range of soil, crop, and growing conditions for which tests are used dictates that no one approach will likely be successful for all situations. Procedures selected for use in a specific location are generally those that have provided the best correlation to K, Ca, and Mg availability for specified crops of that area.

I. MINERALOGY, EQUILIBRIA, AND DYNAMICS THAT MAY AFFECT SOIL TEST RESULTS

A. Potassium

Soil K has been categorized into soluble, exchangeable, fixed, and structural K (Sparks & Huang, 1985). The interrelations among these forms of soil K are illustrated in Fig. 8-3. Procedures that are based on extractable

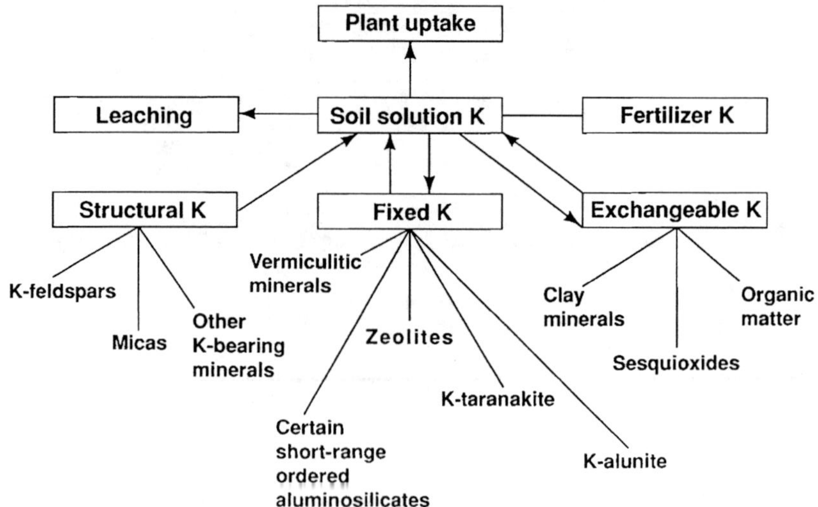

Fig. 8-3. Interrelationship of various forms of soil K (Sparks & Huang, 1985).

K measure soil solution K^+, most of the exchangeable K, and small, but varying, proportions of fixed or structural K. Numerous processes regulate the dynamics of the soil-K-plant system. Measurement of extractable K from a soil sample treated in a specific manner can be expected to serve only as an index to plant-available K.

The direction and rate of equilibrium reactions in the soil largely determine the fate of applied K. Added K can be taken up by plants, maintained in available forms in the soil, converted to less available forms, or leached to lower horizons. More precise fertilizer management recommendations could be made if the various reaction parameters were known for these processes in any given soil. The complexity of the system across different soils makes this a difficult area of research. However, based on results of research with many different soils and a wide range of conditions, it is possible to provide a generalized expression of reaction rates for the possible interconversions among the four major forms of soil K (Table 8–1). Recent detailed reviews of these aspects of soil K availability have been presented by Bertsch and Thomas (1985) and Sparks and Huang (1985).

The following discussion is presented as a general review of the soil-K plant system. It is intended to serve as a basis for understanding the different approaches used in testing soils for K and to help realize the strengths and limitations of various analytical methods.

1. Structural Potassium

The original source of K in soils is from weathering of rocks containing K-bearing minerals. Feldspars and micas are considered to be most important in this regard (Tisdale et al., 1985). The generalized chemical composition and approximate K content of important K-bearing feldspar and mica minerals are presented in Table 8–2.

When primary minerals are physically and chemically altered, secondary layer silicate clays may be formed. These can have K present in the crystal structure or in interlayer positions. Such minerals (e.g., dioctahedral or trioctahedral illite) can influence the supply and availability of K^+ to plants, depending on amounts present, degree of weathering, and past soil management.

2. Potassium on Clay Minerals and Organic Matter

Soil clays and organic matter influence K^+ availability in several ways. Exchange sites of clay minerals attract K^+, providing a sink of readily

Table 8–1. Generalized rate of reaction for conversion of soil K from one form to another (see Fig. 8–3).

Soil K form	Rate of conversion to soil solution K^+†
Structural K (feldspars, micas, etc.)	Slow, geological process (yr)
Fixed K^+	Several hours to several weeks
Exchangeable K^+	Nearly instantaneous to several hours

† Reverse reactions, where they can occur, would be generally similar in rate.

Table 8–2. Some important K-bearing primary minerals (based on Malavolta, 1985; and Rich, 1968).

Mineral	Chemical composition	Potassium content, g kg^{-1}
Feldspar		110–150
Orthoclase	$(K, Na)AlSi_3O_8$	
Microcline	$(Na, K)AlSiO_4$	
Sanidine	$KAlSi_2O_6$	
Micas		
Muscovite	$KAl_2(AlSi_3)O_{10}(OH)_2$	80
Biotite	$K(Mg, Fe^{2+})_3(AlSi_3)O_{10}(OH)_2$	70
Phlogopite	$KMg_3(AlSi_3)O_{10}(OH)_2$	70

exchangeable K in dynamic equilibrium with K^+ in the soil solution. Because of electrical attraction, movement of K^+ in soils by either mass flow or diffusion is not a simple process. Quantities and types of clays (McLean, 1978), soil water and temperature effects (Schaff & Skogley, 1982a), and other factors interact to create a complex situation concerning soil clays and the availability of K^+ to plants.

Three types of adsorption sites for K on clay minerals have been postulated (Fig. 8–4). Those on planar surfaces (p-position) have low K^+ selectivity, those on edges (e-position) have medium K^+ selectivity, and those in interlayer sites (i-position) have high K^+ selectivity. The K present in the i-position is referred to as "fixed" K. Fixed K adds an additional dimension to plant-K availability due to slow rates of release. Micaceous-, illitic-, and smectitic-clays (in decreasing order of K in i-positions) are involved in this relationship. Soils that contain a large proportion of these clays will exhibit varying degrees of K^+ fixation and/or release. Availability of added fertilizer K may be reduced in soils with high-fixation capacity. Under certain conditions, K^+ from interlayer positions may contribute substantially toward meeting crop K demands (Mengel & Kirkby, 1980). Miller (1988) found that wheat (*Triticum aestivum* L.) used nonexchangeable K in the near rhizosphere over extractable and solution K at greater distances in soils containing illitic-type clays.

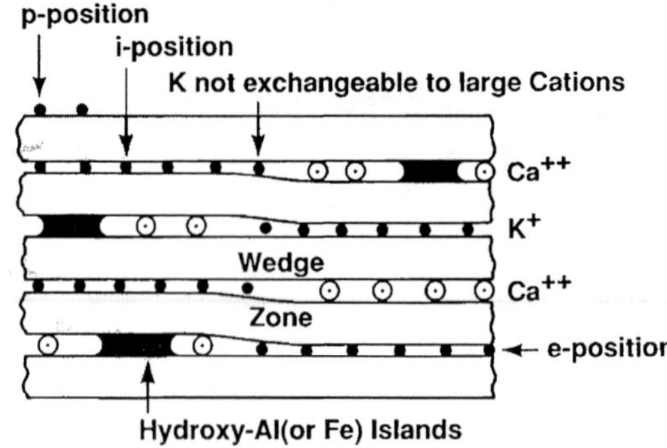

Fig. 8–4. Model of an expandable 2:1 clay mineral with interlayer K^+, wedge zones, and p-, e-, and *i*-positions (Mengel & Kirkby, 1980).

Soil organic matter imparts additional complexity to the K-availability system. Acidic functional groups (carboxyls, phenols, and enols) or organic polymers are sources of negative charge in the soil. The extent of dissociation of these functional groups depends on pH, electrolyte concentration, and cation species, but the charge is high (200–400 $cmol_c\ kg^{-1}$) compared to that of phyllosilicates (1–200 $cmol_c\ kg^{-1}$) (Greenland & Hayes, 1978). The affinity of soil organic matter exchange sites for K^+ is relatively low (Allison, 1973) and certainly much less than when micaceous or illitic clays are involved (Mengel & Kirkby, 1980). Hence, organic matter-derived cation-exchange capacity (CEC) contributes less to soil K^+ relations than its proportionate share of the exchange complex. As a result, it is often overlooked when considering soil factors of importance in K availability. This may be a serious oversight, especially in soils containing several percent organic matter.

Clays and organic materials also form many complexes that can result in altered surface charge properties of both the clay and the organic matter (Allison, 1973). Organic polymers may also be involved in the release of fixed K as a result of their chelating powers (Tan, 1978). Consequently, additions of organic materials to soils and actions and products of roots and microorganisms growing in the soil should not be overlooked as factors that can influence soil-K availability.

3. Soil Solution Potassium

The soil solution is the medium from which plants absorb nutrients. The concentration of K^+ in soil solution is highly important to plant-K availability. The higher the concentration, the greater will be K^+ movement toward plant roots via mass flow, and the greater will be the diffusion gradient (assuming a K^+ depletion zone surrounding absorbing roots). Other factors influence mass flow and diffusion rates. These include soil water content, temperature, and tortuosity of the diffusion path (Bertsch & Thomas, 1985). Nevertheless, K^+ concentrations and gradients develop the primary influences and driving forces for these processes.

A major problem in studying soil-solution nutrient relations has been the difficulty in separating solution and solid phases without altering the nutrient composition of the solution. Adams (1974) listed five categories of methods that have been used in attempting to obtain soil-solution samples. These are: (i) suction or pressure, (ii) displacement, (iii) compaction, (iv) centrifugation, and (v) molecular adsorption. Only the first two methods have been used to any large extent, due to serious deficiencies encountered with the latter three approaches. However, more recently the centrifugation approach has been successfully developed by using heavy organic liquids that are immiscible with water to displace soil solutions under large centrifugal forces (Mubarek & Olsen, 1976).

Most data in the literature concerning soil solution composition have been obtained from either the suction/pressure method or by displacement. Displacement in this terminology involves placing a column of water over

Table 8-3. Soil solution concentration of K^+ in ILD extracts and extractable K^+ as influenced by temperature, soil water tension (ψ), and K additions to Amsterdam silt loam, 0 to 15 cm (Skogley, 1986). Values are means for samples taken 3 and 14 d after K fertilization.

	5 °C		20 °C	
Added K	100 kPa (1 bar)	33 kPa (⅓ bar)	100 kPa (1 bar)	33 kPa (⅓ bar)
kg ha^{-1}		Soil solution K^+, mg L^{-1}		
0	48.9	39.5	60.2	48.5
100	72.3	61.4	84.1	71.6
400	190	162	220	176
		NH_4-OAc-Extractable, cmol$_c$ K^+ kg^{-1}		
0	1.22	1.23	1.31	1.34
100	1.50	1.51	1.51	1.55
400	2.01	2.08	2.08	2.04

a moist column of soil and collecting the desired quantity of displaced liquid at the bottom of the column. In the suction/pressure method, soils are normally wetted to levels above field-normal conditions (e.g., saturated paste extracts) to obviate problems relative to small quantities of extracts. Results from either of these types of extracts will not truly represent quantities and proportions of various ions that would exist under lower soil water contents. Consider, for example, the influence of the "ratio law" on relative concentrations of K^+ and Ca^{2+} as the soil solution becomes more dilute (Thomas, 1974). Even under field conditions, the composition of the soil solution will vary according to soil properties, soil water content, and rate of nutrient extraction by plant roots. Hence, meaningful studies based on soil-solution analysis are difficult to develop, but the nearer one can simulate conditions that exist during plant growth, the more useful the results should be. This is a major advantage of the immiscible liquid displacement (ILD) method.

A large range exists in reported values for K^+ in soil solutions. Most of the data from column displacement or pressure/vacuum extractions fall within about 1 and 265 mg of K^+ L^{-1} for a broad selection of soils and soil conditions from various areas of the world (Nemeth et al., 1970; Adams, 1974; Barber, 1984). Most values for normal agricultural soils of humid regions fall within 2 to 5 mg of K^+ L^{-1}, whereas most from arid region soils are an order of magnitude higher. Results from a study where K^+ contents of ILD extracts from a semiarid Amsterdam silt loam (fine-silty, mixed Typic Haploboroll) were measured are summarized in Table 8-3 (Skogley, 1986). The influence of soil water content (within a range normal for field conditions), temperature, and added fertilizer K are shown, as are values for ammonium acetate-extractable K^+. Converting these latter values to soil test ratings in mg kg^{-1} shows that the soil was one that would test "high" in available K^+ (477–813 mg kg^{-1}) when fertilizer K was not added.

Soil-solution concentrations of K^+ from ILD extraction are at the high end of the range of soil solution K^+ values of normal arid region soils as reported by Barber (1984). This would be expected because there is no dilution of the soil solution in the ILD method. The data illustrate the wide range of values that can occur in response to soil conditions. Even within the wet end of plant-available water (33–100 kPa), the drier the soil, the higher the K^+ concentration. Warmer-extraction temperature also resulted in higher solution K^+ concentrations. Additions of fertilizer K (as KCl) dramatically increased solution K^+. Additions in the range of normal crop K utilization (100–400 kg of K ha^{-1}) resulted in up to 4.1-fold increase in solution K^+. Ammonium acetate-extractable K^+ increased a maximum of 1.7-fold. Ching and Barber (1979), using the column displacement method of soil-solution extraction, reported similar effects of soil temperature and added K on soil-solution K^+ concentration, although their absolute values were much lower.

Results of studies with soil solution provide evidence to help explain why specific soil test methods, based on extractable K, sometimes fail to be strongly related to true nutrient availability and uptake by crops. Extractable K is not nearly as sensitive to changes in conditions as is soil-solution K^+. Furthermore, extractable K does not account for most processes that may become rate-limiting to K availability.

4. Equilibria and Dynamics Affecting Phytoavailability of Potassium

Thermodynamic principles have long been used to obtain exchange coefficients and thermodynamic parameters for various cation-exchange systems (e.g., Gaines & Thomas, 1953). More recently, studies have been conducted using chemical kinetic approaches to determine thermodynamic parameters for K^+ exchange in clays and soils (Sparks & Jardine, 1981; Jardine & Sparks, 1984; Barber, 1985; Sparks & Huang, 1985). Since K^+ movement to plant roots is a dynamic process, thermodynamic equilibrium of K^+ with the soil probably never occurs in the active rhizosphere during crop growth. Hence it is reasoned that evaluation of K^+ availability will be more realistic if it is based on a modelling approach in which kinetic parameters are used, rather than those based primarily on calculations of thermodynamic equilibrium (Barber, 1984).

The root-soil interfacial zone contains only a small amount of a plant's total requirement for K^+. Root volume of annual crops is <1% of soil volume. Thus, the K requirement of the plant obtained from this zone will be small (Barber, 1985). The remainder of the K required by the crop will have to move in the soil to plant root surfaces either in the water being drawn to the root in response to transpiration (mass flow) or by diffusion. If the concentration of K^+ in the soil solution that moves by mass flow is sufficiently high, most or all of the crop K^+ demand could be met in this manner. Any deficit from this process would have to be satisfied by diffusion. Actually, both processes would normally be operating simultaneously during active crop growth. The relative importance of each in supplying K^+ will depend on soil conditions, most of which are in a continual state of change.

Estimates of the relative contribution of mass flow can be made if the total crop K uptake, water use, and soil-solution K^+ concentration are known. Using this approach, Barber (1985) suggested that the soil-solution K^+ concentration would need to be maintained near 100 mg of K^+ L^{-1} to completely satisfy crop K demands through mass flow. Saturation paste extracts or column-extracted soil solution K^+ concentrations range from 2 to 60 mg L^{-1}. Even soil-solution extracts obtained by ILD extraction from a soil "high" in extractable K (see Table 8–3) have concentrations that account for only 30 to 40% of crop K requirements via mass flow (Skogley, 1986). Mass flow could be expected to account for most of the K^+ movement to crop roots only when relatively high rates of fertilizer K are applied to the soil and when most of the water absorbed by plant roots comes from the fertilized zone. Thus, under most cropping conditions, a large fraction of the total crop K demand would have to move to roots in response to a diffusion gradient. When diffusion is impeded for any reason (e.g., cold or dry soil conditions), it could be a rate-limiting process for K^+ availability to crops.

To understand the dynamic soil-K plant system, it is necessary to learn about related mechanisms and conditions in both the soil and the plant. Plant factors include plant age, K status, temperature, transpiration rate, root morphology and growth rate, and K^+-absorption mechanisms of the root. Soil conditions of importance include those factors that influence K^+ diffusion (temperature, water content, tortuosity, and K^+ concentration of soil solution) and exchangeable-K relations (amounts and proportions of K to other cations, soil-K buffer capacity, rate of release to solution phase), as well as soil solution parameters and plant water use related to mass flow. Barber (1984) and his associates have developed models to predict K^+ uptake by corn and soybean [*Glycine max* (L.) Merr.] Their results suggest that root morphology and rate of growth contribute most strongly to total K uptake, followed by soil parameters that control K^+ flux in the soil. These parameters include initial K^+ concentration of the soil solution, soil K^+ buffer power, and the effective diffusion coefficient of K^+ in the soil.

B. Calcium

1. Calcium Sources

The Ca content of earth's crust averages about 36.4 g kg^{-1}. The majority of this Ca exists as difficultly soluble primary minerals. These minerals include the Ca-bearing aluminum silicates such as feldspars, amphiboles, Ca phosphates, and Ca carbonates, the latter being particularly important in calcareous soils (Mengel & Kirkby, 1978). The plagioclase mineral, anorthite ($CaAl_2Si_2O_8$), is the most important primary source of Ca (Tisdale et al., 1985). Other minerals in this group, including impure albite, are of less significance. Pyroxenes (augite) and amphiboles (hornblende) are fairly common Ca minerals in soils. Calcium may also be solubilized from biotite, epidote, apatite, and certain borosilicate minerals.

Soils vary widely in Ca content. Coarse-textured, humid-region soils formed from rocks low in Ca-containing minerals are low in Ca. Humid region soils formed from limestones are frequently acid in the surface layers because of the removal of Ca and other basic cations by leaching. Calcium carbonate, or calcite ($CaCO_3$), is often the dominant source of Ca in soils of semiarid and arid regions. Dolomite [$CaMg(CO_3)_2$] may also be present in association with calcite. Gypsum ($CaSO_4 \cdot 2H_2O$) is present in some soils in semiarid regions.

2. Forms of Calcium

Calcium in soils may be classified as nonexchangeable which includes mineral forms, exchangeable, and soil solution Ca^{2+}. Calcium is usually the most dominant of soil-exchangeable cations. Exchangeable Ca is the major reserve of soil Ca available to plant roots and can range from <25 mg kg^{-1} to more than 5000 mg kg^{-1}. Lower exchangeable Ca concentrations exist mainly in low CEC, acid, humid region soils. Higher Ca levels are found in higher CEC, semiarid and arid region soils that contain $CaCO_3$, and in limed acid soils. The term extractable is preferred instead of exchangeable for Ca removed by dissolution of $CaCO_3$ from alkaline or limed acid soils by acidified extracting solutions. The advent of inductively coupled plasma emission (ICP) spectrometers allows nutrient levels to be analyzed at much higher ranges. As a result, extractable Ca levels >45 000 mg kg^{-1} are sometimes reported.

Calcium is absorbed by plants as Ca^{2+} from soil solution. Rapid equilibrium occurs between exchangeable Ca and soil-solution Ca^{2+}. The amount of Ca^{2+} in soil solution is usually high relative to other cations in nonsodic soils. Adams and Henderson (1962) cited soil solution Ca^{2+} values ranging from 68 to 778 mg kg^{-1} from many experiments.

3. Factors Affecting Uptake

Soil factors of greatest importance that determine Ca availability to plants are listed by Tisdale et al. (1985) as total Ca supply, soil pH, CEC, Ca-saturation percentage of the soil colloids, type of soil colloid, and the ratio of Ca^{2+} to other cations in solution.

a. Calcium Supply. The CaO content of soil ranges from about 55 g kg^{-1} in Aridisols (arid region soils) to 16 g kg^{-1} in Mollisols (grassland prairie soils) and <10 g kg^{-1} in older soils such as Alfisols, Spodosols, Ultisols, and Oxisols (humid region soils) (McLean, 1975). Calcium has an ion diameter of 9.9 × 10^{-9} cm and forms a base on hydrolysis. Calcium moves to the root surface by mass flow in response to the transpiration stream, by root interception of growing roots, or by ion diffusion from higher to lower concentrations (McLean, 1975).

Calcium availability to the plant is largely a factor of the supply in the soil but can be affected by soil properties. In very sandy, acid soils with low CEC, Ca supply can be too low to provide sufficient available Ca to crops. Adams and Moore (1983) reported visual symptoms of Ca deficiency in cotton roots when the soil E and B horizon solution Ca^{2+} concentration ranged between 10.8 and 13.6 mg kg^{-1} and the Ca-saturation level was 170 g kg^{-1} or less. Melsted (1953) observed Ca-deficiency symptoms in corn grown on acid soils containing <400 mg extractable Ca kg^{-1}. He noted that Ca deficiency was not evident until all other nutrient deficiencies were corrected by large applications of soluble fertilizers. Deficiencies were verified by plant Ca concentrations of 2 g kg^{-1} or less.

b. Soil pH. Semiarid and arid region soils are relatively unweathered. Alkaline soils of these regions usually contain an excess of Ca in the plant-rooting depth. Much of this Ca is in the form of $CaCO_3$ that has a solubility of about 5.6 mg Ca L^{-1} in cold water. The majority of crops in these regions are adequately supplied with Ca. Calcium concentration of 0.25 mg L^{-1} was sufficient to obtain maximum growth rates in a nutrient solution at pH 5.6 (Lund, 1970). More than 2.5 mg Ca^{2+} L^{-1} was required at a solution pH of 4.5, and 5 mg of Ca^{2+} L^{-1} was insufficient when pH was 4.0. This is in agreement with Melsted's (1953) earlier field research on pH 4.5 soils. The elongation rate of cotton roots in 0.001 *M* $CaHPO_4$ adjusted to varying pH levels with H_3PO_4 was inhibited only at solution pH <4.25 (Howard & Adams, 1965). Soils approaching this pH level rarely remain in agronomic crop production.

c. Cation Exchange Capacity, Calcium Saturation, and Type of Soil Colloid. Cation exchange capacity is important to soil Ca availability in relation to the Ca saturation of soil colloids and the type of soil colloids. Soil CEC increases with increasing soil pH. This is notable relative to cation saturation percentages. Many crops respond to Ca applications when the degree of Ca saturation of soil CEC falls below 25%. Kaolinite clays are able to satisfy the Ca requirements of most plants at saturation values of only 40 to 50%. Soil clays with two silica to one octahedral layer (2:1), such as montmorillonite, require a Ca saturation of 70% or more before this element is released in sufficient supply to growing plants (Tisdale et al., 1985). Allaway (1945) found that the availability of the replaceable Ca from various colloids as measured by soybean uptake was in the order peat > kaolinite > illite > Wyoming bentonite > Mississippi bentonite. At 40% saturation, kaolinite supplied soybean with more Ca than did Mississippi bentonite at 80% saturation.

d. Cation Ratio. The cation-ratio concept probably originated from New Jersey work that projected an ideal soil as one with the following distribution of exchangeable cations: 65% Ca, 10% Mg, 5% K, and 20% H (Bear et al., 1945). These concentrations provide the ratios: Ca/Mg of 6.5:1, Ca/K of 13:1, and Mg/K of 2:1. Bear et al. (1945) proposed 65% as the Ca saturation of the CEC of the ideal soil for optimum plant growth. Graham (1959) suggested modification of the ideal Ca-saturation ratio to a range of 65 to 85%. Eckert and McLean (1981) reported that German millet (*Setaria italica* L. Beauv.) grew best at Ca saturations below the minimum level of 65%. Geraldson (1957) indicated that excess total salts can cause a Ca deficiency in tomato (*Lycopersicon esculentum* Mill.) even when the Ca ratio is considered high or adequate. He reported that Ca^{2+} uptake by tomato was decreased to the greatest degree by the NH_4^+ ion. Claassen and Wilcox (1974) reported a similar effect of NH_4^+ on Ca^{2+} absorption by corn.

McLean et al. (1983) found poor correlation between yield response of corn, soybean, wheat, and alfalfa to Ca/Mg ratios. The average of 83% saturation of the CEC was common for maximum and minimum yields. Soil pH was better correlated with yields than cation ratios or basic cation saturation ratio (BCSR). Eckert (1987) found that wide variations in BCSR ratios were of little consequence as long as gross imbalances were not created. He found no results in the literature that confirmed the existence of the ideal cation saturation ratio. Liebhardt (1981) also concluded that Ca and Mg recommendations based on cation saturation percentages and the resulting Ca/Mg and Ca/K ratios are not warranted. He suggested maintenance of Ca^{2+} availability to plants by use of a liming program that maintains soil pH between 5.5 and 6.0. Liming to these pH levels generally maintains the Ca + Mg saturation levels between 65 to 75%, alleviates metal toxicities, and increases availability of other fertilizer nutrients.

C. Magnesium

1. Magnesium Sources

The Mg content of most soils generally ranges between 0.5 g kg^{-1} for sandy soils and 5 g kg^{-1} for clay soils (Mengel & Kirkby, 1978). Magnesium is present in relatively easily weatherable ferromagnesian minerals such as biotite, serpentine, hornblende, and olivine. It occurs in secondary clay minerals including chlorite, vermiculite, montmorillonite, and illite. Some soils contain Mg as $MgCO_3$ or $CaMg(CO_3)_2$. In arid or semiarid regions, soils may contain large amounts of Mg as $MgSO_4$.

2. Forms of Magnesium

Soil Mg may be divided into nonexchangeable, exchangeable, and water-soluble forms, all in equilibrium. In a fractionation study of selected temperate and tropical soils, Mokwunye and Melsted (1972) ranked Mg distribution in the following order: primary > acid soluble > exchangeable > organically complexed. Soil Mg is a moderately leachable nutrient, and as with Ca, greater amounts are often found in the subsoil than in upper parts of the profile. Highly leached and weathered soils such as Alfisols, Ultisols, and Oxisols are generally low in Mg compared to the relatively unweathered soils such as Entisols, Vertisols, Inceptisols, Aridisols, and Mollisols.

Magnesium is taken up by plants as Mg^{2+} from soil solution. Expandable clay minerals supply Mg^{2+} to plants from both lattice and interlayer positions (Christenson & Doll, 1973). Exchangeable Mg normally constitutes from 4 to 20% of the CEC (Mengel & Kirkby, 1978), and varies between

17 and 2431 mg L^{-1} in soil solution (Fried & Shapiro, 1961). Most frequently, concentrations exist between 50 and 120 mg L^{-1} (Mengel & Kirkby, 1978).

3. Factors Affecting Uptake

Soil factors that affect Mg^{2+} uptake by plants include Mg^{2+} supply, soil pH, K application rate, soil CEC, percent Mg saturation, soil texture, type of soil colloid, and Mg/cation ratio (Tisdale et al., 1985).

a. Magnesium Supply. The total Mg supply in the Ap layer of five southeastern USA sandy, acid Coastal Plains soils ranged from 89 to 302 mg kg^{-1} (Rice & Kamprath, 1968). The percentage of the total Mg in exchangeable form ranged from 3.8 to 9.4 g kg^{-1}, and generally increased as total Mg increased. Prince et al. (1947) found no correlation between the total Mg in eastern USA soils and their crop-yield potential.

Plant-available Mg is largely in the exchangeable and water-soluble forms (Arnold, 1967). Fox and Piekielek (1984) reported that the exchangeable Mg level recommended for agronomic crop production by different soil-testing services ranged from 25 to 180 mg kg^{-1}. Arnold (1967) reported that soils most susceptible to Mg deficiency are light-textured, acid soils with low-exchangeable Mg levels and soils of high K status regardless of exchangeable Mg level. Reith (1967) found depressed Mg^{2+} supply to plants when K was applied in excess to soils already containing satisfactory or high K contents. A 1.2 ratio of Mg/K was reported by Hooper (1967) as the level needed to obtain a Mg concentration of 2 g kg^{-1} in herbage. Magnesium

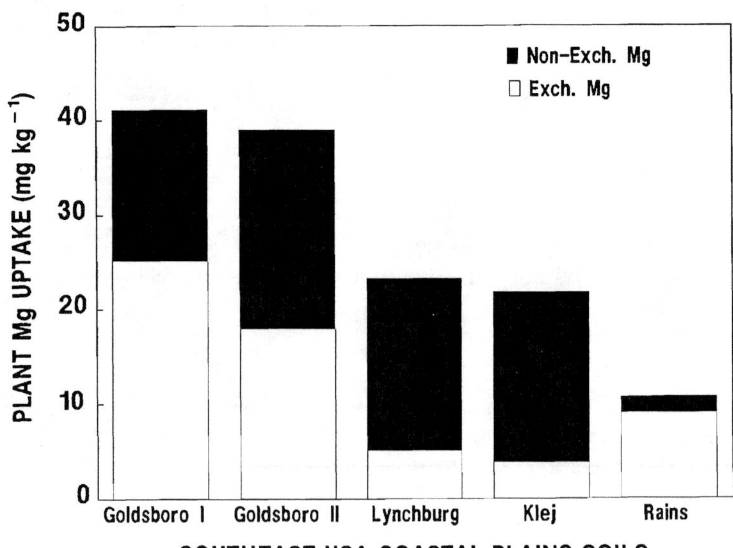

Fig. 8-5. Plant uptake of Mg from exchangeable and nonexchangeable sources. (Data from Rice & Kamprath, 1968.)

uptake was lowered by Ca^{2+} (Farina et al., 1980) and by NH_4^+ (Claassen & Wilcox, 1974).

The soil content of available Mg thought to be deficient or adequate varies. Georgia soils containing 50 mg kg^{-1} or less had inadequate Mg^{2+} for good plant growth (Boswell et al., 1967; Gallaher et al., 1975). According to Reith (1967) in northern Scotland, responses could be expected in slightly acid mineral soils at Mg^{2+} levels below 30 mg kg^{-1} when extracted by tumbling end-over-end in a 1:40 soil:2.5% acetic acid solution for 2 h. Adequate Mg levels could be attained with 160 mg kg^{-1} soluble Mg^{2+} in the soil. Haby et al. (1979) reported no response of Coastal bermudagrass to Mg application when the surface 15 cm of an Alfisol contained 7.3 mg of Mg kg^{-1}, and similar lack of corn response on another Alfisol that contained 36 mg of Mg kg^{-1}. According to Rice and Kamprath (1968), exchangeable Mg in the range of 6.1 to 28 mg kg^{-1} appeared to be equally available for plant uptake because the plants obtained a large part of their Mg from nonexchangeable forms (Fig. 8–5). Charlesworth (1967) related the Mg content of carrot (*Daucus carota* var. *sativus* [Hoffm.] Arcang.) leaves to Mg content within the upper 45 cm of soil.

b. Soil pH. The effect of soil pH on Mg availability is related to several factors. Sudangrass [*Sorghum vulgare sudanense* (Piper) Hitche.] response to Mg was greatest at pH 6.5 on Norfolk loamy sand (fine-loamy, siliceous, thermic Typic Paleudult) and on Hartsells fine sand (fine-loamy, siliceous, thermic Typic Hapludult), but was lower at pH 6.5 than at 5.5 on the Dickson silt loam (fine-silty, siliceous, thermic Glossic Fragiudult) (Adams & Henderson, 1962). Total soil Mg tended to be less at pH 6.5 than at 5.5 on Mg-deficient soils, but was greater at pH 6.5 on Mg-sufficient soils. Christenson et al. (1973) indicated that low pH affected Mg uptake by oat (*Avena sativa* L.) only when the Mg level in the soil was low. Hooper (1967) determined that soil test correlations between several soil-extraction methods and percentage Mg in herbage were consistently greater at pH >6.5. Increasing pH markedly decreased 0.01 *M* $CaCl_2$ extractable Mg while double-acid Mg availability remained essentially unchanged (Farina et al., 1980). In some soils, yields were depressed by high pH levels and Mg content was lower, but Mg applications did not increase yields. Evidence suggested that salt-extractable Mg did not accurately reflect plant availability. According to Jones and Haghiri (1963), increasing pH had a greater effect on Mg^{2+} uptake, whereas the addition of Mg without an increase in soil pH had only a slight effect on uptake of Mg^{2+}.

c. Potassium Supply. Considerable evidence indicates that heavy applications of potash fertilizer or a high level of K^+ in the soil can lead to a low-Mg content in the plant. Doll and Hossner (1964) reported that fertilizer K decreased potato (*Solanum tuberosum* L.) yields at every level of Mg fertilization. Stout and Baker (1981) indicated that the differential adsorption of K was the controlling factor in the uptake of Mg by corn seedlings. Soil K was significantly correlated with Mg concentration in corn plants but explained <40% of the variation in plant uptake of K^+ (Walker & Peck,

1975). Rahmatullah and Baker (1981) reported a correlation between Mg concentration in corn and ½ pMg-pK, an expression of the relative availability of Mg and K by the Baker method (Baker & Amacher, 1981). Data indicated the plant uptake of Mg from these soils was more a function of K availability than of Mg availability. Doll and Hossner (1964) suggested that the suppressing effect of high K on Mg uptake may be due partially to reactions between applied K and soil components. Specifically, Mg^{2+} may be trapped in an inaccessible position within interlayers of collapsed plates of 2:1 lattice clays.

d. Percent Magnesium Saturation. The exchangeable-soil Mg level recommended for agronomic crop production by different soil-testing services ranges from 25 to 180 mg kg^{-1}, or approximately 10% saturation of the soil CEC. Fox and Piekielek (1984) reported that at least 10% saturation was required to obtain 2 g of Mg kg^{-1} in corn silage grown on Ultisols and Alfisols studied. They indicated that 5% saturation of the soil CEC was adequate to produce 1 g of Mg kg^{-1} in the ear leaf and grain yield was not decreased at this level. Stout and Bennet (1983) observed a slight corn grain yield increase to Mg application on an Inceptisol that had an exchangeable Mg level of 3.75% saturation of the CEC. At 5.65% Mg saturation there was no response. Soils that had <4% of the CEC saturated with Mg were Mg deficient for maximum yields of sudangrass and ladino clover (*Trifolium repens* L.) (Adams & Henderson, 1962). Three percent Mg saturation of a high CEC soil was not sufficient for maximum yields of alfalfa (McLean & Carbonell, 1972). They indicated that it was important to have a high level of Mg saturation of the CEC with adequate but not excessive levels of exchangeable K. Martin and Page (1969) reported that at 30, 50, and 100% base saturation, about 3 to 6% exchangeable Mg was associated with Mg-deficiency symptoms and reduced growth of sweet orange [*Citrus sinensis* (L.) Osb.] seedlings. About 7 to 13% exchangeable Mg was associated with mild Mg deficiency symptoms, but plant growth was not reduced. About 4% exchangeable Mg was critical for the Hanford soil (coarse-loamy, mixed, nonacid, thermic Typic Xerorthent). At this level, growth was not reduced but leaf Mg-deficiency symptoms appeared. Three to 4% exchangeable Mg in the Merriam soil (reclassified to Placentia series, a Typic Natrixeralf; L.J. Lund, 1989, personal communication) was sufficiently low to cause severe Mg-deficiency symptoms and reduced plant growth (Martin & Page, 1969).

e. Cation Ratio. Ratios of K/Mg reported as favorable for crop production range from 0.8:1 to 20:1. Ologunde and Sorensen (1982) indicated that as long as the absolute amounts of K and Mg in a sand culture growth medium were adequate to meet the demands of the plant, the K/Mg ratio could vary widely (1–25) without causing any adverse effect on plant dry matter production. The K/Mg ratio was not useful in predicting the amount of growth of sorghum (*Sorghum bicolor* L. Moench) in their study. Mulder (1950) noted in a review paper that deficiency symptoms appeared in apple (*Malus sylvestris* Mill) trees when the K/Mg ratio in the Morgan extract of the top 20 cm of soil was 2:1 to 1.6:1, while under normal trees it was 0.8:1.

Pratt et al. (1957) found no correlation between available Mg and soil K/Mg ratio in the 0 to 15 or 15 to 30 cm soil depths and suggested the K/Mg ratio of the 30 to 91 cm depth provided the best estimate of Mg availability to citrus. Similar results were reported by McCollock et al. (1957) who found a correlation of $r = 0.91$ between citrus leaf Mg and the exchangeable K/Mg ratio of the 46 to 76 cm soil depth. Their data indicated that additional Mg was needed to prevent Mg deficiency when the ratio of soil K/Mg exceeded 0.5. Charlesworth (1967) found the K/Mg ratio was better related to the Mg content of carrot leaves than exchangeable Mg in the 0 to 46 cm of soil depth. Batey (1967) concluded that the effectiveness of the K/Mg ratio depended on the crop, its stage of growth, climate, and soil depth, drainage, and structure. Slow-growing crops may tolerate a wider ratio of K/Mg in the soil than quick-growing, intensive, horticultural crops. He suggested maintaining K/Mg ratios of 2:1 or less in the soil by balancing soil K and K applications with dressings of kieserite ($MgSO_4 \cdot H_2O$) or other Mg-containing fertilizers to avoid Mg deficiency. Soils with low Mg availability (<100 mg kg^{-1}) and a pH of <5.4, or a K/Mg ratio >4 were likely to be Mg deficient.

The Mg/K ratio has been evaluated in several studies (Adams & Henderson, 1962; McLean et al., 1983; Olson et al., 1982). The ratio of exchangeable Mg/K was not a better indicator of available Mg than percent Mg saturation of the soil CEC (Adams & Henderson, 1962). Olson et al. (1982) indicated that the Mg/K ratio varied from 2.2:1 to 6:1 with no significant effect on corn yield or change in available soil K after 7 yr on four major central U.S. Great Plains soils. McLean et al. (1983) evaluated corn, soybean, wheat, and alfalfa relative to Mg/K ratios and found no significant relationship between Mg/K ratio of the soil and yield of any of these crops.

The Mg concentration in ryegrass (*Lolium perenne* L.) was proportional to

$$(\sqrt{{}^{a}Mg})/[(\sqrt{({}^{a}Ca + Mg)}) + (B{}^{a}K)]$$

where ${}^{a}Mg$, ${}^{a}Ca + Mg$ and ${}^{a}K$ were the initial ion activities calculated from the composition of the equilibrium soil solution, and B is a proportionality factor determined by cations in the plant and cation activities in soil solution (Salmon, 1964). The Mg concentration in the grass was correlated ($r = 0.95$) with this ratio and explained some variations in Mg availability on different acid soils.

Hooper (1967) indicated that the Mg/Ca ratio was little better than exchangeable Mg for improving the correlation with Mg uptake. The correlation of Mg/Ca with herbage Mg was better in soils with pH >6.5.

Fox and Piekielek (1984) evaluated the Ca/Mg ratio and reported no reduction in corn yield at ratios from 1.8 to 36.9. Simson et al. (1979) reported that the Ca/Mg ratio could probably go below 0.5 before Mg toxicity or Ca deficiency would affect corn growth. Eckert and McLean (1981) indicated that the balance of cations in the soil was unimportant, except at the extremely wide ratios. At wide ratios, deficiencies of one element were caused by excesses of others, therefore, no best ratio existed for German millet or alfalfa. Olson et al. (1982) concluded that cation balance was not an essential consideration in estimating crop nutrient needs for the yields obtained and soil conditions of their study.

II. SOIL TEST APPROACHES AND METHODS

A. Potassium

1. Rapid Chemical Extractions

Soil-testing laboratories generally rely on rapid methods of extraction and analysis to achieve the short turn-around time expected by clients. An estimate of exchangeable K, which usually includes soil solution K^+, is the standard index of K availability in the USA and Canada. Exchangeable K is related to tissue K concentration (Fig. 8–6) and crop yields (Fig. 8–7) under many conditions. Even when the contribution of nonexchangeable K to plant nutrition is significant, exchangeable K has been closely correlated with uptake of K by plants (Doll & Lucas, 1973). Exchangeable K is determined by a variety of extractants (Fig. 8–8 and Table 8–4), most of which employ NH_4^+ or Na^+ as the cation to replace K^+ on exchange sites. Outlines of procedures used in the north-central and southern regions of the USA have been published (Dahnke, 1988; Johnson et al., 1984).

Determination of total exchangeable K requires exhaustive leaching or washing with 1.0 *M* ammonium acetate, pH 7.0 (Knudsen et al., 1982). This extractant has been used for more than 50 yr (Chapman & Kelley, 1930).

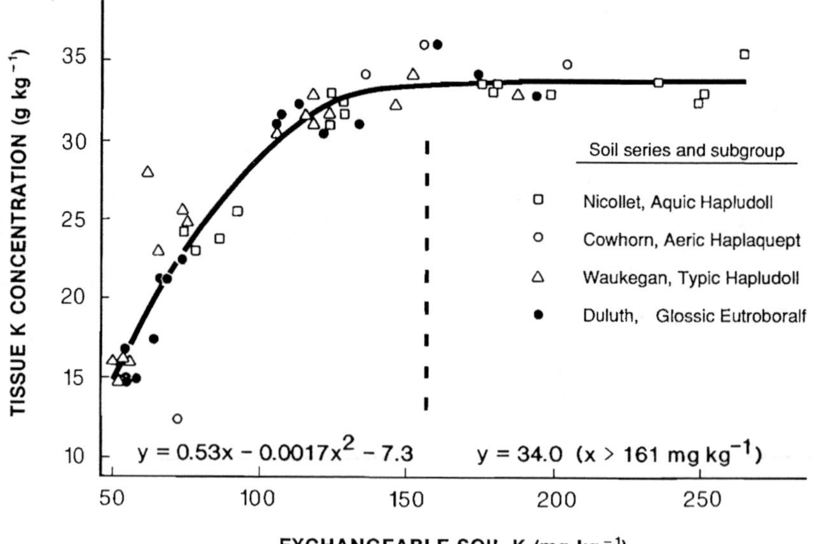

Fig. 8–6. Relationship between exchangeable soil K and tissue K concentration in birdsfoot trefoil (*Lotus corniculatus* L.) seedlings grown in pots for 42 d in the greenhouse (Russelle et al., 1989).

Typical soil-test procedures are designed to minimize extraction time and labor, while attaining adequate precision in results. Extraction at a 1:10 soil/solution ratio with vigorous shaking (at least 200 oscillations per minute) for 5 min is currently recommended for laboratories determining extractable soil K with neutral 1.0 *M* ammonium acetate in north-central USA (Knudsen et al., 1982). Shaking time has a variable impact on soil test results (Fig. 8–9).

Although neutral 1.0 *M* ammonium acetate has long been and is presently the most widely used extractant for K in public soil-testing laboratories, other extractants are gaining in popularity. Many of these solutions facilitate the removal of several elements in one extraction, and are referred to as "universal" extractants. Their use is complemented by the availability of multi-element analytical techniques such as ICP spectrometry.

Because of the chemistry regulating phytoavailability of other nutrient elements, such as P, Zn, and Mn, extracting solution composition and pH have been adjusted for the soils being analyzed; acidic solutions are typically used in neutral to acidic soils and alkaline solutions are usually used in neutral to calcareous soils (Fig. 8–8). Removal of exchangeable soil K is also affected by changes in solution composition, so new methods must be calibrated to crop response and not simply correlated with nutrient extraction by the old method.

Mehlich (1953) proposed the use of the double-acid (Mehlich I) extractant for P, K, Ca, Mg, Na, Mn, and Zn. It is used primarily in states on the Atlantic and Gulf Coasts, where soils are predominantly Ultisols, Inceptisols, and Spodosols (Fig. 8–8). Amounts of K, Ca, and Mg removed by the double acid extractant are closely correlated with, but usually lower than, amounts removed in neutral 1.0 *M* ammonium acetate. The double acid extractant removes excessive amounts of P in calcareous soils and in soils where apatite is the predominant source of P.

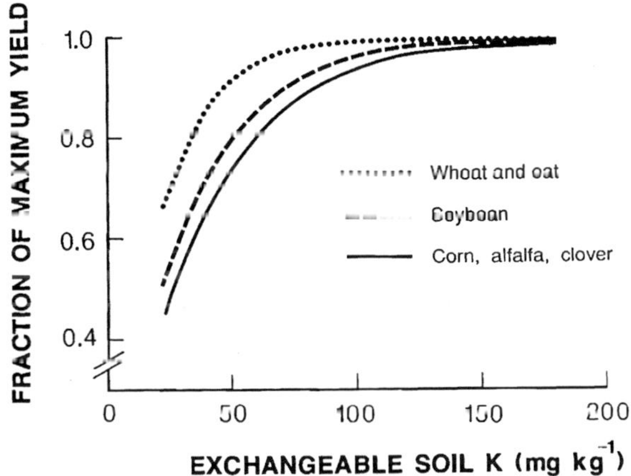

Fig. 8–7. Relationship between exchangeable K in topsoil and yield of several crops on soils with CEC < 12 $cmol_c$ kg^{-1}. (Redrawn from Bray, 1945.)

Mehlich (1978) suggested a new extractant (Mehlich II), which provided improved results for P over a wider range of soils. Extraction of K, Ca, and Mg was also more closely related to the standard ammonium acetate extraction. However, the high concentration of Cl^- (0.21 M) was corrosive to laboratory equipment. No public soil-testing laboratories currently recommend the use of the Mehlich-II extractant.

Because of problems with corrosion and the desire to improve the reliability of Cu extraction, Mehlich (1984) designed an improved extractant by substituting nitrates for chlorides and by adding EDTA. Five states have adopted this extractant, and it is under investigation by several public and

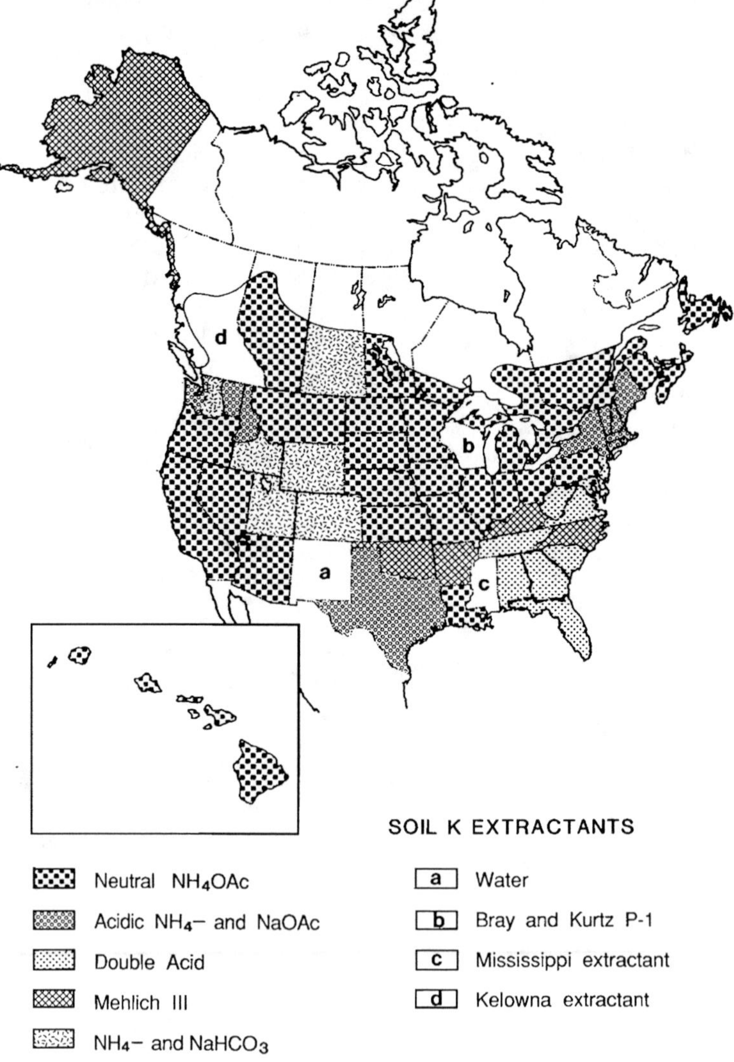

Fig. 8–8. Soil K extractants used or recommended by public soil-testing laboratories in the USA and Canada (from a survey conducted by Russelle).

Table 8-4. A selection of K-extraction methods used in the USA and Canada, 1988.

Method	Components	Concentration, mol L^{-1}	pH	Soil/solution ratio	Extraction time	Reference
Ammonium acetate	CH_3COONH_4	1.0	7.0	1:10	5 min	Knudsen et al., 1982
Baker	I. CH_3COONH_4	1.0	7.0	1:10	30 min	--
	II. KCl	2.5×10^{-4}	7.3	1:10	24 h	Baker & Amacher, 1981
	$MgCl_2$	10.0×10^{-4}				
	$CaCl_2$	50.0×10^{-4}				
	DTPA†	4.0×10^{-4}				
	TEA					
Bray and Kurtz P_1	HCl	0.025	2.6	1:10	5 min	Knudsen & Beegle, 1988
	NH_4F	0.03				
Double acid (Mehlich I)	HCl	0.05	--	1:5	5 min	Mehlich, 1953
	H_2SO_4	0.0125				
Kelowna	CH_3COOH	0.25	3.2	1:10	5 min	van Lierop, 1985
	NH_4F	0.015				
Mehlich III	HNO_3	0.013	2.5	1:10	5 min	Mehlich, 1984
	NH_4F	0.015				
	CH_3COOH	0.2				
	NH_4NO_3	0.25				
	EDTA	0.001				
Mississippi	I. HCl	0.05	--	1:1	10 min	Lancaster, 1980
	II. NH_4F	0.037	4.0	1:4	10 min	
	CH_3COOH	1.57				
	$CH_2(COOH)_2$	0.0625				
	$CH_2CHOH(COOH)_2$	0.0933				
	$AlCl_3$	0.0124				
Morgan	CH_3COOH	0.52	4.8	1:5	30 min	Lunt et al., 1950
	CH_3COONa	0.73				
Modified Morgan (one example)	CH_3COOH	1.25	4.8	1:5	15 min	McIntosh, 1969
	NH_4OH	0.62				
Olsen	$NaHCO_3$	0.5	8.5	1:20	30 min	Olsen et al., 1954
Soltanpour	NH_4HCO_3	1.0	7.6	1:2	15 min	Soltanpour & Schwab, 1977
	DTPA	0.005				
Texas	CH_3COONH_4	1.4	4.2	1:20	60 min	A.B. Onken 1980, unpublished data (listed in Johnson et al., 1984)
	HCl	1.0				
	EDTA	0.025				

† DTPA = diethylenetriaminepentaacetic acid, TEA = triethanolamine, and EDTA = ethylenediaminetetraacetic acid.

private soil-testing laboratories. It is apparently useful over a broad range of soils and climatic conditions (Fig. 8–8) and has obvious advantages as a universal extractant (Hanlon & Johnson, 1984). Potassium removal is about 20% greater than, but highly correlated with the standard ammonium acetate method (J.S. Harrison and V.W. Case, 1988, personal communication). However, salt deposition in nebulizers of certain brands of atomic absorption and ICP spectrometers has occurred with the Mehlich-III extractant.

Onken et al. (1980) reported that P extracted by EDTA solution was significantly related ($R^2 = 0.71$) to grain sorghum yield response in the field. Since that time, 0.025 *M* HEDTA has been added to the 1.4 *M* CH_3COONH_4-1.0 *M* HCl (pH 4.2) extracting solution in Texas. This solution is used as a universal extractant for K, Ca, Mg, and other plant nutrients (Hons et al., 1990; Johnson et al., 1984). The initial problem with salt deposition has been eliminated by modification of the instrument nebulizer (H.D. Pennington, 1988, personal communication).

Soltanpour and Schwab (1977) and Soltanpour and Workman (1979) proposed the use of 1.0 *M* NH_4HCO_3 with 0.005 *M* DTPA, pH 7.6, as an

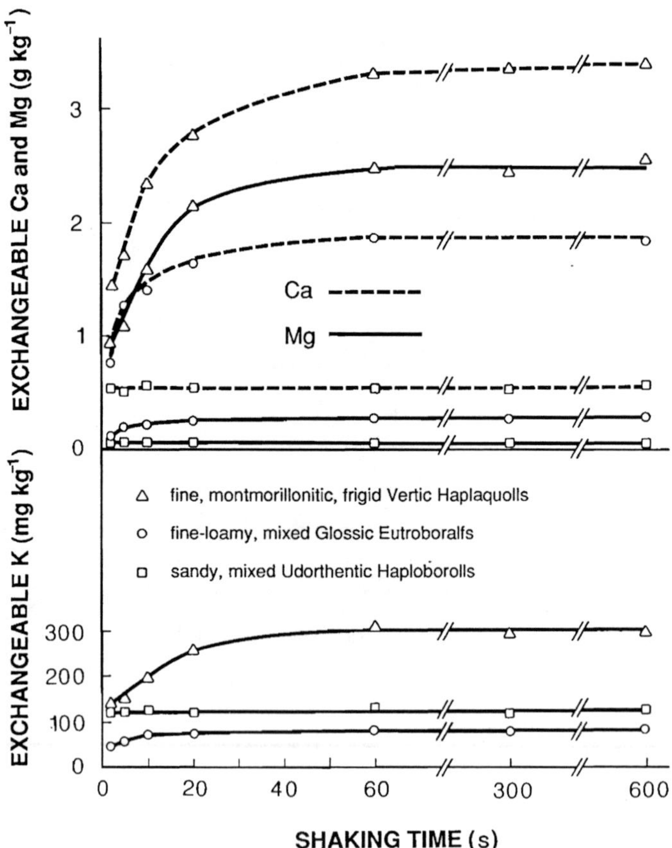

Fig. 8–9. Relationship between shaking time in neutral 1 *M* ammonium acetate and K, Ca, and Mg soil test results (M.P. Russelle and I.T. Aighewi, 1985, unpublished data).

extractant for NO_3-N, P, K, Zn, Fe, Cu, and Mn from alkaline soils. They tested a group of 481 soil samples, having diverse chemical properties. Soil test results with the new extractant were well correlated with standard extraction techniques, although the limits between low, medium, and high soil test levels had to be adjusted for several elements. No change was required for the soil K test. This extractant is presently used in Colorado and Wyoming (Fig. 8–8). Sodium bicarbonate (Olsen et al., 1954) is used to extract alkaline soils in three western states and in Saskatchewan. However, analysis of Ca and Mg in this extractant by atomic absorption spectroscopy is difficult due to the milky precipitate that forms when lanthanum or strontium chloride is added for ionization suppression to increase Ca and Mg sensitivity.

The Morgan extractant (Table 8–4) was designed to simulate the dissolution of soil constituents by dissolved CO_2 (Lunt et al., 1950). It has been used extensively in the northeastern states and to some extent in the Pacific Northwest as a universal extractant for P, K, Ca, Mg, and N. Four states now use the Morgan extractant (Fig. 8–8). Variations of the Morgan solution use NH_4^+ in place of Na^+ to effect more complete displacement of K from clays (McIntosh, 1969). These variations are generally referred to as "modified Morgan" extractants and are used in a few northeastern states. Other extractants currently used include water (New Mexico), Bray and Kurtz P_1 (Wisconsin) (Bray & Kurtz, 1945), the Kelowna extractant (British Columbia), and the Mississippi extractant (Table 8–4).

Dilute or concentrated acids will remove both exchangeable K and a fraction of nonexchangeable K. Hunter and Pratt (1957) demonstrated that soil K extraction with H_2SO_4 could reliably predict K removal by various crops in field and greenhouse conditions. Extracting K by refluxing with HCl was the most precise and fastest means of determining soil K status, including the rate and amount of fixed K release (Singh et al., 1983). However, McCallister (1985) has shown that K extraction with boiling HNO_3 (Wood & DeTurk, 1940; Brown et al., 1973) or sodium tetraphenylboron (Scott et al., 1960; Schulte & Corey, 1965) do not mimic natural feldspar weathering, even though they may provide estimates of plant-available soil K. Hazards associated with use of concentrated acids in routine soil testing have limited the adoption of these methods.

Use of cation exchange resins to remove plant-available soil K has been suggested (e.g., Haagsma & Miller, 1963), but has met with variable success (Skogley & Haby, 1981; Schaff & Skogley, 1982a; Singh et al., 1983). McLean and Watson (1985) have presented an excellent discussion of these and other approaches to soil K extraction. Skogley and Georgitis (1988) reported results of continued research on development of a resin approach as a universal extractant.

2. Incubation and Displacement Techniques

A substantial number of experiments have confirmed that exchangeable K in the topsoil can be used to predict fertilizer K response under many conditions (e.g., Havlin & Westfall, 1985). Under other conditions, however, exchangeable K has been a poor indicator of crop response. For example, Skogley and Haby (1981) found no soil test method for K to be adequate for the cool, dry, high-exchangeable K soils of Montana. Application of KCl significantly decreased the incidence of dryland common root rot (*Fusarium* spp.) in dryland barley (*Hordeum vulgare* L.) in Montana (Garvin et al., 1981). That K_2SO_4 in the same experiment had no effect on this root rot, suggested that disease reduction because of Cl^- could be partially responsible for the observed yield increases attributed to K in these high-exchangeable K Northern Great Plains soils. Evidence from South Dakota confirmed that yield response to KCl fertilizer on high K-testing soils is due, in part, to improved Cl nutrition (Fixen et al., 1986). On low-exchangeable K, sandy east Texas Timberlands soils and Coastal Plains soils of the southeastern USA, lack of initial crop response to applied K was because of rapid release of nonexchangeable K from micaceous minerals in the subsoil (Hons et al., 1976; Yuan et al., 1976). Due to their low clay content, intensive cropping of the deeper sands for several years soon depletes plant-available subsoil K. Barbarick (1985) suggested that alfalfa response to K_2SO_4 on Colorado soils with high-exchangeable K throughout the profile may be because of depression of Na absorption. Only 13% of the variability in K uptake by corn was accounted for by exchangeable K at 51 field sites in Wisconsin (Wietholter, 1983). Considering the complexity of and the interactions among the factors affecting K availability to plants, it is remarkable that simple measures of soil K bear any relationship to K uptake and yield.

Several methods have been proposed to permit more thorough characterization of soil K availability. These methods are not generally used in routine soil testing in the USA and Canada, but may be used in the future to more precisely estimate K-fertilizer requirements for groups of soils and specific crops.

The soil solution is the major pool of K^+ available to plants. Potassium in solution tends to be in equilibrium with K on exchange sites; if addition to (e.g., via fertilization) or removal from (e.g., via absorption by roots) solution occurs, some K^+ will attach to or be released from the exchange sites. Dissolution or formation of hydrous micas will also occur in response to changes in soil solution K^+ concentrations (McLean, 1978; Dubetz & Dudas, 1981; Sparks & Huang, 1985).

Changes in exchangeable K must be balanced by opposite changes in other exchangeable cations, notably Ca and Mg. Because changes in K availability in the soil do not occur independently of other cations, it is appropriate to express the status of K relative to the status of the dominant cations (Schofield, 1947). The activity, or effective concentration, of K^+ in soil solution divided by the combined activity of Ca^{2+} and Mg^{2+} is called the activity ratio (AR). This ratio facilitates comparison of different soils, because the K activity is expressed at comparable $Ca^{2+} + Mg^{2+}$ activity.

Several parameters related to K availability can be derived from the relationship between the AR of K^+ (intensity) and the amount of exchangeable K (quantity) when a soil is equilibrated with different amounts of added K^+ (e.g., Beckett, 1964b; Sparks & Huang, 1985). This relationship is called the quantity/intensity (Q/I) curve. The three most important parameters of this relationship are: (i) the X intercept, which indicates the equilibrium AR for K^+ (AR_e); (ii) the Y intercept, which is an estimate of labile or exchangeable K using this procedure; and, (iii) the slope of the curve at AR_e, which is called the potential buffering capacity (PBC). The greater the PBC, the less change will occur with K addition or removal.

The Q/I approach has been useful in explaining K supply in particular soils (Sparks & Liebhardt, 1981) or in related soils (Zandstra & MacKenzie, 1968), but is not applicable to soils of widely different Ca and Mg contents (Beckett, 1964a; Quemener, 1979). Parra and Torrent (1983) suggested a method of more rapidly determining Q/I relationships. A greatly simplified Q/I analysis provided good estimates of K uptake by white clover (During & Duganzich, 1979).

Goedert and Corey (1973) used a single equilibration with 0.001 *M* $SrCl_2$ to estimate ion activity, exchangeable level, buffering capacity, and AR of K^+, Ca^{2+}, and Mg^{2+}. Wietholter (1983) improved on their approach by using 0.004 *M* $Sr(NO_3)_2$. He used a diffusion-controlled K uptake model to predict K accumulation in field-grown corn and was able to account for 58% of the variability in measured K uptake. Further research on this extractant is underway at the Univ. of Wisconsin.

Several comprehensive models of plant nutrient availability have been developed (e.g., Barber, 1984). Use of such models will likely become more common in areas where important climatic factors are not too variable and where soil factors, especially K buffer power and the effective diffusion coefficient for K^+, can be reasonably estimated from existing databases such as soil survey information. Critical plant parameters, such as root growth rate, nutrient uptake activity, and effective radius, must be determined for different crop species and genotypes. This model was less successful in predicting K uptake by alfalfa seedlings (Aighewi, 1988) than has been reported for corn (Shaw et al., 1983) or soybean (Silverbush & Barber, 1983). Mechanistic models will be able to provide useful predictions only to the extent that we understand the system being modeled and that we provide accurate and precise values for the input variables.

Baker's method of soil testing incorporates estimates of quantity and intensity (Baker & Amacher, 1981). Quantity is determined by extraction with standard solutions (e.g., neutral 1 *M* ammonium acetate for K, Ca, and Mg) or by equilibration with the Baker soil test solution, and thermodynamic analysis is used to calculate availability (intensity). The Baker soil test solution (Table 8–4) was designed to represent optimum concentrations of K^+, Ca^{2+}, and Mg^{2+} in an ideal soil. Reduction of the K^+ concentration in solution after a 24-h incubation indicates that insufficient K is available to maintain the optimum concentration. The Baker method is presently being used only in Maine and in a small lab in Pennsylvania, due to the seeming complexity of calculations and the need, as with all soil-testing procedures, of correlation and calibration data for different soil conditions. The complexity of the calculations should not be a hindrance because they can be made rapidly by computer.

McLean's method uses the classical (Fig. 8–7) or updated response curves for field crops with standard ammonium acetate extraction of exchangeable K, but adjusts the fertilizer K recommendation by fixation factors specific to various groups of soils (McLean et al., 1982; McLean & Watson, 1985). The fixation factor can be determined by equilibrating duplicate soil samples for 2 h with and without a known amount of added K, followed by measurement of exchangeable K. The non-amended sample provides a measure of the existing K level; the decrease in K recovered in the amended sample is an index of the K-fixation capacity. These 2-h values can more accurately reflect field conditions by adjusting them to the fixation measured after 8 wk of incubation (Moorhead & McLean, 1985). The relationships between short- and long-term incubations need be determined only once and can be limited to selected soil groups representative of the region from which samples are received.

McLean's method has the potential of more precisely predicting K fertilizer requirements for individual soils because it does not rely on the standard, but inaccurate, 60% recovery figure used in many recommendations. Soils vary from <20% to >90% fertilizer K recovery. The primary disadvantage is that two extractions per soil sample are needed, requiring significantly more bench space, equipment, reagents, and labor.

3. Electro-ultrafiltration

The technique of electro-ultrafiltration (EUF) was developed more than 60 yr ago by Bechold (cited by Nemeth, 1979). It combined the methods of electrodialysis and ultrafiltration, both of which had been used to separate dissolved nutrients from soil. Most of the current literature on this method is by K. Nemeth and colleagues (e.g., Nemeth, 1979, 1982; Nemeth et al., 1970). In 1981, EUF was being used in more than 20 countries either in research or in soil-testing laboratories (Nemeth, 1982). Since 1974, the Tulln Sugar Factory in Austria has used EUF as a routine method of extracting nutrient elements important to sugarbeet producers. They extract up to 20 000 soil samples per year. The Southern German Sugar Company (Rain, West Germany) has 26 EUF units and has computerized the collection and interpretation of data from ICP spectrometers and various N analyses (J.T. Moraghan, 1988, personal communication).

Like universal extractants, EUF provides a method of extracting several elements simultaneously, including NH_4^+-N, NO_3^--N, P, K, Ca, Mg, Na, S, B, Mn, and Zn. The apparatus consists of three cylindrical compartments or "cells" connected end to end and separated by microfiber filters and platinum electrodes. A 1:10 soil/solution suspension is stirred in the central cell while voltage is applied to the electrodes. Vacuum in the outer cells withdraws water and dissolved constituents from the central cell; water is added continuously to the suspension to maintain a 1:10 ratio. Voltage and suspension temperature are increased during extraction to remove the less labile forms of the elements. Fractions are often collected from the outer cells at 5-min intervals over 35 min, but for routine soil-testing applications, various groups of fractions are combined before analysis, e.g., 0 to 10 min, 10 to 30 min, and 30 to 35 min. Fractions collected on the anode and cathode sides are usually combined before analysis, because mass flow causes incomplete separation of charged species during the initial phases of extraction. The fractions are analyzed by standard techniques for the elements of interest.

An example of a typical K^+ extraction curve is shown in Fig. 8–10. The amount of K^+ recovered after 5 min at 50 V and 20 °C is approximately equal to the K^+ in soil solution (Nemeth, 1979; van Lierop & Tran, 1985). The K^+ recovered in the first 10 min (5 min at 50 V and 5 min at 200 V) is closely correlated with the AR_e, but should not be regarded as a measurement of intensity, because it apparently includes some exchangeable K (Sinclair, 1980). Amounts of K^+ extracted by EUF over a longer period (5 min at 50 V and 25 min at 200 V) were closely correlated with K extracted by ammonium acetate, double acid, and Mehlich-II techniques, but were not as closely related to K^+ removed by $SrCl_2$ (Table 8–5). Potassium extracted during the last 5 min at 400 V and 80 °C reflects the buffering capacity of the soil, with little release from kaolinitic and substantial release from smectitic clays (Nemeth, 1979).

Fertilizer requirements vary with clay content (Fig. 8–11) and the relationship between easily and slowly available K. The values in Fig. 8–11 were developed on central European soils and may not be applicable for soils of different parent materials and development. Certainly, the rates required for clay soils that have low amounts of EUF extractable K^+ would be prohibitively expensive for contemporary North American farms.

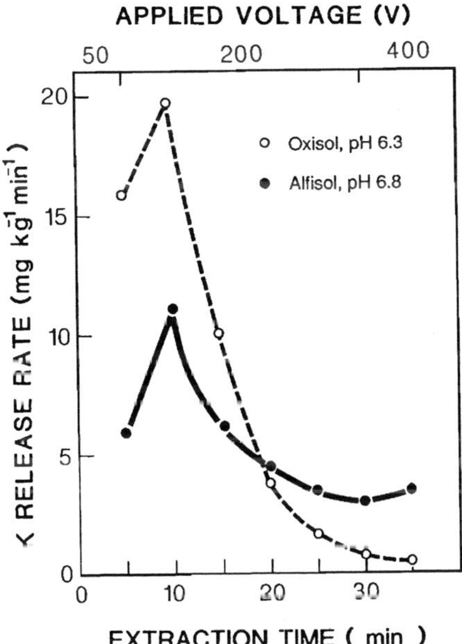

Fig. 8–10. Typical EUF-K release rate curves for two soils of similar exchangeable K content (Oxisol, 262 mg kg^{-1}; Alfisol, 238 mg kg^{-1}), but different K release patterns. (Redrawn from Nemeth, 1979.)

Table 8-5. Relationship between soil K determined after chemical extraction and after EUF extraction (5 min at 50 V and 25 min at 200 V) for 102 soils (van Lierop & Tran, 1985).

Extractant (X)	Potassium concentration range	Estimate of EUF-K (Y)	r^2	SE†
	mg kg^{-1}			
Ammonium acetate 1.0 *M*, pH 7.0	8-250	Y = 0.78X − 13.1	0.85	16.9
Double acid	15-200	Y = 1.0X + 3.2	0.90	13.4
Mehlich II	8-240	Y = 0.80X − 8.6	0.88	16.8
Strontium chloride 0.01 *M*	1-95	Y = 1.51X + 13.7	0.76	21.2

† SE = Standard error of the estimate of Y.

This method of soil nutrient extraction has received little attention in the USA and Canada, although a few researchers have begun to evaluate EUF. Commercially available automated units cost more than $20 000 (U.S. dollars) in 1988. One EUF unit can be used to extract only 8 to 16 samples per day, depending on extraction time. The expense of purchasing several

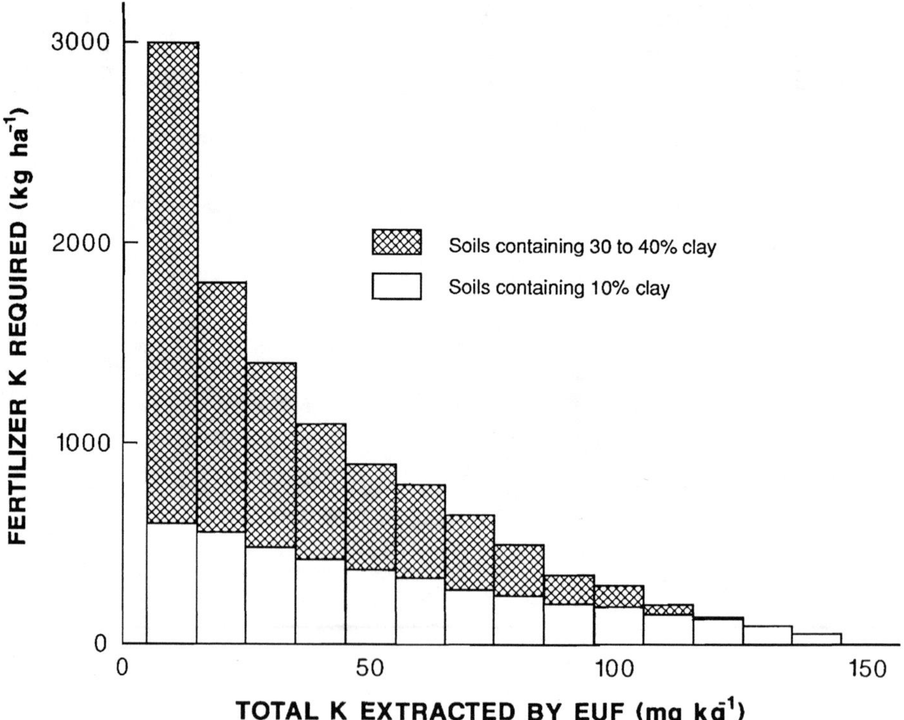

Fig. 8-11. Quantity of fertilizer K needed to increase soil EUF-K values to 150 mg kg^{-1} as related to clay and measured EUF-K content. (Adapted from Nemeth, 1979.)

of these units, the necessity of employing a highly trained, technical expert to supervise the analyses of these chemically complex samples, and the need for extensive calibration data for North American conditions are definite impediments to the expansion of this method. van Lierop and Tran (1985) concluded that little additional information could be obtained by EUF compared to less-expensive and labor-intensive chemical procedures. Others disagree (see Nemeth, 1982). More research is needed to resolve these issues.

4. Sampling and Sample Handling

The effect of soil moisture content on exchangeable-K concentrations has been recognized since at least 1928 (Steenkamp, 1928). The direction of change in exchangeable K upon drying depends on the mineralogy, equilibrium concentration of K^+, and the deviation from the equilibrium concentration at sampling. Haby (1975) analyzed the 0- to 15- and 15- to 30-cm horizons of 18 Montana soils, and found that no change occurred during air drying when exchangeable K in field moist samples was 421 mg kg^{-1} (Fig. 8–12). Fixation occurred during drying at higher K levels, release at lower K levels. If a similar value is characteristic of soils from more humid regions, most would be expected to release K upon drying, because few fields would have soil test levels approaching 1000 kg ha^{-1}.

Dramatic changes in exchangeable K can occur in samples as the moisture content of the air-dry conditions changes with relative humidity (Fig. 8–13; Luebs et al., 1956). The slopes of the curves in Fig. 8–13 range from -360 to -960 mg of K kg^{-1} water at the air-dry moisture content. Small changes

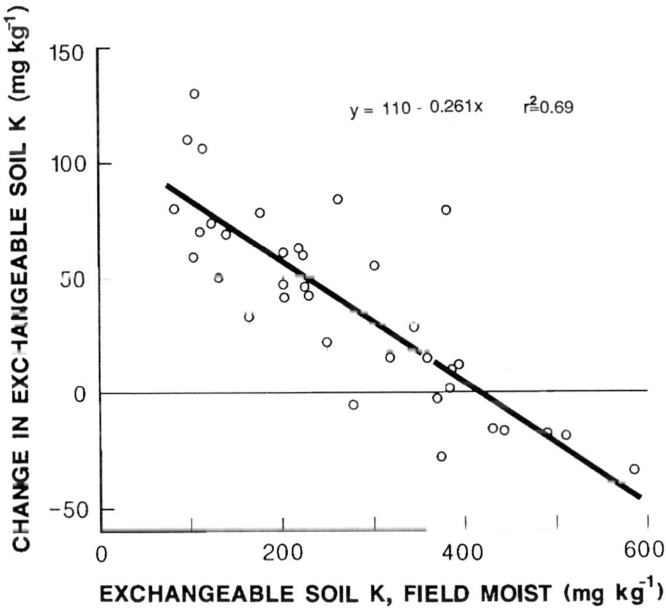

Fig. 8–12. Effect of initial exchangeable K content of field-moist soils of the northern Great Plains on K release or fixation upon air drying at 25 °C. (Adapted from Haby, 1975.)

in relative humidity during drying can alter soil test results, with greater effects occurring in finer-textured soils. The effect of air drying is more pronounced in K-fertilized soil than in nonfertilized soil (Fig. 8–14). Differences between control and fertilized soils are often greater for field-moist samples than for dried samples (Grava et al., 1961). Extractable K was linearly related to soil moisture content in 9 of 10 Montana soils evaluated (Haby et al., 1988).

Although air-drying soils is known to cause unpredictable changes in exchangeable K, field-moist soil samples are not being routinely analyzed by any public laboratory at present. For more than 20 yr, the public laboratory in Iowa used field-moist samples for K analysis, but has begun drying all samples. Sample storage is simplified and subsamples are less variable. Changes in exchangeable K because of drying are smaller in topsoils than in subsoils, because topsoils usually contain less clay. Analytical results of both dry and moist samples generally reflect previous management of topsoils. Drying and wetting, and, in northern climates, freezing and thawing cycles, are also common in topsoils. Therefore, air drying of topsoil samples may not be an unreasonable treatment. Oven-drying soil samples, even

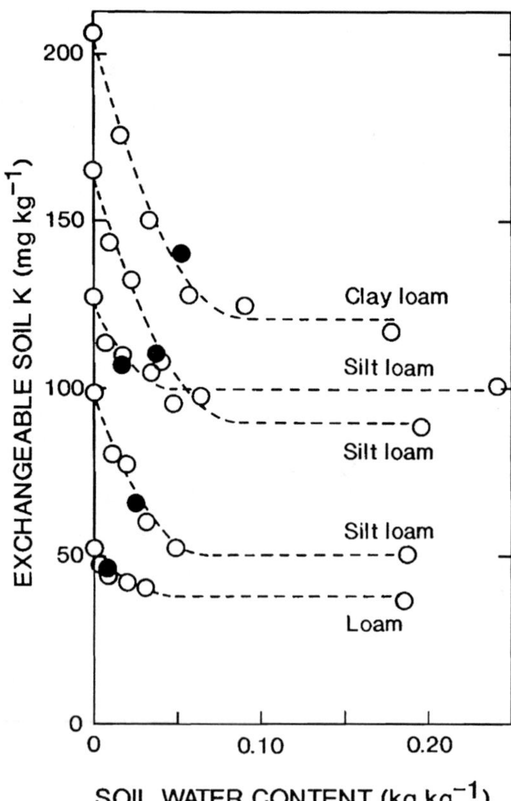

Fig. 8–13. Relationship between exchangeable-K concentration and soil moisture content of several Illinois topsoils (R.A. Bohannon and T.R. Peck, 1957, unpublished data). Closed circles indicate air-dry condition for each soil (25 °C at about 50% relative humidity).

as low as 60 °C, for K analysis should be avoided because this more severe drying causes greater K release (Haby et al., 1988).

Variations in soil test values for K over time should be expected in soils containing micas, vermiculite, or smectites (e.g., Grava et al., 1961; Fig. 8-14). These changes occur because of altered moisture content, freezing and thawing (which are essentially changes in soil moisture content), equilibration of exchangeable and nonexchangeable pools of K after K^+ addition or removal, equilibrium reactions after changes in the concentration of other cations, uptake and release of K^+ from plants and animals (including biocycling of subsoil K to the topsoil [c.f., James et al., 1975]), and leaching. Exchangeable K contents will reflect the net effect of these contrasting mechanisms. Changes in soil solution K^+ will be more extreme than alteration of exchangeable K during periods of plant growth, whereas changes in less labile forms of K could be expected to be rather small over the course of 1 or 2 yr.

Rapid increases in exchangeable K occur after fertilization, but concentrations typically decline during the growing season (Fig. 8-14). Similar decreases in nonfertilized soil also occur, but are less pronounced. In most

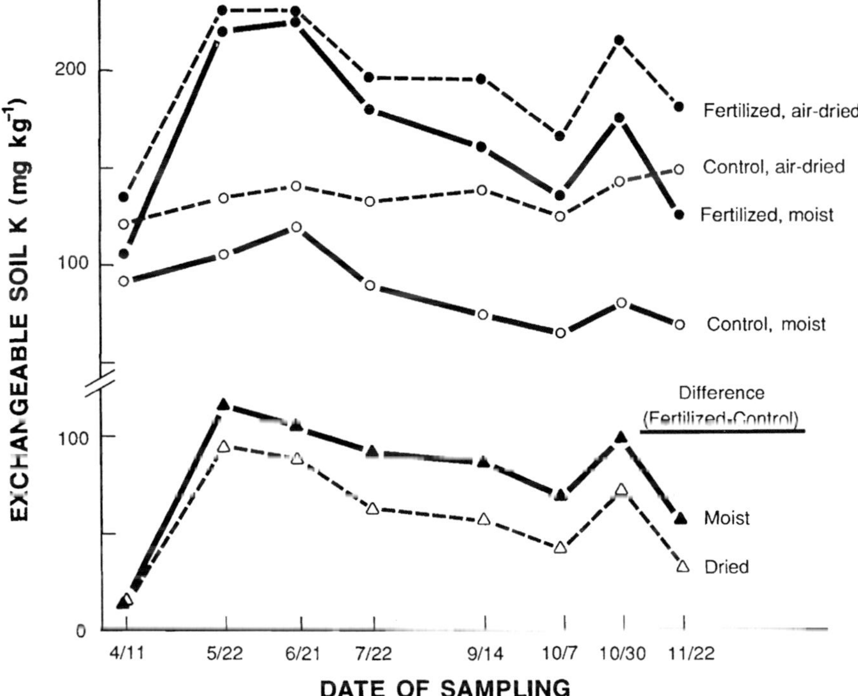

Fig. 8-14. Effect of sampling date and air drying on exchangeable K in the Ap horizon of an Aquic Hapludoll. Fertilized plots had received spring broadcast K applications of 74 kg ha^{-1} yr^{-1} for 3 yr, including the year of sampling, and corn was grown on the plots each year. (Adapted from Grava et al., 1961.)

Canadian provinces and the northern USA, maximum exchangeable-K concentration in unamended soils will generally occur during winter due to freezing. This conclusion has been demonstrated in Illinois (T.R. Peck, 1988, personal communication) and Ohio (Lockman & Molloy, 1984). Because changes in exchangeable K may be appreciable and because we are not yet able to predict these changes (Sparks & Huang, 1985), soil samples should be taken at the same time of year to reduce time-dependent variability.

B. Calcium and Magnesium

The most common extractant for determining what is considered exchangeable Ca and Mg from soils is molar ammonium acetate at pH 7.0 (Lanyon & Heald, 1982). Most determinations of exchangeable soil Ca and Mg are made by high-temperature flame analysis using atomic absorption or ICP emission spectrometry. Ammonium and acetate are volatile in the flame, leaving no salt residue on the burners of these instruments.

Many reports discuss the relationship of a soil-extracted plant nutrient to plant uptake of the particular nutrient under consideration. Much of the crop response-soil test calibration research has been conducted under simulated conditions. Because of the many climatic and soil factors that affect plant availability of Ca and Mg, the amounts of these nutrients removed from a soil by a particular extractant must be correlated with field crop response to applied Ca or Mg in order for that soil extractant to accurately predict the crop's nutrient needs.

Calcium soil test levels and analytical methods used to predict availability of Ca to plants have not received the attention extended to Mg. Scientists and professional agricultural consultants generally agree that Ca availability to plants is not a problem in alkaline soils. Increasing acidity lowers exchangeable soil Ca. It has been well demonstrated that Al or Mn will often become toxic in acid soils before Ca concentrations become deficient. Maintenance of a favorable soil pH by limestone application to acid soils also provides adequate Ca for plant growth.

In contrast to Ca, research on soil test methods for Mg has been extensive for two reasons: (i) to predict crop yield response to applied Mg, and (ii) to predict Mg uptake by the plant to prevent hypomagnesaemia in ruminant animals. Chemical extractants evaluated for exchangeable Mg are many and include use of dilute salt solutions, acidified salt solutions, and dilute acids. MacDonald et al. (1978) correlated molar ammonium acetate-acetic acid, pH 4.8, with Morgan's solution (Table 8–4) to find a method that would be noncorrosive to the flame burner tip. The two methods extracted similar concentrations of Ca and Mg over a soil sample pH range of 4.4 to 7.8. Hooper (1967) found that Morgan's reagent at a soil/extractant ratio of 1:5 (wide ratio) extracted the most Mg. Molar ammonium nitrate at 1:5 soil/extractant and molar ammonium acetate at 1:25 extracted similar amounts of Mg. Morgan's reagent, narrow range (1:2, soil/extractant), shaken for 2 min extracted only two-thirds the Mg as did the wide range.

Calcium chloride removed only about one-half the Mg extracted by Morgan's reagent (wide ratio), ammonium nitrate, and ammonium acetate.

Other methods evaluated for available soil Mg include 0.05 *M* NaCl at a 1:5 ratio (Reith, 1967), neutral *M* ammonium acetate at a 1:5 ratio (Fox & Piekielek, 1984), and 1 *M* sodium acetate at pH 7 and pH 1 (Rice & Kamprath, 1968). Farina et al. (1980) found that 0.01 *M* $CaCl_2$ extracted lower amounts of Mg than did neutral *M* ammonium acetate. These scientists questioned the predictive value of exchangeable Mg as an index of plant availability because drastic decreases in exchangeable Mg were not accompanied by reduced Mg absorption.

III. CALIBRATION AND INTERPRETATION OF SOIL TESTS

There are basically two philosophies of soil test interpretation used by soil-testing organizations to arrive at fertilizer recommendations for K, Ca, and Mg: (i) the nutrient sufficiency level concept and (ii) the nutrient maintenance level concept.

A. Sufficiency Level

In the sufficiency level concept, there are definable levels of individual nutrients in the soil below which crops will respond to added fertilizers with some probability and above which they likely will not respond (Eckert, 1987). It has come from long-term calibration of soil tests with field yield response data establishing ranges of response assured, likely, possible, and unlikely, otherwise expressed as very low, low, medium, and high nutrient levels, respectively (Olsen et al., 1954; Olson et al., 1954). Some measure of the buffering capacities of individual soils is necessary to allow for reasonable soil buildup recommendations (Eckert, 1987), but the main objective of recommendations based on the sufficiency level concept is to "fertilize the plant." The sufficiency level concept of soil test interpretation requires the development of extensive databases from field trials that relate yield response of crops to fertilizer application on soils with different initial soil test nutrient levels. These databases are all too often poorly organized, no longer available for review, or several decades old. Field calibration research is expensive, time consuming, labor intensive, and not always conducive to publication—all of which strongly inhibit the development and maintenance of comprehensive databases. Although criticisms are made about the lack of relevance of calibration equations, they are frequently adequate to sustain high yields and to maximize profit (Olson et al., 1982).

B. Maintenance Level

The maintenance level concept is one of "fertilizing the soil." Adequate nutrient is applied annually to replace that which is expected to be used by the crop, plus some extra to increase the soil test level. Once the soil test reaches the optimum high or very high levels, a maintenance application equal to the amount removed by the crop is applied to prevent the crop from lowering soil nutrient reserves. Soil test calibration for the maintenance level concept may include knowledge of the fixation capacity of a given soil for various nutrients, but is generally based on a standard value for the fixation capacity. Values for crop removal are usually calculated from average elemental concentrations and estimated or measured yields.

Optimum soil test levels can be based simply on extractable concentrations or on the ratio of extractable nutrients to one another. This "conservation" of a soil's nutrient-supplying capacity has strong appeal but discounts the economic aspect to the farmer in those situations where the soil's delivery capacity of a given nutrient may be adequate for top yields for some years to come (e.g., K in Houston Black clay in Texas).

The BCSR concept is used by some who advocate the maintenance approach. In a discussion of the BCSR concept, a distinction must be made between it and basic cation saturation of the soil CEC. With the BCSR concept, attempts are made to determine the amounts of fertilizer K, Ca, and Mg from ratios of these nutrients in the soil sample analyzed without knowledge of the soil CEC. Soil CEC must be known to compute the basic cation saturation percentage.

Numerous experiments during the past 40 yr have demonstrated that use of the BCSR approach alone in making fertilizer recommendations is both scientifically and economically questionable (Hunter, 1949; Key et al., 1955; Geraldson, 1957; Salmon, 1964; McLean & Carbonel, 1972; Claassen & Wilcox, 1974; Simson et al., 1979; Schulte et al., 1981; Eckert & McLean, 1981; Olson et al., 1982; Donohue, 1983; McLean et al., 1983; Fox & Piekielek, 1984). According to Leibhardt (1981), recommendations based on cation saturation percentages of 75% Ca, 15% Mg, and 2 to 5% K and the resulting cation ratios are not warranted, and may result in reduced yields, increased fertilizer costs, or both. Use of the BCSR typically increases fertilizer recommendations compared to other approaches. This method also ignores the selectivity that plants demonstrate during absorption of ions from soil solution.

McLean et al. (1982) have suggested that fertilizer recommendations could be based on a combination of Fisher's (1974) approach and short-term fertilizer recovery tests. This method assumes there is an optimum exchangeable K concentration in soil which varies positively with CEC. Fertilizer recommendations are calculated by (i) determining the difference in exchangeable K of the soil sample and the desired exchangeable K level, and (ii) dividing by the predicted recovery of added K. Potassium recovery varies with soil clay content and mineralogy (McLean, 1978; McLean et al., 1982; Stout, 1982). McLean's analytical method was discussed in section II.A.2.

IV. SOIL TEST LEVELS FOR ADEQUACY

A. Potassium

In the survey of public soil-testing laboratories, respondents were asked for general soil K test levels deemed adequate for maximum yield of different crops. As expected, these minimum adequate levels vary among the laboratories. States and provinces with the highest minimum values for corn (>200 mg of K kg^{-1}) are British Columbia, Manitoba, Missouri, Montana, Ohio, Quebec, and Wisconsin. Those with the lowest minimum values for adequate soil K (<110 mg of K kg^{-1}) include Alabama, Colorado, Delaware, Idaho, Massachusetts, Pennsylvania, South Carolina, and Utah.

Minimum adequate soil test levels vary with crop, soil, and management. In Minnesota, fertilizer K is not recommended for high yields of sugarbeets (*Beta vulgaris* L.), corn, and alfalfa on soils testing >100, 150, and 175 mg kg^{-1}, respectively (e.g., Fig. 8-15; Rehm et al., 1985). In Alabama, minimum adequate soil K test levels for corn must be above the medium level of 40, 60, or 80 mg kg^{-1}, for noncalcareous soils with CEC ranges of 0 to 4.6, 4.6 to 9.0, or >9.0 $cmol_c$ kg^{-1}, respectively (Cope et al., 1981). Addition of K fertilizer to alfalfa and corn is recommended in New York if soil K test levels are below the very high zone that begins at 150 mg kg^{-1} in fine-textured lacustrine sediments, and at 270 mg kg^{-1} in coarse-textured glacial

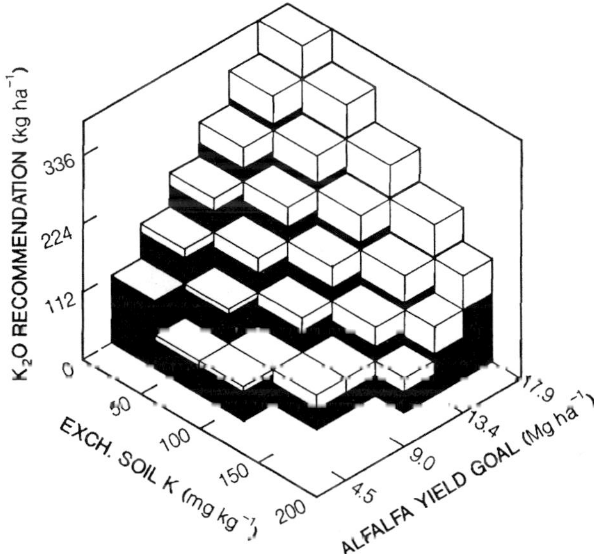

Fig. 8-15. Fertilizer K recommendations for alfalfa in Minnesota for different soil test K levels and yield goals. (Adapted from Rehm et al., 1985.) Dark columns are recommendations for fine- to medium-textured soils; light columns indicate additional fertilizer rates for coarse-textured soils.

till or outwash with low K reserves (Anonymous, 1984). Topdressed K fertilizer is recommended for alfalfa in Arkansas regardless of soil test levels (W.E. Sabbe, 1988, personal communication).

These few examples indicate the wide range of approaches to soil test interpretation and K-fertilizer recommendation provided by public soil-testing laboratories. Private soil-testing laboratories promote still other perspectives on interpretation and calibration. It is no wonder that producers are often confused or suspicious about the use of soil testing to predict fertilizer needs.

About 92% of all K fertilizer used in the USA is applied in states east of 95 °W longitude (USDA, 1984), as expected from the distribution of K-responsive soils (Fig. 8–1). An average rate of only 4.3 kg of K ha^{-1} (total K use divided by cropland area) was applied to cropland in the western states in 1982, in contrast to average rates >50 kg ha^{-1} in 18 of the 31 eastern states. Greatest average application rates were in Hawaii (288 kg of K ha^{-1}), followed by Florida at 159 kg of K ha^{-1}. Obviously, actual application rates vary with crop, soil, and management conditions.

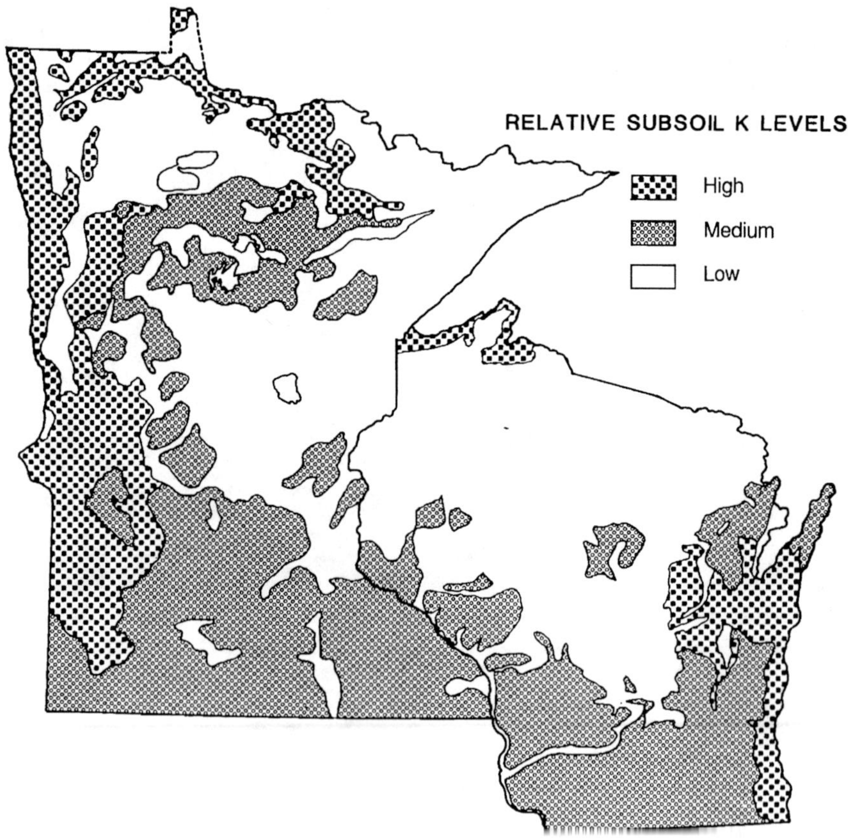

Fig. 8–16. Subsoil K status in Minnesota and Wisconsin. (Adapted from Kelling et al., 1981; and Rehm et al., 1985.)

The impact of subsoil K availability on crop yields has been alternately emphasized and disputed over the years. Seven public laboratories currently modify their recommendations with subsoil K fertility indexes, and a few others are conducting or contemplating further research. An example of subsoil K maps is presented in Fig. 8–16. Subsoil K status maps are usually based on standard exchangeable K analysis of dried samples acquired through extensive sampling of representative soil profiles. This approach could be questioned, because of the variable and often large changes in exchangeable K which occur upon air drying. Some laboratories predict K availability through use of other subsoil characteristics, such as depth, texture, bulk density, and average moisture status.

Although the use of subsoil K-fertility levels to modify K-fertilizer recommendations in intensively managed agriculture is not widespread, it is recognized as being important in many situations. Lack of crop yield response on some sandy Paleudults on the Coastal Plains and Timberlands areas of the southeastern USA and Texas has been reported because of an available supply of subsoil K (Yuan et al., 1976; Hons et al., 1976; Sparks et al., 1980; Woodruff & Parks, 1980). The Fertility Capability Classification System has a modifier, k, which is used to designate soils where K-supplying power is low (Sanchez et al., 1982).

Recent research in Montana has demonstrated that soil characteristics, such as mean annual soil temperature at 50 cm, moist consistence of the Ap horizon, moist and dry consistence of the B horizon, and the clay content of calcic horizons, can be used to identify sites and soils where a response to K can be expected (Schaff & Skogley, 1982b). These characteristics are related to the rate of K diffusion in the soil.

Reasonable success has been achieved in predicting subsoil K-availability indexes with other characteristics, such as clay content, which are typically included in Soil Survey information databases (Aighewi, 1988). These equations cannot be extrapolated beyond the region in which they were developed. It is unlikely that the size of one pool of soil K can be successfully predicted by another, because the relationship among K pools is not consistent in different soils (Fig. 8–17). Climatic factors are extremely influential in the response actually observed in the field, because soil moisture content and temperature greatly affect K diffusion (Schaff & Skogley, 1982a). Clearly, using inherent soil factors alone will not result in consistent predictions of K availability to crops.

B. Calcium

The major areas of concern for Ca supply to plants are in highly leached, acid soils and in soils where excessive levels of other cations such as K^+, Na^+, or NH_4^+ salts have been used, or where Mg in serpentine soils inhibits Ca uptake by plants. In sufficiently acid soils, Ca is primarily supplied as limestone to alleviate actual or potential phytotoxicities of Al^{3+}, Fe^{2+}, Mn^{2+}, and H^+. Only in the absence of toxic amounts of these ions can one determine the need for Ca (Doll & Lucas, 1973). Melsted (1953) observed

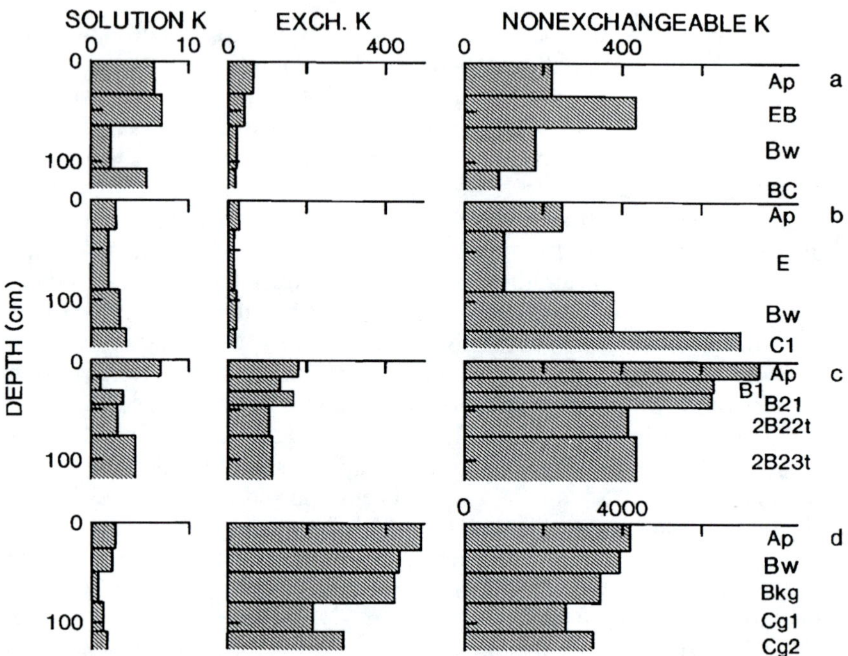

Fig. 8–17. Relationship between genetic horizon and various "pools" of soil K (all in mg kg^{-1}) for four Minnesota soils: (a) Entic Haploboroll; (b) Alfic Udipsamment; (c) Aquollic Hapludalf; and (d) Vertic Haplaquoll (I.T. Aighewi and M.P. Russelle, 1988, unpublished data). Note change in scale for nonexchangeable K for soil d.

Ca-deficiency symptoms on plants growing in soils having pH values below 4.5 and containing <400 mg Ca kg^{-1}. Many crops will respond to Ca applications when the degree of Ca saturation of the exchange capacity falls below 25%. Soybean was reported to suffer Ca deficiency at Ca saturations of 20% or less, while sugarcane in Hawaii grew normally with 12% Ca saturation in volcanic soils (Tisdale et al., 1985). Kaolinitic clays are able to satisfy the Ca requirements of most plants at lower saturation percentages than montmorillonite, which requires a Ca saturation of 70% or more before this element is released in sufficient supply to growing plants.

Although most suggestions are for application of Ca as limestone to correct the soil pH, nearly all laboratories maintain ratings of extractable Ca. Calcium levels at which crops are no longer expected to respond to Ca application vary among laboratories and soil types, and range from 250 mg kg^{-1} in Wisconsin sandy soils to 500 mg kg^{-1} in Wisconsin silty clay soils. Several states use the Ca saturation percentage of the soil CEC to determine adequate Ca availability. Maine defines 75% Ca saturation as adequate.

Most laboratories use the same extractant for Ca as is used for K (Fig. 8–8). The most widely accepted Ca extraction method appears to be 1 *M* ammonium acetate at pH 7 for soils of the USA and Canada. Concentrations of Ca extracted by this method and especially by any acidic extracting solution, will include Ca solubilized from native or applied and unreacted

$CaCO_3$. Extractable Ca would be a more appropriate term to use than exchangeable, especially when acidic extractants are used.

C. Magnesium

Attempts to correlate extraction of soil Mg with crop response rarely include evaluation of subsoil Mg. Elimination of plant access to subsoil Mg is one reason why greenhouse research on soil Mg and crop response is usually more successful than field research in producing statistically significant correlations. Soil material used in greenhouse research is normally collected from the soil surface and is only part of the rooting medium to which plants have access in the field. Significant exchangeable Ca and Mg movement into subsoil to depths of at least 45 cm because of limestone application on acid sandy soils was reported by Haby et al. (1979) and Messick et al. (1984). Acid, sandy Alfisols and Ultisols in which plants might be expected to respond to Mg application, often contain an argillic horizon or zone of clay accumulation. In addition, certain of these soils contain interstratified mica and vermiculite in the fine silt and clay fractions (Hons et al., 1976; Yuan et al., 1976). In the micaceous minerals, Mg can exist in the interlayer positions and can partially substitute for Al in the octahedral layer. This Mg would be considered nonexchangeable. Rice and Kamprath (1968) reported that nonexchangeable Mg was readily soluble in dilute acid extractants and may be important for plant growth. A large part of the Mg taken up by corn plants in four of the five soils they studied was from nonexchangeable forms (Fig. 8-5). They hypothesized that a microzone of H^+ ions exists around plant roots. Since the sandy soils used in their study did not have a high buffer capacity, the H^+ ions exchanged from the roots could be quite effective in releasing Mg from the nonexchangeable form. This mechanism might be responsible for many Coastal Plains and other sandy, acid soils not responding to Mg fertilization even though the exchangeable Mg content in the surface is low. This concept was verified by Christenson and Doll (1973), but would depend to a large degree on the ratio of anions to cations absorbed by the plant, as this ratio affects rhizosphere pH.

Two main methods of reporting soil test Mg levels on which to base fertilizer recommendations for Mg are percent saturation of the soil CEC and exchangeable Mg. Fox and Piekielek (1984) indicated exchangeable Mg was highly correlated ($r = 0.99$) with Mg as a percent saturation of the soil CEC. The general consensus appears to be that 5% Mg saturation of the soil CEC is adequate for optimum yields of most crops. For those crops that require a higher concentration of basic cations, such as alfalfa, and for crops such as corn silage and cool season forages needing higher concentrations of Mg to prevent hypomagnesaemia in ruminant animals, 10% Mg saturation of the soil CEC is suggested to maintain the Mg concentration in the dry matter at 2 g kg^{-1} or above. Where acidified soil extractants are used, such as pH 3, *M* ammonium acetate used in Maine, 15% is the Mg saturation percentage above which Mg application is no longer considered necessary for increased yields. Lancaster (1958) reported that the probability of cotton response to Mg application on heavy clay soils was excellent at 3% Mg saturation of the soil CEC and no response was expected above 6.4% saturation.

The sum of exchangeable cations by routine soil extraction includes soluble and soil solution cations and may not represent the total extraction of the exchange complex. Therefore, the actual soil CEC rather than the sum of exchangeable cations should be used to be most accurate when using percent Mg saturation to predict Mg requirements. Until recently, determination of soil CEC was not adapted as a routine analytical procedure suitable for use in a soil-testing laboratory. Begheyn (1987) has developed a single step CEC analysis procedure that may meet the rapid and routine requirements of soil-testing laboratories.

Fox and Piekielek (1984) reported that the exchangeable Mg level recommended for agronomic crop production by different soil-testing services ranged from 25 to 180 mg kg^{-1}. Research on various crops and soils indicated a range of critical levels for soil exchangeable Mg that varied from about 7 to 35 mg kg^{-1}, depending upon subsoil Mg supply, soil K and Ca content, and fertilizer NH_4 application. Respondents to our survey indicated that the range of exchangeable Mg above which fertilizer Mg is no longer recommended varied from 25 to 60 mg kg^{-1}. This range varied from 30 mg kg^{-1} for sandy soils to 50 mg kg^{-1} for silts and clays, and was highest in the Piedmont soils.

According to Doll and Lucas (1973), guidelines established by the advisory workers in the United Kingdom (N.A.A.S., 1968) agree fairly well with North American observations. These levels of exchangeable Mg and interpretations are as follows:

1. 0–25 mg of Mg kg^{-1}: Deficiency symptoms general in most field crops, vegetables, fruit, and glasshouse crops. Magnesium is certainly advised.
2. 26–50 mg of Mg kg^{-1}: Deficiency expected in sugarbeet, potato, kale (*Brassica oleraceae* DC.), fruit, and glasshouse crops. Magnesium is advised except for cereal crops.
3. 51 to 100 mg of Mg kg^{-1}: Absolute deficiency is not likely in field and most vegetable crops. If symptoms occurred, they were likely induced by other factors such as wide K to Mg ratio in the soil. However, Mg is suggested for fruit, glasshouse crops, and grassland crops. Before recommending Mg for grassland, the economics of the treatment should be compared with the cost of direct feeding of magnesia (MgO) to animals.
4. 101 to 175 mg of Mg kg^{-1}: Standard Mg treatment should be given to glasshouse crops [tomato, cucumber (*Cucumis sativus* L.), pepper (*Capsicum annuum* L. var. *annuum*)]. Treatment of grassland should be required where hypomagnesaemia in animals has been confirmed.
5. 176 to 250 mg of Mg kg^{-1}: Magnesium suggested only for glasshouse crops.

V. PROGNOSIS

Economic and environmental concerns mandate that fertilizer recommendations be as accurate and precise as possible. In the short term, nutrient availability indexes of extractable K and Ca and exchangeable Mg will continue to be used despite their shortcomings. Improved methods will be devised as the mechanistic relationships among soil, plant, and climatic conditions are resolved and true phytoavailability is understood. Recommendations of nutrient inputs will remain imprecise because of the unpredictability of weather, and therefore plant growth. Fundamental advances in understanding nutrient supply, uptake, and analytical methodology made since the second edition of this book in 1973 provide a promising basis for future research.

Development and validation of new methodologies and approaches to soil testing and fertilizer recommendations are time- and resource-consuming. Recognized inefficiencies in current methods make it imperative that this work continue and increase. The excellent cooperation among many public and private soil-testing laboratories in improving soil test correlations, evaluating new extraction techniques, and in sharing discoveries and data should be encouraged, recognized, and supported by administrators. Finally, the extension of this knowledge to producers and supporting industries is crucial—both the potentials and limitations of testing technologies must be understood.

ACKNOWLEDGMENT

The authors appreciate the contribution of time and effort made by those individuals who answered our survey and telephone calls, to Drs. R.A. Bohannon and T.R. Peck for the data in Fig. 8–13, and to Mr. Sung Chou for drafting the maps.

REFERENCES

Adams, F. 1974. Soil solution. p. 441–481. *In* E.W. Carson (ed.) The plant root and its environment. Univ. of Virginia Press, Charlottesville.

Adams, F., and J.B. Henderson. 1962. Magnesium availability as affected by deficient and adequate levels of potassium and lime. Soil Sci. Soc. Am. Proc. 26:65–68.

Adams, F., and B.L. Moore. 1983. Chemical factors affecting root growth in subsoil horizons of Coastal Plain soils. Soil Sci. Soc. Am. J. 47:99–102.

Aighewi, I.T. 1988. Predicting soil potassium supply and modeling its availability to alfalfa (*Medicago sativa* L.). Ph.D. diss. Univ. Minnesota, St. Paul (Diss. Abstr. 8907364).

Allaway, W.H. 1945. Availability of replaceable calcium from different types of colloids as affected by degree of calcium saturation. Soil Sci. 59:207–217.

Allison, F.E. 1973. Soil organic matter and its role in crop production. Elsevier, New York.

Anonymous. 1984. 1985 Cornell recommends for field crops. New York State Coop. Ext. Serv.

Arnold, P.W. 1967. Magnesium and potassium supplying power of soils. p. 39–47. *In* Soil potassium and magnesium. Tech. Bull. 14. Her Majesty's Stationery Office, London.

Ashworth, L.J., Jr., A.G. George, and O.D. McCutcheon. 1982. Disease-induced potassium deficiency and verticillium wilt in cotton. California Agric. 36(9, 10):18–20.

Baker, D.E., and M.C. Amacher. 1981. The development and interpretation of a diagnostic soil testing program. The Pennsylvania State Univ. Agric. Exp. Stn. Bull. 826.

Barbrick, K.A. 1985. Potassium fertilization of alfalfa grown on a soil high in potassium. Agron. J. 77:442–445.

Barber, S.A. 1984. Soil nutrient bioavailability. John Wiley & Sons, New York.

Barber, S.A. 1985. Potassium availability at the soil-root interface and factors influencing potassium availability. p. 309–326. *In* R.D. Munson (ed.) Potassium in agriculture. ASA, CSSA, and SSSA, Madison, WI.

Batey, T. 1967. The ratio of K to Mg in the soil in relation to plant growth. p. 143–146. *In* Soil potassium and magnesium. Tech. Bull. 14. Her Majesty's Stationery Office, London.

Bear, F.E., A.L. Prince, and J.L. Malcolm. 1945. Potassium needs of New Jersey soils. New Jersey Agric. Exp. Stn. Bull. 721.

Beckett, P.H.T. 1964a. Studies on soil potassium. I. Confirmation of the ratio law: Measurement of potassium potential. J. Soil Sci. 15:1–8.

Beckett, P.H.T. 1964b. Studies on soil potassium. II. The 'immediate' Q/I relations of labile potassium in the soil. J. Soil Sci. 15:9–23.

Begheyn, L.T. 1987. A rapid method to determine cation exchange capacity and exchangeable cations in soils, including calcareous, gypsiferous, saline and sodic soils. Commun. Soil Sci. Plant Anal. 18:911–932.

Bertsch, P.M., and G.W. Thomas. 1985. Potassium status of temperate region soils. p. 131–162. *In* R.D. Munson (ed.) Potassium in agriculture. ASA, CSSA, and SSSA, Madison, WI.

Boswell, F.C., O.E. Anderson, and L.S. Jones. 1967. Corn yield responses, soil reaction changes as influenced by lime, magnesium, and nitrogen, when high-analysis fertilizers are used. Georgia Agric. Res. 9(1):3–6.

Bray, R.H. 1945. Soil-plant relations: II. Balanced fertilizer use through soil tests for potassium and phosphorus. Soil Sci. 60:463–473.

Bray, R.H., and L.T. Kurtz. 1945. Determination of total, organic, and available forms of phosphorus in soils. Soil Sci. 59:39–45.

Brown, A.L., J. Quick, and G.J. DeBoer. 1973. Diagnosing potassium deficiency by soil analysis. Calif. Agric. 27(6):13–14.

Chapman, H.D., and W.P. Kelley. 1930. The determination of the replaceable bases and base-exchange capacity of soils. Soil Sci. 30:391–406.

Charlesworth, R.R. 1967. The effect of applied magnesium on the uptake of magnesium by, and on the yield of, arable crops. p. 110–124. *In* Soil potassium and magnesium. Tech. Bull. 14. Her Majesty's Stationery Office, London.

Ching, P.C., and S.A. Barber. 1979. Evaluation of temperature effects on K uptake by corn. Agron. J. 71:1040–1044.

Christenson, D.R., and E.C. Doll. 1973. Release of magnesium from soil clay and silt fractions during cropping. Soil Sci. 116:59–63.

Christenson, D.R., R.P. White, and E.C. Doll. 1973. Yields and magnesium uptake by plants as affected by soil pH and calcium levels. Agron. J. 65:205–206.

Claassen, M.E., and G.E. Wilcox. 1974. Comparative reduction of calcium and magnesium in corn tissue by NH_4-N and K fertilization. Agron. J. 66:521–522.

Cope, J.T., Jr., C.E. Evans, and H.C. Williams. 1981. Soil test fertilizer recommendations for Alabama crops. Alabama Agric. Exp. Stn. Circ. 251.

Dahnke, W.C. (ed.). 1988. Recommended chemical soil test procedures for the North Central region. North Dakota Agric. Exp. Stn. North Central Reg. Publ. 221 (revised).

Doll, E.C., and L.R. Hossner. 1964. Magnesium deficiency as related to liming and potassium levels in acid sandy podzols. Int. Congr. Soil Sci., Trans. 8th, 1964 4:907–912.

Doll, E.C., and R.E. Lucas. 1973. Testing soils for potassium, calcium, and magnesium. p. 133–151. *In* L.M. Walsh and J.D. Beaton (ed.) Soil testing and plant analysis. SSSA, Madison, WI.

Donohue, S.J. 1983. Balancing the soil. . . Does it work? p. 11–13. *In* Virginia Tech. Agron. Tips. Vol. 5. Virginia Polytechnic Institute and State University, Blacksburg.

Dubetz, S., and M.J. Dudas. 1981. Potassium status of a dark brown Chernozem soil after sixty-six years of cropping under irrigation. Can. J. Soil Sci. 61:409–415.

Dumanski, J., L. Brittain, and J.L. Girt. 1982. Spatial association between agriculture and the soil resource in Canada. Can. J. Soil Sci. 62:375–385.

During, C., and D.M. Duganzich. 1979. Simple empirical intensity and buffering capacity measurements to predict potassium uptake by white clover. Plant Soil 51:167–176.

Eckert, D.J. 1987. Soil test interpretations: Basic cation saturation ratios and sufficiency levels. *In* Soil testing: Sampling, correlation, calibration, and interpretation. p. 53–64. SSSA Spec. Publ. 21. SSSA, Madison, WI.

Eckert, D.J., and E.O. McLean. 1981. Basic cation saturation ratios as a basis for fertilizing and liming agronomic crops: I. Growth chamber studies. Agron. J. 75:795–799.

Farina, M.P.W., M.E. Summer, C.O. Plank, and W.S. Letzsch. 1980. Effect of pH on soil magnesium and its adsorption. Commun. Soil Sci. Plant Anal. 11:981–992.

Fisher, T.R. 1974. Some considerations for interpretation of soil tests for phosphorus and potassium. Univ. of Missouri-Columbia Agric. Exp. Stn. Res. Bull. 1007.

Fixen, P.E., R.H. Gelderman, J. Gerwing, and F.A. Cholick. 1986. Response of spring wheat, barley, and oats to chloride in potassium chloride fertilizers. Agron. J. 78:664–668.

Fox, R.H., and W.P. Piekielek. 1984. Soil magnesium level, corn (*Zea mays* L.) yield, and magnesium uptake. Commun. Soil Sci. Plant Anal. 15:109–123.

Fried, M., and R.E. Shapiro. 1961. Soil-plant relationships in ion uptake. Ann. Rev. Plant Physiol. 12:91–112.

Gaines, H.L., and H.C. Thomas. 1953. Adsorption studies on clay minerals. II. A formulation of the thermodynamics of exchange adsorption. J. Chem. Phys. 21:714–718.

Gallaher, R.N., H.B. Harris, O.E. Anderson, and J.W. Dobson. 1975. Hybrid grain sorghum response to magnesium fertilization. Agron. J. 67:297–300.

Garvin, J.P., V.A. Haby, and P.O. Kresge. 1981. Effect of fertilizer N, P, K, Cl, and S on yield, protein percentage, nutrient content and root rot on barley. Proc. Northwest Fert. Conf. 32:87–96.

Geraldson, C.M. 1957. Factors affecting calcium nutrition of celery, tomato, and pepper. Soil Sci. Soc. Am. Proc. 21:621–625.

Goedert, W.J., and R.B. Corey. 1973. Availability indices for soil cations by a single equilibrium with $SrCl_2$. p. 82. *In* Agronomy abstracts. ASA, Madison, WI.

Graham, E.R. 1959. An explanation of the theory and methods of soil testing. Missouri Agric. Exp. Stn. Bull. 734.

Grava, J., G.E. Spalding, and A.C. Caldwell. 1961. Effect of drying upon the amounts of easily extractable potassium and phosphorus in Nicollet clay loam. Agron. J. 53:219–221.

Greenland, D.J., and M.H.B. Hayes. 1978. Soils and soil chemistry. p. 1–27. *In* D.J. Greenland and M.H.B. Hayes (ed.) The chemistry of soil constituents. John Wiley and Sons, New York.

Haagsma, T., and M.H. Miller. 1963. The release of nonexchangeable soil potassium to cation-exchange resins as influenced by temperature, moisture and exchanging ion. Soil Sci. Soc. Am. Proc. 27:153–156.

Haby, V.A. 1975. Evaluation of soil test methods and development of a potassium fertilizer recommendation system for Montana. Ph.D. diss. Montana State Univ., Bozeman (Diss. Abstr. 75-25754).

Haby, V.A., W.B. Anderson, and C.D. Welch. 1979. Effect of limestone variables on amendment of acid soils and production of corn and Coastal bermudagrass. Soil Sci. Soc. Am. J. 43:343–347.

Haby, V.A., J.R. Sims, E.O. Skogley, and R.E. Lund. 1988. Effect of sample pretreatment on extractable soil potassium. Commun. Soil Sci. Plant Anal. 19:91–106.

Hanlon, E.A., and G.V. Johnson. 1984. Bray/Kurtz, Mehlich III, AB/D and ammonium acetate extractions of P, K and Mg in four Oklahoma soils. Commun. Soil Sci. Plant Anal. 15:277–294.

Havlin, J.L., and D.G. Westfall. 1985. Potassium release kinetics and plant response in calcareous soils. Soil Sci. Soc. Am. J. 49:366–370.

Henry, L. 1985. Potassium fertilization in crop production. Agdex 541. Plant Industry Branch, Saskatchewan Agric., Saskatoon, SK, Canada.

Hons, F.M., J.B. Dixon, and J.E. Matocha. 1976. Potassium sources and availability in a deep sandy soil of east Texas. Soil Sci. Soc. Am. J. 40:370–373.

Hons, F.M., L.A. Larson-Vollmer, and M.A. Locke. 1990. NH_4OAc-EDTA extractable phosphorus as a soil test procedure. Soil Sci. 149(5):249–256.

Hooper, L.J. 1967. The uptake of magnesium by herbage and its relationship with soil analysis data. p. 160–173. *In* Soil potassium and magnesium. Tech. Bull. 14. Her Majesty's Stationery Office, London.

Howard, D.D., and F. Adams. 1965. Calcium requirement for penetration of subsoils by primary cotton roots. Soil Sci. Soc. Am. Proc. 29:558–562.

Hunter, A.H., and P.F. Pratt. 1957. Extraction of potassium from soils by sulfuric acid. Soil Sci. Soc. Am. Proc. 21:595–598.

Hunter, A.S. 1949. Yield and composition of alfalfa as affected by variations in the calcium-magnesium ratio in the soil. Soil Sci. 67:53–62.

James, D.W., W.H. Weaver, S. Roberts, and A.H. Hunter. 1975. Potassium in an arid loessial soil: Changes in availability as related to cropping and fertilization. Soil Sci. Soc. Am. Proc. 39:1111–1115.

Jardine, P.M., and D.L. Sparks. 1984. Potassium-calcium exchange in a multireactive soil system: I. Kinetics. Soil Sci. Soc. Am. J. 48:39–45.

Johnson, G.V., R.A. Isaac, S.J. Donohue, M.R. Tucker, and J. Woodruff. 1984. Procedures used by state soil testing laboratories in the southern region of the United States. Agric. Exp. Stn. Southern Coop. Series. Bull. 190. Oklahoma State Univ.

Jones, J.B., and F. Haghiri. 1963. Magnesium deficiency in Columbiana County soils. Ohio Agric. Exp. Stn. Circ. 116.

Kelling, K.A., P.E. Fixen, E.E. Schulte, E.A. Liegel, and C.R. Simson. 1981. Soil test recommendations for field, vegetable, and fruit crops. Univ. of Wisconsin-Extension A2809.

Key, J.L., L. Tokurz, and B.B. Tucker. 1955. Influence of ratio of exchangeable calcium-magnesium on yield and composition of soybean and corn. Soil Sci. 93:265–270.

Knudsen, D., and D. Beegle. 1988. Recommended phosphorus tests. p. 12–15. *In* W.C. Dahnke (ed.) Recommended chemical soil test procedures for the North Central region. North Dakota Agric. Exp. Stn. Bull 499 (revised).

Knudsen, D., G.A. Peterson, and P.F. Pratt. 1982. Lithium, sodium, and potassium. p. 225–246. *In* A.L. Page et al. (ed.) Methods of soil analysis. Part 2. 2nd ed. Agronomy Monogr. 9. ASA and SSSA, Madison, WI.

Lancaster, J.D. 1958. Magnesium status of Blackland soils of northeast Mississippi for cotton production. Mississippi State Univ. Agric. Exp. Stn. Bull. 560.

Lancaster, J.D. 1980. Mississippi soil test methods and interpretation. Mississippi Agric. Exp. Stn. Mimeo.

Lanyon, L.E., and W.R. Heald. 1982. Magnesium, calcium, strontium, and barium. p. 247–262. *In* A.L. Page et al. (ed.) Methods of soil analysis. Part 2. 2nd ed. Agronomy Mongr. 9. ASA and SSSA, Madison, WI.

Liebhardt, W.C. 1981. The basic cation saturation ratio concept of lime and potassium recommendations on Delaware's Coastal Plain soils. Soil Sci. Soc. Am. J. 45:544–549.

Lockman, R.B., and M.G. Molloy. 1984. Seasonal variations in soil test results. Commun. Soil Sci. Plant Anal. 15:741–757.

Luebs, R.E., G. Stanford, and A.D. Scott. 1956. Relation of available potassium to soil moisture. Soil Sci. Soc. Am. Proc. 20:45–50.

Lund, Z.F. 1970. The effect of calcium and its relation to several cations in soybean root growth. Soil Sci. Soc. Am. Proc. 34:456–459.

Lunt, H.A., C.L.W. Swanson, and H.G.M. Jacobson. 1950. The Morgan soil testing system. Connecticut Agric. Exp. Stn. Bull. 541.

MacDonald, G.E., N.H. Peck, and M.T. Vittum. 1978. Relationship between ammonium acetate-acetic acid and Morgan's solution for determining extractable P, K, Ca, and Mg in soils derived from calcareous glacial till. Commun. Soil Sci. Plant Anal. 9:717–728.

Mahler, R.L., A.R. Halvorson, and E.H. Gardner. 1985. Current nutrient status of soils in Idaho, Oregon, and Washington. Pacific Northwest Coop. Ext. Bull. 276.

Malavolta, E. 1985. Potassium status of tropical and subtropical region soils. p. 163–200. *In* R.D. Munson (ed.) Potassium in agriculture. ASA, CSSA, and SSSA, Madison, WI.

Martin, J.P., and A.L. Page. 1969. Influence of exchangeable Ca and Mg and of percentage base saturation on growth of plants. Soil Sci. 107:39–46.

McCallister, D.L. 1985. Electron micrography and microanalysis of naturally and artificially weathered feldspar surfaces. p. 177. *In* Agronomy abstracts. ASA, Madison, WI.

McCollock, R.C., F.T. Bingham, and D.G. Aldrich. 1957. Relationship of soil potassium and magnesium to magnesium nutrition of citrus. Soil Sci. Soc. Am. Proc. 21:85–88.

McIntosh, J.L. 1969. Bray and Morgan soil extractants modified for testing acid soils from different parent materials. Agron. J. 61:259–265.

McLean, E.O. 1975. Calcium levels and availabilities in soils. Commun. Soil Sci. Plant Anal. 6:219–232.

McLean, E.O. 1978. Influence of clay content and clay composition on potassium availability. p. 1–19. *In* G.S. Sekhon (ed.) Potassium in soils and crops. Potash Research Inst. of India, New Delhi.

McLean, E.O., J.L. Adams, and R.C. Hartwig. 1982. Improved corrective fertilizer recommendations based on a two-step alternative usage of soil tests: II. Recovery of soil-equilibrated potassium. Soil Sci. Soc. Am. J. 46:1198–1201.

McLean, E.O., and M.D. Carbonell. 1972. Calcium, magnesium, and potassium saturation ratios in two soils and their effects upon the yield and nutrient contents of German millet and alfalfa. Soil Sci. Soc. Am. Proc. 36:927–930.

McLean, E.O., R.C. Hartwig, D.J. Eckert, and G.B. Triplett. 1983. Basic cation saturation ratios as a basis for fertilizing and liming agronomic crops. II. Field studies. Agron. J. 75:635–639.

McLean, E.O., and M.E. Watson. 1985. Soil measurements of plant-available potassium. p. 277–308. *In* R.D. Munson (ed.) Potassium in agriculture. ASA, CSSA, and SSSA, Madison, WI.

Mehlich, A. 1953. Determination of P, K, Na, Ca, Mg, and NH_4. Soil Test Div. Mimeo. North Carolina Dep. Agric. Raleigh.

Mehlich, A. 1978. New extractant for soil test evaluation of phosphorus, potassium, magnesium, calcium, sodium, manganese and zinc. Commun. Soil Sci. Plant Anal. 9:477–492.

Mehlich, A. 1984. Mehlich 3 soil test extractant: A modification of Mehlich 2 extractant. Commun. Soil Sci. Plant Anal. 15:1409–1416.

Melsted, S.W. 1953. Some observed calcium deficiencies in corn under field conditions. Soil Sci. Soc. Am. Proc. 17:52–54.

Mengel, K., and E.A. Kirkby. 1978. Magnesium. p. 411–423. *In* Principles of plant nutrition. Int. Potash Inst., Worblaufen-Bern, Switzerland.

Mengel, K., and E.A. Kirkby. 1980. Potassium in crop production. Adv. Agron. 33:59–110.

Messick, D.L., M.M. Alley, and L.W. Zelazny. 1984. Movement of calcium and magnesium in Ultisols from dolomitic limestone. Soil Sci. Soc. Am. J. 49:1096–1101.

Miller, R.O. 1988. The bioavailability of potassium on Montana soils as influenced by edaphic and environmental factors. Ph.D. diss. Montana State Univ., Bozeman. (Diss. Abstr. 89-25777)

Mokwunye, A.U., and S.W. Melsted. 1972. Magnesium forms in selected temperate and tropical soils. Soil Sci. Soc. Am. Proc. 36:762–764.

Moorhead, K.K., and E.O. McLean. 1985. Improved corrective fertilizer recommendations based on two-step alternative usage of soil tests: 4. Studies of field plot samples. Soil Sci. 139:131–138.

Mubarek, A., and R.A. Olsen. 1976. Immiscible displacement of soil solution by centrifugation. Soil Sci. Soc. Am. J. 40:329–331.

Mulder, D. 1950. Mg-deficiency in fruit trees on sandy and clay soils in Holland. Plant Soil 2:145–157.

National Agricultural Advisory Service. 1968. Magnesium in agriculture. Advisory paper no. 5. Advisory Serv. Branch, London.

Nemeth, K. 1979. The availability of nutrients in the soil as determined by electro-ultrafiltration (EUF). Adv. Agron. 31:155–188.

Nemeth, K. 1982. Application of electro-ultrafiltration (EUF) in agricultural production. Martinus Nijhoff, Dr. W. Junk Publ. The Hague, Netherlands.

Nemeth, K., K. Mengel, and H. Grimme. 1970. The concentration of K, Ca, and Mg in the saturation extract in relation to exchangeable K, Ca, and Mg. Soil Sci. 109:179–195.

Ologunde, O.O., and R.C. Sorensen. 1982. Influence of concentrations of K and Mg in nutrient solutions on sorghum. Agron. J. 74:41–46.

Olsen, S.R., C.V. Cole, F.S. Watanabe, and L.A. Dean. 1954. Estimation of available phosphorus in soils by extraction with sodium bicarbonate. USDA Circ. 939. U.S. Gov. Print. Office, Washington, DC.

Olson, R.A., K.D. Frank, P.H. Grabouski, and G.W. Rehm. 1982. Economic and agronomic impacts of varied philosophies of soil testing. Agron. J. 74:492–499.

Olson, R.A., M.B. Rhodes, and A.F. Dreier. 1954. Available phosphorus status of Nebraska soils in relation to series classification, time of sampling and method of measurement. Agron. J. 46:175–180.

Onken, A.B., R. Matheson, and E.J. Williams. 1980. Evaluation of EDTA-extractable phosphorus as a soil test procedure. Soil Sci. Soc. Am. J. 44:783–786.

Parra, M.A., and J. Torrent. 1983. Rapid determination of the potassium quantity-intensity relationships using a potassium-selective ion electrode. Soil Sci. Soc. Am. J. 47:335–337.

Pearen, R. 1984. Effect of gypsum amendments on the growth and chemical composition of barley seedlings grown on a solonetzic soil. p. 30–33. Research Highlights. Res. Stn. Lacombe, Alberta.

Penney, D.C. 1985. Potassium fertilizer application in crop production. Agdex 542-9. Print Media Branch, Alberta Agric., Edmonton, AB, Canada.

Pratt, P.F., W.W. Jones, and F.T. Bingham. 1957. Magnesium and potassium content of orange leaves in relation to exchangeable magnesium and potassium in the soil at various depths. Proc. Am. Soc. Hortic. Sci. 70(2):245–251.

Prince, A.L., M. Zimmerman, and F.E. Bear. 1947. The magnesium supplying powers of 20 New Jersey soils. Soil Sci. 63:69–78.

Quemener, J. 1979. The measurement of soil potassium. IPI Res. Topics no. 4. Int. Potash Inst., Bern-Worblaufen, Switzerland.

Rahmatullah, and D.E. Baker. 1981. Magnesium accumulation by corn (*Zea mays* L.) as a function of potassium-magnesium exchange in soils. Soil Sci. Soc. Am. J. 45:899–903.

Rehm, G.W., C.J. Rosen, J.F. Moncrief, W.E. Fenster, and J. Grava. 1985. Guide to computer programmed soil test recommendations for field crops in Minnesota. Univ. of Minnesota Agric. Ext. Serv. AG-BU-0519.

Reith, J.W.S. 1967. Effects of soil magnesium levels and of magnesium dressing on crop yield and composition. p. 97–108. *In* Soil potassium and magnesium. Tech. Bull. 14. Her Majesty's Stationery Office, London.

Rice, B., and E.J. Kamprath. 1968. Availability of exchangeable and nonexchangeable Mg in sandy Coastal Plain soils. Soil Sci. Soc. Am. Proc. 32:386–388.

Rich, C.I. 1968. Mineralogy of soil potassium. p. 79–108. *In* V.J. Kilmer et al. (ed.) The role of potassium in agriculture. ASA, CSSA, and SSSA, Madison, WI.

Russelle, M.P., L.L. Meyers, and R.L. McGraw. 1989. Birdsfoot trefoil seedling responses to soil phosphorus and potassium availability indexes. Soil Sci. Soc. Am. J. 53:828–836.

Salmon, R.C. 1964. Cation activity ratios in equilibrium soil solution and the availability of magnesium. Soil Sci. 98:213–221.

Sanchez, P.A., W. Couto, and S.W. Buol. 1982. The fertility capability soil classification system: Interpretation, applicability and modification. Geoderma 27:283–309.

Schaff, B.E., and E.O. Skogley. 1982a. Diffusion of potassium, calcium, and magnesium in Bozeman silt loam as influenced by temperature and moisture. Soil Sci. Soc. Am. J. 46:521–524.

Schaff, B.E., and E.O. Skogley. 1982b. Soil profile and site characteristics related to winter wheat response to potassium fertilizers. Soil Sci. Soc. Am. J. 46:1207–1211.

Schofield, R.K. 1947. A ratio law governing the equilibrium of cations in the soil solution. Proc. 11th Int. Congr. Pure Appl. Chem. 3:257–261.

Schulte, E.E., and R.B. Corey. 1965. Extraction of potassium from soils with tetraphenylboron. Soil Sci. Soc. Am. Proc. 29:33–35.

Schulte, E.E., K.A. Kelling, and C.R. Simson. 1981. Too much magnesium in soil? Solutions 24(6):106–116.

Scott, A.D., R.R. Hunziker, and J.J. Hanway. 1960. Chemical extraction of potassium from soils and micaceous minerals with solutions containing sodium tetraphenylboron. I. Preliminary experiments. Soil Sci. Soc. Am. Proc. 24:191–194.

Shaw, J.K., R.K. Stivers, and S.A. Barber. 1983. Evaluation of differences in potassium availability in soils of the same exchangeable potassium level. Commun. Soil Sci. Plant Anal. 14:1035–1049.

Silverbush, M., and S.A. Barber. 1983. Prediction of phosphorus and potassium uptake of field-grown soybeans with a mechanistic mathematical model. Soil Sci. Soc. Am. J. 47:262–265.

Simson, C.R., R.B. Corey, ad M.E. Sumner. 1979. Effect of varying Ca:Mg ratios on yield and composition of corn (*Zea mays*) and alfalfa (*Medicago sativa*). Commun. Soil Sci. Plant Anal. 10:153–162.

Sinclair, A.H. 1980. Desorption of cations from Scottish soils by electroultrafiltration. J. Sci. Food Agric. 31:532–540.

Singh, K.D., K.W.T. Goulding, and A.H. Sinclair. 1983. Assessment of potassium in soils. Commun. Soil Sci. Plant Anal. 14:1015–1033.

Skogley, E.O. 1986. Rate limiting processes of phytoavailability of potassium on Montana soils. p. 93–97. *In* J. Havlin (ed.) Proc. Great Plains Soil Fertility Workshop, Denver, CO. 4–5 Mar. Kansas State Univ., Manhattan.

Skogley, E.O., and S.J. Georgitis. 1988. Phytoavailability soil test. p. 164–167. *In* J. Havlin (ed.) Proc. Great Plains Soil Fert. Workshop, Denver, CO. 8–9 Mar. Kansas State Univ., Manhattan.

Skogley, E.O., and V.A. Haby. 1981. Predicting crop responses on high-potassium soils of frigid temperature and ustic moisture regimes. Soil Sci. Soc. Am. J. 45:533–536.

Skogley, E.O., and B.E. Schaff. 1985. Ion diffusion in soils as related to soil physical and chemical properties. Soil Sci. Soc. Am. J. 49:847–850.

Soltanpour, P.N., and A.P. Schwab. 1977. A new soil test for simultaneous extraction of macro- and micro-nutrients in alkaline soils. Commun. Soil Sci. Plant Anal. 8:195–207.

Soltanpour, P.N., and S. Workman. 1979. Modification of the NH_4HCO_3-DTPA soil test to omit carbon black. Commun. Soil Sci. Plant Anal. 10:1411–1420.

Sparks, D.L., and P.M. Huang. 1985. Physical chemistry of soil potassium. p. 201–276. *In* R.D. Munson (ed.) Potassium in agriculture. ASA, CSSA, and SSSA, Madison, WI.

Sparks, D.L., and P.M. Jardine. 1981. Thermodynamics of potassium exchange in soil using a kinetics approach. Soil Sci. Soc. Am. J. 45:1094–1099.

Sparks, D.L., and W.C. Liebhardt. 1981. Effect of long-term lime and potassium applications on quantity-intensity (Q/I) relationships in sandy soil. Soil Sci. Soc. Am. J. 45:786–790.

Sparks, D.L., D.C. Martens, and L.W. Zelazny. 1980. Plant uptake and leaching of applied and indigenous potassium in Dothan soils. Agron. J. 72:551–555.

Steenkamp, J.L. 1928. The effect of dehydration of soils upon their colloid constituents. I. Soil Sci. 25:163–182.

Stout, W.L. 1982. Potassium and magnesium recovery from selected soils of the Allegheny plateau. Soil Sci. Soc. Am. J. 46:1023–1027.

Stout, W.L., and D.I. Baker. 1981. Effect of differential adsorption of potassium and magnesium in soils on magnesium uptake by corn. Soil Sci. Soc. Am. J. 45:996–997.

Stout, W.L., and O.L. Bennet. 1983. Effect of Mg and Zn fertilization on soil test levels, ear leaf composition, and yields of corn in northern West Virginia. Commun. Soil Sci. Plant Anal. 14:601–613.

Tan, K.H. 1978. Effects of humic and fulvic acids on release of fixed potassium. Geoderma 21:67–74.

Thomas, G.W. 1974. Chemical reactions controlling soil solution electrolyte concentrations. p. 483–506. *In* E.W. Carson (ed.) The plant root and its environment. Univ. Press of Virginia, Charlottesville.

Tisdale, S.L., W.L. Nelson, and J.D. Beaton. 1985. Soil and fertilizer potassium. p. 249–250. *In* Soil fertility and fertilizers. Macmillan, New York.

United States Department of Agriculture. 1984. Agricultural statistics 1984. U.S. Gov. Print. Office, Washington, DC.

van Lierop, W. 1985. Comparison of laboratory methods for evaluating plant-available soil phosphorus. *In* The role of soil analysis in resource management. Proc. 9th British Columbia Soils Workshop. B.C. Ministry of Environment, Vancouver.

van Lierop, W., and T.S. Tran. 1985. Comparative potassium levels removed from soils by electro-ultrafiltration and some chemical extractants. Can. J. Soil Sci. 65:25–34.

Walker, W.M., and T.R. Peck. 1975. Effect of potassium upon the magnesium status of the corn plant. Commun. Soil Sci. Plant Anal. 6:189–194.

Whitney, D.A. 1975. Low K soils showing up. Better Crops Plant Food. 2:8–9.

Wietholter, S. 1983. Predicting potassium uptake by corn in the field using the strontium nitrate soil testing method and a diffusion-controlled uptake model. Ph.D. diss. Univ. of Wisconsin-Madison (Diss. Abstr. DA832180).

Wood, L.K., and E.T. DeTurk. 1940. The absorption of potassium in soils from non-replaceable forms. Soil Sci. Soc. Am. Proc. 5:152–161.

Woodruff, J.R., and C.L. Parks. 1980. Topsoil and subsoil potassium calibration with leaf potassium for fertility rating. Agron. J. 72:392–396.

Yuan, T.L., L.W. Zelazny, and A. Ratanaprasatporn. 1976. Potassium status of selected Paleudults in the lower Coastal Plain. Soil Sci. Soc. Am. J. 40.229–233.

Zandstra, H.G., and A.F. MacKenzie. 1968. Potassium exchange equilibria and yield responses of oats, barley, and corn on selected Quebec soils. Soil Sci. Soc. Am. J. 32:76–79.

Chapter 9

Testing Soils for Copper, Iron, Manganese, and Zinc[1]

D. C. MARTENS, *Virginia Polytechnic Institute and State University, Blacksburg*

W. L. LINDSAY, *Colorado State University, Fort Collins*

Recognition that Cu, Fe, Mn, and Zn are essential for plant growth and that Cu, Fe, Mn, and Zn deficiencies occur in the field preceded the development of soil tests for the micronutrient cations. The essentiality of Fe for plant growth was proven in 1844 by Gris (Bonner & Galston, 1952) and that of Mn in 1905 by Bertrand (Stout, 1956). In 1914, Maze provided the evidence that Zn was needed by plants and, in 1928, Sommer and Lipman demonstrated the plant requirement for Cu (Stout, 1956). The criterion for essentiality was based on the inability of plants to complete their life cycles under conditions of insufficient Cu, Fe, Mn, or Zn.

After establishment of the essentiality of Cu, Fe, Mn, and Zn, it was a normal progression to evaluate whether lack of these elements caused abnormal plant growth on problem soils. These micronutrient deficiencies were first identified under field conditions in horticultural crops. Copper deficiency of citrus, or "dieback," and Mn deficiency of tomato (*Lycopersicon esculentum* Mill.) were identified in Florida by Grossenbacher (1916) and Skinner and Ruprecht (1930), respectively. Deficiencies of Fe (Thomas & Haas, 1928) and Zn (Chandler et al., 1932) were diagnosed in Californian citrus. Thereafter, these deficiencies were confirmed in agronomic crops under field conditions. Copper and Mn deficiencies of oat (*Avena sativa* L.) plants were confirmed in Wales (Davies & Jones, 1931) and Florida (Harris, 1947), respectively. A chlorotic condition of sorghum [*Sorghum bicolor* (L.) Moench], which limited grain yields on the Southern Great Plains, was diagnosed as Fe deficiency (Myers & Johnson, 1933). Barnette et al. (1936) reported that Zn application as $ZnSO_4$ increased corn (*Zea mays* L.) grain yield and that

[1] Contribution of the Department of Agronomy, Virginia Polytechnic Inst. and State Univ., Blacksburg, VA 24061, and Colorado State Univ. Exp. Stn., Fort Collins, CO 80523.

application of superphosphate with $ZnSO_4$ rendered the Zn less efficient in correction of the "white bud" in corn.

The need for laboratory procedures to identify soils with inadequate amounts of Cu, Fe, Mn, and Zn for plants became apparent in the 1930s as these deficiencies were confirmed under field conditions. It was recognized early that the plant availabilities of Cu, Fe, Mn, and Zn were governed by soil properties and, therefore, that the total concentrations of these elements in soils would not serve as a suitable indicator of plant-available Cu (Mulder, 1939), Fe and Mn (Mann, 1930), or Zn (Alben & Boggs, 1936). An exception was organic soils with low total concentrations and concomitant low availabilities of these micronutrient cations.

The inconsistent relationship between the micronutrient cation availabilities and total concentrations in most soils led to the conclusion that availability tests for these elements should be based on portions extracted from soil (Hibbard, 1940). The initial soil tests were based on amounts of the micronutrient cations extracted from soils by inorganic reagents or bioassay procedures with the fungus, *Aspergillus niger*. Immediately, problems were encountered on the choice of extractant and the methodology for determining the low concentrations of metals extracted from soil (Hibbard, 1940).

Extraction and determination of low amounts of Cu and Zn were accomplished with an *A. niger* bioassay. Initial attempts to use the bioassay for estimation of plant-available Cu and Zn in soil were unsuccessful. The problem encountered was the nutrient solution contamination supplied the low Cu and Zn requirements for *A. niger* and, consequently, that standard curves were unobtainable. Use of a nutrient solution purification procedure developed by Steinberg (1919), which entailed metal adsorption on $CaCO_3$ added to nutrient solution and separation of the liquid and solid phases by decantation, led to successful use of the bioassay for Cu (Mulder, 1939) and Zn (Bould et al., 1949). Mulder (1939) estimated available Cu in soil by comparing *A. niger* spore color on soil cultures and a series of cultures with increasing increments of Cu. Severe Cu deficiency of oat plants occurred on soils with a level of <0.6 mg of Cu kg^{-1}. Bould et al. (1949) estimated available Zn in soil by comparing the dry weight of *A. niger* tissue on soil cultures and a series of cultures with increasing increments of Zn. They confirmed the occurrence of Zn deficiency in fruit trees by this bioassay method.

Interest in the *A. niger* bioassay for evaluation of available Cu and Zn waned with the development of improved analytical methods for micronutrient cation determination. Furthermore, the bioassay method was not completely adaptable to the soil test requirements, which were summarized by Bray (1948) as follows:

1. The extraction solution and procedure should solubilize a proportionate part of the available forms of a nutrient in soils with variable properties.
2. The amount of the nutrient in the extract should be measurable with reasonable accuracy and speed.
3. The amount extracted should be correlated with crop response to that nutrient under various conditions.

The long incubation period for the bioassay negates its adaptability as a routine soil test procedure. In addition, the procedure does not meet the current goal of simultaneous extraction and determination of Cu, Fe, Mn, and Zn to allow rapid conveyance of the soil test results to crop producers at a reasonable cost (Havlin & Soltanpour, 1981; Selvarajah et al., 1982). Hence, the *A. niger* bioassay will not receive further comment, and the chemical extractants, which are adaptable to the soil test requirements, will receive emphasis herein.

The purpose of this chapter, which is a revision of the one written by Viets and Lindsay (1973), is to review the current status of soil testing for Cu, Fe, Mn, and Zn. Advances in simultaneous extraction and determination of the micronutrient cations as well as improvements in soil test calibration data warrant this revision. Portions of this subject are covered in other chapters (Cox & Kamprath, 1972; Viets & Lindsay, 1973; Mortvedt, 1977; Knezek & Ellis, 1980; Sillanpaa, 1982; Lindsay & Cox, 1985). It is our goal to minimize redundancy with the aforementioned chapters, but for clarity's sake, some overlap is unavoidable.

I. PROCEDURES FOR SOIL TEST DEVELOPMENT

The three sequential steps generally followed to develop micronutrient cation soil tests are extractant selection, greenhouse evaluation, and field calibration. The first step is to select an extractant that will solubilize a proportionate part of labile forms of the micronutrient cations from different soils. The second step is to evaluate if amounts of cations extracted are related to the quantities absorbed by plants from the different soils. To decrease cost and to obtain data more quickly, this step commonly is completed in the greenhouse rather than in the field. It is assumed that, if the amounts of extractable micronutrient cations are unrelated to crop response to their application under controlled greenhouse conditions, then a suitable relationship will not be obtainable in harsher field environments. The third step, field calibration of a soil extractant, is conducted if a suitable relationship is established during the greenhouse evaluation.

Field calibration of a micronutrient cation soil test commonly entails the determination of a critical level, i.e., the soil test value that separates soils into responsive and nonresponsive categories. The critical level is based on yield response to application of the micronutrient under field conditions. Another method used in field calibration research is to determine the insufficient, sufficient, and the transitional zone between insufficient and sufficient levels of the available micronutrient cations in soils. The small yield increases from micronutrient cation applications on many soils led to difficulty in use of the latter method. Statistical procedures using the interaction chi-square technique could be used to identify the boundaries of the transitional zone (Keisling & Mullinix, 1979; Havlin & Soltanpour, 1982).

Economically, it would be desirable to calibrate micronutrient cation soil tests under greenhouse conditions. There are, however, several reasons why critical levels determined for micronutrient cation soil tests in the greenhouse may not be applicable to field conditions. Differences in cation uptake occur in greenhouse as compared with field for the following reasons:

1. Use of higher rates of NH_4^+ fertilizer causes lower levels of pH in pots than in field soils.
2. Use of larger amounts of nutrients in pots leads to higher soluble-salt levels.
3. Greater root-to-soil contact occurs from plant root confinement in pots.
4. Abnormal light, relative humidity, and moisture regimes occur in the greenhouse (Logan & Chaney, 1983; Mortvedt, 1977).

Quite often nutrients are supplied at rates inadequate to obtain maximum plant growth under greenhouse conditions (Terman, 1974). Such insufficiency contributes to variations in plant concentration, uptake, and yield response to micronutrient cation application and, hence, to an unacceptable greenhouse evaluation of a soil test. Also, it is unacceptable to extrapolate critical levels from the vegetative growth stage in the greenhouse to the maturity growth stage under field conditions.

Greenhouse soil tests are commonly conducted with soil from the Ap horizon, whereas in the field, plants absorb nutrients from the Ap and lower horizons. This difference leads to the anomaly illustrated in the research by Iyengar et al. (1981). Zinc application increased the dry weight of corn grown on a Dunmore silt loam (clayey, kaolinitic, mesic Typic Paleudult) in the greenhouse, but not in the field where roots absorbed some Zn from acidic soil below the Ap horizon. In other research, Cu application increased tick trefoil (*Desmodium uncinatum*) and white clover (*Trifolium repens* L.) yields under greenhouse conditions, but only white clover yield was increased by Cu application in the field (Andrew & Thorne, 1962). The difference was attributed to the deep root system of tick trefoil as compared with the shallow root system of white clover. Copper absorption from subsoil by tick trefoil negated a yield response to Cu application.

It is desirable to base the greenhouse evaluation step on increases in plant dry weights from application of the micronutrient cations. Although growth responses to Cu, Fe, and Zn application commonly are obtained with plants grown in the greenhouse, growth responses to Mn application are not always possible. For example, Tierney (1981) collected eight soils from fields where soybean [*Glycine max* (L.) Merr.] plants were severely Mn deficient, but did not obtain a growth response to Mn application in the greenhouse. Unavoidable changes in pe + pH in pots increased Mn availability and, thus, negated soybean yield response to Mn application in the greenhouse. It is for this reason that steps two and three frequently are combined and completed under field conditions for development of a Mn soil test.

In some cases, critical micronutrient cation soil test values obtained in greenhouse research have been used as guides to evaluate whether micronutrient cation deficiencies would occur under field conditions. This use of greenhouse critical levels is open to criticism. Nevertheless, this approach has been necessary on an interim basis during completion of the field calibration research for a soil test.

II. MICRONUTRIENT CATION FORMS SOLUBILIZED BY VARIOUS EXTRACTANTS

Ideally, extractants for Cu, Fe, Mn, and Zn soil tests should be selected from the standpoint of solubilization of a proportionate part of the labile forms in different soils. This selection requires knowledge of the labile micronutrient cation forms in soil. A labile ion is an ion in soil solution or one in the solid phase that may exchange with the same kind of ion in soil solution.

A. Micronutrient Cation Forms in Soils

The micronutrient cations exist in the following forms in soils: (i) as free and complexed ions in soil solution, (ii) as nonspecifically and specifically adsorbed cations, (iii) as ions occluded mainly in soil carbonates and hydrous oxides, (iv) in biological residues and living organisms, (v) in the lattice structure of primary and secondary minerals, and (vi) as precipitates (McLaren & Crawford, 1973; Sims & Patrick, 1978; Lindsay, 1979; Iyengar et al., 1981). The solubilities of Fe and Mn hydrous oxides are sufficiently low that precipitation of these compounds commonly occurs even in slightly acid soils (Lindsay, 1979). In contrast, Cu and Zn hydrous oxides would be unstable in acid soils except where unusually high activities of the two micronutrients occur in soil solution.

Nonspecifically adsorbed Cu^{2+}, Fe^{2+}, Mn^{2+}, and Zn^{2+} occur in soils from formation of an outer-sphere complex with a functional group and the cation, i.e., through bond formation by electrostatic interaction (Sposito, 1981). This bond formation is exemplified by the attraction between these cations and the charge arising from isomorphous substitution in phyllosilicates (Sposito, 1981) or between Mn^{2+} and a functional group of organic matter (Gamble et al., 1976). A nonspecifically adsorbed cation usually is referred to as an exchangeable cation.

Specific adsorption refers to formation of an inner-sphere complex between a micronutrient cation and functional group of soil components (Sposito, 1981). In soils, Cu, Fe, Mn, and Zn cations would be specifically adsorbed by carbonates, hydrous oxides of Al, Fe, and Mn, soil organic matter, and phyllosilicates (Udo et al., 1970; Schnitzer, 1978; Kabata-Pendias & Pendias, 1985).

Occlusion is the formation of a second layer of adsorbed ions over an adsorbed ion such that the originally adsorbed ion no longer is in contact with soil solution and, therefore, cannot dissociate into soil solution. This occlusion occurs as a layer of Al, Fe, or Mn hydroxide precipitates over micronutrient cations, which are complexed by organic matter or adsorbed on carbonate, hydrous oxide, and phyllosilicate surfaces.

Labile micronutrient cations consist of free and complexed cations in soil solution, which provide the intensity of soil to supply nutrients to plants, and the nonspecifically and specifically adsorbed cations, which provide the soil with the capacity to replenish these cations into soil solution. Labile forms are the plant-available sources of the cations. Thus, a soil test extractant should solubilize the labile forms for estimates of plant availability.

Labile micronutrient cations are in metastable equilibria with nonlabile forms in soils. Therefore, portions of labile forms revert to nonlabile forms with time and, conversely, portions of the nonlabile forms revert to the labile forms upon weathering. It is recognized that the labile forms of a cation in soil change rapidly under dynamic soil conditions such as fluctuations in pe + pH. An inherent limitation of soil micronutrient cation soil tests is the attempt to extract labile forms from a dry or moist soil sample, which may differ from labile forms in field soil from the same location during a growing season.

B. Forms Solubilized by Various Extractants

The many soil test extractants that have been developed for solubilizing micronutrient cations fall within four categories: chelating agents, inorganic acids, neutral salts, and reducing agents (Table 9–1). Neutral salt solutions extract nonspecifically adsorbed micronutrient cations that are in a metastable equilibrium with soil solution. The cation in the neutral salt may displace hydrated Al^{3+} that could hydrolyze in soil solution to produce H_3O^+. The H_3O^+ ions then displace specifically adsorbed micronutrient cations by dissociating cation-organic matter complexes and by partially dissolving carbonate and hydrous oxide surfaces. Overall, the amounts of micronutrient cations solubilized with neutral salts are much lower than the quantities extracted with chelating agents, reducing agents, and strong and weak acids (McLaren & Crawford, 1973; Viets & Lindsay, 1973; Sims & Patrick, 1978; Lindsay, 1979; Iyengar et al., 1981).

The quantities of micronutrient cations solubilized from soil by extraction with strong or weak acid depend on the amount of H_3O^+ neutralization in the soil-extraction solution mixture and on the extraction time. A solution with a very low acid concentration (activity) would extract small amounts of micronutrient cations from about the same sources as would neutral salts; proportionally higher amounts would be extracted with increases in acid concentration and extraction time. Extraction with 0.1 *M* HCl commonly has been used for estimation of plant-available Zn. This extractant would solubilize micronutrient cations by dissociating cation-organic matter complexes, by displacing specifically adsorbed cations from carbonate, hydrous oxide, and phyllosilicate surfaces, and by releasing nonlabile occluded and precipitated cations during partial acid decomposition of minerals. The acid also would release octahedrally coordinated Cu, Fe, Mn, and Zn from phyllosilicates as the exposed mineral surfaces undergo acid attack. The amounts released from phyllosilicates would vary with soil mineralogy because some phyllosilicates, e.g., kaolinite, are more resistant to acid attack than are others (Lindsay, 1979).

Table 9-1. Critical levels for Cu, Fe, Mn, and Zn soil tests based on either greenhouse evaluation or field calibration research.

Research	Crop(s)	Soil(s)	Extractant	Critical level	Reference
			Cu soil tests		
Field	Barley and oat	30 soils, pH 5.2–6.7	0.05 *M* EDTA	1.1 mg Cu kg^{-1}	Reith, 1968
Greenhouse	Rice	16 soils, pH 5.5–8.5	0.05 *M* HCl	0.1 mg Cu kg^{-1}	Ponnamperuma et al., 1981
Greenhouse	Corn, soybean, and wheat	15 noncalcareous soils	Mehlich-I	0.26 mg Cu dm^{-3}	Makarim & Cox, 1983
			Mehlich-Bowling	0.62 mg Cu dm^{-3}	
			NH_4HCO_3-DTPA	0.53 mg Cu dm^{-3}	
			Mehlich-III	0.37 mg Cu dm^{-3}	
Field	Soybean and wheat	7 noncalcareous soils	Mehlich-Bowling	0.70 mg Cu dm^{-3}	Makarim & Cox, 1983
			Fe soil tests		
Greenhouse	Sorghum	35 calcareous soils	DTPA-TEA	4.5 mg Fe kg^{-1}	Lindsay & Norvell, 1978
Greenhouse	Sorghum	40 soils, 35 calcareous, and 5 noncalcareous	NH_4HCO_3-DTPA	4.8 mg Fe kg^{-1}	Havlin & Soltanpour, 1981
			DTPA-TEA	4.8 mg Fe kg^{-1}	
			Mn soil tests		
Field	Soybean	25 soils, pH 5.7–7.4	0.033 *M* H_3PO_4	20 mg Mn kg^{-1}	Hoff & Mederski, 1958
			1.0 *M* $NH_4H_2PO_4$	20 mg Mn kg^{-1}	
			Alcoholic hydroquinone	63 mg Mn kg^{-1}	
Field	Oat	25 soils, pH 6.7–7.7	0.033 *M* H_3PO_4	20 mg Mn kg^{-1}	Hammes & Berger, 1960a
			1.5 *M* $NH_4H_2PO_4$	20 mg Mn kg^{-1}	
			0.05 *M* EDTA	50 mg Mn kg^{-1}	
			Hydroquinone-NH_4OAc	65 mg Mn kg^{-1}	
Field	Soybean	17 soils, pH 5.2–7.1	Mehlich-I	pH variable-5.2 mg Mn kg^{-1} at pH 6.0	Cox, 1968
Field	Soybean	Two sandy-textured soils, pH 6.2–7.0	Mehlich-I	2.6 mg Mn kg^{-1}	Shuman et al., 1980
			DTPA-TEA	0.22 mg Mn kg^{-1}	
			NH_4HCO_3-DTPA	0.40 mg Mn kg^{-1}	
Field	Corn	One sandy soil, pH 6.4–7.2	Mehlich-III	pH variable-3.0 mg Mn dm^{-3} at pH 6.4	Mascagni & Cox, 1984

(continued on next page)

Table 9-1. Continued.

Research	Crop(s)	Soil(s)	Extractant	Critical level	Reference
			Mn soil tests		
Field	Soybean	30 soils, pH 5.1–6.9	Mehlich-I	pH variable-4.6 mg Mn kg^{-1} at pH 6.0	Gettier et al., 1985c
Field	Soybean	38 soils, pH 5.5–7.1	Mehlich-I	pH variable-4.7 mg Mn kg^{-1} at pH 6.0,	Mascagni & Cox, 1985
			Mehlich-III	3.9 mg Mn dm^{-3} at pH 6.0	
			Zn soil tests		
Field	Corn	15 soils, pH 4.7–6.2	0.1 *M* HCl	1.0 mg Zn kg^{-1}	Wear & Sommer, 1948
Greenhouse	Corn	42 soils, neutral and calcareous	EDTA-$(NH_4)_2CO_3$	1.4 mg Zn kg^{-1}	Trierweiler & Lindsay, 1969
			Dithizone-NH_4OAc	0.95 mg Zn kg^{-1}	
Greenhouse	Sweet corn	92 soils, most with a pH of 7.0–8.2	DTPA-TEA	0.5 mg Zn kg^{-1}	Brown et al., 1971
			Dithizone-NH_4OAc	0.55 mg Zn kg^{-1}	
Field	Corn	10 soils, pH 5.4–7.2	EDTA-$(NH_4)_2CO_3$	0.8 mg Zn kg^{-1}	Alley et al., 1972;
Field	Corn	34 soils, pH 4.7–8.3	Mehlich-I	0.8 mg Zn kg^{-1}	Cox & Wear, 1977
			0.1 *M* HCl	1.4 mg Zn kg^{-1}	
			DTPA-TEA	0.5 mg Zn kg^{-1}	
Greenhouse	Corn	42 calcareous and non-calcareous soils	DTPA-TEA	0.8 mg Zn kg^{-1}	Lindsay & Norvell, 1978
Greenhouse	Sorghum	35 calcareous and non-calcareous	DTPA-TEA	0.6 mg Zn kg^{-1}	Lindsay & Norvell, 1978
Greenhouse	Green gram	22 soils, pH 7.0–8.7	DTPA-TEA	0.48 mg Zn kg^{-1}	Gupta & Mittal, 1981
			EDTA-$(NH_4)_2CO_3$	0.70 mg Zn kg^{-1}	
			0.1 *M* HCl	2.2 mg Zn kg^{-1}	
Greenhouse	Corn	40 soils, 35 calcareous and 5 noncalcareous	NH_4HCO_3-DTPA	0.9 mg Zn kg^{-1}	Havlin & Soltanpour, 1981
			DTPA-TEA	0.7 mg Zn kg^{-1}	
Greenhouse	Rice	22 soils, pH 5.5–8.5	0.05 *M* HCl	1.0 mg Zn kg^{-1}	Ponnamperuma et al., 1981
Greenhouse	Rice	46 soils, pH 8.8–10.7	DTPA-TEA	0.86 mg Zn kg^{-1}	Singh & Takkar, 1981
			EDTA-$(NH_4)_2CO_3$	1.00 mg Zn kg^{-1}	

The noncrystalline Al, Fe, and Mn hydrous oxides are more susceptible to acid attack with release of occluded, adsorbed, and precipitated micronutrient cations than the crystalline hydrous oxides of these elements. This susceptibility reflects the greater surface area of the noncrystalline compounds. At sufficiently low levels of extraction solution pH, specifically adsorbed and occluded micronutrient cations are released as crystalline primary minerals such as gibbsite, pyrolusite, and strengite undergo partial acid decomposition. Because acids extract portions of nonlabile metal cations, either soil pH or titratable alkalinity often is used in conjunction with an acid extractable cation to improve the prediction of the plant-available cation status in soil (Nelson et al., 1959; Cox, 1968).

The reducing agent, hydroquinone, has been used to solubilize Mn in the development of a Mn soil test. Leeper (1934) extracted easily reducible Mn with a solution of 1.0 *M* NH_4OAc and 0.05% hydroquinone to estimate available soil Mn. The hydroquinone reduces a portion of the Mn in noncrystalline Mn hydrous oxides and thereby liberates Mn into solution (Sherman et al., 1942). There could be some readsorption of Mn by soil at the pH of this soil-extraction solution mixture.

The diethylenetriaminepentaacetic acid-triethanolamine (DTPA-TEA) (0.005 *M* DTPA; 0.01 *M* $CaCl_2$; 0.1 *M* TEA, buffered at pH 7.3) soil test procedure is among the more widely used techniques to identify soils with inadequate levels of available Cu, Fe, Mn, and Zn. Lindsay and Norvell (1978) described the theoretical basis for the procedure. The DTPA molecules form water-soluble complexes with free metal cations and, thereby, decrease the cation activities in solution. In response, cations desorb from soil surfaces or dissolve from labile solid phases to replenish solution cations. Amounts of chelated cations that accumulate in solution during the extraction are a function of both cation activity in the soil solution (intensity factor) and the ability of the soil to replenish those cations (capacity factor). Plant uptake of a micronutrient increases with increases in soil solution cations and the capacity of soil solid phases to replenish cations depleted from soil solution (Lindsay, 1979).

The chelating agent, DTPA, was selected for the soil test because it had the most favorable combination of stability constants for simultaneous complexation of Cu, Fe, Mn, and Zn (Lindsay & Norvell, 1978). Since Fe and Zn deficiencies are prevalent on calcareous soils, the extractant was designed to avoid excessive dissolution of $CaCO_3$ with the release of occluded micronutrient cations, which normally are not available for absorption by plant roots. Excessive dissolution was prevented by inclusion of soluble Ca^{2+} as $CaCl_2$ in the extraction solution and by buffering the solution at pH 7.3 with TEA.

C. Effects of Soil Sample Preparation

Amounts of extractable Cu, Fe, Mn, and Zn in soils are affected by sample drying, grinding force and time, and quantity of sample being ground. The effects of sample preparation procedures on extractable Cu, Fe, Mn, and Zn vary among soils and laboratories. It would be desirable to standardize sample preparation procedures to ensure that comparable levels of extractable Cu, Fe, Mn, and Zn are obtained in different laboratories (Soltanpour et al., 1976).

Drying moist samples of diverse United Kingdom soils decreased the concentrations of easily reducible Mn and increased water soluble plus exchangeable Mn (Goldberg & Smith, 1984). Explanations for the decrease in easily reducible Mn were as follows: (i) dehydration of Mn hydrous oxides, (ii) reduction of Mn in hydrous oxides by organic matter, and (iii) alteration of functional groups that tightly complex Mn. Hammes and Berger (1960b) attributed Mn release on drying of moist soil to chemical oxidation of the organic component of organo-Mn complexes with release of Mn^{2+} to labile water soluble and exchangeable forms.

Drying moist samples affects the amounts of micronutrient cations extracted from soils by currently used soil tests. Air drying increased Mn extracted by 0.033 *M* H_3PO_4 from seven soils (Hammes & Berger, 1960b). Amounts of DTPA-TEA extractable micronutrient cations were increased by air-drying neutral and calcareous soil samples: two- to threefold for Fe and 1.3- to 1.5-fold for Cu, Mn, and Zn (Leggett & Argyle, 1983). The DTPA-TEA extractable Fe and Mn increased over the drying temperature range from 22 to 100 °C, whereas DTPA-TEA extractable Cu and Zn levels changed only slightly over this temperature range. Similar temperature effects occurred for levels of extractable Cu, Fe, Mn, and Zn by the DTPA-TEA and NH_4HCO_3-DTPA (1.0 *M* NH_4HCO_3, 0.005 *M* DTPA, buffered at pH 7.6) procedures for two alkaline soils (Leggett & Argyle, 1983). Although inconvenient, it may be necessary to keep soil moist and to store moist samples under aerobic conditions prior to extraction to minimize changes in labile micronutrient cations (Bartlett & James, 1980).

Differences in grinding force and time as well as amounts of soil being ground cause variation in levels of DTPA-TEA extractable micronutrient cations (Soltanpour et al., 1976). As the amount of soil being ground was increased from 200 to 400 g, at a grinding force of 6.6 kg for either 30 or 120 s, there was a decrease in extractable Fe from two soils (Table 9–2). The extractable Zn in the soils was increased by a longer grinding time from 30 to 120 s and by an increase in grinding force from 3.2 to 6.6 kg. The magnitude of these Zn increases was much less than for Fe. An increase in grinding time increased the extractable Mn levels in the soils, whereas extractable Cu levels were unaffected by differences in grinding time.

The NH_4HCO_3-DTPA procedure was developed for simultaneous extraction of micronutrient cations and macronutrients from neutral and alkaline soils (Soltanpour & Schwab, 1977). Equilibrium is not attained during the 15-min shaking period for the procedure and, therefore, any factor that affects the reaction rate between the NH_4HCO_3-DTPA solution and soil changes the amounts of extractable Cu, Fe, Mn, and Zn (Soltanpour et al., 1979a). Soltanpour et al. (1979a) studied the effect of weight of soil being ground and grinding force and time on NH_4HCO_3-DTPA extractable Cu, Fe, Mn, or Zn. An increase in weight of soil being ground decreased extract-

Table 9-2. Effect of sample size, grinding force, and grinding time on DTPA-TEA extractable Zn, Fe, Mn, and Cu for two soils (Soltanpour et al., 1976).

Micronutrient cation	200 g Soil				400 g Soil			
	3.2 kg force		6.6 kg force		3.2 kg force		6.6 kg force	
	30 s	120 s	30 s	120 s	30 s	120 s	30 s	120 s
	mg kg^{-1}							
	Soil 1							
Zn*	0.25	0.39	0.32	0.66	0.26	0.39	0.28	0.50
Fe*	6.5	24.3	15.4	50.8	6.2	25.1	10.7	37.8
Mn†	5.5	7.0	6.3	8.9	5.9	7.2	6.0	8.2
Cu‡	0.76	0.71	0.73	0.76	0.76	0.73	0.62	0.76
	Soil 2							
Zn*	0.36	0.41	0.32	0.57	0.29	0.39	0.32	0.36
Fe*	10.0	21.1	14.0	44.8	7.6	16.3	10.9	19.1
Mn†	4.6	5.1	4.4	6.8	4.5	4.8	4.5	5.2
Cu‡	0.39	0.56	0.28	0.80	0.34	0.42	0.28	0.28

* Effects of size of soil sample, grinding time, and grinding force were significant at the 0.05 level. Interaction effect of force × time was significant at the 0.10 level.
† Effect of time was significant at the 0.10 level.
‡ No significant effect.

able Fe slightly and did not affect levels of extractable Cu, Mn, and Zn. An increase in extractable Fe and Zn occurred with increases in both grinding force and time, but these grinding parameters did not affect levels of extractable Cu and Mn.

Effects of grinding parameters on extractable micronutrient cations vary with their forms in soils. Grinding to a smaller particle size increases the surface area of Fe hydroxides and oxides exposed to extraction solution and, hence, increases the extractability of Fe as well as occluded Cu, Mn, and Zn in the hydroxides and oxides (Severson et al., 1979; Soltanpour et al., 1979a). In contrast, grinding to smaller particle size has little effect on extractable cations in soil forms that are soluble in the extraction solution (Soltanpour et al., 1979a).

D. Effects of Extraction Parameters

Amounts of extractable Cu, Fe, Mn, and Zn are affected by concentrations of extraction solution components, extraction time, soil-extraction solution ratio, extraction temperature, type of extraction vessel and shaker, and shaker speed. Differences in any one of these extraction parameters could lead to variation in levels of Cu, Fe, Mn, and Zn solubilized from soil. Standardization of extraction parameters as well as soil sample preparation is required to achieve quality control among soil testing laboratories.

Tedious concentration procedures are required for Cu determination by atomic absorption spectrophotometry if solutions contain <0.01 mg of Cu L^{-1} (Mehlich & Bowling, 1975). This difficulty is not encountered for Fe, Mn, and Zn soil tests, except for procedures that solubilize only soil solution or exchangeable micronutrient cations. Similar amounts of Cu were extracted from organic and mineral-organic soils with 0.5 and 1.0 *M* HCl at soil/solution ratios of 1:2.5 and 1:5, respectively. Amounts of extractable Cu at these acid concentrations and soil/solution ratios were sufficiently high for Cu determination by atomic absorption spectrophotometry without Cu concentration. However, of amounts Cu extracted with lower HCl concentrations in the range of 0.075 to 0.2 *M* required Cu concentration prior to its determination by atomic absorption spectrophotometry.

Effects of H_3PO_4 concentration and extraction time on levels of extractable Mn were studied for two soils on which soybean had moderate to severe Mn deficiency (Hoff & Mederski, 1958). Extractable Mn increased with H_3PO_4 strengths from 0.003 to 0.333 *M* and with extraction times from 10 to 120 min. An acid concentration of 0.033 *M* and an extraction period of 60 min were selected because solubilization of unavailable Mn from Mn hydrous oxides would be less likely at this concentration than at higher concentrations and longer extraction times.

Salcedo and Warncke (1979) evaluated the effects of various soil/solution ratio and extraction time combinations on the amounts of Mn extracted with 0.1 *M* HCl, 0.033 *M* H_3PO_4, DTPA-TEA, and 1.0 *M* NH_4OAc from 12 soils. Extraction periods of 10, 30, 60, and 120 min were combined with soil/solution ratios of 1:5, 1:10, 1:25 and, in addition, 1:2 for only the DTPA-TEA extractant. A soil/solution ratio of 1:25 with a shaking time of 120 min maximized extractable Mn by the four procedures. Correlations between Mn uptake by soybean plants grown in the greenhouse on the 12 soils and both DTPA-TEA and NH_4OAc extractable Mn were independent of the soil/solution ratio shaking-time combinations. Overall, the highest correlation with these combinations ($r = 0.92^{**}$, significant at $P = 0.01$) was obtained between Mn uptake by the soybean plants and Mn extracted by 0.033 *M* H_3PO_4 with a 1:5 soil/solution ratio and a 60-min shaking period.

The amount of Zn extracted by shaking 10 g of soil with 30 mL of 0.1 *M* HCl increased with shaking times from 1 to 240 min (Barrows & Drosdoff, 1960). A 2-min extraction period was chosen for the procedure because the amounts of Zn extracted from five soils did not differ greatly over the 1 to 60 min extraction periods. Sorensen et al. (1971) reported that the amounts of Fe, Mn, and Zn displaced by 0.1 *M* HCl from 18 soils increased with extraction times from 15 to 300 min and with increased soil-extraction solution ratios from 1:5 to 1:25. The magnitude of the increases varied widely among the soils for 0.1 *M* HCl extractable Zn, but varied little among the soils for extractable Fe and Mn.

During development of the DTPA-TEA soil test, it was found that extractable Cu, Fe, Mn, and Zn increased with extraction time (Lindsay & Norvell, 1978). The rate of release decreased sharply after the first hour for Cu, Fe, and Zn, and thus, a 2-h extraction time was selected for the procedure. Increases in temperature during the extraction from 15 to 35 °C increased amounts of the extractable cations and indicated the need for temperature control at 25 °C for the procedure. Higher DTPA concentrations increased the amount of extractable Cu, Fe, Mn, and Zn by placing greater stress on the labile micronutrient forms. A DTPA concentration of 0.005 *M* was selected for the test because this concentration provided ample chelating capacity to remove measurable amounts of all four micronutrients and a sufficient excess to prevent competitive secondary interactions among the four metals.

Effects of type of extraction vessel and shaker, shaking speed and length, soil/solution ratio, and filtering time on amounts of DTPA-TEA extractable Cu, Fe, Mn, and Zn were evaluated for Colorado soils (Soltanpour et al., 1976). Extractable Cu, Fe, and Mn increased with soil/solution ratios from 1:2 to 1:6 for 10-g soil samples, whereas extractable Zn was not consistently increased with magnitude of the soil/solution ratios. At a shaker speed of 92 oscillations min^{-1}, amounts of extractable Cu, Fe, Mn, and Zn were higher when shaken in conical flasks than when shaken in rectilinear bottles. Increasing shaker speed increased the levels of metals extracted in rectilinear bottles more than in Erlenmeyer flasks. The difference was attributed to better mixing of soil and DTPA-TEA solution in the flasks. Higher amounts of Cu, Fe, Mn, and Zn were extracted when samples were shaken on a reciprocal shaker than on a rotary shaker. These data reflect the better mixing of soil and DTPA-TEA solution with the reciprocal shaker as compared with the settling of soil to the vessel bottom with use of the rotary shaker. Filtering time of up to 10 min had little effect on levels of extractable Cu, Fe, Mn, and Zn.

III. GREENHOUSE EVALUATION OF SOIL TESTS

Greenhouse evaluation of a soil test usually is undertaken once it is predicted through soil chemical principles that the extraction solution and conditions solubilize a proportionate part of the labile forms of a nutrient from different soils. The purpose of the greenhouse evaluation is to determine whether the soil test warrants field calibration. Field calibration is undertaken if the greenhouse evaluation indicates that the extractant and extraction conditions solubilize amounts of micronutrient cations that are closely related to plant-available forms in soils.

Simple and multiple correlation and regression analyses commonly are used for greenhouse research to evaluate whether the extraction solution and conditions solubilize the plant available forms of a nutrient from soil. One of the following dependent variables is selected for this evaluation: plant nutrient concentration in tissue, nutrient uptake, dry weight, or dry weight change in response to the nutrient application. The independent variable for these analyses usually is the amount of the nutrient solubilized from soil by the extraction solution and conditions. In some cases, independent variables such as soil pH, cation exchange capacity, clay content, and organic C concentrations have been included in the statistical analyses to account for a diversity of properties for soils included in the greenhouse evaluation.

Statistical data from greenhouse evaluation research require careful interpretation to determine whether the relationship is suitable for a soil test to warrant field calibration. With a large number of observations, for example, a highly significant correlation ($\alpha = 0.01$) may be obtained with a relatively low *r* or *R* value, Due to this potential anomaly, a new procedure often is evaluated against a method that has been calibrated for similar soils. Frequently, the 0.1 *M* HCl extractable Zn (Wear & Sommer, 1948) or the DTPA-TEA extractable Cu, Fe, Mn, and Zn (Lindsay & Norvell, 1978) method has been included as a standard test in greenhouse evaluation studies.

Where growth responses are not obtained, correlation analyses may give invalid results because the relationship between the availability of a nutrient to plants and the amount of the extractable nutrient may not be linear in the deficiency range (Trierweiler & Lindsay, 1969). Consequently, there is little justification to extrapolate to the nutrient-deficiency range from regression equations developed with only soils that supply adequate amounts of the nutrient. It is, therefore, appropriate to use a group of soils that supply inadequate and adequate amounts of a nutrient to plants and to evaluate a new soil test procedure against yield response to application of a nutrient. This procedure necessitates control and micronutrient cation treatments to indicate whether soils supply insufficient or sufficient amounts of the micronutrients to plants in the greenhouse evaluation research.

The technique developed by Brown et al. (1962) is highly suitable for use in greenhouse evaluation studies to determine whether a new soil test warrants field calibration. They evaluated the suitability of dithizone-NH_4OAc (1.0 *M* NH_4OAc + 0.0004 *M* dithizone in CCl_4) extractable Zn as a procedure to detect soils with inadequate plant-available Zn. Zinc application increased dry weight of sweet corn on 27 of the 53 soils assayed in their greenhouse study. They presented data in bar graphs with levels of dithizone extractable Zn in ascending order from the origin along the abscissa and visually estimated the critical Zn level, which gave the best separation of soils into Zn deficient and sufficient categories, from the ordinate (Fig.

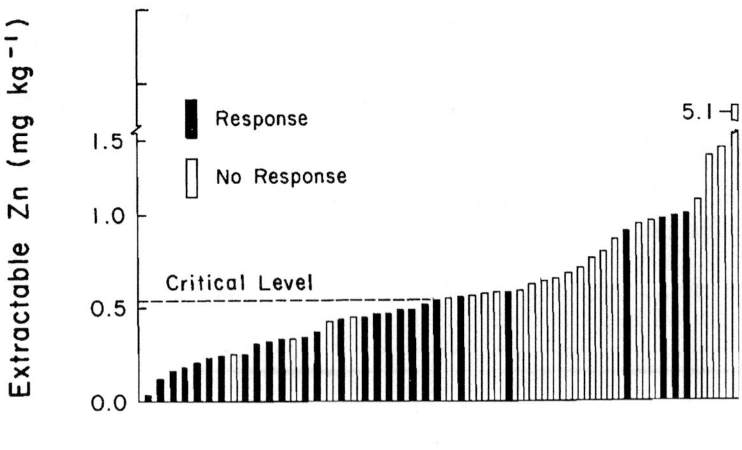

Fig. 9–1. Responses of sweet corn to soil-applied $ZnSO_4$ in relation to dithizone extractable Zn (Brown et al., 1962).

9–1). By this procedure, 84% of the soils below the critical level of 0.55 mg kg^{-1} supplied inadequate Zn, and 76% of the soils above this level supplied adequate Zn. An acceptable separation of nonresponsive and responsive soils may be obtained by this technique even where *r* values are not inordinately high between the extractable nutrient and tissue concentration or uptake of the nutrient from a series of soils (Trierweiler & Lindsay, 1969; Shuman et al., 1980).

The Cate-Nelson (Cate & Nelson, 1971) and linear plateau (Anderson & Nelson, 1975) models are highly suitable for use in greenhouse evaluation research. These models, like the graphic approach (Brown et al., 1962) indicate whether a soil test procedure warrants field calibration and provide a means to determine a critical nutrient level for a soil test. This critical level often is used for comparison purposes by others during completion of greenhouse evaluation research and frequently serves as a temporary critical level while a soil test is undergoing field calibration.

IV. FIELD CALIBRATION OF SOIL TESTS

The same statistical and graphic methods are used to obtain critical levels for a soil test in greenhouse evaluation and field calibration research. Critical levels for field calibration studies have been determined by use of the nutrient quantity extracted either alone or with other soil properties as the independent variable(s) in statistical analyses. The dependent variables used in these analyses have been the critical nutrient level in plant tissue or the yield response to micronutrient application. Critical levels for soil tests also have been obtained through comparison of amounts of an extractable nutrient with the presence or absence of deficiency symptoms. Reliance on plant physiology relationships dictates a need for knowledge in this area during completion of the field calibration research.

A. Differences in Plant Requirements

Caution must be exercised when the ability of soil to supply a micronutrient cation is evaluated by visual observation of deficiency symptoms. It is assumed by this technique that the micronutrient cation deficiency is recognizable, which as discussed below may not be valid for all micronutrient cation deficiencies. Increases in dry weight of soybean from Cu application occurred in greenhouse pots even though deficiency symptoms did not develop on plants (Makarim & Cox, 1983). The first visual symptom of Cu deficiency of field-grown wheat (*Triticum aestivum* L.) often is heads without a full complement of kernels and, when this symptom occurs, Cu application prior to the observation of symptoms increases seed yield (Alloway & Tills, 1984). Cotton (*Gossypium hirsutum* L.) seed yields were increased by Mn application in field experiments where Mn-deficiency symptoms were not observable on plants (Anderson & Boswell, 1968). A corn grain yield increase from Zn application occurred on a Norfolk loamy sand (fine-loamy, siliceous, thermic Typic Paleudults) where plants on the control had mild Zn deficiency early in the growing season, but did not have symptoms late in the growing season (Schnappinger et al., 1969). It is apparent from these findings that reliance on visual observation of deficiency symptoms may lead to incorrect separation of soils into adequate and inadequate micronutrient cation availability groups.

Recognition that critical micronutrient cation concentrations vary with plant part and age and with cultivar is required where critical deficiency levels are used in soil test calibration research. Ohki et al. (1979) determined the critical Mn level for determinate soybean 'Ransom' based on blades from trifoliolate leaves at the R2 growth stage. The critical Mn levels in leaf blades sampled at position one (plant top) down the axis in sequential order to position five were 18, 13, 11, 11, and 10 mg of Mn kg^{-1}. Copper, Fe, Mn, and Zn concentrations from control and Mn treatments on three soils on which Mn application increased seed yields were as follows for uppermost matured trifoliolate leaves at the R1 growth stage: leaflet > entire leaf > petiole (Gettier et al., 1985a, b). Copper, Mn, and Zn concentrations were determined in the hypocotyl-crown, stem and leaves of first and second lateral branches, and the main stem above the branches from three peanut (*Arachis hypogaea* L.) cultivars at various growth stages (Martens et al., 1969). The main stem generally contained the highest Mn and Zn concentrations, whereas the hypocotyl-crown usually had the highest Cu concentrations. Differences among cultivars occurred in Mn and Zn concentrations of similar plant parts at one or more growth stages. All plant portions decreased in concentration of one or more micronutrient during the growing season, except the stemmy portion of the first and second lateral branches.

Translocation of micronutrient cations requires consideration where critical deficiency levels are used in soil test calibration research. Copper translocation in five tropical and five temperate pasture legumes was evaluated in greenhouse research by Andrew and Thorne (1962). More Cu accumulated in roots than in shoots of most species. The species that were most sensitive to Cu deficiency were less able to translocate Cu from root to shoots. Likewise, relatively little translocation of absorbed Cu occurred from roots to shoots in corn plants (Dragun et al., 1976) and citrus trees (Fiskell & Leonard, 1967). Proportionally more absorbed Mn and Zn are translocated from roots to shoots than either Cu or Fe (Logan & Chaney, 1983).

Low amounts of Cu translocation from roots to shoots may lead to low correlation between either plant Cu uptake or concentration and the Cu-supplying power of soil. The critical Cu level in plants is sufficiently low that small experimental errors may lead to low correlations where soils supply insufficient amounts of Cu. This relationship is further justification to calibrate micronutrient cation soil tests on the basis of yield response to Cu application rather than on the basis of critical Cu levels in plant tissue. An alternate calibration procedure is to relate the Cu concentration in roots to soil test values (Fiskell & Leonard, 1967). This procedure was feasible because the Cu concentrations in roots reflected rates of Cu application and because the critical Cu level was higher in roots than in leaves. It is difficult, however, to obtain representative root samples.

Plants differ in sensitivities to Cu (Nelson et al., 1956; Andrew & Thorne, 1962; Reith, 1968), Fe (Olson & Carlson, 1950; Barak & Chen, 1982), Mn (Mascagni & Cox, 1984), and Zn (Massey, 1957; Brown et al., 1964) deficiencies under soil conditions conducive to these abnormalities. Copper deficiencies are common in cereal crops, Fe in peanut and sorghum, Mn in peanut and soybean, and Zn in corn and rice (*Oryza sativa* L.). Because plants vary widely in sensitivities to these micronutrient deficiencies, it is necessary to calibrate a soil test for each plant species or species with equal sensitivities.

B. Seasonal Effects on Plant Requirements

Climatic conditions affect yield response to micronutrient cation application. Surface application of Cu alleviated Cu deficiency of wheat in the greenhouse, but in-field trials on the same soils increases in vegetative growth from surface-applied Cu were not always accompanied by increases in grain yield (Grundon & Best, 1981). The difference between the greenhouse and the field research was attributed to variations in moisture content. Copper application did not increase yields in field experiments where negligible Cu was absorbed from the dry fertilizer zone. Infertile pollen, because of inadequate Cu absorption, led to failure of the wheat to produce grain.

Alloway and Tills (1984) pointed out that Cu deficiency occurs in crops when soil solution is unable to supply adequate Cu for plant uptake and that fast-growing crops require higher levels of soil solution Cu than do slow-growing crops. Therefore, increased availabilities of other nutrients, especially N, and highly productive crop cultivars will create a greater demand on the capacity of the solid phase to replenish Cu into soil solution. Since Cu is not readily translocated in plants, the deficiency develops during periods of rapid growth on soils with low solution Cu because of an insufficient rate of replenishment from the soil.

Excess irrigation water may cause Fe chlorosis in plants on calcareous soils (Lindsay & Thorne, 1954). In contrast, summer drought may cause death of seedlings that are susceptible to Fe chlorosis (Hutchinson, 1970). Iron relationships in soybean plants grown under different moisture regimes were studied by Elgala and Maier (1964). Plants were grown in the greenhouse on soils with 75 and 120% moisture-equivalent treatments. Higher moisture plants were chlorotic from Fe deficiency, and low moisture plants were normal. The lower amount of physiologically active Fe in chlorotic plants paralleled the higher amount of Ca and P in soil solution at the higher moisture content. Elgala and Maier (1964) concluded that the higher Ca and P uptake led to inactivation of Fe in cell walls and, thus, to less active Fe in protoplasm. It also is possible that the higher Ca activity led to more complexation of Ca in cell wall and membrane sites with a concomitant lower amount of transport of Fe into the protoplasm.

Manganese deficiencies commonly occur on poorly drained soils. Soil solution Mn activities in these soils often have a pronounced seasonal variation. Higher Mn activity in soil solution during periods of reducing conditions in winter and spring leads to higher Mn availability. As soils dry in the summer, the oxidizing conditions lead to decreases in soil solution Mn and thus to Mn deficiency. Soybean plants grown during the transition period may contain adequate Mn during early vegetative growth stages and, yet, develop the deficiency as late as 2 wk before flowering (Gettier et al., 1985c). For this reason, Mn concentrations in young plants are poorly related to the development of Mn deficiency later in the growing season.

Zinc-deficiency symptoms are generally most pronounced during a cool, wet spring and often disappear by mid-season (Bauer & Lindsay, 1965). The following explanations were presented for the relationship: (i) Zn uptake increases as the plant root system expands rapidly just prior to mid-season, and (ii) much of the available Zn comes from decomposition of organic residues for which the Zn release rate increases as greater biological activity occurs at the higher temperatures with advance of the growing season. In this research, incubation of soil at 43 °C for 1 to 3 wk increased Zn uptake by corn plants grown in the growth chamber and hence increased shoot and whole plant dry weight. Possible mechanisms proposed for the alteration of Zn availability by incubation at the higher temperature were release of natural chelating agents because of increased microbial growth or an increase in CO_2 pressure with an attendant increase in H_3O^+ activity. Both factors increase the amounts of Zn in soil solution and thereby increase Zn availability.

V. SOIL TESTS FOR INDIVIDUAL MICRONUTRIENT CATION EXTRACTION

The current trend is to develop soil tests for simultaneous extraction and determination of micronutrient cations either alone or with macronutrients. This trend is desirable for rapid conveyance of soil test data to the crop producer at a reasonable cost. Through the years, reliable single micronutrient cation soil tests have been developed that currently are used in some soil-testing laboratories and which frequently are used as standard tests in greenhouse evaluation research.

A. Copper

Reith (1968) completed greenhouse research with 41 soils from northeast Scotland to evaluate the suitability of 0.05 *M* EDTA (pH 7.0) extractable Cu for detection of soils with inadequate available Cu. Although relatively low correlations were obtained between the EDTA extractable Cu with both grain yield ($r = 0.38$*, significant at $P = 0.05$) and dry weight ($r = -0.54$**) response of oat plants to Cu application under greenhouse conditions on the 41 soils, the test gave a good indication of grain yield response of oat and barley (*Hordeum vulgare* L.) plants to Cu application in field experiments on 30 soils. Grain yields were not restricted by Cu deficiency on soils with an EDTA extractable Cu concentration > 1.1 mg kg^{-1}

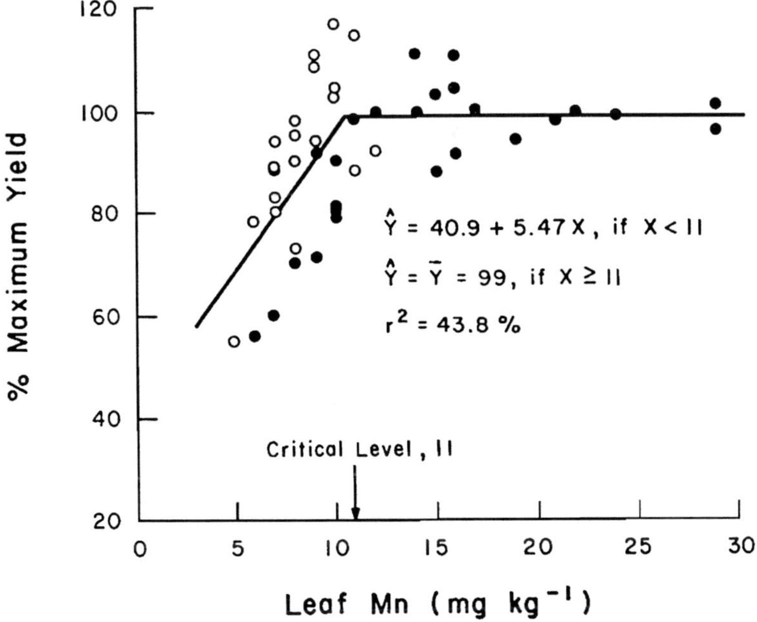

Fig. 9–2. The relationship between yield response and leaf Mn concentration in corn grown on two soils, which are represented by open and solid circles (Mascagni & Cox, 1984).

(Table 9–1). There were large increases in grain yields from Cu application on soils with <0.75 mg kg^{-1} of EDTA extractable Cu, and small increases on soils with EDTA-extractable Cu between 0.75 and 1.1 mg kg^{-1}.

In greenhouse research, the critical level of Cu extracted by the Mehlich-Bowling (0.5 *M* HCl, 0.016 *M* $AlCl_3$) method and growth response of corn, soybean, and wheat to Cu application was determined by the linear plateau model (Anderson & Nelson, 1975) to be 0.62 mg dm^{-3} of soil across the three crops (Makarim & Cox, 1983). The critical Cu level was extrapolated from the abscissa at the intersection of two straight lines, one in the linear portion and the other in the nonlinear portion, for a graph of percent maximum corn yield (ordinate) and extractable Cu (abscissa). Use of this technique is exemplified in Fig. 9–2 for research conducted by Mascagni and Cox (1984). Markarim and Cox (1983) obtained a critical level of 0.70 mg of Cu dm^{-3} of soil for the Mehlich-Bowling procedure by the Cate-Nelson model (Cate & Nelson, 1971) for soybean and wheat grown in 12 field experiments (Table 9–1). A Cate-Nelson plot is illustrated in Fig. 9–3 for research reported by Shuman et al. (1980).

B. Iron

Soil tests were developed for simultaneous extraction of Fe with other micronutrient cations either alone or with macronutrients (Soltanpour & Schwab, 1977; Lindsay & Norvell, 1978; Havlin & Soltanpour, 1981). Calibration data for these tests (Table 9–1) will be covered in a subsequent section.

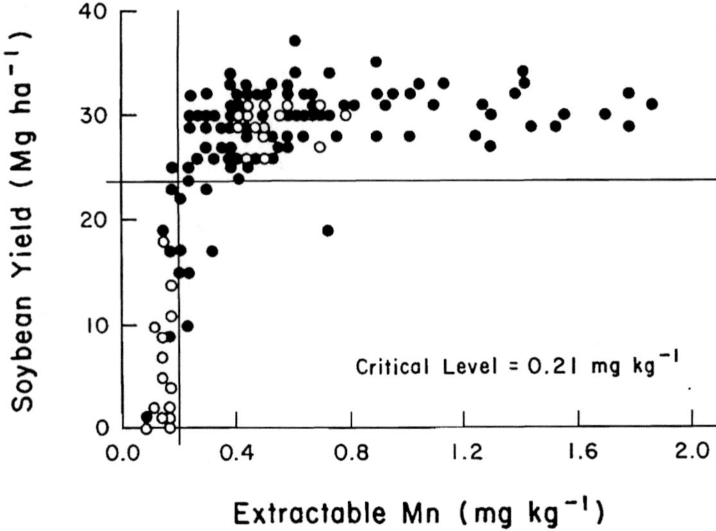

Fig. 9–3. Cate-Nelson plot for soybean seed yield vs. DTPA-TEA extractable Mn for two soils, which are represented by open and solid circles (Shuman et al., 1980).

C. Manganese

Eight soil tests were evaluated for their ability to predict uptake of native Mn by corn plants grown in the greenhouse on 63 Wisconsin soils (Browman et al., 1969). The soil tests evaluated were based on extraction with 1.0 *M* CH_3COONH_4 (pH 7.0); 0.5 *M* $Mg(NO_3)_2$; 0.2% hydroquinone in 1.0 *M* CH_3COONH_4 (pH 7.0); 0.033 *M* H_3PO_4; 3 *M* $NH_4H_2PO_4$; 1.5 *M* $NH_4H_2PO_4$; and EDTA-$(NH_4)_2CO_3$ [0.01 *M* EDTA, 1.0 *M* $(NH_4)_2CO_3$]. Of the soil tests, Mn uptake was most closely related to EDTA (r = 0.60**) and H_3PO_4 (r = 0.58**) extractable Mn. Regression equations were derived by comparing Mn uptake with each soil test in combination with pH and, from these analyses, the best prediction of Mn uptake was provided by 1.0 *M* CH_3COONH_4 extractable Mn and pH (R = 0.73**). It was impossible to determine if the two variables would separate soils into Mn deficient and sufficient categories because a Mn treatment was not included in the greenhouse research.

Relationships between extractable Mn by nine procedures and levels of Mn in soybean tissue were evaluated for plants grown in 25 fields on Ohio soils (Hoff & Mederski, 1958). Soybean plants had Mn-deficiency symptoms on 18 of the 25 soils. Tissue Mn correlated most closely with 1.0 *M* $NH_4H_2PO_4$ (r = 0.90*), alcoholic-hydroquinone (r =0.86*), and 0.033 *M* H_3PO_4 (r = 0.86*) extractable Mn. Hoff and Mederski (1958) calibrated the three soil tests on the basis of presence or absence of Mn-deficiency symptoms of soybean plants at a specific Mn soil test level of (Table 9–1).

Hammes and Berger (1960b) assayed Mn availability to oat plants in the greenhouse on 20 different soils that had been stored in a moist condition prior to potting. Available soil Mn was estimated by extraction with 1.5 *M* $NH_4H_2PO_4$, 0.033 *M* H_3PO_4, and 0.33 *M* H_3PO_4 on moist as well as air-dried samples. Correlations indicated that 0.033 *M* H_3PO_4 extractable Mn from moist samples gave the best relationship with Mn uptake by oat plants ($r = 0.85^{**}$). Since a Mn treatment was not used in the greenhouse research, it was not possible to determine if the H_3PO_4 extractable Mn procedure would separate soils into Mn deficient and sufficient categories.

In subsequent field research, Hammes and Berger (1960a) calibrated Mn soil tests with air-dry soil samples and oat grain yield response to Mn application in 25 field experiments. The soils used in this research were neutral to alkaline in reaction and developed from alkaline, lacustrine parent material. The 0.033 *M* H_3PO_4 and 1.5 *M* $NH_4H_2PO_4$ extractable Mn gave a much better separation of the 25 soils into Mn deficient and sufficient categories than did either 0.05 *M* EDTA or hydroquinone-NH_4OAc extractable Mn. Although they obtained suitable calibration data for all four procedures (Table 9–1), they recommended the H_3PO_4 soil test because of its high reliability and ease of usage. Based on the aforementioned greenhouse and field research and other greenhouse investigations (Randall et al., 1976; Salcedo & Warncke, 1979; Salcedo et al., 1979), the 0.033 *M* H_3PO_4 procedure currently is being used as a soil test for evaluation of the Mn status of soils for Mn-responsive crops in some North Central States (Whitney, 1980).

D. Zinc

Wear and Sommer (1948) provided the first calibration data for a micronutrient cation soil test. Their calibration data were based upon the occurrence of Zn-deficiency symptoms in corn grown in the field on 15 soils with pH levels from 4.7 to 6.2. Zinc-deficiency symptoms were not present in corn grown on seven soils with 0.1 *M* HCl extractable Zn levels > 1.2 mg kg^{-1} and were present on eight soils with extractable Zn levels in the range of 0.5 to 0.9 mg kg^{-1}. From these data, a critical level of 1.0 mg Zn kg^{-1} was selected for the 0.1 *M* HCl procedure (Table 9–1).

Subsequently, Wear and Evans (1968) compared the 0.1 *M* HCl and Mehlich-I (0.05 *M* HCl in 0.0125 *M* H_2SO_4) procedures as predictors of Zn uptake by corn and sorghum grown in the greenhouse on 12 sandy-textured soils with pH levels of 5.7 to 6.8. Correlation coefficients were higher for the Mehlich-I procedure and, therefore, they concluded that this procedure, which is routinely used in many southeastern laboratories, should replace the HCl extractable Zn soil test. The 0.1 *M* HCl, Mehlich-I, and DTPA-TEA procedures were calibrated for corn plants (Table 9–1) in a southeastern region project (Cox & Wear, 1977).

The 0.1 *M* HCl extractable Zn procedure extracts Zn from soil in excess of labile Zn. It dissolves portions of soil $CaCO_3$ and Al, Fe, and Mn hydrous oxides with a release of occluded Zn that, under normal soil conditions, is inaccessible for plant uptake (Trierweiler & Lindsay, 1969; Lauer, 1971). Nelson et al. (1959) developed a procedure to evaluate the Zn status of soils with two variables, 0.1 *M* HCl extractable Zn and titratable alkalinity. The latter is the amount of acid required to acidify soil to pH 5.0. They obtained a good separation of soils with adequate or inadequate Zn for responsive crops even though the titratable alkalinity factor corrects all soils for release of the same concentrations of occluded Zn in acid-soluble components. Different amounts of occluded Zn in acid-soluble components among soils can lead to errors in the use of the technique. Shaw and Dean (1952) obtained a good separation of soils into Zn-deficient and nondeficient categories on the basis of the two variables, dithizone-NH_4OAc (pH 7.0) extractable Zn and soil pH. Their separation was based on the presence or absence of Zn-deficiency symptoms of crops grown on 41 soils from areas of Zn deficiency in the USA.

Trierweiler and Lindsay (1969) modified the EDTA-$(NH_4)_2CO_3$ procedure, which was originally developed by Viro (1955), to buffer at pH 8.6 with $(NH_4)_2CO_3$ and to contain 0.01 *M* EDTA. A good separation of 42 Colorado neutral and high lime soils into adequate and inadequate Zn categories for growth of corn in the greenhouse was obtained by the EDTA-$(NH_4)_2CO_3$ and dithizone-NH_4OAc procedures, but not by the 0.1 *M* HCl plus titratable alkalinity procedure. They used the approach of Brown et al. (1962) to obtain critical levels for the EDTA-$(NH_4)_2CO_3$ and dithizone-NH_4OAc procedures (Table 9–1), except that they divided soils into Zn deficient, Zn deficient with high P, and Zn-sufficient categories (Fig. 9–4). Justification for the modification was that it would be advisable to apply Zn if a soil was sufficiently near the deficient level and if the soil had a relatively high level of available P.

Precautions are necessary for use of the dithizone-NH_4OAc extractable Zn procedure because CCl_4 is hazardous to human health and, therefore, this procedure is less adaptable to routine use than is the EDTA-$(NH_4)_2CO_3$ procedure. The EDTA-$(NH_4)_2CO_3$ method is superior to the acid extractant

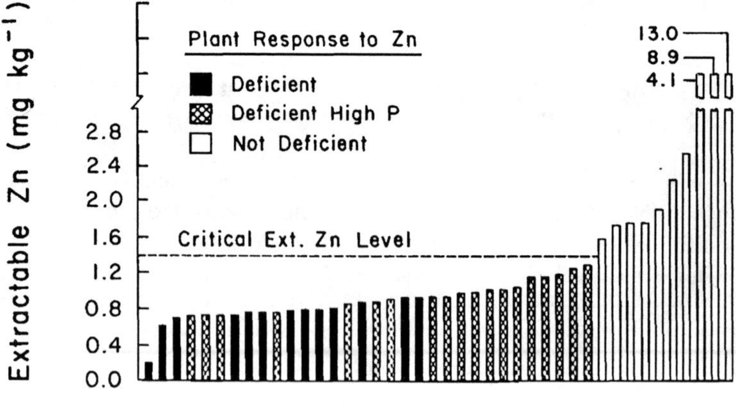

Fig. 9–4. The EDTA-$(NH_4)_2CO_3$ extractable Zn of 42 Colorado soils in relation to Zn response of corn in the greenhouse (Trierweiler & Lindsay, 1969).

because, under conditions of high soil pH, this technique suppresses the carbonate and oxide dissolution and thereby avoids extraction of occluded Zn.

Alley et al. (1972) used the graphic approach developed by Brown et al. (1962) to calibrate the EDTA-$(NH_4)_2CO_3$ extractable Zn procedure with yield response of corn to Zn application. They estimated the critical level for EDTA-$(NH_4)_2CO_3$ extractable Zn to be 0.8 mg kg^{-1} for growth of corn under field conditions (Table 9-1). Zinc application increased corn grain yield on 5 of the 10 soils used in their calibration research.

The dithizone-NH_4OAc and DTPA-TEA extractable Zn procedures were about equally effective in separation of 92 California soils into Zn deficient and sufficient groups based on dry weight response of sweet corn to Zn application in the greenhouse (Brown et al., 1971). Both procedures were superior to the two-variable method, 0.1 *M* HCl and titratable alkalinity, for separating soils into Zn deficient and sufficient categories. An analysis of these data by a binomial statistical test indicated that the DTPA-TEA procedure was superior to the dithizone-NH_4OAc and 0.1 *M* HCl methods for separation of the 92 soils into Zn deficient and sufficient categories (Keisling & Mullinix, 1979). As use of the DTPA-TEA extraction procedure became prominent in the early 1970s, calibration research was discontinued for the EDTA-$(NH_4)_2CO_3$, dithizone-NH_4OAc, and 0.1 *M* HCl procedures, except for green gram (*Phaseolus aureus* Roxb.) and rice (Gupta & Mittal, 1981; Singh & Takkar, 1981).

VI. SOIL TESTS FOR MULTI-NUTRIENT EXTRACTION

Simultaneous extraction of micronutrient cations was initiated with the development of the DTPA-TEA procedure by Lindsay and Norvell (1969). Currently, this procedure is used in many soil-testing laboratories for detection of soils with inadequate amounts of Cu, Fe, Mn, and Zn for normal plant growth. The Mehlich-I (Nelson et al., 1953) and NH_4HCO_3-DTPA (Soltanpour & Schwab, 1977) procedures presently are being used to detect soils that supply inadequate amounts of two or more micronutrient cations and certain macronutrients. It is expected that the Mehlich-III method (0.2 *M* HOAc, 0.25 *M* NH_4NO_3, 0.15 *M* NH_4F, 0.01 *M* HNO_3, and 0.0005 *M* EDTA) will replace the Mehlich-I procedure in the future (Mehlich, 1984).

A. Simultaneous Extraction of Micronutrient Cations

The DTPA-TEA soil test was developed to identify near-neutral and calcareous soils with insufficient available Cu, Fe, Mn, or Zn for maximum crop yields (Lindsay & Norvell, 1978). The soil test successfully separated Colorado soils on the basis of crop response to Fe and Zn fertilization in greenhouse research. Critical Fe and Zn levels were based on visual observation of the soil test levels that separated responsive and nonresponsive soils (Table 9-1). Comparison of visual symptoms of Fe deficiency observed in the field with soil test results confirmed the applicability of the critical soil test level for Fe. Since yield responses to Cu and Mn application were not obtained for the plants grown in the greenhouse, tentative critical DTPA-TEA extractable Cu and Mn levels were estimated to be 0.2 and 1.0 mg kg^{-1}, respectively (Lindsay & Norvell, 1978). The critical Mn level was based on the relationship between DTPA-TEA and hydroquinone-NH_4OAc extractable Mn and the calibration for the latter procedure. The critical Cu level was based on the DTPA-TEA extractable Cu concentration of 0.18 mg kg^{-1} for two soils that supplied inadequate Cu to plants.

There is a need for a procedure to detect soils that supply inadequate amounts of both Cu and Zn for wetland rice because Cu deficiencies have been diagnosed on some soils and Zn deficiencies are widespread (Ponnamperuma et al., 1981). Relationships between Cu, Fe, Mn, and Zn concentrations in rice grown on flooded soils in the greenhouse and DTPA-TEA extractable Cu, Fe, Mn, and Zn in saturated 5-g soil samples incubated at 20 °C for 21 d were investigated by Tiller et al. (1979). A correlation of $r = 0.79^{**}$ was obtained between Zn concentrations in young rice plants grown on 11 soils and levels of DTPA-TEA extractable Zn. Correlations for Cu, Fe, and Mn were nonsignificant.

Ponnamperuma et al. (1981) concluded that the DTPA-TEA procedure required too much labor, time, and cost and, therefore, determined whether a 0.05 *M* HCl extraction method would be as suitable as the DTPA-TEA procedure for detection of soils that supply inadequate Cu and Zn for rice production. Amounts of HCl extractable Zn correlated more closely with tissue Zn levels of rice grown inthe greenhouse on 22 soils ($r = 0.88^{**}$) than did quantities of DTPA-TEA extractable Zn ($r = 0.31$ NS, nonsignificant at the 0.05 level). Likewise, the HCl extractable Cu correlated more closely with Cu concentration ($r = 0.74^{**}$) than did DTPA-TEA extractable Cu ($r = 0.20$ NS). A critical level of 1.0 mg of Zn kg^{-1} was determined for the 0.05 *M* HCl extractable Zn procedure (Table 9-1) by the graphic method first used by Brown et al. (1962). A critical Zn level was not established for the DTPA-TEA procedure. It was not possible to estimate a critical Cu level for the HCl test because Cu application did not increase dry weight of rice plants grown in the greenhouse study. In another study, a correlation of $r = 0.98^{**}$ was obtained between Zn concentration in rice grown in the greenhouse and 0.05 *M* HCl extractable Zn in both dry and moist soil samples (Selvarajah et al., 1982). The pH of the nine soils used in this study ranged from 5.1 to 8.5.

B. Simultaneous Extraction of Micronutrient Cations and Macronutrients

The NH_4HCO_3-DTPA soil test was developed to simultaneously extract plant available P, K, Cu, Fe, Mn, NO_3-N, and Zn (Soltanpour & Schwab, 1977). Two criteria considered for the development of the test were as follows: (i) the soil test procedure should be rapid, reproducible, and economical, and (ii) the extraction of nutrients should be from the labile form that supplies those nutrients to plant roots (Havlin & Soltanpour, 1981). The extractant contains 1.0 *M* NH_4HCO_3 and 0.005 *M* DTPA adjusted to pH 7.6 (Soltanpour & Schwab, 1977). The high correlation between amounts of NH_4HCO_3-DTPA and DTPA-TEA extractable Cu, Fe, Mn, and Zn (r = 0.97–0.99**) coupled with evidence that the DTPA-TEA procedure extracts the labile micronutrient cations from soil (Lauer, 1971; Rule & Graham, 1976) suggests that the NH_4HCO_3-DTPA procedure likewise would extract the labile forms of the micronutrients. Overall, the amount of Fe extracted by the two procedures was about the same, whereas the NH_4HCO_3-DTPA method extracted 0.3, 0.8, and 0.5 mg kg^{-1} more Cu, Mn, and Zn, respectively. Critical levels for NH_4HCO_3-DTPA extractable Cu, Fe, Mn, and Zn were established by regression procedures on the basis of DTPA-TEA extraction data for bean, corn, potato (*Solanum tuberosum* L.), sorghum, sudan [*Sorghum sudanense* (Piper) Stapf], and sorghum-sudan hybrids (Soltanpour & Schwab, 1977).

Improvements in computerized analytical instruments have provided stimulus for simultaneous multi-element determinations from a single solution. Soltanpour et al. (1979b) compared inductively coupled plasma optical emission spectrometry (ICP-OES) with atomic absorption spectrophotometry (AAS) for determination of Cu, Fe, K, Mn, and Zn in NH_4HCO_3-DTPA extracts (Soltanpour & Schwab, 1977). They concluded that the ICP-OES and flame AAS analyses were comparable and, therefore, that ICP-OES, which has the advantage of simultaneous multi-element analysis, could be used for determination of Cu, Fe, K, Mn, and Zn in NH_4HCO_3-DTPA extracts.

Greenhouse studies were carried out to evaluate the ability of the NH_4HCO_3-DTPA extraction procedure to separate soils that supply adequate and inadequate amounts of Fe and Zn to plants (Havlin & Soltanpour, 1981). The NH_4HCO_3-DTPA soil test was as effective as the DTPA-TEA soil test of Lindsay and Norvell (1978) in separating 40 soils into deficient and nondeficient categories. The critical Fe level for sorghum was 4.8 mg kg^{-1} for both tests, whereas the critical Zn levels for corn were 0.9 and 0.7 mg kg^{-1} for the NH_4HCO_3-DTPA and DTPA-TEA procedure, respectively (Table 9–1). The NH_4HCO_3-DTPA soil test is more economical than the DTPA-TEA procedure because the former simultaeously extracts both macronutrients and micronutrient cations. The advantages of the NH_4HCO_3-DTPA procedure are as follows: (i) one extracting solution replaces NH_4OAc for K, $NaHCO_3$ for P, and DTPA; (ii) extraction time is rapid, i.e., 15 min; (iii) rapid analysis is possible with ICP-OES; (iv) results are very reproducible; and (v) there are savings in labor and reagents (Havlin & Soltanpour, 1981).

Baker (1973) developed a multi nutrient soil test procedure based on theoretical ion exchange relationships between soil and equilibration solution that contains nutrients of interest, i.e., Ca, Cu, Fe, K, Mg, Mn, Na, P, S, and Zn. The equilibration solution has a lower DTPA concentration and is less buffered as compared to the DTPA-TEA extractant. This procedure has not been widely adapted for routine soil analysis possibly because of the overall complexity of the method.

Makarim and Cox (1983) calibrated the Mehlich-I (Nelson et al., 1953), Mehlich-Bowling (Mehlich & Bowling, 1975), NH_4HCO_3-DTPA (Soltanpour & Schwab, 1977) and Mehlich-III (Mehlich, 1984) producers for detection of soils with inadequate Cu levels. Organic and mineral soils located in the Atlantic Coastal Plain region were used in this greenhouse research. The percent maximum yield of corn, soybean, and wheat was related to the amounts of extractable Cu by the four procedures with a linear plateau model (Anderson & Nelson, 1975). The critical Cu levels in soil, which were estimated by the intersection of two straight lines in the model, were 0.26, 0.62, 0.53, and 0.37 mg dm $^{-3}$ across crops, for the Mehlich-I, Mehlich-Bowling, NH_4HCO_3-DTPA, and Mehlich-III procedures, respectively (Table 9-1).

The Mehlich-I procedure has been used for decades to predict the need for Ca, K, Mg, and P fertilization in southeastern USA. Cox (1968) refined the procedure to include soil pH with Mehlich-I extractable Mn as an indicator of Mn availability to soybean on 17 Atlantic Coastal Plain soils with pH levels from 5.2 to 7.1. He developed the following regression equation for the 17 field locations: $\Delta Y = 500\,X_1 - 115X_2 - 2405$; where ΔY = soybean yield response to Mn application in kg ha $^{-1}$, X_1 = soil pH, and X_2 = Mehlich-I extractable Mn in mg kg $^{-1}$. The equation accounted for 84% of the variation in soybean seed yield response to Mn application. He developed calibration data from the regression equation that accounted for the inverse relationship between soil pH and Mn availability (Table 9-1). He reasoned that a seed yield response of 67 kg ha $^{-1}$ was required to cover Mn application costs and used this seed yield value to solve the regression equation to obtain the relationship with response probable and improbable at various levels of soil pH and extractable Mn.

Gettier et al. (1985c) modified the calibration for the Mehlich-I procedure on the basis of field research on 30 sites in the Atlantic Coastal Plain region. Manganese application increased soybean seed yields on 17 of the 30 sites. They obtained a prediction equation ($r = 0.68^{**}$) from these sites as follows: $Y = 205.71 - 24.71X_1 + 7.03\,X_2$; where Y = percent maximum yield response to Mn application on the 30 field sites, X_1 = soil pH, and X_2 = Mehlich-I extractable Mn in mg kg $^{-1}$. Their region of uncertain response to Mn application was obtained from solution of the regression equation for 80 to 100% of maximum yield. The critical level for the test was determined similarly with 90% of maximum yield (Table 9-1). The approaches by Gettier et al. (1985c) and Cox (1968) led to relatively similar results, except that the slope of the prediction equation was about 24% less for the equation by Cox (1968). Consequently, fewer incidences of Mn deficiency will be predicted by the calibration for the procedure reported by Gettier et al. (1985c) when the soil pH and Mehlich-I extractable Mn are >6.1 and 5.0 mg kg $^{-1}$, respectively, and more incidences of Mn deficiency will be predicted below these values.

Shuman et al. (1980) used the Cate and Nelson (1971) method for determination of the critical levels of soil tests to detect Mn deficiency of soybean on two soils. The method consists of construction of a graph with yield on the ordinate and extractable Mn on the abscissa. A cross is then placed over data points and moved until the upper left and lower right quadrants have a minimum number of points (Fig. 9-3). The critical level is the amount of extractable Mn where the cross intersects the abscissa. Based on this method, the critical levels (Table 9-1) were 0.40 mg kg^{-1} for NH_4HCO_3-DTPA, 0.22 mg kg^{-1} for DTPA-TEA, 2.6 mg kg^{-1} for Mehlich-I, and 1.8 mg kg^{-1} for Mehlich-II (0.2 *M* NH_4Cl, 0.2 *M* HOAc, 0.015 *M* NH_4F, and 0.012 *M* HCl).

A second technique used by Shuman et al. (1980) to determine the Mn critical level for the soil tests was a regression approach based on the critical deficiency level of 12 mg of Mn kg^{-1} in soybean leaves based on a 10% reduction in yield (Ohki et al., 1979). Linear regression equations were developed for the relationship of leaf Mn (*Y*) and extractable Mn (*X*). The critical Mn level for each soil test was then calculated by substitution of the critical tissue Mn concentration for the Y value in the linear portion of the regression equation. The respective critical levels for the NH_4HCO_3-DTPA and DTPA-TEA extractants of 0.33 and 0.24 mg of Mn kg^{-1} were close to those determined by the Cate-Nelson method, whereas the critical values for the Mehlich-I and Mehlich-II extractants were much lower than those determined by the Cate-Nelson method. The Cate-Nelson approach was assumed by Shuman et al. (1980) to be more accurate than their regression technique, because the latter approach depends on the accuracy of the Mn critical level in tissue.

Quite often soil tests are evaluated on a weight basis by a researcher, whereas analyses are carried out with a volume of sample in soil testing laboratories. Therefore, Mascagni and Cox (1984, 1985) calibrated the Mehlich-I and Mehlich-III procedures on a volume basis for detection of Mn deficiency in soybean and corn (Table 9-1). Their calibration research was based on one location in the Atlantic Coastal Plain region for corn and on 38 experiment years in this region for soybean. The importance of this research is indicated by the fact that future usage of the Mehlich-III procedure for estimation of available Ca, K, Mg, Mn, and Zn is anticipated in southeastern USA.

VII. ADDITIONAL SOIL TEST USES

The micronutrient cation soil tests were developed to identify agricultural soils with inadequate available Cu, Fe, Mn, or Zn for high crop yields. Because the soil tests extract labile forms of the micronutrients, they have additional applications. The soil tests have been used rather extensively in research on downward movement and reversion of micronutrients in soils. Frequently, soil tests are used to evaluate these reactions in soils that received high rates of waste products such as municipal sludge and high Cu manure (Payne et al., 1988; Rappaport et al., 1988).

A. Downward Movement

Less variation in micronutrient cation concentrations often occurs in labile cation forms determined with soil tests than in total concentrations from the same soil. This difference probably reflects less heterogeneity of an element in the labile form than in the nonlabile form in soil. For this reason, soil tests have been used to monitor downward movement of the micronutrient cations under laboratory and field conditions.

Wilson et al. (1981) investigated the amount of leaching from three annual applications of 56 kg of Mn ha^{-1} in an Olustee-Leefield sand (sandy, siliceous, thermic Ultic Haplaquod-loamy, siliceous, thermic Arenic Plintaquic Paleudult) with soil pH values of 6.2 to 6.7. Data from samples obtained at different depths showed that most of the DTPA-TEA and Mehlich-I extractable Mn remained in the upper 30 cm, i.e., in the soybean root zone. These data indicated that low Mn uptake by soybean was not due to leaching of Mn, but rather to formation of insoluble Mn hydrous oxides in this soil.

The Zn distribution in Nebraska sandhill soils where cropping changed from native grass to intensive irrigated corn was studied over a 5-yr period (Rehm et al., 1984). Zinc extracted with 0.1 *M* HCl indicated negligible downward movement of Zn from five annual applications of 13.6 kg of Zn ha^{-1} in a Thurman loamy fine sand (sandy, mixed, mesic Udorthentic Haplustoll). It was concluded that there was little, if any, potential for movement of Zn into the groundwater even though relatively high rates had repeatedly been used on the irrigated sandy soil.

B. Reversion Reactions

There is a tendency for micronutrient cations applied to soils to revert from labile forms to less soluble nonlabile forms with time. Reversion rates of the micronutrient cations to plant-unavailable forms have been evaluated with soil tests. Reversion of broadcast Zn mixed into the surface 20 cm of Ritzville fine sandy loam (coarse-silty, mixed, mesic Calciorthidic Haploxeroll) was studied annually for 6 yr after Zn application (Boawn et al., 1960). There was a decrease in 0.1 *M* HCl extractable Zn over the 6-yr period where 2.2, 4.4, 8.8, and 27.6 kg of Zn ha^{-1} as $ZnSO_4$ had been applied to the soil. After 6 yr, the extractable Zn levels were higher where Zn was applied than in the control treatment and the increases in extractable Zn over the control treatment increased with rate of Zn application. These data showed that applied Zn had a residual effect over several years. Much of the Zn applied as $ZnSO_4$ at rates of 112 and 336 kg of Zn ha^{-1} remained in a dithizone-NH_4OAc extractable form after 3 and 4 yr in a Tulare clay loam (fine montmorillonitic (calcareous), thermic Vertic Haplaquoll) and a peaty muck, respectively (Brown et al. 1962).

Eleven soils were fertilized with Cu, Fe, Mn, and Zn and examined periodically during a 14-wk incubation period for concentrations of the DTPA-TEA extractable micronutrients (Follett & Lindsay, 1971). During the incubation period, DTPA-TEA extractable nutrients declined to 61% of the original value for Cu, 14% for Mn, and 44% for Zn where these metals were supplied as sulfates. Only 20% of the Fe added as $FeSO_4 \cdot 7H_2O$ remained extractable at 1 wk, whereas 70% of the Fe added as FeEDDHA was extractable at 7 wk and 26% at 14 wk. These results are in general agreement with reported residual response from these fertilizers and suggest that the DTPA-TEA soil test may be used to monitor the availability of micronutrient cations in fertilized as well as unfertilized soils. A similar conclusion can be made for NH_4HCO_3-DTPA extractable Fe and Zn based on experimental data from greenhouse research by Havlin and Soltanpour (1984).

Mehlich-I extractable Mn in a Dragston fine sandy loam (coarse-loamy, mixed, thermic Aeric Ochraquult) 1 yr after application of 0, 10, 20, 40, and 50 kg of Mn ha^{-1} as $MnSO_4$ correlated closely with rate of Mn application ($r = 0.99$**) and with soybean seed yields ($r = 0.96$**) (Gettier et al., 1984). Although these data show that there was a residual effect of Mn the second year after application, yields were higher from re-application of Mn even where as much as 60 kg of Mn ha^{-1} had originally been applied to the soil. These data are in general agreement with the finding that Mn fertilizers remain available in highly acid soils but rapidly oxidize and precipitate to less-available forms under near-neutral and alkaline conditions. The data illustrate that good correlations can be obtained between soil test data and crop yields from micronutrient cation application on one soil or a group of soils with similar properties. Often, however, unsuitable relationships are obtained when a soil test is used to predict micronutrient cation availabilities from a group of soils with diverse properties.

VIII. LIMITATIONS OF SOIL TESTS

A. Soil Variability and Sampling

The distribution of micronutrient cation deficiencies often varies widely in fields. These deficiencies frequently occur in areas of <1 ha scattered throughout a field or in extensive areas with mild to severe or almost uniform deficiencies. Variable patterns of these deficiencies sometimes reflect natural heterogeneity in chemical and physical properties of different soils from drainage, P availability, pH, lime content, and other factors. In some cases, the different deficiency patterns within a field are related to variable pH changes from limestone application in soils with different buffer capacities or to variable P availabilities from P application in soils with different phosphate equilibria. Commonly, the micronutrient deficiencies reflect nonuniformity in limestone or P application.

Soil micro-heterogeneity could lead to incorrect interpretation of soil test data. This incorrect interpretation could occur for soils where soil acidity was incompletely neutralized by limestone application because of a short-term equilibration period. In this case, the applied limestone could neutralize an acidic extraction solution and, yet, have little effect on plant uptake of the micronutrient cations.

The first decision in obtaining a sample for evaluation of the micronutrient cation status from a field with much soil variation is whether to obtain a composite sample of each soil type or to obtain a composite sample from the entire field. One composite sample may mask the actual occurrence of the deficiencies. There is not one prescribed methodology for selection of either multiple or single composite samples from a field. Instead, the need for single or multiple composite samples from a field is based on a case-by-case consideration of the importance of a representative sample for a specific area.

B. Soils with Diverse Properties

The DTPA-TEA soil test was developed to detect Cu, Fe, Mn, and Zn deficiencies in alkaline and calcareous soils. Application of this procedure to noncalcareous soils may require inclusion of soil pH along with amounts of DTPA-TEA extractable micronutrient cations in models for prediction of these micronutrient deficiencies. For example, the need for Zn fertilization in corn production on Ontario soils is predicted from the following model (Bates, 1984): Zn-availability index = 203 + 4.5(DTPA extractable Zn) − 50.7(soil pH) + 3.33(soil pH)^2. This equation adjusts the Zn-availability index for soil pH up to approximately 7.0 and provides little adjustment for pH values between 7.0 and 8.0. A Zn-availability index below 15 indicates potential Zn deficiency in corn.

Logically, extractable P, soil pH, and extractable Cu, Fe, or Zn could be included in models for prediction of the micronutrient availabilities, because, at a given level of extractable Cu, Fe, and Zn, these micronutrient availabilities decrease with an increase in both available P and soil pH. A good relationship ($r = 0.945^{**}$) was shown between actual Zn uptake and Zn uptake predicted by the following equation in corn plants grown under greenhouse conditions:

$$Y = 780.2 + 68.8X_1 - 101.3X_2 - 0.4X_3$$

where Y = Zn uptake by corn plants, X_1 = Mehlich-I extractable Zn, X_2 = soil pH, and X_3 = Mehlich-I extractable P (Alley et al., 1972). This equation accurately separated 9 of 10 soils into Zn deficient and sufficient categories. The appeal of this approach is that each of these soil parameters is routinely determined in soil testing laboratories and, hence, recommendations can be computerized with ease.

C. Seasonal Weather Fluctuations

The micronutrient soil test results indicate only whether a micronutrient cation deficiency is likely to occur. The test results do not indicate the actual magnitude in yield increase from application of the micronutrient. An actual yield increase is weather dependent, i.e., a greater yield increase from micronutrient application occurs during a growing season with weather conditions conducive to high yields. In fact, statistical procedures often are not sufficiently sensitive to indicate a micronutrient cation deficiency when small yield increases are obtained because of a drought in non-irrigated crops. This lack of sensitivity leads to an error in soil test calibration, for no response may be obtained on a soil when yields are low, whereas a yield response would have been obtained on the soil under environmental conditions conducive to high yields.

Manganese deficiency of field crops is very weather dependent. Often this deficiency occurs when an inadequate amount of Mn is supplied to plants after Mn is oxidized and forms insoluble hydrous oxides. In some cases, Mn soil test data are used to flag areas of potential Mn deficiency, and foliar Mn is applied to plants when Mn-deficiency symptoms are observable. This method is suitable for correction of Mn deficiency in soybean because the plant has characteristic Mn-deficiency symptoms and relatively high yields can be obtained by correction of Mn deficiency by foliar Mn application (Gettier et al., 1985a). The method is unsuitable, however, in cases of incipient micronutrient deficiencies.

D. Recommendations from Soil Test Results

A major problem with micronutrient cation recommendations is the zone of uncertainty about a soil test critical level where a micronutrient deficiency may or may not occur. In these cases, it is necessary to consider economic factors in development of a recommendation for micronutrient application. The cost/benefit ratio must be considered for various crop and management regimes for these recommendations. The main danger is that individuals responsible for the recommendations may lack the background to recognize the existence of the zone of uncertainty for soil test data and, therefore, will not consider economic factors along with soil test data for fertilizer recommendations.

IX. COMMENTARY

The goal in development of a micronutrient cation soil test is to extract a proportionate part of the labile forms from different soils. This goal can be achieved with present knowledge of soil chemistry as is shown by, for example, the good separation of soils into Cu, Fe, Mn, and Zn deficient and sufficient categories by the DTPA-TEA procedure. Development of the many satisfactory soil tests is an accomplishment of considerable magnitude when one considers the labile micronutrients are extracted from a dry or moist sample under laboratory conditions, and amounts of labile micronutrients vary under diverse environmental conditions while plants are grown on the field sample site.

Although satisfaction can be taken in past achievements, it is recognized that improvements in micronutrient cation soil testing are needed in the future. Improved tests must be developed as advances are attained in soil micronutrient chemistry and plant physiology. More field calibration data are needed for present tests to improve fertilizer recommendations for Cu, Fe, Mn, and Zn. It would be appropriate to direct more attention to standardization of extraction conditions among laboratories to provide the quality control necessary for valid interpretation of the data. More emphasis should be placed on procedures for simultaneous extraction of both macronutrients and micronutrients by the same soil test procedure to provide more rapid transference of fertilizer recommendations to the crop producer at a lower cost and to facilitate the consideration of ion uptake interactions. The accuracy for prediction of a specific nutrient must not be compromised in the development of multi-nutrient soil tests.

It is recognized that a soil test for a micronutrient cation is just one component that can be applied to an overall systems approach to predict whether a Cu, Fe, Mn, or Zn deficiency of a plant will occur on a specific soil. Other logical components of this system would be the plant grown, variable plant response to different weather conditions such as temperature and moisture, and differences in properties from soil to soil that affect crop uptake of the elements. In the future, more attention should be directed toward an overall systems approach rather than to the sole use of a micronutrient cation soil test for prediction of fertilizer needs.

REFERENCES

Alben, A.O., and H.M. Boggs. 1936. Zinc content of soils in relation to pecan rosette. Soil Sci. 41:329–332.

Alley, M.M., D.C. Martens, M.G. Schnappinger, and G.W. Hawkins. 1972. Field calibration of soil tests for available zinc. Soil Sci. Soc. Am. Proc. 36:621–624.

Alloway, B.J., and A.R. Tills. 1984. Copper deficiency in world crops. Outlook Agric. 13:32–42.

Anderson, O.E., and F.C. Boswell. 1968. Boron and manganese effects on cotton yield, lint quality and earliness of harvest. Agron. J. 60:488–493.

Anderson, R.L., and L.A. Nelson. 1975. A family of models involving intersecting straight lines and concomitant experimental designs useful in evaluating response to fertilizer nutrients. Biometrics 31:303–318.

Andrew, C.S., and P.M. Thorne. 1962. Comparative responses to copper of some tropical and temperate legumes. Aust. J. Agric. Res. 13:821–835.

Baker, D.E. 1973. A new approach to soil testing. II. Ionic equilibria involving H, K, Ca, Mg, Fe, Cu, Zn, Na, P, and S. Soil Sci. Soc. Am. Proc. 37:537–541.

Barak, P., and Y. Chen. 1982. The evaluation of iron deficiency using a bioassay-type test. Soil Sci. Soc. Am. J. 46:1019–1022.

Barnette, R.M., J.P. Camp. J.D. Warner, and O.E. Gall. 1936. The use of zinc sulphate under corn and other field crops. Fla. Agric. Exp. Stn. Bull. 292:3–51.

Barrows, H.L., and M. Drosdoff. 1960. A rapid polarographic method for determining extractable zinc in mineral soils. Soil Sci. Soc. Am. Proc. 24:169–171.

Bartlett, R., and B. James. 1980. Studying dried, stored soil samples-some pitfalls. Soil Sci. Soc. Am. J. 44:721–724.

Bates, T.E. 1984. Soil test for zinc availability—revised. p. 52. *In* Land resource science progress report. Univ. of Guelph, ON, Canada.

Bauer, A., and W.L. Lindsay. 1965. The effect of soil temperature on the availability of indigenous soil zinc. Soil Sci. Soc. Am. Proc. 29:413–416.

Boawn, L.C., F.G. Viets, Jr., C.L. Crawford, and J.L. Nelson. 1960. Effects of nitrogen carrier, nitrogen rate, zinc rate, and soil pH on zinc uptake by sorghum, potatoes, and sugar beets. Soil Sci. 90:329–337.

Bonner, J., and A.W. Galston. 1952. Principles of plant physiology. W.H. Freeman and Company, San Francisco.

Bould, C., D.J.D. Nicholas, J.A.H. Tolhurst, T. Wallace, and J.M.S. Potter. 1949. Zinc deficiency of fruit trees in Britain. Nature (London) 164:801–802.

Bray, R.H. 1948. Requirements for successful soil tests. Soil Sci. 66:83–89.

Browman, M.G., G. Chesters, and H.B. Pionke. 1969. Evaluation of tests for predicting the availability of soil manganese to plants. J. Agric. Sci. 72:335–340.

Brown, A.L., B.A. Krantz, and P.E. Martin. 1962. Plant uptake and fate of soil applied zinc. Soil Sci. Soc. Am. Proc. 26:167–170.

Brown, A.L., B.A. Krantz, and P.E. Martin. 1964. The residual effect of zinc applied to soils. Soil Sci. Soc. Am. Proc. 28:236–238.

Brown, A.L., J. Quick, and J.L. Eddings. 1971. A comparison of analytical methods for soil zinc. Soil Sci. Soc. Am. Proc. 35:105–107.

Cate, Jr., R.B., and L.A. Nelson. 1971. A simple statistical procedure for partitioning soil test data into two classes. Soil Sci. Soc. Am. Proc. 35:658–660.

Chandler, W.H., D.R. Hoagland, and P.L. Hibbard. 1932. Little-leaf or rosette of fruit trees, II: Effect of zinc and other treatments. Am. Soc. Hortic. Sci. Proc. 29:255–263.

Cox, F.R. 1968. Development of a yield response prediction and manganese soil test interpretation for soybeans. Agron. J. 60:521–524.

Cox, F.R., and E.J. Kamprath. 1972. Micronutrient soil tests. p. 289–317. *In* J.J. Mortvedt et al. (ed.) Micronutrients in agriculture. SSSA, Madison, WI.

Cox, F.R., and J.I. Wear. 1977. Diagnosis and correction of zinc problems in corn and rice production. North Carolina State Univ. Southern Coop. Ser. Bull. 222.

Davies, D.W., and E.T. Jones. 1931. Gray speck disease of oats. Welsh J. Agric. 7:349–358.

Dragun, J., D.E. Baker, and M.L. Risius. 1976. Growth and elemental accumulation by two single-cross hybrids as affected by copper in solution. Agron. J. 68:466–470.

Elgala, A.M., and R.H. Maier. 1964. Chemical forms of plant and soil iron as influenced by soil moisture. Plant Soil 21:201–212.

Fiskell, J.G.A., and C.D. Leonard. 1967. Soil and root copper. Evaluation of copper fertilization by analysis of soil and citrus roots. J. Agric. Food Chem. 15:350–353.

Follett, R.H., and W.L. Lindsay. 1971. Changes in DTPA-extractable zinc, iron, manganese, and copper in soils following fertilization. Soil Sci. Soc. Am. Proc. 35:600–602.

Gamble, D.S., C.H. Langford, and J.P.K. Tong. 1976. The structure and equilibria of a manganese(II) complex of fulvic acid studied by ion exchange and nuclear magnetic resonance. Can. J. Chem. 54:1239–1245.

Gettier, S.W., D.C. Martens, D.L. Hallock, and M.J. Stewart. 1984. Residual Mn and associated soybean yield response from $MnSO_4$ application on a sandy loam soil. Plant Soil 81:101–110.

Gettier, S.W., D.C. Martens, and T.B. Brumback. Jr. 1985a. Timing of foliar manganese application for correction of manganese deficiency of soybeans. Agron. J. 77:627–630.

Gettier, S.W., D.C. Martens, and T.B. Brumback, Jr. 1985b. Boron, copper, iron and zinc concentrations in soybeans as affected by manganese application. J. Fert. Issues 2:130–135.

Gettier, S.W., D.C. Martens, and S.J. Donohue. 1985c. Soybean yield response prediction from soil test and tissue manganese levels. Agron. J. 77:63–67.

Goldberg, S.P., and K.A. Smith. 1984. Soil manganese: E values, distribution of manganese-54 among soil fractions, and effects of drying. Soil Sci. Soc. Am. J. 48:599–564.

Grossenbacher, J.G. 1916. Some bark diseases of citrus trees in Florida. Phytopathology 6:29–50.

Grundon, N.J., and E.K. Best. 1981. Rainfall pattern and response of wheat to soil dressings of copper sulfate. p. 359. *In* J.F. Loneragan et al. (ed.) Copper in soils and plants. Academic Press, New York.

Gupta, V.K., and S.B. Mittal. 1981. Evaluation of chemical methods for estimating available zinc and response of green gram (*Phaseolus aureus* Roxb.) to applied zinc in non-calcareous soils. Plant Soil 63:477–484.

Hammes, J.K., and K.C. Berger. 1960a. Manganese deficiency in oats and correlation of plant manganese with various soil tests. Soil Sci. 90:239–244.

Hammes, J.K., and K.C. Berger. 1960b. Chemical extraction and crop removal of manganese from air dried and moist soils. Soil Sci. Soc. Am. Proc. 24:361–364.

Harris, H.C. 1947. A nutritional disease of oats apparently due to lack of copper. Science 106:398.

Havlin, J.L., and P.N. Soltanpour. 1981. Evaluation of the NH_4HCO_3-DTPA soil test for iron and zinc. Soil Sci. Soc. Am. J. 45:70–75.

Havlin, J.L., and P.N. Soltanpour. 1982. Greenhouse and field evaluation of the NH_4HCO_3-DTPA soil test for Fe. J. Plant Nutr. 5:769–783.

Havlin, J.L., and P.N. Soltanpour. 1984. Changes in NH_4HCO_3-DTPA extractable zinc and iron as affected by various soil properties. Soil Sci. 137:188–193.

Hibbard, P.L. 1940. The chemical status of zinc in the soil with methods of analysis. Hilgardia 13:1–29.

Hoff, D.J., and H.J. Mederski. 1958. The chemical estimation of plant available soil manganese. Soil Sci. Soc. Am. Proc. 22:129–132.

Hutchinson, T.C. 1970. Lime chlorosis as a factor in seedling establishment on calcareous soils. New Phytol. 69:143–157.

Iyengar, S.S., D.C. Martens, and W.P. Miller. 1981. Distribution and plant availability of soil zinc fractions. Soil Sci. Soc. Am. J. 45:735–739.

Kabata-Pendias, A., and H. Pendias. 1985. Trace elements in soils and plants. CRC Press, Boca Raton, FL.

Keisling, T.C., and B. Mullinix. 1979. Statistical considerations for evaluating micronutrient tests. Soil Sci. Soc. Am. J. 43:1181–1184.

Knezek, B.D., and B.G. Ellis. 1980. Essential micronutrients IV: Copper, iron, manganese, and zinc. p. 259–286. *In* B.E. Davies (ed.) Applied soil trace elements. John Wiley and Sons, New York.

Lauer, D.A. 1971. Evaluation of plant available Zn by the DTPA soil test, 0.1 *M* HCl extraction, and labile Zn measurements. Ph.D. diss. Colorado State Univ., Fort Collins, (Diss. Abstr. 31-6157B).

Leeper, G.W. 1934. Relationship of soils to manganese deficiency of plants. Nature (London) 134:972–973.

Leggett, G.E., and D.P. Argyle. 1983. The DTPA-extractable iron, manganese, copper, and zinc from neutral and calcareous soils dried under different conditions. Soil Sci. Soc. Am. J. 47:518–522.

Lindsay, W.L. 1979. Chemical equilibria in soils. John Wiley and Sons, New York.

Lindsay, W.L., and F.R. Cox. 1985. Micronutrient soil testing for the tropics. Fert. Res. 7:169–200.

Lindsay, W.L., and W.A. Norvell. 1969. Development of a DTPA micronutrient soil test. p. 84. *In* Agronomy abstracts. ASA, Madison, WI.

Lindsay, W.L., and W.A. Norvell. 1978. Development of a DTPA soil test for zinc, iron, manganese, and copper. Soil Sci. Soc. Am. J. 42:421–428.

Lindsay, W.L., and D.W. Thorne. 1954. Bicarbonate ion and oxygen level as related to chlorosis. Soil Sci. 77:271–279.

Logan, T.J., and R.L. Chaney. 1983. Utilization of municipal wastewater and sludge on land—metals. p. 235–323. *In* A.L. Page (ed.) Utilization of municipal wastewater and sludge on land. Univ. of California, Riverside.

Makarim, A.K., and F.R. Cox. 1983. Evaluation of the need for copper with several soil extractants. Agron. J. 75:493–496.

Mann, H.B. 1930. Availability of manganese and of iron as affected by applications of calcium and magnesium carbonates to the soil. Soil Sci. 30:117–142.

Martens, D.C., D.L. Hallock, and M.W. Alexander. 1969. Nutrient distribution during development of three market types of peanuts. II. B, Cu, Mn, and Zn contents. Agron. J. 61:85–88.

Mascagni, Jr., H.J., and F.R. Cox. 1984. Diagnosis and correction of manganese deficiency in corn. Commun. Soil Sci. Plant Anal. 15:1323–1333.

Mascagni, Jr., H.J., and F.R. Cox. 1985. Calibration of a manganese availability index for soybean soil test data. Soil Sci. Soc. Am. J. 49:382–386.

Massey, H.F. 1957. Relation between dithizone-extractable zinc in the soil and zinc uptake by corn plants. Soil Sci. 83:123–129.

McLaren, R.G., and D.V. Crawford. 1973. Studies on soil copper. I. The fractionation of copper in soils. J. Soil Sci. 24:172–181.

Mehlich, A. 1984. Mehlich 3 soil test extractant: A modification of Mehlich 2 extractant. Commun. Soil Sci. Plant Anal. 15:1409–1416.

Mehlich, A., and S.S. Bowling. 1975. Advances in soil test methods for copper by atomic absorption spectrophotometry. Commun. Soil Sci. Plant Anal. 6:113–128.

Mortvedt, J.J. 1977. Micronutrient soil test correlations and interpretations. p. 99–117. *In* T.R. Peck et al. (ed.) Soil testing: Correlating and interpreting the analytical results. ASA Spec. Publ. 29. ASA, CSSA, and SSSA, Madison, WI.

Mulder, E.O. 1939. On the use of micro-organisms in measuring the deficiency of copper, magnesium and molybdenum. Antonie van Leeuwenhoek 6:99–109.

Myers, H.E., and E.W. Johnson. 1933. The cause and control of chlorosis in western Kansas. Trans. Kans. Acad. Sci. 36:106–110.

Nelson, L.G., K.C. Berger, and H.J. Andries. 1956. Copper requirements and deficiency symptoms of a number of field and vegetable crops. Soil Sci. Soc. Am. Proc. 20:69–72.

Nelson, J.L., L.C. Boawn, and F.G. Viets, Jr. 1959. A method for assessing zinc status of soils using acid-extractable zinc and "titratable alkalinity" values. Soil Sci. 88:275–283.

Nelson, W.L., A. Mehlich, and E. Winters. 1953. The development, evaluation and use of soil tests for phosphorus availability. p. 153–188. *In* W.H. Pierre and A.G. Norman (ed.) Soil and fertilizer phosphorus in crop nutrition. Agronomy Monogr. Academic Press, New York.

Ohki, K., F.C. Boswell, M.B. Parker, L.M. Shuman, and D.O. Wilson. 1979. Critical manganese deficiency level of soybean related to leaf position. Agron. J. 71:233–234.

Olson, R.V., and C.W. Carlson. 1950. Iron chlorosis of sorghums and trees as related to extractable soil iron and manganese. Soil Sci. Soc. Am. Proc. 14:109–112.

Payne, G.G., D.C. Martens, E.T. Kornegay, and M.D. Lindemann. 1988. Availability and form of copper in three soils following eight annual applications of copper-enriched swine manure. J. Environ. Qual. 17:740–746.

Ponnamperuma, F.N., M.T. Cayton, and R.S. Lantin. 1981. Dilute hydrochloric acid as an extractant for available zinc, copper and boron in rice soils. Plant Soil 61:297–310.

Randall, G.W., E.E. Schulte, and R.B. Corey. 1976. Correlation of plant manganese with extractable soil manganese and soil factors. Soil Sci. Soc. Am. J. 40:282–287.

Rappaport, B.D., D.C. Martens, R.B. Reneau, Jr., and T.W. Simpson. 1988. Metal availability in sludge-amended soils with elevated metal levels. J. Environ. Qual. 17:42–47.

Rehm, G.W., R.C. Sorensen, and R.A. Wiese. 1984. Soil test values for phosphorus, potassium, and zinc as affected by rate applied to corn. Soil Sci. Soc. Am. J. 48:814–818.

Reith, J.W.S. 1968. Copper deficiency in crops in northeast Scotland. J. Agric. Sci. 70:39–45.

Rule, J.H., and E.R. Graham. 1976. Soil labile pools of manganese, iron, and zinc as measured by plant uptake and DTPA equilibrium. Soil Sci. Soc. Am. J. 40:853–857.

Salcedo, I.H., and D.D. Warncke. 1979. Studies in soil manganese: I. Factors affecting manganese extractability. Soil Sci. Soc. Am. J. 43:135–138.

Salcedo, I.H., B.G. Ellis, and R.E. Lucas. 1979. Studies in soil manganese. II. Extractable manganese and plant uptake. Soil Sci. Soc. Am. J. 43:138–141.

Schnappinger, Jr., M.G., D.C. Martens, and G.W. Hawkins. 1969. Response of corn to Zn-EDTA and $ZnSO_4$ in field investigations. Agron. J. 61:634–636.

Schnitzer, M. 1978. Humic substances: Chemistry and reactions. p. 1–64. *In* M. Schnitzer and S.U. Khan (ed.) Soil organic matter. Elsevier Sci. Publ. Co., New York.

Selvarajah, N., V. Pavanasasivam, and K.A. Nandasena. 1982. Evaluation of extractants for Zn and Cu in paddy soils. Plant Soil 68:309–320.

Severson, R.C., J.M. McNeal, and J.J. Dickson. 1979. Effects of soil preparation on DTPA-extractable elements in soils of the northern great plains. Soil Sci. 128:70–79.

Shaw, E., and L.A. Dean. 1952. Use of dithizone as an extractant to estimate the zinc nutrient status of soils. Soil Sci. 73:341–347.

Sherman, G.D., J.S. McHargue, and W.S. Hodgkiss. 1942. Determination of active manganese in soil. Soil Sci. 54:253–257.

Shuman, L.M., F.C. Boswell, K. Ohki, M.B. Parker, and D.O. Wilson. 1980. Critical soil manganese deficiency levels for four extractants for soybeans grown in sandy soil. Soil Sci. Soc. Am. J. 44:1021–1025.

Sillanpaa, M. 1982. Micronutrients and the nutrient status of soils: A global study. Soils Bull. 48, FAO, Rome.

Sims, J.L., and W.H. Patrick, Jr. 1978. The distribution of micronutrient cations in soil under conditions of varying redox potential and pH. Soil Sci. Soc. Am. J. 42:258–262.

Singh, H.G., and P.N. Takkar. 1981. Evaluation of efficient soil test methods for Zn and their critical values in salt-affected soils for rice. Commun. Soil Sci. Plant Anal. 12:383–406.

Skinner, J.J., and R.W. Ruprecht. 1930. Fertilizer experiments with truck crops. Fla. Agric. Exp. Stn. Bull. 218.

Soltanpour, P.N., A. Khan, and W.L. Lindsay. 1976. Factors affecting DTPA-extractable Zn, Fe, Mn, and Cu from soils. Commun. Soil Sci. Plant Anal. 7:797–821.

Soltanpour, P.N., and A.P. Schwab. 1977. A new soil test for simultaneous extraction of macro- and micro-nutrients in alkaline soils. Commun. Soil Sci. Plant Anal. 8:195–207.

Soltanpour, P.N., A. Khan, and A.P. Schwab. 1979a. Effect of grinding variables on the NH_4HCO_3-DTPA soil test values for Fe, Zn, Mn, Cu, P, and K. Commun. Soil Sci. Plant Anal. 10:903–909.

Soltanpour, P.N., S.M. Workman, and A.P. Schwab. 1979b. Use of inductively-coupled plasma spectrometry for the simultaneous determination of macro- and micronutrients in NH_4HCO_3-DTPA extracts of soils. Soil Sci. Soc. Am. J. 43:75–78.

Sorensen, R.C., D.D. Oelsligle, and D. Knudsen. 1971. Extraction of Zn, Fe, and Mn from soils with 0.1 *N* hydrochloric acid as affected by soil properties, solution:soil ratio, and length of extraction period. Soil Sci. 111:352–359.

Sposito, G. 1981. The operational definition of the zero point of charge in soils. Soil Sci. Soc. Am. J. 45:292–297.

Steinberg, R.A. 1919. A study of some factors in the chemical stimulation of the growth of *Aspergillus niger*. Am. J. Bot 6:330–372.

Stout, P.R. 1956. Micronutrients in crop vigor. J. Agric. Food Chem. 4:1000–1006.

Terman, G.L. 1974. Amounts of nutrients supplied for crops grown in pot experiments. Commun. Soil Sci. Plant Anal. 5:115–121.

Thomas, E.E., and A.R.C. Haas. 1928. Injection method as a means of improving chlorotic orange trees. Bot. Gaz. 86:355–362.

Tierney, C.E. 1981. The effect of the number and the timing of foliar Mn application upon soybean yield. M.S. thesis. Virginia Polytechnic Inst. and State Univ., Blacksburg.

Tiller, K.G., E. Suwadji, and R.S. Beckwith. 1979. An approach to soil testing with special reference to the zinc requirement of rice paddy soils. Commun. Soil Sci. Plant Anal. 10:703–715.

Trierweiler, J.F., and W.L. Lindsay. 1969. EDTA-ammonium carbonate soil test for zinc. Soil Sci. Soc. Am. Proc. 33:49–54.

Udo, E.J., H.L. Bohn, and T.C. Tucker. 1970. Zinc adsorption by calcareous soils. Soil Sci. Soc. Am. Proc. 34:405–407.

Viets, Jr., F.G., and W.L. Lindsay. 1973. Testing soils for zinc, copper, manganese, and iron. p. 153–172. *In* L.M. Walsh and J.D. Beaton (ed.) Soil testing and plant analysis. rev. ed. SSSA, Madison, WI.

Viro, P.J. 1955. Use of ethylenediaminetetraacetic acid in soil analysis: I. Experimental. Soil Sci. 79:459–465.

Wear, J.I., and C.E. Evans. 1968. Relationship of zinc uptake by corn and sorghum to soil zinc measured by three extractants. Soil Sci. Soc. Am. Proc. 32:543–546.

Wear, J.I., and A.L. Sommer. 1948. Acid-extractable zinc of soils in relation to the occurrence of zinc deficiency symptoms of corn: A method of analysis. Soil Sci. Soc. Am. Proc. 12:143–144.

Whitney, D.A. 1980. Micronutrient soil tests—Zinc, iron, manganese and copper. p. 18–21. *In* W.C. Dahnke (ed.) Recommended chemical soil test procedures for the North Region. North Dakota Agric. Exp. Stn. Bull. 499.

Wilson, D.O., F.C. Boswell, K. Ohki, M.B. Parker, and L.M. Shuman. 1981. Soil distribution and soybean plant accumulation of manganese in manganese-deficiet and manganese-fertilized field plots. Soil Sci. Soc. Am. J. 45:549–552.

Chapter 10

Testing Soils for Sulfur, Boron, Molybdenum, and Chlorine[1]

GORDON V. JOHNSON, *Oklahoma State University, Stillwater*

P. E. FIXEN, *South Dakota State University, Brookings*

The nutrient elements S, B, Mo, and Cl are all relatively mobile in soils. This, in general, is due to their existence as relatively large anions (Cl^-) or oxo anions (SO_4^{2-} and MoO_4^{2-}) that form relatively soluble compounds with the common cations abundant in soils, or as in the case of B, chemical stability as a soluble uncharged molecule (H_3BO_3). Plants absorb the nutrients primarily in these chemical forms. The commonly formed alkali metal and alkaline earth compounds of the elements, when present in solid phase, are sufficiently soluble to provide a more than adequate supply of the nutrients for crops.

Except for S, the elements are required in such small amounts by plants that they are commonly categorized as micronutrients. For this reason, crop deficiencies are the exception rather than the rule, even in humid environments where the potential for leaching is high. For different reasons, S deficiencies are also somewhat uncommon and the economic impact of shortages of these elements in crop production is small relative to N, P, and K. Consequently, the evolution of soil testing which has led to reliable procedures used in routine testing for N, P, and K has not occurred to the same degree for S, B, Mo, and Cl. In those environments and cropping systems where these elements are lacking, correctly identifying the deficiency and degree to which it exists has great economic importance and may be aided through soil testing as described in the following sections. These sections will consider in order each element with respect to (i) extraction procedures, (ii) analytical procedures, and (iii) interpretation.

[1]Contribution from the Dep. of Agronomy, Oklahoma State Univ., Stillwater, OK 74078 and the Dep. of Plant Science, South Dakota State Univ., Brookings, SD 57007.

I. SULFUR

A. Extraction Procedures

Successful procedures for estimating available S in soils have centered on extraction of SO_4-S using dilute acids or dilute salt solutions. The principle involved is to remove water soluble, easily exchangeable, and adsorbed or labile S. Attempts to measure mineralizeable S have met the same obstacles as for mineralizable N and hence mineralizable S has not been a component of successful S soil tests. At least 20 different reagents or reagent concentrations have been successful to some degree for extracting available S from soils (Table 10–1). It is obvious that many solutions, including water, have potential for extracting available S from soils. Several factors account for the apparent success or failure of extraction procedures for estimating available S.

In arid and semiarid regions of the world, where leaching has been minimal and soils contain an abundance of simple sulfate salts, procedures which extract water soluble sulfate can easily differentiate nonresponsive soils from soils that may potentially respond to added S. In more humid regions of the world, exchangeable and loosely bound or labile forms of S become increasingly more important sources of S to sustain crop production. Extracting solutions that contain a replacing ion such as phosphate are likely to be more successful in these climates.

Weak electrolyte solutions of a monovalent cation often extract considerable organic matter that may seriously interfere with sulfate analysis.

Table 10–1. Reagents for extracting available S.

Reagent	Reference
H_2O	Bansal et al., 1979
HCl (0.001 *M*)	Bansal et al., 1979
HCl (0.05 *M*)	Islam & Ponnamperuma, 1982
LiCl (0.1 *M*)	Islam & Ponnamperuma, 1982
LiCl (0.01 *M*)	Maynard et al., 1987
NaCl (10 g L^{-1})	Bansal et al., 1979
KCl (0.01 *M*)	Maynard et al., 1987
$CaCl_2$ (0.01 *M*)	Yli-Halla, 1987
$CaCl_2$ (1.5 g L^{-1})	Palaskar & Ghosh, 1985
NH_4OAc (1.0 *M*)	Palaskar & Ghosh, 1985
NH_4OAc (0.5 *M*)	Islam & Ponnamperuma, 1982
NH_4OAc (0.003 *M*)	Maynard et al., 1987
NH_4Cl (0.003 *M*)	Maynard et al., 1987
NH_4Cl (0.01 *M*)	Maynard et al., 1987
$Mg(OAc)_2$ (1.0 *M*, pH 7.0)	Palaskar & GHosh, 1985
$NaHCO_3$ (0.5 *M* $NaHCO_3$)	Tiwari et al., 1983
NaOAc (1.0 *M*) + HOAc (to pH 4.8)	Bansal & Pal, 1987
NH_4OAc (0.5 *M*) + HOAc (0.5 *M*)	Yli-Halla, 1987
KH_2PO_4 (500 mg L^{-1} P)	Bansal et al., 1983
$Ca(H_2PO_4)_2$ (500 mg L^{-1} P)	Islam & Bhuiyan, 1988
$Ca(H_2PO_4)_2$ (500 mg L^{-1} P + 2 *M* HOAc)	Hoeft et al., 1973

This is especially true when turbidimetric analysis with Ba is used. For these extractants, decolorization using charcoal must precede the final analysis. Extractants using Ca or Mg salts will usually provide clear, colorless solutions for analysis. Stepwise procedures for extraction of soluble and soluble plus adsorbed S are detailed elsewhere (Tabatabai, 1982).

B. Analytical Procedures

There are several analytical techniques available for measuring extracted soil S. The most common method in the USA is the turbidimetric analysis after reaction of sulfate with Ba (Johnson, 1987). The foundation of this method is the low solubility of solid $BaSO_4$ that forms under controlled conditions when fine $BaCl_2$ crystals are added to the extract containing sulfate. This method is accurate for mg kg^{-1} levels and, while somewhat cumbersome, usually can be performed with existing equipment (Eik, 1980). Modifications of the turbidimetric method include analysis of the precipitated Ba by atomic absorption analysis (Oeien, 1979) and the use of autoanalyzers to help standardize conditions and speed analysis to as many as 200 samples/d (Wall et al., 1980; Lea & Wells, 1980).

More recent work has demonstrated succesful analysis of S in soil extracts by ion chromatography and inductively coupled plasma (ICP) atomic emission spectrometry (Maynard et al., 1987) Analysis by ICP has been shown to give higher results (Yli-Halla, 1987) than turbidimetric analysis and may include S forms other than sulfate in the extract.

The colorimetric procedures, using sulfate reduction and subsequent reaction with methylene blue (Johnson & Nishita, 1952), continues to be used and improved to provide more rapid and inexpensive analysis (Kowalenko, 1985; Pirela & Tabatabai, 1987).

These procedures all have merit and should be considered when they take advantage of existing equipment and advanced technology to allow rapid accurate analysis.

C. Interpretation

Sulfur soil test interpretations have taken many forms, and include relating critical soil concentration to crop response parameters such as yield, plant S concentration, or N/S ratio. Although crop yield is usually the most important index for assessing adequate nutrition, many studies have depended upon plant response in the greenhouse for evaluating S soil tests. In these studies, plant S concentration and N/S ratios have often been used.

Attempts to calibrate S soil tests have been unsuccessful for many of the same reasons N soil test calibrations have failed. Principle reasons among these is the fact that the primary nutrient form absorbed, sulfate like nitrate is relatively mobile in the soil. Hence, interpretation of the soil test must take into account the production level or yield goal (see chapter 6 in this book, by Dahnke and Johnson). Consequently, greenhouse studies have limited value for identifying a critical soil test level because the S-supplying soil volume is dependent on pot or container size that is usually restrictive relative to field conditions.

Unlike N fertilizer needs, S deficiencies are seldom more than a few kg/ha (or lb/acre) and may be corrected by applying about 1/20 of the N requirement for the crop. For this reason, instead of trying to determine accurately an existing small deficiency of S by use of a soil test, an adequate amount of S (10–15 kg/ha) is often recommended for crops growing in deep, sandy soils, susceptible to S deficiency (Johnson, 1987).

Critical extractable SO_4-S levels vary greatly depending upon the extractant used and whether or not the investigation was performed in the field or greenhouse; choice of crop has minor influence. A critical level of 30 mg/kg of S, extracted by NH_4OAc, has been reported for corn (*Zea mays* L.) (Palaskar & Ghosh, 1985), rice (*Oryza sativa* L.) (Islam & Ponnamperuma, 1982), and cabbage (*Brassica oleracea* L.) (Palaskar & Ghosh, 1981). By comparison, a critical level of about 10 mg/kg has commonly been reported for phosphate-containing extractants (Bansal et al., 1979; Islam & Ponnamperuma, 1982; Bansal et al., 1983; Palaskar & Ghosh, 1985; Islam & Bhuiyan, 1988). Available S in subsoil should also be measured to accurately determine crop needs. When subsoil S has been included critical levels as low as 3.5 mg/kg have been identified using phosphate solutions for extraction (White et al., 1981).

Using S soil tests, which in themselves are difficult to interpret, as a means of predicting response to S fertilizer often fails because of extraneous or incidental S additions to the field. Amounts of S ranging from 10 to 30 kg/ha may be supplied from rainfall annually (Sharpley et al., 1985). Similar levels of S may be supplied from soil organic matter mineralization, manures, other fertilizers, soil ammendments, pesticides, atmospheric SO_2 or particulate deposition.

II. BORON

A. Extraction Procedures

The most widely used and investigated B soil test is the hot water extraction method (Berger & Truog, 1939). Other successful methods including variations on the hot water extraction, are extractions with Morgan's reagent (NaOAc-HOAc) (Lombin, 1985), mannitol (Jin et al., 1988), $CaCl_2$ + mannitol (Cartwright et al., 1983), and hot $CaCl_2$ (Parker & Gardner, 1981; Aitken et al., 1987; Jeffrey & McCallum, 1988).

The hot water extraction method has consistently proven to be a reliable method for estimating available B. The major drawback to this method is that it requires refluxing the boiling soil-water mixture. This requirement has made it difficult to adapt the method to rapid routine analysis for many samples. Refluxing the boiling sample in weak $CaCl_2$ solutions (0.01–0.02 *M*) usually provides a clear, colorless extract without altering the amount of B extracted. This is a significant improvement when colorimetric analysis is used because it eliminates the need to add charcoal as a decolorizing agent. In addition to saving a step, it also removes a source of negative error in the extraction (Gupta, 1979; Parker & Gardner, 1981).

Additional improvements in the hot water extraction in recent time have been the consideration of lengthening the extraction period and substitution of plastic containers for glass-refluxing equipment. Extending the boiling time from 5 to 10 min gives more consistent results (Odom, 1980) without complicating the procedure. Boiling the soil-water mixture in plastic pouches, instead of refluxing with glass, offers a major advancement in the method with the advantage of lower equipment cost, less labor, and improved sample turn-around time (Mahler et al., 1984). Similar advantages have been obtained by performing the extraction in plastic tubes heated in an aluminum block (Nilsson, 1986).

B. Analytical Procedures

Colorimetric analysis using either curcumin or azomethine-H have been the most common methods of measuring extracted soil B. Continued improvements in these methods have resulted in eliminating interferences, improving sensitivity, and adaptation to automated analysis (Wikner & Uppstroem, 1980; Parker & Gardner, 1981; Porter et al., 1981). A standard procedure for these methods of analysis and the preceding soil extraction are given in detail elsewhere (Bingham, 1982).

Recent technological advances in analysis of elements by emission spectrometry have resulted in this becoming an attractive analytical tool for soil testing. Both ICP and direct current plasma (DCP) have been found acceptable alternatives to the colorimetric procedures (Gestering & Soltanpour, 1981; Nilsson & Jennische, 1986; Jeffrey & McCallum, 1988). Using either ICP or DCP instruments, the analysis is fast because soil extracts may be analyzed directly. Emission analysis is relatively free of interferences and provides excellent detection limits and precision.

C. Interpretation

Boron nutrient requirements are dependent on the crop to be grown and its anticipated yield. Requirements for most crops, however, are usually met when soils contain only 0.5 to 1.0 mg kg^{-1} of hot water soluble B, and levels above about 5 mg kg^{-1} are likely to be toxic (Ponnamperuma et al., 1981). For this reason, deficiencies are easily corrected by adding small amounts (0.5–1.0 kg/ha) of B and care must be taken to avoid adding too much and causing a toxic condition.

III. MOLYBDENUM

Molybdenum availability in soils has not been extensively researched, in large part because deficiencies are scarce. Secondly when deficiencies have been suspected, factors other than soil test availability indexes have shown to be just as useful for predicting response to Mo additions. While this latter conditional application of the soil test result is not unusual among micronutrient soil tests, having to measure other soil parameters does sometimes detract from soil test application.

A. Extraction Procedure

The most commonly used extractant has been ammonium oxalate (Grigg, 1953). However, even this extractant has not been consistently successful and, in spite of early promise, it has not been better for diagnosing Mo deficiency than soil type (Grigg, 1960). In more recent work (Lombin, 1985), the Mo extracted by ammonium oxalate was only poorly related ($R^2 = 0.64$) to Mo uptake by peanut (*Arachis hypogaea* L.) in the greenhouse when soil organic matter was included as an independent variable. An intensive study involving several soils and crops in a large geographical area of the southeastern USA found acid oxalate extractable Mo poorly related to plant parameters associated with available soil Mo (Mortvedt & Anderson, 1982). Correlations of soil Mo and plant Mo were not statistically significant for forage legumes. Relative forage yield and plant Mo concentration were highly correlated to soil pH, however. The correlation with these crop parameters was only slightly improved by including both soil pH and soil Mo. Resulting R^2 values were only about 0.5 indicating poor predictive ability of the soil tests. Although the acid ammonium oxalate procedure has not been a successful extractant from the standpoint of identifying Mo-deficiency conditions, it still appears to hold the most promise for measuring available Mo.

B. Analytical Procedures

Analytical methods must be sensitive to measure extracted Mo since the concentrations will usually be in the range of about 10 to 50 μg/L. Successful methods include graphite furnace-atomic absorption spectrometry (Mortvedt & Anderson, 1982) and direct-current plasma emission spectrometry (Pierzynski et al., 1986).

C. Interpretation

Critical levels of extractable available Mo remain unknown for lack of suitable extracting reagents or Mo-deficient soils. Identifying critical levels is made more difficult by the fact that plant response in marginally deficient soils appears to be influenced more by liming than application of Mo (Mortvedt & Anderson, 1982).

IV. CHLORINE

Chlorine has been recognized as an essential plant nutrient for many years. However, because the element is so commonly present in the environment, deficiencies in the field are rare, if not absent altogether. Interpretation of field responses to chloride fertilizers are confounded by the prophylactic properties of the element (Christensen et al., 1981). Nevertheless, recent work presents strong evidence for chloride insufficiencies in the Northern Great Plains of the USA (Fixen et al., 1986b). Low rainfall and little or no need for K fertilizer (KCl) are somewhat unique features of this large landlocked geographical area.

A. Extraction Procedures

Chloride can be extracted from soils using water (Fixen et al., 1988) or any weak electrolyte since the common Cl salts are highly water soluble. Choice of extractant may be influenced by analytical method and compatibility with simultaneous extraction of other elements. Care must be taken to avoid contamination from dust, water, glassware, filter paper, paper bags, perspiration, and many common cleaning agents (Fixen et al., 1988).

B. Analytical Procedures

Several methods of analysis are available for quantifying extractable Cl. Procedures for colorimetric analysis using mercury (II) thiocyanate; potentiometric analysis using a Cl^- electrode; and analysis by ion chromatrography have been described in detail (Fixen et al., 1988). Titration of Cl in the soil extract using $AgNO_3$ has also been used successfully (Fixen et al., 1986a).

C. Interpretation

The critical level of extractable Cl is still poorly defined. However, soil chloride levels above about 45 kg/ha in the top 60 cm of soil have been indicated as adequate for near maximum yield of hard red spring wheat (*Triticum aestivum* L.). Continued research is needed in this area and will undoubtedly better define the needs for soil testing and fertilizer application of this micronutrient.

REFERENCES

Aitken, R.L., A.J. Jeffrey, and B.L. Compton. 1987. Evaluation of selected extractants for boron in some Queensland soils. 1987. Aust. J. Soil Res. 25:263–273.

Bansal, K.N., D.P. Motiramani, and A.R. Pal. 1983. Studies on sulfur in vertisols. I. Soil and plant tests for diagnosing sulfur deficiency in soybean [*Glycine max* (L.) Merr.]. Plant Soil 70:133–140.

Bansal, K.N., and A.R. Pal. 1987. Evaluation of a soil test method and plant analysis for determining the sulfur status of alluvial soils. Plant Soil 98:331–336.

Bansal, K.N. Sharma, D.N., and Singh, D. 1979. Evaluation of some soil test methods for measuring available sulfur in alluvial soils of Madhya Pradesh, India. J. Indian Soc. Soil Sci. 27:308–313.

Berger, K.C., and E. Truog. 1939. Boron determination in soils and plants. Ind. Eng. Chem. Anal. Ed. 11:540–545.

Bingham, F.T. 1982. Boron. p. 501–538. *In* A.L. Page et al. (ed.) Methods of soil analysis. Part 2. 2nd ed. Agronomy Monogr. 9. ASA and SSSA, Madison, WI.

Cartwright, B., K.G. Tiller, B.A. Zarcinas, and L.R. Spouncer. 1983. The chemical assessment of the boron status of soils. Aust. J. Soil Res. 21:321–332.

Christensen, N.W., R.G. Taylor, T.L. Jackson, and B.L. Mitchell. 1981. Chloride effects of water potential and yield of winter wheat infected with take-all root rot. Agron. J. 73:1053–1058.

Eik, K. 1980. Recommended sulfate-sulfur test. p. 25–28. *In* W.C. Dahnke (ed.) Recommended chemical soil test procedures for the North Central Region. North Dakota Agric. Exp. Stn. Bull. 499.

Fixen, P.E., G.W. Buchenau, R.H. Gelderman, T.E. Schumacher, J.R. Gerwing, F.A. Cholick, and B.F. Farber. 1986a. Influence of soil and applied chloride on several wheat parameters. Agron. J. 78:736–740.

Fixen, P.E., R.H. Gelderman, and J.L. Denning. 1988. Chloride tests. p. 26–28. *In* Recommended chemical soil test procedures for the North Central Region. North Dakota Agric. Exp. Stn. North Central Reg. Publ. 221 (revised).

Fixen, P.E., R.H. Gelderman, J. Gerwing, and F.A. Cholick. 1986b. Response of spring wheat, barley, and oats to chloride in potassium chloride fertilizer. Agron. J. 78:664–668.

Gestring, W.D., and P.N. Soltanpour. 1981. Boron analysis in soil extracts and plant tissue by plasma emission spectroscopy. Commun. Soil. Sci. Plant Anal. 12:733–742.

Grigg, J.L. 1953. Determination of the available molybdenum of soils. N. Z. J. Sci. Technol. 34:405–414.

Grigg, J.L. 1960. The distribution of molybdenum in the soils of New Zealand. I. Soils of the North Island. N. Z. J. Agric. Res. 3:69–86.

Gupta, U.C. 1979. Some factors affecting the determination of hot-water-soluble boron from Podzol soils using azomethine-H. Can. J. Soil Sci. 59:241–247.

Hoeft, R.G., L.M. Walsh, and D.R. Kenney. 1973. Evaluation of various extractants for available soil sulfur. Soil Sci. Soc. Am. Proc. 37:401–404.

Islam, M.M., and F.N. Ponnamperuma. 1982. Soil and plant tests for available sulfur in wetland rice soils. Plant Soil 68:97–113.

Islam, M.M., and N.I. Bhuiyan. 1988. Evaluation of various extractants for available sulfur in wetland rice (*Oryza sativa*) soils of Bangladesh. Indian J. Agric. Sci. 58:603–606.

Jeffrey, A.J., and L.E. McCallum. 1988. Investigation of a hot 0.01 *M* calcium chloride soil boron extraction procedure followed by ICP-AES analysis. Commun. Soil Sci. Plant Anal. 19:663–673.

Jin, J., D.C. Martens, and L.W. Zelazny. 1988. Plant availability of applied and native boron in soils with diverse properties. Plant Soil 105:127–132.

Johnson, C.M., and H. Nishita. 1952. Microestimation of sulfur in plant materials, soils, and irrigation waters. Anal. Chem. 24:736–742.

Johnson, G.V. 1987. Sulfate: Sampling, testing, and calibration. p. 89–96. *In* J.R. Brown (ed.) Soil testing: Sampling, correlation, calibration, and interprettion. SSSA Spec. Publ. 21. SSSA, Madison, WI.

Kowalenko, C.G. 1985. A modified apparatus for quick and versatile sulfate sulfur analysis using hydriodic acid reduction. Commun. Soil Sci. Plant Anal. 16:289–300.

Lea, R., and C.G. Wells. 1980. Determination of extractable sulfate and total sulfur from plant and soil material by an autoanalyzer. Commun. Soil Sci. Plant Anal. 11:507–516.

Lombin, G. 1985. Micronutrient soil tests for the semi-arid savannah of Nigeria: Boron and molybdenum. Soil Sci. Plant Nutr. 31:1–11.

Mahler, R.L., D.V. Naylor, and M.K. Fredrickson. 1984. Hot water extraction of boron from soils using sealed plastic pouches. Commun. Soil Sci. Plant Anal. 15:479–492.

Maynard, D.G., Y.P. Kalra, and F.G. Radford. 1987. Extraction and determination of sulfur in organic horizons of forest soils. Soil Sci. Soc. Am. J. 51:801–805.

Mortvedt, J.J., and O.E. Anderson. 1982. Forage legumes: Diagnosis and correction of molybdenum and manganese problems. Univ. of Georgia Southern Coop. Ser. Bull. 278.

Nilsson, L.G., and P. Jennische. 1986. Determination of boron in soils and plants. Chemical methods and biological evaluation. Swed. J. Agric. Res. 16:97–103.

Odom, J.W. 1980. Kinetics of the hot water soluble boron soil test. Commun. Soil Sci. Plant Anal. 11:759–765.

Oeien, A. 1979. Determination of easily soluble sulfate and total sulfur in soil by indirect atomic absorption. Acta Agric. Scand. 29:71–74.

Palaskar, M.S., and A.B. Ghosh. 1981. Evaluation of some soil test methods for available sulfur using cabbage as a test crop. J. Nucl. Agric. Biol. 10:88–91.

Palaskar, M.S., and A.B. Ghosh. 1985. An appraisal of some soil test procedures for diagnosing sulfur availability to maize (*Zea mays* L.) grown on alluvial soils. Fert. News 30:25–30.

Parker, D.R., and E.H. Gardner. 1981. The determination of hot-water-soluble boron in some acid Oregon soils using a modified azomethine-H procedure. Commun. Soil Sci. Plant Anal. 12:1311–1322.

Pierzynski, G.M., S.R. Crouch, and L.W. Jacobs. 1986. Use of direct-current plasma spectrometry for the determination of molybdenum in plant tissue digests and soil extracts. Commun. Soil Sci. Plant Anal. 17:419–428.

Pirela, H.J., and M.A. Tabatabai. 1987. Determination of sulfate in soils by reduction with tin and phosophoric acid. Commun. Soil Sci. Plant Anal. 18:1131–1141.

Ponnamperuma, F.N., M.T. Clayton, and R.S. Lantin. 1981. Dilute hydrochloric acid as an extractant for available zinc, copper, and boron in rice soils. Plant Soil 61:297–310.

Porter, S.R., S.C. Spindler, and A.E. Widdowson. 1981. An improved automated colorimetric method for the determination of boron in extracts of soils, soilless peat-based composts, plant materials and hydroponic solutions with azomethine-H. Commun. Soil Sci. Plant Anal. 12:461–473.

Sharpley, A.N., S.J. Smith, R.G. Menzel, and R.L. Westerman. 1985. The chemical composition of rainfall in the Southern Plains and its impact on soil and water quality. Oklahoma Agric. Exp. Stn. Tech. Bull. T-162.

Tabatabai, M.A. 1982. Sulfur. p. 501–538. *In* A.L. Page et al. (ed.) Methods of soil analysis. Part 2. 2nd ed. Agronomy Monogr. 9. ASA and SSSA, Madison, WI.

Tiwari, K.N., V. Nigam, and A.N. Pathak. 1983. Evaluation of some soil test methods for diagnosing sulfur deficiency in rice in alluvial soils of Ultar Pradesh. J. Indian Soc. Soil Sci. 31:245–249.

Wall, L.L., C.W. Gehrke, and J. Suzuki. 1980. An automated turbidimetric method for total sulfur in plant tissue and sulfur in soils. Commun. Soil Sci. Plant Anal. 11:1087–1103.

White, P.J., M.J. Whitehouse, L.A. Warrel, and P.R. Burrill. 1981. Field calibration of a soil sulfate test on sward lucerne on the eastern Darling Downs, Queensland. Aust. J. Exp. Agric. Anim. Husb. 110:303–310.

Wikner, B., and L. Uppstroem. 1980. Determination of boron in plants and soils with a rapid modification to the curcumin method utilizing different 1,3-diols to eliminate interferences. Commun. Soil Sci. Plant Anal. 11:105–126.

Yli-Halla, M. 1987. Assessment of extraction and analytical methods in estimating the amount of plant available sulfur in the soil. Acta Agric. Scand. 37:419–425.

Chapter 11

Testing Soils for Toxic Metals[1]

JEFFERY A. RISSER, *University of Maine, Orono*

DALE E. BAKER, *Pennsylvania State University, University Park*

The toxic metals Cd, Cr, Hg, Ni, and Pb present a wide range of properties, characteristics, and problems with respect to soil testing and plant analysis. Cadmium, Hg, and Zn form the group IIb elements. These elements have filled inner d orbitals and outer s orbitals. Oxidation states in excess of two are not observed. Chromium and Ni are transition metals; Pb is in group IVa with Si, Ge, and Sn. A discussion of these toxic elements as a group in soils is difficult because of their chemical differences. However, their origin and distribution in soils, uptake by plants and toxicity to plants, animals and humans allows their collective discussion. All of these toxic metals are economically important and are used by people in substantial quantities, providing the possibility of soil and plant contamination from anthropogenic activities such as mining, smelting, and waste disposal, in addition to contributions from mineral weathering of parent materials.

Many of the issues and concerns involved with toxic and potentially toxic materials in the soil, plant, animal, and human environment have been discussed by Baker and Chesnin (1975). Cadmium, Hg, and Pb are toxic to animals and humans, making their behavior in soils and plants of principle importance in relation to food chain contamination (Baker et al., 1979). Chromium can be toxic to plants in its common oxidation states, Cr(III) and Cr(VI) (Mortvedt & Giordano, 1975; Bartlett & James, 1979). The Cr(III) cation is normally the form found in plants and is essential for human nutrition (Huffman & Allaway, 1973). Chromium(VI) is anionic, mobile in soils (Artiole & Fuller, 1979) and waters, and toxic to aquatic life, microorganisms, and animals (Ross et al., 1981; Northeast Regional Res. Publ., 1985). Hexa-

[1] Contribution from the Dep. of Plant and Soil Sci., Univ. of Maine, Orono, ME 04469 and the Dep. of Agronomy, Pennsylvania State Univ., University Park, PA 16802. Maine Agric. Exp. Stn. Publ. 1288, Univ. of Maine, Orono, ME.

valent chromium is also a suspected human carcinogen. The relative contribution of Cr(III) and Cr(VI) to plant Cr is not known, but is presumably dependent on soil conditions. Elevated levels of Ni are of concern in soils primarily for phytotoxicity reasons (Soane & Saunder, 1959; Webber, 1972). Nickel is a required trace element for animal and human nutrition, and most background levels in soils and plants are of little concern for human health (Welch & Cary, 1975) since inhibition of plant growth and development by excessive Ni places a limit on Ni entering the food chain.

One or more availability indices is usually recommended to evaluate potential plant uptake of the toxic metals, although availability and plant uptake can change with environmental and biological conditions. Analysis of soil solution or soil water extracts for the toxic metals in the absence of waste additions or enrichments generally result in concentrations below the μg/kg range. Although these levels may be of chemical interest, both analytical difficulties and lack of correlation with plant uptake data may limit the agronomic value of soil solution analysis for these metals. Exceptions may exist for Cr(VI) and Hg for which sensitive methods are available and for which potential movement and toxicity in groundwater are of concern.

There are many methods available for analysis of the toxic metals (American Public Health Assoc., 1985; Assoc. of Official Analytical Chemists, 1980; Page et al., 1982; USEPA, 1979b). Because levels of Cd, Cr, Hg, Ni, and Pb are often low in soils, care in sampling and handling soil is required. An aluminum sampling tube or one constructed of another material, if tests prove negative for the metals of interest, may be used to obtain soil samples. Selected composite samples should be mixed in an acid washed container and placed in polyethylene bags (Baker & Amacher, 1982). For Cr(VI) analysis, soils should not be dried (Bartlett & James, 1979). For other metals, soils may be air dried in a clean environment if the test method is not sensitive to the drying process. Samples to be analyzed for Hg must be dried below 50 °C and must not be exposed to laboratory atmospheres or other areas polluted with Hg, which can be absorbed by soils (Andersson, 1979). Samples should be sieved through a 20-mesh nylon or polyethylene sieve and returned to their original bag (Baker & Amacher, 1982).

Behavior of these metals may be differentiated in soils because Hg will be less prone to adsorption by soil minerals owing to the stability and solubility of its chloride and hydroxide complexes, although organic soil components may retain Hg (Andersson, 1979). Lead and Cr(III) are readily precipitated. Chromium(VI) may be formed or persist at low levels despite being metastable over much of the soil system redox range, and Cd, owing to its bioactivity, complexation with organic matter and mineral surfaces and formation of stable ion-pairs is most sensitive to changes in the chemical environment of the soil (Pickering, 1980). A more general review of these and other elements, referred to as heavy metals, in soils has recently been published (Alloway, 1990). The numerous differences among these toxic metals necessitates some discussion of each metal individually, but where practical they will be considered as a group.

I. CONSIDERATIONS WHEN TESTING SOILS FOR TOXIC METALS

A. Source of Metal

When testing soils for toxic metals one should consider the source of the element. Although a single source at any one location is unlikely, the dominance of one source is quite possible. Background levels may be dominated by soil mineral or parent material composition, while industrial wastes or other sources may govern toxic metal characteristics in soils with elevated levels. For example, Cd levels may be elevated as a result of long-term P fertilization, but are dependent upon the phosphate rock source (Mulla et al., 1980; Mortvedt, 1987). If minerals or parent materials are the major source of a toxic metal to a well-developed soil, we may expect that a steady state has been achieved, which may continue as long as new environmental or cultural conditions are not imposed (e.g., tillage operations, lime application to change rather than maintain pH or irrigation). Examples of mineral contributions to soil toxic metals include soils developed over serpentine deposits that are high in Cr and Ni (Soane & Saunder, 1959; Proctor & Woodell, 1975), or cinnabar deposits (and mine tailings) contributing to high Hg background levels (Harsh & Doner, 1981; Lindberg et al., 1979). Other sources of Hg to areas with elevated levels or to the global Hg pool may include smelters, coal-fired plants (Crockett & Kinnison, 1979), urban areas, waste disposal sites (Andersson, 1979), and geothermal areas (Phelps & Buseck, 1980). Zinc deposits and areas surrounding Zn mine sites are often high in the associated element Cd (Baker & Bowers, 1988). Some shale-derived soils (Lund et al., 1981; Peterson & Alloway, 1979) and phosphate mine sites (Hutchison & Wai, 1979) also contain sufficient Cd to cause elevated levels of plant Cd. Cadmium migration under prevailing soil conditions has been demonstrated at smelter sites with long histories of metal additions (Kuo et al., 1983), and migration may also be inferred from an inability to completely recover Cd after 11 yr of sludge addition (Hinesly et al., 1982a). Galena is the major Pb-bearing mineral, and is common enough to constitute a significant mineral source of Pb to soils (Colbourn & Thornton, 1978; Palmer & Kucera, 1980). Lead-enriched fuels and paints have also contributed to Pb in soils of many areas.

When mineral forms of a toxic element predominate in soils, soil chemistry may be clarified because the chemistry of minerals is relatively well known, and mineralogical methods to determine the contribution of mineral forms to the total soil content are available. The behavior of the fraction of a toxic metal no longer in its mineral form presents a more difficult, but typical, problem in soils. The initial identification of an elements source is important because some soils with a high ore content may have amounts of toxic metals higher than that of soils contaminated with the same metals in other forms, but produce plants with similar metal content. A statistical method to distinguish between contaminated and noncontaminated soils (Davies, 1983) may be useful in identifying problem sites. Fractionation of metal forms in soils, as well as documentation of source differences and relative effectiveness of extractants, has been reported by several authors (Cd, Pb [Farrah & Pickering, 1978; Miller & McFee, 1983; Harrison et al., 1981]); (Hg [Harsh & Doner, 1981; Lindberg et al., 1979; Hogg et al., 1978b]); (Cr, Ni [Shewry & Peterson, 1976]); (Cd, Ni [Soon & Bates, 1982; Hickey & Kittrick, 1984]).

1. Elemental and Ionic Forms

Although not a common form for toxic elements in soil or soil amendments (with the exception of Hg^0 vapor), if an element is added or still present in the elemental form, considerable chemical literature regarding solubility and possible reactions and their rates at various pH and redox levels becomes useful. Studies using laboratory columns and metal salts to examine the mobility of Cd and Ni (Tyler & McBride, 1982b) and Pb and Cd (Singh & Sekhon, 1983) in soils have shown all these metals to be mobile (Pb $<$ Ni $\leq$ Cd). These observations agree with hydroxide precipitation or oxide adsorption of the metals (Tyler & McBride, 1982b), although many mechanisms may be active in the adsorption of metals by soils (Harter, 1979).

2. Organic Forms

The common practice of adding municipal and industrial sludge materials to soils often results in toxic metals in an organic form being added to soils. One of the larger sources of high Cr organic wastes is the leather tanning industry, which may be producing 281 000 Mg yr^{-1} according to an estimate by Dawson (1978).

Although specific characterization of organic-toxic metal complexes is difficult, some generalizations regarding the behavior of these organic materials are possible. The soil microbial population is responsible for organic-toxic metal material decomposition and constitute a major influence on the chemistry of toxic metals in soil. Some organic-metal complexes (such as methyl mercury) may be directly absorbed by plants and microorganisms and may make a significant contribution to the toxic metal content of biological systems (Czuba et al., 1981). Humic materials can decrease ionic activities by complexing with metals and possibly reduce plant uptake of toxic metals (Tyler & McBride, 1982a; Brady & Pagenkopf, 1978). In addition, the existence of soluble toxic metal chelates in soil may decrease (Elliott & Denneny, 1982) or increase (Chubin & Street, 1981) metal sorption depending on soil conditions (pH, mineralogy, and oxides of Fe and Al), and presumably affect metal mobility in soils (Stevenson & Welch, 1979; Welch & Lund, 1987). Although high soil pH may reduce plant uptake of toxic metals (Adriano et al., 1982), the related increased retention of toxic metals by soil metal oxides (Kuo et al., 1985; Abd-Elfatt & Wada, 1981) is of equal or greater importance. Mercury appears to be an exception as it has a particular affinity for, and is retained by, soil organic matter (Zyrin et al., 1981) even at low pH (Andersson, 1979).

The contribution of organic toxic metal complexes to the total soil load of a toxic metal is difficult to determine and may vary with soil and organic type (Petruzzelli et al., 1981). Within a short time following waste application, estimates of the organic portion of the soil toxic metal load may be made from a knowledge of the organic content and organic decomposition rate of the waste material in soils (Wiseman & Zibilske, 1988). In the absence of this information, or after longer periods following application, estimates of organic forms of toxic metals can be made either by selectively extracting the metal from organic materials or by breaking down organic materials in soils and determining the quantity of released metals. The selective extraction method may either employ a metal that is strongly complexed by organic materials and competes for binding sites with toxic metals, or an organic chelate that strongly binds the toxic metal of interest and can be employed to compete with the organic materials in soil for the metal. The chelate extraction method may be employed using two theoretically different but functionally similar procedures. These approaches are discussed in the methods section.

The organic matter destruction approach encompasses many different analytical methods but all involve chemically or physically altering the soil organic matter. High temperature ashing or low temperature ashing in an enriched O_2 environment effectively destroys the organic materials but may also oxidize, dehydrate, or otherwise change the mineral components of soils. The chemical approaches usually employ oxidizing agents such as hydrogen peroxide or hypochlorite to accomplish organic matter destruction, making them similar to the dry oxidation methods. Methods of altering or selectively extracting organic materials in soils are also available and have been described by Schnitzer (1978, 1982). An extractant that is not specific for organic metals may be used prior to and following organic matter destruction; the difference being attributed to organically bound metals.

These methods may be useful to define the organically bound pool of metals in a soil, but the distinction between the native metal-organic pool and the pool because of waste addition cannot normally be made. Because of microbial and plant modification of the organic matter pool, this distinction is no longer meaningful unless a synthetic organic component remains important among the compounds formed biologically in the soil and original sludge material. This may be the case for chlorinated hydrocarbons, and other refractory organic molecules.

The net rate of release of organically bound toxic metals is dependent on many factors, among which are their toxicity to soil microbes, the availability of additional organic materials with which they may become complexed, temperature and other environmental and cultural conditions in the soil environment. The fate of these released toxic metals is not clear, but possibilities include movement in the soil profile, increased plant and microbial uptake, sorption by clays or metal oxides, and "reversion" into either mineral or organic forms (Giordano et al., 1979).

3. Air and Water-Borne Depositions

Air and water-borne deposits may be in the organic or elemental form, but are most likely to be metal oxides or other mineral compounds. Deposits left by air and water streams have distinct downstream patterns, generally with the concentration inversely proportional to distance from the source. If remedial action is required, definition of the areas and estimation of the severity of toxic metal effects are necessary. The materials deposited from plumes are generally sorted by size, and the toxic metal concentration pattern can be uneven and without distinct boundaries. The chemical form(s) of the deposited toxic metal is (are) likely to be uniform over the contaminated area, with differences among sites because of environmental and particle-size differences. An example of a nonhomogenous pattern may be found in Pb distribution along highways (Fergusson et al., 1980). Additional considerations may also apply in some cases. For example, air-borne Hg deposits that result from mercury volatilization (Hg^0) may be more easily revolatilized and thus have greater effects on organisms than equal amounts of particulate mineral forms.

B. Distribution of Metals in a Geographic Region

1. Discrete Area

Discrete areas of toxic metal concentration, with relatively sharp boundaries, are generally characteristic of recent or intentional additions. Organic forms of toxic metals in sludges are most often intentionally applied to a defined and often previously surveyed land area. Toxic metal distribution from these means normally provides adjacent areas without metal applications that may be available for comparison. Although areas adjacent to waste application sites frequently differ in cultural histories, application sites may be compared to surrounding areas, monitored as a unit, and possibly require remedial action or isolation. Waste application sites generally have had uniform past management practices, and allow more options for future supervision. When sampling soils from a discrete area contaminated with toxic metals, a small survey of the area is probably sufficient to establish the current conditions, and subsequent sampling and monitoring can probably be done on a small number of composite samples (provided the area proves uniform). Depending upon the extent of contamination and the potential for further environmental degradation, remedial action may involve soil amendments such as organic matter and lime to reduce metal availability, dedication of the site to waste disposal or other low-risk use, or isolation of the site from humans or animals at risk from exposure to the toxic metals. Among the options available for sites that do not need to be isolated are selection of crop species, cultivars, cultural practices, and the ultimate disposition of the crop to minimize health impacts. An alternative to isolation is removal of the site from agricultural production, and dedication to forestry or park woodland resource.

2. Contamination Halos

In cases where distinct boundaries of toxic metal accumulation are not observed, presumably resulting from anthropogenic sources or dispersion of natural concentrations arising from igneous intrusion or mineral deposition, additional problems are presented. For either anthropogenic or natural sources of a toxic metal a decrease in metal concentration away from the source, and in a direction and pattern that results from the force dispersing the metal must be expected. This plume shadow area may require extensive sampling to describe its size, shape, and concentration distribution of metals. When the source of metals are known, the sampling may be simplified by knowledge of the methods of distribution: wind, water, gravitational, or other important means under the prevailing conditions. For those sources that have not been identified, their identification is often more difficult, as long periods of time and major changes in environmental conditions may have occurred over the period of plume formation. Definition of the extent and increment of added metals, especially if the source of the metal has not been identified, may require many samples and significant analytical expense, although more limited efforts may also be useful (Miller & McFee, 1983). The deposited plume may cover large areas and include an array of environmental conditions, cultural practices, natural or synthetic disturbances and topographical features. In general, more extensive monitoring, and a greater number of rehabilitational alternatives are required when a concentration gradient, halo, or plume pattern is present.

C. Nature of Toxic Hazard

1. Direct Soil Ingestion

Human and animal consumption of metals can occur directly from soil materials ingested intentionally or as a consequence of food contamination. The importance of direct ingestion of toxic metals in soil materials is greatest for those elements acutely toxic to, or chronically accumulated by, animals and humans. In contrast, toxic metals that are selectively taken up by plants are most likely to accumulate and concentrate relative to other elements in the food chain. Toxic elements accumulated by plants in small amounts, but present in soils at potentially dangerous levels, may pose their greatest risk from direct ingestion. Of the five toxic metals considered here, Pb is the element most likely to pose such a threat. Soil particulates acquired through direct soil ingestion or dust inhalation by grazing animals (Thorton, 1974) or playing children probably constitutes the greatest health hazard from soils high in Pb. Roadside and inner city area soil and dust have been shown to be particularly high in Pb and Cd (Harrison et al., 1981). Surface contamination of plants by soil materials or aerosol particulates may constitute a major source of Pb in plant materials (Palmer & Kucera, 1980), and thus in the food chain.

Human consumption of soil is most frequently a problem for young children who play in soil and accidentally or intentionally ingest soil materials. Garden soils in urban or other high metal areas may present health problems as a result of several pollutants at one site (Preer, 1985). Among the factors contributing to metals in garden soils are their proximity to the source, e.g., leaded fuels, lead paint chips, industrial aerosols, and the addition of metals in domestic or industrial sludges and composted organic materials containing high metal materials such as Pb-rich newsprint (Elfving et al., 1979). Combinations of these sources may provide the same individuals with a consistent supply of metal-enriched vegetables and the possibility of direct ingestion of soil materials.

2. Plant Uptake and Food Chain Contamination

Elements or materials that are selectively accumulated or particularly hazardous to humans, but less so to plants, present their greatest health risk through the food chain. Cadmium and Hg fit the pattern of food chain contamination well. Their potential toxicity to humans is great in comparison to their effects on plants. Cadmium is probably of most widespread concern, as it is readily accumulated from soils by plants and may thereby enter the food chain (Chang et al., 1983) and accumulate in animals (Williams et al., 1978; Baker et al., 1979). Levels of plant Cd that cause concern in the food chain usually occur before plant toxicity responses are observed (Kelly et al., 1979; Miles & Parker, 1979b). Cadmium concentrations are generally greater in root than vegetative and leafy plant parts, which in turn are greater than seed portions (Jastrow & Koeppe, 1980; Peterson & Alloway, 1979), except in accumulator species such as Swiss chard (*Beta vulgaris* L.) (Kuo et al., 1985). Concentrations in plants may also be dependent on variety or hybrid within a species, form of Cd in the soil (Hinesly et al., 1982a, b; Street et al., 1977) and soil temperature (Giordano et al., 1979). Presumably plants can selectively partition Cd out of new generation materials (Chang et al., 1982). The extent of partitioning of Cd out of seed portions of crops may be of significance for the important grain crops of the world (Jastrow & Koeppe, 1980). This does not mean that levels of Cd high enough to be of health concern cannot occur in grain and seed portions of plants (Williams et al., 1978; Re et al., 1983).

Mercury may be accumulated by plants either by root uptake or by shoot and leaf absorption of Hg^0 volatilized from soil or present in polluted air (Lindberg et al., 1979). The relative contributions of these modes of uptake are variable, but in certain instances both may be important. Plant roots seem to retain Hg obtained from soil, while aboveground plant content arises largely from atmospheric Hg (Hg^0 vapor or particulates). However, mushrooms (*Agaricus bisporus*) seem to accumulate Hg and present potential problems since high Hg levels may be found in compost used by the mushroom industry (Loughton & Frank, 1974). The concentration of Hg in plant roots and tops depends both on soil and atmospheric Hg sources and climatic conditions (Elsokkary, 1982). Cappon (1987) reported several forms of Hg in vegetables that vary in relative proportion among species. Factors affecting Hg^0 volatilization from soil include temperature (Lindberg et al., 1979) and biological activity. Soil Hg content does not appear to be a good indicator of volatility (Rogers & McFarlane, 1979). Mercury seems to be strongly retained by both organic and mineral components of the soil (Kromer et al., 1981), with movement more readily occurring in the vapor phase or with root growth rather than in the solution phase (Hogg et al., 1978a, b).

Chromium added to soils in inorganic forms, especially Cr(VI), is more readily extracted by water (Grove & Ellis, 1980) and accumulated by plants than Cr applied in organic wastes (Mortvedt & Giordano, 1975). This is probably because of greater proportions of the more soluble Cr(VI) (McGrath, 1982). Field studies of short duration with Cr applied in organic wastes (Stomberg et al., 1984; Hemphill et al., 1985) or inorganic additions (Rehab & Wallace, 1978a) failed to show increased plant uptake of Cr. In most cases, but with exceptions (Czuba & Hutchinson, 1980), the Pb and Cr found in plants is largely accumulated and partitioned in plant roots (Shewry & Peterson, 1974; Cary et al., 1977; Huffman & Allaway, 1973). In the case of Pb, elevated root levels may be exaggerated because of surface contamination by soil materials. Exceptions to root partitioning of Pb may occur on organic soils (Czuba & Hutchinson, 1980).

3. Plant Toxicity

A third form of heavy metal hazard is plant toxicity. Among the toxic metals considered here, Cr and Ni are the most phytotoxic under prevalent soil conditions. The hexavalent forms of Cr are believed to be deleterious to humans, most mobile in soils because of their anionic form, and probably the form most readily accumulated by plants. The difficulty of demonstrating the form most readily absorbed by plants is related to the interacting complexities of the solubility, redox, pH and microenvironment of the plant root, which determine plant availability of Cr. Combining redox and pH values into a single term (pe + pH) as advocated by Lindsay (1979) may help overcome some of these difficulties. Elevated levels of plant Ni are generally not a human nutrition problem because phytotoxicity prohibits excessive plant content, although tolerance by some plant species have been observed for soils enriched with Ni and Cr (Shewry & Peterson, 1976). When equal soil concentrations of the metals Cd, Cr, Ni, and Pb are compared (Dijkshoorn et al., 1979), Ni and Cr seem to be the most phytotoxic, although Mitchell et al. (1978) reported that Cd is more phytotoxic than Ni. The extent of metal accumulation in plants and the order of metal phytotoxicity based on plant content is species dependent, but when comparisons are based on natural or expected abundances of the elements in soils, Ni more often limits plant growth.

Suggested tolerance levels of toxic metals in soils, plants, animal feeds, the human diet, and drinking water are presented in Table 11-1. The maximum concentration of toxic metals in wastes presumed to be acceptable for application of waste materials to soils may also be found in the Northeast Regional Publ. (1985).

Table 11-1. Suggested tolerance levels for toxic metals in soils, plants, animal feeds, foods, and primary drinking water.†

Toxic metal	Maximum cumulative soil levels‡	Plant concentration	Animal feeds	Dietary tolerance	Primary drinking waters§
	kg ha^{-1}	mg kg^{-1}		μg d^{-1}	mg L^{-1}
Cd	2.5–5.0	1	0.5	≈65	0.010–0.005¶
Cr	112–672	2	3000		0.05
Hg	0.9–2.2	0.04	2		0.002
Ni	11.2–67.2	3	50		
Pb	112–672	10	30		0.05

† Northeast Regional Research Publ. (1985) except as noted.
‡ Maximum loading dependent upon soil texture.
§ USEPA (1975).
¶ Commission of the European Communities (1978).

II. SELECTING APPROPRIATE SOIL TEST METHODS

Many factors may be active in the process of retention, and conversely the release of a metal from soil. Among the important factors are ion competition for metal retention sites, metal species and concentration, soil chemical and mineralogical composition and surface area, dissolution of metal hydroxides by proton addition, formation of stable solution complexes and the kinetics of dissolution and precipitation (Harding & Healy, 1979; Pickering, 1980). Additionally, displacement depends on the concentrations and relative affinities for the soil surface of the ions involved and the pH and ionic strength dependence of complex formation (Pickering, 1980).

Dilute extractants such as water and neutral salt solutions are usually too weak to remove a substantial amount of a metal, or to provide an estimate of the buffering or replenishment ability of a soil for the element of interest. However, they may provide an estimate of the availability and perhaps of the amount retained on exchange sites. Conversely, strong extractants such as mineral acids may change soil conditions greatly and dissolve soil minerals so that the extracted amount of a metal has little relationship to that portion of soil metal removed by plants (Palmer & Kucera, 1980). Information provided by both types of analyses may be helpful to evaluate the metal status in soils. To more accurately predict plant and soil response to toxic metals, several levels of metal addition and extracts with a range of displacement abilities can be employed (McBride et al. 1981).

For those nutrients and metals that are complexed by chelating agents, soil extraction by a strong chelate may remove a larger quantity of metal than a crop, but still closely represent plant uptake of metals from soils. The amount of metal removed from a soil by a chelator is dependent on soil and extractant characteristics, many of which can be measured or controlled. Chelate soil tests have been employed to evaluate plant availability for all of the toxic metals, but with varying success. For Cd, Pb (Sadiq, 1983) and Cd, Ni, and Pb (Baker, 1973; Baker & Amacher, 1981) encouraging results have been found, especially when additional soil characteristics are considered. The mineral stability and slow kinetics of Cr dissolution and the relative instability of Hg chelates lessens the utility of chelate soil tests for these metals. The difference between the chelate-extracted quantity and plant uptake might in part be accounted for by soil conditions, root-soil contact, and extent of root exploitation of the soil profile. Plant roots generally exploit a smaller portion of the soil volume than an extraction solution, but the portion of the soil in close contact to plant roots may be substantially depleted. These factors could be especially important when comparisons are made among species and across sites since both soil and crop conditions may contribute to differences in plant uptake.

Metals associated with organic materials may be extracted with the same techniques used for other soil materials, or an approach using a strongly bound metal to compete for organic-binding sites may be employed. The competing metal approach most often employs Cu, a metal known to bind strongly to organic materials, and thereby displace a suite of metals previously present on the mineral and organic matter surfaces. To the extent that Cu or any other selected metal is specific for organic surfaces, this approach is both logical and useful. Cadmium, Ni, and Zn extracted with 0.125 *M* Cu(II) acetate was shown to be significantly correlated with soil organic matter content (Soon & Bates, 1982).

The USEPA (1979a) recommended a strong acid extraction for sediments. This method has been used to compare untreated soils with comparable soil samples from locations treated with sewage sludge or other wastes. The method provides results for total sorbed metals, which includes the nonvolatile metals within the nonsilicate fraction of the soil. By comparing untreated and treated soils, a measure of the total loadings of Cd, Ni, Pb, Cr, and other nonvolatile metals is obtained. This method with some revisions has more recently been made an official method for sediments, soils, and sludges (USEPA, 1986).

Other commonly used methods for Cd, Pb, and Ni include dilute HCl (0.1 *M* HCl) and double acid (0.05 *M* HCl and 0.0125 *M* H_2SO_4) extractants (Cox & Wear, 1977; Korcak & Fanning, 1978). A rapid method for detection of Pb-contaminated soils is also available (Preer & Murchison, 1986). Methods for soil Cr determinations generally may be divided into soluble Cr(VI), organic Cr forms and total Cr tests. For Hg, total, soluble and perhaps volatilizable fractions (Cannon & Dudas, 1982; Dudas & Cannon, 1983) are the options for testing soils. Lead isotope ratios may be used as an aid in identifying the sources and relative contribution of anthropogenic Pb to soil (Gulson et al., 1981).

There are many instrumental techniques available for analysis of toxic and trace metals. Among the most commonly employed methods are flame and furnace atomic absorption spectrophotometry (AAS), inductively coupled Ar plasma spectroscopy (ICP), and cold vapor analysis for Hg (American Public Health Assoc., 1985; Assoc. of Official Analytical Chemists, 1980; Page et al., 1982; USEPA, 1979b). Anodic stripping voltammetry and polarographic techniques are sensitive methods for quantification of many elements, and thus may be considered as optional methods of toxic metal analysis (Street & Peterson, 1982).

Soil and mineral standard reference materials are available from the National Bureau of Standards, U.S. Dep. of Commerce. These materials are essential to assure the quality of results produced by analytical methods for total element content of samples. Reference materials for use with soil test methods to determine available or extractable metals are generally not commercially obtainable. Any time a test method is used frequently in a laboratory, one or more soils should be selected and developed as an in-house reference material by repeated analyses for use in checking accuracy and precision of the soil test. In addition to laboratory check samples, a periodic exchange of samples among several laboratories employing the method of interest should be used to check for systematic errors and calculation or interpretation differences. For analysis of toxic and other metals, laboratories are urged to follow the USEPA (1979c) quality assurance/quality control procedures.

A summary of some of the extractants used to test soils for toxic metal content and influence on plant composition is presented in Table 11–2. These citations are not meant to be comprehensive, but illustrative, and additional examples are readily available in Page et al. (1982), the text and bibliography of this review or in current issues of many environmental journals.

III. SOIL TESTING METHODS

A. Chelate Extraction Method

The approach pioneered by Norvell and Lindsay (1969) and applied in soil testing by Soltanpour (1985) and Lindsay and Norvell (1978) involves the use of chelates in the 0.01 *M* range as extractants, with resulting extraction levels correlated to biological response in the soil system. Lindsay's method remains one of the more commonly employed, although that of Soltanpour (1985) is probably quicker and more easily adapted to routine testing. The approach pioneered by Baker (Baker, 1973; Baker & Amacher, 1981) involves lower levels of chelate intended to cause less perturbation of the system and invoke a small exchange of metals with the soil surface. Baker's method (Baker & Amacher, 1981) and the approach of Fujii et al. (1983) both employ iterative calculation of metal ion activity values in soil. Both methods include the effects of pH and other metals on the stability of the chelate complex and the ion activity, theoretically making them applicable to a wider range of soil, environmental and plant conditions. Fujii et al. (1983) employed a second extract to accurately fix $CdCl_2$ activity. All chelate approaches may be thought of as extractions to a fixed metal-chelating sink, meaning that the extractant capacity and metal-chelate affinity for all soils is fixed upon selection of a chelate and its concentration. The primary differences between these methods are in subsequent calculations and interpretation. The method of Baker has been applied extensively for mine soil and

Table 11-2. Summary of selected extractants for toxic metals in soils and plants for which correlation is reported and citation.

Extractants	Elements	Plants	Reference
NTA, H_2O, $HCl + AlCl_3$, EDTA, DTPA, CH_3COOH, CH_3COONH_4, $HNO_3 + HCl$, $(COOH)_2 + (COOHN_4)_2$	Cd, Cu, Ni, and Zn	Swiss chard (*Beta vulgaris* L.)	Haq et al., 1980
1 *M* KCl, 2 *M* HNO_3	Cr	Corn roots (*Zea mays* L.)	Hafez et al., 1979
0.42 *M* CH_3COOH, 0.002 *M* $CaCl_2$, 0.05 *M* $CaCl_2$, 0.05 *M* EDTA	Cd	Sorption isotherms	Jarvis & Jones, 1980
Total analysis	Hg	Rice roots (*Oryza sativa* L.)	Morishita et al., 1982
Total, weak acid, CH_3COONH_4	Cr, Ni	Many	Proctor & Woodell, 1975
Bray P_1	Pb	Soybean [*Glycine max* (L.) Merr.]	Miller et al., 1975
$CH_3COONH_4 + CH_3COOH$ (pH 4.8)	Pb	Six vegetables	Nicklow et al., 1983
Total (added as sulfate salts)	Cd, Ni	Cotton (*Gossypium* sp.)	Rehab & Wallace, 1978b
Total (x-ray fluorescence spectroscopy)	Pb	Five vegetables	Spittler & Feder, 1979
DTPA	Cd, Cu, Pb, and Zn	Various	Miles & Parker, 1979a; Baker & Bowers, 1988
0.5 *M* $NaHCO_3$, 1 *M* CH_3COONH_4, total analysis	Hg	Alfalfa (*Medicago sativa* L.)	Lindberg et al., 1979

spoil reclamation (Baker & Buck, 1988); land application of sewage sludges (Shipp & Baker, 1975) and more recently for "clean-up" and human food chain Cd evaluations for a USEPA super-fund site (Baker & Bowers, 1988). The method has been described (Baker & Amacher, 1981), but it requires appropriate computer software for calculating and interpreting the laboratory results.

The method of Norvell and Lindsay (1969) is described in the following section. The methods cited above and other methods employing a variety of organic chelates may be used, provided data for interpretation of the results are available. Buffering the extract solution to pH 7.3 with triethanolamine (TEA) reduces the dissolution of minerals and release of trace metals by proton exchange. Calcium chloride is included to minimize effects of $CaCO_3$ dissolution in soils containing free carbonates. All glassware used for the extraction, filtration, and storage of extracts should be acid soaked and thoroughly rinsed in high-quality deionized water.

1. Reagents and Procedures

Diethylenetriaminepentaacetic acid (DTPA) extracting solution—To prepare 1 L of this solution, dissolve 14.9 g of reagent-grade TEA, 1.97 g of DTPA, and 1.47 g of $CaCl_2 \cdot 2H_2O$ in about 200 mL of deionized water. After dissolution, dilute to 900 mL and adjust the pH to 7.3 with 1 *M* HCl while stirring, then complete the dilution to 1 L. The resulting solution is 0.005 *M* DTPA, 0.01 *M* $CaCl_2$, and 0.1 *M* TEA.

To 10 g of air-dried soil in a 125-mL flask, add 20 mL of the DTPA solution. Cover with plastic stoppers and shake at two cycles s^{-1} for exactly 2 h. Gravity filter the suspension through Whatman no. 42 or equivalent filter paper. The shaking time is critical as the extraction is not complete, and metals will continue to dissolve if left for greater periods. Standards should be prepared in the extracting solution by adding appropriate volumes of 1000 mg L^{-1} of stock standards prior to final dilution of the extracting solution. Analyze the filtrate for Cd, Ni, and Pb or other elements of interest by AAS or ICP.

2. Interpretation

The results of trace metal analysis are most commonly expressed as mg kg^{-1} on a soil dry weight basis. This is equivalent to ppm in the soil, and can be obtained by multiplying the solution concentration by 2. If other units are used, especially volume-based units, then the methods used to obtain those units and conversion factors to express results on a weight of soil basis should be included. Suggested approaches for reporting soil test results have been published (Dahnke, 1988).

B. Dilute HCl Method

Dilute HCl as a soil extractant, relating soil-extractable Zn levels and titratable alkalinity to available Zn was described by Nelson et al. (1959).

They reported that for a range of calcareous and high pH noncalcareous soils the combination of extractable Zn and alkalinity predicted the crop Zn status. Although similar success has not been demonstrated for toxic metals, a dilute HCl extract may correlate well with the total metal content of the soil, and under some conditions, plant uptake. As with total metal analyses, other soil parameters must be considered to estimate plant available amounts of toxic metals. Among these parameters are organic matter content, clay content, soil pH and silicate and nonsilicate minerals present.

1. Reagents and Procedures

Prepare 0.1 *M* HCl solution by diluting 1 volume of redistilled 6 *M* HCl with 59 volumes of deionized water.

Add 20 mL of 0.1 *M* HCl to 2.00 g of soil in a 30- to 50-mL centrifuge tube. Cover, shake for 5 min at two or greater cycles s^{-1}, uncover, and centrifuge until a clear supernatant is obtained. Decant the supernatant and repeat the extraction two or more times until the equilibrium solution pH is <2.0. Adjust the combined extracts to 100 mL final volume and analyze the mixed solution for Cd, Cr, Hg, Ni, and Pb or other metals of interest using AAS, ICP, or cold vapor (for Hg) detection. Standards should be prepared in the extracting solution by adding appropriate volumes of 1000 mg L^{-1} stock standards prior to final dilution of the HCl to 0.1 *M*.

2. Interpretation

There are many variations of the 0.1 *M* HCl method and unless a specific method is carefully followed, the results of analyses may not be comparable. The critical factors are: the final solution pH is below 2.0, a minimum of 15 min soil-solution contact time is obtained, and that at least two cycles s^{-1} shaking is observed. This method has been reported as a single extraction method, but on high pH soils cation exchange converts the extractant to the equivalent of approximately 0.05 *M* $CaCl_2$. A final pH of 2.0 must be obtained.

C. Double Acid Method

The double-acid method has been developed for use on soil with low CEC, low pH, and low levels of organic matter. Comparisons of results for the double-acid extract, the dilute HCl method and the DTPA method have been made (Wear & Evans, 1968). When the method is applied to soils that fit the conditions under which it was developed then the results may be equally as useful as other methods of predicting metal availability. Additional soil parameters, as for the other toxic metal tests, should also improve the reliability of this test.

1. Reagents and Procedures

Dilute 16 mL of 6 *M* redistilled HCl and 1.4 mL of concentrated sulfuric acid to 2 L with deionized water.

To 5 g of air-dry soil, add 25 mL of the extraction solution in an acid-washed 50-mL flask and shake for 15 min at three cycles s^{-1}. Filter the solution through Watman no. 42 or equivalent filter paper. Standards should be prepared in the extracting solution by adding appropriate volumes of 1000 mg L^{-1} stock standards prior to final dilution of the extracting solution. Analyze the filtrate for Cd, Cr, Hg, Ni, and Pb or other elements of interest using AAS, ICP, or cold vapor (for Hg) detection.

2. Interpretation

Comparison of the double-acid extract with DTPA have shown each to provide superior correlations with plant uptake under some conditions, but found DTPA to be superior over a wider range of soil conditions (Korcak & Fanning, 1978).

Some of the problems and potential sources of variation in estimating metal available to plants from soil tests are discussed by Barry and Clark (1978). For toxic metals, the extreme natural variation and the additional variation that can be introduced by human activity may require a 50% or more increase over background levels to detect metal enrichment in the soils of an area. For Hg, a budgeting approach that attempts to account for inputs to and losses from a soil may be more sensitive to changes in metal content than soil analysis (Andersson, 1979).

Each soil test is most useful under conditions where the extractable amount of a metal has been related to crop uptake under a particular set of soil, crop, and environmental conditions. These conditions exist where extensive calibration work has been completed for the soil test and regional crops and soils. A test may become more useful if regression equations are available to relate both the extractable level of a metal and the soil pH to crop response. For chelate soil tests, the inclusion of changes in metal availability with pH allow the application of chelate extractions over wider ranges of soil conditions (Korcak & Fanning, 1978). Under more complex, or less studied soil and climatic conditions, one of the chelate soil tests that includes extractable levels of other metals and soil cations may be more useful because it enables more of the potentially important soil factors to be included in an estimate of plant-available metals. Disadvantages of the chelate soil tests include a potentially long equilibration time, greater number of analyses if the interactions of other metals are to be included in the predictive results, and possibly extensive calculations to complete an estimate of the contribution by the soil environment to the availability of the toxic metal of interest.

D. Total Sorbed Metals (From USEPA, 1986)

1. Reagents and Procedures

Concentrated HNO_3, and HCl, hydrogen peroxide (30%) and American Society for Testing and Materials (ASTM) type II water are required reagents. The acids and oxidant should be analyzed separately to determine the levels of impurities. If the method blank is less than the method detection limit the acids and oxidant may be used. Reagent grade acids are generally acceptable for flame atomic absorption spectroscopy. However, ultrapure acids may be needed for graphite furnace atomic absorption spectroscopy.

Add 10 mL of 1:1 HNO_3 to 2 g of air-dry soil, in a 150-mL pyrex beaker and cover with a watch glass. Place the samples on a hot plate and cautiously heat to 95 °C and reflux for 15 min without boiling. Cool the beaker, add 5 mL of concentrated HNO_3, replace the watch glass, and reflux for 30 min. Repeat the last step. With the watch glass covering three-fourths of the beaker, evaporate the solution to 5 mL without boiling. Cool the sample, add 2 mL of type II water and 3 mL of 30% H_2O_2. With the beaker covered, heat the sample gently to start the peroxide reaction. If the effervescence becomes excessively vigorous, remove the sample from the hot plate and cool. Continue to add 30% H_2O_2 in 1-mL increments, followed by gentle heating, until effervescence is minimal. Next, add 5 mL of concentrated HCl and 10 mL of type II water, return the covered beaker to the hot plate and reflux for 15 min without boiling. After cooling, filter the sample through Whatman no. 41 filter paper (or equivalent) and dilute to 50 mL with type II water.

For samples to be analyzed by graphite furnace atomic absorption spectroscopy, do not add concentrated HCl to the sample but continue heating the HNO_3-peroxide digestate until the volume has been reduced to 5 mL. After cooling, filter the sample with Whatman no. 41 filter paper (or equivalent) and dilute to 50 mL. Standards should be prepared in the appropriate aced mixture for the analysis method by diluting 1000 mg L^{-1} stock standards to the desired final concentrations.

2. Interpretation

This method is not suitable for Hg. For Cd, Ni, Pb, Cr, As, and Se and other trace metals in soils, the method provides a measure of the total concentrations that are exchangeable or adsorbed by soil components. The method does not include trace metals associated with silicates. Therefore, the use of the procedure to measure total metals is not recommended. However, this method is the official procedure used by USEPA for metals in soils and it gives a reliable measure of the amounts of the metals added to soils as nonsilicates from industrial sources.

E. Total Content Analysis of Soil for Chromium, Nickel, and Other Metals

A carbonate fusion (Lim & Jackson, 1982) or lithium metaborate method may be used for total analysis of Cr and Ni. Many sample decomposition methods and potential problems of elemental analysis are described by Bock (1979). Fusion methods are usually unacceptable unless shown otherwise for the volatile elements Cd, Hg, and Pb. For these elements, a low temperature wet digestion method generally employing NHO_3-$HClO_4$ may be used (Lim & Jackson, 1982). For digestion of samples for Pb analysis, the use of H_2SO_4 should be avoided to preclude $PbSO_4$ precipitation (Burau, 1982).

F. Readily Available Chromium Test

1. Reagents and Procedures

Prepare 0.01 *M* KH_2PO_4 by dissolving 1.36 g of KH_2PO_4 in deionized water. Diphenylcarbazide reagent is prepared by dissolving 0.4 g of *S*-diphenylcarbazide in 100 mL of ethanol, and mixing this solution with a solution prepared from 120 mL of 85% H_3PO_4 diluted to 400 mL with water.

Three grams of fresh, moist field soil are extracted with 25 mL of the KH_2PO_4 solution by shaking for 5 min in a 50-mL centrifuge tube. One mL of the diphenylcarbazide reagent is added to 8 mL of the centrifuged soil extract (diluted if necessary), mixed, and transmittance determined at 540 nm after 20 min. Readily available Cr on a dry weight basis is calculated from the moisture content of a separately dried soil.

2. Interpretation

Interpretation of soil tests for Cr requires consideration of soil redox reactions and of the factors that affect the distribution and transformations of Cr species in soils such as soil Mn coatings, soil organic matter, and soil pH (Bartlett & James, 1979, 1988; James & Bartlett, 1983).

G. Total Soil Mercury

A wet digestion of soil or sediment in a reflux flask using H_2SO_4 and HNO_3 is the approach Stewart and Bettany (1982) recommend. Availability indices are not well developed, but solution extractable amounts and plant uptake or specially constructed incubators capable of measuring Hg^0 volatilization are reasonable possibilities (Rogers & McFarlane, 1979; Lindberg et al., 1979). Alternatively, a method for estimating an allowable soil Hg level has been developed by Bashor and Turri (1986) and may serve as a guide for other researchers.

IV. INTERPRETING SOIL TEST RESULTS

Among the many soil tests that have been used to quantify metals in soils, the selection of a particular test should be based on the usefulness of the information to be obtained. One of the principle factors affecting usefulness and a major justification for selecting a particular test is the availability of good correlation data, both for the soils of the area and for the crop being grown. When such data are available, the selection of that test is usually indicated. If sufficient data is to be collected to allow recorrelation of field results to an alternate test, or cross correlation among several soil tests selected for comparison then other tests may be considered. The use and interpretation of soil tests require knowledge of the extractant, the soil on which it is used and crop to be grown.

Each soil-plant system studied with respect to toxic metals should be independently analyzed and also compared to similar systems (Pickering, 1980). Ultimately, some efforts must be made to standardize methods and select necessary measurements on the soil-plant system to allow description and prediction of the systems behavior. In the interim, well-documented methods should be employed, and whenever possible, the methods selected should be from among a short list with proven value.

REFERENCES

Abd-Elfattah, A., and K. Wada. 1981. Adsorption of lead, copper, zinc, cobalt and cadmium by soils that differ in cation-exchange materials. J. Soil Sci. 32:271–283.

Adriano, D.C., A.L. Page, A.C. Chang, and A.A. Elseewi. 1982. Co-recycling of sewage sludge and fly-ash: Cadmium accumulation by crop. Environ. Tech. Lett. 3:145–150.

Alloway, B.J. (ed.). 1990. Heavy metals in soils. John Wiley & Sons, New York.

American Public Health Association. 1985. Standard methods for the examination of water and wastewater. 16th ed. Am. Public Health Assoc., Washington, DC.

Andersson, A. 1979. Mercury in soils. Review. *In* J.O. Nriagu (ed.) The biogeochemistry of mercury in the environment. Top. Environ. Health 3:79–112.

Artiole, J., and W.H. Fuller. 1979. Effects of crushed limestone barriers on chromium attenuation in soils. J. Environ. Qual. 8:503–510.

Association of Official Analytical Chemists. 1980. Official methods of analysis. 13th ed. AOAC, Washington, DC.

Baker, D.E. 1973. A new approach to soil testing: II. Ionic equilibria involving H, K, Ca, Mg, Mn, Fe, Cu, Zn, Na, P and S. Soil Sci. Soc. Am. Proc. 37:537–541.

Baker, D.E., and M.C. Amacher. 1981. The development and interpretation of a diagnostic soil testing program. Pennsylvania State Univ. Agric. Exp. Stn. Bull. 826.

Baker, D.E., and M.C. Amacher. 1982. Nickel, copper, zinc, and cadmium. p. 323–336. *In* A.L. Page et al. (ed.) Methods of soil analysis. Part 2. 2nd ed. Agronomy Monogr. 9. ASA and SSSA, Madison, WI.

Baker, D.E., M.C. Amacher, and R.M. Leach. 1979. Sewage sludge as a source of cadmium in soil-plant-animal systems. Environ. Health Perspect. 28:45–49.

Baker, D.E., and M.E. Bowers. 1988. Human health effects of cadmium predicted from growth and composition of lettuce in gardens contaminated by emissions from zinc smelters. p. 281–295. *In* D.D. Hemphill (ed.) Trace substances in environmental health-XXII. Univ. of Missouri, Columbia.

Baker, D.E., and J.K. Buck. 1988. Using computerized expert systems, unique soil testing methods, and monitoring data in land management decisions. Mine Drainage and Surface Mine Reclamation. Vol. II: Mine Reclamation, Abandoned Mine Lands and Policy Issues. USDI Bur. of Mines Circ. 9184:246–256.

Baker, D.E., and L. Chesnin. 1975. Chemical monitoring of soils for environmental quality and animal and human health. Adv. Agron. 27:305–374.

Barry, S.A.S., and S.C. Clark. 1978. Problems of interpreting the relationship between the amounts of lead and zinc in plants and soil on metalliferous wastes. New Phytol. 81:773–783.

Bartlett, R., and B. James. 1979. Behavior of chromium in soils. III. Oxidation. J. Environ. Qual. 8:31–35.

Bartlett, R., and B. James. 1988. Mobility and bioavailability of chromium in soils. p. 267–304. *In* J.O. Nriagu and E. Nieboer (ed.) Chromium in the natural and human environments. John Wiley & Sons, New York.

Bashor, B.S., and P.A. Turri. 1986. A method for determining an allowable concentration of mercury in soil. Arch. Environ. Contam. Toxicol. 15:435–438.

Bock, R. 1979. Decomposition methods in analytical chemistry. Halsted Press, div. of John Wiley and Sons, New York.

Brady, B., and G.K. Pagenkopf. 1978. Cadmium complexation by soil fluvic acid. Can. J. Chem. 56:2331-2336.

Burau, R.G. 1982. Lead. p. 347-365. *In* A.L. Page et al. (ed.) Methods of soil analysis. Part 2. 2nd ed. Agronomy Monogr. 9. ASA and SSSA, Madison, WI.

Cannon, K., and M.J. Dudas. 1982. Seasonal pattern of gaseous mercury emanations from forested soils in Alberta. Note. Can. J. Soil Sci. 62:541-543.

Cappon, C.J. 1987. Uptake and speciation of mercury and selenium in vegetable crops grown on compost-treated soil. Water, Air, Soil Pollut. 34:353-361.

Cary, E.E., W.H. Allaway, and O.E. Olson. 1977. Control of chromium concentrations in food plants. I. Absorption and translocation of chromium by plants. J. Agric. Food Chem. 25:300-304.

Chang, A.C., A.L. Page, and K.W. Foster, and T.E. Jones. 1982. A comparison of cadmium and zinc accumulation by four cultivars of barley grown in sludge-amended soils. J. Environ. Qual. 11:409-412.

Chang, A.C., A.L. Page, J.E. Warneke, M.R. Resketo, and T.E. Jones. 1983. Accumulation of cadmium and zinc in barley grown on sludge-treated soils: A long-term field study. J. Environ. Qual. 12:391-397.

Chubin, R.G., and J.J. Street. 1981. Adsorption of cadmium on soil constituents in the presence of complexing ligands. J. Environ. Qual. 10:225-228.

Colbourn, P., and I. Thornton. 1978. Lead pollution in agricultural soils. J. Soil Sci. 29:513-526.

Commission of the European Communities. 1978. Criteria (does/effect relationships) for cadmium. Report of a working group of experts prepared for the commission of the European communities, directorage-general for social affairs, health and safety directorate. Pergamon Press, Elmsford, NY.

Cox, F.R., and J.I. Wear (ed.). 1977. Diagnosis and correlation of zinc problems in corn and rice production. Southern Coop. Ser. Bull. 222 for Coop. Reg. Res. Proj. S-80.

Crockett, A.B., and R.R. Kinnison. 1979. Mercury residues in soil around a large coal-fired power plant. Environ. Sci. Technol. 13:712-715.

Czuba, M., H. Akagi, and D.C. Mortimer. 1981. Quantitative analysis of methyl- and inorganic-mercury from mammalian, fish and plant tissue. Environ. Pollut. Ser. B. 2:345-352.

Czuba, M., and T.C. Hutchinson. 1980. Copper and lead levels in crops and soils of the Holland marsh area—Ontario. J. Environ. Qual. 9:566-575.

Dahnke, W.C. (ed.). 1988. Recommended chemical soil test procedures for the North Central region. North Dakota Agric. Exp. Stn. North Central Reg. Publ. 221 (revised).

Davies, B.E. 1983. A graphical estimation of the normal lead content of some British soils. Geoderma 29:67-75.

Dawson, R. 1978. Leather tanning industry: Sludge problems ahead. Sludge 1:24-27.

Dijkshoorn, W., L.W. van Broekhoven, and J.E.M. Lampe. 1979. Phytotoxicity of zinc, nickel, cadmium, lead, copper and chromium in three pasture plant species supplied with graduated amounts from the soil. Neth. J. Agric. Sci. 27:241-253.

Dudas, M.J., and K. Cannon. 1983. Seasonal changes in background levels of mercury in surface horizons of forested soils in Alberta. Note. Can. J. Soil Sci. 63:397-400.

Elfving, D.C., C.A. Bache, and D.J. Lisk. 1979. Lead content of vegetables, millet and apple trees grown on soils amended with colored newsprint. J. Agric. Food Chem. 27:138-140.

Elliott, H.A., and C.M. Denneny. 1982. Soil adsorption of cadmium from solutions containing organic-ligands. J. Environ. Qual. 11:658-663.

Elsokkary, I.H. 1982. Contamination of soils and plants by mercury as influenced by the proximity industries in Alexandria, Egypt. Sci. Total Environ. 23:55-60.

Farrah, H., and W.F. Pickering. 1978. Extraction of heavy metal ions sorbed on clays. Water, Air, Soil Pollut. 9:491-498.

Fergusson, J.E., R.W. Hayes, T.S. Yong, and S.H. Thiew. 1980. Heavy-metal pollution by traffic in Christchurch, New Zealand. Lead and cadmium content of dust, soil, and plant samples. N. Z. J. Sci. 23:293-310.

Fujii, R., L.L. Hendrickson, and R.B. Cory. 1983. Ionic activities of trace metals in sludge-amended soils. Sci. Total Environ. 28:179-190.

Giordano, P.M., D.A. Mays, and A.D. Behel, Jr. 1979. Soil temperature effects on uptake of cadmium and zinc by vegetables grown on sludge-amended soil. J. Environ. Qual. 8:233-236.

Grove, J.H., and B.G. Ellis. 1980. Extractable Chromium as related to soil pH and applied chromium. Soil Sci. Soc. Am. J. 44:238–242.

Gulson, B.L., K.G. Tiller, K.J. Mizon, and R.H. Merry. 1981. Use of lead isotopes in soils to identify the source of lead contamination near Adelaide, South Australia. Environ. Sci. Technol. 15:691–696.

Hafez, A.A.R., H.M. Reisenauer, and P.R. Stout. 1979. Solubility and plant uptake of chromium from heated soils. Commun. Soil Sci. Plant Anal. 10:1261–1270.

Haq, A.U., T.E. Bates, and Y.K. Soon. 1980. Comparison of extractants for plant-available zinc, cadmium, nickel and copper in contaminated soils. Soil Sci. Soc. Am. J. 44:772–777.

Harding, I.H., and T.W. Healy. 1979. Cadmium uptake on organic and inorganic constituents of soil. Prog. Water Technol. 11:265–273.

Harrison, R.M., D.P.H. Laxen, and J.S. Wilson. 1981. Chemical associations of lead, cadmium, copper and zinc in street dusts and roadside soils. Environ. Sci. Technol. 15:1378–1383.

Harsh, J.B., and H.E. Doner. 1981. Characterization of mercury in a river-wash soil. J. Environ. Qual. 10:333–337.

Harter, R.D. 1979. Adsorption of copper and lead by Ap and B2 horizons of several northeastern United States soils. Soil Sci. Soc. Am. J. 43:679–683.

Hemphill, D.D., Jr., V.V. Volk, P.J. Sheets, and C. Wickliff. 1985. Lettuce and broccoli response and soil properties resulting from tannery waste application. J. Environ. Qual. 14:159–163.

Hickey, M.G., and J.A. Kittrick. 1984. Chemical partitioning of cadmium, copper, nickel, and zinc in soils and sediments containing high levels of heavy metals. J. Environ. Qual. 13:372–376.

Hinesly, T.D., D.E. Alexander, K.E. Redborg, and E.L. Ziegler. 1982a. Differential accumulations of cadmium and zinc by corn hybrids grown on soil amended with sewage sludge. Agron. J. 74:469–474.

Hinesly, T.D., K.E. Redborg, E.L. Ziegler, and J.D. Alexander. 1982b. Effect of soil cation exchange capacity on the uptake of cadmium by corn. Soil Sci. Soc. Am. J. 46:490–497.

Hogg, T.J., J.R. Bettany, and J.W.B. Stewart. 1978a. The uptake of 203-Hg-labeled mercury compounds by brome-grass from irrigated undistrubed soil columns. J. Environ. Qual. 7:445–450.

Hogg, T.J., J.W.B. Stewart, and J.R. Bettany. 1978b. Influence of the chemical form of mercury on its adsorption and ability to leach through soils. J. Environ. Qual. 7:440–445.

Huffman, E.W.D., and W.H. Allaway. 1973. Chromium in plants: Distribution in tissues, organelles and extract and availability of bean leaf Cr to animals. J. Agric. Food Chem. 21:982–986.

Hutchison, F., and C.M. Wai. 1979. Cadmium, lead, and zinc in reclaimed phosphate mine waste dumps in Idaho. Bull. Environ. Contam. Toxicol. 23:377–380.

James, B.R., and R.J. Bartlett. 1983. Behavior of chromium in soils: V. Fate of organically complexed Cr(III) added to soil. J. Environ. Qual. 12:169–172.

Jarvis, S.C., and L.H.P. Jones. 1980. The contents and sorption of cadmium in some agricultural soils of England and Wales. J. Soil Sci. 31:469–479.

Jastrow, J.D., and D.E. Koeppe. 1980. Uptake and effects of cadmium in higher plants. p. 607–638. *In* J.O. Nriagu (ed.) Cadmium in the environment, Part 1. Ecological cycling. Environmental science and technology. John Wiley and Sons, New York.

Kelly, J.M., G.R. Parker, and W.W. McFee. 1979. Heavy-metal accumulation and growth of seedlings of five forest species as influenced by soil cadmium level. J. Environ. Qual. 8:361–364.

Korcak, R.F., and D.S. Fanning. 1978. Extractability of cadmium, copper, nickel and zinc by double acid versus DTPA and plant content at excessive soil levels. J. Environ. Qual. 7:506–512.

Kromer, E., G. Friedrich, and P. Wallner. 1981. Mercury and mercury compounds in surface air, soil gas, soils and rocks. J. Geochem. Explor. 15:51–62.

Kuo, S., P.E. Heilman, and A.S. Baker. 1983. Distribution and forms of copper, zinc, cadmium, iron, and manganese in soils near a copper smelter. Soil Sci. 135:101–109.

Kuo, S., E.J. Jellum, and A.S. Baker. 1985. Effects of soil type, liming, and sludge application on zinc and cadmium availability to Swiss chard. Soil Sci. 139:122–130.

Lim, C.H., and M.L. Jackson. 1982. Dissolution for total elemental analysis. p. 1–12. *In* A.L. Page et al. (ed.) Methods of soil analysis. Part 2. 2nd ed. Agronomy Monogr. 9. ASA and SSSA, Madison, WI.

Lindberg, S.E., D.R. Jackson, J.W. Huckabee, S.A. Janzen, M.J. Levin, and J.R. Lund. 1979. Atmospheric emission and plant uptake of mercury from agricultural soils near the Almaden mercury mine. J. Environ. Qual. 8:572–578.

Lindsay, W.L. 1979. Chemical equilibria in soils. John Wiley and Sons, New York.

Lindsay, W.L., and W.A. Norvell. 1978. Development of a DTPA test for zinc, iron, manganese and copper. Soil Sci. Soc. Am. J. 42:421–428.

Loughton, A., and R. Frank. 1974. Mercury in mushrooms (*Agaricus bisporus*). p. 347–355. *In* K. Mori (ed.) Mushroom Sci. (Part 1) Proc. Int. Sci. Conf. on Cultivation of Edible Fungi, Tokyo. Mushroom Res. Inst., 8 Hiraicho, Kiryu, Japan.

Lund, L.J., E.E. Betty, A.L. Page, and R.A. Elliott. 1981. Occurrence of naturally high cadmium levels in soils and its accumulation by vegetation. J. Environ. Qual. 10:551–556.

McBride, M.B., L.D. Tyler, and D.A. Hovde. 1981. Cadmium adsorption by soils and uptake by plants as affected by soil chemical properties. Soil Sci. Soc. Am. J. 45:739–744.

McGrath, S.P. 1982. The uptake and translocation of tri- and hexavalent chromium and effects on the growth of oat in flowing nutrient solution and in soil. New Phytol. 92:381–390.

Miles, L.J., and G.R. Parker. 1979a. DTPA soil extractable and plant heavy metal concentrations with soil-added Cd treatments. Plant Soil 51:59–68.

Miles, L.J., and G.R. Parker. 1979b. The effect of soil-added cadmium on several plant species. J. Environ. Qual. 8:229–232.

Miller, J.E., J.J. Hassett, and D.E. Koeppe. 1975. The effect of soil properties and extractable lead levels on lead uptake by soybeans. Commun. Soil Sci. Plant Anal. 6:339–347.

Miller, W.P., and W.W. McFee. 1983. Distribution of cadmium, zinc, copper and lead in soils of industrial northwestern Indiana. J. Environ. Qual. 12:29–33.

Mitchell, G.A., F.T. Bingham, and A.L. Page. 1978. Yield and metal composition of lettuce and wheat grown on soils amended with sewage sludge enriched with cadmium, copper, nickel, and zinc. J. Environ. Qual. 7:165–171.

Morishita, T., K. Kishino, and S. Idaka. 1982. Mercury contamination of soils, rice plants, and human hair in the vicinity of a mercury mine in Mie Prefecture, Japan. Soil Sci. Plant Nutr. 28:523–534.

Mortvedt, J.J. 1987. Cadmium levels in soils and plants from some long-term soil fertility experiments in the United States of America. J. Environ. Qual. 16:137–142.

Mortvedt, J.J., and P.M. Giordano. 1975. Response of corn to zinc and chromium in municiple wastes applied to soil. J. Environ. Qual. 4:170–174.

Mulla, D.J., A.L. Page, and T.J. Ganje. 1980. Cadmium accumulations and bioavailability in soils from long-term phosphorus fertilization. J. Environ. Qual. 9:408–412.

Nelson, J.L., L.C. Boawn, and F.G. Viets, Jr. 1959. A method for assessing zinc status of soils using acid-extractable zinc and "titratable alkalinity" values. Soil Sci. 88:275–283.

Nickow, C.W., P.H. Comas-Haezebrouck, and W.A. Feder. 1983. Influence of varying soil lead levels on lead uptake of leafy and root vegetables. J. Am. Soc. Hortic. Sci. 108:193–195.

Northeast Regional Research Publication. 1985. Criteria and recommendations for land application of sludges in the Northeast. Pennsylvania Agric. Exp. Stn. Bull. 851.

Norvell, W.A., and W.L. Lindsay. 1969. Reactions of EDTA complexes of Fe, Zn, Mn, and Cu with soils. Soil Sci. Am. Proc. 33:86–91.

Page, A.L., R.H. Miller, and D.R. Keeney (ed.). 1982. Methods of soil analysis. Part 2. 2nd ed. Agronomy Monogr. 9. ASA and SSSA, Madison, WI.

Palmer, K.T., and C.L. Kucera. 1980. Lead contamination of sycamore and soil from lead mining and smelting operations in eastern Missouri. J. Environ. Qual. 9:106–111.

Peterson, P.J., and B.J. Alloway. 1979. Cadmium in soils and vegetation. Review. *In* M. Webb (ed.) Chemistry, biochemistry and biology of cadmium. Top. Environ. Health 2:45–92.

Petruzzelli, G., G. Guidi, and L. Lubrano. 1981. Influence of organic matter on lead adsorption by soil. Note. Z. Pflanzenernaehr. Bodenkd. 144:74–76.

Phelps, D., and P.R. Buseck. 1980. Distribution of soil mercury and the development of soil mercury anomalies in the Yellowstone geothermal area, Wyoming. Econ. Geol. 75:730–741.

Pickering, W.F. 1980. Cadmium retention by clays and other soil or sediment components. p. 365–397. *In* J.O. Nriagu (ed.) Cadmium in the environment. Part 1. Ecological cycling. Environmental science and technology. John Wiley and Sons, New York.

Preer, J.R. 1985. Potential food-chain transfer of toxic trace substances in urban soils. p. 30. *In* Agronomy abstracts. ASA, Madison, WI.

Preer, J.R., and G.B. Murchison, Jr. 1986. A simplified method for detection of lead contamination of soil. Environ. Pollut. Ser. B. 12:1–13.

Proctor, J., and S.R.J. Woodell. 1975. The ecology of serpentine soils. Adv. Ecol. Res. 9:255–366.

Re, M., M.G. Garagiola, E. Crovato, and R. Cella. 1983. Cadmium distribution within corn plants as a function of cadmium loading of the soil. Environ. Res. 30:44–49.

Rehab, F.I., and A. Wallace. 1978a. Excess trace metal effects on cotton: 4. Chromium and lithium in Yolo loam soil. Commun. Soil Sci. Plant Anal. 9:645–651.

Rehab, F.I., and A. Wallace. 1978b. Excess trace metal effects on cotton: 6. Nickel and cadmium in Yolo loam soil. Commun. Soil Sci. Plant Anal. 9:779–784.

Rogers, R.D., and J.C. McFarlane. 1979. Factors influencing the volatilization of mercury from soil. J. Environ. Qual. 8:255–260.

Ross, D.S., R.E. Sjogren, and R.J. Bartlett. 1981. Behavior of chromium in soils: IV. Toxicity to microorganisms. J. Environ. Qual. 10:145–148.

Sadiq, M. 1983. Complexing of lead by DTPA in calcareous soils. Water, Air, Soil Pollut. 20:247–255.

Schnitzer, M. 1978. Humic substances: Chemistry and reactions. p. 1–64. *In* M. Schnitzer and S.U. Khan (ed.) Soil organic matter. Elsevier North-Holland, New York.

Schnitzer, M. 1982. Organic matter characterization. p. 581–594. *In* A.L. Page et al. (ed.) Methods of soil analysis. Part 2. 2nd ed. Agronomy Monogr. 9. ASA, Madison, WI.

Shewry, P.R., and P.J. Peterson. 1974. The uptake and transport of chromium by barley seedlings (*Hordeum vulgare* L.). J. Exp. Bot. 25:785–797.

Shewry, P.R., and P.J. Peterson. 1976. Distribution of chromium and nickel in plants and soil from serpentine and other sites. J. Ecol. 64:195–212.

Shipp, R.S., and D.E. Baker. 1975. Pennsylvania's sewage sludge research and extension program. Compost Sci. 16:6–8.

Singh, B., and G.S. Sekhon. 1983. Leaching of zinc, lead and cadmium in columns of calcareous soils. Z. Pflanzenernaehr. Bodenkd. 146:531–538.

Soane, B.D., and D.H. Saunder. 1959. Nickel and chromium toxicity of serpentine soils in southern Rhodesia. Soil Sci. 88:322–330.

Soltanpour, P.N. 1985. Use of ammonium bicarbonate DTPA soil test to evaluate elemental availability and toxicity. Commun. Soil Sci. Plant Anal. 16:323–338.

Soon, Y.K., and T.E. Bates. 1982. Chemical pools of cadmium, nickel and zinc in polluted soils and some preliminary indications of their availability to plants. J. Soil Sci. 33:477–488.

Spittler, T.M., and W.A. Feder. 1979. Study of soil contamination and plant lead uptake in Boston urban gardens. Commun. Soil Sci. Plant Anal. 10:1195–1210.

Stevenson, F.J., and L.F. Welch. 1979. Migration of applied lead in a field soil. Environ. Sci. Technol. 13:1255–1259.

Stewart, J.W.B., and J.R. Bettany. 1982. Mercury. p. 367–384. *In* A.L. Page et al. (ed.) Methods of soil analysis. Part 2. 2nd ed. Agronomy Monogr. 9. ASA and SSSA, Madison, WI.

Stomberg, A.L., D.D. Hemphill, Jr., and V.V. Volk. 1984. Yield and elemental concentration of sweet corn grown on tannery waste-amended soil. J. Environ. Qual. 13:162–166.

Street, J.J., W.L. Lindsay, and B.R. Sabey. 1977. Solubility and plant uptake of cadmium in soils amended with cadmium and sewage sludge. J. Environ. Qual. 6:72–77.

Street, J.J., and W.M. Peterson. 1982. Anodic stripping voltametry and differential pulse polarography. p. 133–148. *In* A.L. Page et al. (ed.) Methods of soil analysis. Part 2. 2nd ed. Agronomy Monogr. 9. ASA and SSSA, Madison, WI.

Thornton, I. 1974. Biogeochemical and soil ingestion studies in relation to the trace-element nutrition of livestock. p. 151–154. *In* Hoekstra et al. (ed.) Trace element metabolism in animals. Univ. of Wisconsin, Madison

Tyler, L.D., and M.B. McBride. 1982a. Influence of Ca, pH and humic-acid on Cd uptake. Plant Soil 64:259–262.

Tyler, L.D., and M.B. McBride. 1982b. Mobility and extractability of cadmium, copper, nickel and zinc in organic and mineral soil columns. Soil Sci. 134:198–205.

U.S. Environmental Protection Agency. 1975. Water programs; national interim primary drinking water regulations. Fed. Reg. 40:59570.

U.S. Environmental Protection Agency. 1979a. Criteria for classification of solid waste disposal facilities and practices; final, interim final, and proposed regulations. Fed. Reg. 44:53438–53464.

U.S. Environmental Protection Agency. 1979b. Methods for chemical analysis of water and wastes. USEPA Publ. 600/4-79-020. USEPA, Cincinnati.

U.S. Environmental Protection Agency. 1979c. Handbook for analytical quality control in water and water laboratories. USEPA Publ. 600/4-79-019. USEPA, Cincinnati.

U.S. Environmental Protection Agency. 1986. Acid digestion of sediment, sludge and soils. *In* Test methods for evaluating solid waste SW-846. USEPA, Cincinnati.

Wear, J.I., and C.E. Evans. 1968. Relationship of zinc uptake by corn and sorghum to soil zinc measured by three extractants. Soil Sci. Soc. Am. Proc. 32:543–546.

Webber, J. 1972. Effects of toxic metals in sewage on crops. Water Pollut. Contr. 71:404–413.

Welch, J.E., and L.J. Lund. 1987. Soil properties, irrigation water quality, and soil moisture level influences on the movement of nickel in sewage sludge-treated soils. J. Environ. Qual. 16:403–410.

Welch, R.M., and E.E. Cary. 1975. Concentration of chromium, nickel, and vanadium in plant materials. J. Agric. Food Chem. 23:479–482.

Williams, P.H., J.S. Shenk, and D.E. Baker. 1978. Cadmium accumulation by meadow voles (*Microtus pennsylvanicus*) from crops grown on sludge-treated soil. J. Environ. Qual. 7:450–454.

Wiseman, J.T., and L.M. Zibilske. 1988. Effects of sludge application sequence on carbon and nitrogen mineralization in soil. J. Environ. Qual. 17:334–339.

Zyrin, N.G., B.A. Zvonarev, L.K. Sadovnik, and N.I. Voronova. 1981. Distribution of mercury in soils of the North Ossetain Plains. Sov. Soil Sci. 13:44–52.

Chapter 12

Testing Soils for Salinity and Sodicity

J. D. RHOADES, *U.S. Salinity Laboratory, Riverside, California*

S. MIYAMOTO, *Texas A&M University, Agricultural Research Center, El Paso, Texas*

The diagnosis, management, and need for reclamation of salt-affected soils are evaluated by measurements of their salinity and sodicity. *Soil salinity,* which refers to the presence of excessive levels of dissolved inorganic solutes, is generally assessed by the electrical conductivity of the saturation extract (EC_e), since EC is a practical index of total ionized solute concentration. *Soil sodicity,* which refers to the excessive presence of Na, is generally expressed in terms of the exchangeable sodium percentage (ESP) or the sodium adsorption ratio of the saturation extract (SAR_e). The latter is more conveniently and accurately determined and these two parameters are closely related.

In this chapter, we present procedures for testing soils for salinity and sodicity; in addition, guidelines are given for diagnostic purposes. Detailed analytical methods for determining the concentrations of the individual salt constituents in waters and soil extracts are described in Rhoades (1982a); and descriptions of alternative methods for determining salinity in the field are described in Rhoades (1984) and Rhoades and Oster (1986).

I. LABORATORY METHODS FOR SALINITY AND SODICITY APPRAISAL

A. Soil Sampling, Storage, and Preparation

1. Soil Sampling

Soil salinity is among the most variable properties of soils, and its variation within a field is generally greater than analytical errors. Thus, the reliability of analytical data for salinity appraisal is often limited by sampling error. Kelley (1922) described it realistically: "It is evident that the analysis of a single soil drawn from one place in the area studied has very little

 Soil Testing and Plant Analysis, 3rd ed.—SSSA Book Series, no. 3.

value. . . If similar variation exists in alkali soils generally, it may be safe to conclude that the analysis of samples such as commonly submitted by practical farmers is a waste of time.''

The method of appropriate sampling is dictated largely by the purpose for collecting soil samples and the variation existing within the sampling area. In some cases, we may only want to identify the salinity level at a selected point in a field. If this is the objective, statistical considerations may be academic. In many cases, soil samples submitted for routine analysis come from an area characterized as a ''saline area'' or a ''sodic spot,'' covering a fraction of a hectare to many hectares within a cultivated field. Such an area may consist of a single soil type or several types, and salinity and sodicity typically vary greatly within such areas. For instance, salinity readings in surface-irrigated orchards in the El Paso Valley of Texas varied two- to threefold within salt-affected areas of <1 ha (Miyamoto & Cruz, 1986). Even greater variation in salinity was observed in a 0.8 ha fallow field near Bet Dagan (Bresler et al., 1984). The variation in sodicity is usually smaller than that of salinity in irrigated fields (e.g., Miyamoto & Cruz, 1986), but can be large in nonirrigated soils as documented by Sayegh et al. (1958) in the Baker Valley of Oregon. The spatial variations in salinity and sodicity are site specific, and their approximate ranges for applicable soil types, or series, or fields must be known before soil samples can be collected appropriately. This information can be estimated from the much more easily acquired field measurements of bulk soil electrical conductivity (EC_a) made using portable instrumentation.

Once the variability is known, the question is reduced to sampling adequacy. The frequency distribution of salinity in an area consisting of multiple soil types or series often presents a skewed distribution (e.g., Wagenet & Jurinak, 1978; Miyamoto & Cruz, 1986). However, in an area consisting of a single soil type, both salinity and sodicity (on a depth-averaged basis) generally follow an approximately normal distribution (Miyamoto & Cruz, 1986, 1987). Therefore, the number of soil samples (N) required to meet a desired statistical accuracy can be estimated by

$$N = t_{\alpha}^{2}\mathrm{SD}^{2}/(d\,\bar{x})^{2} \qquad [1]$$

where t_{α} is the normalized deviate for a given confidence level of α (e.g., $t = 1.96$ at a confidence level $\alpha = 5\%$), SD is the standard deviation, and d is the deviate range from the true mean, $\bar{x}$. The value for N is site-dependent, and in torrifluvents ranges from 5 to 20 samples per hectare of border or basin irrigated land consisting of a single soil type when d is taken as 15% of the true mean at a 5% confidence level (e.g., Miyamoto, 1988). The numbers of samples required to characterize salinity or sodicity at individual soil depths are usualy greater than those required for depth-averaged sampling, owing to greater variation in the former case. Once the number of samples required is determined, soil samples can be collected within an area consisting of a single soil type at random sites or at sites following determined spatial schemes.

When detailed soil maps are not available and soil characteristics are poorly known, an equal-spaced grid pattern can be used for sampling (e.g., Petersen & Calvin, 1965). The data collected in this fashion can be used to describe the spatial distribution of salinity or sodicity over the area through kriging (Hajrasuliha et al., 1980). Such information is useful, for instance, for planning reclamation of saline or sodic soils or for making alternative management decisions. However, this type of spatial analysis generally demands numerous samples and is not usually included in routine appraisals. If the purpose is to obtain the value of the mean salinity of an area, a systematic grid sampling is not nearly as efficient as is a soil type-based sampling system (Sayegh et al., 1958; Miyamoto, 1988).

Soil salinity and sodicity typically vary with depth. Under irrigated conditions, salinity generally increases approximately exponentially with depth in sandy soils and approximately linearly in clay soils to the bottom of the root zone. In clay soils having water infiltration problems or shallow water tables, it is common to observe higher levels of salinity at shallow depths. Soil samples should be collected from that part of the profile from which the plant is extracting significant amount of water by observed horizons in the profile or by arbitrary depth-increments, such as every 30 cm, in the absence of such horizons. The samples collected by arbitrary depth-increments generally show greater spatial variability than those collected by horizons, especially when the thickness of the horizons vary with location within the sampling area.

Salinity distribution patterns are also influenced by the methods of irrigation and cultivation (e.g., Bernstein & Fireman, 1957; Ayers & Westcot, 1985; Yaron et al., 1973). Since high-salt accumulations often occur in the ridges of furrow-irrigated crop beds and at the wetting front surrounding drip emitters, these depths should be avoided or else sampled separately when assessing the salinity hazard to established crops. (These high-salt bands present at the soil surface or the extremities of the crop root zone do not generally harm established crops). The near-surface depth of soil above and surrounding the seeding depth may be collected to assess relevent salt conditions for seedling emergence and establishment (Miyamoto et al., 1985b).

Equipment conventionally used for general soil sampling purposes are satisfactory for sampling saline or sodic soils. Tube samplers are most convenient when composite samples are to be taken. Large diameter samplers generally yield smaller variation in salinity measurements than do small tube samplers (Hassan et al., 1983). It is preferable to sample when the soil is reasonably dry and after all loose plant material and debris are removed from the surface. If a salt crust is present it should be sampled separately, if desired, but in no case allow it to be included in the major soil sample. Care should be taken to prevent dislodgement of an upper layer while collecting the deeper layers when using augers or tube samplers to obtain successively deeper samples. To avoid contamination and to facilitate mixing of the samples, it is advisable to pass each depth-sample through a 0.6 to 0.8 cm screen and thence into a separate plastic bucket. For routine analyses, 200 g (clay soils) to 400 g (sandy soils) of soil are usually sufficient. The samples should be transferred into plastic, air-tight bags with tags for labeling (preferably placed inside, if the samples are sufficiently dry). All soil samples should be coded for reference. Pertinent field information should be recorded, such as irrigation methods, irrigation water quality, crop condition, soil moisture level, time since last irrigation, and soil permeability (leaching fraction) to help in the interpretation of analytical results.

2. Sample Storage and Preparation

The method of sample storage depends on the duration of the intended storage as well as on whether analyses of pH and Ca and carbonate concentrations are desired. For short-term storage, samples can be stored in a cool room or, if available, in a refrigeration unit without drying. Prior to analysis, samples may be air dried to facilitate mixing to obtain a homogeneous sample. For long-term storage, the samples should be first air dried. Prolonged air drying should be avoided, especially under a hot and dry climate, as salts can precipitate and become difficult to dissolve and extract. Likewise, oven drying of soil samples should be avoided. Salts can be extracted without difficulty if samples are processed as soon as they become dry enough to conveniently grind and mix.

Excessive grinding of samples should be avoided. Reduce the clods and large aggregates to <4 mm using a wooden or rubber roller, mallet, or any effective method that will not pulverize the soil particles. Weigh and discard the >4-mm material. Place the sieved soil on a plastic sheet; mix by repeatedly pulling opposite corners of the sheet to the center of the sheet. Determine moisture content on an aliquot of the sample, and store the rest in a sealed plastic container.

B. Saturation Extract and Other Aqueous Extracts

1. Principles

It is desirable to know the composition of solutes existing in the soil water at field water contents. However, because present methods of obtaining soil water samples at field water contents are not practical for routine purposes, soil solution extracts are generally used. Because the absolute and relative concentrations of various solutes in the extract are influenced by the soil/water ratio (Reitemeier, 1946), the ratio must be specified and it is desirable that it be standardized to obtain results that can be applied and interpreted uniformly. Soil salinity is conventionally defined and measured on aqueous extracts of saturated soil pastes (U.S. Salinity Laboratory Staff, 1954). This soil/water ratio is the lowest practical ratio for which enough extract for analysis can be readily removed from the soil with low pressure or vacuum. It is generally related in a predictable manner to field soil water content. It is quite reproducible, once trained in its preparation. Normal deviations in the ratio do not significantly affect salinity appraisal. Almost all of the salinity crop-tolerance data used to diagnose salinity status is expressed in terms of the electrical conductivity of the saturation extract and relations used to predict ESP from SAR are also based on the saturation extract (U.S. Salinity Laboratory, 1954; Sposito & Mattigod, 1979; Rhoades, 1982b; Jurinak et al., 1984).

Solutions from more dilute extraction ratios, 1:1, 1:5, etc., are easier to obtain than saturation, but they are not related in as predictable a manner to field soil water contents. Errors from peptization, hydrolysis, cation exchange, and mineral dissolution are enhanced with such extracts. This is particularly true in gypsiferous soils where the dissolution of gypsum ($CaSO_4 \cdot 2H_2O$) significantly affects the concentration of dissolved salts. When relative changes in solute concentrations are to be monitored, these dilute extraction ratios may be used to advantage, but they are not advised for diagnosis purposes.

Once soil extract samples are obtained, laboratory chemical analyses are carried out to determine the EC and the SAR and, if needed, the concentrations of other solutes in the extract.

2. Saturation Extract

Weigh 200 to 400 g of air-dry soil of known water content into a plastic container having a snaptight lid. Weigh the container plus contents. Add distilled water to the soil with stirring until it is nearly saturated. Allow the mixture to stand covered for several hours to permit the soil to imbibe the water, and then add additional water to achieve a uniformly saturated soil-water paste. At this point, which is generally reproducible to within C = 5%, the soil paste glistens as it reflects light, flows slightly when the container is tipped, slides freely and cleanly off a spatula, and consolidates easily by tapping or jarring the container after a trench is formed in the paste with the side of the spatula. After mixing, allow the sample to stand for at least 2 h (preferably overnight) and then recheck it for saturation. Free water should not collect on the soil surface, nor should the paste have stiffened markedly or lost its glisten. If the paste is too wet, add additional dry soil of known weight to the paste mixture. Upon attainment of saturation, reweigh the container plus contents. Record the increase in weight, which is the amount of water added. Calculate the saturation water percentage from the weight of oven dry soil and the sum of the weights of water added and that initially present in the air-dry sample; SP = 100 (g of water/g of oven-dry soil). After allowing the saturated soil paste to stand 2 h or more, transfer it to a Buchner or Richards-type (1949) filter funnel fitted with highly retentive filter paper, such as Whatman No. 42. Apply vacuum, and collect the filtrate in a test tube or bottle. If the initial filtrate is turbid, refilter or discard it. Terminate the filtration when air begins to pass through the filter. Before storage, add one drop of 0.1% $(NaPO_3)_6$ solution[1] for each 25 mL of extract; alternatively dilute 1:1 with distilled water. This last step can be omitted if the effect of $CaCO_3$ precipitation upon the composition is negligible.

[1] Sodium hexametaphosphate [$(NaPO_3)_6$] solutions, 0.1%: Dissolve 0.1 g of $(NaPO_3)_6$ in water and dilute the solution to 100 mL.

3. Other Aqueous Extracts

To prepare aqueous extracts of soil/water ratios of 1:1 and 1:5, etc. weigh a sample of air-dry soil of appropriate size, and transfer it to a flask or bottle. Add the required amount of distilled water, stopper the container, and shake in a mechanical shaker for 1 h. If a mechanical shaker is not available, shake the container vigorously by hand for 1 min at least four times at 30-min intervals. Filter the suspension using highly retentive filter paper, such as Wattman No. 42. (Discard or refilter the initial filtrate if it is turbid.) Add 0.1% $(NaPO_3)_6$ solution[1] at a ratio of one drop for each 25 mL of extract.

4. Comments

In the preparation of the saturation paste, it is recommended to record the soil water content at saturation—the so-called *saturation percentage*. This information is useful to estimate soil texture (see Soil Survey Staff, 1951) and the salinity of the soil water at different field moisture levels. Given knowledge of the soil bulk density (ρ_b), the electrical conductivity of soil water (EC_w) at volumetric content (θ_w) may be estimated as: $EC_w = \rho_b EC_e SP/100 \theta_w$ (after Rhoades et al., 1989c). Special precautions should be taken to prepare a saturated soil paste with peat and muck soils or very fine or very coarse-textured soils. If possible, peat and muck soils should not be allowed to dry following collection because their saturation water content changes upon drying. Peat and muck, especially if coarse or woody, require at least an overnight imbibition period to obtain a definite endpoint for the saturation point. After the first wetting, pastes of these soils usually stiffen upon standing. Adding water and remixing then give a mixture that usually retains the characteristics of a saturated paste. One may also experience a similar problem in preparing the saturation paste of Na-rich soils high in clay content. With fine-textured soils, water should be initially added, with a minimum of mixing, to bring the sample nearly to saturation and then allow time for the water to be imbibed. This minimizes the formation of unwetted clumps during the subsequent stirring, speeds the mixing process, and helps attain a more definite endpoint. Care should be taken not to overwet coarse-textured soils. The presence of free water on the surface of the paste after standing is an indication of oversaturation. Even small amounts of free water can lead to significant errors in saturation paste water contents for very coarse-textured soils. Additional soil (of known weight) can be added to the paste in the case of oversaturation to correct this problem. This paste should be remixed and reequilibrated.

Alternative methods of preparing the saturated soil paste have been described by Longenecker and Lyerly (1964), who proposed wetting the soil sample on a capillary saturation table; Beatty and Loveday (1974), who recommended predetermining the amount of water at saturation on a separate soil sample using a capillary wetting technique; and Allison (1973), who recommended slowly adding soil to water (oversaturation method). Similar results are obtained with these methods. The choice of method is primarily one of personal preference. Sonneveltand van den Ende (1971) recommend use of a volume-based extract made up of one part soil and two parts water.

Sodium hexametaphosphate is added to the extract to inhibit the precipitation of $CaCO_3$. The amount of $(NaPO_3)_6$ solution added increases the Na concentration by about 0.5 mg/L, or 0.02 $mmol_c$/L, which is inconsequential compared with the possible loss of $CaCO_3$. A 1:1 dilution with distilled water is an alternative to the use of $(NaPO_3)_6$ to prevent $CaCO_3$ precipitation during storage. Thymol can be added to the paste to minimize the effect of microbial activity on saturation extract composition during equilibration (Carlson et al., 1971). The extracts should be stored at about 4 °C until analyzed.

C. Determination of Salinity: Electrical Conductivity of Extracts

For this determination, use a direct readout, temperature compensating conductivity meter.

1. Apparatus

a. Conductivity meter.
b. Conductivity flow cell with automatic temperature compensation.

2. Reagents

a. Standard KCl solutions, 0.010 and 0.100 N solution: For 0.010 N solution (1.412 dS/m at 25 °C) dissolve 0.7456 g of KCl in distilled water, and add water to make 1 L at 25 °C. For 0.100 N solution (12.900 dS/ m at 25 °C), use 7.456 g of KCl.

3. Procedure

Rinse and fill the conductivity cell with standard KCl solution. Adjust the conductivity meter to read the standard conductivity. Rinse and fill the cell with the extract and read EC, corrected to 25 °C, directly from the digital display.

4. Comments

Because of marked differences in the equivalent weights, equivalent conductivities, and proportions of major solutes in soil extracts and water samples, the relationships between EC and salt concentration or between EC and osmotic potential are only approximate. They are still quite useful, however. These relationships at 25 °C are:

a. Total cation (or anion) concentration, $mmol_c$/L ≈ 10 x EC, in dS/m.
b. Salt concentration, mg/L ≈ 640 x EC, in dS/m
or Salt concentration, mg/L ≈ 740 x EC, in dS/m (gypsiferous soils)
c. Osmotic potential, bars ≈ 0.4 x EC, in dS/m.

The electrical conductivity of an extract is generally lower than that of the soil solution because of dilution. In the case of the saturation extract, the water content of the paste averages two (ranging from 1.8–2.5) times field capacity and four (ranging from 3–5) times permanent wilting percentage, except in sandy soils. The electrical conductivities of the soil solutions at these two important soil moisture levels are greater than EC_e roughly by the same factors, except in gypsiferous soils. The conductivity of aqueous extracts (EC), including the saturation extract, can be related to that of soil solution (EC_{sw}) at known or given gravimetric soil water contents (*wf*) using Eq. [2].

$$EC_{sw} = \alpha(w/wf)EC \qquad [2]$$

where w is the gravimetric soil water content at extraction, and α the correction factor. The value for α depends on ion composition, soil bulk density (see p. 304) and, to a lesser extent, the concentration of dissolved salts, and usually ranges from 0.7 to 1.0 (dS/m). The error in estimating EC_{sw} increases with increasing w/wf and with the increasing content in the soil of the less-soluble salt species. In general, the error becomes unacceptably high when the extraction is made at 1:1 or at higher water content, unless α is determined either empirically or theoretically (e.g., McNeal et al.,1970).

D. Determination of Sodicity: Sodium Adsorption Ratio of Saturation Extract

Since the determination of ESP requires several cumbersome procedures subject to numerous errors (Thomas, 1982), soil sodicity hazard is now commonly defined and evaluated in terms of the SAR_e. The ESP and SAR_e are quantitatively related (Rhoades, 1982b; Jurinak et al., 1984); they are nearly the same in value over the range 0 to 30 (U.S. Salinity Laboratory Staff, 1954).

Determine Ca, Mg, and Na using an atomic absorption spectrometer (or by other suitable methods such as described in *Methods of Soil Analysis*, Part 1. 2nd ed. Agronomy 9, Am. Soc. of Agron.). If large numbers of samples are to be routinely analyzed, it is extremely helpful to have this unit equipped for automatic sample transport, sequencing, siphoning, reading, and recording.

1. Apparatus

a. Atomic absorption spectrometer.
b. Sampling and sequencing system.
c. Acetylene gas (C_2H_2), commercial grade.
d. Volumetric flasks and sample vials.

2. Reagents

a. Suppressant solution for Ca and Mg: Add 29.0 g of La_2O_3, 250 mL of concentrated hydrochloric acid (HCl), and enough distilled water to make up to 500-mL volume. Add a sufficient amount of this solution to the aliquot and diluent (distilled water) to achieve 10% (by volume) of this $LaCl_3$ solution in the final solution.

b. Suppressant solution for Na: Add 6.358 g of LiCl and make to 1 L in distilled water (0.15 *N*). Add enough of this solution to the aliquot and diluent (distilled water) to give 10% (by volume) of this LiCl solution in the final solution.

c. Standard cation solutions: Ca (0–0.4 $mmol_c/L$), Mg (0–0.1 $mmol_c/L$), and Na (0–1.0 $mmol_c/L$).

3. Procedure

Adjust the atomic absorption spectrometer controls and settings for each cation as recommended by the manufacturer. Set the atomic absorption spectrometer readout to read the upper and lower standard solutions. Then initiate the transport/readout system, which automatically positions a sequence of samples, siphons and aspirates the samples in the air-acetylene flame, and reads and records (as a digital printout) the concentration of the cation in the aspirated solution. Standard solutions are analyzed every 20 samples to ensure stability of instrument calibration during the automated run. Two-hundred samples can be processed per hour without attendance after the automatic sampling process is initiated. Alternatively, the same sequence of operations can be performed by hand.

4. Calculations

Concentrations of cations in the saturation extract in $mmol_c/L$ = (atomic absorption spectrometer readout, $mmol_c/L$, of aspirated sample) × (analytical dilution factor). The dilution factor must include the 1:1 predilution made at sampling time to prevent $CaCO_3$ precipitation during storage, if appropriate.

The SAR is calculated as

$$SAR = Na/\sqrt{(Ca + Mg)/2} \quad [3]$$

where the total cation concentrations in the saturation extract are in $mmol_c/L$.

5. Comments

The SAR of an extract decreases with increasing soil water content at extraction due to dilution effects as well as dissolution of Ca-bearing soil minerals such as calcite and gypsum. Thus, the SAR determined in the extracts of low dilution such as the saturation extract is generally more reliable than those determined in the extracts of higher dilution. When there is no mineral dissolution, the SAR decreases in proportion to the square root of the dilution factor. The decrease in SAR in soil systems is less than this estimate, because the exchange reaction buffers the change (e.g., Oster & McNeal, 1971).

The ESP of typical soils is approximately equal to SAR_e at values below about 30. However, the ESP of gypsiferous soils is usually greater than SAR_e, partly because the measured soluble Ca and Mg concentrations include Ca- and Mg-sulfate ion pairs having no net electrical charge (e.g., Rao et al., 1968). In gypsiferous soils, Ca-sulfate ion pairs can constitute up to about one-third of the total dissolved Ca. When SO_4 concentrations are known, the SAR_e can be corrected numerically for the formation of ion pairs to yield a better estimate of ESP (Sposito & Mattigod, 1979; Oster & Sposito, 1980); however, this is seldom done routinely.

For more detailed testing needs, it is sometimes desirous to determine the compositional makeup of the other solutes in the saturation extract.

Alkalinity (CO_3 + HCO_3), EC, and pH determinations should be made immediately on fresh extracts or on the solutions treated with hexametaphosphate. The cations can be determined in any sequence. The preferred sequence of anion determinations, to minimize $CaCO_3$ precipitation problems and estimate aliquot size for the more time-consuming and difficult-to-analyze anions, especially SO_4, is alkalinity, then Cl, NO_3, and SO_4. Finally, B is determined. Boron is of negligible concentration compared to the major cations and anions, but it is toxic to many plants, even in small concentrations.

There are many satisfactory analytical methods for determining individual solute concentrations in extracts. Methods in common use in laboratories having modern, but not fully automated, instrumentation are described in Rhoades (1982a). Methodology more suited to laboratories without such conveniences are described in Bower and Wilcox (1965).

II. INSTRUMENTAL FIELD METHODS OF SALINITY APPRAISAL

A. Saturation Paste Conductivity

1. Principles

The EC_e may be estimated from measurement of the electrical conductivity of the saturated soil-paste (EC_p) and estimates of saturation percentage. The measurement of EC_p and the estimate of SP are made using an EC-cup of known geometry and volume. The method is suitable for both laboratory and field applications, especially the latter, because the apparatus is inexpensive, simple, and rugged and because the determination of EC_p can be made much more quickly than EC_e.

Rhoades et al. (1989b) have shown that the following relation describes the electrical conductivity of saturated soil pastes,

$$EC_p = \left[\frac{(\theta_s + \theta_{ws})^2\, EC_e\, EC_s}{(\theta_s)\, EC_e + (\theta_{ws})\, EC_s}\right] + (\theta_w - \theta_{ws})\, EC_e \qquad [4]$$

where EC_p and EC_e are as defined previously, θ_w and θ_s are the volume fractions of total water and solids in the paste, respectively, θ_{ws} is the volume fraction of water in the paste that is coupled with the solid phase to provide an electrical pathway through the paste (a series coupled pathway), and EC_s is the average specific electrical conductivity of the solid particles. The difference $(\theta_w - \theta_{ws})$ is θ_{wc}, which is the volume fraction of water in the paste that provides a continuous pathway for electrical current flow through the paste (a parallel pathway to θ_{ws}). Assuming the average particle density (ρ_s) of mineral soils to be 2.65 g/cm^3 and the density of saturation soil-paste extracts (ρ_w) to be 1.00, θ_s and θ_w are directly related to SP as follows:

$$\theta_w = SP/\left[\frac{100}{\rho_w \rho_s} + SP\right] \qquad [5]$$

$$\theta_s = 1 - \theta_w. \qquad [6]$$

As shown by Rhoades et al. (1989b, c), saturation percentage of mineral soils, generally, can be adequately estimated in the field for purposes of salinity appraisal from the weight of the paste-filled cup. Figure 12-1 may be used for this purpose; for details of the relations inherent in this figure see Wilcox (1951).

The EC_e can be determined for any soil solely from measurement of EC_p and SP (using Eq. [4]-[6]), if values of ρ_s, θ_{ws}, and EC_s are known. These parameters can be adequately estimated, as demonstrated by Rhoades et al. (1989b, d), for typical arid land soils of the southwestern USA. ρ_s may be assumed to be 2.65 g/cm^3. EC_s may be estimated from SP as: EC_s = 0.019 (SP) − 0.434. The difference $(\theta_w - \theta_{ws})$ may be estimated from SP as: $(\theta_w - \theta_{ws}) = 0.0237\ (SP)^{0.6657}$.

2. Apparatus

For this determination, use any suitable conductivity meter and cup-type conductivity cell.

a. Conductivity meter, temperature compensating type.
b. Conductivity cell of 50 cm^3 volume, such as the "Bureau of Soils" cup (U.S. Salinity Laboratory, 1954).
c. Portable balance capable of weighing accurately to the nearest 1 g.

3. Reagents

a. Standard KCl solutions, 0.010 and 0.100 *N* solution: For 0.010 *N* solution (1.412 dS/m at 25 °C), dissolve 0.7456 g of KCl in distilled water, and add water to make 1 L at 25 °C. For 0.100 *N* solution (12.900 dS/m at 25 °C), use 7.456 g of KCl.

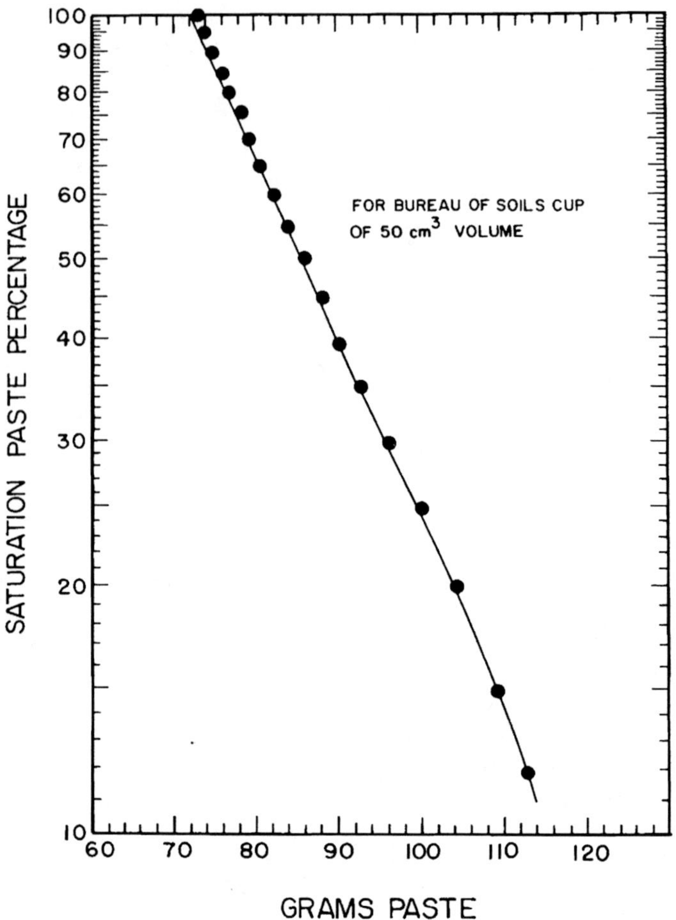

Fig. 12–1. Theoretical relation between saturation percentage (SP) and weight (in grams) of 50 cm^3 of saturated soil paste, assuming a particle density of 2.65 g/cm^3.

4. Procedure

Rinse and fill the conductivity cup with standard KCl solution. Adjust the conductivity meter to read the standard conductivity. Rinse and fill the cup with the saturated soil-paste; tap the cup to dislodge any air entrapped within the paste. Level off the paste with the surface of the cup. Weigh the cup plus paste; subtract the cup tare weight to determine the grams of paste occupying the cup. Obtain the SP value from Fig. 12–1 corresponding to this weight. Connect the cup electrodes to the conductivity meter and determine the EC_p, corrected to 25 °C, directly from the meter display. Obtain EC_e from Fig. 12–2 from EC_p using the curve corresponding to the SP value.

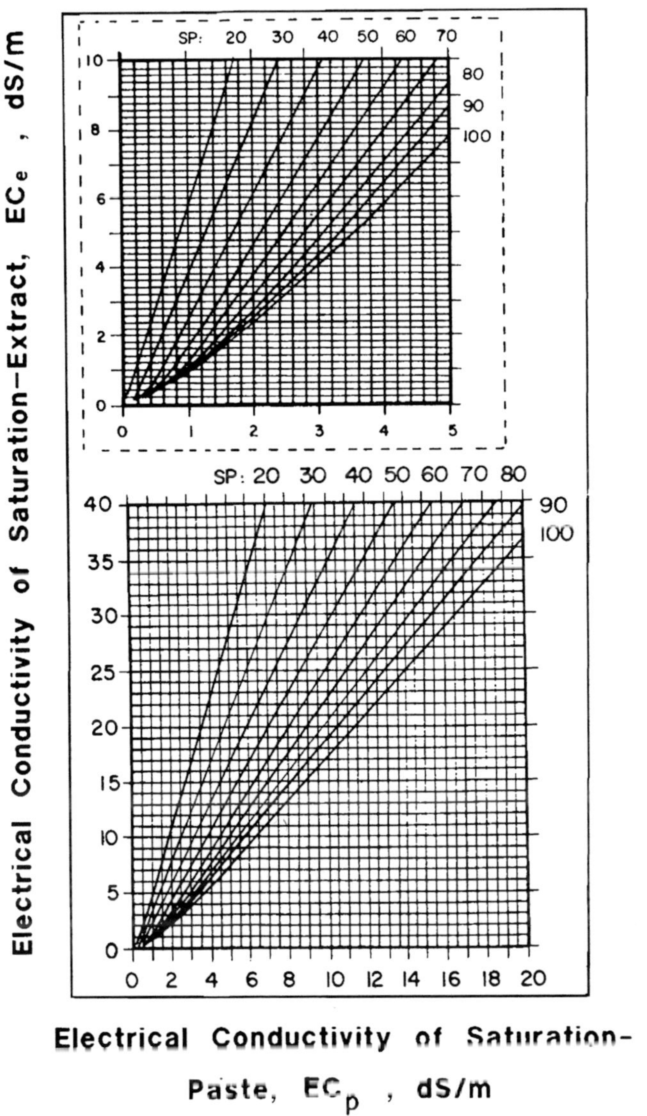

Fig. 12-2. Relations between electrical conductivity of saturated soil-paste (EC_p), electrical conductivity of saturation extract (EC_e), and saturation percentage (SP), for representative arid-land soils.

5. Comments

Sensitivity analyses and tests have shown that the estimates used in this method are generally adequate for salinity appraisal purposes of typical mineral arid-land soils of the southwestern USA. For organic soils or soils of different mineralogy or magnetic properties, these estimates may be inappropriate.

For such soils, appropriate values for ρ_s, EC_s, and θ_{ws} will need to be determined using analogous techniques to those Rhoades et al. (1989b) used. The accuracy requirements of these estimates may be evaluated using the relations given in Rhoades et al. (1989d).

The curves relating EC_p, EC_e, and SP were developed by solving Eq. [4] using the quadratic formula as follows:

$$EC_e = (-b \pm \sqrt{b^2 - 4ac})/2a \qquad [7]$$

where $a = [\theta_s (\theta_w - \theta_{ws})]$, $b = [(\theta_s + \theta_{ws})^2(EC_s) + (\theta_w - \theta_{ws})(\theta_{ws}EC_s) - (\theta_s) EC_p]$ and $c = -(\theta_{ws})(EC_s)(EC_p)$.

B. Bulk Soil Electrical Conductivity Sensors

In situ or remote devices capable of measuring electrical conductivity of the bulk soil can be used advantageously for purposes of soil salinity appraisal, monitoring, and mapping. Three kinds of portable sensors are now available, each with its own advantages and limitations: (i) four-electrode sensors, (ii) electromagnetic induction sensors, and (iii) time domain reflectometry (TDR) insertion parallel guide electrodes. All three measure the electrical conductivity of the bulk soil (EC_a), which depends upon its soil water content and salt concentration.

1. Principles

Because most soil minerals are insulators, electrical conduction in moist, saline soils is primarily through the large water-filled pores, which contain the dissolved salts (electrolytes). There is also a relatively small contribution of exchangeable cations (associated with the solid phase) to electrical conduction in soils, the so-called surface conduction (EC_s), because these electrolytes are more limited in their amounts and mobilities. The value of EC_s is assumed, for practical purposes, to be essentially constant for any given saline soil. EC_s is coupled in series with the electrolyte present in the water films associated with the solid surfaces and in the small water-filled pores that bridge adjacent particles to provide a secondary pathway for current flow in moist soils. This pathway acts in parallel with the major, continuous flow pathway (large water-filled pores). The relative flow of current in the two pathways depends on the solute concentration of the soil water, the magnitude of EC_s, and the contents of water in the two different categories of pores.

A mathematical description of the above model of electrical current flow in soils is given in Eq. [8] after Rhoades et al. (1989c):

$$EC_a = \left[\frac{(\theta_s + \theta_{ws})^2 \, EC_{ws} \, EC_s}{(\theta_s) \, EC_{ws} + (\theta_{ws}) \, EC_s}\right] + (\theta_w - \theta_{ws}) \, EC_{wc} \qquad [8]$$

where EC_a, θ_s, θ_w, and EC_s are as previously defined, θ_{ws} and $(\theta_w - \theta_{ws})$ are the volumetric soil water contents in the series-coupled pathway (the fine water-filled pores) and the separate continuous liquid pathway (large water-filled pores), respectively, and EC_{ws} and EC_{wc} are the specific electrical conductivities of the soil water that is in the two corresponding pathways, respectively.

The relation between EC_{ws}, EC_{wc}, and EC_e is (after Rhoades et al., 1989c):

$$(EC_{wc}\,\theta_{wc} + EC_{ws}\,\theta_{ws})/\rho_b = EC_e\,SP/100$$

where ρ_b is the bulk density of the soil. For practical purposes of salinity appraisal, it is assumed that $EC_{wc} \approx EC_{ws}$ and, therefore, that $(EC_w\,\theta_w) \approx (EC_{wc}\,\theta_{wc} + EC_{ws}\,\theta_{ws})$. Data exist to support the general validity of this assumption for typical field soils (Rhoades et al., 1989d).

The other relations used in the practical application of EC_a measurements to appraise soil salinity are (after Rhoades et al., 1990):

$$SP = 0.76\,(\%C) + 27.25, \quad [10]$$

$$\rho_b = 1.73 - 0.0067\,(SP), \quad [11]$$

$$\theta_s = \rho_b/2.65, \quad [12]$$

$$\theta_{wfc} = SP \cdot \rho_b/200, \quad [13]$$

$$\theta_w = \theta_{wfc} \cdot FC/100, \quad [14]$$

$$\theta_{ws} = 0.639\,\theta_w + 0.011, \text{ and} \quad [15]$$

$$EC_s = 0.019\,SP - 0.434 \quad [16]$$

where %C is clay percentage as estimated by "feel" methods, θ_{wfc} is the estimated volumetric water content at field capacity, and FC is the percent water content of the soil relative to that at field capacity, as estimated by the feel method. Use of the above relations permits the EC_e to be estimated in the field sufficiently accurately for salinity appraisal purposes from the measurement of EC_a and the estimates of %C and θ_{wfc} made by feel methods. That such procedures are generally adequate for typical arid land mineral soils of the southwestern USA has been demonstrated by Rhoades et al. (1990).

2. Apparatus

a. Four-Electrode Sensors

A combination electric current source and resistance meter, four metal electrodes, and connecting wire are needed for large soil volume (surface ar-

Fig. 12–3. Photograph of four electrodes positioned in a surface array and a combination electric generator and resistance-meter.

ray) measurements (Fig. 12–3). A four-electrode salinity probe, in which the electrodes are built into the probe [(Rhoades & van Schilfgaarde, 1976)] is needed for small soil volume measurements (Fig. 12–4). The current source-meter unit may be either a hand-cranked generator (Fig. 12–3) or a battery-powered type (Fig. 12–4). Units designed for geophysical purposes generally read in ohms and should measure from 0.1 to 1000 ohms for general soil

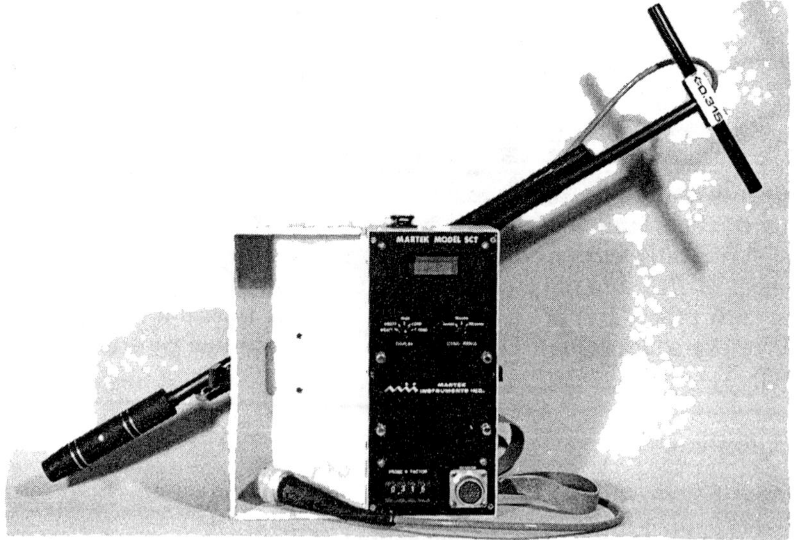

Fig. 12–4. Photograph of commercial four-electrode conductivity probe and generator-meter.

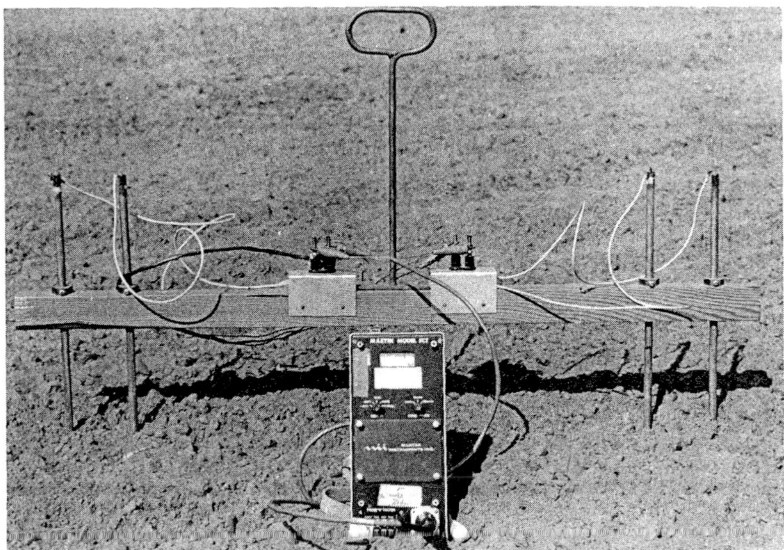

Fig. 12-5. Photograph of a "fixed-array" four-electrode apparatus and commercial generator-meter.

salinity measurement needs. Units specifically designed for use with the four-electrode salinity probe are much smaller and more convenient (Austin & Rhoades, 1979). A commercial unit, Martek·SCT[2], reads directly in EC_a corrected to 25 °C (Fig. 12-4).

Electrodes are made of stainless steel, copper, brass, or almost any other corrosion-resistant metal. Array electrode size is not critical, except that the electrode must be small enough to be easily inserted into the soil, to not tip over and maintain firm contact with the soil, when inserted to a depth of 5 cm or less. Electrodes 1.0 to 1.25 cm in diameter by 45 cm long are convenient for most array purposes, although smaller electrodes are preferred for determination of EC_a within shallow depths (<30 cm). Any flexible, well-insulated, multi-stranded, 12- to 18-gauge wire is suitable for connecting the array electrodes to the meter.

For survey or traverse work, the array electrodes may be mounted in a board with a handle (see Fig. 12-5) so that soil-resistance measurements can be made quickly for a given inter-electrode spacing (Rhoades & Halvorson, 1977). These "fixed-array" units save the time involved in positioning the electrodes. For most purposes, an inter-electrode spacing of 30 or 60 cm is adequate and convenient (wider spacings require lengthy, cumbersome units).

b. Electromagnetic Induction Sensors

The basic principle of operation of the EM soil electrical conductivity meter is shown schematically in Fig. 12-6. A transmitter coil located in one end of the instrument induces circular eddy current loops in the soil. The

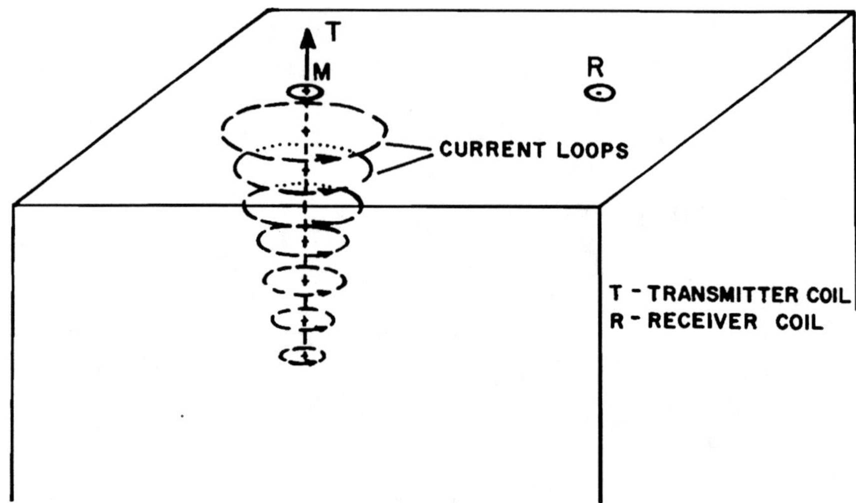

Fig. 12–6. Diagram showing the principle of operation of electromagnetic induction soil conductivity sensor.

magnitude of these loops is directly proportional to the conductivity of the soil in the vicinity of that loop. Each current loop generates a secondary electromagnetic field that is proportional to the value of the current flowing within the loop. A fraction of the induced electromagnetic field from each loop is intercepted by the receiver coil and the sum of these signals is amplified and formed into an output voltage that is linearly related to a depth-weighted soil EC_a.

Figure 12–7 shows the commercially available EM soil salinity sensor (Geonics EM-38[2]) being held in the vertical and horizontal (coils) positions. This device has an inter-coil spacing of 1 m, operates at a frequency of 13.2 kHz, is powered by a 9-V battery, and reads *depth-weighted* EC_a directly. The coil configuration and inter-coil spacing were chosen to permit measurement to effective depths of approximately 1 and 2 m when placed at ground level in a horizontal and vertical configuration, respectively. The device contains appropriate circuitry to minimize instrument response to the magnetic susceptibility of the soil and maximize response to EC_a.

c. *Time Domain Reflectometry Sensors*

With the use of TDR sensors, the apparent dielectric constant of the soil, ϵ, is obtained by measuring the transit time, t, of a voltage pulse applied to the parallel transmission line (dual-rod sensor) of length L embedded in the soil of electrical conductivity EC_a, and applying the relation:

$$\epsilon = (ct/2L)^2 \quad [17]$$

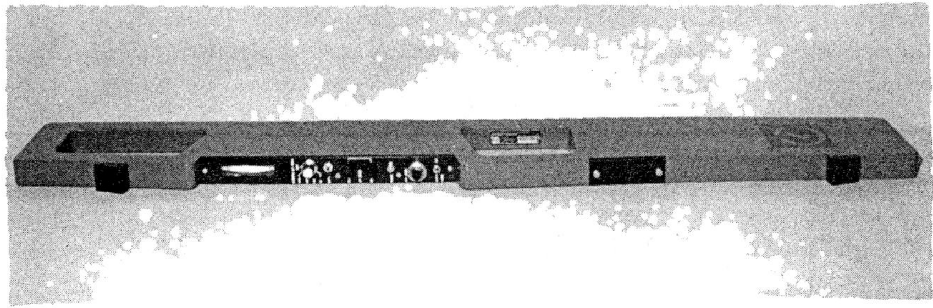

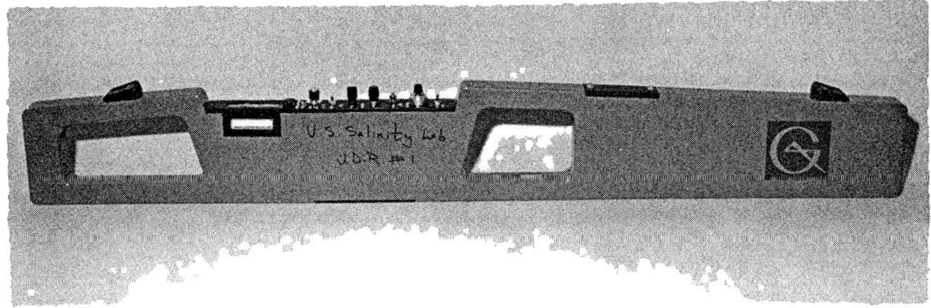

Fig. 12-7. Photograph of electromagnetic induction soil conductivity sensor.

where c is the velocity of light in vacuum. The signal is attenuated in proportion to EC_a so that the transmitted voltage, V_T, is reduced to V_R according to:

$$V_R = V_T \exp(-2\alpha L). \quad [18]$$

The attenuation coefficient, α, increases linearly with EC_a as:

$$\alpha = 60\pi EC_a/(\epsilon)^{1/2}. \quad [19]$$

Figure 12-8 shows a TDR-insertion sensor (homemade unit) and a commercially available TDR tester (Tektronix 7603 oscilloscope-7512 TDR sampler[2]).

Recent laboratory studies using this equipment have shown that EC_a and EC_w are well correlated with V_R/V_T (Dalton et al., 1984). The TDR method offers the distinct advantage of measuring both water content and soil electrical conductivity simultaneously; however, the practical attributes of this method have not yet been fully evaluated.

[2]Mention of trademark or proprietary products in this manuscript does not constitute a guarantee or warranty of the product by the USDA or by the Texas Agric. Exp. Stn., and does not imply its approval to the exclusion of other products that may also be suitable.

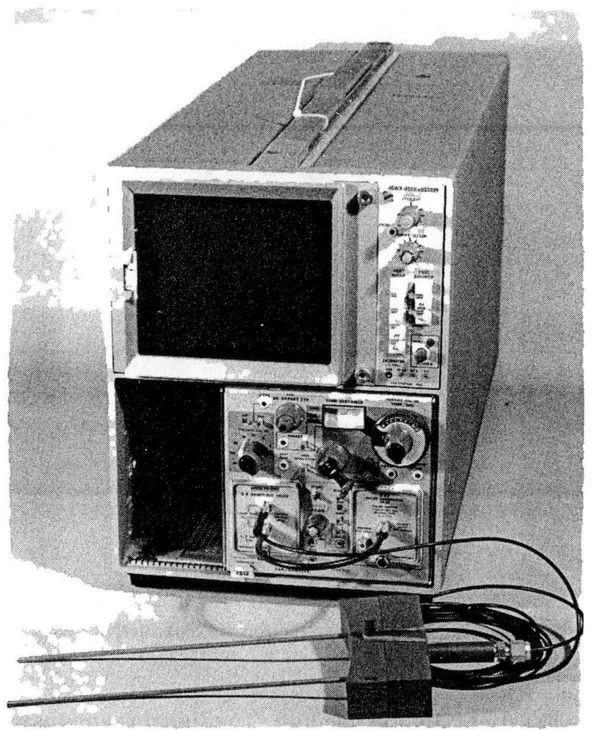

Fig. 12–8. Photograph of time domain reflectometry (TDR) probe and meter.

3. Procedures

a. Large Volume Measurements

For the purpose of determining soil salinity of entire root zones, or some fraction thereof, it is desirable to make the measurement when the current flow is concentrated within this soil depth. This is accomplished with the four-electrode equipment by selecting the appropriate spacing between the two current (outer) electrodes that are inserted into the soil surface to a depth of about 5 cm. In this arrangement, four electrodes are placed in a straight line. With conventional geophysical resistivity measurements the electrodes are equally spaced in the so-called *Wenner array* (Rhoades & Ingvalson, 1971). With the Martek SCT meter, each of the inner pair of electrodes is placed inward from its closest outer-pair counterpart, a distance equal to 10% of the spacing between the outer pair. The inner pair is used to measure the potential while current is passed between the outer pair. The effective depth of current penetration for either configuration (in the absence of appreciable soil layering) is equal to about one-third the outer-electrode spacing, y, and the average soil salinity is measured to approximately this depth (Rhoades & Ingvalson, 1971; Halvorson & Rhoades, 1976). Thus, by varying the spacing between current electrodes, one can measure average soil salinity to different depths and within different volumes of soil. Another advantage of this method is the relatively large volume of soil measured as compared with soil samples. The volume of measurement is about $(\pi y/3)^3$. Hence, effects of small-scale variations in field-soil salinity on sampling requirements can be minimized by these large-volume measurements.

For measurements taken in the Wenner array (electrodes equally spaced) using geophysical-type meters that measure resistance, the soil electrical conductivity is calculated, in dS/m, from:

$$EC_a = 159.2\, f_t/a\, R_t \qquad [20]$$

where a is the distance between the electrodes in cm, R_t is the measured resistance in ohms at the field temperature t, and f_t is a factor[3] to adjust the reading to a reference temperature of 25 °C. For measurements made with the Martek SCT meter, a factor is supplied in chart form for each spacing of outer electrodes; this factor is dialed into the meter and the correct soil EC_a reading is displayed in the meter readout.

Large volumes of soil can also be measured with the electromagnetic induction technique. The volume and depth of measurement can be increased by increasing the spacing between coils, reducing the current frequency, and positioning the coils so that their axis is vertical to the soil surface plane. The effective depths of measurement of the Geonics EM-38 device are about 1 and 2 m when it is placed on the ground and the coils are positioned horizontally and vertically, respectively. The EM-38 device does not integrate soil EC_a linearly with depth. The 0 to 0.30, 0.30 to 0.61, 0.61 to 0.91, and 0.91 to 1.22 m depth intervals contribute about 43, 21, 10, and 6%, respectively, to the EC_a reading of the EM unit when positioned on homogeneous ground in the horizontal position (Rhoades & Corwin, 1981). Thus, the weighted bulk soil electrical conductivity read by the EM device in this configuration (EM_{HO}) is given by:

$$EM_{HO} = 0.43EC_{0\text{-}0.3} + 0.21EC_{0.3\text{-}0.6} + 0.10EC_{0.6\text{-}0.9} + 0.06EC_{0.9\text{-}1.2} + 0.2EC_{>1.2} \qquad [21]$$

where the subscript designates the depth interval in meters.

It is desirable to determine soil EC_a by depth intervals for calculating soil salinity within various parts of the root zone as needed for making assessments and management decisions. Since the proportional contribution of each soil depth interval to EC_a, as measured by the EM unit, can be varied by raising it aboveground to higher heights, it is possible to calculate the EC_a depth relation from a succession of EM measurements made at various heights aboveground (Rhoades & Corwin, 1981). The EC_a values of each soil-depth interval are simply correlated with the succession of EM readings (0–4) as:

[3] $f_t = (0.0004)(T^2) - (0.043)(T) + 1.8149$; based on data given on p. 90 in U.S. Salinity Laboratory Staff (1954).

Table 12-1. Equations for predicting EC_a within different soil-depth increments from electromagnetic measurements made with the EM-38 device placed on the ground in the horizontal (EM_H) and vertical (EM_V) configurations.

Depth, cm	Equations for electrical conductivity†
	for $EM_H \leq EM_V$
0–30	$\hat{EC}_a = 3.023\ \hat{EM}_H - 1.982\ \hat{EM}_V$
0–60	$\hat{EC}_a = 2.757\ \hat{EM}_H - 1.539\ \hat{EM}_V - 0.097$
0–90	$\hat{EC}_a = 2.028\ \hat{EM}_H - 0.887\ \hat{EM}_V$
30–60	$\hat{EC}_a = 2.585\ \hat{EM}_H - 1.213\ \hat{EM}_V - 0.204$
60–90	$\hat{EC}_a = 0.958\ \hat{EM}_H + 0.323\ \hat{EM}_V - 0.142$
	for $EM_H > EM_V$
0–30	$\hat{EC}_a = 1.690\ \hat{EM}_H - 0.591\ \hat{EM}_V$
0–60	$\hat{EC}_a = 1.209\ \hat{EM}_H - 0.089$
0–90	$\hat{EC}_a = 1.107\ \hat{EM}_H$
30–60	$\hat{EC}_a = 0.554\ \hat{EM}_H + 0.595\ \hat{EM}_V$
60–90	$\hat{EC}_a = -0.126\ \hat{EM}_H + 1.283\ \hat{EM}_V - 0.097$

† $\hat{EC}_a$, $\hat{EM}_H$, and $\hat{EM}_V$ are the fourth roots of EC_a, EM_H, and EM_V.

$$EC_{a,\,0\text{-}0.3} = \beta_0 EM_0 \times \beta_1 EM_1 + \beta_2 EM_2 + \beta_3 EM_3 + \beta_4 EM_4 \quad [22a]$$

$$EC_{a,\,0.3\text{-}0.6} = \gamma_0 EM_0 + \gamma_1 EM_1 + \gamma_2 EM_2 + \gamma_3 EM_3 + \gamma_4 EM_4. \quad [22b]$$

The values of the coefficients of Eq. [22] reported by Rhoades and Corwin (1981) are generally applicable, though exceptions have been found.

Another series of equations and coefficients have been derived to obtain EC_a within soil-depth intervals from just two measurements made with the magnetic coils of the EM instrument positioned at ground level, first horizontally and then vertically (Corwin & Rhoades, 1982; Rhoades et al.,1989a). For the depth increment x1 to x2, the equations are of the form:

$$EC_{a,\,x1\text{-}x2} = k_H\ EM_H - k_V\ EM_V + k \quad [23]$$

where EM_V and EM_H are the apparent bulk soil electrical conductivities measured electromagnetically at the soil surface in the vertical and horizontal positions, respectively; x1–x2 is the soil-depth increment in centimeters and k_H, k_V, and k are empirically determined coefficients for the depth increment. Equation [23] is more easily solved than Eq. [22] and is almost as accurate. Values of the coefficients are given in Table 12–1, after Rhoades et al. (1989a).

b. Small Volume Measurements

Sometimes information on salinity distribution within a localized volume of the whole root zone is desired, such as that within the seed bed or under

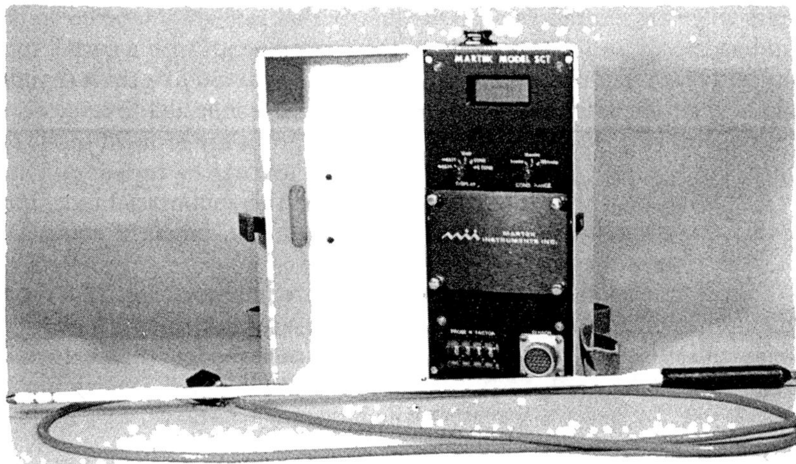

Fig. 12-9. Photograph of commercial seed-bed four electrode conductivity probe and generator-meter.

the furrows. For such conditions, the four-electrode salinity probe (Rhoades & van Schilfgaarde, 1976) and burial-type probe (Rhoades, 1979) are recommended. The seed bed probe (see Fig. 12-9) is designed to be directly inserted into the soil. In the larger probes (see Fig. 12-4 and 12-10), four

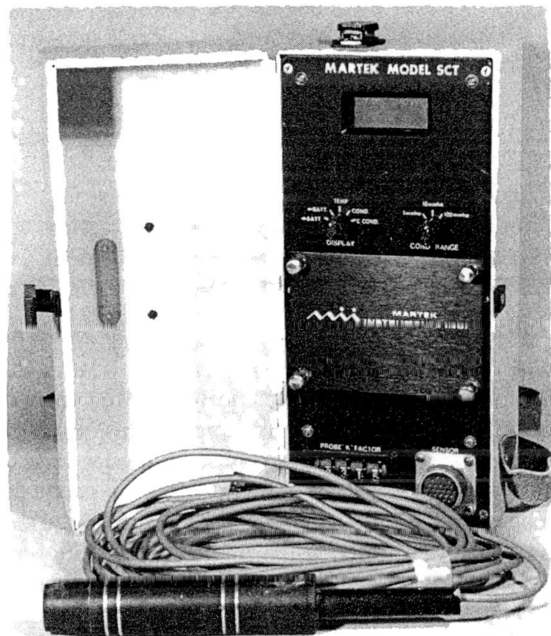

Fig. 12-10. Photograph of burial-type four electrode conductivity probe and generator-meter.

annular rings are molded in a plastic matrix that is slightly tapered so that it can be inserted into a hole made to the desired depth with a coring tube. In the portable version (Fig. 12–4), the probe is attached to a shaft (handle) through which the electrical leads are passed and connected to a meter. In the burial unit (Fig. 12–10), the leads from the probe are brought to the soil surface. The volume of sample under measurement can be varied by changing the spacing between the current electrodes. The commercial unit, Martek SCT[2], has a spacing of 6.6 cm and measures a soil volume of about 2350 cm^3.

To determine soil EC_a with the four-electrode probe (Fig. 12–4), core a hole in the soil to the desired depth of measurement using a Lord[2] soil sampling tube (or sampler of similar diameter). Insert the four-electrode probe into the soil and record the resistsance, or the displayed EC_a depending on the meter used. When using meters which display resistance, EC_a in dS/m is calculated as:

$$EC_a = k\, f_t/R_t \qquad [24]$$

where k is an empirically determined geometry constant (cell constant) for the probe in units of 1000 cm^{-1}, R_t is the resistance in ohms at the field temperature, and f_t is a factor to adjust the reading to a reference temperature of 25 °C (see footnote 3).

4. Calculations

The EC_w is calculated from the solution of Eq. [8] and [10] to [16] using the quadratic formula:

$$EC_w = (-b \pm \sqrt{b^2 - 4ac})/2a \qquad [25]$$

where $a = -[(\theta_s)(\theta_w - \theta_{ws})]$, $b = [(\theta_s EC_a) - (\theta_s + \theta_{ws})^2 (EC_s) - (\theta_w - \theta_{ws})(\theta_{ws} EC_s)]$, and $c = [(\theta_w)(EC_s)(EC_a)]$. Then EC_e can be solved from Eq. [19]. Alternatively obtain EC_e, given measurements of EC_a and reasonable estimates of %C and θ_w, using Fig. 12–11a–l.

5. Comments

Sensitivity analyses and tests have shown that the estimates used in this method are generally adequate for salinity appraisal purposes of typical mineral, arid-land soils of the southwestern USA (Rhoades et al., 1989a, 1990). For organic soils or soil of different mineralogy or magnetic properties, these estimates may be inappropriate. For such soils, appropriate estimating procedures will have to be developed using analogous techniques to those used by Rhoades et al. (1990). The accuracy requirements of these estimates may be evaluated using the relations given in Rhoades et al., 1989d).

As seen in Fig. 12–11a–l, water content (as well as salinity) affects soil electrical conductivity, and determinations are made preferably when the soil is near field capacity. However, measurements and salinity appraisals can

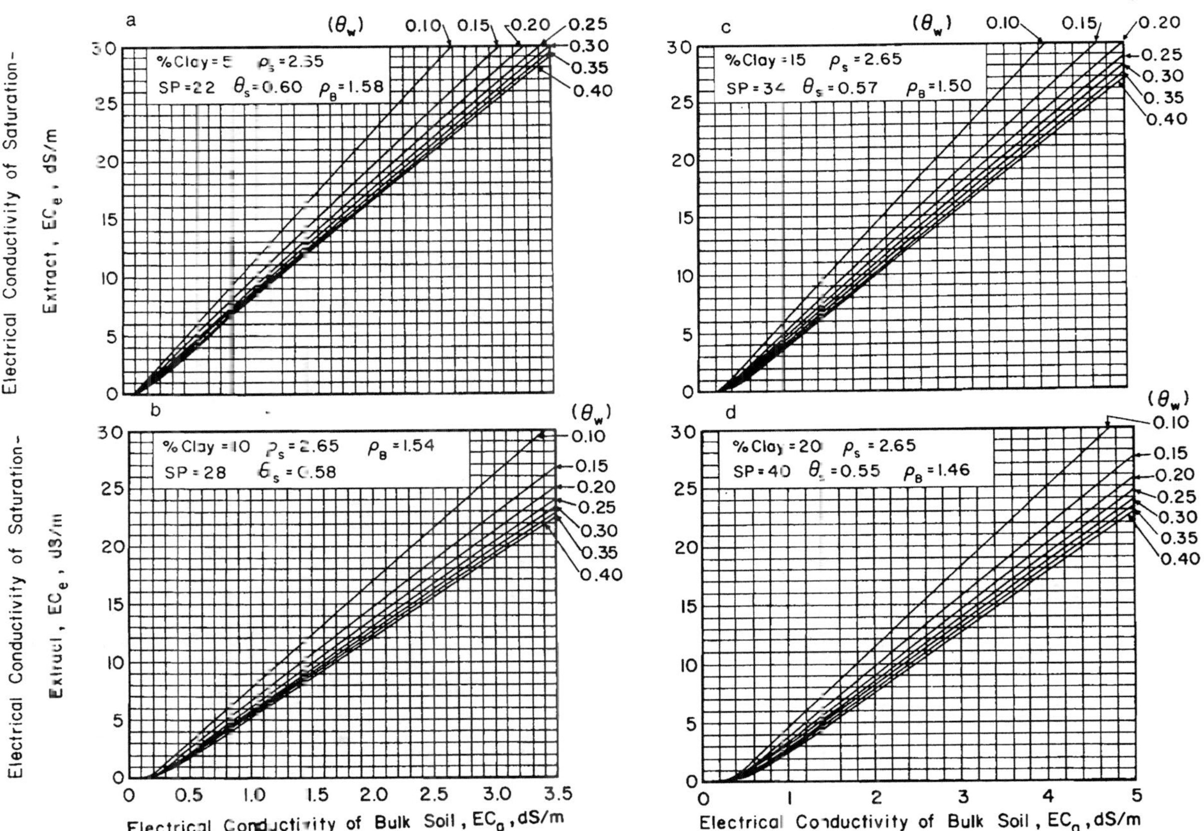

Fig. 12–11a–1. Relations between electrical conductivity of bulk soil (EC_a), electrical conductivity of saturation-extract (EC_e), soil volumetric water content (θ_w), and soil clay content (% clay), for representative arid-land soils.

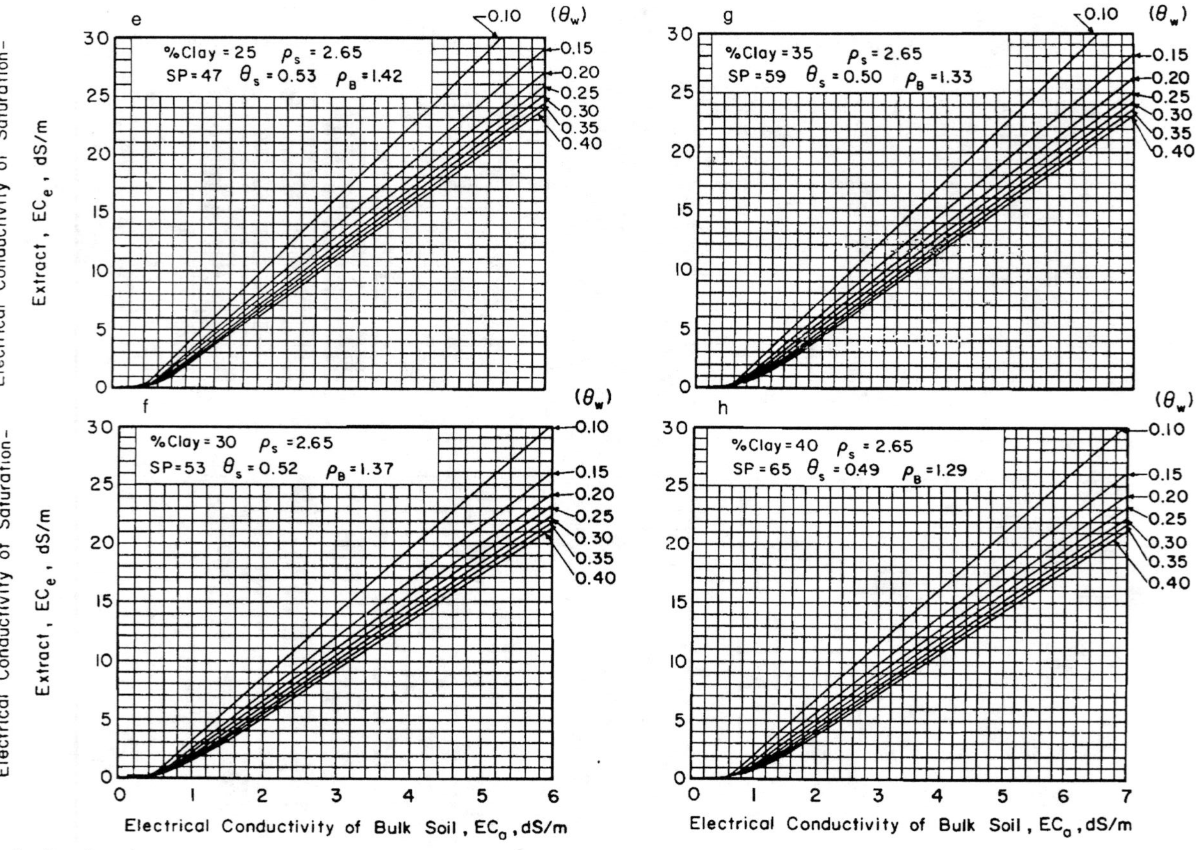

Fig. 12-11a-1. Continued.

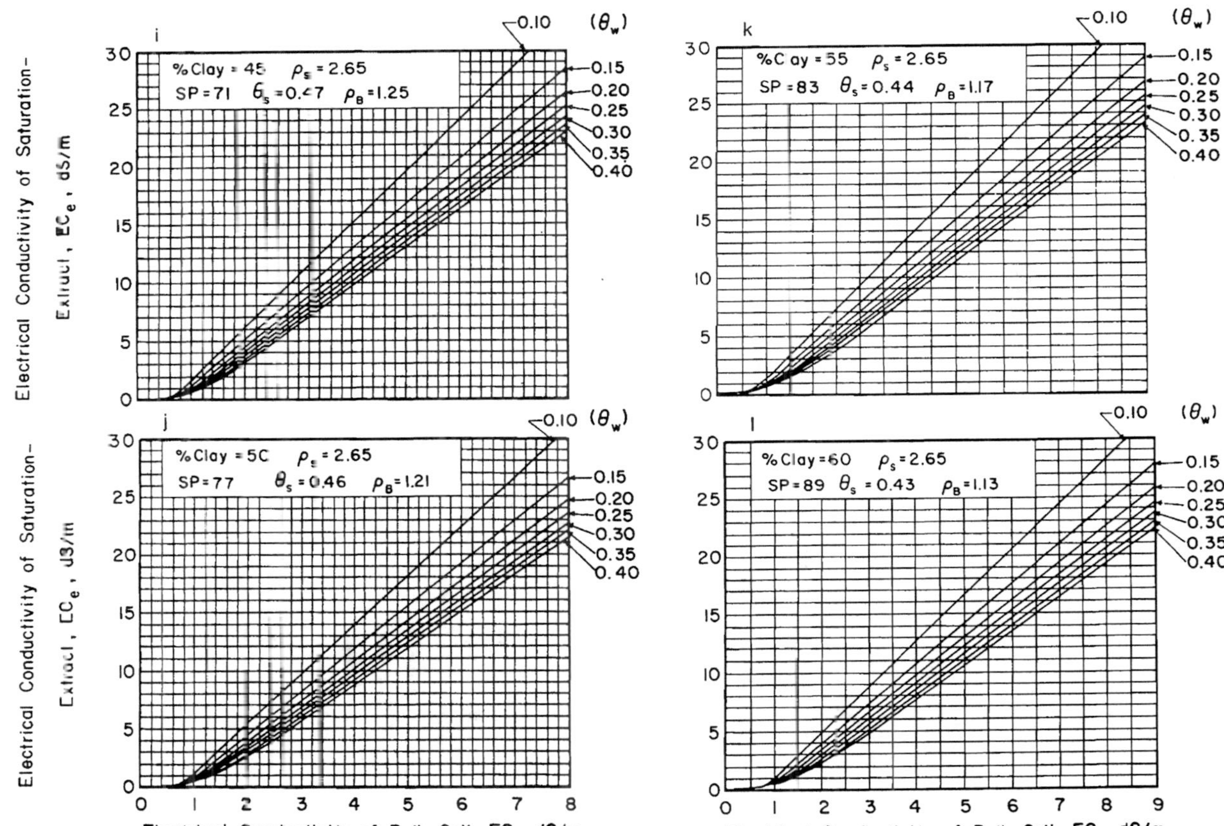

Fig. 12-11a-l. Continued.

be made at lower water contents as described above. However, a certain minimum water content is required in the soils for the measurements of EC_a and the model calculations to be valid; this water content is about 10% on a gravimetric basis, though it may be somewhat higher for very sandy soils.

The ratio SP/100 in Eq. [9] may be replaced by the ratio (θ_e/ρ_p), where ρ_p is the bulk density of the saturated paste and θ_e is the total volumetric content of water in the saturated paste. It should be noted that $(EC_e\ \theta_e)$ is not equivalent to $(EC_w\ \theta_w)$ because different amounts of soil are involved in the two measurements. The relation between these two products is given in Eq. [9]. θ_e is related to SP as follows:

$$\theta_e = SP/[(\rho_e\ 100/\rho_s) + SP] \qquad [26]$$

where ρ_e is the density of the saturation extract (~ 1.00 g/cm^3). ρ_p (soil dry weight basis) is related to SP as follows:

$$\rho_p = 100[(100/\rho_s) + (SP/\rho_e)]. \qquad [27]$$

These relations are described in more detail elsewhere (Rhoades et al., 1989b, 1990).

If devices are available to measure θ_w, or if other more appropriate values for any of the other estimated parameters are available, then, of course, they should be used in place of the estimates obtained by the methods given here. If more accurate measurements of EC_e or EC_w are required than can be obtained by the estimation procedures provided, quantitative measurements of θ_w, EC_s, and ρ_b, should be made using appropriate methods.

The depth-weighted value of EC_e, as calculated from the value of EM_{HO} measured by the EM-38 unit (see Eq. [21]), may be appropriate to use as a single index of soil salinity in some cases, as it roughly corresponds to the water-extraction behavior of plants. Irrigated crops tend to remove the soil approximately in the proportions 40:30:20:10 by successively deeper quarter-fractions of their root zone, which is about 1 m in depth for many crops, and to respond to water uptake-weighted salinity (Bernstein & Francois, 1973; Rhoades & Merrill, 1976).

III. DIAGNOSIS GUIDELINES

A. Salinity Hazard

Excess salinity (essentially independent of its composition) in the root zone adversely affects the growth of established plants by a general reduction in growth rate. Salt increases the energy that must be expended by the plant to extract water from the soil and to make the biochemical adjustments necessary to grow under stress (Maas, 1984, 1985). This energy is diverted from the processes that lead to normal growth and yield. Hence, the potential adverse effects of salinity on crop production should be assessed using information of salinity levels in the regions of the soil where the roots are actively extracting water. A problem is likely if the level of salinity in the vicinity of the active rootlets exceeds the tolerances of the crops for growth and yield.

To assess the likelihood of a salinity problem for conventionally irrigated and established crops, first estimate the depth-average salinity of the major root zone (i.e., where the majority of water extraction occurs). When high frequency or drip irrigation is used, estimate the water-uptake-weighted mean by using an appropriate soil water extraction pattern. Secondly, determine the mean and the extremes of salinity over the area of concern. If there are sufficient number of measurements (at least 10), compute the standard deviation. If no salt damage is permitted in any part of the area, the maximum salinity observed in the area should be taken as the soil salinity level for the field. If salt problems are permitted to occur in about 15% of the area, the sum of mean and standard deviation (SD) should be used as the salinity level. If the SD is small (e.g., <15% in coefficient of variability, CV), the sample mean may be used as the soil salinity level. Thirdly, compare the value of the soil salinity level obtained with that of the salt tolerance of the crop(s) to be grown. The salt tolerances of crops based on growth and yield are given in Table 12-2, after Maas and Hoffman (1977) and Maas (1985), in terms of their threshold values and percentage decreases in yield per unit increase of soil salinity above the threshold (where the unit of soil salinity is the electrical conductivity of the extract of a saturated soil paste in dS/m). Such salt-tolerance data cannot provide accurate, quantitative crop yield losses from salinity for every situation, since actual response to salinity varies with growing conditions. Such conditions include climate, irrigation method, agronomic management, and crop variety. Such salt-tolerance data are useful, however, to diagnosis the *likelihood* of salinity problems and predict how one crop might fare relative to another one under similar salinity conditions. For detailed information on the tolerance of specific crops to salinity refer to Francois and Maas (1978).

Improvement in diagnosis can be made by using computed average salinity of the soil solution per se rather than that of the saturation extract (see section II.C.4). Salinity of the saturation extract does not take into account the effect of soil water depletion on increasing salinity of the soil water between irrigations (Rhoades et al., 1981; Miyamoto et al., 1985a). With the use of high-frequency irrigation methods, the use of the saturation extract is even less appropriate as a measure of soil water salinity (Rhoades, 1982b). Soil solution-based salinity appraisal is also preferred for sandy soils, especially those irrigated with saline waters. The EC_e values of such soils are considerably less than those of medium-textured soils similarly treated (e.g., Longenecker, 1973). In fact, it is common to underrate the salt hazard of sandy soils irrigated with saline waters. The use of soil solution-based salinities necessitates the conversion of crop salt-tolerance data from EC_e to EC_{sw} (see Eq. [9]).

The salt-tolerance data of Table 12-2 apply most directly to flood-irrigated crops and typical irrigation management, they are less applicable

Table 12–2. Relative salt tolerances of agricultural crops. After Maas and Hoffman (1977).

Crop	Scientific name	Threshold salinity of saturation extract	Decrease in yield at soil salinities above the threshold
		dS/m	% (dS/m)
	Sensitive crops		
Bean	*Phaseolus vulgaris* L.	1.0	19
Carrot	*Daucus carota* L.	1.0	14
Strawberry	*Fragaria* sp.	1.0	33
Onion	*Allium cepa* L.	1.2	16
Almond	*Prunus dulcis* (Mill.)	1.5	19
Blackberry	*Rubus* sp.	1.5	22
Boysenberry	*Rubus ursinus*	1.5	22
Plum, prune	*Prunus domestica* L.	1.5	18
Apricot	*Prunus armeniaca* L.	1.6	24
Orange	*Citrus sinensis* (L.)	1.7	16
Peach	*Prunus persica*	1.7	21
Grapefruit	*Citrus paradisi* Macfad.	1.8	16
	Moderately sensitive crops		
Turnip	*Brassica rapa*	0.9	9.0
Radish	*Raphanus sativus* L.	1.2	13
Lettuce	*Lactuca sativa* L.	1.3	13
Clover, berseem	*T. alexandrinum* L.	1.5	5.7
Clover, strawberry	*T. fragiferum* L.	1.5	12
Clover, red	*T. pratense* L.	1.5	12
Clover, alsike	*Trifolium hybridum* L.	1.5	12
Clover, ladino	*Trifolium repens* L.	1.5	12
Foxtail, meadow	*Alopecurus pratensis* L.	1.5	9.6
Grape	*Vitaceae*	1.5	9.6
Orchardgrass	*Dactylis glomerata* L.	1.5	6.2
Pepper	*Capsicum annuum* L.	1.5	14
Sweet potato	*Ipomoea batatas* (L.)	1.5	11
Broadbean	*Vicia faba* L.	1.6	9.6
Corn	*Zea mays* L.	1.7	12
Flax	*Linum usitatissimum* L.	1.7	12
Potato	*Solanum tuberosum* L.	1.7	12
Sugarcane	*Saccharum officinarum* L.	1.7	5.9
Cabbage	*B. oleracea capitata*	1.8	9.7
Celery	*Apium graveolens* L.	1.8	6.2
Corn (forage)	*Zea mays* L.	1.8	7.4
Alfalfa	*Medicago sativa* L.	2.0	7.3
Spinach	*Spinacia oleracea* L.	2.0	7.6
Trefoil, big	*Lotus uliginosus*	2.3	19
Cowpea (forage)	*Vigna unguiculata*	2.5	11
Cucumber	*Cucumis sativus* L.	2.5	13
Tomato	*Lycopersicon Lycopersicum*	2.5	9.9
Broccoli	*Brassica oleracea botrytis*	2.8	9.2
Vetch, common	*Vicia angustifolia*	3.0	11
Rice, paddy	*Oryza sativa* L.	3.0	12
Squash, scallop	*Cucurbita pepo melopepo*	3.2	16
	Moderately tolerant crops		
Wildrye, beardless	*E. triticoides* Buckl.	2.7	6.0
Sudangrass	*Sorghum sudanense*	2.8	4.3
Wheatgrass, std. crested	*Agropyron desertorum*	3.5	4.0

(continued on next page)

Table 12-2. Continued.

Crop	Scientific name	Threshold salinity of saturation extract	Decrease in yield at soil salinities above the threshold
		dS/m	% (dS/m)
	Moderately sensitive crops		
Fescue, tall	*Festuca elatior*	3.9	5.3
Beet, red	*Beta vulgaris* L.	4.0	9.0
Hardinggrass	*Phalaris tuberosa*	4.6	7.6
Squash, zucchini	*C. pepo melopepo*	4.7	9.4
Cowpea	*Vigna unguiculata* (L.)	4.9	12
Soybean	*Glycine max* (L.)	5.0	20
Trefoil, birdsfoot	*Lotus corniculatus* L.	5.0	10
Ryegrass, perennial	*L. perenne* L.	5.6	7.6
Wheat, durum	*T. durum* Desf.	5.7	5.4
Barley (forage)	*Hordeum vulgare* L.	6.0	7.1
Wheat	*Triticum aestivum* L.	6.0	7.1
Sorghum	*Sorghum bicolor* (L.)	6.8	16
	Tolerant crops		
Date palm	*Phoenix dactylifera* L.	4.0	3.6
Bermudagrass	*Cynodon dactylon* (L.)	6.9	6.4
Sugarbeet	*Beta vulgaris* L.	7.0	5.9
Wheatgrass, fairway crested	*A. cristatum*	7.5	6.9
Wheatgrass, tall	*A. elongatum*	7.5	4.2
Cotton	*Gossypium hirsutum* L.	7.7	5.2
Barley	*Hordeum vulgare* L.	8.0	5.0

to high-frequency forms of irrigation, such as drip irrigation. Sprinkler-irrigated crops may suffer additional damage from foliar salt uptake and "burn" caused by contact with the spray. The available data-base for predicting yield losses from foliar spray effects is limited (Maas, 1984, 1985). The degree of foliar injury depends not only upon salinity of the irrigation water but also upon weather conditions, the size of sprinkler droplets, crop type, and growth stage. The tolerances of crops to foliar-induced salt damages does not usually coincide with that of root-induced damage. In the case of sprinkler irrigation, it is necessary to evaluate irrigation water salinity in addition to soil salinity.

Salinity also adversely affects crop establishment, especially when furrow-irrigation is used. In fact, obtaining a good crop stand is among the hardest tasks associated with crop production in saline areas. Once the crop is established, management risks are substantially reduced. The reason for the high risk at the seedling establishment stage is partly related to the lower salt tolerance of seedlings compared to established plants. However, the major reason for salt damage to seedlings is the exposure to pronounced salt accumulation occurring at or near the soil surface in the immediate vicinity of the seed or small plant. The salinity of the soil solution at the near-surface of furrow irrigated beds can easily reach that of sea water in a matter of several weeks (e.g., Miyamoto et al., 1985b). Salt concentrations in crop beds vary markedly with depth and time. Therefore, it is essential to have a clear specification of the depth and time of sampling when interpreting such salinity data.

Suppose that soil samples were collected at seedling depths (excluding the surface salt crust) immediately before or after time of seeding. The salinity readings can then be used to assess potential effects on seed germination. Since many of the published data of salt effects on seed germination use salinity of incubating solutions, salinity of the extract may have to be converted to that of the soil solution (Eq. [9]).

For many crops, stand problems begin after the seed has germinated. Hypocotyls emerging from the seed may have trouble passing through the soil layer above, which is often high in salts because of salt deposition occurring there during water evaporation. During this emergence process, hypocotyl mortality can occur, especially with crops sensitive to foliar salt damage. The levels of salinity that cause hypocotyl mortality vary widely. The levels are as low as 5 dS m^{-1} in EC_e of the top 5 mm of soil for guayule (*Parthenium argentatum* A. Gray), 10 dS m^{-1} for carrot (*Daucus carota* ssp. *sativus*) and more than 40 dS m^{-1} for tomato [*Lycopersicon lycopersicum* (L.)] (Miyamoto, 1986). One approach to handle this problem is to remove the surface crust of salts by mechanical means prior to seedling emergence (Miyamoto et al., 1986).

Emerged seedlings also undergo mortality when seedling roots are exposed to the highly saline zone typically present in the ridge region of furrow-irrigated beds or when substantial rain leaches the surface accumulated salts back into the seedling zone (Bernstein, 1974). Mortality of emerged seedlings also often occur when seedling leaves are exposed to saline splatters caused by light showers on salted soil surfaces (Miyamoto et al.,1986). However, rigid criteria to diagnose the extent of seedling mortality from those processes have not yet been fully developed.

Aside from the evaluation of salt damage to crops, salinity data obtained at different locations and depths in the field are useful for assessing irrigation efficiency and uniformity of water infiltration, and for estimating the leaching fraction (Rhoades, 1980).

B. Sodicity Hazard

The suitability of soils for cropping depends appreciably on their ability to conduct water and air (permeability) and on aggregate properties that control the friability of the seed bed (tilth). In contrast to saline soils, sodic soils have reduced permeabilities and poorer tilth. Sodicity adversely affects these physical properties of the soil and hence its suitability as a medium for crop growth. The direct effect of Na per se on crop growth is limited to a few crop species and is discussed later as a case of specific ion hazard (see section III.C).

The ESP has, in the past, been used to assess adverse Na effects. A classic criterion used for defining a sodic soil is an ESP value of ≥ 15 (U.S. Salinity Laboratory Staff, 1954). However, the presence of sufficient electrolyte can counteract the adverse effects of Na. Thus, the problems traditionally associated with sodicity should be assessed only in conjunction with a consideration of the accompanying level of salinity.

A short-range adhesive force (commonly referred to as van der Waals force) is primarily responsible for the flocculation of clay particles. The adsorption of Na molecules on clay surfaces enlarges the thickness of the diffuse-double layer existing around clay particles, thus increasing the repulsive force between adjacent particles of like charge. A consequence is dispersion of clay particles and deterioration of soil structure. With increasing ESP (or SAR) beyond about 15, Na enters the interlayer positions between the parallel platelets of smectitic clay particles and brings about swelling (Shainberg & Letey, 1984). Increasing electrolyte concentration reduces the thickness of the diffuse layer inside the shear plane by forcing more ions into it by mass action and counteracts the dispersive and swelling effects of loosely absorbed Na ions.

These underlying mechanisms largely dictate the effects of sodicity and salinity on soil physical properties of practical importance. Hydraulic conductivity, for instance, generally decreases with increasing sodicity or decreasing salinity. When ESP is $\lesssim 15$, the slaking of aggregates and dispersion of clay particles and accompanying loss of continuity of water conducting pore space are the principal causes of reduced infiltration rate and hydraulic conductivity. As ESP exceeds 15, the swelling factor comes into greater play in soils, especially in those containing smectitic clays. The magnitude of reduction in hydraulic conductivity depends on soil properties. When the electrolyte concentration of percolating solution is 10 $mmol_c/L$, ESP that causes a 15 to 25% reduction in hydraulic conductivity ranges from 5 to 25 in soils of mixed mineralogy (McNeal & Coleman, 1966a, b; Quirk & Schofield, 1955; Frenkel et al., 1978). Hydraulic conductivity of soils often decreases even with ESPs of 5 when electrolyte concentration is lower than about 3 to 5 $mmol_c/L$ (Shainberg, 1984). Under unsaturated flow conditions, sodicity causes lesser effects on hydraulic conductivity (e.g., Russo & Bresler, 1977). An approximate conversion between electrolyte concentration and EC was given earlier (see section I.C.4).

Water infiltration into soils is sensitive to ESP and electrolyte concentration as has been shown by the lysimeter data of Oster and Schroer (1979) and Miyamoto (1989). When water infiltrates into the soil surface, the soil solution of the topsoil is essentially that of the infiltrating water while the exchangeable Na percentage is essentially that pre-existent in the soil (since ESP is buffered against rapid change by the soil CEC). All water entering the soil must pass through the surface; hence, the stability of the topsoil aggregates influences the water entry rate of the soil. Representative threshold values of SAR ($\sim$ESP) and the electrical conductivity of infiltrating water for maintenance of soil permeability can be estimated from Fig. 12-12 (Rhoades, 1982b). Because significant differences exist in the Na-salinity response among soils, this relation should only be used as an approximate guideline. Reductions in water infiltration under rainfall or sprinkler and surface irrigation may be even more of a problem than is indicated in Fig. 12-12 because of enhanced particle dispersion and aggregate slaking caused

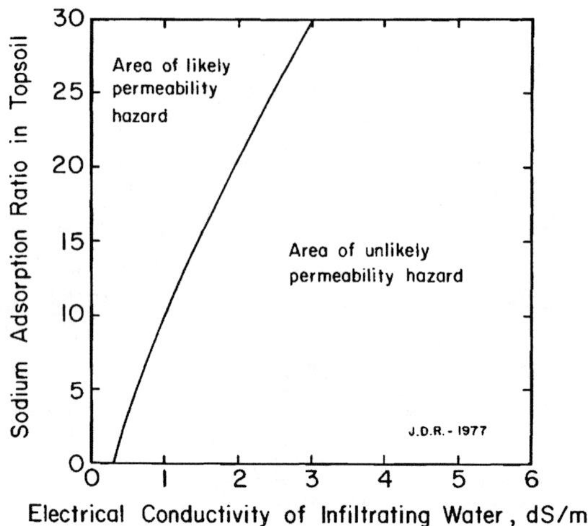

Fig. 12–12. Threshold value of sodium adsorption ratio (SAR) of topsoil and electrical conductivity of infiltrating water for maintenance of soil permeability.

by the energy of falling and flowing water (Oster & Schroer, 1979; Miyamoto, 1982; Shainberg, 1984). The dispersed particles form a "washed-in" layer of low permeability (Shainberg, 1984) and a crust through which seedlings have difficulty in penetrating during emergence.

Sodic soils can be improved with amendments such as gypsum, sulfuric acid, organic matter, and sulfur (Rhoades, 1982b; Stroehlein et al., 1978; Miyamoto et al., 1975; Stromberg & Tisdale, 1979). Generally, fast-acting amendments are expensive; slower ones are less so.

C. Specific Ion and Toxicity Effects

Certain salt constituents are specifically toxic to some crops. Boron is highly toxic to many crops when present in the soil solution at concentrations of only a few milligrams per liter (Maas, 1984; Keren et al., 1985; Bingham et al., 1985). In some woody crops, Na and chloride may accumulate in the tissue to toxic levels (Bernstein, 1974; Maas, 1984, 1985). These toxicity problems are not generally major ones. The effects of salinity and toxic solutes on the physiology and biochemistry of plants are discussed in more detail by Maas and Nieman (1978) and Maas (1984).

Sodic soil conditions may induce Ca and various micronutrient deficiencies by the associated high pH and bicarbonate conditions repressing their solubilities and thereby lowering their concentrations and, in the case of "heavy metals," due to precipitation of hydrous oxides.

IV. SUMMARY

Soluble salts in soils can be determined or estimated from measurements made (i) on aqueous extracts of soil samples, (ii) on saturated soil pastes, and (iii) on the field soil. The latter measurements can be made using four-electrode probes, TDR probes or electromagnetic (EM) sensors. The appropriate method of measuring soil salinity should be selected for the specific condition and purpose. If only a measure of total soluble electrolyte level in the soil is needed, EM or four-probe devices are recommended. When determination of a particular solute is needed, then collection and extraction of soil samples is required. Soil sample extracts give relative comparisons only, because the soils are adjusted to unnaturally high water contents during extraction. A combination of the various methods minimizes the need for sample collection and chemical analysis, especially when monitoring salinity changes with time and characterizing large feld or project situations. For this purpose, use of EM and four-probe field techniques are recommended with supplemental use of the other methods as needed. If soil samples are used in the field and only salinity is needed, the measurement of the saturated soil-paste is recommended.

The ESP of soils can be adequately estimated from the more rapid and simpler measurement of the SAR of the saturation extract. The sodicity hazard of a soil should be assessed from the combined values of estimated ESP, electrolyte concentration, and pH taking into consideration soil texture and clay mineralogy, along with the irrigation and tillage practices. The focus of attention should be the near-surface soil, in this regard.

While salinity and sodicity are the major factors limiting crop growth in most salt-affected soils, specific ion toxicities may be important in some cases. Special consideration should be given to Na and chloride, where woody, perennial crops are being grown and to B for susceptible crops.

REFERENCES

Allison, L.E. 1973. Oversaturation method for preparing saturation extracts for salinity appraisal. Soil Sci. 116:65–69.

Austin, R.S., and J.D. Rhoades. 1979. A compact, low-cost circuit for reading four-electrode salinity sensors. Soil Sci. Soc. Am. J. 43:808–810.

Ayers, R.S., and D.W. Westcot. 1985. Water quality for agriculture. FAO Irrigation and Drainage Paper 29. Revised. FAO, Rome.

Beatty, H.J., and J. Loveday. 1974. Soluble cations and anions. p. 108–117. *In* J. Loveday (ed.) Methods for analysis of irrigated soils. Tech. Commun. 54, Commonwealth Bureau of Soils, Commonwealth Agricultural Bureaus, Farmham Royal, Bucks, England.

Bernstein, L. 1974. Crop growth and salinity. p. 39–54. *In* Jan van Schilfgaarde (ed.) Drainage for agriculture. Agronomy Monogr. 17. ASA, Madison, WI.

Bernstein, L., and M. Fireman. 1957. Laboratory studies on salt distribution in furrow-irrigated soil with special reference to pre-emergence period. Soil Sci. 83:249–263.

Bernstein, L., and L.E. Francois. 1973. Leaching requirement studies: Sensitivity of alfalfa to salinity of irrigation and drainage waters. Soil Sci. Soc. Am. Proc. 37:931–943.

Bingham, F.T., J.D. Rhoades, and R. Keren. 1985. An application of the Maas-Hoffman salinity response model for boron toxicity. Soil Sci. Soc. Am. J. 49:672–674.

Bower, C.A., and L.V. Wilcox. 1965. Soluble salts. p. 933–951. *In* C.A. Black et al. (ed.) Methods of soil analysis. Agronomy Monogr. 9. ASA, Madison, WI.

Bresler, E., G. Dagan, R.J. Wagenet, and A. Laufer. 1984. Statistical analysis of soil salinity and texture effects on spatial variability of soil hydraulic conductivity. Soil Sci. Soc. Am. J. 48:16–25.

Carlson, R.M., R. Overstreet, and D.V. Naylor. 1971. Effect of microbial activity on saturation extract composition. Hilgardia 40:553–564.

Corwin, D.L., and J.D. Rhoades. 1982. An improved technique for determining soil electrical conductivity depth relations from above ground electromagnetic measurements. Soil Sci. Soc. Am. J. 46:517–520.

Dalton, F.N.,W.N. Herklerath, D.S. Rawlins, and J.D. Rhoades. 1984. Time domain reflectrometry: Simultaneous measurement of the soil water content and electrical conductivity with a single probe. Science 224:989–990.

Francois, L.E., and E.V. Maas. 1978. Plant response to salinity: An indexed bibliography. ARM-W-6. USDA, Washington, DC.

Frenkel, H., J.O. Goertzen, and J.D. Rhoades. 1978. Effects of clay type and content, exchangeable sodium percentage, and electrolyte concentration on clay dispersion and soil hydraulic conductivity. Soil Sci. Soc. Am. J. 42:32–39.

Hajrasuliha, S., N. Baniabbassi, J. Metthey, and D.R. Nielsen. 1980. Spatial variability of soil sampling for salinity studies in southwestern Iran. Irrig. Sci. 1:197–208.

Halvorson, A.D., and J.D. Rhoades. 1976. Field mapping soil conductivity to delineate dryland saline seeps with four-electrode technique. Soil Sci. Soc. Am. J. 40:571–575.

Hassan, H.M., A.W. Warrick, and A. Amoozegar-Fard. 1983. Sampling volume effects on determining salt in a soil profile. Soil Sci. Soc. Am. proc. 47:1265–1267.

Jurinak, J.J., C. Amrhein, and R.J. Wagenet. 1984. Sodic hazard: The effect of SAR and salinity in soils and overburden materials. Soil Sci. 137:152–159.

Kelley, W.P. 1922. Variability of alkali soil. Soil Sci. 14:177–189.

Keren, R., F.T. Bingham, and J.D. Rhoades. 1985. Plant uptake of boron as affected by boron distribution between the liquid and the solid phases in soil. Soil Sci. Soc. Am. J. 49(4):297–302.

Longenecker, D.E. 1973. The influence of soil salinity upon fruiting and shedding, boll characteristics, fiber quality and yields of two cotton species. Soil Sci. 115:94–302.

Longenecker, D.E., and P.J. Lyerly. 1964. Making soil pastes for salinity analysis: A reproducible capillary procedure. Soil Sci. 97:268–275.

Maas, E.V. 1984. Salt tolerance of plants. p. 57–75. *In* B.R. Christie (ed.) Handbook of plant science in agriculture. Vol. 2. CRC Press, Boca Raton, FL.

Maas, E.V. 1985. Salt tolerance in plants. Appl. Agric. Res. 1:12–26.

Maas, E.V., and G.J. Hoffman. 1977. Crop salt tolerance—Current assessment. J. Irrig. Drainage Div., ASCE 103(IR2):115–134.

Maas, E.V., and R.H. Nieman. 1978. Physiology of plant tolerance to salinity. p. 277–299. *In* G.A. Jung (ed.) Crop tolerance to suboptimal land conditions. ASA Spec. Publ. 32. ASA, CSSA, and SSSA, Madison, WI.

McNeal, B.L., and N.T. Coleman. 1966a. Effect of solution composition on soil hydraulic conductivity. Soil Sci. Soc. Am. Proc. 30:308–312.

McNeal, B.L., and N.T. Coleman. 1966b. Effect of solution composition on the swelling of extracted soil clays. Soil Sci. Soc. Am. Proc. 30:313–317.

McNeal, B.L., J.D. Oster, and J.T. Hatcher. 1970. Calculation of electrical conductivity from solution composition data as an aid to in-situ estimation of soil salinity. Soil Sci. 110:405–414.

Miyamoto, S. 1989. Cause and remedies of slow water infiltration. p. 27–37. *In* E. Herrera (ed.) Proc. Western Pecan Conf. 5–7 Mar., Las Cruces, NM. New Mexico State Univ., Las Cruces.

Miyamoto, S. 1988. Sampling irrigated soils for salinity appraisal. Texas Agric. Exp. Stn. Tech. Bull. 1570.

Miyamoto, S., and I. Cruz. 1986. Spatial variability and soil sampling for salinity and sodicity appraisal in surface-irrigated orchards. Soil Sci. Soc. Am. J. 50:1020–1026.

Miyamoto, S., and I. Cruz.1987. Spatial variability of soil salinity in furrow-irrigated Torrifluvents. Soil Sci. Soc. Am. J. 51:1019–1025.

Miyamoto, S., G.R. Gobran, and K. Piela. 1985a. Salt effects on seedling growth and in uptake of three pecan rootstock cultivars. Agron.J. 7:383–388.

Miyamoto, S., K. Piela, and J. Petticrew. 1985b. Salt effects on germination and seedling emergence of several vegetable crops and guayule. Irrig. Sci. 6:159–170.

Miyamoto, S., K. Piela, and J. Petticrew. 1986. Seedling mortality of several crops induced by root, stem or leaf exposure to soils. Irrig. Sci. 7:97–106.

Miyamoto, S., J. Ryan, and J.L. Stroehlein. 1975. Potentially beneficial uses of sulfuric acid in southwestern agriculture. J. Environ.Qual. 4:431–437.

Oster, J.D., and B.L. McNeal. 1971. Computation of soil solution composition variation with water content for desaturated soils. Soil Sci. Soc. Am. Proc. 35:436–442.

Oster, J.D., and F.W. Schroer. 1979. Infiltration as influenced by irrigation water quality. Soil Sci. Soc. Am. J. 43:444–447.

Oster, J.D., and G. Sposito. 1980. The Gapon coefficient and the exchangeable sodium percentage-sodium adsorption ration relation. Soil Sci. Soc. Am. J. 44:258–260.

Petersen, R.G., and L.D. Calvin. 1965. Sampling. p. 54–72. *In* C.A. Black et al. (ed.) Methods of soil analysis. Part 1. Agronomy Monogr. 9. ASA, Madison, WI.

Quirk, J.P., and R.K. Schofield. 1955. The effect of electrolyte concentration on soil permeability. J. Soil Sci. 6:163–178.

Rao, T.S., A.L. Page, and N.T. Coleman. 1968. The influence of ionic strength and ion-pair formation between alkali-earth metals and sulfate on Na-divalent cation exchange equilibria. Soil Sci. Soc. Am. Poc. 32:639–643.

Reitemeier, R.F. 1946. Effect of moisture content on the dissolved and exchangeable ions of soils of arid regions. Soil Sci. 61:195–214.

Rhoades, J.D. 1979. Inexpensive four-electrode probe for monitoring soil salinity. Soil Sci. Soc. Am. J. 43:817–818.

Rhoades, J.D. 1980. Determining leaching fraction from field measurements of soil electrical conductivity. Agric. Water Manage. 3:205–215.

Rhoades, J.D. 1982a. Soluble salts. p. 167–178. *In* A.L. Page et al. (ed.) Methods of soil analysis. Part 2. Agronomy Monogr. 9. 2nd ed. ASA and SSSA, Madison, WI.

Rhoades, J.D. 1982b. Reclamation and management of salt-affected soils after drainage. p. 123–197. *In* Proc. 1st Annu. Western Provincial Conf. Rationalization of Water and Soil Res. Manage., Lethbridge, AB. 29 Nov.–2 Dec. Alberta Agric., Lethbridge, AB.

Rhoades, J.D. 1984. Principles and methods of monitoring soil salinity. p. 130–142. *In* Soil salinity and irrigation—processes and management. Vol. 5. Springer Verlag New York, New York.

Rhoades, J.D., and D.L. Corwin. 1981. Determining soil electrical conductivity—depth relations using an inductive electromagnetic soil conductivity meter. Soil Sci. Soc. Am. J. 45:255–260.

Rhoades, J.D., D.L. Corwin, and G.J. Hoffman. 1981. Scheduling and controlling irrigations from measurements of soil electrical conductivity. p. 106–115. *In* Proc., ASAE, Irrigation Scheduling Conf., Chicago. 14 Dec. ASAE, St. Joseph, MI.

Rhoades, J.D., D.L. Corwin, and P.J. Shouse. 1988. Use of instrumental and computer assisted techniques to assess soil salinity. p. 50–103. *In* Symp. Proc. of Int. Symp. in Solonetz Soils, Osijek, Yugoslavia. Agricultural Faculty, Univ. of Osijek, Osijek, Yugoslavia.

Rhoades, J.D., and A.D. Halvorson. 1977. Electrical conductivity methods for detecting and delineating saline seeps and measuring salinity in Northern Great Plains soils. ARS W-42. U.S. Gov. Printing Office, Washington, DC.

Rhoades, J.D., and R.D. Ingvalson. 1971. Determining salinity in field soils with soil resistance measurements. Soil Sci. Soc. Am. Proc. 35:54–60.

Rhoades, J.D., S.M. Lesch, P.J. Shouse, and W.J. Alves. 1989a. New calibrations for determining soil electrical conductivity—Depth relations from electromagnetic measurements. Soil Sci. Soc. Am. J. 53(1):74–79.

Rhoades, J.D., N.A. Manteghi, P.J. Shouse, and W.J. Alves. 1989b. Estimating soil salinity from saturated soil-paste electrical conductivity. Soil Sci. Soc. Am. J. 53:428–433.

Rhoades, J.D., N.A. Manteghi, P.J. Shouse, and W.J. Alves. 1989c. Soil electrical conductivity and soil salinity: New formulations and calibrations. Soil Sci. Soc. Am. J. 53:433–439.

Rhoades, J.D., and S.D. Merrill. 1976. Assessing the suitability for irrigation: theoretical and empirical approaches. FAO Soil Bull. 31:69–109.

Rhoades, J.D., and J.D. Oster. 1986. Solute content. p. 985–1006. *In* A. Klute (ed.) Methods of soil analysis. Part 1. 2nd ed. Agronomy Monogr. 9. ASA and SSSA, Madison, WI.

Rhoades, J.D., P.J. Shouse, W.J. Alves, N.A. Manteghi, and S.M. Lesch. 1990. Determining soil salinity from soil electrical conductivity using different models and estimates. Soil Sci. Soc. Am. J. 54:46–54.

Rhoades, J.D., and J. van Schilfgaarde. 1976. An electrical conductivity probe for determining soil salinity. Soil Sci. Soc. Am. J. 40:647–651.

Rhoades, J.D., B.L. Waggoner, P.J. Shouse, and W.J. Alves. 1989d. Determining soil salinity from soil and soil-paste electrical conductivities: Sensitivity analysis of models. Soil Sci. Soc. Am. J. 53:1368–1374.

Richards, L.A. 1949. Filter funnels for soil extracts. Agron. J. 41:466.

Russo, D., and E. Bresler. 1977. Effect of mixed Na/Ca solution on the hydraulic properties of unsaturated soils. Soil Sci. Soc. Am. J. 41:713–717.

Sayegh, A.H., L.A. Alban, and R.G. Petersen. 1958. A sampling study in a saline and alkali area. Soil Sci. Soc. Am. Proc. 22:252–254.

Shainberg, I. 1984. The effect of electrolyte concentration on the hydraulic properties of sodic soils. p. 49–64. *In* I. Shainberg and J. Shalhevet (ed.) Soil salinity under irrigation. Springer-Verlag New York, New York.

Shainberg, I., and J. Letey. 1984. Response of soils to sodic and saline conditions. Hilgardia 52:1–57.

Soil Survey Staff. 1951. Soil survey manual. USDA Handb. 18. U.S. Gov. Print. Office, Washington, DC.

Sonnevelt, C., and J. van den Ende. 1971. Soil analysis by means of a 1:2 volume extract. Plant Soil 35:505–516.

Sposito, G., and S.V. Mattigod. 1979. On the chemical foundation of the sodium adsorption ratio. Soil Sci. Soc. Am. J. 41:323–329.

Stroehlein, J.L., S. Miyamoto, and J. Ryan. 1978. Sulfuric acid for improving irrigation waters and reclaiming sodic soils. Univ. of Arizona, Tucson. Agric. Eng. Soil Sci. Bull. 78-5.

Stromberg, L.K., and S.L. Tisdale. 1979. Treating irrigated arid-land soils with acid forming sulfur compounds. Tech. Bull. 24. The Sulfur Inst., Washington, DC.

Thomas, G.W. 1982. Exchangeable cations. p. 159–165. *In* A.L. Page et al. (ed.) Methods of soil analysis. Part 2. 2nd ed. Agronomy Monogr. 9. ASA and SSSA, Madison, WI.

U.S. Salinity Laboratory Staff. 1954. Diagnosis and improvement of saline and alkali soils. USDA Handb. 60. U.S. Gov. Print. Office, Washington, DC.

Wagenet, R.J., and J.J. Jurinak. 1978. Spatial variability of soluble salt content in a Mancos shale watershed. Soil Sci. 126:342–349.

Wilcox, J.V. 1951. A method for calculating the saturation percentage from the weight of a known volume of saturated soil paste. Soil Sci. 72:233–237.

Yaron, B., D. Shimski, and J. Shalhevet. 1973. Patterns of salt distribution under trickle irrigation. *In* Physical aspects of soil water and salt in ecosystems. Ecol. Stud. 4:389–394.

Chapter 13

Testing Artificial Growth Media and Interpreting the Results

DARRYL D. WARNCKE, *Michigan State University, East Lansing*

Testing containerized growth media provides many challenges and opportunities. Since the mid-1960s, growth media used for producing floral, vegetable, and ornamental plants in containers have changed from soil-based mixes to peat- and bark-based mixes. These changes in growth media composition have necessitated changes in testing procedures from those developed for soil to ones more suited for mixes of peat, bark, sand, or vermiculite. In the USA, a gradual change has occurred from the soil-oriented Spurway procedure to water-extraction methods. Use of the saturated media extract has increased among university and commercial testing laboratories. In the European countries, water extract methods have been in use for a longer period as have artificial growth media.

The many artificial growth media being used have been developed for improvement of drainage and aeration with nutrient-supplying abilities being considered second (Boodley & Sheldrake, 1972; Lucas & Rieke, 1968; Matkin et al., 1957; Whitcomb, 1984). Even though the peat, composted bark and vermiculite have nutrient-holding capabilities the nutrients are held less tightly than by mineral soils. Table 13-1 shows that a high percentage of N, P, and K added to two peat-based growth media remained recoverable by water extraction. The presence of expanded vermiculite in the medium did reduce the water-extractable K and NH_4 (Bunt, 1988). By being able to extract with 0.03 *M* acetic acid, 100% of the N, P, and K added to a sphagnum peat, Nowosielski and Beresniewic (1957) demonstrated that the nutrients not recoverable by water extraction are not bond tightly by the substrates. All P and K added to a shredded pine bark was also 100% recoverable. Nitrogen recovery decreased to 30% after 12 wk of composting due to immobilization but increased to more than 100% after 6 mo of composting.

In soils, the capacity factor is extremely important for the continual supply of nutrients to plant roots. However, the importance of the intensity factor overshadows the capacity factor when plants are intensively grown in soilless media with frequent watering; periodically with a nutrient solution. As the content of soil in a growth media decreases, the importance of

Table 13-1. Percentage of added nutrients recovered in 1 L of leachate from peat-sand (75–25%) and peat-vermiculite (50–50%) composts. (Bunt, 1988).†

Element	Peat-sand	Peat-vermiculite
NH_4-N	81	33
NO_3-N	87	75
P	60	43
K	70	45

† Pot capacity was 1 L.

the capacity factor decreases while that of the intensity factor increases. The relative roles of the capacity and intensity factors are, therefore, an important consideration in developing an appropriate testing methodology for artificial growth media.

I. HANDLING AND PREPARING GROWTH MEDIA FOR TESTING

The prime objective in testing a growth medium, whether soil or soilless, is to characterize the nutrient environment as closely as possible to the way the plant root system experiences it. Improper handling of a growth medium sample can greatly increase the difficulty of attaining this goal. Field soil samples are usually dried, ground, and sieved prior to testing, but it is recognized that drying, especially at temperatures >38 °C, may change extractable nutrient levels.

Grinding of samples usually helps facilitate a uniform sample for testing. However, some components of artificial media, such as expanded vermiculite, polystyrene beads, and calcined clay, do not grind well. The grinding and sieving process will give a uniform sample, but may also systematically alter both the chemical and physical properties of the original medium that could result in testing a material whose composition is different from the medium in which plants are grown. Grinding a medium that contains slow-release fertilizers generally increases the test levels. For these reasons, grinding of soilless growth media is not recommended. Sieving is also questionable as it may result in separation of constituent materials and the testing of a medium different from that used. Recognizing that artificial growth media are not homogenous, the best approach to collecting and testing a representative sample is to use as large a sample as practical.

Should the sample to be tested be measured by weight or by volume? Mehlich (1972, 1973) has presented a good discussion of this matter for field soils. Prime consideration must be given to the fact that plants grow in a volume of growth medium. The bulk density of growth media vary considerably, ranging from <0.2 to >0.8 Mg m^{-3}. Cation-exchange capacity (CEC) values of low-density materials such as peat, composted barks, and vermiculite appear quite high when expressed on a weight basis. However, when expressed on a volume basis the CEC values are actually similar to or less than those of mineral soils (Brown & Pokorny, 1975; Bunt, 1988; Lucas, 1983; Whitcomb, 1984). For example, on a weight basis the CEC of sphagnum peat, pine bark, and a loam soil are 1000, 440, and 120 mmol kg^{-1}, whereas on a volume basis the respective CEC values are 80, 110, and 150 mol m^{-3}. Since many testing methods specify a given sample/solution ratio, measuring the sample by weight or volume will influence the test result and subsequent interpretation. Consider two growth media, A and B. Medium A has a density of 0.4 Mg m^{-3} and medium B has a density of 0.8 Mg m^{-3}. Ten millimole of nitrate-N (NO_3-N) is mixed with 1 L of each medium. If each medium is extracted on a volume/volume (v/v) basis with a sample/solution ratio of 1:5, the extract from each will contain 2 mmol NO_3-N L^{-1}. However, when extracted on a volume/weight (v/w) basis at a ratio of 1:5, the extract of medium A will contain 5 mmol NO_3-N L^{-1} whereas medium B will contain 2.5 mmol NO_3-N L^{-1}. Hence, samples measured on a weight basis will necessitate conversion of the results to a volume basis or require separate sets of interpretation guidelines based on medium density. In considering these factors, measuring and testing of growth media should be conducted on a volume basis.

Handling a container growth medium in a moist state provides the opportunity to test the medium in a condition similar to how it will be or was used. However, some changes may occur if the sample is maintained in the moist state for several days while being shipped to a lab for testing. Most concern is for N conversions that may occur, resulting in more or less available N. Furthermore, the distribution of NH_4 and NO_3 forms may change. The occurrence of more NO_3 would increase the soluble salt level by bringing more cations into solution. However, Markus and Steckel (1980) found the NO_3 and NH_4 content did not change in moïst media held in airtight containers at 7 or 22°C for 3 and 6 d. Hence, the concern for N conversions may not be as serious as commonly thought. If a growth medium contains slow-release fertilizer, maintaining it in a moist state for several days may result in some inflation of the test results due to continued nutrient release. But drying and rewetting these media may actually result in greater nutrient release from the slow-release fertilizers. And complete drying of growth media may alter the ionic equilibrium (Jackson, 1958). Some soluble nutrients may precipitate in forms not readily solubilized when rewetted during extraction. Peat-based media rewet slowly after being dried and some researchers have found use of a surfactant helpful in rewetting them (White et al., 1975). Even then complete wetting may not occur during the extraction times used in many testing labs. Markus (1986) extracted higher amounts of P, K, Ca, and Mg from moist samples compared with samples that were air dried. Hence, maintaining media samples moist prior to testing appears to be best.

II. ANALYTICAL APPROACHES

With frequent watering and fertilization of container-grown plants, the importance of exchangeable cations and immobilized nutrients in the growth media is diminished. Analytical procedures must reflect the nutrient environment of the root system. The concentration (intensity) and balance of the nutrients in the solution phase is important in these weakly buffered systems. While an analytical method must provide accuracy and reproducibility, Bunt (1986) has pointed out that for use in a service lab a method should also: (i) allow for handling several samples per day; (ii) incorporate simplicity to be independent of individual operator skill; and (iii) be suitable for analyzing a diverse range of materials. Furthermore, for extension purposes the analytical procedure must provide results related to plant growth responses and be backed up with interpretation and recommendation guidelines.

Slow-release fertilizers are frequently used to help maintain adequate nutrient concentrations in artificial growth media and this provides an additional challenge for analysis. Ideally, the procedure will be able to determine the nutrient status of a medium containing slow-release fertilizer without stimulating additional nutrient release.

Several different analytical methods are being used for analysis of artificial growth media used for container-grown plants. Bunt (1986) classified these into three categories: (i) suspensions extracted with water, acids, or salts; (ii) saturated media extracts; and (iii) displaced soil (media) solutions.

A. Acid Extracts

1. Suspensions

Since 1940, perhaps the most popular system for analyzing greenhouse soils and soil mixes in the USA has been the method developed by C.H. Spurway during the late 1930s and early 1940s (Spurway, 1943; Spurway & Lawton, 1949). The Spurway method originally used 13 mL of 0.018 M acetic acid to extract plant-available nutrients from 4 cm^3 of mineral soil. Later, some labs modified the method by: (i) using a different medium/solution ratio; (ii) weighing out the sample; and (iii) using other extractants for N and the cations. Surveys of soil testing labs in the mid-1970s indicated the modified Spurway system was being used widely for analysis of both soil-based and soilless growth media. The Spurway system has the advantage of considerable background calibration data and interpretation guidelines for a wide spectrum of plant species. Hence, extension horticulturists have felt comfortable with data generated via this methodology. Several samples can be handled quickly, but, on the other hand, the medium is dried prior to being analyzed and, in some labs, several extractions are used for a complete analysis. The small sample size makes representative sampling of heterogenous growth media difficult. When the Spurway method is used to extract media containing slow-release fertilizer, the test values are inflated rather than reflecting the current available nutrient status. Several other acid and salt extractions have been evaluated by Markus and Steckel (1980). The Mehlich I (double acid) extracted the largest quantity of nutrients, but no one extractant was superior to the others as a diagnostic tool. Several of the extractants correlated well with each other in the extraction of various nutrients, but correlation with nutrient uptake by plants was not good for any of the extractants studied. Sartain (1983) found that Mehlich I analyses were less variable than other methods, but again nutrient uptake by tomato (*Lycopersicon esculentum* Mill.) did not correlate with nutrient levels extracted by Mehlich I.

B. Water Extracts

1. Suspensions

Water extraction is used in western Europe where soilless potting media are used predominantly (Bunt, 1986). Saturated media extracts and displaced solutions are used in research, but water suspension extracts are used by service laboratories. Johnson (1980) found water extracts at various medium/solution ratios, ranging from 1:1.5 to 1:6, gave more consistent results than did chemical extractants. Based on Johnson's studies, the Agricultural Development and Advisory Service (ADAS) of the United Kingdom routinely analyzes all potting media using a 1:6 (v/v) medium to pure water ratio.

The development of water extract procedures has faced two main concerns: (i) whether to use moist or dry samples, and (ii) whether to prepare extracts on a weight (w/v) or volume (v/v) basis. Some concerns regarding air-drying samples prior to extraction are presented in section I. Extraction of moist samples presents the problem of variation in the initial moisture content. Sonneveld and van den Ende (1971) adjusted all mineral soil samples to field capacity before making a one part soil to two parts water (v/v) extract. Conductivity and soluble nutrient levels obtained with this approach agreed much better with saturation extract data than data obtained with 1 part soil to 25 parts water (w/v). The variability in moisture-holding capacity is greater among artificial growth media than among mineral soils. Hence, making suspension extracts without regard for the initial moisture content would give variable results. Johnson (1980) found that by using a wide medium to water ratio, i.e., 1:6, this problem could be overcome. However, as the extraction ratio increases the resulting conductivity values and nutrient levels become less agreeable with saturation extracts (Sonneveld et al., 1974). Dutch researchers have shown variability in the moisture content of growth media is minimized by adjusting the moisture tension to pF 1.5 (about 32 cm of water). They reported this adjustment can be made visually with experience, but periodic checks on a sand box are desirable. Another source of variability with volume extracts is the quantity packed into a unit volume. In preparation for their 1:1.5 (v/v) extracts, Sonneveld et al. (1974) and Johnson (1980) used a constant pressure of 0.1 kg cm^{-2}. Thus, by standardizing the moisture tension and the pressure used in measuring a specified volume, their analytical results correlated very well with those in the medium solution extracted with a hydraulic press; $r^2 = 0.96$ or better (Table 13-2). Nutrient concentrations in the 1:1.5 extract were 20 to 33% the concentration in the press extract.

Table 13-2. Regression equations for relationships between analytical data of press extract (x) and 1:1.5 volume extract (y) Sonneveld et al. (1974).

Determination	Regression equation	*r*
Conductivity	y = 0.311 x − 0.05	0.969
Cl	y = 0.218 x + 0.17	0.975
N	y = 0.256 x + 0.02	0.967
NO_3	y = 0.239 x + 0.18	0.961
NH_4	y = 0.280 x + 0.17	0.986
PO_4	y = 0.270 x + 4.04	0.972
K	y = 0.302 x + 0.17	0.982
Mg	y = 0.298 x − 0.75	0.957

2. Saturated Media Extract

Under conditions of equilibrium, the composition of the solution phase of the growth medium fully characterizes the root environment. Lagerwerff (1958) documented this statement by demonstrating that nutrient uptake was similar from the solution and adsorbed phases. In weakly buffered growth media, the concentration of nutrients in solution will be of greater importance than in well-buffered mineral soils. The saturation extract method was first developed for determining total soluble salt levels in soils (Richards, 1954), but is useful in determining specific soluble nutrient concentrations in weakly buffered artificial growth media. Geraldson (1957, 1967) found that saturation extracts reflected well the available nutrient status of very weakly buffered coarse sands in Florida. Following Geraldson's lead, Lucas et al. (1972) studied the feasibility of using a saturated medium extract (SME) procedure for routine analysis of potting media. Results obtained with the SME method agreed well with those obtained with the modified Spurway system (Table 13-3).

The SME method used by the Michigan State University Soil Testing Laboratory is described in North Central Regional Bull. 221 (Warncke, 1988). Basically, 400 cm^3 of growth medium (just as it comes to the lab) is mixed with additions of pure water until it is just saturated. After equilibrating for 1.5 h, pH is determined in the saturated medium and the solution is extract-

Table 13-3. Correlation of test variables determined by the SME method and modified Spurway extraction procedure.†

Test	*r* Value	Extractant
Soluble salts	0.99	0.018 *M* HOAc
NO_3-N	0.97	0.018 *M* HOAc
P	0.90	1.0 *M* NH_4OAc
K	0.90	1.0 *M* NH_4OAc
Ca	0.57	1.0 *M* NH_4OAc
Mg	0.59	1.0 *M* NH_4OAc
Na	0.35	1.0 *M* NH_4OAc
Zn	0.94	0.1 *M* HCl
Mn	0.98	0.1 *M* HCl
Cu	0.70	1.0 *M* HCl

† Samples of growth media were collected from several greenhouses in Michigan. Original paper by Lucas et al. (1972) does not indicate sample numbers or replications.

Table 13–4. Influence of increasing water volume on the soluble salt and nutrient levels in the saturation extract.

Water volume†	Soluble salts	NH_4-N	P	K	Ca	Mg
L	relative value‡					
0.175	1.18	1.29	1.13	1.14	1.34	1.18
0.200	1.15	1.22	1.04	1.07	1.25	1.16
0.225	1.00	1.00	1.00	1.00	1.00	1.00
0.250	0.97	0.87	0.87	0.93	0.94	0.93
0.300	0.79	0.71	0.65	0.79	0.81	0.80

† Volume of water mixed with 0.40 L of Redi-earth. Saturation was attained near 0.225 L of water.

‡ Actual test values equal to 1.00 were: 1.70 dS m^{-1}; 7 mmol NO_3-N L^{-1} 0.23 mmol P L^{-1}; 0.36 mmol K L^{-1}; 1.7 mmol Ca L^{-1}; and 4.87 mmol Mg L^{-1}.

ed with a vacuum filter, and all subsequent analyses are conducted on the filtrate. This approach overcomes several problems associated with analyzing artificial growth media: (i) large samples can be analyzed without preparatory drying and handling which minimizes the concern for segregation or heterogeneity of components; (ii) since only gentle mixing is involved, medium containing slow-release fertilizer can be analyzed with only minimal and insignificant effects on test values (Warncke, 1986); (iii) only one set of interpretation guidelines is needed; and (iv) calculation of nutrient balance is possible.

Results obtained by the SME method have been more variable than with other methods (Holcomb & White, 1979, Sartain, 1983; Warncke, 1983). Much of the variability is associated with the difficulty of accurately mixing the medium to the point of saturation. With field soil, the criteria for determining the endpoint are relatively clear: (i) the soil just beings to flow; (ii) the soil surface glistens; and (iii) the soil slides cleanly off a spatula. Many potting media will also mix easily and flow slightly at the saturation endpoint. As the content of fibrous peat, coarse bark, polystyrene beads, and similar materials increases, however, the difficulty of accurately determining the endpoint increases. Table 13–4 illustrates how missing the saturation endpoint affects the relative nutrient concentrations in the extract. Deviation of the relative nutrient values from those obtained at saturation was less by adding excess water than by undersaturating the medium, except for P. When expressed as a percentage of the total soluble salts, Warncke (1986) has shown variability of test results is much less than when absolute nutrient concentrations are considered.

Michigan State University started using the SME method for routine analysis of artificial growth media in 1974. Since then, many service laboratories across the USA and Canada have adopted the method. Admittedly, the SME method is time consuming and does not lend itself to doing 40 000 samples per year, as is done by some advisory centers in western Europe. However, some labs in the USA are using the SME method and handling 3000 to 4000 samples per year.

Table 13–5. Summary of nutrient levels in saturation extracts with various extractants.†

Nutrient	Extractant				
	Water	1*M* NH_4OAc	Bray 1	0.02 *M* HOAc	0.1 *M* HOAc
	mmol L^{-1}				
P	0.74	0.64	1.61	0.84	1.13
K	2.63	4.40	--	2.61	2.86
Ca	5.55	20.27	--	8.10	13.12
Mg	4.08	14.29	--	5.91	8.67
Zn	0.007	0.011	0.006	0.012	
Mn	0.014	0.052	0.033	0.021	0.038

† Average values for nine growth media.

There is some concern as to how well nutrient concentrations in water extracts relate to overall nutrient availability in growth media. Warncke (1986) compared SME with other common extractants on nine commercial growth media. Results in Table 13–5 show water-extractable P and K levels compared closely with the active Spurway extractant (0.018 *M* acetic acid). Some additional P and K was extracted with the Bray and Kurtz P_1 and ammonium acetate extractants. The water-extractable cation levels being proportionate to NH_4 acetate extractable levels indicates the SME water-soluble nutrient concentrations reliably reflect the plant-available nutrient status of growth media.

3. Displaced Solution

Parker (1921) demonstrated that the equilibrium soil solution could be accurately displaced by adding water or another liquid to a soil column. The water added to the top of the column acts as a plunger to force the soil solution out the bottom. The nutrient concentration in the displaced solution was found to be inversely proportional to the moisture content of the soil. As Nelson and Faber (1986) increased the medium moisture tension from 0 to 15 kPa, the NO_3-N concentration in the displaced solution increased from 38 to 148 mmol L^{-1}. Burd and Martin (1923) demonstrated that an accurately displaced solution is truly representative of the soil solution. Uniform packing of columns is key to obtaining a true displaced solution. Growth media composed of several components are more difficult to pack to prevent channeling than soil. To deal with this potential problem, Faber and Nelson (1984) packed 1000 cm^3 of medium, 100 cm^3 at a time, into a 5-cm diam. and 60-cm tall column, and then displaced the medium solution using an aqueous solution of 5% potassium thiocyanate in 50% ethanol. Any intermixing of the displacing solution with the medium solution was detected by testing the displaced solution with 0.5% ferric chloride. Formation of the bright-red ferric thiocyanate indicated intermixing had occurred. The column displacement methodology has proven to be useful in research; however, the detail of the procedure precludes its use for routine diagnostic analyses.

4. Pour-Through

A variation of the displaced extract and saturated media extract is the Pour-Through (PT) method (Wright, 1986). The objective of the PT method is to displace the medium solution, which is in equilibrium, by adding water to the container. The growing plant is not disturbed with this approach. Water must be added to the center of the container to avoid channeling down along the sides. Adding excess water can cause dilution and misleading results. Yeager et al. (1983) determined that 40 to 100 mL of water can be added to a 1-gal container without affecting the nutrient concentration in the extract. As with true displaced extracts, the moisture content of the medium at the time of extraction influences the final nutrient concentrations. A standard initial moisture content is essential for development of interpretation guidelines. Wright (1986) suggests the medium moisture content be near container capacity, however, this may be too saturated. Extracting a set time after watering is desirable to allow for equilibration of the water with the medium. The PT method lends itself nicely to monitoring pH and nutrient levels of media in large (1 gal size or larger) containers such as those used for nursery stock. For pot and bedding plants grown in smaller containers or cell-paks accurately obtaining an adequate volume of extract without dilution is difficult.

Holcomb et al. (1982) have demonstrated that medium solution can be extracted from smaller intact pots or containers by directly subjecting them to a 0.103 MPa (15 psi) vacuum. Nutrient concentrations in these extracts correlate well with those obtained with the SME method. The relationships for conductivity values (dS m^{-1}) and extractable K (mmol of K L^{-1}) were: $C_{se} = 1.55\ C_{ve} + 0.61$ and $K_{se} = 1.60\ K_{ve} + 0.002$ where se = saturation extract and ve = vacuum extract. Extracting small containers of media by the PT or vacuum approach at a specific length of time (12–24 h) after watering improves the consistency of results.

III. TESTING CONSIDERATIONS

Producing plants in containers requires intensive management and monitoring of the growth medium pH, soluble salts content, and essential nutrient concentrations. The appropriate levels of each is influenced by kind of growth medium, plant species being grown, and growth stage.

A. Growth Medium pH

Changes in the availability of essential plant nutrients in growth media with changes in pH are well documented in the literature. In moderately weathered mineral soils, the optimum pH range for overall nutrient availability is 6.5 to 6.8 (Truog, 1948). However, in highly weathered and leached mineral soil, the most suitable pH may be lower. In organic (peat) soils, Lucas and Davis (1961) found the best pH range for overall nutrient availability

Table 13-6. Effect of mineral soil content on interpretation of growth media pH. Adapted from Peterson (1984).

Interpretation	Soil mix (>20% mineral soil)	Soilless mix (<20% mineral soil)
	pH	
Very low	5.0–5.4	4.4–4.7
Low	5.5–5.9	4.8–5.1
Slightly low	6.0–6.4	5.2–5.4
Optimum	6.5–6.8	5.5–6.0
Slightly high	6.9–7.2	6.1–6.5
High	7.3–7.4	6.6–6.9
Very high	7.5–7.6	7.0–7.3
Extremely high	7.7–8.4	7.4–7.8

to be 5.5 to 5.8. Since many artificial media are peat-based, the most suitable pH should be closer to 5.5 than to 6.5. Working with one such peat-based growth medium, Peterson (1981) demonstrated that overall nutrient availability was maximal between pH 5.3 and 5.5. Markus (1986) found extractable Cu and Mn were maximal between pH 5.5 and 6.0. As growers have switched from soil to peat-based media, problems have surfaced because they have not recognized that the optimum pH may also have changed. Table 13-6 provides an interpretation guide of pH for growth media with high and low mineral soil content. Optimal pH values for growth media containing composted barks vary with the bark source, softwood or hardwood (Nelson, 1987; Whitcomb, 1984).

The pH of a saturated growth medium reflects closely the pH within the root zone. pH values determined using one part medium to two parts pure water may be 0.5 or more pH units higher than that determined in the saturated medium. Since much of this difference may be due to salt effect, pH values using 0.01 *M* calcium chloride relate more closely to pH's determined by placing the pH electrodes in the saturated medium. Hipp et al. (1979) found pH values of solutions obtained by placing porous ceramic cups into pots of medium agreed quite well with those in saturated medium and could be expressed by $Y = 0.99X + 0.20$ ($r = 0.96$) where Y is the solution pH and X is the saturated medium pH. Further study revealed that pH values obtained by placing pH electrodes in direct contact with the medium surface in a container, with plants growing, gave results that were less acceptable than those obtained with the ceramic cups. Measuring the pH in solution extracts must be done carefully because the solution extract from most container growth media is weakly buffered, and pH values may be easily altered by contamination.

B. Conductivity (Soluble Salts)

The value of conductivity (soluble salts) measurements as a diagnostic aid in obtaining good plant growth has been demonstrated over the years by several researchers (Merkle & Dankle, 1944; McCall et al., 1961; Mascianica, 1983). Soluble salts are determined by a measure of electrical conductance of a solution and is a summation of contributions from all the ions present, both cations and anions. The degree of conductance is influenced by the number of ions per unit volume of the solution and the velocities at which these ions move under the influence of applied electromotive force. The total soluble salt content of an extract solution is measured with a conductivity meter or solu-bridge. At 25 °C, a 0.010 *M* potassium chloride solution gives a conductivity reading of 1.41 dS m^{-1}. Soluble salts increase the osmotic potential of the medium solution, thereby affecting water and nutrient uptake by plant roots. Higher soluble salt levels are permissible in peat-containing media because they hold more water per unit volume than mineral soils (Fig. 13–1).

The soluble salt content of a growth medium is affected by the amount of ionized salts it contains. Fertilizers differ in their "salt index" depending on the degree of dissociation and subsequent effect on the osmotic pressure of a solution. Table 13–7 lists the salt indice and relative salinity of commonly used fertilizer materials.

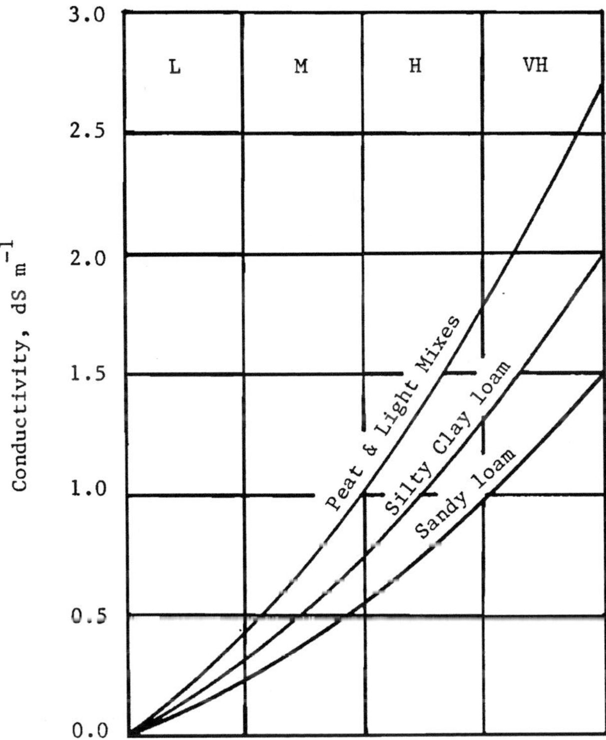

Fig. 13–1. Conductivity (soluble salt) ratings by the 1:2 (v/v) method for three growth media. Low (L): Needs additional fertilizer. Salt effect on plant is negligible. Medium (M): Can be fertilized lightly if in the lower end of the range, but is satisfactory near the top. High (H): Satisfactory in the lower end of the range, but germination and seedling injury may occur at the top. Very High (VH): Do not allow media to become dry or add fertilizer. Water enough to cause some leaching if near the top of the range. Leach extensively if over this range. (Modified from "Crop fertilization based on North Carolina soil tests." Agronomic Div., North Carolina Dep. of Agric., Raleigh.

Table 13-7. Effect of fertilizers on the salinity of growth media. Adapted from Bunt (1988).

Fertilizer	Elemental content	Salt index†	Total nutrients‡	Relative salinity§
	%		%	
Sodium nitrate, $NaNO_3$	16.5 N	100.0	16.5	100.0
Ammonium nitrate, NH_4NO_3	35 N	104.7	35.0	49.4
Ammonium sulphate, $(NH_4)SO_4$	21 N	69.0	21.0	53.7
Ammonia solution, NH_3	82 N	47.1	82.0	9.4
Calcium nitrate, $Ca(NO_3)_2$	11.9 N	52.5	28.8	30.1
Urea, $CO(NH_2)_2$	46 N	75.4	46	26.7
Diammonium phosphate, $(NH_4)HPO_4$	21 N, 23 P	34.2	44	12.7
Monoammonium phosphate, $NH_4H_2PO_4$	12 N, 27 P	29.9	39	12.7
Superphosphate (triple), $Ca(H_2PO_4)\cdot H_2O$	19.6 P	10.1	19.6	8.5
Potassium chloride, KCl	49.8 K	116.3	49.8	38.5
Potassium nitrate, KNO_3	13 N, 38 K	73.6	51	23.6
Potassium sulphate, K_2SO_4	45 K	46.1	45	17.0
Calcium carbonate, $CaCO_3$	40 Ca	4.7	40	1.9
Calcium sulphate, $CaSO_4$	23 Ca	8.1	23	5.8
Magnesium oxide, MgO	60 Mg	1.7	60	0.5
Magnesium sulphate, $MgSO_4$	16 Mg	44.0	16	44.5
Dolomite, $CaCO_3$ + $MgCO_3$	24 Ca, 12 Mg	0.8	36	0.4

† The "salt index" was calculated from the increase in osmotic pressure by equal weights of fertilizers relative to sodium nitrate.

‡ Total nutrients have been recalculated from the sum of the N, P, K, Ca, and Mg as usually stated in the fertilizer analysis, e.g., monoammonium phosphate = 12 N + 27 P = 39.

§ Relative salinity has been calculated from the increase in osmotic pressure per unit of plant nutrient relative to sodium nitrate.

The most common methods for measuring soluble salt content of growth media is to mix one part of media with two or five parts pure water. The 1:2 ratio is preferred, but 1:5 is necessary when the medium holds large amounts of water. These ratios have been used on a v/v and w/v basis. Unless the sample is completely air dried, the resulting soluble salt value will be influenced by the moisture content of the medium on either basis. A lack of standardization has, at time, led to some confusion (Kirven, 1986). Waters et al. (1970) found the 1:2 v/v method gave values that related better to plant growth than did the 1:2 w/v method. This is reasonable because plants grow in a volume of medium. With the density of growth medium varying considerably, a given weight of a heavy medium will give a much smaller volume than will a light-weight medium. In an analysis of 27 growth media, Waters et al. (1970) showed that soluble salt levels determined by saturation extraction (SME method) were superior in relating to plant growth to those obtained by the 1:2 v/v method. Table 13-8 shows guidelines for soluble salt levels obtained by the three methods used most frequently.

A concern with the 1:2 and 1:5 ratios is the solubilization of slowly soluble compounds, e.g., calcium sulfate or slow release fertilizers. Solubiliza-

Table 13-8. Soluble salt guidelines for growth media by three test methods. (Warncke & Krauskopf, 1983).

Saturated media extract	One part growth media to:† Two parts water	Five parts water	Interpretation
	dS m^{-1}		
0.0–0.74	0.0–0.24	0.0–0.12	Very low salt levels. Indicates very low nutrient status.
0.75–1.99	0.25–0.49	0.13–0.34	Suitable range for seedlings and sensitive plants.
2.00–3.49	0.50–0.99	0.35–0.64	Desirable range for most established plants. Upper range may reduce growth of some salt-sensitive plants and seedlings.
3.50–5.00	1.00–1.49	0.65–0.89	Slightly higher than desirable. Loss of vigor in upper range. Okay for high nutrient-requiring plants.
5.00–6.00	1.50–1.99	0.90–1.10	Reduced growth and vigor. Wilting and marginal leaf burn.
6.00+	2.00+	1.10+	Severe salt symptoms—wilting. Crop failure.

† Measured out on a v/v basis.

tion of these materials will give an inflated test value for soluble salts. Hence, use of the saturation extract approach is encouraged whenever possible.

C. Elemental Variables

1. Nitrogen, Ammonium, and Nitrate

The quantity and form of N present in growth media may significantly affect the quality and rate of plant growth. Excess N will produce rapid soft growth that results in poor quality. Most containerized plants grow best with most of the N in the NO_3 form, and excess NH_4 tends to depress the uptake of other cations, especially Ca. Ammonium toxicity may become apparent in some plants when NH_4 composes more than 30% of the soluble N fraction or makes up more than 3% of the total soluble salts (Bunt, 1988).

Solution NO_3-N content correlates well with leaf N content and plant growth (Gilliam & Wright, 1977). However, the most suitable concentration of NO_3-N in solution varies with composition of the medium. Prasad (1980) found that media containing barks and wood shavings tend to retain N, but peat-based media do not. Hence, soluble NO_3-N levels need to be higher in bark-based media to produce equal growth to peat-based media (Prasad et al., 1981a, 1983).

Generally, it is suggested that samples of moist growth media be extracted immediately or be refrigerated to prevent changes in the N status. However, as pointed out in section I, in working with peat-vermiculite mixes Markus and Steckel (1980) found no change in the NO_3 and NH_4 status when the media was held in air-tight containers for 3 and 6 d at both 7 and 23 °C.

Table 13–9. Correlation coefficients between extractable P by three methods and P uptake by cyclamen grown in three media. Adapted from Prasad et al. (1983).

	Medium					
	Bark		Peat		Peat + soil	
Method	T max†	R^2‡	T max	R^2	T max	R^2
SME	0.52 (16)	0.949	1.10 (34)	0.872	0.19 (6)	0.880
1:1.5	0.13 (4)	0.929	0.39 (12)	0.864	0.10 (3)	0.987
Spurway	2	0.948	4	0.863	2	0.910

† T max = Test value at which maximum P uptake occurred. For SME and 1:1.5 T max is in mmol L^{-1} (mg P L^{-1}). For Spurway T max is in mg kg^{-1}.
‡ R^2 = correlation coefficient between medium test values and P uptake.

2. Phosphorus

Extractable P varies with the composition of the growth media. Fibrous peats retain little P initially, but P retention increases as the peat decomposes (Lucas, 1983). Mineral soils, composted barks and wood shavings retain P quite well. When equal amounts of P are equilibrated with mixes containing the various components extractable P is highest in peat-based media; but all of them supply enough P for optimum plant growth (Prasad et al., 1983; Warncke, 1976).

Prasad et al. (1983) found that P uptake by cyclamen (*Cyclamen persicum* Mill.) and poinsettia (*Euphorbia pulcherrima* Willd.) grown in peat, bark, or peat + soil medium correlated well with P extracted by the SME, 1:1.5, and Spurway methods. However, the level of extractable P at maximum P uptake varied with the extraction method (Table 13–9). Some of the differences are accounted for by the Spurway values being reported as mg of P/kg of medium, whereas the SME and 1:1.5 values are mmol of P L^{-1} (mg of P L^{-1}) extract. The test value required for maximum P uptake was higher for peat than for bark or peat + soil. Hence, the optimum P test value varies with the extraction method and composition of the growth medium.

3. Potassium

In many potting media being used today, K, being held rather loosely on the limited exchange sites, is easily displaced by the divalent cations, Ca and Mg. Prasad et al. (1981e) found water and ammonium acetate extractable K to be linearly related for several growth media of varying composition. Even though ammonium acetate extracted more K, the correlation between extractable K and plant uptake was similar for both extractants. Hence, the K content of the solution phase characterizes the K environment of the root quite well.

Retention of K among the various growth media components is less variable than P retention. Peats and expanded vermiculite retain K slightly better than wood wastes (Prasad, 1980). Both vermiculite and wood materials supply significant amounts of K for plant growth (Markus & Flannery, 1983; Prasad et al., 1981e).

4. Calcium and Magnesium

Plant availability of Ca and Mg in artificial growth media has received little study. Since divalent cations are bound more tightly within growth media than monovalent ions, water extracts do not reflect the available status of Ca and Mg as well as for K. Exchangeable Ca and Mg would be more reflective of the availability within the root environment. Verloo (1980) found that adequate Ca helped keep humic complexes flocculated and thereby maintained good physical condition of organic materials.

5. Sodium and Chloride

Determination of the Na and Cl content of containerized growth media is important because of their potential adverse effects on plant growth both directly and indirectly. Both contribute to the total soluble salt content of the medium solution. Sodium held on the exchange sites presents less potential consequence, although it may displace other essential cations and also compete for uptake sites on plant roots. Individually, Cl is of more concern because of its toxic effects at high concentrations. In solution culture, Geraldson (1970) found that vegetable plant growth was not adversely affected as long as the Na or Cl each accounted for $<10\%$ of the total soluble salts in solution.

6. Micronutrients

Commercially prepared peat-based growth media are quite low in available micronutrients unless they have been added to the media. A summary of saturation extracts of peat-based media by the Michigan State University Soil Test Laboratory revealed that Zn and Mn concentrations in saturation exctracts were below 0.004 mmol L^{-1} (0.25 mg L^{-1}) and 0.006 mmol L^{-1} (0.35 mg L^{-1}) in 74 and 82% of the samples, respectively. Composted barks are reported to contain adequate Zn and Mn, but are low in Cu. With micronutrient concentrations in saturation extracts being quite low and the range being narrow, development of interpretation guidelines has been difficult. When Berghage et al. (1987) substituted various acids, salts, or chelates for water in the SME method, they were able to significantly increase the levels of Zn, Mn, and Fe in the extract. 0.005 *M* of diethylenetriaminepentaacetic acid (DTPA) was selected for modification of the SME method because it had minimal effects on the other variables and uptake of Zn, Mn, Fe, and Cu related well to extracted levels (Berghage, 1986). The DTPA improved the extraction of the cationic micronutrients in peat, peat/pine bark, and soil-based media. DTPA used as an extractant of micronutrients seems reasonable since Verloo (1980) has shown peat and peat decomposition products complex micronutrients. Using Mehlich I and DTPA extractants at medium to extractant ratios of 1:5 and 1:4 (v/v), Markus et al. (1981) increased the amount of Zn, Mn, Fe, and Cu extracted from peat-vermiculite media. Extracted Zn and Mn levels correlated well with concentrations in tomato leaf tissue.

Table 13-10. Estimated optimum P values based on maximum plant dry weight of cyclamen for three methods of extracting P from peat, bark, and peat + soil (Prasad et al., 1983).

Test method	Substrate	R^2†	Optimum level‡
1:1.5	Peat	0.422	0.26–0.28 (8.1–8.8)
SME	Peat	0.504	0.79–0.86 (24.5–26.7)
Spurway	Peat	0.559	3.2–3.5
1:1.5	Bark	0.432	0.12–0.13 (3.7–4.0)
SME	Bark	0.546	0.45–0.48 (13.9–15.0)
Spurway	Bark	0.525	1.4–1.6
1:1.5	Peat + soil	0.420	0.045–0.061 (1.4–1.9)
SME	Peat + soil	0.464	0.074–0.119 (2.3–3.7)
Spurway	Peat + soil	0.438	0.9–1.2

† Correlation coefficient between plant dry weight and extractable P.
‡ SME and 1:1.5 values are mmol L^{-1} (mg L^{-1}). Spurway values are mg kg^{-1}.

IV. INTERPRETATION OF TEST RESULTS

Methods for extraction of nutrients are easily developed. However, the success of an extractant for nutrient evaluation depends on a direct and predictable correlation with plant uptake, growth, or quality. This relationship can then be used in the development of good interpretation guidelines that are essential for an extraction method to be used as a useful diagnostic tool.

Prasad et al. (1983), studying the effects of nutrient levels on growth of cyclamen and poinsettia, found that test values from three separate methods correlated equally well with nutrient uptake. Regression equations for each of the three extracting methods (1:1.5; SME; and Spurway) were quadratic for N, P, and K. Plants grown in bark and similar mixes that retain N take up less N than plants grown in peat (Prasad et al., 1981b, 1983). Although there were differences in N uptake, the desirable N levels in the peat and bark media were similar, with the bark values being slightly higher. Nitrogen levels that produce the best growth and quality may vary greatly with plant species. The optimum N concentrations in the extract by the 1:1.5 method were found to be near 11.7, 4.4, and 17.2 mmol L^{-1} for mums, (*Chrysanthemum morifolium* Ramat.), verbena [*Aloysia triphylla* (L'Hér.)] and tomato, respectively.

As illustrated in Table 13-10, P levels desirable for optimum plant growth vary considerably with the composition of the growth media (Prasad et al., 1981c, d, 1983). Across all three test methods compared, maximum P uptake occurred with quite low P concentrations in the extract when mineral soil was included in the growth media. Similarly, maximum P uptake was achieved from bark-based media at much lower test values than from peat-based media. Apparently, some of the plant-available P is held by soil and bark in forms not readily extracted with water or weak acetic acid (Spurway extractant). However, these forms are in direct equilibrium with the water-soluble P. In these studies, the R^2 values were lower when based on plant dry weight than on P uptake.

Table 13-11. Standard values for total soluble salts content and nutrient levels in extracts of growth media obtained by the 1:1.5 and SME methods.†

Analytical value	Unit	Optimum		Very high	
		1:1.5	SME	1:1.5	SME
Conductivity‡	$dS\ m^{-1}$	1.3–1.8	2.0–3.5	3.6+	5.0+
NO_3-N	$mmol\ L^{-1}$	3.7–5.4	7.1–13.2	9.0+	21.4+
P	$mmol\ L^{-1}$	0.48–0.68	0.23–0.42	1.13+	0.61+
K	$mmol\ L^{-1}$	1.5–2.1	4.0–6.0	3.5+	9.0+
Ca	$mmol\ L^{-1}$	--	2.5–5.0	--	12.5+
Mg	$mmol\ L^{-1}$	0.65–0.90	1.5–3.0	1.5+	7.25+
Na	$mmol\ L^{-1}$	--	<3.0	--	13.6+
Cl	$mmol\ L^{-1}$	<3.3	<2.5	6.7+	5.7+

† Data compiled from Bik and Boertje (1975), Peterson (1984), Sonneveld et al., (1974), and Warncke and Krauskopf (1983).

‡ Specific conductivity of the extract at 25 °C.

In contrast to N and P, the relationship between K uptake and test values are similar for peat- and bark-based growth media. Several test methods predict K uptake equally well (Prasad et al., 1981e, 1983). Potassium test values for optimum uptake and plant growth were near 3.0 and 4.1 mmol L^{-1}, and 1.0 mg kg^{-1}, respectively, for the 1:1.5, SME, and Spurway methods.

Limited studies at the Horticulture Research Center in Levin, New Zealand indicate that a factor of 2.5 to 3.0 can be used to convert Spurway test values to 1:1.5 (Dutch Method) test values (Prasad et al., 1983) on peat and bark media. Studies by Bik and Boertje (1975) indicate that the relationship between conductivity values (dS m^{-1}) determined by the 1:1.5 and SME methods can be expressed as $EC_{(1:1.5)} = 0.384\ EC_{(SME)} + 0.012$. They also found that the relationship between NO_3-N test values (mmol L^{-1}) by the two methods was represented by $N_{(1:1.5)} = 0.38\ N_{(SME)} + 1.14$. These equations illustrate that nutrients are more concentrated in saturation extracts due to a narrower ratio between medium and extractant compared with the 1:1.5 method. The standard values given in Table 13–11 also bear out this general relationship between test values obtained by the two methods. The 1:1.5 standard values are the guidelines commonly used by many of the testing laboratories in western European countries. These guidelines agree reasonably well with those developed by Verdure (1980) after 4 and 6 wk of plant growth. The SME standard values are used by many testing laboratories in the USA. Relative agreement between guidelines for the two methods is quite good, except for P.

Geraldson (1967, 1970) has shown nutrient balance in solution culture and in saturation extracts of weakly buffered sand soil to be of equal or greater importance than the quantities of nutrients present. He found a good nutrient balance as a percentage of total soluble salts to be: NO_3-N, 8 to 10; NH_4-N, <3; K, 11 to 13; Ca, 14 to 16; Mg, 4 to 6; Na, <10; and chloride, <10. Since most artificial growth media are weakly buffered, this nutrient balance logically applies for saturation extracts of these media. Expressing the nutrient contents as a percentage of total salts is helpful in as-

Table 13-12. Pour through soil solution elemental and soluble salt levels associated with vigorous growth of *Ilex crenata* and other nursery species growing in a pine bark medium (Wright, 1986).

Element	Concentration in leachate	
	mmol L^{-1}	mg L^{-1}
NO_3-N	5.3–7.1	(75–100)
P	0.32–0.48	(10–15)
K	0.77–1.28	(30–50)
Ca	0.25–0.37	(10–15)
Mg	0.42–0.62	(10–15)
Soluble salts	0.6–2.0 dS m^{-1}	

sessing the most limiting nutrient or in the case of NH_4 and chloride, whether the potential for toxicity exists. Holcomb and White (1979) have shown that when plants are constantly fertilized through the watering system K is adequate at 3% of the total soluble salts. Recent studies by Biernbaum (1988, unpublished data) and George (1989) also indicate that top-quality plants can be grown with lower nutrient levels when subirrigation is used.

Nursery species that grow less rapidly than annual plant species have been shown to do well at lower media test levels. Table 13–12 presents the nutrient levels in a PT extract that Wright (1986) feels are adequate for good growth of a range of nursery stock. Nitrogen and P values agree with those in Table 13–11 and K falls in line as a percentage of total salts with the findings of Holcomb and White (1979). However, Ca and Mg values are much lower. Apparently, divalent cations do not reach the same equilibrium concentration in solution with the PT method as when media is mixed in the SME method.

Interpretation guidelines for the micronutrients are limited. Berghage et al. (1987) indicate that in 0.005 *M* DTPA saturation extracts B, Mn, Zn, and Fe concentrations of 0.065, 0.29, 0.21, and 0.27 mmol L^{-1} (0.7, 16, 14, and 15 mg L^{-1}) respectively, may be near optimum. However, there was considerable variation in response to micronutrient levels in studies with mums, poinsettias, and marigolds (*Tagetes erecta* L.).

V. SUMMARY AND FUTURE CONSIDERATIONS

Unlike naturally occurring mineral soil, artificial growth media for producing containerized plants can be uniform across large geographical areas. Even though some growers continue to prepare their own special medium, the trend is toward uniform media prepared by large manufacturers. Uniformity of growth media cries out for a standard well-defined testing methodology that is applicable across large geographical areas. Currently, water extracts are the most popular with the 1:1.5 (Dutch) and 1:6 (v/v) methods being widely used in European countries. The SME method has received increased use in the USA. Variations of these methods such as the

PT method seem appropriate for special situations. The modified Spurway method continues to be used by some labs because they have a wealth of supportive data to aid in interpretation of results.

If the SME method is to receive strong support in the future, it will require generating an interpretative base for a large variety of plant types. A means will also need to be found for consistently mixing growth media to the point of saturation. Variation in judgment of the saturation point is currently the major weakness with the SME method.

Micronutrient testing of artificial growth media needs additional study. Regardless of test methodology, little information is available to assist in determining whether test values are low, optimum, or high. Recent developments with a DTPA modification of the SME procedure appears to have laid the groundwork for future progress in this area.

With increasing use of fairly uniform growth media, uniform reporting of results as well as establishment of uniform interpretation guidelines is essential. While nutrient concentrations in saturation extracts are acceptable indexes of nutrient availability, reporting soluble nutrient concentrations in a volume of growth medium may be more meaningful to growers.

Testing of artificial growth media has progressed significantly in the last 10 to 15 yr. To keep pace with developments in production management, continual research and refinement of testing methodology of growth media is essential for testing laboratories to service the greenhouse and nursery industries.

REFERENCES

Berghage, R.D. 1986. Micronutrient testing of greenhouse growth media. M.S. thesis. Michigan State Univ., East Lansing.

Berghage, R.D., D.M. Krauskopf, D.D. Warncke, and I. Widders. 1987. Micronutrient testing of plant growth media: Extractant identification and evaluation. Commun. Soil Sci. Plant Anal. 18:1089–1109.

Bik, A.R., and G.A. Boertje. 1975. Fertilizing standards for potting composts based on the 1:1½ volume extraction method of soil testing. Acta Hortic. 50:153–156.

Boodley, J.W., and R. Sheldrake, Jr. 1972. Cornell peat-lite mixes for commercial plant growing. New York College Agric. Cornell Univ. Info Bull. 43.

Brown, E.F., and F.A. Pokorny. 1975. Physical and chemical properties of media composed of milled bark and sand. Am. Soc. Hortic. Sci. 100:191–121.

Bunt, A.C. 1986. Problems in the analysis of organic and light-weight potting substrates. Hortic. Sci. 21:229–231.

Bunt, A.C. 1988. Media and mixes for container-grown plants. Unwin Hyman Ltd., London.

Burd, J.S. and J.C. Martin. 1923. Water displacement of soils and the soil solution. J. Agric. Sci. 13:254–295.

Feber, W.R., and P.V. Nelson. 1984. Evaluation of methods for bulk solution collection from container root media. Commun. Soil Sci. Plant Anal. 15:1029–1040.

George, R.K. 1989. Flood subirrigation systems for greenhouse production and the potential for disease spread. M.S. thesis. Michigan State Univ., East Lansing.

Geraldson, C.M. 1957. Soil soluble salts: Determination of and association with plant growth. Proc. Fla. State Hortic. Soc. 70:121–217.

Gerladson, C.M. 1967. Evaluation of the nutrient intensity and balance system of soil testing. Proc. Soil Crop Sci. Fla. 27:59–67.

Geraldson, C.M. 1970. Intensity and balance concepts as an approach to optimal vegetable production. Commun. Soil Sci. Plant Anal. 1:187–196.

Gilliam, C.H., and R.D. Wright. 1977. Effect of four nitrogen levels on soil, soil solution, and tissue nutrient levels in three container-grown *Ilex* cultivars. J. Am. Soc. Hortic. Sci. 102(5):662–664.

Hipp, B.W., D.L. Morgan, and D. Hooks. 1979. A comparison of techniques for monitoring pH of growing medium. Commun. Soil Sci. Plant Anal. 190:1233–1238.

Holcomb, E.J., and J.W. White. 1979. Using plant uptake to determine optimum values for soil tests. J. Am. Soc. Hortic. Sci. 104:365–368.

Holcomb, E.J., J.W. White, and S.A. Tooze. 1982. A vacuum extraction procedure for N, K and conductivity evaluation of potted crop media. Commun. Soil Sci. Plant Anal. 13:879–889.

Jackson, M.L. 1958. Soil chemical analysis. Prentice-Hall, Englewood Cliffs, NJ.

Johnson, E.W. 1980. Comparison of methods of analysis for loamless composts. Acta Hortic 99:197–204.

Kirven, D.M. 1986. Horticultural testing: Is our language confusing? Hortic. Sci. 21:215–217.

Lagerwerff, J.V. 1958. Comparable effects of absorbed and dissolved cations on plant growth. Soil Sci. 86:63–69.

Lucas, R.E. 1983. Organic soils (Histosols): Formation, distribution, physical and chemical properties and management for crop production. Michigan State Univ. Res. Rep. 435.

Lucas, R.E., and J.F. Davis. 1961. Relationships between pH values of organic soils and availability of 12 plant nutrients. Soil Sci. 92:172–182.

Lucas, R.E., and P.E. Rieke. 1968. Peats for soil mixes. Int. Peat Congr. Proc. 3rd. 3:261–263.

Lucas, R.E., P.E. Rieke, and E.C. Doll. 1972. Soil saturated extract method for determining plant nutrient levels in peats and other soil mixes. Int. Peat Congr. Proc. 4th. 3:221–230.

Markus, D.K. 1986. Spurway/acid extraction procedures. Hortic. Sci. 21:217–222.

Markus, D.K., and R.L. Flannery. 1983. Macronutrient status in potting mix substrates and in tissue of azaleas from the effects of slow-release fertilizer. Acta Hortic. 133:179–190.

Markus, D.K., and J.E. Steckel. 1980. Periodical analysis of artificial rooting media and tomato leaf analysis from New Jersey greenhouses. Acta Hortic. 99:205–217.

Markus, D.K., J.E. Steckel, and J.R. Trout. 1981. Micronutrient testing in artificial mix substrates. Acta Hortic. 125:219–235.

Mascianica, M.P. 1983. Relationships between electrical conductivity and fertilizer levels in an azalea line growing medium. Commun. Soil Sci. Plant Anal. 14:575–584.

Matkin, O.A., P.A. Chandler, and K.R. Baker. 1957. Components and development of mixes. p. 86–107. *In* K.F. Baker (ed.) The U.C. system for producing healthy container-grown plants. California Agric. Exp. Stn. Ext. Serv. Manual 23.

McCall, W.W., R.F. Stinson, and R.S. Lindstrom. 1961. The effect of salinity upon the growth of container-grown greenhouse roses. Michigan Agric. Exp. Stn. Q. Bull. 44:61–66.

Mehlich, A. 1972. Uniformity of expression soil test results: A case of calculating results on a volume basis. Commun. Soil Sci. Plant Anal. 3:417–424.

Mehlich, A. 1973. Uniformity of soil test results as influenced by volume weight. Commun. Soil Sci. Plant Anal. 4:475–486.

Merkle, F.G., and G.C. Dankle. 1944. The soluble salt content of greenhouse soils as a diagnostic aid. J. Am. Soc. Agron. 36:10–19.

Nelson, R.V. 1987. Greenhouse operation and management. Reston Publ. Co., Reston, VA.

Nelson, P.V., and W.R. Faber. 1986. Bulk solution displacement. Hortic. Sci. 21:225–227.

Nowosielski, O., and A. Beresniewicz. 1957. Nutrient recovery from sphagnum peat, brown coal and bark substrates as measured by a soil testing universal procedure. Acta Hortic. 37:1945–1956.

Parker, F.W. 1921. Methods of studying the concentrations and composition of the soil solution. Soil Sci. 12:209–232.

Peterson, J.C. 1984. Current evaluation ranges for the Ohio state floral crop growing medium analysis program. Ohio Florists Assoc. Bull. 654:7–8.

Peterson, J.C. 1981. Effects of pH upon nutrient availability in a commercial soilless root medium utilized for floral crop production. p. 16–19. *In* Ornamental plants-1981: A summary of research. Ohio Agric. Res. Dev. Ctr. Rep.

Prasad, M. 1980. Retention of nutrients by peats and wood wastes. Sci. Hortic. 12:203–209.

Prasad, M., T.M. Spiers, and I.C. Ravenwood. 1981a. Soil testing of horticultural substrates (i) Evaluation of 1:1.5 water extract for nitrogen. Commun. Soil Sci. Plant Anal. 12(9):811–824.

Prasad, M., T.M. Spiers, I.C. Ravenwood, and R.W. Johnston. 1981b. Soil testing of horticultural substrates (ii) Desirable nitrogen values for the 1:1.5 water extract. Commun. Soil Sci. Plant Anal. 12(9):825–838.

Prasad, M., T.M. Spiers, and I.C. Ravenwood. 1981c. Soil testing of horticultural substrates (iii) Evaluation of 1:1.5 water extract and Olsen's exract for phosphorus. Commun. Soil Sci. Plant Anal. 12(9) 839–852.

Prasad, M., T.M. Spiers, I.C. Ravenwood, and R.W. Johnston. 1981d. Soil testing of horticultural substrates (iv). Desirable phosphorus values for the 1:1.5 water and Olsen's extract. Commun. in Soil Sci. Plant Anal. 12(9):853–866.

Prasad, M., T.M. Spiers, I.C. Ravenwood, and R.W. Johnston. 1981e. Soil testing of horticultural substrates (v) Evaluation of 1:1.5 water extract and ammonium acetate extract for potassium and desirable potassium values. Commun. Soil Sci. Plant Anal. 12(9).

Prasad, M., R.E. Widmer, and R.R. Marshall. 1983. Soil testing of horticultural substrates for cyclamen and poinsettia. Commun. Soil Sci. Plant Anal. 14:553–573.

Richards, L.A. 1954. Diagnosis and improvement of saline and alkaline soils. USDA Agric. Handb. 60. U.S. Gov. Print. Office, Washington, DC.

Sartain, J.B. 1983. Analysis of potting media as influenced by extractants and media:solution ratios. Soil Crop Sci. Soc. Fla Proc. 42:165–168.

Sonneveld, C., and J. van den Ende. 1971. Soil analysis by means of a 1:2 volume extract. Plant Soil 35:505–516.

Sonneveld, C., J. van den Ende, and P.A. van Dijk. 1974. Analysis of growing media by means of a 1:1½ volume extract. Commun. Soil Sci. Plant Anal. 5:183–202.

Spurway, C.H. 1943. Soil fertility control for greenhouse. Michigan State College Agric. Exp. Stn. Spec. Bull. 325.

Spurway, C.H., and K. Lawton. 1949. Soil testing. Michigan Agric. Exp. Stn. Bull. 132.

Truog, E. 1948. Lime in relation to availability of plant nutrients. Soil Sci. 65:1–7.

Verdure, M. 1980. Standards of interpretation of analytical results for horticultural composts. Acta Hortic. 99:231–236.

Verloo, M.G. 1980. Peat as a natural complexing agent for trace elements. Acta Hortic. 99:51–56.

Warncke, D.D. 1976. Saturation extractable nutrients levels of six greenhouse soil mixes equilibrated with four rates of fertilizer. p. 155. *In* Agronomy abstracts. ASA, Madison, WI.

Warncke, D.D. 1983. Analysis of greenhouse growth media: A saturation extract method. p. 1–4. *In* Benchmarks. Vol. 2, No. 2. W.R. Grace, Cambridge, MA.

Warncke, D.D. 1986. Analyzing greenhouse growth media by the saturation extraction method. Hortic. Sci. 21:223–225.

Warncke, D.D. 1988. Recommended test procedure for greenhouse growth media. p. 34–37. *In* W.C. Dahnke (ed.) Recommended chemical soil test procedures. North Dakota Agric. Exp. Stn. North Central Reg. Publ. 221.

Warncke, D.D., and D.M. Krauskopf. 1983. Greenhouse growth media: Testing and nutrition guidelines. Michigan State Univ. Coop. Ext. Serv. Bull. E-1736.

Waters, W.F., W. Llewellyn, and J. NeSmith. 1970. The chemical, physical and salinity characteristics of 27 soil media. Proc. Fla. State Hortic. Soc. 83:482–488.

Whitcomb, C.E. 1984. Plant production in containers. Lacebark Publ., Stillwater, OK.

White, J.W., R.J. Thomas, R.F. Fletcher, and M.N. Long. 1975. Analytical method for peat substrates. Acta Hortic. 50:157–163.

Wright, R.D. 1986. The pour-through nutrient extraction procedures. Hortic. Sci. 21:227–229.

Yeager, T.H., R.D. Wright, and S.J. Donahue. 1983. Comparison of pour-through and saturated pine bark extract N, P, K, and pH levels. J. Am. Soc. Hortic. Sci. 108:112–114.

Chapter 14

Principles and Practices in Plant Analysis[1]

ROBERT D. MUNSON, *University of Minnesota, St. Paul*

WERNER L. NELSON, *Potash and Phosphate Institute, West Lafayette, Indiana*

The principles and practices of plant analysis have evolved over many years and changed as our scale of observation and depth of knowledge have improved. Early plant researchers identified the chemical elements found in plants. Initially, they concentrated on determining the elements essential for growth, development, and maturation of plants. An element is considered essential if a plant cannot complete its life cycle and produce viable seed for reproduction of the next generation without it. De Saussure (in France) conducted research on plant analysis at the earliest development level. The last element determined to be essential for plants in general was Cl (Broyer et al., 1954), which exists in the soil and plant as the chloride (Cl^-) ion. Field research has confirmed deficiencies of and responses to Cl^- in various parts of the world, including the Northern Great Plains and Pacific northwest in the USA (Jackson, 1986; Fixen, 1985; Fixen et al., 1986; Goos, 1986). In connection with Cl^- toxicities, crops such as corn (*Zea mays* L.) can essentially exclude it (Parker et al., 1985), while some varieties of soybean [*Glycine max* (L.) Merr.] are adversely affected by higher levels (Parker et al., 1983). Marschner (1986) has indicated that the discovery of the essentiality of other micronutrients ranged from Fe in 1860 to Mo in 1938. Cobalt was confirmed as an essential element for the symbiotic fixation of N in 1960 (Ahmed & Evans, 1960). Subsequent research has shown nickel (Ni) to be essential to that process (Dalton et al., 1985) and selenium (Se) to be beneficial (Evans, 1989).

Earlier reviews of the development of plant analysis have been written by Goodall and Gregory (1947) and Ulrich (1948). Chapman (1966) edited a comprehensive book that included use of plant analysis for diagnostic criteria for crops. Australian workers have published an interpretation manual on plant analysis (Reuter & Robinson, 1986).

[1]Contribution of the Minnesota Exp. Stn. Paper No. 17,827.

 Soil Testing and Plant Analysis, 3rd ed.—SSSA Book Series, no. 3.

In terms of instrumentation for rapid analysis, it has evolved through the emission spectrograph, atomic absorption, x-ray florescence, mass spectrograph, electron microprobe, ion chromatograph, and the inductively coupled plasma (ICP) spectrograph. Near infrared reflectance instruments can be used to analyze some elements in forages, but lacks accuracy for others (Clark et al., 1987). Laboratories are now highly automated. Many elements can be determined simultaneously, on newer ICP instruments. The analytical instruments are linked to computers so that the results can be simultaneously printed for the researcher or customers' reports. These reports can either be mailed to the customer or sent electronically for more rapid transmission of results. The latter are printed out on site for further analysis and interpretation.

Plant analysis in a narrow sense is the determination of the elemental composition of plants or a portion of the plant for elements essential for growth. The concentration or extractable fraction of an element is determined from a sample taken at a specific time or stage of physiological or morphological development. The concentration is usually expressed on a dry matter basis. In a broader sense, plant analysis could include determination of specific organic compounds or substances such as amino or organic acids, plant growth substances, or hormones. It can also include determining elements that are detrimental to growth or animals and humans through our food chain. In this chapter, we deal with plant analysis in its narrower sense.

Plant analysis has become an integral part of most agronomic experimentation and is used to document and verify treatment effects. It is also used at the farm level by dealers and crop consultants to monitor production fields. Many private and public laboratories extend their services to dealers and farmers alike, and field use has greatly increased to confirm deficiency symptoms, diagnose problems, and search for the critical path to improved yields and profitability in crop production. This chapter covers the basic principles and relationships involved in plant analysis.

I. BASIC PRINCIPLES

Plant analysis is based on the principle that the concentration of an element or nutrient within the plant or one of its parts is an integral value of all of the factors that have interacted to affect plant growth, including the availability of the element. From an essential element standpoint, other than C, O, and H, which are derived from the air and soil water, we are dealing primarily with N, P, K, Ca, Mg, S, B, Cl, Cu, Fe, Mn, Mo, and Zn for most crops, with the addition of Co and Ni for symbiotic N_2 fixing bacteria associated with legumes. Sodium and Si have been shown to be essential for some crops, while in other cases they have been highly beneficial and provided economic crop responses. For example, in studies on wild rice (*Oryza sativa* L.) yield increases of nearly 50% have been found in response to Si application (Bloom & Meyer, 1988).

Much research has involved determining the optimum concentration of essential and beneficial elements for a given crop under experimental conditions. Realistically, under the usual two-dimensional experimentation, yield vs. one nutrient variable, seldom are all of the elements at or near optimum levels. If one element is found limiting, the sufficiency of others cannot really be determined until the limiting element is brought to sufficiency. Also, other nonessential elements, such as Al, above specific concentrations can adversely affect root and shoot growth, influencing the availability and uptake of other elements, and yields (Ohki, 1987). Excess concentrations of essential elements can also become detrimental to growth and lead to yield decreases.

The steps in plant analysis involve: (i) Collecting a representative sample or a portion of the crop, sampled at a time or stage of development that will be useful in reflecting its true nutrient status. (ii) Handling the sample so that the analysis will provide an accurate measure of the crop's elemental status. (iii) Using the best method and instrumentation for chemically analyzing the elements of interest or their fractions. (iv) Having enough knowledge so that the results of the analyses can be properly interpreted from a research or advisory standpoint. (v) Making recommendations to economically improve the nutrient status and productivity of the crop.

Steenbjerg (1951) discussed various yield curves and the relationships that are found with plant analysis. In the following sections, basic principles and specific examples of some relationships will be shown.

A. Yield in Relation to Nutrient Concentration and Supply

In developing the basic relationships among or between growth rate, dry matter accumulation (yield) and nutrient supply or concentration in the crop, usually one, two, or at most three elements are varied in an experiment. All others are considered to be present in the adequate amounts range. It should be remembered that when one element is added in increasing amounts, one or more other elements are usually being applied at the same time. Therefore, systems are seldom as well controlled as we would like in most studies.

A typical model or curve of relationships involving crop yield and nutrient concentration with increasing nutrient supply from the soil and fertilizer is shown in Fig. 14-1. Nutrients other than the one being considered are assumed to be in adequate supply. If one were to view the entire yield curve on a soil essentially devoid of the limiting element, the curve would be sigmoid with increasing nutrient supply. In field experiments, usually only segments of continuous curves are found in a given year due to the varying levels of the limiting nutrient in or released by the soil. The concept and segments of the curves as presented are similar to those Macy (1936) proposed. The concept as presented does not consider economics, but indicates interrelationships. Also, remember that curves or regressions are estimated with error. Therefore, the points on the curves, such as optimum yield, critical percentage, critical level, or optimum are estimated with associated errors, which indicates ranges rather than specific values.

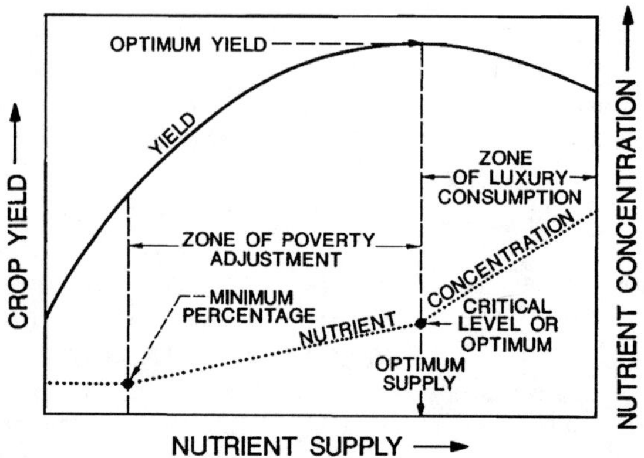

Fig. 14-1. Schematic graph of the manner in which nutrient concentration and crop yield varies with the supply of nutrient (Brown, 1970).

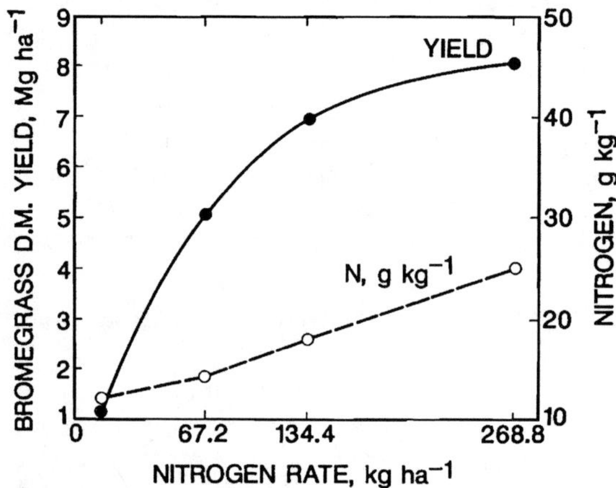

Fig. 14-2. Yield of first cutting of bromegrass and its N concentration as related to N applied as ammonium nitrate (Russell et al., 1954).

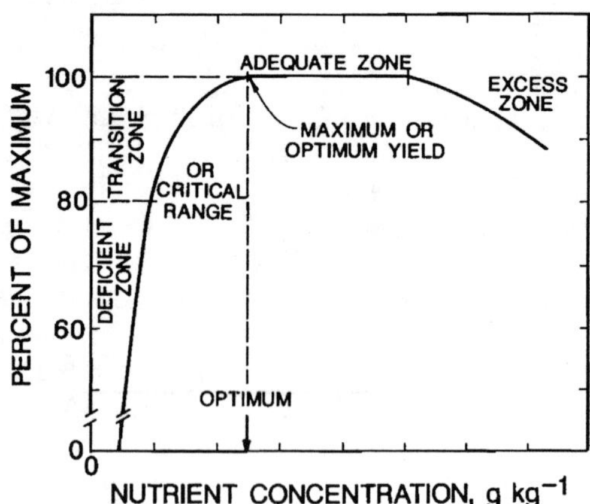

Fig. 14-3. Schematic showing the relationship between percentage of maximum yield of growth rate and nutrient concentration of a specific plant part sampled at a given stage of development (Ulrich & Hills, 1967; Dow & Roberts, 1982).

Results from a field study in which N rates were applied to a single cutting of smooth bromegrass (*Bromis inermis* Leyss.) are shown in Fig. 14–2. It appears that the plant analysis range includes the minimum percentage and zone of poverty adjustment, approaching the critical N concentration that produced the highest or optimum yield. The objective of evaluating results of such studies is to determine the critical or optimum concentration so that plant analysis can be used to monitor the sufficiency of the nutrient in terms of need or supply. Unfortunately, the zone of luxury consumption or concentration could not be evaluated from this example.

B. Yield or Growth Relationships with Nutrient Concentration

Graphing the yield and nutrient concentration is the approach often used to determine the critical level or concentration of an element in a given crop. Ulrich (1949, 1961, 1976) and his colleagues (Ulrich & Hills, 1967), as well as others, (Ohki, 1976, 1984a, b) have used this technique for many crops and found it particularly useful. The technique is based on the fact that if a plant is deficient in an element, growth rates and yields will be decreased. By adding increasing amounts of the nutrient applied, the concentration of the element in the plant or plant part increases until an optimum level is reached. Growth or yield using this approach is expressed as a percentage of the maximum. A schematic of this type of relationship for an element is shown in Fig. 14–3. Note how rapidly growth or yield declines when concentrations are in the deficient zone or range.

The zone between the deficient and optimum concentration in Ulrich's approach is referred to as the *transition zone*. Ulrich indicated that the transition zone starts with concentrations that produce about a 20% reduction in growth or yield and continues to those that produce the optimum or 100% of the maximum. Dow and Roberts (1982) have referred to this zone as the *critical range* within which the researcher may select the yield reduction and nutrient concentration considered acceptable. Ulrich and his colleagues set the critical level at concentration that produced a 10% reduction in growth or yield. Ware et al. (1982) found that it made a great deal of difference in the critical level determined, depending upon if a graphic analysis or a Mitscherlich model were used. Usually the concentrations determined by the Mitscherlich model were significantly greater than those determined by graph analysis.

In general, the yield reduction that a researcher is willing to accept in setting the critical concentration is subjective and could depend upon several factors. These factors include the sophistication of the production system, economic value of the crop, environmental constraints, and the risk and uncertainty associated with the system. Parker and Walker (1986) in connection with peanut (*Arachis hypogaea* L.) production and Mn concentrations concluded that "if concentrations of Mn in peanut leaves is only high enough to produce 90% of maximum yield and corrective treatments are not made, growers can suffer severe economic losses because peanut is a high income field crop.'' The same could be said for many other crops and production systems if economic considerations alone were the only consideration. Roberts and Dow (1982) have suggested critical range values that produce 95 to 100% of maximum yield for irrigated potato (*Solanum tuberosum* L.), a high value crop. They refer to concentrations above those necessary for maximum yield as adequate, which would correspond to the sufficiency range or optimum range used by others (Planck, 1979; Embleton et al., 1976). In our example in Fig. 14-3, the optimum was selected as being that associated with the maximum yield. However, under a high-risk production system, such as tropical pastures, a researcher might accept a critical concentration associated with 80% of maximum yield (Lanyon & Smith, 1985).

The reason we have accepted the optimum concentration or yield at 100% of maximum relates to the percentage sufficiency concept and nutrient interactions. Based on the percentage sufficiency concept, if two essential elements were at concentrations in the plant that would only allow a yield of 90% for each element, all others being at the optimum or 100%, the highest rate of growth or yield that could be achieved would be 81% of maximum ($0.9 \times 0.9 = 0.81 \times 100 = 81\%$). Assuming that there are 13 mineral elements that come from the soil and one accepts a critical concentration for each at a level that will produce a yield that is 90% of maximum, theoretically using 0.9 multiplied by itself 12 times for the 13 elements, the highest yield that could be achieved would be just over 25% of the maximum potential. Even if all elements were at concentrations that would produce 95% of maximum, based on the concept, one could only achieve 51% of the potential yield level. It is our view that the validity of this concept has not yet been adequately tested or verified, in spite of some practical evidence that would support it.

Hylton et al. (1967) have shown that the critical level of one element can shift rather widely if another element can substitute or interfere with the uptake of the first element. For example, with Italian ryegrass, they found that the critical level for K increased from 8 to 35 g/kg (0.8–3.5%), depending on the concentration of Na in the blade tissue. Ulrich and Hills (1967) and Hills and Ulrich (1976) point out the importance of this interrelationship for sugarbeet (*Beta vulgaris* L.). Sodium concentration influenced the range of K concentrations at which leaf petioles or blades show deficiency symptoms. Reneau et al. (1983) discussed the interrelationships between yield, K, and Mg concentrations in forage sorghum [*Sorghum bicolor* (L.) Moench] with P and K fertilization. Munson (1968) discussed the interrelationship of cations in plants and the manner in which nutrient additions, nutrient sources, and several other factors can influence yields and nutrient concentrations.

The adequate zone in Fig. 14-3 conforms to the sufficiency zone or optimum range terms used by others. It is a range that indicates nutrient concentrations that are at or above those needed for maximum yields under the conditions of the experiment, but with no reduction in yield. Some might refer to concentrations above the optimum levels as an indication of luxury consumption, while others would only indicate adequacy or sufficiency as

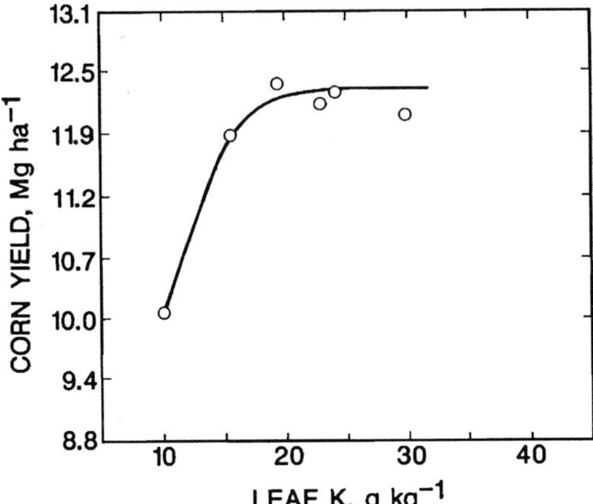

Fig. 14–4. Relationship of corn yield to leaf K concentrations sampled at tasselling stage of development (Overdahl & O'Leary, 1981).

long as yields were not reduced. When concentrations of one element increase above adequacy, one always needs to evaluate other elements to determine if they indeed might be limiting or are present in excess. When yields are decreased by higher concentrations of one nutrient, the level would be in the excess zone or at a level some would refer to as toxicity or excessive.

The relationship between irrigated corn yield and leaf K concentrations sampled at tassel is shown in Fig. 14–4. As K concentrations of leaf K were increased from 10.5 to 19.3 g/kg, grain yields were increased 2320 kg/ha (37 bu/acre), reaching a yield of 12.4 Mg/ha (198 bu/acre). No further increase in yield was observed, even though leaf K concentrations were increased to nearly 30 g/kg. This may indicate the possibility of other limiting elements or production constraints. With irrigation the minerals in some coarse-textured soils, such as those used in this study, release significant amounts of K, even though the K soil test is relatively low. However, these soils will not produce optimum yields unless additional K is applied (Overdahl & O'Leary, 1981), even with irrigation.

Chapman (1967) presented a yield-nutrient concentration relationship curve. A modification of that curve and the terminology is shown in Fig. 14–5. Note that the deficiency range is divided into severe and low and the optimum yield and optimum concentration are associated with the beginning of the sufficiency range. The lower portion of the curve may be found with an element such as a micronutrient, when the initial deficiency is severe and increased availability greatly increases dry matter yields, causing a dilution of its measured concentration.

In actual practice, it is of little value to sample a crop with different treatments for comparative purposes until plants have developed sufficiently to cause differential concentrations among the treatments. Data of Viets et al. (1954) indicate that the N concentrations in corn leaves sampled 17

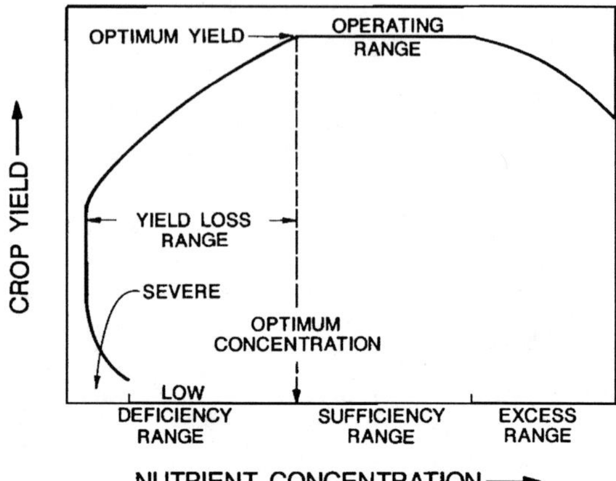

Fig. 14–5. Schematic of yield or growth of a crop as related to nutrient concentrations with interpretive ranges (Chapman, 1967).

June did not differ over N treatments, since all were at or above the critical level (Fig. 14–6). However, samples taken 9 July were sufficiently different in concentrations so that a meaningful relationship could be determined. With that sampling it appeared that even the highest N concentration was below that necessary for the optimum yield. Because of experiences such as these, sampling at a specific stage of developmet is usually suggested. For example, for corn leaf samples, it is recommended that the ear leaf or the leaf opposite and below the ear is sampled when the crop is 75% silked or entering the reproducive development stage.

Loué (1963) found the relationship between corn grain yield and leaf K shown in Fig. 14–7. He suggested that rather than referring to a specific critical level, it would be better to refer to a critical zone, which he determined to be between 17 and 20 g/kg (1.7 and 2.0%) K. It is presumed that if the level fell below 17 g/kg, the sample would be deficient.

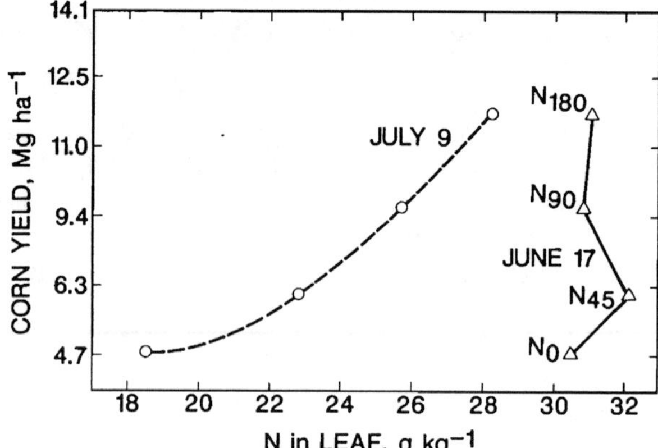

Fig. 14–6. The relationship of corn grain yield and leaf N composition as influenced by rates of N and date of sampling (Viets et al., 1954).

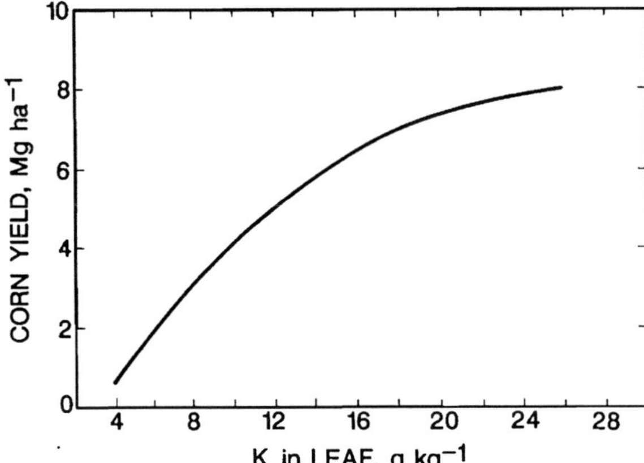

Fig. 14-7. The relationship of corn grain yield and K composition of the ear leaf sampled at silking (Loué, 1963).

Rehm et al. (1983) have conducted P studies on corn. They used regression analysis relating the percentage of maximum yield of silage or grain to either ear leaf P or P concentration in whole plants at maturity. Using the respective quadratic regression equations and solving for optimum ear leaf P concentration for maximum yield, the optimum levels were found to be 2.63 g of P/kg for silage and 2.64 g of P/kg for grain. When whole plant analysis were used, the optimum P concentrations were 2.85 g/kg for silage production and 3.46 g/kg for grain production. They noted substantial year-to-year variations in the affects of applied P on leaf P and in the mature crop.

Because in field situations one seldom deals with one element at a time and percentage of maximum yield is the criterion used to determine the optimum concentration, it could be beneficial to evaluate the leaf concentrations of essential nutrients found in corn from experiments that approach current maximum yields. Roy L. Flannery (1987, personal communication) of the New Jersey Agricultural Experiment Station provided the plant analyses of corn leaves sampled at early tassel from five experiments conducted on the same experimental site for 5 yr, with grain yields ranging from 17.9 to 21.2 Mg/ha (Table 14-1). It is assumed that these values are within the

Table 14-1. Corn yields and elemental ear leaf concentrations sampled at early tassel for 5 yr. (R. Flannery, 1987, personal communication).

Grain yield	Percent max.	Elemental concentration-ear leaf N	P	K	Ca	Mg	S	Mn	Zn	Cu	B
Mg/ha		g/kg						mg/kg			
21.2	100	33.8	3.2	23.9	5.9	1.8	2.8	41	28	10	21
19.6	92	31.8	4.0	22.8	5.3	1.5	2.0	43	39	11	14
19.4	91	31.6	4.1	26.1	4.8	1.8	1.9	39	26	9	7
18.1	86	30.4	3.4	26.6	6.1	1.9	2.3	37	28	12	12
17.9	84	30.2	3.2	26.3	5.8	1.8	2.2	36	27	10	10
						Mean values					
19.2		31.4	3.6	25.1	5.6	1.8	2.2	39	30	10	13

sufficiency range for each element because of the high yields achieved. Walworth et al. (1988) have used these values in developing DRIS norms, which will be discussed in a later section of this chapter.

Flannery (1982) provided the analyses of trifoliate soybean leaves sampled at early flowering for soybean that yielded 7.3 Mg/ha (113 bu/acre). This yield approaches current experimental maxima for soybean. The results were as follows: N, 58.2 g/kg; P, 4.3 g/kg; K, 22.3 g/kg; Ca, 9.1 g/kg; Mg, 3.2 g/kg; S, 2.4 g/kg; B, 42 mg/kg; Cu, 12 mg/kg; Fe, 126 mg/kg; Mn, 35 mg/kg; and Zn, 50 mg/kg. These values should be within the sufficiency range for the elements determined.

C. Nutrient Concentrations and Physiological Maturity

In establishing sampling procedures for plant analysis, the researcher and user must be aware that the concentrations of elements or soluble fractions change rather rapidly with time and physiological maturity. The major reason for this is rapid dry matter accumulation and dilution of the nutrient concentrations, even though the uptake of some nutrients continue until near maturity. Tyler and Lorenz (1962) have shown that potato petiole nitrate-N (NO_3-N) and phosphate-P (PO_4-P) increase with rates of application and decrease with seasonal sampling and development stage. From such information, one can extend the seasonal sampling period and determine whether the level of nutrition of a given field is in the sufficient, intermediate, or deficient ranges.

Dow and Roberts (1982) have proposed and Roberts and Dow (1982) have discussed an approach using seasonal critical nutrient range (CNR). In potato rate of P studies on two soil types, yields and petiole samples were taken for each P rate at five sampling dates during the growing season. From these data, Dow and Roberts regressed percentage petiole-P on sampling dates in days after formation of 2-cm tubers to determine two regression equations which encompassed yields between 95 and 100% of maximum. These equations were used to establish the seasonal P deficient, CNR, and adequate P levels (Fig. 14–8). Early in the season, the CNR for petiole-P was from 3.8 to 4.5 g/kg (0.38–0.45% P), while for the last sampling the range was from 1.4 to 1.7 g/kg.

Lutrick et al. (1986) used a graphic analysis to monitor seasonal NO_3 levels in cotton (*Gossypium hirsutum* L.), breaking the ranges into deficient, adequate, and excess. Either approach should be applicable for any nutrient or combination of nutrients, to determine adequacy anytime during the growing season. The critical range should provide a more certain guide than that used by Westermann and Kleinkopf (1985a), which sets a level of total P concentration in the tops and active leaves of potato at >2.2 g/kg as being adequate for dry matter production and tuber growth rate. Westermann and Kleinkopf (1985b) used a similar approach for N and concluded that a fourth petiole NO_3-N concentration of 15 000 mg/kg was a concentration adequate for optimum dry matter production.

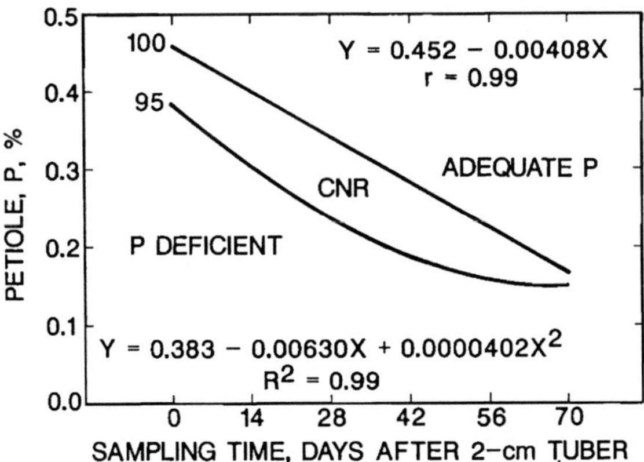

Fig. 14-8. Use of 95 and 100% of maximum potato yield regression equations to establish seasonal petiole P deficient, critical nutrient range, and adequate P levels sampled after 2-cm tuber formation (Roberts & Dow, 1982).

Many researchers have studied the changes in nutrient concentrations of different plant parts of crops over the growing season, but few have used modeling or regression techniques, such as applied by Roberts and Dow (1982). The latter approach is sound if elemental concentrations are increasing or decreasing. Sayer (1955) conducted extensive research on concentrations of elements in various portions of corn inbreds and hybrids at different stages of maturity. More recent research on a given hybrid was conducted in Iowa under low and higher fertility levels with respect to P and K based on the research of Nael El-Hout (R. Killorn & R.D. Voss, 1985, personal communication). Leaves, sheaths, stalk, tassel, lower leaves, husks, shank, silks, cobs, and grain were analyzed from developmental stage V4 to R1 and every 11 d after that for 55 days. The concentrations of each portion was determined for each sampling date, as well as to total amounts of N, P, K, S, Ca, and Mg in each portion and the total crop. These are probably the most thorough set of data that exist for an intermediate yield level of 10.16 Mg/ha. Others have analyzed portions of corn plants samples at different times over the season, including Hanway (1962), Gorsline et al. (1965), Flannery (1986a), and Karlen et al. (1988). Using the dry matter and amounts of nutrients in the aerial portion of the crop, calculations from Flannery's research indicates the following changes in concentrations for nutrients at 32 d after emergence to Day 146: N, 42.6 to 12.2 g/kg; P, 4.64 to 2.16 g/kg; K, 55.9 to 11.0 g/kg; Ca, 3.55 to 1.87 g/kg; Mg, 3.55 to 1.38 g/kg; and S, 2.13 to 1.27 g/kg.

Grain

Some researchers have used postharvest analysis of grain as a means to determine whether a crop was grown under optimum levels of nutrition. Pierre et al. (1977a, b) studied the relationship between the N concentration of the corn grain and yield expressed as a percentage of maximum as a measure of N sufficiency. They used both regression and graphic techniques. They found that when corn yields approached maximum yields based on N-fertility studies, the postharvest N concentration of the grain was 15.2 g/kg by graphic analysis or 15.4 g/kg based on regression analysis. These concentrations would be viewed as the critical levels at which N would not have been the limiting factor to yield. It does not guarantee that some other nutrient or management practice might not have been limiting. They used this technique to estimate a N requirement index (NRI) need for different situations to increase yields to either the maximum or economic optimum. Based on their analysis, it would take an average of 34 kg of N to produce a megagram of grain when yields were 45% of maximum, but about 75 kg of N when yields were at 90% of maximum.

Randall et al. (1987) studied increasing rates of N using two commercial adapted corn hybrids where the previous crop was corn or soybean. The N concentrations of the grain for the two hybrids were different even though yield increases from additional N were small or negligible. One hybrid following corn yielding at 99.6% of maximum had a N grain concentration of 11.8 g/kg. Following soybean with corn yields of 97 and 98% of maximum, the two hybrids had grain N concentrations of 12.3 and 13.6 g/kg, well below the levels suggested by Pierre and his colleagues. Following soybean, the yields of the two hybrids were 1.38 and 1.59 Mg/ha greater than following corn.

Ray Lockman (1986, personal communication) has analyzed field samples of corn grain taken from an Illinois farmer's (Herman Warsaw) test area that had yielded 20.446 Mg/ha (326 bu/acre) or 17.28 Mg/ha of dry matter. That yield approaches current maximum yields being grown. On a dry matter basis, the nutrient concentrations of the grain were as follows: N, 16 g/kg; P, 3.4 g/kg; K, 3.7 g/kg; Ca, 0.4 g/kg; Mg, 1.1 g/kg; S, 1.3 g/kg; Cu, 3.8 mg/kg; Fe, 34 mg/kg; Mn, 8.9 mg/kg; and Zn, 30 mg/kg. From these values, using the dry matter yield of the grain, one can estimate the amounts of nutrients removed from the field.

Dumenil (1974), working on soil productivity in Iowa, related P and K concentrations in the grain of corn, oat, and soybean to increasing soil test values. With P, as the soil test values were increased from 5 to 50 mg of P/kg, the P concentration of corn grain increased from 1.9 to 3.6 g/kg, oat from 2.5 to 4.5 g/kg, and soybean from 4.5 to 7.3 g/kg. With increasing soil K levels from 25 to 350 mg of K/kg, the K concentration of corn grain increased from 3.0 to 5.1 g/kg and soybean increased from 16.9 to 25.3 g/kg. No attempt was made to relate the concentrations to the yields in these studies. However, it should be noted that Sale and Campbell (1986) found that the yield of soybean and oil ($r = 0.97$), protein ($r = -0.94$), and K concentration ($r = 0.98$) were closely related. Bean yields increased until the K concentration of the mature beans was >18 g/kg. Results from field studies in which yields were from over 4.4 and 5.8 Mg/ha, the K concentrations of the bean of soybean were 18 and 17.9 g/kg, respectively (see Table 2, Bundy & Oplinger, 1984). Goos et al. (1982) used grain protein of winter wheat as a postharvest measure of N sufficiency for maximum yield of grain. They determined the critical grain protein level to be 115 g/kg and determined that the transitional zone between deficient and sufficiency ranged from 111 to 120 g/kg.

D. Nutrient Concentrations and Variety or Hybrid

Many researchers have observed that genetics exerts a high degree of control over nutrient uptake and concentrations in plants, the actual concentration reflecting genetics, cultural practices, and environmental conditions. Genetics establishes the boundaries of physiological and metabolic potential. Manipulation of cultural and management practices, nutritional regime, and environmental modification determines to what extent the genetic potential is fulfilled. Munson (1970, 1974) has reviewed genetics and nutrient concentrations for many crops. Munson and Nelson (1973) included a section on the subject. Parks (1985) has specifically evaluated K variability with varieties and hybrids for many crops. Carlone and Russell's (1987) evaluation of 28 corn genotypes developed and used during different eras in corn production have found marked differences in their response to plant densities and N rates. While N concentrations were not studied per se, this represents a classic paper for those interested in genotypes and interactions with nutrient response.

Some of the differences among varieties or hybrids relate the differential capacities of roots to absorb or take up nutrients. Baligar and Barber (1979) found that four parental corn genotypes from Indiana and four from Florida and their single crosses exhibited different appearing root systems, causing differential rates of nutrient uptake in short-term studies. The single crosses exhibited heterosis in growth of shoots and roots. The ranges in I_{max} values among the Florida genotypes were much greater for K, Ca, Mg, and P in terms of pmol (cm/s), the values being 0.75 to 2.96, 0.32 to 1.08, 0.04 to 0.28, and 0.18 to 0.89, respectively, than for the Indiana genotypes. The latter genotypes showed ranges of 1.30 to 2.06, 0.04 to 0.23, 0.12 to 0.19, and 0.09 to 0.19 for the same elements.

Elliott and Lauchli (1985), in studying 15 different corn inbreds, found significant differences among genotypes for rates of P utilization, but not for P absorption by roots. The lack of P use was thought to be limited by partitioning of P between free orthophosphate and that organically combined. However, in these studies, all genotypes exhibited a P-induced Fe deficiency caused by an apparent inhibition of Fe transport from the roots to the shoots. This brings up the direct and indirect effects of applying an increasing rate of a nutrient or nutrients to soils and the resultant impact on the availability of nutrients other than those being applied.

Bruetsch and Estes (1976) evaluated 12 corn genotypes of differing relative maturities, with final yields ranging from 10.9 to 14.3 Mg/ha, for variations in dry matter, and nutrient accumulation. They found that there was significant variability among genotypes with respect to dry matter and nutrient accumulation. Nutrients that showed genotype differences in dry weight production per gram of nutrient were N, P, K, Ca, and Mg. The relationship between P concentration and relative maturity of the genotype was most consistent, with the higher P concentrations being in the earlier maturing genotypes. However, in terms of efficiency of dry matter produced per unit of P absorbed, the later-maturing hybrids ranked highest.

Malzer et al. (1984) under high yield conditions evaluated nine different corn hybrids for grain yield, leaf nutrients, grain and stover nutrients and total uptake. In most cases, the results indicated a high probability of differences among the hybrids for the criteria measured. Different hybrids also appeared to respond differently depending upon the previous crop (corn after corn or corn after soybean).

Mackay and Barber (1986) studied the N uptake and root growth in response to N to explain the differences between two corn genotypes, Pioneer 3732 and B73 × Mo17, on two soil types. They found that in contrast with the P 3732 hybrid, by midsilk, N increased the root length of B73 × Mo17 on both soils by 35 and 37% and showed extended root growth after midsilk. They proposed root growth after midsilk as one mechanism to explain the advantage of that hybrid under higher N rates. They did not evaluate the effects of N on the leaf concentrations or uptake of other elements.

Ebelhar et al. (1987) studied the response of prolific (Pr), semiprolific (SPr) and nonprolific (NPr) corn genotypes to N and K rates at two plant populations. They found that increasing N rates resulted in greater K uptake and that N rate in fact had a greater effect on K uptake than K rate. This caused differential leaf concentrations of Ca, Mg, and K among the genotypes. The Pr and SPr genotypes showed greater increases of Ca, Mg, and K in response to N rates than did the NPr genotype. There was a marked genotype effect of K rate on K uptake with the Pr and SPr lines taking up more K than the NPr line. Maximum yields were obtained with leaf N concentrations of 28 to 30 g/kg for yields that ranged from 12.1 to 13.6 Mg/ha, depending upon the year. Increased N concentrations increased Ca+Mg/K ratios for the Pr genotype, but did not change those of the NPr. Earlier, Anderson et al. (1984) studied the effects of N on several factors related to the prolificacy of genotypes and found differences among the lines with respect to N concentrations in grain, with the NPr being the highest, 15.7 to 16.0 g/kg, and the Pr being the lowest, 12.6 to 13.7 g/kg.

Differences in nutrient concentrations between and within species and cultivars have been studied by many. Some of these include studies on corn by Baker et al. (1970) and Gallaher et al. (1972); forage legumes by Loper and Smith (1961); soybean by Karlen et al. (1982), Gettier et al. (1985), Sale and Campbell (1986, 1987), and Silberbush and Barber (1985); barley (*Hordeum vulgare* L.) by Dick et al. (1985); tomato (*Lycopersicon esculentum* Mill.) by Carpena et al. (1988); cotton by Anderson and Harrison (1970); and peanut by Coffelt and Hallock (1986).

E. Interrelationships with Other Factors

Availability of an element has a dominant effect on the concentration of that element found in a plant or plant part being analyzed, but it can also have a dominant effect on the uptake and concentration of one or more other elements. An earlier example was given for N and K, as well as for the relationships between K and Na in some species. If a nutrient such as N is applied and the availability of another essential element is limiting, initially, there may be a decrease in the concentration of the limiting element, and its concentration will drop to the minimum necessary to support growth. By the same token, because of the effects of soil moisture (Mackay & Barber, 1985a, b; Kuchenbuch et al., 1986) and temperature (Kuchenbuch & Barber, 1988) on nutrient availability and uptake, they can exert influences on the concentration of elements in plant tissue. The drought of 1988, for example, produced K deficiency symptoms on soils that under more normal soil moisture would not have shown apparent symptoms. Diagnosticians need to be aware of the differences within a season and among seasons even at the same experimental site. These differences may complicate interpretation unless adequate ranges are used. Changes in management practices such as increasing plant densities per unit area, changing tillage or increasing soil compaction, or adjusting soil pH may cause nutrient concentrations to change depending upon their effects upon nutrient availabilities.

F. Nutrient Use Efficiencies

A researcher can determine how efficiently a crop is using a nutrient internally by measuring the total uptake of the nutrient and determining how much yield, either dry matter or harvested portion, is produced per unit of nutrient. This requires total dry matter measurements and plant analysis. These values will not indicate how efficient a particular crop is in recovering either soil or fertilizer nutrients unless other comparisons are made. For corn, which produced 2-yr-average-yield over 20 Mg/ha, R.L. Flannery (1985, personal communication) found that average grain output/uptake ratio based on total aerial or aboveground uptake were 48.8 kg of grain/kg of N, 240.8 kg of grain/kg of P, 50.6 kg grain/kg of K. Rhoads and Stanley (1981) found grain output to uptake ratios of 62.2 for N, 295 for P, and 50.8 for K for corn that yielded 12.7 Mg/ha. For corn yields of 13.5 Mg/ha, Karlen and Camp (1985) found grain to uptake efficiencies of 59.4 kg of grain/kg of N, 221 kg of grain/kg of P, and 50.7 kg of grain/kg of K. The efficiencies for P amoung the three studies varied the most, but those for K were amazingly similar. The standard deviations for this small sample were 38.3 for P, 7.68 for N, and 0.1 for K. It is easy to take the reciprocal of the above efficiency values, multiply them by 1000 and obtain estimates of the total uptake per metric ton of grain production.

For soybean that produced 6787 kg/ha, Flannery (1986b) found that for the output per unit of highest seasonal nutrient uptake, efficiencies were 11.05 kg of bean/kg of N, 105.1 kg of bean/kg of P, 16.86 kg of bean/kg of K, 36.78 kg of bean/kg of Ca, 107.7 kg of bean/kg of Mg, and 218.9 kg of bean/kg of S. The above values and those for corn provide an indication of the internal efficiency with which these crops convert nutrients taken up into grain yield under high-yield conditions. From an energy standpoint, the relationship found between corn and soybean produced per unit of K uptakes K is interesting. Usually it is considered to take three times the energy to produce a unit of soybean as a unit of corn, because of the greater oil and protein contents of bean. If one divides the mean output per internal unit of input of K for corn from the above example (50.7 kg of grain/kg of K) by three, you obtain 16.9 kg of grain/kg of K, which is the same value found for soybean (16.86 kg of beans/kg of K). This may indicate the nutritional importance and stability of K in energy relations in both crops.

The recovery of one nutrient from the soil or a fertilizer application depends upon the availability of a second nutrient. For example, K uptake and efficiency from either soil or K applications often depend upon the rate of N used (Ebelhar et al. 1987). Likewise, if K availability or supply is low, N uptake and utilization may be very low. Similar interrelatioships are observed between N and P. In the Iowa studies of Nael El-Hout, mentioned earlier, increased application of P and K, with the same rate of N, increased yield by 74% (5.7–9.93 Mg/ha); N uptake by 68%; P uptake by 253%; K uptake by 162%; S uptake by 39%; Ca uptake by 46%; and Mg uptake by 20%. It may well be that nutrients other than P and K became limiting at higher rates of these nutrients. With the environmental issues that confront us, there will be increased emphasis on the efficiency of nutrient recovery and more stress will be placed on determining limiting factors, and on varieties or hybrids that increase the output per unit of soil or applied nutrient. It is unlikely that simply looking at leaf concentrations will produce the answers needed for the future. Plant analysis can, however, be used to monitor crops throughout the growing season for nutrient adequacy.

II. PRACTICES OF PLANT ANALYSIS

Research has established critical or optimal levels of nutrients for crops, as well as levels considered to be deficient, low, sufficient, or excess. These will be discussed in detail for the various crops in subsequent chapters. It is interesting that variations in these values exist among states and experiment stations. Perhaps this is not surprising when one considers the weather, climate, and soil differences, as well as genetic differences in specific crops that exist. In each case, one assumes that data were properly evaluated in determining the established values. Optimal values will stand the test of time and experimentation and practical use in the field by practicing agronomists, soil scientists, and crop consultants.

The availability of plant analysis facilities has increased markedly. Most land grant universities have laboratories that are used in both research and extension and are available to farmers and dealers for a fee. Also, many private laboratories have agronomists or crop consultants that work directly with dealers to serve their customers.

A. Experimental Techniques

Every year many soil fertility experiments are conducted in which the nutrient availability is varied, both in terms of nutrient soil test levels and fertilizer application. Plants are sampled and analyzed at appropriate times to relate these values to growth and yield. These are used to improve our knowledge base through calibration and correlation studies. If and when enough new data become available, researchers may adjust the critical or optima levels or ranges that are currently being used. With such input, there is a continual shifting and updating of information based on the latest results that may come with newer varieties or hybrids and cultural or management practices. The trend to analyze for all of the essential elements, as well as for heavy metals or those that may cause health and nutritional problems broadens the usefulness of the information collected.

B. Verification of Specific Symptoms

Nearly every crop consultant or practicing agronomist has used plant analysis to verify the adequate nutritional status of crops or identify an apparent nutrient deficiency. While nutrient deficiencies should be rare in high production systems, nevertheless, they are observed from time to time. If young plants are sampled and deficiencies found, often there is still time to make corrective applications of nutrients, heading off the possibility of large economic losses. Sometimes the apparent symptoms of an excess of one element will look like the deficiency of another element. For ecample, excess B in corn produces symptoms that appear similar to K deficiency. Plant analysis provides direction for proper action.

C. Field Surveys

Field plant analysis surveys are used by researchers, and extension and industry personnel to provide information about possible deficiency of a specific element or elements and to define the area over which the deficiency exists. Such surveys usually involve an indicator crop, which may be particularly sensitive or has a high requirement for the element in question. Some industry people have used this technique within a given trade area and may refer to it as a 'Blitz'. Both soil sampling and testing are combined with the plant sampling and analyses for complimentary information. Researchers in many states have also used these techniques to define the boundaries or soil associations in which specific problems exist. Lanyon and Griffith (1988) cite data, for example, showing that the ratios of K fertilizer applied to K removals for alfalfa (*Medicago sativa* L.) and corn grain in the USA are 0.28 and 2.8, respectively. Such discrepancies need to be carefully studied to evaluate soil test levels, the capacities of the two crops to recover and respond to soil and fertilizer K, and the amounts needed for increases or maintenance of yield.

D. Analyses of Normal and Abnormal Plants

Sometimes no apparent deficiency symptom is showing in plants, but a comparison of different portions of a field make it obvious that some areas are not growing normally. In such situations, it is helpful to be careful in your evaluation. It is important to dig up plants and examine roots, look for insect feeding, nematode damage, or disease symptoms, as well as examining the soil for compaction or surface sealing. Such conditions can lead to abnormal growth and reduced uptake or concentration of elements in a crop. Both plant and soil samples taken from the respective areas will aid in assessing corrective action. Soil samples can sometimes identify problems other than fertility as the cause of plant stress.

Apparent deficiency of N in a legume crop is often due to a deficiency of elements other than N that are essential for the N_2 fixation process. Elements that could be affecting N_2 fixation are P, K, S, Cu, Mo, Ni, or Co.

E. Evaluation of Crops on a Given Soil

Many times, farmers and their advisors will develop fertility programs and want to determine if there may be a need for further adjustments. Even though yields may be well above average, plant analysis may indicate the next limiting elements or an imbalance. A Diagnosis and Recommendation Integrated System (DRIS) analysis (mentioned below) of the plant analysis data could indicate a need to make shifts in the fertilizer program. An Illinois farmer, Herman Warsaw, has achieved corn grain yields on a test area of 23.2 Mg/ha. Periodically, he took plant samples to determine if elements were within the sufficiency range. For a 3-yr yield of more than 19.5 Mg/ha (312 bu/acre) from Warsaw's farm, R.B. Lockman (1986, personal communication) of Agrico Chemical Company Testing Laboratory in Ohio and cooperators provided ear leaf analyses data sampled at tassel which had average elemental values as follows: N, 33.7 g/kg; P, 3.77 g/kg; K 27 g/kg; Ca, 5.1 g/kg; Mg, 1.4 g/kg; S, 2.7 g/kg; B, 7.7 mg/kg; Cu. 5.7 mg/kg; Fe, 106 mg/kg; Mn, 48 mg/kg; and Zn, 27 mg/kg. Analyses were also determined on the mature grain. The average values based on dry matter were as follows: 15.7 g of N/kg, 3.2 g of P/kg, 3.1 g of K/kg, 0.3 g of Ca/kg, 1.2 g of Mg/kg, 1.3 g S/kg, 1.6 mg of B/kg, 3.1 mg of Cu/kg, 38 mg of Fe/kg, 8 mg of Mn/kg, and 41 mg of Zn/kg. These latter values readily convert to kg/Mg or g/Mg to estimate nutrient removals by the 19.5 Mg/ha grain yield.

F. Crop Logging

In crop logging, periodic analysis of seasonal samples of high value crops, such as sugarcane (*Saccharum officianarum* L.), sugarbeet, and pineapple [*Ananas comosus* (L.) Merr.], is used so that nutrient additions can be made if an element or elements drop below the optimum or sufficiency levels

(Clements, 1961; Su, 1969). A detailed record is kept of weather conditions, pest infestations, and other growth-influencing factors. Baver (1961) demonstrated how crop logging and soil testing could be combined with other management practices to develop improved recommendation for sugarcane. By closely monitoring the crop to pick up stress conditions under intensive management, nutrients can be applied through irrigation systems or foliarly by ground or aerial applicators. Jones and Bowen (1981) compared crop log and DRIS diagnoses (see below) for sugarcane and found that it depended upon the element being considered which approach gave the most accurate diagnosis. However, nearly identical diagnosis would have been made by either approach for P, Ca, Mg, Mn, S, and Si. As production of field crops becomes more sophisticated, it may be that crop logging based on local values will come into wider use. Some dealerships and farmers are already using crop consultants and scouting to pick up problems earlier and to improve the timeliness of corrective actions.

G. Diagnosis and Recommendation Integrated System

The DRIS, as a system in crop production initially designed by Beaufils (1961), should be applied to both plant and soil analyses. The system was developed to determine (i) the concentrations and interrelations of essential elements in plants and (ii) the soil nutrient levels and conditions and cultural practices associated with optimum plant growth and the highest attainable yields. The system was first applied to rubber trees (*Hevea brasiliensis* L.) in Vietnam. It is based on the premise that with optimum concentrations of one or more elements, the yield level achieved may be high, low, or intermediate, depending upon one or more other factors affecting crop growth. These include availability of other nutrients, soil characteristics, water availability, temperature, plant population, weeds, insects, or diseases and cultural practices. Both the individual elemental concentrations and ratios of the elements, one to another, or their multiples can be used, and these values are divided into those from high- and low-yielding subpopulations. The high- and low-yielding subpopulations are used to determine the means or norms, coefficients of variation, variance, and standard deviations. All expressions for which a significant variance ratio (SA/SB) between the high (A) and low (B) yield subpopulations are found, are retained because these are the expressions which should be useful to determine if a nutrient(s) is/are approaching optimum levels or are deficient or excess. Few researchers have combined both the soil and plant approach to DRIS. Only recently has an effort been made to apply it to soil nutrient parameters with modest success for lower-yielding populations of soybean in the southeastern USA. (Evanylo et al., 1987; Evanylo & Sumner, 1987).

M.E. Sumner, a former colleague of Beaufils from South Africa, introduced the DRIS approach in the USA in the mid-1970s. Since then, DRIS norms have been developed for many crops, including corn or maize (Beaufils, 1973; Sumner, 1977a, b, d, 1979, 1981; Escano et al., 1981; Elwali et al., 1985; Elwali & Gascho, 1988; Walworth et al., 1988); soybean (Sumner,

1977b; Hanson, 1981; Beverly et al., 1986; Evanylo & Sumner, 1987; Hallmark et al., 1988; Hallmark, 1988); wheat (*Triticum aestivum* L.) (Sumner, 1977c, 1981; Amundson & Koehler, 1987); sugarcane (Jones & Bowen, 1981; Elwali & Gascho, 1983; Elwali & Gascho, 1984); alfalfa (Kelling et al., 1985–1986; Russelle & Sheaffer, 1986; Walworth et al., 1986); coastal bermudagrass (*Cynodon dactylon* L.) (Tarpley et al., 1985; Robinson & Tarpley, 1986); potato (Beaufils, 1973; MacKay et al., 1987; Evanylo & Zehnder, 1988); tobacco (*Nicotiana tabacum* L.) (Evanylo et al., 1988a, b); sweet cherry [*Prunus avium* (L.)]; and filbert (*Corylus avellana* L.) (Righetti et al., 1988a, b). The DRIS has also been applied to fir trees [*Abies fraseri* (Pursh) Poir.] (Hockman et al., 1989) for evaluating nutrient balance for Christmas tree quality.

The DRIS approach as applied to plant analysis allows the researcher or diagnostician to place a relative ranking of the element from the most to the least deficient and calculate an index based on the sum of the positive and negative values found for the various nutrients. The DRIS analysis has been found in some cases to be more sensitive than the critical or sufficiency level in identifying the need for higher levels of one or more nutrients or nutrient balance problems. Some public and private laboratories are new using graphic printouts as well as numerical values to indicate the degree of deficiencies or excess of nutrients, such as were proposed by Tarpley et al. (1985).

Because DRIS uses ratios of nutrients, dry matter dilution due to crop maturity is minimized and time of sampling should be less important (Sumner, 1977a). Apparently, the results are somewhat dependent upon the crop. Evanylo et al. (1988a) found that they could make precise diagnoses for tobacco with tissue sampled either early or late if many nutrients are considered simultaneously. Amundson and Koehler (1987) indicated that for wheat, sampling at early heading provided the most correct diagnoses. Hallmark et al. (1988) found that physiological age was important and that more accurate diagnoses were made when plants sampled at the same growth stage were used as the database. Also, there is evidence that for properly developed norms, the portion of the plant sampled becomes less important. However, one should sample the portion most likely to reflect the nutritional status of the crop.

Initially, it was suggested that DRIS norms for different crops established at one geographic location should be applicable to other regions. Results of studies on corn, soybean, alfalfa, wheat, and potato have indicated that norms developed locally or regionally produced more accuracy in diagnosing deficiencies than those from other regions (Escano et al., 1981; Beverly et al., 1986; Walworth et al., 1986; Amundson & Koehler, 1987; MacKay et al., 1987). Within a region, the degree of accuracy has proven to be rather high in predicting likely responses to nutrients based on experimental results (Tarpley et al., 1985).

Researchers are beginning to test the validity of early sampling using both critical levels and DRIS to see if they can detect seasonal nutrient shortages and increase nutrient efficiencies by delayed applications (Elwali & Gascho, 1988). It is expected that research of this type will increase, especially with elements that may be related to environmental problems. Also, researchers are comparing norms generated from a few high-yielding experiments provided from a relatively small database, with those from a broader or worldwide database (Walworth et al., 1988). From the initial comparison, it appears that more diagnostic accuracy may be obtained from norms developed from the high-yielding database. Combinations of DRIS and modified-DRIS (M-DRIS) norms using nutrient ratios and dry matter concentrations from full-bloom trifoliate leaf samples from soybean yielding >3.5 Mg/ha (52 bu/acre) indicated that M-DRIS norms were especially helpful in correctly diagnosing nutrients that were not deficient (Hallmark, 1988). The DRIS norms were most useful in correctly diagnosing situations where P and K were most deficient.

Even though some claim that there is little physiological basis for some of the ratios used in DRIS (Smith, 1986), it is a system that provides the possibility of bringing all of the elements involved in nutrition together and evaluating them simultaneously, with yield level being part of the process. Hence, it should provide an improved opportunity to look at nutrient balance. While DRIS is not perfect, it often provides a better estimate for predicting nutrient shortages and responses than the critical level or sufficiency concepts. The latter estimates identify the most-limiting nutrient or nutrients, which must be supplied before the sufficiency of other elements can properly be evaluated. The DRIS is designed to overcome this limitation, using the ratio norms to place an order on the second, third, or fourth most limiting nutrients.

Karlen et al. (1988) applied DRIS norms to corn results in which Flannery had produced 19.3 Mg/ha (308 bu/acre) of grain and found that all but four of the indices were within +/− 18 for most of the samplings. These indices apparently indicate relatively good nutrient balance. It is probable that the use of DRIS will continue to increase. If adequate calibration data are available and properly evaluated, the sufficiency level and DRIS approaches should in the long-run provide similar answers to diagnosticians' questions.

Some experiment stations now have copyright computer disketts available for DRIS analysis of various crops, enhancing the availability and use of the programs. The University of Georgia is one such institution. Many soil and plant analysis laboratories now provide a DRIS analysis when returning the plant analysis results to the customer.

H. Determination of Total Nutrient Uptake of High-Yielding Crops

One of the major functions of plant analysis is to determine the uptake and use of various nutrients by different crops over the season and at maturity. Knowing the amounts of nutrients taken up at various times during the season is important because uptake is rate driven, with peak periods of nutrients required, depending upon the developmental stages of growth. These peak demands for high yields may not be met by many soils unless greater-than-normal levels of nutrients are present or special applications of some are provided. The total uptake during different periods helps determine the quantities that must be provided by the soil and fertility program, and the amounts that will be removed by the harvested crop. The latter values, long-term, help determine the maintenance amounts of nutrients to maintain or increase yields and fertility of soils.

Rhoads and Stanley (1981, 1984); Flannery (1986a, b); and Karlen et al. (1985, 1987, 1988) have determined the nutrient uptake of high-yielding corn. Karlen et al. (1988) found that for a corn crop yielding 19.3 Mg/ha (308 bu/acre), the greatest amounts of nutrients per hectare in the aerial portion of the crop were: N, 386 kg; P, 70 kg; K, 370 kg; Ca, 59 kg; Mg, 44 kg; S, 40 kg; B, 0.13 kg; Cu, 0.14 kg; Fe, 1.9 kg; Mn, 0.8 kg; and Zn, 0.8 kg. Flannery (1986a) found that average rates of nutrient uptake for N, P, and K during a 13-d period between the 12th leaf and early tassel stages of growth were 12.38 kg ha^{-1} d^{-1} of N, 1.39 kg ha^{-1} d^{-1} of P, and 14.24 kg ha^{-1} d^{-1} of K. During that 13-d period, the crop accumulated about 42% of its N, 26% of its P and >50% of its K. Karlen et al. (1987) found that peak accumulation rates for N, P, and K were 10, 1.6, and 28 kg ha^{-1} d^{-1}, respectively, at a growth stage associated with 550 growing degree units for corn that yielded 14 Mg/ha.

Henderson and Kamprath (1970), Hanway and Weber (1971), Terman (1977), Karlen et al. (1982), and Flannery (1986b) have studied dry matter and nutrient accumulation during the growth and development of soybean. The bean yields ranged from 4.77 (Terman) to 5.38 (Henderson and Kamprath) to 6.79 (Flannery) Mg/ha in the respective studies. In Flannery's study, the highest total aerial uptake of nutrients were as follows: N, 614 kg; P, 64.6 kg; K, 402 kg; Ca, 185 kg; Mg, 63 kg; and S, 31 kg. The peak rate of uptake of N was 12.8 kg ha^{-1} d^{-1} during a 21-d period between pod development and soft seed. Forty-four percent of the total N was taken up during that period. For P, the highest rate of uptake occurred during the same period as the N, and was 1.35 kg ha^{-1} d^{-1}, with the same percentage of the total taken up as for N. The highest rate of K uptake occurred between full bloom and pod development, from the 67th to 82nd day following emergence, and was 8.92 kg of K ha^{-1} d^{-1}. Between the 51st and 82nd day, more than 213 kg of K or nearly 55% of crops total K was taken up.

Total nutrient uptake values have been determined for wheat yielding more than 11 Mg/ha (Brown, 1986), lowland rice grain (*Oryza sativa* L.) yielding 9.8 Mg/ha (DeDatta & Mikkelsen, 1985), alfalfa yielding from 17.9 to more than 20 Mg/ha (Tesar, 1981; Lanyon et al., 1983), potato crops yielding 79 Mg/ha (Roberts & McDole, 1985), cassava (*Manihot esculenta* Crantz) with a fresh root yield of 52.1 Mg/ha (Howeler, 1985), and cauliflower (*Brassica oleracea* L.) yielding 17.9 Mg/ha of curd (Welch et al., 1987). Practitioners desiring these data should check those papers for details.

For high-yielding crops, the uptake per megagram of crop produced or the nutrient removal per megagram of crop harvested will often be nearly constant. There is less variation in the elemental content of the harvested grain than in portions that may be left in the field. For example, Bundy and Oplinger (1984) found that when yields of soybean were increased from 4.4

Table 14-2. Concentrations of nutrients in soybean seed at two yield levels and row widths (Bundy & Oplinger, 1984).

Soybean yield	Row width	Nutrient concentration in seed									
		N	P	K	Ca	Mg	S	Zn	B	Mn	Cu
Mg/ha	cm	g/kg						mg/kg			
5.846	20.3	58.5	6	17.9	1.9	2.4	3.1	41	26	25	10
4.435	76.2	58.3	6	18.0	2.0	2.3	3.0	43	27	25	11

to more than 5.8 Mg/ha by narrowing the row spacing, the elemental concentrations of the different nutrients in the seed for the two yield levels remained nearly constant (Table 14-2), but the removal of nutrients was, of course, greater for the higher yield level.

Gill and Kamprath (E.J. Kamprath, 1987, personal communication) conducted research on rice in Sumatra and found that with increasing rates of applied K, the K concentration of the grain only ranged from 2.9 to 3.1 g/kg, the higher value being on the control plots, while the grain dry matter yield was increased 1.65 Mg/ha. A high correlation was found between the rice yield and the K concentration in the straw at harvest (r = 0.90).

For forages, such as alfalfa, it is relatively easy to evaluate the nutrient requirements per unit of production or efficiencies of nutrients. In a study in which rates of K and irrigation levels were used to increase yields, at the two highest levels of production, Sheaffer (1984) found the results shown in Table 14-3. Even though there was more than a 4.7 Mg/ha differential in dry matter yield, the nutrient concentrations for the two yield levels did not differ significantly. By taking the reciprocal of the values presented, the internal efficiencies for the different nutrients are determined.

III. THE FUTURE

While plant analysis has not proven to be the panacea that some had hoped in pointing the way to higher yields or solving soil fertility problems, it is nevertheless an important tool in our diagnostic arsenal. In many cases where the answers to soil fertility problems are fairly simple, it is extremely effective. However, in cases where problems are more complex, application of DRIS, which allows one to evaluate the relationships among nutrients, may provide a better approach. If DRIS can provide better answers than the critical level or sufficiency range, it will probably become the system or

Table 14-3. Alfalfa nutrient uptake per Mg of dry matter (DM) produced at two yield levels (Sheaffer, 1984).

Alfalfa yield-DM	Nutrient uptake of dry matter								
	N	P	K	Ca	Mg	Zn	B	Cu	Mn
Mg/ha	kg/Mg								
13.7	29	2.2	29.0	15.5	2.55	0.02	0.05	0.01	0.055
9.96	31	2.15	24.5	16.0	2.75	0.02	0.055	0.01	0.070

approach of choice. The DRIS norms,once developed, can easily be entered into computers so that the analysis is readily available to both laboratories and individuals.

Extension specialists and agents, as well as, dealers and consultants are becoming more computer literate and sophisticated in their operations. They are now using plant analysis more at all production levels. They can have information available at the flick-of-a-switch and can access communication networks that were only dreamed of a few short years ago. While some farm-supply consulting groups have chosen to back off from their initial goal of putting many crop consultants in the field, indications are that the number of private consultants will continue to increase. This will be in response to farmer demand for special attention in their specific production systems, seasonal monitoring, and improvement of nutrient use efficiency for rational optimum yields. More attention will be given to matching genotypes, soil fertility, and cultural management programs. This approach will help balance economic and environmental considerations.

The challenge for researchers and producers is to determine how to produce a bigger package or yield in a given time frame, for a single crop or with multiple cropping. As has been shown through the application of plant analysis in high-yield research, such systems place special nutrient stress on soils in terms of rates, processes, and nutrient availabilities during peak periods of uptake. In the future, large gains will be made in crop production. Study of quantum mechanics and randomness of our system indicate that striking accomplishments are likely in many fields, including agriculture, especially as new genotypes become available from biotechnology. The soil fertility and plant nutrition researchers must be part of that technology to analyze and test the new lines that result from that effort. Plant analysis will play a key role in future studies.

REFERENCES

Ahmed, S., and H.J. Evans. 1960. Cobalt: A micronutrient element for the growth of soybean plants under symbiotic conditions. Soil Sci. 90:205–210.

Amundson, R.L., and F.E. Koehler. 1987. Utilization of DRIS for diagnosis of nutrient deficiencies in winter wheat. Agron. J. 79:472–476.

Anderson, O.E., and R.M. Harrison. 1970. Micronutrient variation within cotton leaf tissue as related to variety and soil location. Commun. Soil Sci. Plant Anal. 1(3):163–172.

Anderson E.L., E.J. Kamprath, and R.H. Moll. 1984. Nitrogen fertility effects on accumulation, remobilization, and partitioning of N and dry matter in corn genotypes differing in prolificacy. Agron. J. 76:397–404.

Baker, D.E., A.E. Jarrell, L.E. Marshall, and W.I. Thomas. 1970. Phosphorus uptake from soil by corn hybrids selected for high and low phosphorus accumulation. Agron. J. 62:103–106.

Baligar, V.C., and S.A. Barber. 1979. Genotype differences of corn in ion uptake. Agron. J. 71:870–873.

Baver, L.D. 1961. A research background for recommendations for sugarcane fertilization. (Reprint.) Int. Sugar J. 63:195–199, 227–230, 259–262, 295–296.

Beaufils, E.R. 1961. Les desequilibres chimiques chez l'*Hevea brasiliensis*. La methode dite du diagnostic physiologique. Ph.D. diss. Sorbonne, Paris.

Beaufils, E.R. 1973. Diagnostic and recommendation integrated system (DRIS). A general scheme for experimentation and calibration based on principles developed from research in plant nutrition. Soil Sci. Bull. 1. Univ. of Natal, Pietermaritzburg, S. Africa.

Beverly, R.B., M.E. Sumner, W.S. Letzsch, and C.O. Planck. 1986. Foliar diagnosis of soybeans by DRIS. Commun. Soil Sci. Plant Anal. 17(3):237–256.

Bloom, P., and M. Meyer. 1988. Silicon and nitrogen nutrition of wild rice. p. 26–39. *In* Minnesota wild rice research 1987. Minnesota Agric. Exp. Stn. Misc. Publ. 54.

Brown, B.D. 1986. Nutrient uptake of high yielding wheat. Better Crops Plant Food 70:27.

Brown, J.R. 1970. Plant analysis. Missouri Agric. Exp. Stn. Bull. SB881.

Broyer, T.C., A.B. Carlton, C.M. Johnson, and P.R. Stout. 1954. Chlorine-A micronutrient element for higher plants. Plant Physiol. 29:526–532.

Bruetsch, T.F., and G.O. Estes. 1976. Genotype variation in nutrient uptake efficiency in corn. Agron. J. 68:521–523.

Bundy, L.G., and E.S. Oplinger. 1984. Narrow row spacings increase soybean yields and nutrient removal. Better Crops Plant Food 68:16–17.

Carlone, M.R., and W.A. Russell. 1987. Response to plant densities and nitrogen levels for four maize cultivars from different eras of breeding. Crop Sci. 27:465–470.

Carpena, O., A. Masaguer, and M.J. Sarro. 1988. Nutrient uptake by cultivars of tomato plants. Plant Soil 105:294–296.

Chapman, H.D. (ed.). 1966. Diagnostic criteria for plants and soils. Univ. of Calif., Div. of Agric., Berkeley.

Chapman, H.D. 1967. Plant analysis values suggestive of nutrient status of selective crops. p. 77–92. *In* G.W. Hardy et al. (ed.) Soil testing and plant analysis. Part II. Plant analysis. SSSA Spec. Publ. 2. SSSA, Madison, WI.

Clements, H.F. 1961. Crop logging of sugar cane in Hawaii. p. 131–137. *In* W. Reuther (ed.) Plant analysis and fertilizer problems. Publ. 8. Am. Inst. of Biol. Sci., Washington, DC.

Clark, D.H., H.F. Mayland, and R.C. Lamb. 1987. Mineral analysis of forages with near infrared reflectance spectroscopy. Agron. J. 79:485–490.

Coffelt, T.A., and D.L. Hallock. 1986. Soil fertility responses of Virginia-type peanut cultivars. Agron. J. 78:131–137.

Dalton, D.A., H.J. Evans, and F.H. Hanus. 1985. Stimulation by nickel of soil microbial urease activity, and hydrogenase activities in soybeans grown in low-nickel soil. Plant Soil 88:245–258.

DeDatta, S.K., and D.S. Mikkelsen. 1985. Potassium nutrition of rice. p. 665–699. *In* R.D. Munson (ed.) Potassium in agriculture. ASA, Madison, WI.

Dick, A.C., S.S. Malhi, P.A. O'Sullivan, and D.R. Walker. 1985. Chemical composition of whole plant and grain yield of nutrients in grain of five barley cultivars. Plant Soil 86:257–264.

Dow, A.I., and S. Roberts. 1982. Proposal: Critical nutrient ranges for crop diagnosis. Agron. J. 74:401–403.

Dumenil, L. 1974. What affects P and K soil fertility levels? EC-904. *In* 26th Ann. Fertilizer and Agchemical Dealers Conf. Iowa State Univ., Ames.

Ebelhar, S.A., E.J. Kamprath, and R.H. Moll. 1987. Effects of nitrogen and potassium on growth and cation composition of corn genotypes differing in average ear number. Agron. J. 79:875–881.

Elliott, G.C., and A. Lauchli. 1985. Phosphorus efficiency and phosphate-iron interaction in maize. Agron. J. 77:399–403.

Elwali, A.M.O., and G.J. Gascho. 1983. Sugarcane response to P, K and DRIS corrective treatments on Florida Histisols. Agron. J. 75:79–83.

Elwali, A.M.O., and G.J. Gascho. 1984. Soil testing, foliar analysis, and DRIS as guides for sugarcane fertilization. Agron. J. 76:466–470.

Elwali, A.M.O., and G.J. Gascho. 1988. Supplemental fertilization of irrigated corn guided by foliar critical nutrient levels and diagnosis and recommendation integrated system norms. Agron. J. 80:243–249.

Elwali, A.M.O., G.J. Gascho, and M.E. Sumner. 1985. DRIS norms for 11 nutrients in corn leaves. Agron. J. 77:506–508.

Embleton, T.W., W.W. Jones, and R.G. Platt. 1976. Leaf analysis as a guide to citrus fertilization. p. 4–9. *In* H.M. Reisenauer (ed.) Soil and plant-tissue testing in California. Div. Agric. Sci. Univ. Calif. Bull. 1879.

Escano, C.R., C.A. Jones, and G. Uehara. 1981. Nutrient diagnosis in corn grown on Hydric Dystrandepts. II. Comparison of two systems of tissue diagnosis. Soil Sci. Soc. Am. J. 45:1140–1144.

Evans, H.J. 1989. Hydrogen recycling in legume nodules: Molecular biology and role of trace elements. p. 11. *In* Agronomy abstracts. ASA, Madison, WI.

Evanylo, G.K., J.H. Grove, and J.L. Sims. 1988a. Effect of sampling period on nutrient diagnostic indices for burley tobacco. Agron. J. 80:615–619.

Evanylo, G.K., J.L. Sims, and J.H. Grove. 1988b. Nutrient norms for cured burley tobacco. Agron. J. 80:610–614.

Evanylo, G.K., and M.E. Sumner. 1987. Preliminary development and testing of DRIS soil norms for soybean production. Commun. Soil Sci. Plant Anal. 18(12):1355–1377.

Evanylo, G.K., M.E. Sumner, and W.S. Letzsch. 1987. Utilization of the boundary line approach in the development of soil nutrient norms for soybean production. Commun. Soil Sci. Plant Anal. 18(12):1379–1401.

Evanylo, G.K., and G.W. Zehnder. 1988. Potato growth and nutrient diagnosis as affected by systemic pesticide and physiological growth stage. Commun. Soil Sci. Plant Anal. 19:1731–1745.

Fixen, P.E. 1985. Response of spring grains to chloride fertilization. p. 13–24. *In* Soils, Fertilizer and Agricultural Pesticides Short Course Proc. Minneapolis. Dec. Agric. Ext. Serv., Univ. of Minnesota, St. Paul.

Fixen, P.E., G.B. Farber, R.H. Gelderman, and J.R. Gerwing. 1986. Role of Cl in maximum yield environments: I. Evidence of yield response and Cl requirements. p. 41–51. *In* T.L. Jackson (ed.) Chloride and crop production. Spec. Bull. 2. Potash and Phosphate Inst., Atlanta.

Flannery, R.L. 1982. Maximum yield soybean experiment in 1982. Soils and crops Res. Center Rep., Rutgers Univ., NJ.

Flannery, R.L. 1986a. Plant food uptake in a maximum yield corn study. Better Crops Plant Food 70:4–5.

Flannery, R.L. 1986b. Plant food uptake in a maximum yield soybean study. Better Crops Plant Food 70:6–7.

Gallaher, R.N., W.L. Parks, and L.M. Josephson. 1972. Effect of levels of soil potassium, fertilizer potassium and season on yield and ear leaf potassium content of corn inbreds and hybrids. Agron. J. 64:645–647.

Gettier, S.W., D.C. Martens, and S.J. Donohue. 1985. Soybean yield response prediction from soil test and tissue manganese levels. Agron. J. 77:63–67.

Goodall, D.W., and F.G. Gregory. 1947. Chemical composition of plants as an index of their nutritional status. Imp. Bur. Hortic. Plantation Crops. Tech. Commun. 17. Ministry of Agric., London.

Goos, R.J. 1986. Potassium chloride fertilization and common root rot of barley—an update. p. 89–94. *In* Proceedings of the 16th North Central Extension-Industry Soil Ferility Workshop. Proc. Bridgeton, MO. 29–30 Oct. Potash and Phosphate Inst., Manhattan, KS.

Goos, R.J., D.G. Westfall, A.E. Ludwick, and J.E. Goris. 1982. Grain protein content as an indicator of N sufficiency for winter wheat. Agron. J. 74:130–133.

Gorsline, G.W., D.E. Baker; and W.I. Thomas. 1965. Accumulation of eleven elements by field corn (*Zea mays* L.). Pennsylvania Agric. Exp. Stn. Bull. 725.

Hallmark, W.B. 1988. Comparison of DRIS and M-DRIS for diagnosing P and K deficiencies in soybeans. Better Crops Plant Food 72(3):20–21.

Hallmark, W.B., C.J. deMooy, H.F. Morris, J. Pesek, K.P. Shao, and J.D. Fontenot. 1988. Soybean phosphorus and potassium deficiency detection as influenced by plant growth stage. Agron. J. 80:586–591.

Hanson, R.G. 1981. DRIS evaluation of N, P, K status of determinant soybeans in Brazil. Commun. Soil Sci. Plant Anal. 12(9):933–948.

Hanway, J.J. 1962. Corn growth and composition in relation to soil fertility: III. Percentages of N, P, and K in different plant parts in relation to stages of growth. Agron. J. 54:222–229.

Hanway, J.J., and C.R. Weber. 1971. N, P, and K percentages in soybeans [*Glycine max* (L.) Merrill] plant parts. Agron. J. 63:286–290.

Henderson, J.B., and E.J. Kamprath. 1970. Nutrient and dry matter accumulation by soybeans. North Carolina Agric. Exp. Stn. Bull. 197.

Hills, F.J., and A. Ulrich. 1976. Plant analysis as a guide for mineral nutrition of sugar beets. p. 18–21. *In* H.M. Reisenauer (ed.) Soil and plant-tissue testing in California. Div. Agric. Sci., Univ. of California Bull. 1879.

Hockman, J.N., J.A. Burger, and D.Wm. Smith. 1989. A DRIS application to Fraser fir Christmas trees. Commun. Soil Sci. Plant Anal. 20:305–318.

Howeler, R.H. 1985. Potassium nutrition of cassava. p. 819–841. *In* R.D. Munson (ed.) Potassium in agriculture. SSSA, Madison, WI.

Hylton, L.O., A. Ulrich, and D.R. Cornelius. 1967. Potassium and sodium interrelations in growth and mineral contents of Italian ryegrass. Agron. J. 59:311-315.

Jackson, T.L. (ed.). 1986. Chloride and crop production. Potash and Phosphate Inst., Atlanta.

Jones, C.A., and J.E. Bowen. 1981. Comparative DRIS and crop log diagnosis of sugarcane tissue analyses. Agron. J. 73:941-944.

Karlen, D.L., and C.R. Camp. 1985. Row spacing, plant population, and water management effects on corn in the Atlantic Coastal Plain. Agron. J. 77:393-398.

Karlen, D.L., C.R. Camp, and J.P. Zublena. 1985. Plant density, distribution, and fertilizer effects on yield and quality of irrigated corn silage. Commun. Soil Sci. Plant Anal. 16(1):55-70.

Karlen, D.L., R.L. Flannery, and E.J. Sadler. 1988. Aerial accumulation and partitioning of nutrients by corn. Agron. J. 80:232-242.

Karlen, D.L., P.G. Hunt, and T.A. Matheny. 1982. Accumulation and distribution of K, Ca, and Mg by selected determinate soybean cultivars grown with and without irrigation. Agron. J. 74:347-354.

Karlen, D.L., E.J. Sadler, and C.R. Camp. 1987. Dry matter, nitrogen, and potassium accumulation rates by corn on Norfolk loamy sand. Agron. J. 79:649-656.

Kelling, K.A., E.E. Schulte, and T. Erickson. 1985-1986. Adapting DRIS for alfalfa: What are the diagnostic norms? Better Crops Plant Food 70:18-20.

Kuchenbuch, R.O., and S.A. Barber. 1988. Significance of temperature and precipitation for maize root distribution in the field. Plant Soil 106:9-14.

Kuchenbuch, R.O., N. Claassen, and A. Jungk. 1986. Potassium availability in relation to soil moisture. I. Effect of soil moisture on potassium diffusion, root growth and potassium uptake of onion plants. Plant Soil 95:221-231.

Lanyon, L.E., J.E. Baylor, and W.K. Waters. 1983. Understanding alfalfa nutrient uptake. Better Crops Plant Food 67:12-13.

Lanyon, L.E., and W.K. Griffith. 1988. Nutrition and fertilizer use. p. 333-372. *In* Alfalfa and alfalfa improvement. ASA, CSSA, and SSSA, Madison, WI.

Lanyon, L.E., and F.W. Smith. 1985. Potassium nutrition of alfalfa and other forage legumes: Temperate and tropical. p. 861-893. *In* R.D. Munson (ed.) Potassium in agriculture. ASA, CSSA, and SSSA, Madison, WI.

Loper, G.M., and D. Smith. 1961. Changes in micronutrient composition of herbage of alfalfa, medium red clover, ladino clover, and bromegrass with advance in maturity. Res. Rep. 8. Wisconsin Agric. Exp. Stn., Madison, WI.

Loué, A. 1963. A contribution to the study of the inorganic nutrition of maize, with special attention to potassium. Fertilité 20 (Nov.-Dec.).

Lutrick, M.C., H.A. Peacock, and J.A. Cornell. 1986. Nitrate monitoring for cotton lint production on a typic Paleudult. Agron. J. 78:1041-1046.

Mackay, A.D., and S.A. Barber. 1985a. Soil moisture effects on root growth and phosphorus uptake by corn. Agron. J. 77:519-523.

Mackay, A.D., and S.A. Barber. 1985b. Soil moisture effect on potassium uptake by corn. Agron. J. 77:524-527.

Mackay, A.D., and S.A. Barber. 1986. Effect of nitrogen on root growth of two corn genotypes in the field. Agron. J. 78:699-703.

Mackay, D.C., J.M. Carefoot, and T. Entz. 1987. Evaluation of the DRIS for assessing the nutritional status of potato (*Solanum tuberosum* L.). Commun. Soil Sci. Plant Anal. 18:1331-1353.

Macy, P. 1936. The qualitative mineral nutrient requirements of plants. Plant Physiol. 11:749-764.

Malzer, G.L., J. Geadelmann, and T. Graff. 1984. High corn yield experiments on coarse textured soil of Minnesota. Agric. Exp. Stn. Misc. Publ. 2(rev.).

Marschner, H. 1986. Mineral nutrition of higher plants. Academic Press, Orlando, FL.

Melsted, S.W., H.L. Motto, and T.R. Peck. 1969. Critical plant nutrient composition values useful in interpreting plant analysis data. Agron. J. 61:71-20.

Munson, R.D. 1968. Interaction of potassium and other ions. p. 321-353. *In* V.J. Kilmer et al. (ed.) The role of potassium in agriculture. SSSA, Madison, WI.

Munson, R.D. 1970. Plant analysis: Varietal and other considerations. p. 84-104. *In* F. Greer (ed.) Proceedings from a symposium on plant analysis. Int. Minerals and Chemicals Corp., Skokie, IL.

Munson, R.D. 1974. Plant breeding and nutrient concentration or uptake: A perspective. Mimeo. Crop Sci. Soc. Am. Symp., Chicago. Potash Inst. of North Am., Lafayette, IN.

Munson, R.D., and W.L. Nelson. 1973. Principles and practices in plant analysis. p. 223–248. *In* L.M. Walsh and J.D. Beaton (ed.) Soil testing and plant analysis (rev.). Madison, WI.

Ohki, K. 1976. Manganese deficiency and toxicity levels for 'Bragg' soybeans. Agron. J. 68:861–864.

Ohki, K. 1984a. Manganese deficiency and toxicity effects on growth, development, and nutrient composition in wheat. Agron. J. 76:213–218.

Ohki, K. 1984b. Zinc nutrition related to critical deficiency and toxicity levels for sorghum. Agron. J. 76:253–256.

Ohki, K. 1987. Aluminum stress on sorghum growth and nutrient relationships. Plant Soil 98:195–202.

Overdahl, C.J., and M. O'Leary. 1981. Corn yields from varying levels of potash p. 27. *In* A report on field research in soils. Soil Ser. 109.

Parker, M.B., T.P. Gaines, and G.J. Gascho. 1985. Chloride effects on corn. Commun. Soil Sci. Plant Anal. 16(12):1319–1333.

Parker, M.B., G.J. Gascho, and T.P. Gaines. 1983. Chloride toxicity of soybeans grown on Atlantic coastal Flatwood soils. Agron. J. 75:439–443.

Parker, M.B., and M.E. Walker. 1986. Soil pH and manganese effects on manganese nutrition of peanut. Agron. J. 78:614–620.

Parks, W.L. 1985. Interaction of potassium with crop varieties or hybrids. p. 535–558. *In* R.D. Munson (ed.) Potassium in agricultre. ASA, CSSA, and SSSA, Madison, WI.

Pierre, W.H., V.D. Jolley, J.R. Webb, and W.D. Schrader. 1977a. Relation between corn yield, expressed as percentage of the maximum, and the N percentage in the grain. I. Various N-rate experiments. Agron. J. 69:215–220.

Pierre, W.H., L. Dumenil, and J. Henao. 1977b. Relationship between corn yield, expressed as a percentage of the maximum, and the N percentage of the grain. II. Diagnostic use. Agron. J. 69:221–226.

Planck, C.O. 1979. Plant analysis handbook for Georgia. Univ. of Georgia Coop. Ext. Serv. Bull. 735.

Randall, G.W., P.L. Kelly, and P.M. Russelle. 1987. Rotation nitrogen study. p. 117–122. *In* A report on field research in soils. Univ. of Minnesota Agric. Exp. Stn. Misc. Publ. 2(rev.).

Rehm, G.W., R.C. Sorensen, and R.A. Wiese. 1983. Application of phosphorus, potassium, and zinc to corn grown for grain or silage: Nutrient concentration and uptake. Soil Sci. Soc. Am. J. 47:697–700.

Reneau, R.B., Jr., G.D. Jones, and J.B. Friedericks. 1983. Effect of P and K on yield and chemical composition of forage sorghum. Agron. J. 75:5–8.

Reuter, D.J., and J.B. Robinson (ed.). 1986. Plant analysis: An interpretation manual. Inkata Press Proprietary, Melbourne and Sidney, Australia.

Rhoads, F.M., and R.L. Stanley, Jr. 1981. Fertilizer scheduling, yield, and nutrient uptake of irrigated corn. Agron. J. 73:971–974.

Rhoads, F.M., and R.L. Stanley, Jr. 1984. Yields and nutrient utilization efficiency of irrigated corn. Agron. J. 76:219–223.

Righetti, T.L., O. Alkoshab, and K. Wilder. 1988a. Diagnostic biases in DRIS evaluations on sweet cherry and hazelnut. Commun. Soil Sci. Plant Anal. 19(13):1429–1447.

Righetti, T.L., O. Alkoshab, and K. Wilder. 1988b. Verifying critical values from DRIS norms. Commun. Soil Sci. Plant Anal. 19(13):1449–1466.

Roberts, S., and A.I. Dow. 1982. Critical nutrient ranges for petiole phosphorus levels of sprinkler-irrigated Russet Burbank potatoes. Agron. J. 74:583–585.

Roberts, S., and R.E. McDole. 1985. Potassium nutrition of potatoes. p. 799–818. *In* R.D. Munson (ed.). Potassium in agriculture. ASA, CSSA, and SSSA, Madison, WI.

Robinson, D.L., and M.L. Tarpley. 1986. DRIS proves useful for diagnosing nutrient deficiencies in coastal bermudagrass. Better Crops Plant Food 70:8–9.

Russell, J.S., C.W. Bourg, and H.F. Rhoades. 1954. Effect of nitrogen fertilizer on the nitrogen, phosphorus, and cation contents of bromegrass. Soil Sci. Soc. Am. Proc. 18:292–296.

Russelle, M.P., and C.C. Sheaffer. 1986. Use of the diagnosis and recommendation integrated system with alfalfa. Agron. J. 78:557–560.

Sale, P.W.G., and L.C. Campbell. 1986. Yield and composition of soybean seed as a function of potassium supply. Plant Soil 96:317–325.

Sale, P.W.G., and L.C. Campbell. 1987. Differential responses to K deficiency among soybean cultivars. Plant Soil 104:183–190.

Sayer, J.D. 1955. Mineral nutrition of corn. p. 293–314. *In* G.F. Sprague (ed.) Corn and corn improvement. Agronomy 5. ASA, Madison, WI.

Sheaffer, C.C. 1984. Potash for irrigated alfalfa boosts yield and reduces stand loss. Better Crops Plant Food 68:8–9.

Silberbush, M., and S.A. Barber. 1985. Root growth, nutrient uptake and yield of soybean cultivars grown in the field. Commun. Soil Sci. Plant Anal. 16:119–127.

Smith, F.W. 1986. Interpretation of plant analysis concepts and principles. p. 1–12. *In* D.J. Reuther and J.B. Robinson (ed.) Plant analysis: An interpretation manual. Inkata Press Proprietary, Melbourne and Sidney, Australia.

Steenbjerg, F. 1951. Yield curves and chemical plant analyses. Plant Soil 3:97–109.

Su, N.R. 1969. Research on fertilization of pineapple in Taiwan and some associated cultural practices. Spec. Publ. 1. Soc. Soil Scientists and Fertilizer Technologists of Taiwan, Taipei.

Sumner, M.E. 1977a. Application of Beaufils diagnostic indices to maize data published in the literature irrespective of age and conditions. Plant Soil 46:359–369.

Sumner, M.E. 1977b. Preliminary N, P, and K foliar diagnostic norms for soybeans. Agron. J. 69:226–230.

Sumner, M.E. 1977c. Preliminary NPK foliar diagnostic norms for wheat. Commun. Soil Sci. Plant Anal. 8:149–167.

Sumner, M.E. 1977d. Effect of corn leaf sampled on N, P, K, Ca and Mg content and calculated DRIS indices. Commun. Soil Sci. Plant Anal. 8:269–280.

Sumner, M.E. 1979. Interpretation of foliar analysis for diagnostic purposes. Agron. J. 71:343–348.

Sumner, M.E. 1981. Diagnosing the sulfur requirement of corn and wheat using foliar analysis. Soil Sci. Soc. Am. J. 45:87–90.

Takkar, P.N., M.S. Mann, R.L. Bansal, N.S. Randhawa, and H. Singh. 1976. Yield and uptake response of corn to zinc as influenced by phosphorus fertilization. Agron. J. 68:942–946.

Tarpley, M.L., D.L. Robinson, B.K. Gustavson, and M.M. Eichhorn. 1985. The DRIS for interpretation of coastal bermudagrass analysis. Commun. Soil Sci. Plant Anal. 16(12):1335–1348.

Terman, G.L. 1977. Yields and nutrient accumulation by determinant soybeans as affected by applied nutrients. Agron. J. 69:234–238.

Tesar, M.B. 1981. High yield alfalfa research. *In* Weed, Seed, and Fertilizer Conf. Proc., East Lansing, MI. Michigan State Univ., East Lansing.

Tyler, K.B., and O.A. Lorenz. 1962. Diagnosing nutrient needs in vegetables. Better Crops Plant Food 46(3):6–13.

Ulrich, A. 1948. Plant analysis—methods and interpretation of results. p. 157–198. *In* H.B. Kitchen (ed.) Diagnostic techniques for soils and crops. Am. Potash Inst. Washington, DC.

Ulrich, A. 1949. Critical nitrate levels of sugar beets estimated from analysis of petiole and blades, with special reference to yields and sucrose concentrations. Soil Sci. 69:291–309.

Ulrich, A. 1961. Plant analysis in sugar beet nutrition. p. 190–211. *In* W. Reuter (ed.) *In* Plant analysis and fertilizer problems. Am. Inst. of Biol. Sci., Washington, DC.

Ulrich, A. 1976. Plant tissue analysis, plant analysis as a guide in fertilizing crops. p. 1–4. *In* H.M. Reisenauer (ed.) Soil and plant-tissue testing in California. Univ. of California Bull. 1879.

Ulrich. A., and F.J. Hills. 1967. Principles and practices of plant analysis. p. 11–24. *In* Soil testing and plant analysis. Part II. SSSA Spec. Publ. Ser. 2. SSSA, Madison, WI.

Viets, F.G., Jr., C.E. Nelson, and C.L. Crawford. 1954. The relationship among corn yields, leaf composition and fertilizers applied. Soil Sci. Soc. Am. Proc. 18:297–301.

Walworth, J.L., M.E. Sumner, R.A. Isaac, and C.O. Plank. 1986. Preliminary DRIS norms for alfalfa in the southeastern United States and a comparison with midwestern norms. Agron. J. 78:1046–1052.

Walworth, J.L., H.J. Woodward, and M.E. Sumner. 1988. Generation of corn tissue norms from a small, high-yield data base. Commun. Soil Sci. Plant Anal. 19(5):563–577.

Ware, G.O., K. Ohki, and L.C. Moon. 1982. The Mitscherlich plant growth model for determining the critical nutrient deficiency levels. Agron. J. 74:88–91.

Welch, N.C., K.B. Tyler, and D. Ririe. 1987. Split nitrogen applications best for cauliflower. Calif. Agric. 41(6):21–22.

Westermann, D.T., and G.E. Kleinkopf. 1985a. Phosphorus relationships in potato plants. Agron. J. 77:490–494.

Westermann, D.T., and G.E. Kleinkopf. 1985b. Nitrogen requirements of potatoes. Agron. J. 77:616–621.

Chapter 15

Sampling, Handling, and Analyzing Plant Tissue Samples

J. BENTON JONES, JR., *University of Georgia, Athens*

VERNON W. CASE, *University of Kentucky, Lexington*

Plant analysis (sometimes referred to as leaf analysis) is the determination of the total elemental content of a specified plant part. The emphasis in this chapter will be on the determination of those elements required for plant growth. Interpretation is normally based on the use of a "critical value" or "sufficiency range" (Smith, 1962) comparison between the elemental concentration found and a known norm (Goodall & Gregory, 1947; Chapman, 1966; Reuter & Robinson, 1986; Adriano, 1986; Martin-Prevel et al., 1987). An alternative method of interpretation is Diagnosis and Recommendation Integrated System (DRIS), which interprets the ratios of elements (N/P, K/Ca, and K/Mg) as indicators of elemental status (Beaufils, 1973; Sumner, 1977, 1982).

Most growers primarily use a plant analysis for diagnosing suspected elemental insufficiencies, while its most significant, yet little used application, is for evaluating the soil/plant elemental status. This is partially reflected in the relatively few plant tissue samples assayed for growers, about 500 000, in the USA each year (Jones, 1985). Tissue testing, an elemental assay of extracted cell sap by means of quick chemical tests in the field, seems to be gaining an interest at levels equal to that observed several decades ago.

A plant analysis is carried out in a series of steps as shown in Fig. 15–1. The results obtained are no better than the care taken in collecting, handling, preparing, and analyzing the collected tissue. An error made in one of these steps can result in an erroneous interpretation leading to recommendations that may be either unnecessary, costly, or even damaging to the crop. Therefore, it is important for those employing either a plant analysis or tissue test to follow the proper sampling, preparation, and analysis procedures. This chapter deals with the procedures required to successfully conduct a plant analysis or tissue test.

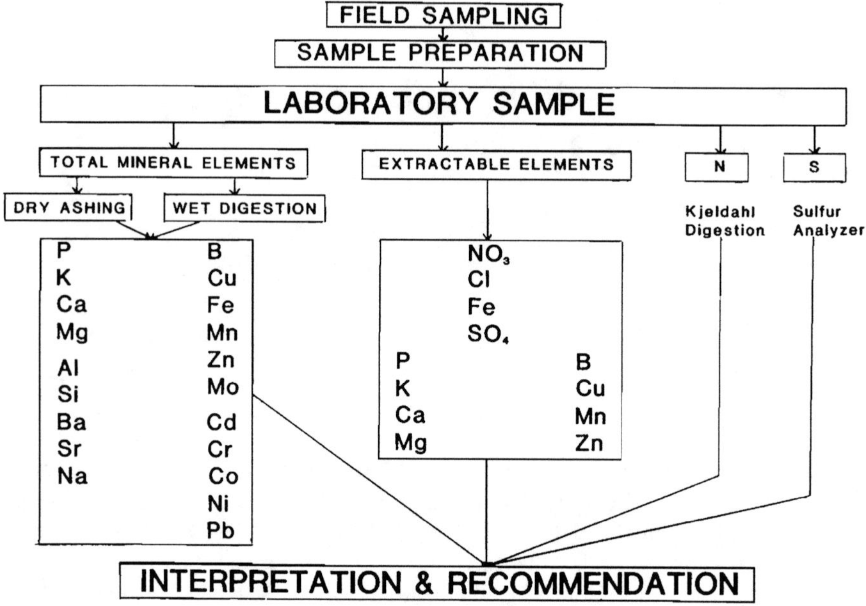

Fig. 15-1. Sequence of procedures for conducting a plant analysis.

I. SAMPLING

The validity and usefulness of the determined elemental content of a collected plant tissue sample hinge upon an intelligent and realistic approach to the problem of how to obtain a reliable sample. If the sample taken is not representative of the general population, all the careful and costly work put into the subsequent analysis will be wasted because the results will be invalid. To obtain a representative sample from a particular plant species is a complex problem, and expert knowledge is required before it can be attempted.

A. Elemental Heterogeneity

The elemental content of a plant is not a fixed entity, but varies from month to month, day to day, and even from hour to hour on the same day as well as differing among the various parts of the plant itself (Goodall & Gregory, 1947; Jones, 1970). A specific plant part at a definite location on the plant obtained at a definite stage of growth (on the basis of physiological age) constitute the sampling parameters. In general, tissues that are either physiologically young and undergoing rapid change in elemental content, or those past full maturity should not be sampled.

The plant part selected and the time of sampling must correspond to the best relationship that exists between its elemental content and yield, or the physical appearance of the plant (Bates, 1971). Frequently, no single time for sampling of a particular plant part is ideal for evaluating every element. Therefore, several plant parts at different growth stages may need to be taken. Comparisons of analyses between leaves and petioles, stems and leaves, and upper and lower plant parts may assist in evaluating a plant analysis (Bates, 1971). Jones (1967) noted that the determination of the homogeneity, or the lack of it, may be a useful technique when diagnosing certain suspected elemental deficiencies in some crops. For example, corn (*Zea mays* L.) plants deficient in K contain less K in their lower leaves than in their upper leaves. When plants contain sufficient K, the reverse is true. It was also noted that differences in concentation of B and Zn among the upper and lower leaves of corn decrease as the plant approaches B or Zn deficiency (Jones, 1967). However, the practical application of this technique of comparison of analysis results between plant parts has yet to develop into a practical system of plant analysis interpretation.

B. Statistical Considerations

Once it has been determined what plant part is to be sampled to represent the plant's elemental status, the number of plants to sample for adequate representation must be decided. What constitutes an adequate number has been determined to some degree from the results of previous research.

Plants growing adjacent to each other can differ considerably in their elemental content. Lilleland and Brown (1943), when studying the P nutrition of peach (*Prunus persica* L.) trees, found that the composition of morphologically homologous leaves taken from adjacent trees receiving the same fertilizer treatment differed considerably. This was also the experience that Thomas (1945) found with apple (*Malus sylvestris* Mill.) trees, and Steyn (1959) with citrus trees and pineapple [*Ananas comosus* (L.) Merr.] plants.

Elemental content variations within single plant parts as well as from plant to plant must be considered. Steyn (1959) has shown that there are relatively small variations in the N, P, K, Ca, Mg, Fe, Mn, Zn, and Cu concentration among the selected sampling material on a single citrus tree or a single pineapple plant. It is reasonable to assume that this could be the case for other plants as well. Therefore, provided the sampled material is carefully selected, a relatively small sample could adequately represent the elemental content of a single plant.

When considering the variation in element content from plant to plant, the situation can be entirely different. If large variations exist, then intensive sampling is required to obtain sufficient plant tissue for representing the element content of the plant sampled. Steyn (1961) conducted a statistical sampling study on citrus trees and pineapple plants in which adjacent plants in blocks were intensively sampled under rigorously controlled conditions. Some of the analysis results from citrus trees are shown in Table 15–1. Similar trends were also observed for pineapple plants.

Table 15-1. Minimum number of trees (in blocks of 16 trees) to sample to show a 5% level significance in differences (D) exceeding D% of the mean values (Steyn, 1961).

Item	N	P	K	Ca	Mg	Fe	Mn	Zn	Cu
	Element								
	g kg^{-1}					mg kg^{-1}			
Mean	26.8	1.17	4.8	36.3	4.8	80.0	23.4	11.0	5.9
CV, %	5.3	3.6	15.8	7.2	18.3	8.0	16.9	20.0	18.3
	Number of trees								
D=10%	3	2	23	5	31	6	26	37	31
D=20%	1	1	6	2	8	2	7	10	8

It was found that, when the citrus trees were in poor condition, much more intensive sampling than that shown in Table 15-1 was necessary for adequate representation. In this study, variation in N and P content was usually considerably less than that observed for the other elements. Potassium and Mg showed the greatest degree of variation, followed by some of the micronutrients, such as Cu and Zn. When the concentration of the element was at a deficient level, the variation was exceptionally large. This is evident for Zn as shown in Table 15-1 where virtually all the trees in the block of 16 would have to be sampled for adequate representation of plant Zn status. Similar results were obtained in the pineapple sampling study. Therefore, if all the essential elements are to be determined in a single sample, a requirement to adequately interpret a plant analysis is to follow an intensive sampling procedure. As was observed for citrus, many leaf tissue samples would be required if pineapple plants under nutrient element stress were being evaluated for their elemental content.

Colonna (1970), working in a homogeneous coffee (*Coffea arabica* L.) plantation, recommended sampling two leaves per tree from 40 randomly selected trees per hectare in 1 ha. In fertilizer trials, he found it necessary to sample five to six replicates with 20 to 25 coffee plants per treatment to obtain useful plant analysis data. Similar sampling studies for the more commonly grown annual field crops have yet to be done to establish the sampling intensity required to ensure reasonable analytical reliability.

C. Tissue Sampling Techniques

Kenworthy (1969), Chapman (1964), and Jones et al. (1971) have summarized sampling techniques that have been generally accepted for interpreting plant analyses. A list of recommended sampling procedures is given in Table 15-2. If other sampling procedures are used, interpretation of the plant analyses data may differ. Since there is a substantially large potential for errors to occur due to improper sampling technique, only thoroughly trained and experienced technicians should be responsible for collecting tissue samples in the field. The number of plants to sample in a particular situation depends on the general condition of the plants, soil homogeneity, and the purpose for which the analysis result will be used. To ensure representa-

Table 15–2. Suggested sampling procedures for field and vegetable crops, fruit and nuts, and ornamentals (Jones et al., 1971).

Crop	Stage of growth	Plant part to sample	Number of plants to sample
Field crops			
Corn	Seedling stage (<12 in.)	All the aboveground portion	20–30
	Prior to tasselling	The entire leaf fully developed below the whorl	15–25
	From tasselling and shooting to silking	The entire leaf at the ear node (or immediately above or below it)	15–25
Soybean or other bean	Seedling stage (<12 in.)	All the aboveground portion	20–30
	Prior to or during flowering*	Two or three fully developed leaves at the top of the plant	20–30
		*Sampling after pods begin to set not recommended	
Small grain (including rice)	Seedling stage (<12 in.)	All the aboveground portion	50–100
	Prior to heading	The fourth uppermost leaves	50–100
		Sampling after heading is not recommended	
Hay, pasture or forage grasses	Prior to seed head emergence or at the optimum stage for best quality forage	The fourth uppermost leaf blades	40–50
Alfalfa	Prior to or at 1/10 bloom stage	Mature leaf blades taken about one-third of the way down the plant	40–50
Clover and other legumes	Prior to bloom	Mature leaf blades taken about one-third of the way down the plant	40–50
Sugarbeets	Mid-season	Fully expanded and mature leaves midway between the younger center leaves and the oldest leaf whorl on the outside	40–50
Tobacco	Before bloom	Uppermost fully developed leaf	8–12
Sorghum-milo	Prior to or at heading	Second leaf from top of plant	15–25
Peanuts	Prior to or at bloom stage	Mature leaves from both the main stem and either cotyledon lateral branch	40–50
Cotton	Prior to or at first bloom or when first squares appear	Youngest fully mature leaves on main stem	30–40
Vegetable crops			
Potato	Prior to or during early bloom	Third to sixth leaf from growing tip	20–30
Head crops (cabbage, etc.)	Prior to heading	First mature leaves from center of whorl	10–20
Tomato (field)	Prior to or during early fruit set	Third or fourth leaf from growing tip	20–25

(continued on next page)

Table 15-2. Continued.

Crop	Stage of growth	Plant part to sample	Number of plants to sample
Vegetable crops			
Tomato (Greenhouse)	Prior to or during fruit set	Young plants: Leaves adjacent to second and third clusters	20–25
		Older plants: Leaves from fourth to sixth clusters	20–25
Bean	Seedling stage (<12 in.)	All the aboveground portion	20–30
	Prior to or during initial flowering	Two or three fully developed leaves at the top of the plant	
Root crops (carrots, onions, & beets, etc.)	Prior to root or bulb enlargement	Center mature leaves	20–30
Celery	Mid-growth (12–15 in. tall)	Petiole of youngest mature leaf	15–30
Leaf crops (lettuce, spinach, etc.)	Mid-growth flowering	Youngest mature leaf from the top of the plant	35–60
Peas	Prior to or during initial flowering	Leaves from the third node down	30–60
Sweet corn	Prior to tasselling	The entire fully mature leaf below the whorl	
	At tasselling	The entire leaf at the ear node	20–30
Melons (water, cucumber, & muskmelon)	Early stages of growth prior to fruit set	Mature leaves near the base portion of plant on main stem	20–30
Fruits and nuts			
Apple, apricot, almond, prune, peach, pear, cherry	Mid-season	Leaves near base of current year's growth or from spurs	50–100
Strawberry	Mid-season	Youngest fully expanded mature leaves	50–75
Pecan	6 to 8 wk after bloom	Middle pair of leaflets from mid-portion of terminal growth	30–45
Walnut	6 to 8 wk after bloom	Middle pair of leaflets from mature shoots	30–35
Lemon, lime	Mid-season	Mature leaves from last flush or growth on nonfruiting terminals	20–30
Orange	Mid-season	Spring cycle leaves, 4 to 7 mo old from nonbearing terminals	20–30
Grapes	End of bloom period	Petioles from leaves adjacent to fruit clusters	60–100

(continued on next page)

Table 15-2. Continued.

Crop	Stage of growth	Plant part to sample	Number of plants to sample
Fruits and nuts			
Raspberry	Mid-season	Youngest mature leaves on lateral or "primo" canes	20–40
Ornamentals and flowers			
Ornamental trees, shrubs	Current year's growth	Fully developed leaves	30–100
Turf	During normal growing	Leaf blades; clip by hand to avoid contamination with soil or other material	¼ liter of material
Roses	During flowering production	Upper leaves on the flowering stem	20–30
Chysanthe-mums	Prior to or at flowering	Upper leaves on flowering stem	20–30
Carnations	Unpinched plants	Fourth or fifth leaf pairs from base of plant	20–30
	Pinched plants	Fifth and sixth leaf pairs from top of primary laterals	20–30
Poinsettias	Prior to or at flowering	Most recently mature fully expanded leaves	15–20

tion, sampling as many plants as practical is recommended, collecting samples during a particular time of day and under calm climatic conditions.

As a general rule, mature leaves exposed to full sunlight just below the growing tip on main branches or stems are usually preferred, taken just prior to or at the time the plants begin their reproductive stage of growth. In some situations, sampling may be necessary at earlier periods in the plant's growth cycle with the same maturity.

There are as many instructions on what *not* to sample as there are on what to sample. For example, do not collect tissue that is covered with soil or dust, or from plants damaged by insects, mechanically injured, or diseased. Dead plants should not be sampled, or dead tissue included as a part of a collected sample. Whole plants or plants beyond full maturity should not constitute the sample or a portion of the sample. In addition, sampling is not recommended when plants are under moisture or temperature stress. Seeds are not normally useful for assessing the nutrient element status of plants, except possibly for the element N. In some instances, seed analyses have been of value in determining the Mo and Zn supply for young plants developing from that seed (Reisenauer, 1956; Shaw et al., 1954).

Plants under stress due to a possible elemental deficiency or imbalance should be sampled only at the initiation of the stress. After a period of stress, plants develop unusual element concentrations in their tissues that can lead to a misinterpretation of a plant analysis result. When visual symptoms occur, or a deficiency is suspected, the analysis of the same plant part from adjoining normal plants in the same field or area can aid in the interpretation (Munson & Nelson, 1973). However, if the plants being compared differ in their vigor and stage of development, the same plant part at the same stage of development may not exist. Therefore, comparing the analysis results between two sets of tissues may be confusing. It is also advisable to collect soil samples from the same area where plants have been selected for sampling. Comparing analysis results between and among soil tests and plant analysis results can assist in the interpretation.

Normally, after pollination and as plants begin setting and developing fruit or seed, the elemental content of the vegetative portions of the plant begin to change substantially, making a plant analysis interpretation difficult if leaf tissue is collected at this time. Therefore, sampling after pollination is not recommended for most grain and fruit crops.

Standardizing sampling techniques cannot be over emphasized, since criteria for elemental analysis interpretation have been established for specific plant conditions. Therefore, for elemental concentration determinations to be meaningful, it is essential to adhere to the given sampling procedures designed for the plant species and element(s) to be assayed.

II. TISSUE PREPARATION

After collection, the plant tissue must be: (i) cleaned to remove surface contamination; (ii) oven dried to stop enzymatic reactions and remove moisture; (iii) reduced in particle size to homogenate and obtain a suitable laboratory sample; and (iv) finally dried to a constant weight. Two additional steps may be needed: (i) storage and transport of the fresh material prior to cleaning (decontamination) and oven drying; and (ii) storage of the dried and powdered tissue prior to analysis.

A. Decontamination

Exposed plants are always covered with a thin film of dust that is difficult to remove by mechanical wiping or brushing. Failure to remove this dust normally affects only the determinations for Fe and Al unless the dust cover is considerable or of a specific composition. Leaf material may be contaminated with spray residues, whether applied for nutritional, insecticidal, or fungicidal purposes (Ashby, 1969). Therefore, these sources of contamination that will affect the analytical result must be removed when the leaf material is fresh. Steyn (1959) and Wallace et al. (1980) found that the only satisfactory way of removing contamination is by washing the tissues in a 0.1 to 0.3% detergent solution followed by rinsing in pure water. Sonneveld and van Dijk (1982) recommend dipping tissue for 15 s into an ample volume of solution containing 1% Teepol (detergent) and 0.1 *M* HCl, followed by rinsing in pure water. The washing procedure must be done quickly to avoid long contact between the solution and tissue as there is the danger that some elements, such as K and Cl, will be removed by leaching (Bhan et al., 1959). Smooth leaves can be easily wiped with a damp cloth or washed to decontaminate, while, leaves that are pubescent or not smooth surfaced are difficult, if not impossible, to decontaminate completely. The effect that washing can have on the elemental content of orange [*Citrus sinensis* (L.) Osbeck], corn, and apple leaves are shown in Table 15-3.

Unless leaf tissue is visually coated with dust or other foreign substances, washing to remove these substances is usually not necessary. The only exception would be when Fe (Wallace et al., 1982) as well as Al and Si are determined elements. If the plant material is not washed, the method of organic matter destruction (procedures that are discussed in more detail later) may also affect the analysis for these three elements.

B. Oven Drying

After washing (decontamination), the plant tissue must be dried as rapidly as possible to minimize chemical and biological change. If drying is unduly delayed, considerable loss in dry weight may occur due to respiration (Lockman, 1970), while proteins will be broken down to simpler nitrogenous compounds with the potential release of ammonia. Also, a too high drying temperature can affect the dry weight (Grant & MacLaughlin, 1968). Dry weight preservation is essential since element content is expressed on a dry weight basis of prepared tissue.

The requirements that must be satisfied when drying plant material are that the temperature is sufficiently high to destroy the enzymes responsible for decomposition and sufficient for moisture removal yet below the temperature of thermal decomposition. According to Tauber (1949), enzymes present in plant tissue are rendered inactive at temperatures above 60 °C. Therefore, air drying plant tissue prior to more than 1 or 2 d of storage, particularly if the tissue was previously washed, may not be sufficient to prevent enzymatic decomposition from occurring. Therefore, tissue should be dried quickly in a dust-free, forced-draft electrical oven as soon after collection as possible. Steyn (1959) determined the extent of losses over 8 wk when citrus and pineapple leaves were dried in a forced-draft oven set at 50, 65, and 105 °C, respectively. The samples were dried for 24 h and weighed. The results are shown in Table 15-4. Drying at 50 °C was not sufficient as there was enough retained moisture for enzymatic action to occur during storage. Drying at 65 °C resulted in a certain amount of thermal decomposition, but enzymatic action was successfully stopped even after 8 wk storage.

During the particle-size reduction process, previously oven-dried plant tissue will absorb varying amounts of moisture that must be removed before weighing the tissue for analysis. Because of the difference in physical state between plant tissue powder and the unground material, a similar drying experiment as that described above was carried out by Steyn (1959) on plant tissue powder.

Steyn (1959) found much greater thermal decomposition when leaf powder was dried at the higher temperatures than when the plant tissue was

Table 15-3. Element concentration in orange, corn, and apple leaves as affected by washing.

Element	Orange leaves†			Corn leaves				Apple leaves¶	
	Unwashed	Detergent washed	Detergent plus acid washed	Unwashed‡	Wiped with deterdent solution†	Unwashed§	Washed in distilled water§	Unwashed	Washed with detergent
	$g\ kg^{-1}$								
Ca	39.7	39.7	39.6	8.7	8.5	4.8	4.5	--	--
K	10.7	10.8	20.4	21.7	21.7	12.2	12.6	17.8	18.0
Mg	4.2	4.1	4.2	1.6	1.6	3.9	4.1	2.5	2.3
N	25.3	25.6	25.5	--	--	29.3	31.1	22.2	21.9
P	1.5	1.5	2.7	--	--	2.2	2.2	--	--
	$mg\ kg^{-1}$								
B	367.0	368.0	369.0	17.0	17.0	11.0	10.0	28.0	27.0
Cu	5.6	5.1	5.0	29.5	28.8	9.0	10.0	--	--
Fe	186.0	61.0	61.0	136.0	134.0	96.0	85.0	77.0	72.0
Mn	182.0	94.0	92.0	--	--	73.0	64.0	38.0	36.0
Mo	--	--	--	--	--	1.1	1.2	--	--
Zn	123.0	68.0	65.0	23.1	30.5	22.0	22.0	14.1	15.6

† Labanauskas (1968). ‡ Baker et al (1964). § Jones (1963). ¶ Ashby (1969).

Table 15–4. Loss of weight of fresh leaves at constant temperature (Steyn, 1959).†

	Loss in wt., %					
	Citrus leaves			Pineapple leaves		
Time, h	50 °C	65 °C	105 °C	50 °C	65 °C	105 °C
5	41.9‡	57.7‡	58.5§	24.0§	34.1§	90.5¶
10	56.7	57.9	58.5	58.5	90.1	90.7
24	57.2	58.0	58.7	86.4	90.2	90.7
48	57.5	58.2	58.9	88.7	90.3	90.9
72	57.6	58.2	59.0	89.0	90.4¶	91.3
96	57.7	58.2	59.1	89.1	90.6	91.6
168	57.8‡	58.2§	58.5¶	89.5§	90.9¶	92.0¶
14 d	57.8	58.2	59.5	89.7	90.9	92.1
30 d	58.3	58.2	59.5	91.0	90.9	92.1
60 d	58.8	58.3	59.5	92.0	90.8	92.1

† Difference >0.4% is significant.
‡ No scorching, fresh leaf color.
§ Slight scorching, dull-brown mottling.
¶ Severe scorching, brown throughout.

dried in its original physical state at the same temperatures, indicating the sensitivity of the finely divided state of the powder to drying temperature compared to the original material (Table 15–5). Consequently, the prevailing tendency of workers to dry ground plant tissue at 105 °C prior to analysis must be questioned. As shown in Table 15–5, a 2% difference in the dry weight of citrus leaf powder was obtained between the 65 and 105 °C drying temperature in a 24-h period. For the pineapple leaf powder, the difference in dry weight amounted to more than 5.5% in the same 24-h period. Whereas a certain degree of variation in oven temperature when drying fresh plant material may be permissible, oven temperature must be strictly controlled for tissue powder which is more susceptible to thermal decomposition.

Steyn's results would suggest that oven drying plant tissue powder at 65 °C for 24 h removes moisture without thermal decomposition. However, Bowen (1967) recommended drying a hygroscopic kale powder sample at 90 °C for 20 h to obtain the dry weight. Evidently, the ideal drying temperature for powdered samples is dependent on the source of the plant material and exists between 65 to 90 °C. Drying fresh and powdered plant tissue at 80 °C

Table 15–5. Loss of weight of leaf powder at constant temperature (Steyn, 1959).

	Loss in wt., %†			
	Citrus leaf powder		Pineapple leaf powder	
Time, h	65 °C	105 °C	65 °C	105 °C
2	4.09	5.79‡	--	--
4	4.25	5.92	2.33	4.81‡
8	4.35	5.94	2.40	6.03
16	4.40	6.30	2.40	7.00
24	4.42	6.36	2.42	7.99
48	4.47	6.48	3.01‡	9.35§
96	4.64	6.68	--	--
168	4.64	7.16§	3.08	12.30

† Difference > 0.2% is significant. ‡ Severely, scorched. § Blackish brown.

may be the best compromise temperature, although testing may be necessary to determine the appropriate drying temperature to minimize thermal decomposition.

For plant tissue high in soluble sugars, moisture removal is best done by either freeze drying or vacuum drying (Horwitz, 1980).

Plant tissue may be quickly and satisfactorily dried in a microwave oven (Carlier & van Hee, 1971; Shuman & Rauzi, 1981), although the procedure is somewhat tedious and not suited for drying large quantities of material at one time.

C. Particle-Size Reduction

Particle-size reduction of the dried plant material before analysis provides both a suitable form of the sample for manipulation in the laboratory as well as one of uniform composition. Because of the laborious nature of either hand cutting or crushing of dried tissue, particularly when the sample size is large, mechanical devices are normally used. These mechanical devices use either a cutting action, such as with the Wiley or hammer-type mills, or a crushing action obtained when using ball mills, or by abrasion in cyclone or UDY mills. In most mills, particles of the contact surfaces such as Cu and Zn from brass fittings, Fe from steel, Al from Al-made parts, Na and Zn from plastic seals, and Zn from rubber seals will be added to the tissue sample during the milling process. The extent of these elemental additions will depend on the condition of the mill and the length of contact time. When plant tissue samples are to be assayed for their Fe content, either hand cutting or crushing in an agate mortar are probably the only suitable techniques for reducing the particle size. No matter what device is used, test samples should be assayed to determine what elemental additions will occur during the particle-size reduction procedure.

Hood et al. (1944) determined the extent of elemental contamination occurring during the milling of plant samples. The types of milling equipment tested were a (i) Wiley mill, (ii) hammer mill, (iii) jar mill with flint, and (iv) porcelain mill with mullite balls. They found that with all four commonly used mechanical mills some degree of contamination with one or more elements (Al, Cu, Fe, and Zn) occurred. A similar study by Grier (1966) found Wiley-type mills in wide use. He recommended the use of stainless steel as the best substance for the cutting and sieving surfaces to minimize elemental contamination.

Steyn (1959) conducted a similar investigation using a Vetter agate ball mill. A composite citrus leaf sample was washed, dried, and divided into two sets of triplicate subsamples. One subsample was crushed in an agate mortar and the other was ground for 2 h in an agate ball mill. Each subsample was assayed separately. Because no significant contamination was evident in the analysis results (Table 15–6), Steyn concluded that grinding plant tissue in agate is preferred to other methods, particularly when the micronutrients are to be included in the assay.

Table 15-6. Nutrient concentration of citrus leaves after crushing and mechanical grinding in agate (Steyn, 1959).

Element	Method of grinding	
	Crushing	Mechanical grinding
	g kg^{-1}	
N	24.6	24.3
P	1.21	1.22
K	5.6	5.6
Ca	36.7	36.6
Mg	5.8	5.7
	mg kg^{-1}	
Fe	65.0	64.0
Mn	32.0	33.0
Zn	15.0	15.0
Cu	3.4	3.3

The significance of fineness of grinding using a Wiley mill has been investigated. Nelson and Boodley (1965) obtained separation of terminal and basal leaf particles during grinding and weighing. They recommended fine grinding (<20 mesh) and control of electrostatic charge which frequently arises during grinding. Smith et al. (1968) noted variation in the element concentration for various sample size fractions. For example, Zn tended to be concentrated in the larger sized fractions of barley (*Hordeum vulgare* L.) straw and alfalfa (*Medicago sativa* L.) tissue. Jones (1963) found that fractions of ground, unwashed corn leaves screened to pass 20-, 40-, 60-, and 100-mesh sieves varied in composition for the elements Fe and Zn. For example, the tissue passing the 100-mesh sieve contained four times as much Fe as the whole sample. But, this could have been due to contamination since the finer fraction probably contains more dust particles than the coarser fractions.

Fineness can be of considerable importance since it relates to sample homogeneity. Particle-size reduction to pass a 20-mesh screen is sufficient if 0.5 g or larger aliquots of plant tissue are to assayed. If < 0.5 g aliquots are taken, then particle-size reduction to pass a 40-mesh screen is necessary Particle-size reduction that results in a wide range of particle sizes poses problems in homogeneity if the prepared tissue sample segregates during handling and weighing.

Adherence of fine particles to the cutting surface in a Wiley-type cutting mill can be partially overcome by attaching a vacuum system to the mill (Graham, 1972) or by using pulsing air (Ulrich, 1984) as well as by controlling static electricity (Nelson & Boodley, 1965).

Similar studies on particle-size distribution have not been done using ball mills or cyclone (UDY) mills for particle-size reduction. However, plant tissue processed through these types of mills are usually finer and more uniform in particle size than that obtained by cutting in Wiley-type mills.

Another complication is that some types of plant tissues are not easy to grind in a Wiley-type mill due to either the presence of pubescence (such

Table 15-7. Decomposition of fresh citrus leaves on storage (Steyn, 1959).

Storage, d	Loss in dry wt., %		
	In sealed polyethylene	Open to atmosphere	In freezer at −5°C
2	1.8	1.4	0.0
4	2.7	1.7	0.0
7	5.5	2.0	0.3
14	9.7	2.8	0.4

as on apple leaves), or when the tissue is coarse and fibrous (continuing the grinding process until the entire sample passes through the mill) or the tissue is highly deliquescent. Therefore, special care is required when reducing the particle size for all of these types of tissues to ensure homogeneity of the prepared sample.

D. Keeping Quality of Plant Tissue

In dealing with large numbers of plant tissue samples, a considerable delay may occur between the time a sample is collected and when it can be decontaminated and oven dried. Therefore, the need to properly store freshly collected tissues to minimize respiratory losses.

Steyn (1959) compared loss in dry weight when fresh citrus leaves were stored in sealed polyethylene bags in the open laboratory atmosphere with that in sealed polyethylene bags put into a freezer set at −5 °C. He found that the fresh citrus leaves did not decompose significantly in 1 or 2 d if they are air dried prior to transport to the laboratory (Table 15-7). However, the air-dried leaves were difficult to clean properly to remove surface dust contamination, whereas the fresh leaves stored under refrigerated conditions could be cleaned easily and efficiently. Therefore, if there will be a considerable time lag after sampling and before transport to the laboratory, plant materials are best transferred to the laboratory under refrigerated conditions.

Steyn (1959) has further demonstrated that oven-dried and ground plant material should not be stored on an oven shelf for longer than 8 wk before analysis. On the other hand, this same plant material can be stored indefinitely in a sterilized, sealed bottle in a freezer set at −5 °C. The U.S. National Institute of Standards and Technology stores their standard reference plant materials following sterilization by means of gamma radiation in sealed polyethylene bags.

E. Recommended Procedure for the Handling and Preparation of Plant Tissue Samples for Analysis

At each step in the preparative phase, an error made may be manifested in the elemental analysis result. Although any one individual error may not be large, the cumulative effect of several errors may easily result in errors of 10% or more in the final analytical result. Since the whole preparation process is so highly susceptible to error, considerable care is required throughout or the final analytical result will be of little value.

1. Tissues not to be Washed

Place the collected plant tissue in an open-to-the atmosphere container (e.g., a cotton bag) and transport to the laboratory as quickly as possible. When it arrives at the laboratory, transfer to an oven set at a temperature between 65 and 80 °C.

2. Tissue to be Washed

Place the collected plant tissue in a polyethylene bag. If transport to the laboratory is to be the next day, keep the bags containing the tissue under refrigerated conditions. Upon arrival at the laboratory, remove one bag at a time from the refrigerator and wash each tissue sample by sponging with a piece of cotton wool moistened in a 0.1% detergent (Teepol or Dreft) solution, followed by rinsing in two lots of pure water, or in flowing pure water. Place the sample in a clean cotton bag and suspend inside a forced-draft oven set between 65 to 80 °C.

3. Grinding and Storage

After drying for 48 h, grind the entire sample to pass a 20-mesh screen or powder in an agate mechanical ball mill. Place the ground or powdered tissue powder in a clean bottle and dry for an additional 24 h at 65 °C to remove any moisture added during the particle-size reduction step. After drying, seal the bottle and place it in a cool dry place. For long-term storage, store under refrigerated conditions until the analysis can be done.

F. Alternate Procedure for the Handling of Plant Tissue Samples

Good washing, drying, or refrigeration facilities are not often available on or near the collection site. In this situation, the following is recommended:

a. Gather the tissue samples as described in Table 15–2, or follow the sampling directions given by the assaying laboratory.
b. Spread the collected plant sample out in a dust-free area for 1 or 2 d to air dry. This is essential for those types of tissues that have a high water content.
c. After 1 or 2 d of air drying, place the plant sample loosely in a labeled paper or cloth bag. Avoid packing tightly. Do not put undried plant tissue in plastic bags that will not be refrigerated immediately.
d. Place the loosely filled paper or cloth bags in a sturdy container, with crumpled newspaper between samples, and transport them to the laboratory as quickly as possible.

Although this sample handling procedure will not enable the laboratory to account properly for contamination from dust or spray residues, the assay procedure will ensure that tissue samples will reach the laboratory in good physical condition.

III. LABORATORY ANALYSIS

There are various assay methods suitable for the determination of the total elemental content of plant tissue. Most involve destruction of the tissue's organic component, thereby converting the elements to a soluble form for analysis. Some have proposed releasing the elements by extracting them from the dried green tissue, but these techniques have been developed for only a few crops and elements, and the procedure is discussed in detail in the next section. Significant analytical advancements have occurred in the last several decades with the introduction of rapid multielement analyzers that use a single prepared sample for the assay of most of the essential elements. These analyzers offer high speed, excellent sensitivity, and precision. In today's analytical laboratory, those having a quality assurance program in force have markedly improved the reliability of obtained analytical results.

A. Methods of Organic Matter Destruction

Probably no other aspect of plant tissue preparation prior to elemental analysis has stirred as much controversy as how best to destroy the organic matter portion of tissue. The best treatises on the subject are the books written by Gorsuch (1970) and Bock (1978) who describe in considerable detail the advantages and difficulties associated with each organic matter destruction procedure. In addition, the review articles by Tolg (1974) and Gorsuch (1976) provide considerable useful information.

There are essentially two decomposition procedures, wet acid digestion and high temperature dry oxidation, frequently referred to as wet and dry ashing, respectively. The appropriate method will depend on elements sought in the analysis, nature of the plant material, difficulty with dissolution, acceptable degree of recovery, required sensitivity, and the capability of a laboratory to manage a particular method as described by Munter et al. (1984).

Wet oxidation is the destruction of organic matter by high temperature acid digestion. The common acids used are H_2SO_4, HNO_3, and $HClO_4$, usually in some combination of two or all three. Nitric acid is usually included in most digestion mixtures, with the addition of H_2SO_4 to raise the digestion temperature, or the addition of $HClO_4$ or 30% H_2O_2 to speed and complete the digestion. Tolg (1974) lists the characteristics of various acid mixtures as given in Table 15–8.

Sulfuric acid as a component in the digestion mixture is not recommended when digesting plant tissue high in Ca. Relatively insoluble $CaSO_4$ may be formed which will lower the Ca determination and possibly other elements by coprecipitation. Lindner and Harley (1942) used hot H_2SO_4 with repeated additions of 30% H_2O_2 until the digestion was complete. Wolf (1982) suggests that this procedure is best suited for small sample aliquots (0.10–0.25 g), and for tissue easily oxidized and relatively low in Cl content. Nitrogen as well as most of the other mineral elements (P, K, Ca, and Mg) can be determined in their digests. Parkinson and Allen (1975) used the H_2SO_4-30%

Table 15-8. Digestion reagents for use in methods for wet acid oxidation (Tolg, 1974).

Digestion reagents	Applicability to organic matrix	Remarks
H_2SO_4/HNO_3	Vegetable origin	The most used: Danger of volatilization of As, Hg, and Se
H_2SO_4/H_2O_2	Vegetable origin	Pb loss on co-precipitation with $CaSO_4$: Loss of Ge, As, Ru and Se
HNO_3	Biological origin	Easily purified reagent; digestion temp. 250 °C; short digestion time; soluble metal-nitrates
$HClO_4$	Biological origin	Catalyst: $(NH_4)_2MoO_4$, etc.
$H_2SO_4/HClO_4$	Biological origin	Suitable only for small samples; danger of explosion
$HNO_3/HClO_4$	Protein, carbohydrate (no fat)	Less explosive; no loss of Pb
$H_2SO_4/HNO_3/HClO_4$	Universal (also fat and carbon black)	No danger with exact temperature control; As, Sb, Au, and Fe, are volatile, digest under reflux conditions

H_2O_2 mixture plus Se and $LiSO_4 \cdot H_2O$ to digest tissue for the determination of N, P, K, Ca, Mg, Cu, Fe, Mn, and Zn. Cresser and Parsons (1979) used a H_2SO_4-$HClO_4$ digestion mixture for preparing plant tissue for the analysis of N, P, K, Ca, and Mg.

To shorten the digestion time, $HClO_4$ is added to the digestion mixture. However, $HClO_4$, when hot, is a strong oxidant that can react with explosive force when brought into contact with easily oxidizable compounds, especially if the digestion mix approaches dryness. Therefore, extreme care is required when using this acid (Horwitz, 1980), and current USA regulations only permit the use of $HClO_4$ in specially designed fume hoods which continuously wash fumes released in the digestion process.

Today, the wet oxidation procedure is frequently done in digestion tubes inserted into ports of a temperature-controlled digestion block that may be obtained commercially, or constructed by the analyst (Gallaher et al., 1975). With proper control devices, the digestion temperature and time can be carefully controlled, greatly simplifying the digestion procedure. A block-digestion procedure using a mixture of HNO_3 and $HClO_4$ has been described by Zasoski and Burau (1977), $HClO_4$ and H_2O_2 as the digestion mixture by Adler and Wilcox (1985), and a method using HNO_3 alone by Havlin and Soltanpour (1980) and Zarcinas et al. (1987). Huang and Schulte (1985) used a HNO_3-30% H_2O_2 combination for digesting small aliquots of tissue in Folin tubes inserted into a digestion block. Several of these procedures are described in Table 15-9. If a digestion block is not available, most of these procedures can be done in covered beakers placed on a hot plate.

White and Douthit (1985) devised a procedure digesting plant tissue in a mixture of HNO_3 and 30% H_2O_2 heated in a microwave oven, while Ogner (1983) used ultraviolet (UV) radiation as the source of heat. There are several commercial digestion systems that combine microwave heating with sealed digestion vessels for automated rapid wet digestion of plant tissue.

Table 15–9. Acid digestion procedures for organic matter destruction in plant tissue.

Acid digestion procedure (HNO_3 and $HClO_4$)

Weigh 0.5 g of 20-mesh dried plant tissue into a beaker or digestion tube. Add 2.5 mL conc. HNO_3. Cover the beaker with watch glass or place funnel into the mouth of digestion tube. Let it stand overnight. Place covered beaker on hot plate or digestion tube into a port of the digestion block. Set the temperature of plate or block at 80 °C and digest for 1 h. Remove beaker or digestion tube from plate or block and let cool.
Add 2.5 mL of $HClO_4$.
Place covered beaker on hot plate or tube back into digestion block and digest at 180 to 200 °C for 2 to 3 h, or until digest is clear. Remove the glass cover from the beaker or funnel from digestion tube and heat at 80 °C until fumes of $HClO_4$ have dissipated.
Remove beaker from hot plate or digestion tube from the digestion block and let cool.
Add pure water to digest to bring the level to 10 mL.
The digest is ready for elemental assay.

Acid digestion procedure (HNO_3 alone)

Weigh 0.5 g of dried 20-mesh plant tissue into a beaker or digestion tube. Add 5.0 mL conc. HNO_3. Cover beaker with watch glass or place funnel into the mouth of digestion tube. Let it stand overnight.
Place covered beaker on hot plate or digestion tube into a port of the digestion block. Set temperature of plate or block at 125 °C and digest for 4 h.
Remove beaker from hot plate or digestion tube from the digestion block and let cool.
Add pure water to digest to bring the level to 10 mL.
The digest is ready for elemental assay by ICP-AES only.

Wet digestion procedure (HNO_3 and 30% H_2O_2)

Weigh 0.5 g of dried 20-mesh plant tissue into a beaker or digestion tube. Add 5.0 mL of conc. HNO_3. Cover beaker with watch glass or place funnel into the mouth of digestion tube. Let it stand overnight.
Place covered beaker on hot plate or digestion tube into a port of the digestion block. Set temperature of plate or block at 125 °C and digest for 1 h. Remove beaker from hot plate or digestion tube from block and let cool.
Repeat cooling and 30% H_2O_2 additions until the digest is clear. Add more HNO_3 as needed to keep digest from going dry.
When the digest is colorless, reduce the temperature of the hotplate or digestion block to 80 °C, remove the watch glass from the beaker or funnel from the digestion tube and let the digest go almost to dryness. The residue should be either white or colorless. If not, repeat the higher temperature digestion with additional H_2O_2 treatment. Remove the beaker from the hot plate or digestion tube from the digestion block and let cool.
Add 1:10 (HNO_3) or (HCl) to bring the final volume to 10 mL.
The clear solution is ready for elemental assay.

Wet digestion procedure (H_2SO_4 and 30% H_2O_2)

Weigh 0.5 g of dried 20-mesh plant tissue into a digestion tube. Add 3.5 mL conc. H_2SO_4 and let it stand for 30 min. Add 3.5 mL 30% H_2O_2.
Place a funnel in the digestion tube, and the tube into a port in the digestion block set at 250 °C.
Heat for 30 min. Remove tube from the digestion block and let the tube cool. Add 1 mL 30% H_2O_2 until the digest is clear upon cooling. When clear after cooling, dilute to 20 mL with pure water.
The digest is ready for elemental assay.

All the prepared digest can be directly assayed by ICP-AES, and all but the HNO_3 alone procedure by FIA, without further manipulation.
Sulfate sulfur can be determined for all the acid digested tissue except for that prepared in H_2SO_4 + 30% H_2O_2.
The mineral elements can be determined in the H_2SO_4 + 30% H_2o_2 digest plus ammonium.
Boron is usually determined best in tissue samples prepared by dry ashing.

Wet oxidation under pressure can be performed by placing sample and digestion reagents either into a Parr Bomb (Vigler et al., 1980; Okamoto & Fuwa, 1984), in sealed ampules placed into an autoclave at 125 °C under pressure (Sung et al., 1984; Knapp, 1985; Knapp & Grillo, 1986), or with 6 *M* HCl at 80 °C in tightly capped polyethylene bottles (Kuennen et al., 1982). These techniques are used for volatile element containment or for tissues difficult to digest completely.

If B is one of the elements to be determined in the plant tissue digest, the dry ash procedure is recommended (Wikner, 1986) since this element can be partially or almost entirely lost by volatilization during wet oxidation (Feldman, 1961). Boron may be retained during wet oxidation for tissue high in Ca. To avoid B additions from the glassware used, the digestion vessels must be heat acid washed and relatively free from scratches.

The critical considerations for organic matter destruction by high temperature oxidation are: (i) the nature of the ashing vessel, (ii) placement in the muffle furnace, (iii) ashing temperature, and (iv) time. Although the shape and size of the ashing vessel is not generally specified, high-walled vessels large enough to keep the sample depth in the vessel minimal are best. A lid on the ashing vessel can minimize loss of ash and contamination from the muffle furnace (Munter et al., 1984), but is not recommended unless a constant flow of air is passed through the muffle furnace. Silica crucibles are one of the best vessels for dry-ashing plant tissue. Pyrex glass high-form beakers and well-glazed porcelain high-form crucibles are also suitable ashing vessels, although the glass beakers may add B and Na to the solubilized ash, and Al from the use of porcelain crucibles.

A critical initial requirement to prevent flaming of tissue in the ashing vessel requires raising the muffle furnace temperature slowly and keeping the furnace door closed. Furnaces with tightly fit doors, or too many ashing vessels placed in the furnace may cause incomplete oxidation due to low O_2 supply.

Keeping the ashing vessel from contact with a heated muffle furnace floor and away from the door is equally important. Frequently, the obtained ash in vessels placed near the furnace door may look darker than those positioned in the center due to less oxidation (caused by cooler temperatures). Therefore, the analyst should determine if vessel position is a factor for the muffle furnace being used, making adjustments as required to ensure even and complete organic matter destruction for every vessel placed in the muffle furnace.

Baker et al. (1964) ashed plant tissues at 458 ± 5 °C for 8 h in a muffle furnace lined with stainless steel which was necessary to eliminate Al and Zn contamination from the walls. Jones and Warner (1969) ashed at 500 °C for 4 h which proved satisfactory for most plant tissues. Cholak and Story (1941) found dry ashing at 500 °C satisfactory for preparing various biological substances without losses of the elements they measured.

According to Gorsuch (1959, 1970, 1976), as long as the ashing temperature does not exceed 500 °C, volatilization losses should not occur for most elements. There is danger from a too low ashing temperature due to incomplete organic matter destruction which prevents complete recovery of elements from the organic matrix. Isaac and Jones (1972) ashed plant tissues at temperatures from 400 to 700 °C, and found only the elements Al, B, Cu, Fe, K, and Mn affected by ashing temperature. However, they also observed that when the ashing temperature was 500 °C, the known values for 13 elements (Al, B, Ba, Ca, Cu, Fe, K, Mg, Mn, Mo, P, Sr, and Zn) in a corn leaf tissue standard were obtained. Munter et al. (1984) reported similar results when ashing at the 500 °C temperature, although they emphasized the need for careful temperature control of the muffle furnace. The volatile elements such as As, Cd, Hg, and Pb may be lost during dry ashing, although retention seems to be enhanced for tissues that have high Ca (7.0%) contents.

The length of ashing time necessary to completely oxidize the organic components in plant tissue will vary depending on the type of the tissue. Complete oxidation is more difficult to achieve for tissues high in sugar or oil content. Usually, 4 to 8 h at the ashing temperature is sufficient. If crucible covers (with adequate air-flow in the muffle furnace) are used, an additional 2 h may be needed (Munter et al., 1984) for complete oxidation.

Ashing aids may be used to assist in the decomposition of the plant tissue organic matter, particularly for those tissues high in either sugar or oil content. Gorsuch (1970) recommended 10 mL of 10% H_2SO_4 or 10 mL of 7% $Mg(NO_3)_2 \cdot 6\ H_2O$ per 5-g sample as suitable ashing aids. The AOAC Manual (Horwitz, 1980) procedure for dry ashing calls for HNO_3 as an ashing aid by wetting an obtained ash with dilute HNO_3 acid, taking to dryness and placing in the muffle furnace for another hour. Since the need for an ashing aid is more important for highly carbonaceous tissues, the analyst should experiment with the tissue samples of interest to determine if an aid is needed. If organic matter destruction is complete, then the remaining ash should appear "white" and relatively free from particles of unoxidized C. If a clean white ash is not obtained, then an ashing aid may be needed to assist in the oxidation process.

The procedure used to solubilize the ash may also affect the elemental analysis. The common practice is to dissolve the ash at room temperature in a sufficiently concentrated acid, either HNO_3 or HCl, or a combination of both (aqua regia) to make the final dissolved acid solution 1 *N* in concentration. Acid dissolution at room temperature, however, may not completely release Al, Cr, and Fe (Dalquist & Knoll, 1978). Munter and Grande (1981) suggest adding a weighed aliquot of 2 *N* HCl to the ash, heat on a hot plate until fumes of the acid evolve, let cool, and then bring back to the original weight by the addition of pure water. By this technique, they were able to obtain the certified value for Al and Fe, as well as other elements, using *NIST Plant Tissue Standard Reference Materials (SRMs).* The *AOAC Manual* (Horwitz, 1980) calls for solubilizing by HCl treatment of the ash on a hot plate. The procedure for ashing and solubilization of the ash is given in Table 15-10.

Complete solubilization of the heavy metal micronutrients, particularly Zn, in plant ash containing sizable quantities of Si (such as rice plant ash), make the dry-ashing technique unsuitable for organic matter destruction

Table 15-10. Organic matter destruction in plant tissue by high temperature oxidation.

Dry ashing procedure
Weigh 0.5 g of dried 20-mesh plant tissue into a 15 mL high form porcelain or quartz crucible.
Place crucible in a rack. Place the rack in a cool muffle furnace. Set furnace temperature to reach set temperature in about 2 h. After 4 to 8 h of muffling at 500 °C, remove the crucible rack from the furnace and let it cool.
Add 10 mL of dilute acid (300 mL of HCl and 100 mL of HNO_3 in 1 L of pure water) to dissolve the ash. Crucible and contents may be heated to assist in dissolving ash by weighing crucible, heating, and bringing back to original weight after cooling. Allow suspended material to settle to the bottom to the crucible.
The clear solution is ready for elemental assay, with or without further dilution.

unless the silica is removed by treating the ash with HF. Therefore, wet digestion is the recommended procedure for high Si-containing tissues.

It has been frequently observed that differences in elemental content will be obtained when the same tissue is either wet digested or dry ashed. The two commonly observed differences occur for the elements Al and Fe (B may or may not be lost during wet ashing). There may be two possible explanations for these differences. Either the insoluble oxides of Al and Fe are formed when dry ashing, and therefore, not brought into solution when the ash is acid solubilized, or during wet ashing, Al and Fe in dust-contaminated tissue are brought into solution. In either case, higher Al and Fe values usually are obtained when tissue is wet digested.

Low-temperature dry ashing is another technique used to retain the more volatile elements (such as As, Hg, and Se) in samples during the destruction of organic matter. Gleit and Holland (1962) introduced the basic procedure, ashing in the presence of electronically excited O_2 (Gleit, 1963). The ashing temperature is from 100 to 150 °C with periods as much as 5 d required to completely destroy the organic matter. Therefore, the procedure is not well suited for high-volume rapid analysis requirements.

In most instances, the method of organic matter destruction is selected on the basis of personal preference without regard as to type of plant material or elements to be determined. It is difficult to find sufficient fault or advantage with either method that would consistently designate one superior to the other. Therefore, the analyst should compare analysis results of plant tissue that has been prepared by both methods of organic matter destruction, and on this basis, choose the method that gives the desired result.

B. Methods of Extraction

Extraction procedures for evaluating the Ca (Gallaher & Jones, 1976); Fe (Machold & Stephen, 1969; Katyal & Sharma, 1980; Chaney, 1984); N and NO_3 (Baker & Smith, 1969); and S and SO_4 (Spencer et al., 1978) status of plants have been proposed. Grunau and Swiader (1986) used ion chromatography to measure the Cl, NO_3, SO_4, and PO_4 in water extracts of several dried vegetable leaves, with good results reported except for PO_4. Ulrich et al. (1959) used 2% acetic acid to extract P and K from sugarbeet

Table 15-11. Extraction procedure for NO_3 in plant tissue.

Baker and Smith (1969) method

1. Weigh 400 mg of oven-dried ground (80 °C, 20-mesh) tissue into shaking bottle.
2. Add 40 mL extracting solution [0.025 *M* $Al_2(SO_4)_3$ containing 10 μg mL^{-1} NO_3-N and 1 mL L^{-1} preservation].
3. Shake for 15 min.
4. Filter and save filtrate for NO_3-N assay.

Heanes (1982) method

1. Weigh 400 mg of oven-dried ground (80 °C, 20-mesh) tissue into shaking bottle.
2. Add 50 mg of oxidized activated charcoal (AC).
3. Add 40 mL 0.025 *M* $Al_2(SO_4)_3$.
4. Shake for 30 min.
5. Filter into shaking bottle containing 500 mg of oxidized AC.
6. Shake filtrate plus AC for 30 min.
7. Filter and save filtrate for NO_3-N assay.

(*Beta vulgaris* L.) petioles. Sahrawat (1980, 1987), Hunt (1982), and Miyazawa et al. (1984) used dilute HCl as an extraction reagent to determine Ca, Mg, K, Mn, Cu, Zn, Fe, and P content in dried plant tissue. Baker and Greweling (1967) developed an extraction procedure that gave results comparable to those for dry-ashed samples for the elements Ca, Mg, K, Mn, Cu, and Zn. Nicholas (1957) also obtained good correlations between results obtained by extraction with total for the elements Ca, K, Mn, and P. The determination of B by extraction using HCl-HF has been suggested by van der Lee et al. (1987).

The most frequent chemical form obtained by extraction is NO_3-N. A widely used method of extraction first described by Baker and Smith (1969) and then modified by Heanes (1982) is given in Table 15-11. Mills (1980) suggests water as a satisfactory extraction reagent. A good review of the techniques and procedures for extraction and determination of NO_3-N has been written by Keeney and Nelson (1982).

All of these extraction procedures are laboratory conducted tests using oven-dried and milled tissue, and should not be confused with procedures called *tissue-tests,* tests conducted in the field on extracted sap from fresh tissue. Tissue testing will be discussed in another section of this chapter.

C. Methods of Elemental Determination

Advances in analytical chemistry since 1970 have significantly improved the ease and speed for the determination of elements found in plant tissue ash or digests. For most of the elements, frequently referred to as the *mineral elements,* the more traditional wet chemistry procedures have given way to various instrumental procedures that employ either emission or absorption spectrometry.

The classical colorimetric procedures have been described in detail by Piper (1942), Jackson (1958), Johnson and Ulrich (1959), Greweling (1976), and Chapman and Pratt (1982). The last two references describe flame emission and atomic absorption spectrometry procedures as well. Many of these colorimetric and flame emission procedures have been automated by employing an AutoAnalyzer® as described by Isaac and Jones (1970), and Steckel and Flannery (1971) for the determination of Ca, K, Mg, and P, and for B as described by Basson et al. (1969). Some of these same colorimetric procedures can be employed using automated flow injection analysis (Ranger, 1981).

Flame emission spectrometry for the determination of K and Na (Marrodineanu, 1970) as well as Cu, Fe, and Mn (Berneking & Schrenk, 1957; Pickett & Koirtyohann, 1969; Keliher et al., 1984), and atomic absorption spectrometry (whose acronym is AA) for the determination of Ca, Mg, Cu, Fe, Mn, and Zn are procedures that have been described by Greweling (1976), and in review articles by Isaac and Kerber (1971), Isaac (1980), Baker and Suhr (1982), and Ure (1983), and the book by Christian and Feldman (1970). Although both instrumental methods are still in wide use today, they are slow and cumbersome when compared to the more recently developed spectrometer techniques. Having a narrow dynamic reading range of up to two decades, considerable sample manipulation may be required to bring elemental concentrations within the calibration range of the spectrometer. In addition, only one element at a time can be determined.

Direct reading emission spectrometry, using a progression of excitation sources from AC and DC arcs (Cholak & Story, 1941; Mitchell, 1956, 1964; Thompson & Bankston, 1969), AC spark (Jones, 1976), the inductively coupled plasma (Jones, 1977; Dalquist & Knoll, 1978; Munter & Grande, 1981; Soltanpour et al., 1982; Zarcinas, 1984; Zarcinas et al. (1987), and the DC plasma (DeBolt, 1980), has been a major analytical technique for the assay of most elements found in plant tissue ash or digests. With one pass of the prepared sample through the excitation source, most of the plant essential elements (P, K, Ca, Mg, B, Cu, Fe, Mn, and Zn, and S for vacuum spectrometers) as well as nonessential elements (Al, As, Cd, Co, Cr, Ni, Pb, and Se) can be determined either simultaneously, in <60 s, or sequentially in $<$ 120 to 180 s. Plasma spectrometry is also relatively free from matrix and spectral interferences, has excellent sensitivity (usually $<$ 1 mg/kg) and a wide concentration reading range of several decades (between three to five). Inductively coupled agron plasma emission spectrometry is frequently referred to by its acronym, ICP or ICAP. There are two recent books on ICP, one by Walsh (1983) and the other edited by Montaser and Golightly (1987).

The spectrometer receiving and recording an ICP generated emission can have either an air or vacuum light path. Evacuated spectrometers can detect emission lines in the UV region of the spectrum (most useful for the elements B, P, and S). The spectrometer design may be either a sequential monochromator which records one emission line at a time, or a polychromator that has a detector for each element so that elemental determinations are made simultaneously (Nygaard & Sofera, 1988).

Sample preparation is relatively simple. First the organic matter is destroyed, and then the dissolved ash or digest solution is directly introduced into the ICP. This is in marked contrast to the more traditional wet chemical procedures and other single element analyzers that require considerable manipulation of the plant ash or digest solution prior to the determination of its elemental content. Since these latter procedures require substantial amounts of effort and time to perform an analysis, they are costly and subject to potential errors.

Another method of elemental determination in plant tissue has been x-ray fluorescence, procedures that have been described by Dixon and Wear (1964), Alexander (1965), Jenkins and Hurley (1966), Kubota and Lazar (1971), Murray (1975), and Jones (1982). However, the method is strongly affected by the matrix influences and is not a widely used technique for plant analysis.

It is difficult to specifically select any one method or technique as superior to another. There have been various studies conducted that have tried to make this distinction (Kenworthy et al., 1956; Bowen, 1967; Brech, 1968; Jones, 1969; Jones & Isaac, 1969). Others have studied the variance associated with various methods of analysis (Baker et al., 1964; Carpenter et al., 1968; Jones & Warner, 1969; Munter et al., 1984). However, the emphasis today is on quality assurance and performance testing as the means of assuring reliable performance for the analytical procedure used, laboratory management procedures that are discussed in more detail later. Unfortunately in most instances, method selection is based on what instrumentation and facilities are available, rather than a decision based on best method for the element(s) determined and form of the matrix.

1. Nitrogen

Total and approximate total N in plant tissue can be determined by essentially two analytical procedures. The Dumas (Bremner & Tabatabai, 1971; Edeling, 1968) technique for total N involves the conversion of organic N to molecular N. Although automated Dumas instruments are readily available, such as the LECO FP-228 (Sweeney & Rexroad, 1987) and are fairly easy to use, their high cost and low analytical capacity (<10 samples/h) limits their use for routine analytical purposes. In addition, since usually <100 mg samples are assayed, the plant tissue must be finely ground to obtain a reasonably homogenous sample. Nitrogen content determination by the Dumas procedure as compared to that obtained by Kjeldahl digestion, the most widely used N determination method, usually gives slightly (1–4%) higher results.

The original Kjeldahl digestion procedure was developed by Johan Kjeldahl in Denmark in the late 1800s. The first published procedure appeared in 1883 (Morries, 1983), making it the oldest analytical method still in use today. The method is in two steps; first, high temperature (330–450 °C) digestion in concentrated H_2SO_4 in the presence of a catalyst (Cu, Hg, or Se, or the combination of Cu-TiO_2) which converts organic N to inorganic NH_4), which combines with SO_4 to form $(NH_4)_2SO_4$; and the second step, either alkaline distillation of NH_3 and determination of NH_4 by acidimetric titration (Munsinger & McKinney, 1982), or determination of NH_4 by a colorimetric procedure (Warner & Jones, 1970; Crooke & Simpson, 1971; Uhl et al., 1971; Isaac & Johnson, 1976; Smith, 1980; Wang & Oien, 1986), or NH_4 determination by specific ion electrode (Gallaher et al., 1976). Numerous procedures have been proposed on how to conduct the Kjeldahl digestion step designed to simplify it as well as improve precision and accuracy. Nelson and Sommers (1980) have reviewed and evaluated many of the various modifications proposed for the Kjeldahl procedure as well as the NH_4 determination.

Depending on sample size, the Kjeldahl digestion procedure has been classed as being either macro- (1.0 g or greater), semimicro- (1.0–0.5 g), micro- (<0.5 g), with the size of the digestion/distillation apparatus scaled accordingly. Precision declines with decreasing sample size (<0.5 g) due to the lack of homogeneity among plant tissue particles, and, therefore, the need for a more finely ground sample (< 20 mesh) to ensure adequate homogeneity (Batey et al., 1974). The trend from macro- to micro-Kjeldahl digestion is an attempt to reduce the required laboratory space and equipment needed as well as to reduce reagent use (Campbell, 1986).

Today, the digestion block, which can be either constructed by the analyst (Gallaher et al., 1975) or obtained commercially seems to be the procedure of choice (Munsinger & McKinney, 1982). The digestion is carried out in a digestion tube set in the heated block, with tube size dictated by sample size (Nelson & Sommers, 1973; Noel & Hambleton, 1976). The NH_4 formed is determined by titration after being volatilized by steam distillation into a trapping solution (Campbell, 1986). There is also a fully automated Kjeldahl apparatus which is in fairly wide use (Anonymous, 1980).

Nitrogen in plant tissue as either NO_3 or NO_2 is not completely recovered in the Kjeldahl digestion unless converted to NH_4 by pretreatment of the sample with either reduced Fe under acidic conditions (Cataldo et al., 1974) or pretreatment in a moisture-free environment with salicylic acid or thiosulfate (Dalal et al., 1984). The Kjeldahl procedure, with and without NO_3 or NO_2 recovery, and NH_4 determination by alkaline distillation is given in the AOAC Manual under "Fertilizers" (Horwitz, 1980). Data from Whitehead and Olson (1942) suggest that some of the NO_3 in plant material is reduced and included in the Kjeldahl procedure that does not normally include NO_3 in inorganic fertilizer materials. These researchers added 4.2 to 16.6 mg of NO_3-N as KNO_3 solution to varying aliquots (0.5–2.0 g) of oat (*Avena sativa* L.) straw that contained 3.3 to 13.3 mg of N based on Kjeldahl method without added reducing agents. About 40% to more than 70% of the added NO_3-N was recovered. The amount of NO_3 added and type of catalyst had more of an effect than amount of sample used.

Some have used either $HClO_4$ (Batey et al., 1974) or 30% H_2O_2 (Lindner & Harley, 1942; Wolf, 1982; Parkinson & Allen, 1975) in place of a catalyst so that the obtained digest can be used for N determination as well as other elements, such as P, K, Ca, and Mg plus the micronutrients. However, Nelson and Sommers (1973) found that H_2O_2 additions can reduce N recoveries by about 15%.

The addition of either K_2SO_4 or Na_2SO_4, but mostly K_2SO_4, to the digestion mixture will increase the temperature of the digestion from 330 °C with pure H_2SO_4 to 350 °C, or higher, which in turn speeds the digestion

Table 15-12. Standard Kjeldahl digestion procedure (for determination in Kjeldahl flask or digestion tube).

Weigh 500 mg of dried ground plant tissue (80°C and 40 mesh) into a Kjeldahl flask or digestion tube.
Add 5.0 g of digestion mixture [100:1:1000 $CuSO_4 \cdot 5H_2O$:Se:K_2SO_4], or [1:60:1670:($CuSO_4$):TiO_2:K_2SO_4].
Add 10 mL conc. H_2SO_4.
Place Kjeldahl flask on digestion rack, or place funnel in the neck of the digestion tube and place the tube into a port of the digestion block. Heat to rolling boil in Kjeldahl flask or at 360 to 410 °C for digestion block.
Continue to heat for 60 min after clearing.
Discontinue heating and let Kjeldahl flask or digestion tube and contents cool.
Dilute to appropriate volume with pure water and determine NH_4 content in digest.

time and increases N recovered. The amount of either sulfate salt added to the acid is 0.3 to 0.5 g/mL of H_2SO_4. If the addition of K_2SO_4 to the digestion mixture results in solidification during the digestion step, some N will be lost by volatilization.

Either Cu or Hg, or Se alone, or the combination Cu-TiO_2 (Kane, 1986, 1987) have been used as catalysts in the Kjeldahl digestion procedure. Mercury is the catalyst most frequently specified for the N determination in plant tissue and is the one recommended in the official AOAC procedure (Horwitz, 1980) as well as that given by the American Association of Cereal Chemists (Christensen, 1983). If Hg is the catalyst selected, then S_2O_3 must be added to the digest in the distillation step to break the Hg-NH_4 complex that is formed. Mercury as well as Se are elements of concern today due to their potential health hazard. Therefore, their recommended use as catalysts has been discouraged and replaced by either Cu (Rexroad & Cathey, 1976) or Cu-TiO_2 (Kane, 1986, 1987).

Digestion time is an important consideration for the complete conversion of organic N to NH_4 to occur. When the digestion mixture clears, an additional period of two to three times that required for clearing is needed to obtain near complete conversion of organic N to NH_4. At clearing, about 90 to 92% of the organic N in most plant samples has been converted with the additional digestion time needed to obtain most of the remaining 7 to 8% N.

The determination of N in plant tissue by Kjeldahl digestion gives something less than total N content. Therefore, the use of the word "Kjeldahl" to designate "Total N" is not correct. Also, the word Kjeldahl is not sufficient to be used alone without some explanation of how the analysis was performed in terms of sample size, apparatus, and catalyst used as well as whether NO_3 was recovered (Jones, 1987). A generic description of the Kjeldahl digestion step is given in Table 15-12. Verification of the Kjeldahl N determination is important as recommended by Hach and Brayton (1985).

Currently, there are several non-Kjeldahl methods for N determination in plant tissue that are being investigated. One method is by direct distillation, placing a plant tissue sample into alkali and steam distilling for a specified time (Anonymous, 1981). The amount of amino-N converted to NH_4 and measured is compared with a predetermined Kjeldahl N value by means of a calibration curve. An essential requirement for the procedure is precise control of the distillation time and temperature if reliable results are to be obtained.

Another is a chemiluminescent N system based on the conversion of bound N to NO which is converted to NO_2 in the presence of ozone (Anonymous, 1982). Jacques and Peterson (1987) found the technique has excellent sensitivity, but precision is significantly affected by sample size and the combustion boat design.

Near infrared reflectance (NIR) is the procedure that looks promising for determination of plant N (Dorsheimer & Isaac, 1982; Mundel & Schaalje, 1988). A beam of infrared radiation is focused on a finely ground dried plant tissue sample and the reflected radiation measured. The technique is fast and nondestructive; however, the instrument is expensive and the calibration procedure requires the use of standards of the same plant species as those to be assayed. Isaac and Johnson (1983) compared N determinations in corn leaves obtained by NIR with that obtained by Kjeldahl digestion, a comparison they described as quite good, but not sufficient to recommend the NIR technique. However, Blakeney et al. (1988) obtained agreement between N determined by NIR and Kjeldahl digestion using whole rice (*Oryza sativa* L.) stem samples finely ground in a cyclone mill.

Sulfur

Total S in plants can be determined by several techniques with comparable results. One procedure follows wet oxidation of a plant tissue sample in a mixture of HNO_3 and $HClO_4$, converting organic S to SO_4-S. The SO_4 content in the digest is then determined by either the $BaSO_4$ turbidity method or by one of several colorimetric procedures (Beaton et al., 1968). The $BaSO_4$ turbidity method has been adapted for use with an AutoAnalyzer® (Wall et al. 1980).

Automated combustion using a Leco Sulfur Analyzer is another method for determining total plant S. A prepared plant tissue sample is placed into the induction furnace of the analyzer and heated to 1350 °C in a stream of O_2. Plant S is oxidized to SO_2 which is either trapped in an indicator solution and the amount of SO_2 evolved determined by a back titration (Jones & Issac, 1972), or the SO_2 is passed through an infrared analyzer (Hern, 1984). The titration technique requires removing interfering Cl and N prior to S analysis which is done by mixing MgO with the plant tissue and ashing at 500 °C for 2 h.

Sulfur can also be determined by x-ray emission spectrometry (Alexander, 1965; Kubota & Lazar, 1971; Murdock & Murdock, 1977) using a dried ground plant tissue sample. The analysis technique is relatively easy since little sample preparation is required, but the method does require the use of an expensive analytical instrument. In addition to S, several other elements can be determined by this instrumental procedure (Maclauchlan et al., 1987), a method that is sensitive to the physical and chemical characteristics of the sample itself. Although x-ray emission spectrometry has had a fairly long history of use, it has not been widely used as a major instrumental procedure for routine assay of plant tissue for its elemental content.

D. Quality Assurance

Quality assurance in an analytical laboratory is an important management tool which is designed to ensure reliable performance, and is becoming a national policy for many types of testing laboratories (Aldehnoff & Ernest, 1983). Criteria for implementation have been established by the Association of Official Analytical Chemists (Garfield, 1984). The elements of a quality assurance program are: (i) administration, (ii) personnel management, (iii) management of equipment and supplies, (iv) records maintenance, (v) sample analysis, (vi) proficiency testing, (vii) audit procedures, and (viii) design and safety of facilities. Application of quality assurance criteria on an actual analytical procedure are discussed in detail in the books by Dux (1986) and Taylor (1987).

The basis for reliable performance in an analytical laboratory has been titled "Good Laboratory Practices," the subject of a *Federal Register* entry in 1979 (Anonymous, 1979), and is described in some detail by Fischbeck (1980). The Council of Independent Laboratories (Anonymous, 1976), the Environmental Protection Agency (Booth, 1979), and the National Bureau of Standards (Berman, 1980) have recently published proceedings of symposia on the subject of laboratory performance.

Horwitz (1982), based on his many years of association with the Association of Official Analytical Chemists, has discussed the practical limits of acceptable variability for methods of analysis, focusing on the important aspects of reliability, reproducibility, repeatability, systematic error of bias, specificity, and limit of reliable measurement. The impact of these aspects on any analytical procedure varies considerably in terms of sample size, determinations made, concentrations of the analyte, and the characteristics of analytical instrument used. An additional criteria is the ruggedness factor that sets the limits for each step in the analytical procedure, and when exceeded, will invalidate the obtained results. Examples would be how much variation in temperature or time could be tolerated for a given procedure without a significant change in the result occurring, thereby making the obtained result invalid. Unfortunately, most plant analysis procedures have not been so described. Examples would be limit criteria for sample preparation procedures such as moisture removal criteria, ashing and digestion temperatures, and length of time; and procedures described earlier in this chapter but without carefully defined limits in some instances.

The analyst today has a wide range of instrumentation from which to choose. Some of the factors that affect the choice made are described by McLaughlin et al. (1979). A similar evaluation of analytical technique has been developed by Hislop (1980) as shown in Table 15–13.

Practical considerations, such as instrument availability, and its purchase price and operating cost frequently are governing factors that determine the choice of analytical procedure rather than Hislop's criteria. In

Table 15-13. Analytical criteria for selecting technique of analysis (Hislop, 1980).

Accuracy	Elemental coverage
Precision	Single or multielement
Limit of detection	Determine chemical form

addition, the skill and experience of the analyst may also govern the selection of a particular method or instrument. Those in search of a suitable analytical procedure may find the articles by Morrison (1979) and Stika and Morrison (1981) useful as they compare the relative sensitivity and precision of various methods. For many of the elements, sensitivity is not a significant factor when assaying a plant tissue digest or plant ash for its elemental content since many elemental contents are quite high. Therefore, the choice of analytical procedure could then be based on precision alone. This suggests that different instrument techniques may have to be selected depending on the needs in terms of sensitivity and precision requirements.

The issues of sensitivity, precision, and accuracy are important considerations for the analyst and should be significant determining factors in method selection. Sensitivity has been variously defined as two parameters, the minimum level of detectability, and the ability to significantly distinguish between two analytical values. The detection limit is the smallest concentration that can be identified as being greater than zero and is usually defined as either 2 or 3 SD above a background determination. Sensitivity in terms of level distinction is confounded with precision.

Precision is a measure of the degree of variability associated with an obtained analytical result that is determined by repeated assays of the same sample carried through all the steps from sample preparation to the final result. Poor precision may be the compounded result of accumulated errors made over the entire analytical procedure, and therefore, not the result of a single factor. Hislop (1980) has written an excellent article on the requirements for obaining accurate and precise analytical results. Horwitz (1982) also has evaluated various analytical procedures by assigning levels of performance by analytical technique, basing this judgment on years of experience in the determination of elements and substances in various materials. Dux (1986) describes how precision determinations can be made a part of a quality assurance program.

Accuracy is the ability of the method to obtain the "true" value and is dependent on the use of reliable standards and matrix matching. Several kinds of standards are needed in the laboratory, one set to monitor analytical procedures, and another set to calibrate instruments and standardize reagents. Prepared and certified standard solutions and reagents can be obtained from many chemical supply houses. Standard References Materials (Uriano, 1979) are available from the U.S. National Institute for Standards and Technology (Alvarez, 1980). Currently, there are four plant tissues (tomato leaves, citrus leaves, pine needles, and spinach), and one animal tissue (bovine liver) available as SRMs. Unfortunately, not every element of interest to the plant nutritionist is certified in all these SRMs, which does limit their usefulness. However, SRMs should be used when ever possible for verification of accuracy as outlined in the handbook for SRM users (Taylor, 1985).

In the normal laboratory routine, a standard can be used not only to monitor an analytical procedure, but when placed into a sequence of unknowns, it serves the dual purpose as marker and standard. By noting the position and value of the marker-standard at the end of an analytical run, the analyst can determine if an unknown had been skipped or duplicated as well as evaluating if a shift in the calibration has occurred. An evaluated calibration shift may then become the basis for adjustment for results obtained between each marker-standard.

These factors and others form the basis for an established quality assurance program as has been defined by Garfield (1984), Dux (1986), and Taylor (1987). Without such a program of laboratory and analytical management, performance in terms of reliable analytical results cannot be obtained. A recent study reported by Munter et al. (1984) points to certain areas of concern when conducting an elemental assay of plant tissue. They found that both preparation procedures and instrument calibration techniques should be standardized to minimize variation. A similar study by Sterrett et al. (1987) uncovered errors in the determination for various elements in plant tissue. Both these papers point to the need for careful evaluation of each step in the analytical process on the basis of quality assurance parameters.

IV. TISSUE TESTING

A distinction is made between plant (leaf) analysis and tissue testing. The former is a laboratory analysis on dried and ground plant tissue sample, while a tissue test is an assessment of the elemental content of sap taken from fresh tissue with procedures that are usually carried out in the field using special testing techniques.

Krantz et al. (1948) give instructions for the field testing of corn (*Zea mays* L.), and cotton (*Gossypium hirsutum* L.), and soybean [*Glycine max* (L.) Merr.] plants using sap pressed from fresh tissue for the semiquantitative determination for the ions NO_3^-, PO_4^{-3}, and K^+. Wickstrom (1967) has also written procedures that use such tissue tests for field diagnosis. Syltie et al. (1972) have given procedural details for field-conducted tissue tests on corn and soybean for the elements N, P, Mg, and Mn. They give instructions for preparation of the reagents as well as a description of techniques for conducting the tests. Scaife and Stevens (1983) have found the use of "Merckoquant" test strips suitable for NO_3^- determination in the field assessment of the N status of various vegetable crops.

Iron is an element that can be determined by a tissue test conducted in the field, a procedure first developed by Bar-Akiva et al. (1978) and modified by Bar-Akiva (1984). Peroxidase activity is measured by floating leaf discs in a reactive solution with the development of a blue color indicating adequate Fe in the plant tissue.

The ability to perform tissue tests in the field is considered by some to be a significant advantage in terms of immediate test results and low cost compared to that required for a laboratory conducted plant analysis. It should be remembered that most of the tests themselves are not entirely quantitative, but provide the tester with a qualitative "yes or no" evaluation; that is, the element or chemical form is either present or not present at the desired concentration level. Considerable practical experience using these test procedures and repeated observations are required before one can feel confident when making an interpretation based on such a test result. Tissue tests and their interpretation are considered by some to be more of an "art" than a strict quantitative analytical science. However, it should be remembered that the test procedures themselves are based on sound analytical chemistry, but their utilization and interpretation requires skill that can be gained only by repeated practical experience. Combining field observations with soil and plant tissue quick test procedures has been coined "The Diagnostic Approach," a procedure of observation, testing, and evaluation that has been discussed in detail in a special issue of *Better Crops* (Armstrong, 1984).

V. SUMMARY

Considerable care is required when sampling, handling, and assaying plant tissue if a plant analysis is to be a useful diagnostic tool. Failure to follow prescribed procedures anywhere in the process can lead to erroneous results, and then, faulty conclusions. Sampling may still be the weakest link in the plant analysis technique, and those who collect plant tissue samples must be properly instructed and trained. Although improvements in various aspects of the plant analysis technique are still needed, the procedures for proper cleaning, drying, milling, and storing plant tissue have been fairly well established. By using multielement analytical instruments, such as ICP emission spectrometer, or the flow injection analyzer (FIA), an analyst can quickly and easily assay a plant ash or digest for most of the essential elements. Analytically, there is need to standardize the Kjeldahl digestion procedure that will accurately define the N forms included. Analysts should be using known plant tissue standards such as SRMs from the U.S. National Institute of Standards and Technology or other similar sources for instrument calibration and accuracy assessment. Laboratories should have a quality assurance program in force that follows the current guidelines established by the Association of Official Analytical Chemists.

If the collection, preparation, and elemental determination is done properly, then the plant analysis result can effectively evaluate the nutrient status of the plant. The analysis result can either be used to identify suspected elemental deficiencies, serve to monitor the soil/plant nutrient element status, or provide the basis for formulating lime and fertilizer treatments necessary to ensure high yield and quality.

REFERENCES

Adler, P.R., and G.E. Wilcox. 1985. Rapid perchloric acid methods for analysis of major elements in plant tissue. Commun. Soil Sci. Plant Anal. 16:1153–1163.

Adriano, D. 1986. Trace elements in the terrestrial environment. Springer-Verlag New York, New York.

Aldehnoff, G.A., and L.A. Ernest. 1983. Quality assurance program at a municipal wastewater treatment plant laboratory. J. Water Pollut. Control Fed. 55:1132–1137.

Alexander, G.V. 1965. An x-ray fluorescence method for the determination of calcium, potassium, chlorine, sulfur, and phosphorus in biological tissues. Anal. Chem. 37:1671–1674.

Alvarez, R. 1980. NBS plant tissue standard reference materials. J. Assoc. Off. Anal. Chem. 63:806–808.

Anonymous. 1976. Quality control systems-requirements for a testing laboratory. Am. Counc. of Independent Lab., Washington, DC.

Anonymous. 1979. Nonclinical laboratory studies, good laboratory practices. Federal Register, Friday, 22 Dec. Part II. Vol. 43. No. 247. Dep. of Health, Education, and Welfare, FDA. U.S. Gov. Print. Office, Washington, DC.

Anonymous. 1980. Kjel-Foss Automatic Type 16200. Bull. A/S, N. Foss Electric, Hillerod, Denmark.

Anonymous. 1981. Rapid determination of protein content in grains by using the Kjeldahl Auto System DD. Tecator Appl. Note An 33/81, Herndon, VA.

Anonymous. 1982. Model 703C chemiluminescent nitrogen system. Antek Instruments, Houston, TX.

Armstrong, D. (ed.). 1984. The diagnostic approach to maximum economic yields. Potash and Phosphate Inst., Atlanta.

Ashby, D.L. 1969. Washing techniques for the removal of nutrient element deposits from the surface of apple, cherry and peach leaves. J. Am. Soc. Hortic. Sci. 94:266–268.

Baker, A.S., and R. Smith. 1969. Extracting solution for potentiometric determination of nitrate in plant tissue. J. Agric. Food Chem. 17:1284–1287.

Baker, D.E., G.W. Gorsline, C.G. Smith, W.I. Thomas, W.E. Grube, and J.L. Ragland. 1964. Technique for rapid analyses of corn leaves for eleven elements. Agron. J. 56:133–136.

Baker, D.E., and N.H. Suhr. 1982. Atomic absorption and flame emission spectrometry. p. 13–28. *In* A.L. Page et al. (ed.) Methods of soil analysis, Part 2. 2nd ed. Agronomy Monogr. 9. ASA and SSSA, Madison, WI.

Baker, J.H., and T. Greweling. 1967. Extraction procedure for quantitative determination of six elements in plant tissue. J. Agric. Food Chem. 15:340–344.

Bar-Akiva, A. 1984. Substitutes for benzidine as N-donors in the peroxidate assay for rapid diagnosis of iron in plants. Commun. Soil Sci. Plant Anal. 15:929–934.

Bar-Akiva, A., D.N. Maynard, and J.E. English. 1978. A rapid tissue test for diagnosing iron-deficiencies in vegetable crops. Hortic. Sci. 13:284–285.

Basson, W.D., R.G. Bohner, and D.A. Stanton. 1969. An automated procedure for the determination of boron in plant tissue. Analyst 94:1135–1141.

Bates, T.E. 1971. Factors affecting critical nutrient concentrations in plants and their evaluation: A review. Soil Sci. 112:116–130.

Batey, T., M.S. Cresser, and I.R. Willett. 1974. Sulphuric-perchloric acid digestion of plant material for nitrogen determination. Anal. Chim. Acta 69:484–487.

Beaton, J.D., G.R. Burns, and J. Platou. 1968. Determination of sulphur in soils and plant material. Tech. Bull. 14. The Sulphur Inst., Washington, DC.

Beaufils, E.R. 1973. Diagnosis and recommendation integrated system (DRIS). Soil Sci. Bull. 1. Univ. of Natal, South Africa.

Berman, G.A. (ed.). 1980. Testing laboratory performance: Evaluation and accreditation. Natl. Bureau of Standards, Washington, DC.

Berneking, A.D., and W.G. Schrenk. 1957. Flame photometric determination of manganese, iron, and copper in plant material. J. Agric. Food Chem. 5:742–745.

Bhan, K.C., A. Wallace, and D.R. Hunt. 1959. Some mineral losses from leaves by leaching. Proc. Am. Soc. Hortic. Sci. 73:289-293.

Blakeney, A.B., G.D. Batten, and P.E. Bacon. 1988. Tissue test of rice plant nitrogen. Int. Rice Res. News 13:27–28.

Bock, R.A. 1978. Handbook of decomposition methods in analytical chemistry. Int. Textbook Co., Glasgow, Scotland.

Booth, R.L. (ed.). 1979. Handbook for analytical quality control in water and wastewater laboratories. Analytical Quality Control Lab., Natl. Environ. Res. Ctr., Cincinnati.

Bowen, H.J.M. 1967. Comparative elemental analyses of a standard plant material. Analyst 92:124–131.

Brech, F. 1968. Comparison of optical emission and atomic absorption methods for the analysis of plant tissues. J. Assoc. Offic. Anal. Chem. 51:132–136.

Bremner, J.M., and M.A. Tabatabai. 1971. Use of automated combustion techniques for total carbon, total nitrogen, and total sulfur analysis of soils. p. 1–16. *In* L.M. Walsh (ed.) Instrumental methods for analysis of soils and plant tissue. SSSA, Madison, WI.

Campbell, D.C. 1986. Micro-Kjeldahl analysis using 40-table block digestor and steam distillation. J. Assoc. Off. Anal. Chem. 69:1013–1016.

Carlier, L.A., and L.P. van Hee. 1971. Microwave drying of lucerne and grass samples. J. Sci. Food Agric. 22:306–307.

Carpenter, P.N., A. Ellis, H.E. Young, and T.E. Byther. 1968. A critical evaluation of results from spectrographic analysis of plant tissue. Maine Agric. Exp. Stn. Tech. Bull. 30.

Cataldo, D.A., L.E. Schrader, and V.L. Youngs. 1974. Analysis of digestion and colorimetric assay of total nitrogen in plant tissue high in nitrate. Crop Sci. 14:854–856.

Chaney, R.L. 1984. Diagnostic practices to identify iron deficiency in higher plants. J. Plant Nutr. 7:47–67.

Chapman, H.D. 1964. Foliar sampling for determining the nutrient status of crops. World Crops 16:35–46.

Chapman, H.D. (ed.). 1966. Diagnostic criteria for plants and soils. Univ. California Div. Agric. Sci., Riverside, CA.

Chapman, H.D., and P.F. Pratt. 1982. Methods of analysis for soils, plants, and waters. Div. of Agric., Univ. of California, Berkeley, CA. Priced Publ. 4034.

Cholak, J., and R.V. Story. 1941. Spectrochemical determination of trace metals in biological material. J. Opt. Soc. Am. 31:730–738.

Christensen, E.A. (ed.). 1983. Approved methods of the American Association of Cereal Chemists. Vol. II. Sections 46:09–13. Am. Assoc. Cereal Chem., St. Paul.

Christian, G.D., and F.J. Feldman. 1970. Atomic absorption spectroscopy applications in agriculture, biology and medicine. Wiley-Interscience, New York.

Colonna, J.P. 1970. The mineral diet of excelsior coffee plants. Natural variability of the mineral foliar composition on a homogenous plantation. Cashiers Office. Res. Sci. Tech. Outre-Mer Ser. Biol. 13:67–80.

Cresser, M.S., and J.W. Parsons. 1979. Sulfuric-perchloric acid digestion of plant material for the determination of nitrogen, phosphorus, potassium, calcium and magnesium. Anal. Chem. Acta 109:431–436.

Crooke, W.M., and W.E. Simpson. 1971. Determination of ammonium in Kjeldahl digests of crops by an automated procedure. J.Sci. Food. Agric. 22:9–10.

Dalal, R.C., K.L. Sahrawat, and R.J.K. Myers. 1984. Inclusion of nitrate and nitrite in the Kjeldahl nitrogen determination of soils and plant materials using sodium thiosulfate. Commun. Soil Sci. Plant Anal. 15:1453–1461.

Dalquist, R.L., and J.W. Knoll. 1978. Inductively coupled plasma atomic emission spectroscopy: Analysis of biological materials and soil for major, trace, and ultra-trace elements. Appl. Spectro. 32:1–30.

DeBolt, D.C. 1980. Multielement emission spectroscopic analysis of plant tissue using DC agron plasma source. J. Assoc. Off. Anal. Chem. 63:802–805.

Dixon, J.B., and J.I. Wear. 1964. X-ray spectrographic analysis of zinc, manganese, iron, and copper in plant tissue. Soil Sci. Soc. Am. Proc. 28:744–746.

Dorscheimer, W.T., and R.A. Isaac. 1982. Application of NIR analysis. Am. Lab. 14:58–63.

Dux, J.P. 1986. Handbook of quality assurance for the analytical chemistry laboratory. Van Nostrand Reinhold Co., New York.

Edeling, M.E. 1968. The Dumas method for nitrogen in feeds. J. Assoc. Off. Anal. Chem. 51:766–770.

Feldman, C. 1961. Evaporation of boron from acid solutions and residues. Anal. Chem. 33:1916–1920.

Fischbeck, R.D. 1980. Good laboratory practice regulations. Am. Lab. 12:125–130.

Gallaher, R.N., C.O. Wilson, and J.G. Futral. 1975. An aluminum block digestor for plant and soil analysis. Soil Sci. Soc. Am. Proc. 39:803–806.

Gallaher, R.N., and J.B. Jones, Jr. 1976. Total extractable, and oxalate calcium and other elements in normal and mouse ear pecan tree tissues. J. Am. Soc. Hortic. Sci. 101:692–696.

Gallaher, R.N., C.O. Weldan, and F.C. Boswell. 1976. A semi-automated procedure for total nitrogen in plant and soil samples. Soil Sci. Soc. Am. Proc. 40:887–889.

Garfield, F.M. 1984. Quality assurance principles for analytical laboratories. Assoc. Off. Anal. Chem., Arlington, VA.

Gleit, C.E. 1963. Electronic apparatus for ashing biologic specimens. Am. J. Med. Electron. 2:112–118.

Gleit, C.E., and W.D. Holland. 1962. Use of electrically excited oxygen for the low temperature decomposition of organic substances. Anal. Chem. 34:1454–1457.

Goodall, D.W., and F.G. Gregory. 1947. Chemical composition of plants as an index of their nutritional status. Imp. Bur. Hortic. Plant Crops (GB) Tech. Commun. 17.

Gorsuch, T.T. 1959. Radio-chemical investigation on the recovery for analysis of trace elements in organic and biological materials. Analyst 84:135–173.

Gorsuch, T.T. 1970. Destruction of organic matter Int. Series of Monographs in Analytical Chemistry. Vol. 39. Pergamon Press, New York.

Gorsuch, T.T. 1976. Dissolution of organic matter. p. 491–508. *In* P.D. LaFluer (ed.) Accuracy in trace analysis: Sampling, sample handling, analysis. Vol. 1. Spec. Publ. 422. Natl. Bureau of Standards, Washington, DC.

Graham, R.D. 1972. Vacuum attachment for grinding mills. Commun. Soil Sci. Plant Anal. 3:167–173.

Grant, C.L., and D.J. MacLaughlin, Jr. 1968. Preparation of biological samples and standards for spectrochemical analysis. Abstr. Appl. Spectr. 22(4):365.

Greweling, T. 1976. Chemical analysis of plant tissue. Search, Agron. 6, no. 8. Cornell Univ., Ithaca, NY.

Grier, J.D. 1966. Preparation of plant material for plant analysis. J. Assoc. Off. Anal. Chem. 49:292–298.

Grunau, J.A., and J.M. Swiader. 1986. Application of ion chromatography to anion analysis in vegetable leaf extracts. Commun. Soil Sci. Plant Anal. 17:3321–335.

Hach, C.C., and S.V. Brayton. 1985. Primary standards. The bases for accuracy in the Kjeldahl protein method. Tech. Center for Applied Analytical Chemistry. HACH Co., Loveland, CO.

Halvin, J.L., and P.N. Soltanpour. 1980. A nitric acid plant tissue digest method for use with inductively-coupled plasma spectrometry. Commun. Soil Sci. Plant Anal. 11:969–980.

Heanes, D.L. 1982. Determination of nitrate-N in plants by an improved extraction procedure adapted for ultraviolet spectrophotometry. Commun. Soil Sci. Plant Anal. 13:805–818.

Hern, J.L. 1984. Determination of total sulfur in plant materials using an automated sulfur analyzer. Commun. Soil Sci. Plant Anal. 15:99–107.

Hislop, J.S. 1980. Choice of the analytical method. p. 747–767. *In* P. Bratter and P. Schramel (ed.) Trace element analytical chemistry in medicine and biology. DeGruyter, Berlin.

Hood, S.L., R.Q. Parks, and C. Hurwitz. 1944. Mineral contamination resulting from grinding plant samples. Ind. Eng. Chem. Anal. Ed. 16:202–205.

Horwitz, W. (ed.). 1980. Official methods of analysis of the Association of Official Analytical Chemists, 13th ed. Assoc. Off. Anal. Chem., Arlington, VA.

Horwitz, W. 1982. Evaluation of analytical methods used for regulation of foods and drugs. Anal. Chem. 54:67A–76A.

Huang, C.L., and E.E. Schulte. 1985. Digestion of plant tissue for analysis by ICP emission spectroscopy. Commun. Soil Sci. Plant Anal. 16:943–958.

Hunt, J. 1982. Dilute hydrochloric acid extraction of plant material for routine cation analysis. Commun. Soil Sci. Plant Anal. 13:49–55.

Isaac, R.A. 1980. Atomic absorption methods for analysis of soil extracts and plant tissue digests. J. Assoc. Off. Anal. Chem. 63:788–796.

Isaac, R.A., and W.C. Johnson. 1976. Determination of total nitrogen in plant tissue. J. Assoc. Off. Anal. Chem. 59:98–100.

Isaac, R.A., and W.C. Johnson. 1983. Determination of protein nitrogen in plant tissue using near infrared spectroscopy. J. Assoc. Off. Anal. Chem. 66:506–509.

Isaac, R.A., and J.B. Jones. Jr. 1970. Auto-analysis systems for the analysis of soil and plant tissue extracts. p. 57–64. *In* Advances in automated analysis. Technicon Congr. Proc. Technicon Corp., Tarrytown, NY.

Isaac, R.A., and J.B. Jones, Jr. 1972. Effects of various dry ashing temperatures on the determination of 13 nutrient elements in five plant tissues. Commun. Soil Sci. Plant Anal. 3:261-269.

Isaac, R.A., and J.D. Kerber. 1971. Atomic absorption and flame photometry: Techniques and uses in soil, plant and water analysis. p. 17-37. *In* L.M. Walsh (ed.) Instrumental methods for analysis of soils and plant tissue. Soil Sci. Soc. Am., Madison, WI.

Jackson, M.L. 1958. Soil chemical analysis. Prentice-Hall, Englewood Cliffs, NJ.

Jacques, D.I., and J.C. Peterson. 1987. A comparison of the Antek chemiluminscent system and Kjeldahl procedure for determination of total nitrogen in plant tissue. J. Plant Nutr. 10:1683-1388.

Jenkins, R., and P.W. Hurley. 1966. Plant mineral analysis by x-ray fluorescence spectrometry. Analyst 91:395-397.

Johnson, C.M., and A. Ulrich. 1959. II. Analytical methods for use in plant analysis. California Agric. Exp. Stn. Bull. 766.

Jones, A.A. 1982. X-ray fluorescence spectrometry. p. 85-121. *In* A.L. Page et al. (ed.) Methods of soil analysis. Part 2. 2nd ed. Agronomy Monogr. 9. ASA and SSSA, Madison, WI.

Jones, J.B., Jr. 1963. Effect of drying on ion accumulation in corn leaf margins. Agron. J. 55:579-580.

Jones, J.B., Jr. 1967. Interpretation of plant analysis for several agronomic crops. p. 48-49. *In* G.W. Hardy et al. (ed.) Soil testing and plant analysis. Part II. SSSA Spec. Publ. 2. SSSA, Madison, WI.

Jones, J.B., Jr. 1969. Elemental analysis of plant tissue by several laboratories. J. Assoc. Offic. Anal. Chem. 59:900-903.

Jones, J.B., Jr. 1970. Distribution of 15 elements in corn leaves. Commun. Soil Sci. Plant Anal. 1:27-34.

Jones, J.B., Jr. 1976. Elemental analysis of biological substances by direct reading spark emission spectroscopy. Am. Lab. 8:15-20.

Jones, J.B., Jr. 1977. Elemental analysis of soil extracts and plant tissue ash by plasma emission spectroscopy. Commun. Soil. Sci. Plant Anal. 8:345-365.

Jones, J.B., Jr. 1985. Recent survey of number of soil and plant tissue samples tested for growers in the United States. *In* J.R. Brown (ed.) Tenth soil-plant analysts workshop. Counc. Soil Testing and Plant Analysis, Athens.

Jones, J.B., Jr. 1987. Kjeldahl nitrogen determination-what's in a name. J. Plant Nutr. 10:1675-1682.

Jones, J.B., Jr., and R.A. Isaac. 1969. Comparative elemental analyses of plant tissue by spark emission and atomic absorption spectroscopy. Agron. J. 61:393-394.

Jones, J.B., Jr., and M.H. Warner. 1969. Analysis of plant-ash solutions by spark emission spectroscopy. p. 152-160. *In* E.L. Grove and H.D. Perkins (ed.) Developments in applied spectroscopy. Vol. 7A. Plenum Press, New York.

Jones, J.B., Jr., R.L. Large, D.P. Pfleiderer, and H.S. Klosky. 1971. How to properly sample for a plant analysis. Crops Soils 23:15-18.

Jones, J.B., Jr., and R.A. Isaac. 1972. Determination of sulfur in plant material using a LECO sulfur analyzer. J. Agric. Food Chem. 20:192-194.

Kane, P.F. 1986. $CuSO_4$ as Kjeldahl digestion catalysts in manual determination of crude protein in animal feeds. J. Assoc. Off. Anal. Chem. 69:664-666.

Kane, P.F. 1987. Comparison of HgO and $CuSO_4/TiO_2$ as catalysts in manual Kjeldahl digestion for determination of crude protein in animal feed. Collaborative Study. J. Assoc. Off. Anal. Chem. 70:907-911.

Katyal, J.C., and B.D. Sharma. 1980. A new technique of plant analysis to revolve iron chlorosis. Plant Soil 55:105-109.

Keeney, D.R., and D.W. Nelson. 1982. Nitrogen-inorganic forms. p. 643-698. *In* A.L. Page et al. (ed.) Methods of soil analysis. Part 2. 2nd ed. Agronomy Monogr. 9. ASA and SSSA, Madison, WI.

Keliher, P.N., W.J. Boyko, J.M. Patterson III, and J.W. Hershy. 1984. Emission spectrometry. Anal. Chem. 56:133R-155R.

Kenworthy, A.L. 1969. Fruit, nut, and plantation crops deciduous and evergreen. A guide for collecting foliar samples for nutrient element analysis. Hortic. Rep. 11. Michigan State Univ., East Lansing, MI.

Kenworthy, A.L., E.J. Miller, and W.T. Mathis. 1956. Nutrient-element analysis of fruit tree leaf samples by several laboratories. Proc. Am. Soc. Hortic. Sci. 67:16-21.

Knapp, G. 1985. Sample preparation techniques—An important part in trace element analysis for environmental research and monitoring. Int. J. Environ. Anal. Chem. 22:71–83.

Knapp, G., and A. Grillo. 1986. A high pressure asher for trace analysis. Am. Lab. 18(3):76–79.

Krantz, B.A., W.L. Nelson, and L.F. Burkhart. 1948. Plant-tissue tests as a tool in agronomic research. p. 137–156. *In* H.B. Kitchen (ed.) Diagnostic techniques for soils and crops. Am. Potash Inst., Washington, DC.

Kubota, J., and V.A. Lazar. 1971. X-ray emission spectrograph: Techniques and uses for plant and soil studies. p. 67–83. *In* L.M. Walsh (ed.) Instrumental methods for analysis of soils and plant tissue. SSSA, Madison, WI.

Kuennen, R.W., K.A. Wolnik, F.L. Fricke, and J.A. Caruso. 1982. Pressure dissolution and real matrix calibration for multielement analysis of raw agricultural crops by inductively coupled plasma atomic emission spectrometry. Anal. Chem. 54:2146–2150.

Labanauskas, C.K. 1968. Washing citrus leaves for leaf analysis. Calif. Agrig. 22:12–14.

Lilleland, O., and J.G. Brown. 1943. Phosphate nutrition of fruit trees. Proc. Am. Soc. Hortic. Sci. 41:1–10.

Lindner, R.C., and C.P. Harley. 1942. A rapid method for the determination of nitrogen in plant tissue. Science 96:565–566.

Lockman, R.B. 1970. Plant sample analysis as affected by sample decomposition prior to laboratory processing. Commun. Soil Sci. Plant Anal. 1:13–19.

Machold, O., and W.W. Stephen. 1969. Functions of iron porphyrin in chlorophyll biosynthesis. Phytochemistry 8:2189–2192.

Maclauchlan, L.E., J.H. Borden, M.R. Cackette, and J.M. D'Auria. 1987. A rapid, multi sample technique for detection of trace elements in trees by energy-dispersive x-ray fluorescence spectroscopy. Can. J. For. Res. 27:1124–1130.

Martin-Prevel, P., J. Gagnard, and P. Gautier (ed.). 1987. Plant analysis as a guide to the nutrient requirements of temperate and tropical crops. Lavosier Publ., New York.

Marrodineanu, R. (ed.). 1970. Analytical flame spectroscopy. Springer-Verlag New York, New York.

Mclaughlin, R.D., M.S. Hunt, D.L. Murphy, and C.R. Chen. 1979. Instrumentation in environmental monitoring. p. 224–234. *In* D.L. Hamphill (ed.) Trace elements in environmental health-XIII. Univ. of Missouri, Columbia.

Mills, H.A. 1980. Nitrogen specific ion electrodes for soil, plant, and water analysis. J. Assoc. Off. Anal. Chem. 63:797–801.

Mitchell, R.L. 1956. The spectrographic analysis of soils, plants and related materials. Tech. Commun. No. 44. Commonwealth Bureaux of Soils, Harpenden, Herts, England.

Mitchell, R.L. 1964. The spectrochemical analysis of soils, plants and related materials. Tech. Commun. No. 44A. Commonwealth Agric. Bureaux, Franham Royal, Bucks, England.

Miyazawa, M., M.A. Pavan, and M.F.M. Block. 1984. Determination of Ca, Mg, K, Mn, Cu, Zn, Fe, and P in coffee, soybean, corn, sunflower and pasture grass leaf tissues by HCl extraction method. Commun. Soil Sci. Plant Anal. 15:141–147.

Montaser, A., and D.W. Golightly. 1987. Inductively coupled plasmas in analytical atomic spectrometry. VCH Publ., New York.

Morries, P. 1983. A century of Kjeldahl (1883–1983). J. Assoc. Publ. Anal. 21:53–58.

Morrison, G.H. 1979. Elemental trace analysis of biological materials. Crit. Rev. Chem. 8:287–320.

Mundel, H.H., and G.B. Schaalje. 1988. Use of near infrared reflectance spectroscopy to screen soybean lines for plant nitrogen. Crop Sci. 28:157–162.

Munsinger, R.A., and R. McKinney. 1982. Modern Kjeldahl systems. Am. Lab. 14:76–79.

Munson, R.D., and W.L. Nelson. 1973. Principles and practices in plant analysis. p. 223–248. *In* L.M. Walsh and J.D. Beaton (ed.) Soil testing and plant analysis (rev. ed.). SSSA, Madison, WI.

Munter, R.C., and R.A. Grande. 1981. Plant tissue and soil extract analysis by ICP-atomic emission spectrometry. p. 653–672. *In* R.M. Barnes (ed.) Developments in atomic plasma spectrochemical analysis. Heyden and Sons, London.

Munter, R.C., T.L. Halverson, and R.D. Anderson, 1984. Quality assurance of plant tissue analysis by ICP-AES. Commun. Soil Sci. Plant Anal. 15:1285–1322.

Murdock, A.,and O. Murdock. 1977. Analysis of plant material by x-ray fluorescence spectrometry. X-ray Spectros. 6:215–217.

Murray, S.A. 1975. *In vivo* elemental analysis by x-ray fluorescence in heat-treated bushbean. Commun. Soil Sci. Plant Anal. 6:27–32.

Nelson, D.W., and L.E. Sommers. 1973. Determination of total nitrogen in plant material. Agron. J. 65:109–112.

Nelson, D.W., and L.E. Sommers. 1980. Total nitrogen analysis of soil and plant tissues. J. Assoc. Off. Anal. Chem. 63:770–778.

Nelson, P.B., and J.W. Boodley. 1965. An error involved in the preparation of plant tissue for analysis. Proc. Am. Soc. Hortic. Sci. 86:712–716.

Nicholas, D.J.D. 1957. An appraisal of the use of chemical tissue tests for determining the mineral status of crop plants. p. 119–139. *In* T. Wallace (ed.) Plant analysis and fertilizer problems. Inst. de Recherches pour les Huiles et Oleagineux, Paris.

Noel, R.J., and L.G. Hambleton. 1976. Collaborative study of a semi-automated method for the determination of crude protein animal feeds. J. Assoc. Off. Anal. Chem. 59:134–140.

Nygaard, D.D., and J.J. Sofera. 1988. New developments in polychromator plasma emission spectrometers. Am. Lab. 20:30–36.

Ogner, G. 1983. Digestion of plants and organic soils using nitric acid, hydrogen peroxide and UV radiation. Commun. Soil Sci. Plant Anal. 14:936–943.

Okamoto, K., and K. Fuwa. 1984. Low-contamination digestion bomb method using a Teflon double vessel for biological materials. Anal. Chem. 56:1756–1760.

Parkinson, J.A., and S.E. Allen. 1975. A wet oxidation procedure suitable for the determination of nitrogen and mineral nutrients in biological material. Commun. Soil Sci. Plant Anal. 6:1–11.

Pickett, E.E., and S.R. Koirtyohann. 1969. Emission flame photometry—A new look at an old method. Anal. Chem. 41:28A–30A.

Piper, C.S. 1942. Soil and plant analysis. Hassell Press, Adelaide, Australia.

Ranger, C.B. 1981. Flow injection analysis: Principles, techniques, application and design. Anal. Chem. 53:20A–27A.

Reisenauer, H.M. 1956. Molybdenum content of alfalfa in relation to deficiency symptoms and response to molybdenum fertilization. Soil Sci. 81:237–242.

Reuter, D.J., and J.B. Robinson (ed.). 1986. Plant analysis: An interpretation manual. Inkata Press, Melbourne, Australia.

Rexroad, P.R., and R.D. Cathey. 1976. Pollution-reduced Kjeldahl method for crude protein. J. Assoc. Off. Anal. Chem. 59:1213–1217.

Sahrawat, K.L. 1980. A rapid nondigestion method for determination of potassium in plant tissue. Commun. Soil Sci. Plant Anal. 11:753–757.

Sahrawat, K.L. 1987. Determination of calcium, magnesium, zinc and manganese in plant tissue using a dilute HCl extraction method. Commun. Soil Sci. Plant Anal. 18:947–962.

Scaife, A., and K.L. Stevens. 1983. Monitoring sap nitrate in vegetable crops: Comparison of test strips with electrode methods and affects of time of day and leaf position. Commun. Soil Sci. Plant Anal. 14:761–771.

Shaw, E., R.G. Menzel, and L.A. Dean. 1954. Plant uptake of zinc^{65} from soils and fertilizers in the greenhouse. Soil Sci. 77:205–214.

Shuman, G.E., and F. Rauzi. 1981. Microwave drying of rangeland forage samples. J. Range Manage. 34:426–428.

Smith, J.H., D.L. Carter, M.J. Brown, and C.L. Douglas. 1968. Differences in chemical composition of plant sample fractions resulting from grinding and screening. Agron. J. 60:149–151.

Smith, P.F. 1962. Mineral analysis of plant tissues. Ann. Rev. Plant Physiol. 13:81–108.

Smith, V.R. 1980. A phenol-hypochlorite manual determination of ammonium nitrogen in Kjeldahl digests of plant tissue. Commun. Soil Sci. Plant Anal. 11:709–722.

Soltanpour, P.N., J.B. Jones, Jr., and S.M. Workman. 1982. Optical emission spectroscopy. p. 29–66. *In* A.L. Page et al. (ed.) Methods of soil analysis. Part 2. 2nd ed. Agronomy Monogr. 9. ASA and SSSA, Madison, WI.

Sonneveld, C., and P.A. van Dijk. 1982. The effectiveness of some washing procedures on the removal of contaminates from plant tissue of glasshouse crops. Commun. Soil Sci. Plant Anal. 13:487–496.

Spencer, K., Jr., R. Freney, and M.B. Jones. 1978. Diagnosis of sulphur deficiency in plants. p. 507–513. *In* A.R. Ferguson et al. (ed.) Proc. 8th Int. Colloquium Plant Analysis and Fertilizer Problems. New Zealand Dep. of Sci. Int. Res., Wellington, NZ.

Steckel, J.E., and R.L. Flannery. 1971. Simultaneous determination of phosphorus, potassium, calcium and magnesium in wet digestion solutions of plant tissue by AutoAnalyzer. p. 83–96. *In* L.M. Walsh (ed.) Instrumental methods for analysis of soils and plant tissue. SSSA, Madison, WI.

Sterrett, S.B., C.B. Smith, M.P. Mascianica, and K.T. Demchak. 1987. Comparison of analytical results from plant analysis laboratories. Commun. Soil Sci. Plant Anal. 18:287–299.

Steyn, W.J.A. 1959. Leaf analysis. Errors involved in the preparative phase. J. Agric. Food Chem. 7:344–348.

Steyn, W.J.A. 1961. The errors involved in the sampling of citrus and pineapple plants for leaf analysis purposes. p. 409–430. *In* Walter Reuther (ed.) Plant analysis and fertilizer problems. Am Inst. Biol. Sci., Washington, DC.

Stika, K.M., and G.H. Morrison. 1981. Analytical methods for the mineral content of human tissues. Fed. Proc. 40:2115–2120.

Sumner, M.E. 1977. Use of the DRIS system in the foliar diagnosis of crops at high levels. Commun. Soil Sci. Plant Anal. 8:251–268.

Sumner, M.E. 1982. The diagnosis and recommendation integrated system (DRIS). *In* Proc. Soil and Plant Analysis Seminar. Counc. Soil Test. and Plant Analysis, Athens.

Sung, J.F.C., A.E. Nevussu, and F.B. Dewalle. 1984. Simple sample digestion of sewage and sludge for multi-element analysis. J. Environ. Sci. Health A19:959–972.

Sweeney, R.A., and P.R. Rexroad. 1987. Comparison of LECO FR-228 "Nitrogen determinator" with AOAC copper catalyst Kjeldahl method for crude protein. J. Assoc. Off. Anal. Chem. 70:1028–1030.

Syltie, P.W., S.W. Melstead, and W.M. Walker. 1972. Rapid tissue tests as indication of yield, plant composition; and fertility of corn and soybeans. Commun. Soil Sci. Plant Anal. 3:37–49.

Tauber, H. 1949. Chemistry and technology of enzymes. John Wiley and Sons, New York.

Taylor, J.K. 1985. Handbook for SRM users. NBS Spec. Publ. 260–100. U.S. Dep. of Commerce, Natl. Bureau of Standards, Gaithersburg, MD.

Taylor, J.K. 1987. Quality assurance of chemical measurements. Lewis Publ. Chelsea, MI.

Thomas, W. 1945. Foliar diagnosis. Soil Sci. 59:353–374.

Thompson, G., and D. Bankston. 1969. A technique for trace element analysis of powdered material using d.c. arc and photoelectric spectrometry. Spectrochim. Acta 24B:335–350.

Tolg, G. 1974. The basis of trace analysis. p. 698–710. *In* E. Korte (ed.) Methodium chimicum. Vol. 1. Analytical methods. Part B. Micromethods, biological methods, quality control, automation. Academic Press, New York.

Uhl, D.E., E.B. Lancaster, and C. Vojnovich. 1971. Automation of nitrogen analysis of grain and grain products. Anal. Chem. 43:900–994.

Ulrich, J.M. 1984. Pulsing air enhances Wiley mill utility. Commun. Soil Sci. Plant Anal. 15:189–190.

Ulrich, A., D. Ririe, F.J. Hills, A.G. George, and M.D. Morse. 1959. I. Plant analysis: A guide for sugar beet fertilization. Calif. Agric. Exp. Stn. Bull. 766.

Ure, A.M. 1983. Atomic absorption and flame emission spectrometry. p. 1–54. *In* K. Smith (ed.) Soil analysis. Marcel Dekker, New York.

Uriano, G.A. 1979. The NBS Standard Reference Materials Program: SRM's today and tomorrow. ASTM Standardization News 7:8–13.

Van der Lee, J.J., I. Walinga, P.K. Mangeki, V.J.G. Houba, and I. Novozamsky. 1987. Determination of boron in fresh and in dried plant material by plasma emission spectrometry after extraction with HF-HCl. Commun. Soil Sci. Plant Anal. 18:789–802.

Vigler, M.S., A.W. Varnes, and H.A. Strecker. 1980. Sample preparation techniques for AA and ICP spectroscopy. Am. Lab. 12:31–34.

Wall, L.L., C.W. Gehrke, and J. Suzuki. 1980. An automated turbidimetric method for total sulfur in plant tissue and sulfate sulfur in soils. Commun. Soil Sci. Plant Anal. 11:1087–1103.

Wallace, A., J. Kinnear, J.W. Cha, and E.M. Romney. 1980. Effect of washing procedures on mineral analysis and their cluster analysis for orange leaves. J. Plant Nutr. 2:1–9.

Wallace, A., R.T. Mueller, J.W. Cha, and E.M. Romney. 1982. Influence of washing of soybean leaves on identification of iron deficiency by leaf analysis. J. Plant Nutr. 5:805–810.

Walsh, J.N. 1983. A handbook of inductively coupled plasma spectrometry. Blackie, London.

Wang, L., and A. Oien. 1986. Determination of Kjeldahl nitrogen and exchangeable ammonium in soil by the indolphenol method. Acta. Agric. Scand. 36:60–70.

Warner, M.H., and J.B. Jones, Jr. 1970. A rapid method for nitrogen determination in plant tissue. Commun. Soil Sci. Plant Anal. 1:109–114.

White, R.T., Jr., and G.E. Douthit. 1985. Use of microwave oven and nitric acid-hydrogen peroxide digestion to prepare botanical materials for elemental analysis by inductively coupled agron plasma emission spectroscopy. J. Assoc. Off. Anal. Chem. 68:766–769.

Whitehead, E.I., and O.E. Olson. 1942. Effects of nitrates on determination of protein nitrogen by Kjeldahl method. J. Assoc. Off. Anal. Chem. 25:769–822.

Wickstrom, G.A. 1967. Use of tissue testing in field diagnosis. p. 109–112. *In* G.W. Hardy et al. (ed.) Soil testing and plant analysis. Part II. SSSA Spec. Publ. 2. SSSA, Madison, WI.

Wikner, B. 1986. Pretreatment of plant and soil samples. A problem in boron analysis. Part I. Plants. Commun. Soil Sci. Plant Anal. 17:1–25.

Wolf, B. 1982. A comprehensive system of leaf analysis and its use for diagnosing crop nutrient status. Commun. Soil Sci. Plant anal. 13:1035–1059.

Zarcinas, B.A. 1984. Analysis of soil and plant material by inductively coupled plasma-optical emission spectrometry. Div. Rep. 70. Commonwealth Scientific Industrial Research Organication (CSIRO), Australia.

Zarcinas, B.A., B. Cartwright, and L.P. Spouncer. 1987. Nitric acid digestion and multi-element analysis of plant material by inductively coupled plasma spectrometry. Commun. Soil Sci. Plant Anal. 18:131–147.

Zasoski, R.J., and R.G. Burau. 1977. A rapid nitric-perchloric acid digestion method for multielement tissue analysis. Commun. Soil Sci. Plant Anal. 3:425–436.

Chapter 16

Plant Analysis as an Aid in Fertilizing Sugarbeet

ALBERT ULRICH, *University of California, Berkeley*

F. JACKSON HILLS, *University of California, Davis*

Plant analysis, in its simplest terms, is a study of the relationship of the nutrient content of plants to their growth. Through research and experience, reference concentrations of mineral nutrients in specific plant parts are determined and used as guides to indicate how well plants are supplied with these nutrients at a certain time of sampling. Such reference concentrations provide a tool to assist the agronomist in evaluating nutrient disorders and improving fertilizing practices.

A basic concept of plant analysis is that the concentration of a nutrient within the plant at any particular moment is an integrated value resulting from all the growth factors that have influenced the concentration up to the time of sampling. This is a straightforward diagnostic procedure as the plant is "asked" about its nutrient problems rather than using the indirect approach of soil analysis to interpret plant growth.

Research and practical experience show that plant analysis can serve as an effective guide in the fertilization of sugarbeet (*Beta vulgaris* L.) (Ulrich, 1948a, b; Kirie et al., 1954; Hills et al., 1982). This summary presents the concepts important to the use of plant analysis and suggests ways of using it to improve the management of sugarbeet fertilization. Practically, plant analysis in sugarbeet culture is most useful in controlling the use of N fertilizer to provide sufficient N early in the growing season to develop large roots (storage roots) followed by N depletion prior to harvest to obtain roots high in sucrose concentration and low in N compounds harmful to the recovery of sucrose in processing. This procedure optimizes net income for the grower and processor and minimizes the pollution of groundwater by leaching nitrates.

 Soil Testing and Plant Analysis, 3rd ed.—SSSA Book Series, no. 3.

I. THE CRITICAL CONCENTRATION

A. Estimation

The critical concentration is a useful reference point in using plant analysis as a tool for determining the nutrient status of sugarbeet plants. It is defined, for a given form of nutrient and specific plant part, as the concentration at which the growth rate of the plant begins to decline significantly. It is best estimated through the use of solution culture but soil pot culture or field experiments have also been used (Ulrich, 1950, 1952). Solution culture has the advantage of permitting the addition of a nutrient in quantitative doses so that some plants at harvest can be simultaneously extremely deficient, mildly deficient, and amply supplied with the nutrient being studied. The responses of plants to nutrient additions are largely independent of their source, and it makes little difference to the plant whether its roots are in culture solution or in a soil. The symptoms and nutrient concentrations of the affected leaves are for all practical purposes identical in both growth media. It follows that the critical value or values associated with deficiency symptoms have nearly universal application (Ulrich, 1948a; Table 16–1).

In each nutrient calibration, a series of plants in pots is supplied with increasing amounts of the nutrient under study, maintaining all other nutrients at adequate levels (Ulrich, 1950; Rosell & Ulrich, 1964). In such a series only one nutrient should be limiting growth at a time, otherwise, if a second nutrient becomes deficient, the concentration of the first nutrient will increase greatly and lead to an erroneous estimate of its critical concentration.

Plants are leaf-sampled and harvested when the lower one-third or one-half of the nutrient series show deficiency symptoms (Fig. 16–1). A calibration curve is then constructed, relating nutrient concentration in a specific plant part to growth, with growth usually expressed as a percentage of the treatments giving maximum growth. A schematic diagram of a calibration curve involving four zones is shown in Fig. 16–2. In the first of these zones, the zone of deficiency, plant growth increases sharply as more nutrient is absorbed, but there is little change in the concentration of the nutrient in the plant part analyzed. Within the second zone, the transition zone, both nutrient concentration and growth increase as more nutrient is absorbed. The third, or adequate zone, is that region of the curve where each addition of the nutrient raises the nutrient concentration without a corresponding increase in growth. The fourth zone, the toxic zone, is where growth decreases as nutrient concentration increases.

The critical concentration lies within the transition zone and is associated with a reduction in growth, usually 10%, or the breaking point of the curve, or the midpoint of the transition zone. The critical concentration will differ somewhat depending on which growth criterion is used and the magnitude of the nutrient concentration range within the transition zone. The most useful calibration curve is the one in which the range in values from deficiency to adequacy is large and the transition zone is sharp, i.e., there is a narrow range in nutrient concentration between plants that are deficient

Table 16-1. A plant analysis guide for sugarbeet.

Nutrient	Plant part	Concentration of element† Critical value field samples‡§	Range showing deficiency symptoms¶	Range without deficiency symptoms#
Boron	Blade	21	12–40	35–200
Calcium	Petiole	1 g/kg	0.4–10 g/kg	2–25 g/kg
	Blade	5 g/kg	1–4 g/kg	4–15 g/kg
Chlorine	Petiole	4 g/kg	0.1–0.4 g/kg	8–85 g/kg
Copper	Blade	--	<2	<2
Iron	Blade	55	20–55	60–140
Magnesium	Petiole	--	0.1–0.3 g/kg	1–7 g/kg
	Blade	--	0.25–0.5 g/kg	1–25 g/kg
Manganese	Blade	10	4–20	25–360
Molybdenum	Blade	--	0.01–0.15	0.20–20.0
Nitrogen (NO_3^--N)	Petiole	1 000	70–500	350–35 000
	Storage root	1 000	70–500	800–4 000
Phosphorus ($H_2PO_4^-$-P)	Petiole	750	150–400	750–4 000
	Blade	--	250–700	1 000–8 000
	Seedling:			
	Petiole	1 500	500–1 300	1 600–5 000
	Blade	3 000	500–1 700	3 500–14 000
	Cotyledon	1 500	200–700	1 600–13 000
Potassium (Na >1.5%)	Petiole	10 g/kg	2–6 g/kg	10–110 g/kg
	Blade	10 g/kg	3–6 g/kg	10–60 g/kg
Potassium (Na <1.5%)	Petiole††	--	5–20 g/kg	25–90 g/kg
	Blade	10 g/kg	4–5 g/kg	10–60 g/kg
Sodium	Petiole	--	--	0.2–90 g/kg
	Blade	--	--	0.2–37 g/kg
Sulfur (SO_4^{2-}-S)	Blade	250	50–200	500–14 000
Zinc	Blade	9	2–13	10–80

† All concentrations are for the element on a dry wt. basis and are mg/kg (ppm), except when noted as g/kg [Note, $(g/kg)10^{-1} = \%$).

‡ The critical concentration is that nutrient concentration at which plant growth begins to decrease in comparison with plants above the critical concentration.

§ All critical concentrations except for P in seedlings and for N in roots are based on a sample of leaves that have just fully expanded (Fig. 16–3).

¶ Leaf material for chemical analysis must be collected shortly after appearance of leaf symptoms, otherwise deficient plants may accumulate nutrients in the leaf without restoring chlorotic tissues to normal. Use a color atlas (Ulrich & Hills, 1969) to help identify what deficiency, if any, has occurred.

The upper value reported is the highest value observed to date for normal plants. Abnormally high values are often associated with other nutrient deficiencies; for example, blades low in Fe may contain up to 40 g of Ca/kg.

†† Because of the influence of Na on K content of petioles, blades must be used for K analysis when petioles contain <15 g of Na/kg.

and those that are just well supplied with the nutrient in question. When the transition zone is sharp, any of the above criteria for establishing the critical concentration will lead to essentially the same value.

Calibration curves can be constructed for nutrient concentrations in various parts of a plant (petioles and blades of leaves of different ages, roots, and midstems) and for various forms of a nutrient (e.g., total N, NO_3^--N, total P, and soluble $H_2PO_4^-$-P). The plant part selected to serve as a reliable indicator of nutrient status should give comparable results for all sam-

Fig. 16-1. Sugarbeet plants growing in culture solution to which increasing amounts of Zn were added (Rosell & Ulrich, 1964). To establish a calibration curve (Fig. 16-2) plants must be harvested when there are wide differences in growth and nutrient concentration.

pling dates and should be relatively easy to sample. The form of the nutrient determined should result in a calibration curve with a relatively sharp transition zone, a broad range in nutrient concentration between deficiency and abundance, and a relatively constant critical concentration for a considerable range of sampling dates. Sugarbeet leaves usually are the most appropriate part to sample. Sharper calibration curves are usually obtained with recently matured leaves, i.e., leaves that have just attained maximum size (Fig. 16-3) and with mobile rather than nonmobile forms of a nutrient. The petiole has been found to be most satisfactory for NO_3^--N, $H_2PO_4^-$-P, and Cl^-, while the blade is more satisfactory for SO_4^{2-}-S and the other essential nutrients.

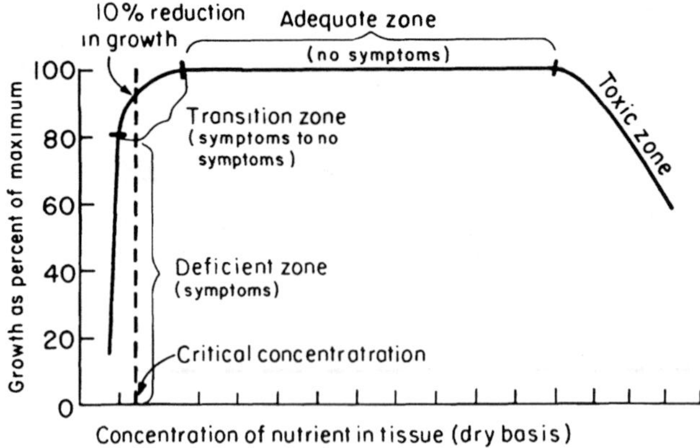

Fig. 16-2. The critical concentration is taken at the point where growth is 10% less than the maximum. Symptoms generally appear below the critical concentration and fail to appear above it. Symptoms may or may not appear in the transition zone. The sharper the transition zone, the more useful the calibration for diagnostic purposes.

Fig. 16-3. Select a recently matured leaf, one that has just reached full size. There are four or five such leaves, similar to the two being held (A) on each plant. Separate blade from petiole where they join (B).

Dry material is preferable to fresh as concentrations tend to be less variable, laboratory work can be more routinely scheduled, and material can be easily stored for additional analyses in the future. The chief disadvantage of reporting results on a dry weight basis occurs through the dilution of nutrients by increasing dry matter accumulation in plant tissues as they mature. This diluting action, however, may be minimized at each sampling date by the selection of plant material of the same physiological age. For this reason, plant material from the youngest "mature" leaves (youngest fully developed leaves) is selected for analysis whenever possible.

The establishment of critical concentration values of NO_3^--N and $H_2PO_4^-$-P in petioles, K in blades and petioles, and for SO_4^{2-}-S and Zn in blades of recently matured sugarbeet leaves is illustrated in Fig. 16-4, 16-5, 16-6, 16-7, and 16-8. Similarly, critical concentrations have been determined for other nutrients as indicated in Table 16-1.

B. Interpretation

When monitoring plants in the field, the likelihood of a growth response from the addition of fertilizer will depend upon whether the nutrient concentration of the plants sampled is above or below the critical level. When the nutrient concentration is above the critical level and remains there throughout the entire growth period, there is little chance of a response in growth from the addition of more nutrient. Conversely, when the nutrient concentration falls below the critical level the chance of a growth response under field conditions becomes much greater as the nutrient concentration in the plants decreases. The magnitude of the response will depend on the relative adequacy of other growth factors, upon the stage of growth when

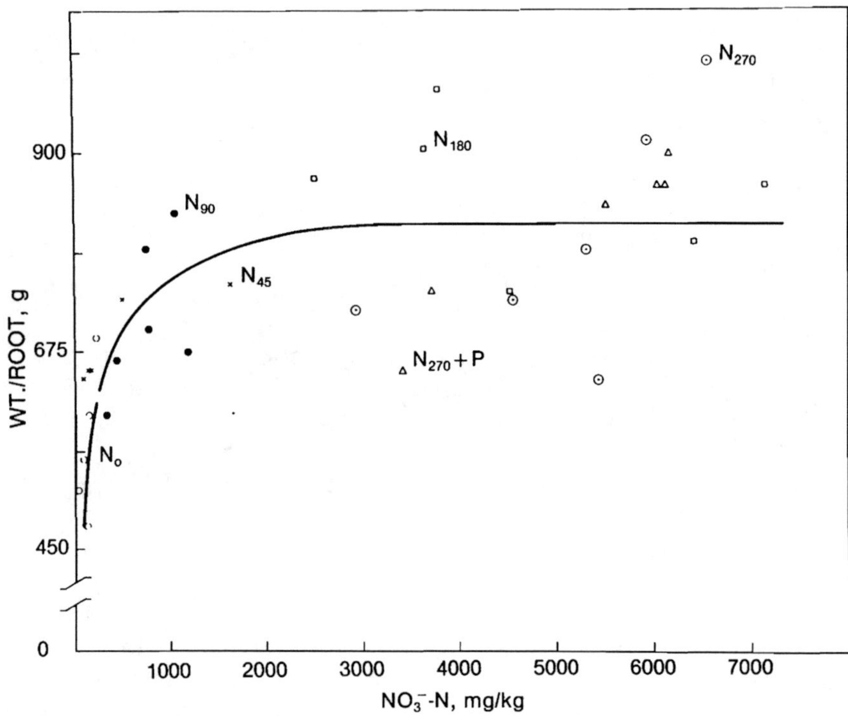

Fig. 16–4. Relation of beet-root fresh weight to the NO_3^--N concentration (dry basis) of petioles of recently matured leaves from a field experiment near Woodland, CA. Subscripts indicate kilograms of applied N/ha and P indicates 97 kg of applied P/ha. The critical N concentration at the breaking point of the curve is about 1000 mg of NO_3^--N/kg. Below the critical concentration growth is retarded, but above it growth is maintained at maximum as long as there are no other factors limiting growth (Ulrich, 1950).

the deficiency occurs and the length of time it persists. When other growth factors are sufficient, addition of the deficient nutrient will result in a relatively large increase in yield. When another factor or set of factors becomes limiting almost immediately, additions of the required nutrient will produce a relatively small increase in yield. Similarly, the chance of a measurable growth response in the field will decrease as the duration of the deficiency decreases. If the deficiency first appears late in the growth season, there is little chance for a significant yield increase.

When a sample consists of plant material taken from a large area within a field, the leaves taken may differ considerably in nutrient concentration. Consequently, when the average concentration of the sample is at or near the critical level, the material collected will represent both deficient and nondeficient plants. Under these conditions, yields will be raised signfiicantly by fertilization only when the proportion of deficient to nondeficient plants is sufficiently large to produce a measurable increase in yield. Thus, the critical level associated with a significant yield increase will be higher under most

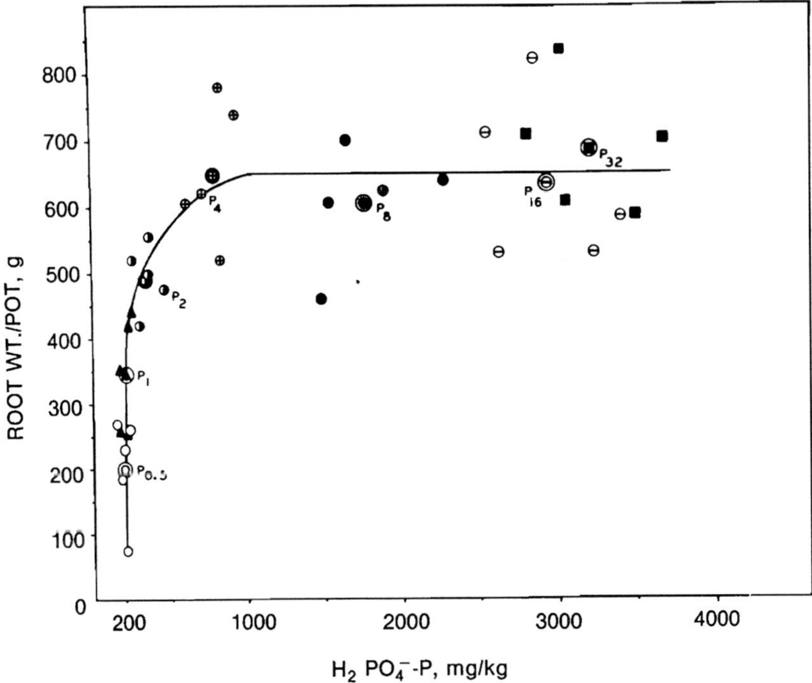

Fig. 16–5. Relation of beet-root fresh wt. to $H_2PO_4^-$-P conc. (dry basis) in petioles of recently matured leaves in a pot experiment. The critical conc. is approximately 750 mg/kg. The subscripts are increments of 434 mg of P/pot (49 kg/ha), e.g., P_2 = (434)2 = 868 mg of P/pot, the circled points are average values for a treatment level.

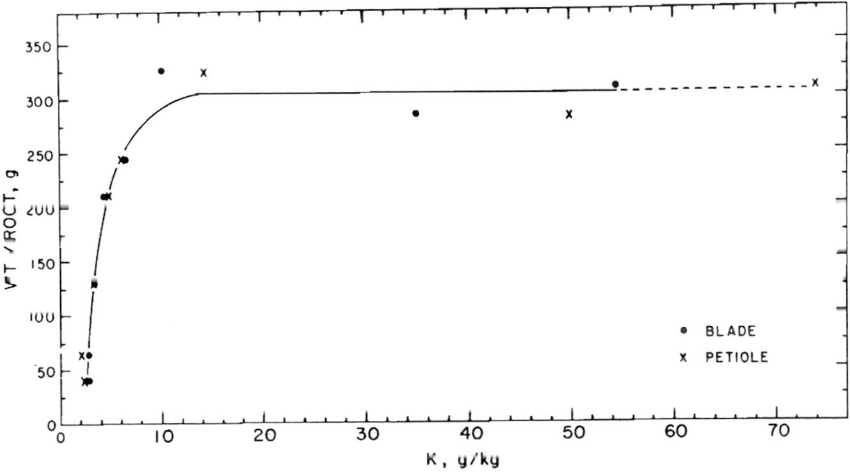

Fig. 16–6. Relation of beet-root fresh wt. to K conc. in petioles and blades (dry basis) of recently matured leaves of plants in culture solution receiving increasing amounts of K and a constant but adequate amount of Na. The critical K value is approximately 10 g/kg (1.0%) for blades, regardless of Na content, and for petioles only when Na conc. is >15 g /kg (1.5%). Each symbol is an average of five replications.

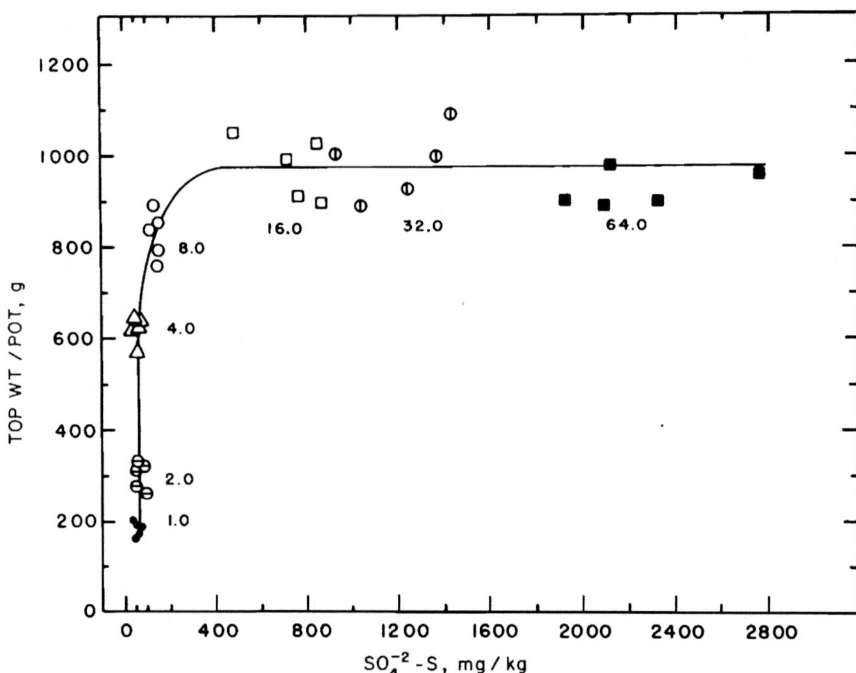

Fig. 16–7. Relation of fresh wt. tops to SO_4^{-2}-S conc. (dry basis) of recently matured blades of plants in culture solution. Numbers adjacent to symbols are mg SO_4^{-2}-S/L initially in the culture solution. The critical conc. at the breaking point of the curve is approximately 250 mg/kg.

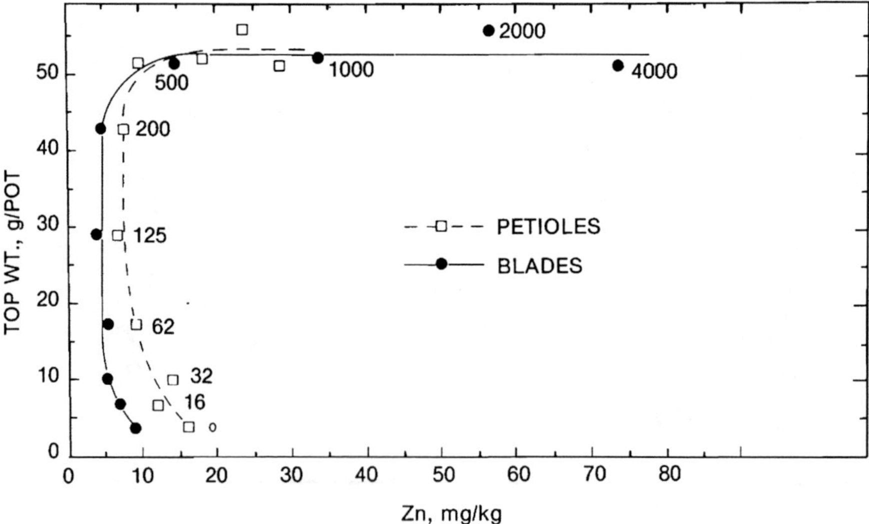

Fig. 16–8. Relation of dry wt. of tops to Zn conc. (dry basis) of blades and petioles of recently matured leaves (Rosell & Ulrich, 1964). Each symbol is the mean of five replications and the numbers adjacent are micrograms Zn as $ZnSO_4$/20 L of initial culture solution. Blades are preferred for analysis because of the broad range in conc. from deficient to adequate. The critical value for blades is approximately 10 mg/kg.

field conditions than when samples are taken from solution or pot cultures, or from small, uniform field plots.

C. A Constant or Changing Critical Value?

Nutrient concentrations in leaves that are increasing in age tend to reflect the characteristics of leaves of that particular age, whereas properly sampled leaves of the same physiological age tend to reflect the changing nutrient status of the plants as influenced by the balance between nutrients supplied from the soil and nutrients required for plant growth. For sugarbeet, where vegetative growth is indeterminate, if the leaf sampled is always of the same physiological age, critical concentrations have not been found to vary enough, regardless of the stage of growth of the plant, to affect their usefulness as a measure of the nutritional status of the plants at the time of sampling (Ulrich, 1964). We have found the concept of the constant critical concentration for several nutrients to be a satisfactory reference point for evaluating the nutrient status of plants throughout the varied growing seasons in California (Ririe et al., 1954), for other states (Robins et al., 1956; Carter et al., 1971), and for other countries (F.J. Hills, 1965–1985, unpublished data).

An exception to the constant critical value is for the evaluation of the nutrient status of seedling plants before a large mature true leaf has developed. In this case, it is advisable to determine the critical concentration for various nutrients for a specific seedling plant part as was done by Sipitanos and Ulrich (1969) for $H_2PO_4^-$-P in various parts of sugarbeet seedlings (Table 16–1).

D. Effect of Other Factors

Variety (cultivar) of sugarbeet apparently has little or no effect on the critical concentration. So far, we have not observed a signifiant difference in the critical concentration among commercial varieties. Effects of variety on the critical level, if any, are expected to be small compared to the wide range in nutrient concentrations between plants with and without an adequate supply of a given nutrient.

Concern is frequently expressed that the moisture regime under which plants are growing may have serious effects on nutrient concentrations and critical concentrations. Loomis and Worker (1963) found that moisture stress had little effect on the concentration of NO_3^--N in mature petioles compared to plants recently irrigated and not exhibiting moisture stress systems. We have observed the same effect (Table 16–2). Even when plants were severely stressed, i.e., wilted to the point where many old leaves were lost, there was little or no reduction in the NO_3^--N concentration in mature petioles. Thus, it appears that moderate changes in weather or the time of day that samples are collected will have little influence on nutrient concentration, particularly from the standpoint of making decisions as to whether plants are deficient in a certain nutrient.

Table 16–2. Effect of moisture stress on the concentration of NO_3^--N in petioles of recently matured sugarbeet leaves, Kern County, California (F.J. Hills, 1963, unpublished data).

Days from last irrigation	Observed plant stress to soil moisture	NO_3^--N, mg/kg (dry basis)
	22 July	
2†	None	3030
20‡	Moderate	3270
		SE = 590
	31 July	
11†	Slight	4370
29‡	Severe	4330
		SE = 850

† Sixteen plots last irrigated 20 July. ‡ Sixteen plots last irrigated 2 July.

Nutrient concentration, however, can be markedly changed by other conditions that affect plant growth. For example, petioles of mature leaves of sugarbeet infected with the beet yellows virus may be higher in NO_3^--N concentration than healthy plants (Table 16–3). This difference is not indicative of a different nutrient requirement in the diseased plants, but is the result of NO_3^--N accumulation in the slow-growing plants.

When there are differences in nutrient concentration associated with variety, weather, moisture stress, and disease they are likely to be related to the kind and extent of root development, or to other factors affecting growth, rather than to differences in internal requirements (Ulrich, 1948a, b).

Table 16–3. The effect of variety and virus infection on the concentration of NO_3^--N in petioles of recently matured sugarbeet leaves, Broom's Barn Exp. Stn., England (F.J. Hills, 1970, unpublished data).

Variety	NO_3^--N, mg/kg (dry basis)		Variety means
	Healthy	Diseased†	
	27 July (V × D = NS)		
US H7A	2880	6410	4650
US H9B	3120	4680	3900
Bush Mono	3660	5240	4450
Disease means	3220	5440**	NS‡
	24 Aug. (V × D = NS)		
US H7A	3060	4460	3760
US H9B	3570	4640	4120
Bush Mono	2970	4590	3780
Disease means	3210	4560**	NS

** Significantly greater than healthy ($P < 0.01$).
† Inoculated with the beet yellows virus on 8 June, soon after thinning.
‡ NS = Not significant.

II. USE OF PLANT ANALYSIS

Experience in the interpretation of nutrient concentrations in plant samples, as related to critical levels, is essential for giving advice concerning fertilizer practice. Optimum economic yield is the usual objective, but fertilizing sugarbeet for maximum top and root production usually does not mean maximum sucrose production. An adequate supply of N is required early in the growing season to ensure vigorous top and root growth, but if storage roots are to be high in sucrose concentration, the plants must be N deficient for a period prior to harvest to retard vigorous growth, especially top growth, and allow sucrose to accumulate in the storage roots. The field experiment summarized in Fig. 16-9 illustrates this. Optimum sugar yield,, a function of root yield and sucrose concentration, was produced at a fertilizer rate that nearly optimized root yield, but this rate was considerably less than the rate required for maximum crop growth (roots plus tops) and was not the rate giving the highest root sucrose concentration (Hills et al., 1978).

Fertilizing to just achieve optimum sugar yield has other advantages. It seldom contributes to NO_3^- pollution of groundwater and, in addition, a considerable amount of N is incorporated into beet tops where, like a green manure crop, a portion is available to the next crop to be grown on the field (Hills et al.,1978; Abshahi et al., 1984). Table 16-4 gives data from the same experiment as Fig. 16-9 and shows that 223 kg of N/ha was removed in the

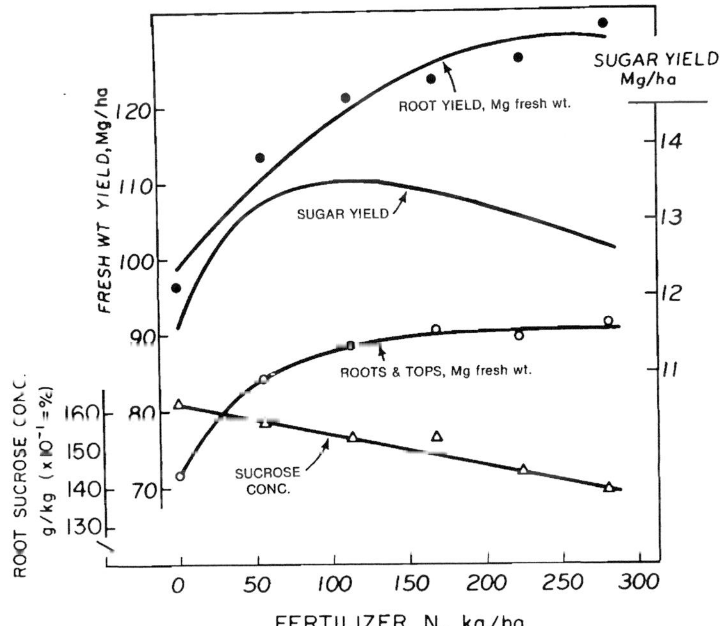

Fig. 16-9. Response of sugarbeet to fertilizer N (Hills et al., 1978). The response functions are: root + tops, Y = 98.66 + 0.24X − 0.00047X^2, R^2 = 0.96; root yield, Y = 90.88 − 19.21 (0.981^X), R^2 = 0.99, % sucrose, Y = 16.2 − 0.0079X, r^2 = 0.95. Sugar yield is computed from root yield and sucrose concentration.

Table 16-4. Response of sugarbeet to fertilizer N and resulting N uptake. (Values are predicted from regressions [Hills et al., 1978].)

Fertilizer N applied	Sugar yield	Fresh wt. yield			N uptake		
		Tops	Roots	Total	Tops	Roots	Total
kg/ha	Mg/ha				kg/hg		
0	11.6	27.0	71.7	98.7	79	79	158
56	13.3	29.3	84.3	113.6	93	101	194
112	13.6	42.8	88.6	131.4	105	118	223
168	13.4	62.1	90.1	152.2	114	132	246
224	13.1	85.4	90.6	176.0	112	141	263
280	12.7	111.9	90.8	202.7	127	147	274

crop that required 112 kg of fertilizer N/ha to give optimum sugar yield. Of the 223 kg of N/ha taken up, only 47% came from the applied fertilizer. Thus, when carefully fertilized, a sugarbeet crop may be viewed as a scavenger crop in terms of its extensive use of soil N (Hills et al., 1983). Figure 16-10 shows the time course of the concentration of NO_3^--N in sugarbeet petioles for certain N rates of the experiment of Fig. 16-9 and Table 16-4. The petioles of sugarbeet fertilized for optimum sugar yield indicated a N deficiency (<1000 mg of NO_3^--N/kg) for about 8 wk prior to a fall harvest. This, and much additional experience, indicates that plants should be deficient for at least 4 wk and as long as 10 wk before harvest.

The predictability of a response to fertilizer added in response to a low concentration of a nutrient in plant tissue decreases as the time elapsed from

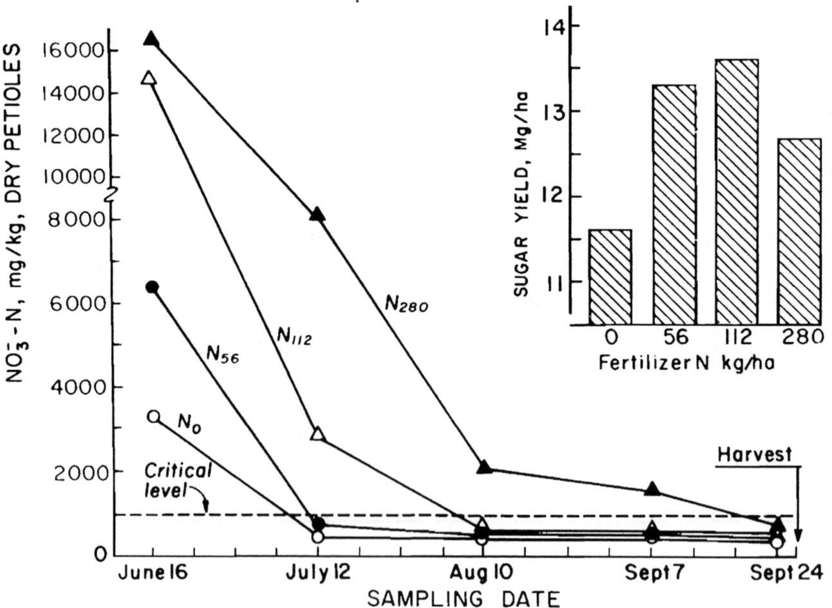

Fig. 16-10. Petiole nitrate depletion curves for selected fertilizer N rates and related sugar yields of the experiment of Fig. 16-9. For the optimum rate of N fertilization (112 kg of N/ha) plants were N deficient for about 8 wk prior to harvest.

sampling to harvest increases. This is because climate and other factors intervening between sampling and harvest may greatly alter soil nutrient supply or plant nutrient demands. For best results in developing a plant analysis program, sampling should be done frequently to establish the proper time and frequency of sampling. This approach is much preferred to depending on the results of a single sampling, as is too often done. A single mid-season sample will not detect an early P deficiency that disappears as soil temperatures warm or a post mid-season release of N mineralized from the organic residue of a previous crop.

Following are ways in which plant analysis can be used to improve crop production.

A. Evaluation of Fertilizer Programs

To determine how well a fertilizer program is meeting the needs of a sugarbeet crop, a series of samples must be taken throughout the growing season. One set of samples should be taken early to detect early season deficiencies of certain nutrients, particularly P (Hills et al.,1970). Subsequent samplings should be at 2- to 4-wk intervals (Fig. 16–10 and 16–11). The minimum number of samplings should be four: one at thinning time, one at early mid-season, one at late mid-season, and one just prior to harvest. By comparing the analyses of these samples to critical nutrient concentrations, it is possible to determine how well the fertilizer program is meeting the needs of the crop. In this manner, impending nutrient deficiencies can be detected

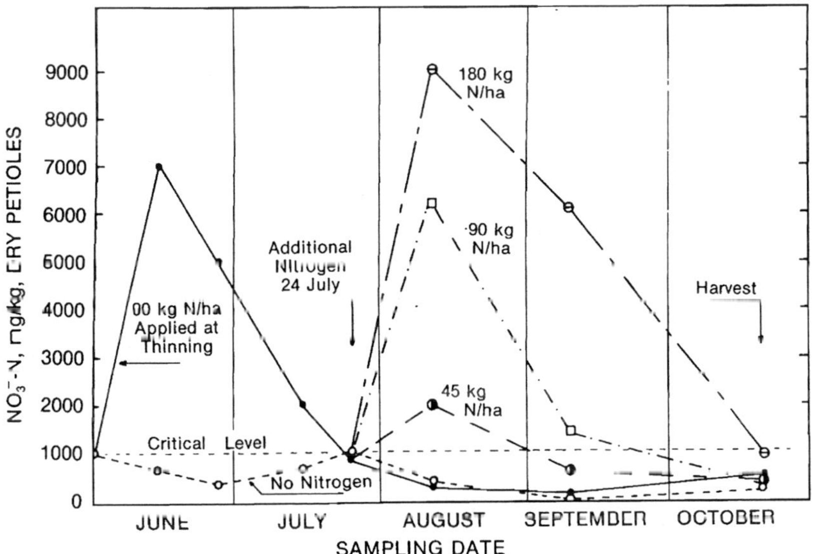

Fig. 16–11. Controlled feeding. The study showed that 90 kg N/ha at thinning was not enough. A second 90 kg of N/ha in mid-July gave the highest sugar yield. Sugar yields (Mg/ha) at harvest for the various N rates (kg/ha) were: Zero N, 7.5; 90 N, 8.2; 90 N + 45 N, 8.8; 90 N + 90 N, 9.4; 90 + 180 N, 9.2 (Ulrich et al., 1959).

and corrected before they occur and appropriate adjustments in a fertilizer program can be made. This system, particularly if it can be applied to more than the sugarbeet crop in a rotation, offers an excellent way to optimize fertilization efficiency. If an impending deficiency is discovered early enough in the growing season, emergency applications of fertilizers may be made to correct the deficiency. However, with more experience and consistent use of plant analysis, emergency applications of fertilizers will become a rarity or completely unnecessary. Adjustments in the fertilizer program often will be limited to increasing the use of one nutrient or lessening that of another, or even adding other elements just becoming short in supply. Also, changes in timing or method of fertilization might be necessary to satisfy more fully the fertilizer needs of the crop.

After alterations in fertilization or cultural practices have been made, the nutrient changes occurring within the crop should again be observed through plant analysis. If the new program leads to better nutrition of the crop and higher sugar yields, it is continued; if not, the program is changed again, thus preventing deficiencies and serious losses in crop yield.

B. Controlled Feeding

Fertilizing crops with N on a demand basis is desirable from the standpoint of maximizing net returns and minimizing potential contamination of groundwater. This can be accomplished for the sugarbeet crop by applying a minimum rate of fertilizer N initially, then waiting until results of plant analysis show a need for more N (Hills et al., 1982). Nitrogen-deficient plants often produce more sucrose than N-sufficient plants unless deficiencies occur more than 10 wk prior to harvest. With longer periods of deficiency, more sucrose is produced by N-sufficient plants due to their more rapid root growth. Thus, an objective of N fertilizer management should be to allow plants to become N deficient prior to harvest. If an error is to be made it appears desirable to make it on the side of underfertilization, since periods of deficiency of up to 10 wk usually can be tolerated without reducing sugar yield (Hills et al., 1963).

The adequacy of an initial fertilizer application can be ascertained from petiole analyses by taking samples from recently matured leaves at 2-wk intervals starting about 2 wk after thinning (Fig. 16–11). Indications of impending deficiencies of more than 10 wk prior to harvest can be estimated based on local experience, or as calculated by Carter et al. (1971) or Hills et al.(1982). Such deficiencies can be corrected by additions of from 45 to 65 kg of N/ha.

C. Diagnosis of Abnormal Growth

Frequently, plants in all or part of a field show abnormal growth characteristics. A knowledge of nutrient deficiency symptoms may lead to a tentative diagnosis, which can be confirmed or discarded by analyses of comparable petioles and blades from abnormal and normal appearing plants (Ulrich &

Table 16-5. Leaf analyses of vigorous and poorly growing sugarbeet reveal S deficiency (Ulrich et al., 1959).

Growth of sugarbeets	Leaf analysis, dry basis			
	Petioles			Blades
	NO_3^--N	$H_2PO_4^-$-P	K	SO_4^{2-}-S
	mg/kg		g/kg	mg/kg
Vigorous	2520	2120	31.5	1880
Poor	9000	2990	38.0	155

Hills, 1969). Plants should be sampled soon after deficiency symptoms appear. After a prolonged period of deficiency, the concentration of certain nutrients will increase in physiologically mature tissue even though the symptoms remain. Table 16–5 illustrates how plant analysis was used to detect a S deficiency in sugarbeet. A large area in a field contained plants that appeared deficient. However, analyses indicated that N, P, and K were all above their critical levels but that SO_4^{2-}-S in blade tissue was deficient. Fertilizer treatments applied to the plants in the affected area confirmed the diagnosis.

The critical nutrient concentrations and nutrient concentrations associated with deficiency symptoms given in Table 16–1 are helpful in diagnosing abnormal plant growth.

D. Guide to Harvest

When there is a choice of fields to be harvested, the results of a systematic plant analysis program may serve as a guide to the scheduling of beet harvest (Duckworth & Hills, 1952). Those fields with the longest period of N deficiency would be scheduled for an early harvest and those high in this nutrient would be delayed until depleted or held as long as possible before harvesting.

Root pulp being analyzed for sucrose can be analyzed for NO_3^- through the use of the NO_3^- electrode or semiquantitatively through the use of the diphenylamine reagent. These tests were first implemented by John T. Alexander (Holly Sugar Company) following a field demonstration by Albert Ulrich as to the use of the diphenylamine reagent to locate NO_3^- in roots and leaves of sugarbeet. They have been useful in revealing areas of low sucrose concentration due to excessive nitrate uptake prior to harvest (Hills & Ulrich, 1971). Sugar companies have found these tests useful in evaluating how well crops in an area have been fertilized with N to meet desirable quality standards. When pulp NO_3^--N values at harvest are coupled with petiole NO_3^--N values for the entire growing season, the grower learns when and for how long the beets were high or low in NO_3^--N before harvest. This information will help in meeting the quality standards required for processing beets efficiently.

E. Nutrient Survey

Plant analysis can be used to survey the nutrient status of crops in a farming area (Duckworth & Hills, 1952). This can result in the detection of deficiencies, improvement of fertilizer practices, and the identification of fields suitable for fertilizer trials. To be fully effective, the plant nutrient survey should include a series of samplings during the growing season, as outlined earlier under the evaluation of fertilizer programs.

III. SAMPLING PROCEDURE

To serve as a measure of variability within a field, a minimum of two samples, but preferably four, should be taken from each field. The number of leaves for each sample depends on the variability between leaves but usually from 25 to 50 leaves are sufficient to estimate reasonably the true mean of the nutrient concentration in the plants represented by the sample (Ulrich & Hills, 1952).

One system to follow in sampling a field is to divide it into imaginary quarters (Hills et al., 1982). The sampler walks across the center of each quarter at right angles to the plant rows and collects from 25 to 50 leaves per quarter of the field. Samples should be placed in an ordinary paper bag of convenient size and taken to the laboratory for processing as soon as possible. Delays of more than 48 h at ordinary temperatures should be avoided, and storage in closed compartments of automobiles during warm weather should not be allowed. If it is necessary to store samples prior to drying, they should be kept at a temperature of 5 °C. At this temperature, concentrations of NO_3^--N, $H_2PO_4^-$-P, and K on an air-dry basis change very little in sugarbeet petioles over a 120-h period (Ulrich, 1948a).

For micronutrient analyses, plant material must be free of dust. Leaves should be washed for 30 s in a bath containing a detergent or a detergent and 0.1 *M* HCl, followed by two successive rinses in distilled water.

A sample may be reduced in size by cutting it into small pieces (about 10 mm) thoroughly mixing the pieces, and placing a subsample (about 100 g of fresh material) in a cut-down paper bag or other suitable container and drying it overnight in an oven, preferably with a forced draft at 70 °C. After drying, the samples are ground to pass a 30- or 40-mesh screen and stored in small glass or plastic vials. Following this procedure, samples have been kept in moisture-tight containers for many years without a significant change in their mineral composition.

IV. ANALYTICAL METHODS

The methods selected for analyzing plant material must be both accurate and rapid. There is little merit in analyzing a few samples accurately during the course of a day, when there are hundreds of samples to be analyzed, nor is there much value in analyzing hundreds of samples rapidly when the analytical error is equal to or greater than the sampling error of the material analyzed. Whenever feasible, the analytical error should be far less than the error associated with the plant material sampled (Ulrich & Hills, 1952).

The time lapse between collection of samples and reporting of results must be as brief as possible if a current season crop is to benefit. Information for efficient fertilizer practices can be provided conveniently within 3 to 7 d by the laboratory methods proposed by Johnson and Ulrich (1959) or by more recent methods involving atomic absorption spectroscopy, atomic fluorescence spectroscopy, inductively coupled plasma (ICP) emission spectrophometry, and the specific ion electrode. As a check on the analyst and the analytical method, it is important to carry along with the unknowns a reference plant sample of known nutrient concentration established by many analyses.

V. LIMITATIONS AND USEFULNESS

The concentration of a mineral nutrient in the tissue of a specified plant part, when referred to a well-established critical concentration, provides a simple way of determining whether the plants represented by the sample are adequately supplied with nutrients at the time of sampling. Such analyses, however, usually reveal only one deficiency at a time. A second nutrient, or even a third nutrient, may be in short supply but, due to reduced growth caused by the primary deficiency, most other nutrients will accumulate in the tissue. When the primary deficiency has been corrected, the increased growth will often decrease the concentration of a second, nutrient which most likely will become deficient rather soon. Conversely, in the case of a P deficiency, NO_3^- uptake has been observed to be depressed, and the depression reversed by the addition of P to the soil or a nutrient solution that lacked P (Sipitanos & Ulrich, 1969; Hills et al., 1970).

A nutrient level below the critical concentration at a single sampling date indicates a deficiency, but later the same plants may show an increase in nutrient uptake. For this reason, multiple sampling dates are preferable to a single sampling date. In the interpretation of the results for multiple sampling dates, the sharper the drop in nutrient concentration and the earlier in the season it occurs, the greater the degree of deficiency. But even when the results of several sampling dates indicate that plants have been deficient for a considerable period, the exact degree of deficiency is not known (Hills et al., 1963). At harvest, however, the degree of deficiency for a given field can be estimated in terms of the maximum yield for the area, and fertilization of the next crop is then adjusted to this finding. For example, a field that produced only 25% of the potential yield, with all other growth factors adequate, is much more deficient and will require more fertilizer than one that produced 90% of the potential for the area.

One of the greatest values of plant analysis is the prevention of deficiencies rather than their correction after they occur. Trends in leaf concentration followed over a period of years can be a valuable guide in delineating the kinds, amounts, frequency, and methods of fertilization for efficient crop production on a given field. In this way, a nutrient deficiency can be predicted or at least detected as soon as it occurs, and corrective measures can be taken before serious crop losses take place.

Plant analysis is often useful in revealing the subtle influence of climate on the supply and demand for nutrients. In a favorable climate, the need for nutrients, particularly for N, increases dramatically during the grand growth period and then declines prior to harvest. During periods of rapid growth, soil N supplies may not be adequate to meet the needs of the crop, but later as the growth rate declines, soil NO_3^--N may exceed demand just when it should be deficient prior to harvest. Such excesses can occur from nitrification of organic material in peat soils, previous crop residues, or heavy applications of manure. However, low night temperature shortly before harvest, will frequently increase sucrose concentrations even in the presence of high petiole NO_3^--N values.

Plant analysis results have also revealed that low soil temperatures induce P deficiency of seedlings on some soils relatively high in P. This deficiency is overcome by P fertilization prior to planting and later by an increase in soil temperature.

Growth rates and the need for nutrients are also greatly influenced by day length, day and night temperature, light intensity, and rainfall, which may move NO_3^- in surface soil into the root zone and thereby lower sucrose concentrations of beets as new growth occurs (Stout, 1964). Comparing plant analysis results, soil N values, and root pulp NO_3^--N tests with sugar produced on each field within and among districts will be useful to the grower, processor, and agronomist in assessing growth problems and fertilizer practices.

REFERENCES

Abshahi, A., F.J. Hills, and F.E. Broadbent. 1984. Nitrogen utilization by wheat from residual sugarbeet fertilizer and soil incorporated sugarbeet tops. Agron. J. 76:954–958.

Carter, J.N., M.E. Jensen, and S.M. Bosma. 1971. Interpreting the rate of change in nitrate-nitrogen in sugarbeet petioles. Agron. J. 63:669–674.

Duckworth, W.R., and F.J. Hills. 1952. Possibilities of improved nitrogen fertilization of sugar beets through the use of leaf analysis. Proc. Am. Soc. Sugar Beet Technol. 7:252–254.

Hills, F.J., F.E. Broadbent, and M. Fried. 1978. Timing and rate of fertilizer nitrogen for sugarbeets related to nitrogen uptake and pollution potential. J. Environ. Qual. 7:368–372.

Hills, F.J., F.E. Broadbent, and O.A. Lorenz. 1983. Fertilizer nitrogen utilization by corn, tomato, and sugarbeet. Agron. J. 75:423–426.

Hills, F.J., G.V. Ferry, A. Ulrich, and R.S. Loomis. 1963. Marginal nitrogen deficiency of sugar beets and the problem of diagnosis. J. Am. Soc. Sugar Beet Technol. 12:476–484.

Hills, F.J., R. Sailsbery, and A. Ulrich. 1982. Sugarbeet fertilization. Univ. of California, Div. of Agric. Sci. Bull. 1891.

Hills, F.J., R.L. Sailsbery, A. Ulrich, and K.M. Sipitanos. 1970. Effect of phosphorus on nitrate in sugar beet. Agron. J. 62:91–92.

Hills, F.J., and A. Ulrich. 1971. Nitrogen nutrition. p. 111–135. *In* R.T. Johnson et al. (ed.) Advances in sugar beet production. Iowa State Univ. Press, Ames.

Johnson, C.M., and A. Ulrich. 1959. Analytical methods for use in plant analysis. California Agric. Exp. Stn. Bull. 766:25–78.

Loomis, R.S., and G.F. Worker, Jr. 1963. Response of sugar beet to low soil moisture at two levels of nitrogen nutrition. Agron. J. 55:509–515.

Ririe, D., A. Ulrich, and F.J. Hills. 1954. The application of petiole analysis to sugar beet fertilization. Proc. Am. Soc. Sugar Beet Technol. 8(1):48–57.

Robins, J.S., C.E. Nelson, and C.E. Domingo. 1956. Some effects of excess water application on utilization of applied nitrogen by sugar beets. Proc. Am. Soc. Sugar Beet Technol. 9:180–188.

Rosell, R., and A. Ulrich. 1964. Critical zinc concentrations and leaf minerals of sugar beet plants. Soil Sci. 97:152–167.

Sipitanos, K.N., and A. Ulrich. 1969. Phosphorus nutrition of sugarbeet seedlings. J. Am. Soc. Sugar Beet Technol. 15(4):332–336.

Stout, M. 1964. Redistribution of nitrate in soils and its effects on sugarbeet nutrition. J. Am. Soc. Sugar Beet Technol. 13:68–80.

Ulrich, A. 1948a. Plant analysis—methods and interpretation of results. p. 157–168. *In* H.B. Kitchen (ed.) Diagnostic techniques for soils and crops. Am. Potash Inst., Washington, DC.

Ulrich, A. 1948b. Plant analysis as a guide to the nutrition of sugar beets in California. Proc. Am. Soc. Sugar Beet Technol. 5:364–377.

Ulrich, A. 1950. Critical nitrate levels of sugar beets estimated from analysis of petioles and blades with special reference to yields and sucrose concentrations. Soil Sci. 69:291–309.

Ulrich, A. 1952. Physiological bases for assessing the nutritional requirements of plants. Ann. Rev. Plant Physiol. 3:207–228.

Ulrich, A. 1964. The relative constancy of the critical nitrogen concentration of sugar beet plants. p. 331–391. *In* C. Bould (ed.) Plant analysis and fertilizer problems. IV. Am. Soc. Hortic. Sci., Michigan State Univ., East Lansing.

Ulrich, A., and F.J. Hills. 1952. Petiole sampling of sugar beet fields in relation to their nitrogen, phosphorus, potassium and sodium status. Proc. Am. Soc. Sugar Beet Technol. 7:32–45.

Ulrich, A., and F.J. Hills. 1969. Sugar beet nutrient deficiency symptoms, a color atlas and chemical guide. Univ. of California, Div. of Agric. Sci., Berkeley.

Ulrich, A., F.J. Hills, D. Ririe, A.G. George, and M.D. Morse. 1959. Plant analysis, a guide for sugar beet fertilization. California Agric. Exp. Stn. Bull. 766:1–24.

Chapter 17

Plant Tissue Analysis of Sugarcane

JOHN E. BOWEN, *University of Hawaii, Honolulu, Hawaii*

Sugarcane (*Saccharum officinarum* L.) is a highly efficient plant photosynthetically, producing up to 40 t of sugar/ha in Hawaii where it is normally grown as a 24-mo crop (Clements, 1964a, 1980; Clements et al., 1952). These excellent yields result from optimization of the myriad factors that affect the crop's growth and development.

The grower cannot control such parameters as minimum and maximum daily temperatures, and intensity and duration of sunlight, of course. He only has partial control of soil moisture if the field is not irrigated.

A grower does have significant control over another important parameter, however; i.e., crop nutrition and availability of essential nutrients in the soil. His obvious goal is to maintain soil fertility at a level that allows maximum potential crop productivity.

Excessive nutrient availability, on the other hand, can be detrimental to profitable yields in any of three ways: (i) some essential nutrients; e.g., B and Mn, are toxic to sugarcane when too much is available, (ii) excesses of some nutrients; e.g., N, interfere with the "ripening" of sugarcane and thus decrease sugar yields; and (iii) overfertilization, a common cause of excessive availability, erodes profits.

It is sometimes a fine line indeed between adequate and excessive availability of a nutrient, especially a micronutrient. As you might anticipate, much research has been done to develop techniques and procedures to assist sugarcane growers in coping with this problem. The most accurate and successful ones to date are those based upon analysis of the plant tissues to assess the crop's nutritional status. Further improvement in usefulness is gained when environmental factors that affect nutrient uptake, accumulation, and utilization are integrated into the tissue analysis program.

I. HISTORY

The concept of analyzing plant tissues to evaluate a crop's nutritional status and determine its future fertilizer requirements is certainly not new,

 Soil Testing and Plant Analysis, 3rd ed.—SSSA Book Series, no. 3.

dating instead from the mid-19th century work of Liebig (Samuels, 1969). It was not until the 1930s, however, that Lagatu and Maume provided the scientific basis for what today we know as "foliar diagnosis" or "plant tissue analysis" (Lagatu & Maume, 1932, 1933a, b). The idea is that certain tissues of a growing plant can be chemically analyzed to measure their essential nutrient contents and, therefore, to measure indirectly the supply of nutrients available in the soil. As we now realize, the assumption that the plant nutrient content is a direct function of nutrient availability in the soil is not always true. When the availability of a nutrient is inadequate, its concentration in the plant tissues drops below a previously defined optimum amount or "critical level." The latter represents a marginal insufficiency that causes a slight reduction in growth without producing visible deficiency symptoms or decreasing crop yield.

II. REVIEW OF TISSUE ANALYSIS METHODS USED FOR SUGARCANE

A. Methods Based on "Critical Level" Concept

In 1936, Halais (1950, 1955, 1957) began experimenting in Mauritius with a foliar diagnostic technique for sugarcane patterned after methods developed for other crops by Lagatu and Maume (1932, 1933a, b). Halais correlated tissue analysis data for N, P, and K with field responses to fertilizers. This led to establishment of critical levels for these nutrients. Nitrogen, P, and K fertilizer recommendations could then be made by comparing the tissue concentrations of these nutrients to the critical levels. The Halais method became known as the top visible dewlap (TVD) sampling procedure and is still used in some regions today, especially those where sugarcane is grown as a 11- or 12-mo crop (Samuels, 1969). The TVD leaf is the youngest leaf with a visible dewlap, often the third leaf from the spindle.

The TVD samples are typically taken 4 to 6 mo after the cane is ratooned; the plant crop is not usually sampled with this diagnostic method. This timing usually corresponds to the peak of the "grand growth period" or "boom stage." Two samples are taken during the 12-mo crop cycle with a minimum interval of 1 mo between them. A sample consists of TVD leaves removed from 50 to 60 plants along a diagonal through the field.

These leaves are weighed, and then four discs are punched from each blade, halfway between the midrib and the margin. Alternatively, the entire middle one-third of the blade, excluding the midrib, may be used. The selected tissue is then dried, ground, and analyzed for its mineral nutrient content.

B. Crop Logging

Clements started studying the effects of climate on sugarcane growth and yield in Hawaii in 1940 (Clements, 1940, 1980). He quickly showed that the effects of climatic, physiological, and nutritional factors on a 24-mo sugarcane crop were closely inter-related. Clements' tissue analysis methods were later adopted by the Hawaiian sugar industry and are now called *crop logging*. Crop logging has undergone frequent revision and extensive improvement in the succeeding decades but its scientific basis remains unchanged (Clements, 1940, 1959, 1964b, 1980; Sund & Clements, 1974). This foliar diagnostic method is often acknowledged as the most comprehensive method used today for evaluating and guiding the growth and development of sugarcane (Samuels, 1959, 1969; Jones & Bowen, 1981).

Plant tissue samples are taken every 35 d throughout the 24-mo crop cycle. However, crop logging has also been adapted to the more common 12-mo crop. In the latter case, the cane is sampled at 14-d intervals rather than the originally suggested 35-d ones (Sund & Clements, 1974; Clements, 1980).

Regardless of the frequency of sampling, though, the elongating sheaths of leaves 3, 4, 5, and 6 (spindle leaf = 1) are removed from the primary stalks of five plants. It is very important that primary stalks *only* be selected because physiological age of the tissue affects nutrient composition. These sheaths are used for fresh weight and tissue moisture content determinations, for total sugar content analysis, and for all other nutrient analyses except N. The latter is measured in leaf punches taken from the blades of these same leaves. (For a discussion of the leaf-numbering system used in crop logging, see Clements & Ghotb, 1969.)

Numerous other data are obtained from the three samples taken during the midpoint or "boom stage" of the crop cycle, usually when the cane is 8- to 12-mo old. The three "boom stage" samples are analyzed for phosphate and the micronutrients. (Micronutrients as used here includes all essential nutrients other than N, P, and K, a convention adopted from the Hawaiian sugar industry.)

Six to 10 crop log samples are taken during an 11- to 12-mo crop, and about 22 samples during a 24-mo crop.

Crop logging differs from many other foliar diagnostic procedures in that it integrates essential nutrient contents of sugarcane tissues with plant growth, sugar production, and various climatic factors known to affect these parameters (Clements, 1940, 1959, 1964a, 1980). The crop logging system was developed in Hawaii under conditions that differ significantly from other regions of the world where sugarcane is grown. Primarily, this is 24 mo crop in Hawaii, irrigated in some areas and subjected to as much as 250 cm of rainfall annually in others, and grown from sea level to 600 m elevation. However, crop logging has been modified locally for use under each of these diverse conditions. This system for monitoring crop development has also been adapted for application in areas beyond Hawaii (e.g., regions of Iran, Iraq, Brazil, and Ecuador).

Researchers at the Hawaiian Sugar Planters' Association (HSPA) were also working on sugarcane tissue analysis, but not foliar analysis, during this period (Samuels, 1959, 1969). The stalk-logging technique, devised by Burr, dictates that 8 to 10 internode samples be taken monthly, starting when the 24-mo crop cane is 6- to 10-mo old and continuing until the last fertilizer has been applied. The stalk is cut below the point of attachment of leaf 10, retaining any older leaves that may be present. Five such stalks constitute a sample. The stalks are then trimmed so that only internodes 8, 9, and 10 remain. These are sliced, dried, ground, and analyzed for N, P, and K (Burr, 1955).

Ewart, also working at the HSPA, developed the self-adjusting stalk indices method. This technique, like that of Burr (1955), was predicated upon the hypothesis that the stalk would be a more reliable indicator of the nutritional status of the plant than would the leaf blades and sheaths. A sample consists of all stalks from pre-selected small, medium, and large sugarcane stools that have been preferentially fertilized. Similar samples are taken from the adjacent commercial field. The stalk tissues are analyzed for N, P, K, and Ca (Samuels, 1959, 1969). Neither this method (Samuels, 1959, 1969) nor that devised by Burr (1955) has gained much acceptance among the world's sugarcane growers, however.

The TVD method developed by Halais (1950, 1955, 1957) was subsequently modified by Evans to include adjustments for crop age at sampling and the amount of rainfall prior to sampling (Evans, 1955). Evans also established TVD critical levels for some of the micronutrients (Evans, 1955).

Numerous other techniques, many of which are regional modifications and adaptations of these methods, are also in use; e.g., the Jamaican and Puerto Rican versions of the TVD method (Samuels, 1959, 1969).

The Jamaican method, as developed by Innes and Chinloy (1955), requires sampling of the blade of the third fully-expanded leaf when the crop is 4- to 5-mo old. The crop field is sampled as is a microplot previously established within the crop field. These tissues are analyzed for N, P, and K. The microplot has received more N fertilizer than the surrounding field. Leaf N concentrations in the crop cane are compared to those from the microplot cane. Nitrogen fertilizer recommendations are made, based upon relative increases in N contents. Potassium and P needs are determined by comparing current data to those from earlier crops.

The Puerto Rican method is similar to crop logging in that leaves 4, 5, and 6 are sampled (Samuels, 1959, 1969). However, the blades and sheaths are combined for N, P, and K analyses, whereas tissue moisture content is determined in the sheaths alone. Samples are usually taken only in cane that is 3-mo old or less. If older cane is sampled, then corrections are made for crop age and tissue moisture content. Fertilizer recommendations are then made, based upon previously determined standard values.

The methods developed in Jamaica (Innes & Chinloy, 1955), Mauritius (Halais, 1950, 1955, 1957), and Puerto Rico (Samuels, 1959, 1969) are particularly well suited to sugarcane when it is grown as a 12-mo crop but have little applicability to longer crop cycles. The provision for early sampling allows corrective fertilizer applications to be made to the current crop. However, no corrections are made for crop age and tissue moisture content in the early samples despite the fact that these parameters are known to have significant effects upon uptake and accumulation of N and K in sugarcane leaf tissues (Clements, 1957, 1964a, 1980).

C. Diagnosis and Recommendation Integrated System

All methods described thus far use the critical level approach to interpret the crop's nutritional status. However, Beaufils (1973) and Beaufils and Sumner (1976) have introduced a totally different concept to plant tissue analysis. Their concept is to use ratios of tissue nutrient concentrations to calculate indices for diagnosis of deficiencies and imbalances. This technique is called the Diagnosis and Recommendation Integrated System (DRIS) (Beaufils, 1973; Beaufils & Sumner, 1976).

The advantages of DRIS are many, according to its developers (Beaufils, 1973; Beaufils & Sumner, 1976; Sumner, 1979). Briefly stated, DRIS interrelates all plant, soil, and environmental factors, and the effect of each upon crop yield. It is irrelevant to users of DRIS whether any specific factor is known to have an effect upon yield. All factors are considered, based upon the premise that any specific factor could become limiting under certain currently undefined conditions.

An index, as a function of yield, is derived for each cultural and environmental factor (Beaufils, 1973; Sumner & Beaufils, 1979). The more negative an index value for a nutritional parameter is, the more that nutrient is limiting crop yield. These indices also permit the ranking of yield-limiting factors in order of their effect upon crop yield.

Jones and Bowen (1981) compared crop logging with DRIS under a wide range of conditions in Hawaii. They found that the latter provided 3 to 5% fewer incorrect diagnoses of nutrient insufficiencies. Thus, DRIS may slightly improve diagnostic accuracy, but this benefit must be weighed against the complexities of DRIS, and the time required to learn the technique and interpret the data.

III. CROP LOGGING

The remainder of this discussion shall be focused on crop logging because this method of tissue analysis has played such a prominent role in the Hawaiian sugarcane industry for more than 40 yr. And, since Hawaii's cane sugar industry leads the world's producers in terms of yields every year (Blume, 1985; Smith, 1978; USDA, 1987). Crop logging can also be said to have had a major impact on world sugar production as well. Further, crop logging is used extensively in those areas that historically have produced the greatest yields of cane and sugar. Indeed, the introduction of crop logging coincides with these sharp increases in yields.

Remember, crop logging was originally developed for a 24-mo sugarcane crop and has subsequently been adapted to 12-mo crops in some areas. These adaptations also will be discussed where appropriate.

Portions of a completed crop log sheet for a 24-mo crop in an unirrigated field are shown in Fig. 17-1, 17-2, 17-3, and 17-4. The top section of Fig. 17-1 is a record of average maximum and minimum daily temperatures in degrees Fahrenheit, and also of sunlight in (g cal/cm^2/d). These data are

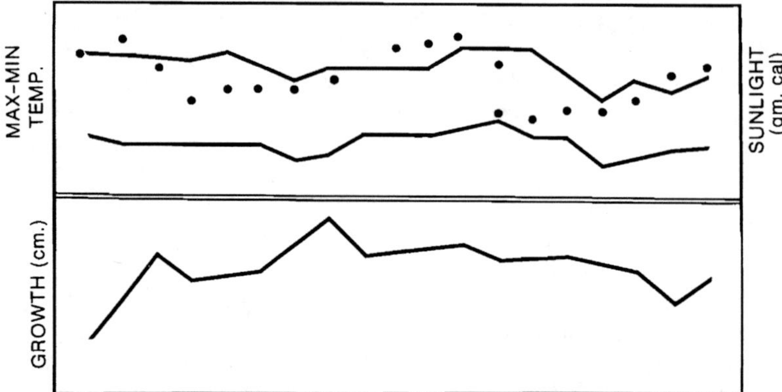

Fig. 17–1. A completed crop log for a 24-mo, unirrigated sugarcane crop in Hawaii. The lower bottom curve depicts crop growth as manifested by fresh weight of the stalks, in grams. The top and middle curves show maximum and minimum daily temperatures, respectively.

for record purposes only, though, because the grower obviously cannot alter either temperature or sunlight.

The lower section of Fig. 17–1 shows crop growth; i.e., total fresh weight in grams of sheaths 3, 4, 5, and 6, divided by five (five plants per sample). Some growers prefer to measure stalk elongation and, in that case, those data are recorded here instead of the fresh weights.

Actual leaf blade N contents (expressed as percentage of dry wt.) are recorded next and are shown as points connected by a solid line (Fig. 17–2). The "x's" are the normal nitrogen index (NNI) values. The NNI is the N level in leaf blade tissues from a previously high-yielding crop whose age and moisture content were identical to those of the present sample.

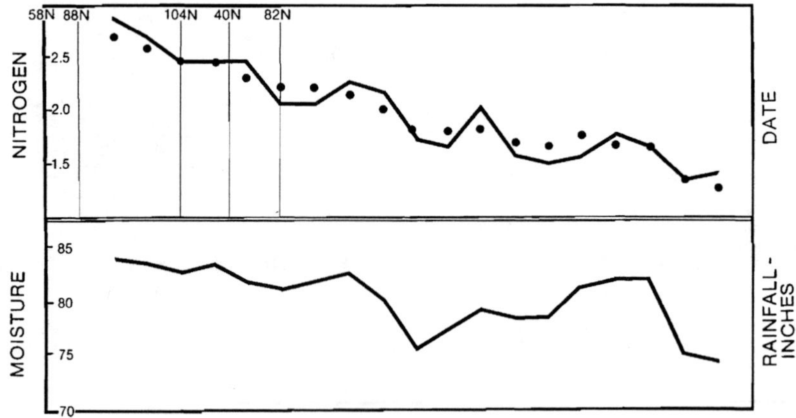

Fig. 17–2. A completed crop log for a 24-mo, unirrigated sugarcane crop in Hawaii. Actual leaf blade N contents (expressed as percentage of dry wt.) are recorded in the upper section and are shown as points connected by a solid line. The "dots" are the NNI values. The NNI is the N level in leaf blade tissues from a previously high-yielding crop whose age and moisture content were identical to those of the present sample. The sheath moisture content, expressed as percentage of fresh weight, is shown in the lower section.

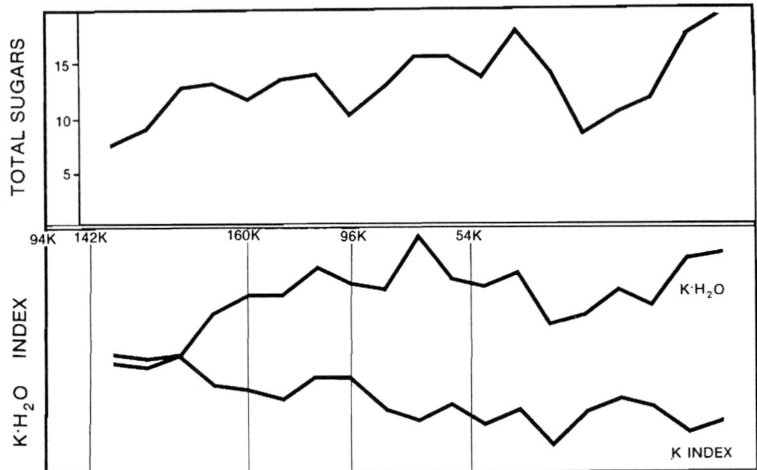

Fig. 17–3. The total soluble sugar content or "primary index" of the elongating leaf sheaths is shown in the top section of this portion of the crop log sheet. (Total soluble sugars = sugars in a water extract from the dried tissue, after inversion with invertase.) Sheath K contents are presented in the lower section as both the K index and the K·H_2O index. The K index is the K content of leaf sheaths [percentage soluble sugar-free (s-f) dry wt.]. Divide the y-axis values by 100 to obtain the actual K index value. The K·H_2O index is the K content of the sheaths expressed as a percentage of the tissue moisture and multiplied by 1000.

Actual crop age in months, measured from the date of planting or ratooning, and sampling dates are normally shown just above the N curves but these have been omitted from Fig. 17–2 to improve clarity. The numbers denote dates and amounts (in lb/acre) of N fertilizer applications.

The fourth section (the lower section in Fig. 17–2) records sheath moisture, rainfall, and irrigation applications if applicable.

The total soluble sugar content or primary index of the elongating leaf sheaths is shown in the next section (Fig. 17–3). (Total soluble sugars = sugars in a water extract from the dried tissue, after inversion with invertase.)

Sheath K contents are presented as both the K index and the K·H_2O index (Fig. 17–3). The K index is the K content of leaf sheaths [percentage soluble sugar-free (s-f) dry wt.]. Divide the y-axis values by 100 to obtain the actual K index value; e.g., a K index of 380 is actually 3.80% s-f dry wt.

The K·H_2O index is the K content of the sheaths expressed as a percentage of the tissue moisture and multiplied by 1000. Thus, a K·H_2O index of 425 is actually 0.425. The K·H_2O index, like that for N, is quite complex. It, too, will be discussed later.

All data in the lower left section of Fig. 17–4 are obtained from the boom stage samples taken when the cane is 8- to 12-mo old.

Three different P indices are used in crop logging; i.e., the standardized phosphorus index (SPI), the Fifth Internode P content, and the amplified phosphorus index (API). The calculation and application of these will be discussed later also.

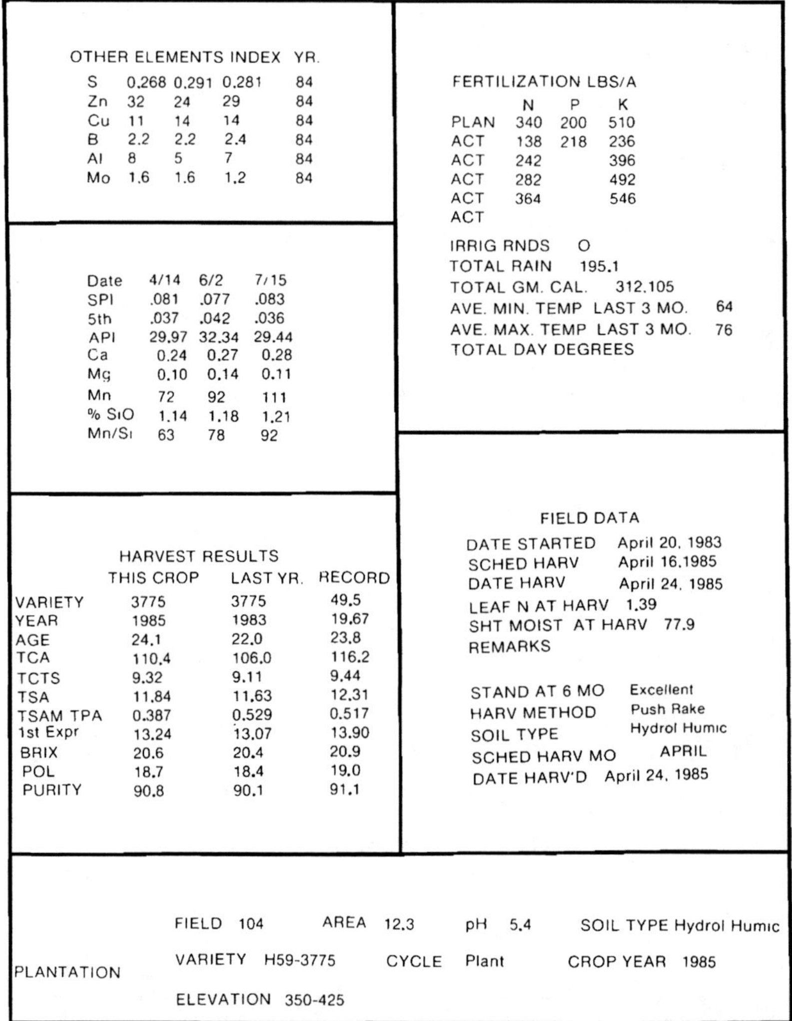

OTHER ELEMENTS INDEX YR.

S	0.268	0.291	0.281	84
Zn	32	24	29	84
Cu	11	14	14	84
B	2.2	2.2	2.4	84
Al	8	5	7	84
Mo	1.6	1.6	1.2	84

FERTILIZATION LBS/A

	N	P	K
PLAN	340	200	510
ACT	138	218	236
ACT	242		396
ACT	282		492
ACT	364		546
ACT			

IRRIG RNDS O
TOTAL RAIN 195.1
TOTAL GM. CAL. 312.105
AVE. MIN. TEMP LAST 3 MO. 64
AVE. MAX. TEMP LAST 3 MO. 76
TOTAL DAY DEGREES

Date	4/14	6/2	7/15
SPI	.081	.077	.083
5th	.037	.042	.036
API	29.97	32.34	29.44
Ca	0.24	0.27	0.28
Mg	0.10	0.14	0.11
Mn	72	92	111
% SiO	1.14	1.18	1.21
Mn/Si	63	78	92

HARVEST RESULTS

	THIS CROP	LAST YR.	RECORD
VARIETY	3775	3775	49.5
YEAR	1985	1983	19.67
AGE	24.1	22.0	23.8
TCA	110.4	106.0	116.2
TCTS	9.32	9.11	9.44
TSA	11.84	11.63	12.31
TSAM TPA	0.387	0.529	0.517
1st Expr	13.24	13.07	13.90
BRIX	20.6	20.4	20.9
POL	18.7	18.4	19.0
PURITY	90.8	90.1	91.1

FIELD DATA

DATE STARTED April 20, 1983
SCHED HARV April 16,1985
DATE HARV April 24, 1985
LEAF N AT HARV 1.39
SHT MOIST AT HARV 77.9
REMARKS

STAND AT 6 MO Excellent
HARV METHOD Push Rake
SOIL TYPE Hydrol Humic
SCHED HARV MO APRIL
DATE HARV'D April 24, 1985

PLANTATION
FIELD 104 AREA 12.3 pH 5.4 SOIL TYPE Hydrol Humic
VARIETY H59-3775 CYCLE Plant CROP YEAR 1985
ELEVATION 350-425

Fig. 17–4. This section of the crop log sheet contains data on micronutrient concentrations in the sheaths of "boom-stage" samples, planned and actual fertilizer applications, field data pertaining to cultural practices, and a historical record of yields from that field.

Calcium and Mg contents of the elongating sheaths, as percentages of the soluble s-f dry wt., are also shown in this lower left section. Comparison of actual Ca and Mg levels with the critical levels (Table 17–1) will guide your decision whether to apply these nutrients to the growing crop or to wait until the next crop is established.

The S content of the sheaths is reported as percentage s-f dry wt.; concentrations of other micronutrients are in mg/kg s-f dry wt., or ppm.

If the tissue level falls below the critical level for one or more of the micronutrients (Table 17–1), the usual practice would be to incorporate the

Table 17-1. Comparison of critical levels of micronutrients in elongating leaf sheaths of sugarcane (Clements, 1980; Bowen, 1983).

	Critical level in elongating sheaths		
		Bowen	
Nutrient	Clements	TCH†	TSH†
Ca‡	0.17	0.20–0.25	0.20
Mg‡	0.08	0.10	0.10
S‡	0.22	0.30	0.30
B§	2.0	2.0	2.0
Zn§	10	10	10
Cu§	10	5	8
Mn§	10	10	10
Si‡	--	2.20	--
Mn/Si_2¶	--	--	50

† TCH = tonnes cane per hectare; TSH = tonnes sugar per hectare.
‡ Percent soluble sugar-free dry wt.
§ Soluble sugar-free dry wt. (mg/kg).
¶ Mn (mg/kg s-f dry wt.)/SiO_2 (percentage s-f dry wt.).

appropriate fertilizer into the soil when the next crop is started. Rarely is a micronutrient fertilizer applied to a growing crop. The most frequent exception to this, however, is S that can indeed be applied successfully to established crops.

The Mn/SiO_2 ratio is accorded great significance in crop logging. Although SiO_2 is not known to be essential for sugarcane, it can often alleviate Mn toxicity.

Certain elements essential to optimal growth and development of sugarcane are sometimes present in soil in excessive quantities and are thus toxic to the plants. This is most likely to be a problem with micronutrients; e.g., B and Mn, but maximum levels of tolerance have yet to be established.

Still other elements are not essential but are nevertheless markedly toxic; e.g., Al, the concentration of which in the sheaths is given in the "Other Element Index" section of Fig. 17-4. This, too, is expressed in mg/kg s-f dry wt. or ppm.

Fertilization data for N, P, and K, both planned application rates and the actual amounts applied, are reported in the "Fertilizer Lb/A" section. The actual values shown for N and K should agree with the totals that can be obtained from the N and K sections of the crop log above. All phosphate fertilizer is applied at the start of the crop, though, so the quantity of phosphate shown in this section represents a single application at the time of planting or ratooning.

There is also space in this section for a summary of total rainfall received and irrigation rounds applied (if any) during the crop; total (g cal)/m^2 of light received; average minimum and maxmum daily temperatures for the last 3 mo prior to harvest; and total "day-degrees."

The concept of "day-degree" requires some explanation. It is defined as the excess of degrees (in °F) more than 70 °F each day. Thus, if the maximum temperature for a given day was 82 °F, then 12 "day-degrees" would be recorded.

Miscellaneous field cultivation and harvest data are given in the "Field Data" section (Fig. 17–4). These data are part of the complete crop record and also contribute to an accurate interpretation of the tissue analysis data.

The lower left section headed "Harvest Results" (Fig. 17–4) contains information on various yield parameters from the current crop, the preceding crop and the record-yielding crop from that field.

Other pertinent data are recorded at the extreme bottom of the crop log sheet; i.e., field number, elevation and area; variety or cultivar being grown; soil type and pH; and crop cycle (plant, first ratoon, and second ratoon).

Crop logging, when applied to a 24-mo crop, permits correction of N and K deficiencies at any time until the crop reaches an age of 14 mo. No fertilizer is applied after this point in the crop cycle because it interferes with proper crop ripening. To alleviate this limitation on the applicability of crop logging, Sund and Clements (1974) recommended that samples be taken at 14-d intervals when working with a 12-mo crop. The sampling procedure was not modified, though.

Many regions in which sugarcane is grown as a 12-mo crop are characterized by saline soils and drought. Water table variations are often shown on crop logs to warn the grower of impending problems with salinity. Salinity is the most dominant negative growth factor in these areas and is thus monitored closely (Sund & Clements, 1974).

The sheath moisture level is regarded as the single most useful index in many regions (Sund & Clements, 1974). In recognition of this fact, these data are plotted for the 12-mo crop, along with normal moisture levels for the cultivar being grown. The normal moisture level is defined as the sheath moisture level at the same age in excellent crops, expressed as a percentage of fresh weight.

Experience has taught that sheath moisture, and tissue N and P concentrations are the factors most likely to pose problems for the cane farmer in areas where crop logging has been adapted to the 12-mo crop (Sund & Clements, 1974). Therefore, these parameters are measured biweekly while other nutrient levels are analyzed only occasionally.

A. Data Interpretation

Inherent to proper interpretation of any tissue analysis data is a thorough knowledge of the amount of each nutrient that must be present in the cane tissues at any given stage of growth if maximum yields are to be obtained; i.e., the critical level for each nutrient. The concentration of each nutrient in the index tissue (elongating sheaths in crop logging) is measured and compared with its concentration in samples from the highest-yielding fields *cultivated under the same conditions*. The latter is important; data obtained under different environmental or cultivation conditions cannot be compared accurately.

This standard of comparison obtained from tissue samples from the highest-yielding fields is the critical level for that nutrient. It is presumed that the nutrient concentrations in tissues from record-yielding fields will be those that must be maintained in subsequent crops in those fields if similar high yields are to be obtained (Clements et al., 1952; Clements, 1980). If the nutrient concentration selected as the critical level is not accurately correlated with the highest-yielding crops, then the standard of comparison is incorrect for future crops; i.e., future yields will be less than the maximum previously realized.

Critical levels for seven essential micronutrients, for apparently nonessential but beneficial Si, and for the Mn/SiO_2 ratio are presented in Table 17-1 (Bowen, 1983; Clements, 1967, 1980; Clements et al., 1974). The critical levels are virtually the same whether one uses tonnes cane per hectare (TCH) or tonnes sugar per hectare (TSH) except in the case of Cu. A tissue concentration of 5 mg Cu/kg s-f dry wt. is associated with maximum cane tonnage but 8 mg Cu/kg s-f dry wt. is required to produce maximum sugar yields. Cane yields decrease as the Cu content of the sheaths increases above 5 mg/kg so maximum sugar production is achieved at a point where cane tonnage is actually less than the maximum. In these experiments, the tonnes cane/tonnes sugar (TCTS) ratio was 10.11 when the sheath Cu content was 5 mg/kg. At a sheath Cu content of 8 mg/kg when sugar production was maximal, the TCTS ratio dropped to a desirable 6.59 (Bowen, 1983).

Much progress has been made in understanding critical levels since 1980. We now know, for instance, that myriad factors bear upon critical levels and tissue nutrient concentrations in addition to nutrient availability in the rhizosphere per se. Crop age and tissue moisture content strongly influence the critical levels for N and K, so much so in fact that these parameters must be considered in every interpretation and application of an N or K analysis (Clements, 1980).

Clements (1980) showed that climatic and ecological factors influence N uptake and accumulation in sugarcane tissues. The significance of these effects is so great that they preclude detection of a direct correlation between N availability in the soil, tissues N levels, and yield. This absence of direct correlations among these parameters may seem initially to be a failure of tissue analysis. However, Clements (1957, 1959, 1980; Clements & Moriguchi, 1942) used this situation to demonstrate that factors other than nutrient availability influence uptake and accumulation. The concept of the standardized N index (SNI) arose from this early work.

Six factors, other than availability, were shown to affect the accumulation of N in sugarcane: sheath moisture content, crop or plant age, minimum and maximum daily temperatures, soil moisture content and light intensity/duration. These six factors account for almost 70% of the variability in tissue N levels without the available N being included (Clements, 1957, 1980; Clements & Moriguchi, 1942; Clements et al., 1952).

Clements et al. (1952) hypothesized that the interpretation of tissue N contents could be greatly facilitated if a relatively simple mathematical equation could be derived to standardize tissue N levels for all conditions under which sugarcane is grown.

Further statistical analysis of crop log data showed that, of the six factors affecting tissue N contents listed above, the dominant two were crop or plant age at sampling and sheath moisture contents. This discovery led to the derivation of a regression equation for the normal nitrogen index (NNI).

The NNI values are calculated with this equation from previous data obtained locally from high-yielding crops. A table is constructed that includes all possible sheath moisture values between 73 and 90%, at intervals of 0.5%. Crop ages in the range of 2 to 28 mo are included, at intervals of 0.5 mo. The theoretical N content of the elongating leaf blades is entered into the table for every combination of sheath moisture and crop age. These tabular N values are called the NNI values because they represent the tissue N content that must be maintained at any given combination of sheath moisture and crop age if yields similar to the earlier record ones are to be obtained.

The actual N content of a particular sample whose age and sheath moisture content are already known is compared to the NNI value from the table. The actual N content should be close to the NNI value. Excess actual N means that fertilizer has been applied injudiciously and that it may not be possible to "ripen" the cane properly. An actual N content below the NNI value indicates that the crop has inadequate N available and that fertilization is likely needed (Clements, 1980; Clements et al., 1952).

The concentration of K in the leaf sheaths, like that of N in the blades, is also affected by factors other than K availability in the soil. Using multiple regression analyses, Clements (1959, 1970, 1980) showed that the single most important factor influencing K accumulation in the elongating sheaths was sheath moisture content. There was no statistically significant correlation between the actual sheath K contents expressed as a percentage of the s-f dry wt. and the ultimate crop yield. However, when these same K contents were expressed as a percentage of the sheath moisture, a very meaningful relationship between K levels and ultimate crop yield was apparent.

The $K \cdot H_2O$ index is calculated with this equation:

$$K \cdot H_2O \text{ index} = \frac{[(\text{K, \%s-f dry wt.})(100 - \text{\%sheath moisture})]}{\text{\%sheath moisture}} .$$

Calculation and application of $K \cdot H_2O$ values are expedited by using a table of pre-calculated values. $K \cdot H_2O$ index values are tabulated for all possible combinations of sheath moistures from 73 to 90%, at 0.5% intervals, and actual tissue K contents ranging from 1.1 to 3.0% s-f dry wt. A $K \cdot H_2O$ index value of 0.425 is normal under most conditions. (In actual practice, the $K \cdot H_2O$ index values are usually multiplied by 1000. Thus, 0.425 would be reported as 425.) Above 0.425, too much K is present and below this index value, inadequate K is available.

Phosphorus accumulation depends upon factors other than P availability (Clements, 1955, 1958, 1980) and requires use of three separate P indices.

Analyses for P are run on the elongating leaf sheath samples and also on the fifth internode samples. The latter is obtained from the fifth internode of the stalk below the point of attachment of the oldest living leaf (Clements, 1958; Clements & Ghotb, 1969).

A standard phosphorus index (SPI) value is calculated using the P content of the sheath samples (percentage s-f dry wt.), the sheath moisture content and the total soluble sugar content of the sheath sample. This SPI value is compared to the SPI value calculated for an assumed moisture content of 81.0% and total sugar content of 10.2%.

Later work by Clements (1955, 1958, 1980) showed that yet another parameter more accurately portrayed the P status of the crop; i.e., the amplified phosphorus index (API):

$$\text{API} = (\text{SPI}) \times (\text{5th internode P})$$

after both the SPI and 5th internode P values have been converted to whole numbers.

The API has consistently proven to be the most useful of the three P indices in assessing the P nutritional status of the crop (Clements, 1958, 1980). Optimal API values vary between plant and ratoon crops. However, 3200 or more is generally considered sufficient for a plant crop and 3600 or greater for a ratoon. The API values are used as guides for phosphate requirements for the next crop only, however. All phosphate is applied at the time of planting or ratooning in Hawaii so there is no opportunity to correct a P insufficiency during the crop cycle.

B. Other Factors Affecting Interpretation

Nitrogen, P, and K are not the only essential nutrients whose uptake and accumulation depend heavily upon factors other than their respective availabilities in the soil, though. Micronutrient accumulation in the elongating sheaths strongly depends upon crop age at the time of sampling, for example (Bowen, 1975).

Sheath Ca and Mg concentrations correlate negatively and nonlinearly with age. Copper and B accumulations are not affected by crop age, however. Thus, the importance of sampling elongating sheaths, or other tissues as the case may be, that are the same age physiologically becomes very obvious, at least for Ca and Mg.

Sugarcane cultivars vary widely in the amounts of essential nutrients required for optimal growth and maximal yields (Bowen, 1973a). Thus, a critical level is not likely to be applicable to all cultivars under all conditions (Table 17–2).

Consider Zn as an example. Cultivar H53-263 requires much less Zn for optimal growth than does cv. H57-5174. The latter readily develops symptoms of Zn deficiency in the field whereas the former rarely does so. Research has revealed that cv. H53-263 roots have a sixfold greater affinity for Zn than those of cv. H57-5174 (Bowen, 1973a). Thus, cv. H53-263 can use the

Table 17-2. Critical levels for micronutrients in three Hawaiian sugarcane cultivars (Bowen, 1983). (TSH used as the yield parameter.)†

Nutrient	H59-3775	Cv. H49-5	H53-263
Ca‡	0.20	0.22	0.22
Mg‡	0.10	0.10	0.10
S‡	0.27	0.30	0.30
B§	2.0	2.0	2.0
Zn§	10.0	10.0	7.0
Cu§	7.5	8.0	8.0
Mn§	10.0	10.0	10.0
Mn/SiO_2¶	50–75	50–75	50–75

† TSH = tonnes sugar per hectare.
‡ Percent soluble sugar-free dry wt.
§ Soluble sugar-free dry wt. (mg/kg).
¶ Mn (mg/kg s-f dry wt.)/SiO_2 (percentage s-f d ry wt.).

Zn that is available more efficiently than H57-5174. This explains, at least in part, why H53-263 can thrive under conditions that would be Zn-deficient for H57-5174 (Bowen, 1973a). Similar differences among cultivars of sugarcane have been found for other essential nutrients (e.g., N and Fe) (Clements, 1980).

Both synergistic and antagonistic interactions between essential nutrients affect their uptake, translocation, and utilization in plant tissues, and therefore affect interpretation of tissue analysis data (Bowen, 1973a, 1983). These problems are especially serious with micronutrients.

For example, synergistic interactions have been observed in the uptake of Ca-S, Ca-Zn, Ca-Cu, Ca-B, Mg-S, Mg-Zn, S-Zn, S-Cu, S-B, S-Mn, Zn-B, and Cu-B (Bowen, 1981). Antagonistic interactions were found between Ca-Mn, Mg-Cu, Mg-B, and Zn-Cu (Bowen, 1981). Thus, a low tissue content of Cu, for instance, may not indicate inadequate Cu availability. Rather, it may mean that excessive Zn is present in the soil, thus interfering with uptake and accumulation of Cu in the sheath tissues.

Silicon is a seemingly nonessential nutrient for sugarcane (Clements, 1967; Clements et al., 1974), although this fact has certainly not been unequivocally established (Elawad et al., 1982, b; Gascho, 1978; Gascho & Andreis, 1974). Silicon does significantly increase accumulation of S and Mg and decrease that of Mn, Cu, and B under some conditions, however (Bowen, 1981; Clements, 1980). Again, an apparent deficiency of Mn or Cu, for example, may indicate high SiO_2 levels in the soil, not Mn or Cu deficiency per se (Clements, 1967; Clements et al., 1974).

IV. TOXICITIES

Toxicities of essential nutrients are now receiving much attention from researchers, particularly micronutrient toxicities (Bowen, 1983). Boron, Cu, and Mn, for example, are markedly toxic to sugarcane and to many other plants when available in excess. There is often a narrow margin between sufficiency and toxicity for B. The critical level for B in elongating sheaths is 2.0 mg B/kg s-f dry wt. Increasing the tissue B level to 3.0 mg/kg s-f dry wt. decreases sugar tonnage by 24% (Bowen, 1970, 1983).

Toxicities occur with other essential nutrients also. Sulfur, for example, has a critical level of 0.30% s-f dry wt. If the sheath S level increases to 0.35% significant decreases in both cane and sugar tonnages are likely to occur (Bowen, 1983).

Excess available N also has a negative effect on sugar yield. One means used to properly "ripen" sugarcane is to induce a N stress or deficiency. When the plant is thus stressed, vegetative growth ceases and sugar rapidly accumulates. If too much N is available during this ripening phase of the crop cycle, though, sugarcane will continue to grow vegetatively at the expense of sugar accumulation in the stalks (Clements, 1959, 1980). This situation is not a true toxicity but the result for the grower is still lower sugar yields.

Nutrient excesses are readily controlled in those cases where the excess is caused by overfertilization. Close monitoring and accurate interpretation of tissue analysis data will assist you in determining the correct amounts of fertilizer to apply as well as the optimum time to apply it.

The toxicity problem is considerably more severe when a micronutrient is naturally present in the soil in amounts detrimental to sugarcane growth. Liming to increase the soil pH is one possible way to render some micronutrients less readily available and thus to lower the tissue concentrations.

Antagonistic interactions may also offer a possible solution to specific problems. For instance, a moderate application of Cu would decrease tissue Zn levels.

A long-range solution to an inherent soil toxicity is the selection of sugarcane cultivars that can tolerate the excess nutrient in question. This is currently an active research field for many crops in the tropics and subtropics. If it is too expensive or otherwise not feasible to correct a detrimental soil condition, the alternative is to modify the crop genetically so that it can adapt to the adverse soil condition.

V. PRACTICAL APPLICATION OF SUGARCANE TISSUE ANALYSIS METHODS

Regardless of just what measures you take to optimize the nutritional status of the sugarcane crop, do not overlook the fact that maximal yields ultimately depend upon a balanced supply of the 17 currently recognized essential nutrients. (It was suggested in 1972 that chlorine be added to this list of essential nutrients for sugarcane [Bowen].) Maintenance of this balance can be greatly facilitated by plant tissue analysis but numerous pitfalls await the unwary practitioner (Bowen, 1978).

Think of plant tissue analysis in terms of two separate operations: (i) sampling of crop fields and preparation of the plant material for the laboratory, and (ii) chemical analysis of the plant material. The latter will usually be done by an independent laboratory and thus will be beyond your control. So, we will concentrate upon the former to make certain that preventable error is not creeping into your results and costing you money through lowered yields or excessive fertilization.

The most critical step in successful crop control through plant tissue analysis is the selection of an index tissue for each nutrient. What tissue are you going to sample? Use one of the established methods such as crop logging or a modified TVD method to avoid this problem.

Critical levels also must be determined accurately before you can expect meaningful results from plant tissue analysis. Again, this has already been done for you if you use one of the established methods.

Let me inject a note of caution here, however. Insufficient concentrations of a specific nutrient in the tissue do not necessarily mean that inadequate amounts of that nutrient are present in the soil. For example, poor root development due to disease, nematodes, insects, or chemical toxicities can manifest itself in reduced nutrient uptake. Therefore, when the nutrient contents of the index tissue drop below the critical levels, the reasons for this should be investigated before any corrective action is taken.

In the field, the weakest link in the tissue analysis procedural chain is selection of the plants to be sampled, both in regards to sites of the sampling stations within the field and the individual plants and stalks chosen. Improper technique at this stage means that subsequent chemical analyses and fertilizer recommendations will be inaccurate. Sampling, therefore, receives much care and attention.

Set forth a few general guidelines in regard to site selection. First, topography or terrain is a major determinant. A uniform field requires fewer sampling stations than does a field dominated by steep slopes, inadequately drained areas and rocky outcroppings. Second, mixed cultivars within a field can necessitate multiple sampling sites because sugarcane cultivars manifest nutritional variabilities. Third, if the date of planting differs significantly in different parts of the field, at least one sample should be taken for each plant age. Lastly, areas of a field that historically show poor growth and low yields, regardless of the reason, should be sampled separately.

We could list numerous other variables that would require separate sampling sites or a smaller area per site. Rather than do that, sample site selection should be done by an experienced person who has detailed knowledge of the field in question.

It is not practical to offer guidelines as to how large an area a single sample site can represent because there are simply too many variables involved. Perhaps it might be useful to mention that one sample station per 10 ha is common in a uniform cane field. Avoid field edges and irrigation ditches when taking plant tissue samples to minimize "edge effects" so familiar to statisticians.

Although there is some disagreement about the need for doing this, take your plant samples within 3 h after sunrise to avoid possible diurnal fluctuations in moisture and nutrient composition. Tissue levels of N and K, for example, are significantly affected by tissue moisture contents as discussed above.

The practical importance of early morning sampling has not been sufficiently evaluated experimentally to prove that it is necessary. But, to be safe, sampling should continue to be done early in the day until it is shown to be unnecessary.

If a lengthy time delay is anticipated between sampling and return of the tissue to the laboratory for fresh weight determinations, seal the cut material in plastic bags to retard moisture loss.

After the plant material has been cut, it must be cleaned, dried, and prepared for chemical analysis. These steps also involve several potential pitfalls. First, surface contaminants such as loose soil and dusts are removed by shaking the sample.

Next, cleaning is needed. One method is to wash the sample quickly in flowing tap water with gentle scrubbing to loosen adherent surface films, and then pass the tissue through a dilute phosphate-free soap solution. The importance of using phosphate-free soap is obvious if reliable P analyses are to be run. Finally, the tissue is rinsed in distilled water and blotted dry.

Actually, the best method for cleaning tissue surfaces is widely debated among tissue analysis practitioners. Recommendations vary from the washing just described, to brushing samples individually with a camel hair brush, to wiping with a damp cloth. Wiping alone is clearly quite insufficient for removing contaminants from sugarcane blades and sheaths, though.

Failure to remove dusts, soil, and pesticide residues will likely have little effect upon apparent N, P, and K concentrations in the sample, but these contaminants will cause erroneously high micronutrient levels.

On the other hand, excessive washing can also introduce error because K and other highly soluble nutrients can be leached from the tissue. Losses of K from sugarcane sheaths are commonly $<2\%$ under the washing regime outlined above, though.

Sugarcane tissues should be quickly dried after sampling to minimize physical and biochemical changes that would affect the accuracy of the analytical data. For example, respiration continues in tissue samples after they have been cut from the plant until they are dried. Since respiratory enzymes convert sugars to CO_2 and water, failure to dry samples promptly can cause a loss in their weights and sugar contents. Also, protein metabolism is altered in collected samples. This may lead to false N readings.

The drying temperature must be sufficiently high to destroy enzymes, but not so high as to cause tissue loss by burning and charring. Overnight drying in a forced draft oven at 70 to 75 °C works well.

It is unlikely that all moisture is removed even under these conditions, however, so the "dry weight" of the sample is probably not really that at all. Some water will be retained because it is difficult, if not impossible, to remove it all without destroying some of the tissue with high heat. Although the samples are not actually being dried completely, we hopefully are drying them to a *constant* moisture content. Continue to refer to the tissue as being "dried," to avoid confusion.

The dried tissue is then normally delivered to the laboratory for chemical analysis. The data are most useful to the grower if he receives it within a few days after sampling. A short "in-laboratory" time permits the farmer to use the data to correct deficiencies in the growing crop; e.g., to apply N fertilizer if the actual N level falls below the NNI value.

Some sources of potential error in the sampling process have been suggested. There are also many such sources in the analytical and interpretive phases of tissue analysis, just as there are in making decisions about necessary measures to correct nutritional problems that are detected. Numerous research projects throughout the sugarcane-growing regions of the world are currently being addressed toward elucidating the optimal procedures for applying plant tissue analysis to this crop.

Sugarcane tissue analysis is a dynamic area of research and application. Thus, it behooves the user to be alert for changes and updates in the analytical and interpretive procedures.

REFERENCES

Beaufils, E.R. 1973. Diagnosis and Recommendation Integrated System (DRIS). A general scheme for experimentation and calibration based on principles developed from research in plant nutrition. Soil Sci. Bull. 1. Univ. Natal, South Africa.

Beaufils, E.R., and M.E. Sumner. 1976. Application of the DRIS approach for calibrating soil and plant factors in their effects on yield of sugarcane. Proc. South Afr. Sugar Tech. Assoc. 50:118–124.

Blume, H. 1985. Geography of sugar cane. Verlag Dr. Albert Bartens, Berlin, West Germany.

Bowen, J.E. 1970. Boron deficiency and toxicity in sugarcane. Sugar Cane Pathol. Newsl. 4:51–52.

Bowen, J.E. 1972. Essentiality of chlorine for optimum growth of sugarcane. Proc. 14th Congr., Int. Soc. Sugar Cane Technol. 1971:1102–1112.

Bowen, J.E. 1973a. Kinetics of zinc absorption by excised roots of two sugarcane clones. Plant Soil 39:125–129.

Bowen, J.E. 1973b. Effect of micronutrient deficiencies on macronutrient accumulation in sugarcane. Trop. Agric. (Trinidad) 50:129–137.

Bowen, J.E. 1975. Micro-nutrient composition of sugar-cane sheaths as affected by age. Trop. Agric. (Trinidad) 52:131–137.

Bowen, J.E. 1978. Plant tissue analysis: Costly errors to avoid. Crops Soils 1978:6–9.

Bowen, J.E. 1981. Micro-element nutrition of sugar-cane. II. Interactions in micro-element accumulation. Trop. Agric. (Trinidad) 58:215–220.

Bowen, J.E. 1983. Micro-element nutrition of sugar-cane. 3. Critical micro-element levels in immature leaf sheaths. Trop Agric. (Trinidad) 60:133–138.

Burr, G.O. 1955. Plant analyses as indices of nutrient availability. Hawaii. Plant. Rec.. 55:113–128.

Clements, H.F. 1940. Integration of climatic and physiologic factors with reference to the production of sugarcane. Hawaii. Plant. Rec. 44:201–233.

Clements, H.F. 1955. The absorption and distribution of phosphorus in sugarcane. Hawaii. Plant. Rec. 55:17–32.

Clements, H.F. 1957. Crop logging of sugarcane: The standard nitrogen index. Hawaii Agric. Exp. Stn. Tech. Bull. 35.

Clements, H.F. 1958. Recent developments in the crop logging of sugarcane—Phosphorus and calcium. Hawaii Agric. Exp. Stn. Prog. Notes 114.

Clements, H.F. 1959. Sugarcane nutrition and culture. Sugar Cane Res. Inst., Lucknow, India.

Clements, H.F. 1964a. Interaction of factors affecting yield. Ann. Rev. Plant Physiol. 15:409–442.

Clements, H.F.1964b. Foundations for objectivity in tissue diagnosis as a guide to crop control. Plant analysis and fertilizer problems IV. Am. Soc. Hortic. Sci., Alexandria, VA.

Clements, H.F. 1967. Effects of silicates on the growth and leaf freckle of sugarcane in Hawaii. Proc. 12th Congr., Int. Soc. Sugar Cane Technol. 1965:197–215.

Clements, H.F. 1970. Crop logging of sugarcane: Nitrogen and potassium requirements and interactions using two varieties. Hawaii Agric. Exp. Stn. Tech. Bull. 81.

Clements, H.F. 1980. Sugar crop logging and crop control: Principles and practices. Univ. Press of Hawaii, Honolulu.

Clements, H.F., and A. Ghotb. 1969. The numbering of leaves and internodes for sugarcane nutrition studies. Proc. 13th Congr., Int. Soc. Sugar Cane Technol. 1968:569–584.

Clements, H.F., and S. Moriguchi. 1942. Nitrogen and sugarcane: The nitrogen index and certain quantitative field aspects. Hawaii. Plant. Rec. 46:163–190.

Clements, H.F., E.W. Putman, R.H. Suehisa, G.L.N. Yee, and M.L. Wehling. 1974. Soil toxicities as causes of sugarcane leaf freckle, macademia leaf chlorosis (Keaau), and Maui sugarcane growth failure. Hawaii Agric. Exp. Stn. Tech. Bull. 88.

Clements, H.F., G.T. Shigeura, and E.K. Akamine. 1952. Factors affecting the growth of sugarcane. Hawaii Agric. Exp. Stn. Tech. Bull. 18.

Elawad, S.H., G.J. Gascho, and J.J. Street. 1982a. Response of sugarcane to silicate source and rate. I. Growth and yield. Agron. J. 74:481–484.

Elawad, S.H., J.J. Street, and G.J. Gascho. 1982b. Response of sugarcane to silicate source and rate. II. Leaf freckling and nutrient content. Agron. J. 74:484–487.

Evans, H. 1955. Studies on the mineral nutrition of sugarcane in British Guyana. II. The mineral status of sugarcane as revealed by foliar diagnosis. Trop. Agric. (Trinidad) 32:295–322.

Gascho, G.J. 1978. Response of sugarcane to calcium silicate slag. I. Mechanisms of response in Florida. Soil Crop Sci. Soc. Fla. Proc. 37:55–58.

Gascho, G.J., and H.J. Andreis. 1974. Sugarcane response to silicate slag applied to organic and sand soils. Proc. 15th Congr., Int. Soc. Sugar Cane Technol. 1974:543–551.

Halais, P. 1950. Foliar diagnosis, a new guide to fertilization of sugarcane in Mauritius. Proc. 7th Congr., Int. Soc. Sugar Cane Technol. 1950:218–232.

Halais, P. 1955. Foliar diagnosis. Annu. Rep. Mauritius Sugar Ind. Res. Inst., p. 41–43.

Halais, P. 1957. Foliar and internode diagnosis. Annu. Rep. Mauritius Sugar Ind. Res. Inst., p. 48–52.

Innes, R.F., and T. Chinloy. 1955. Experiences with crop control in Jamaica. Hawaii. Plant. Rec. 55:149–165.

Jones, C.A., and J.E. Bowen. 1981. Comparative DRIS and crop log diagnosis of sugarcane tissue analysis. Agron. J. 73:941–944.

Lagatu, H., and L. Maume. 1932. Application du diagnostic foliare: il suggere controle et limite le redressement alimentaire d'une vigne mal nourrie. C.R. Acad. Sci. (Paris) 194:812–814.

Lagatu, H., and L. Maume. 1933a. Composition comparee, chez la vigne, de feuilles homologues prises, respectivement, sur des souches fructiferes et sur des souches privees de leur grappes. C.R. Acad. Sci. (Paris) 196:1168–1170.

Lagatu, H., and L. Maume. 1933b. Composition comparee de la materiere seche de feuilles homologues des rameaux naturellement steriles d'une vigne. C.R. Acad. Sci. (Paris) 196:1445–1447.

Samuels, G. 1959. The relative merits of various methods of foliar diagnosis for sugarcane. Proc. 10th Congr., Int. Soc. Sugar Cane Technol. 1959:529–537.

Samuels, G. 1969. Foliar diagnosis for sugarcane. Agric. Res. Publ., Rio Piedras, Puerto Rico.

Smith, D. 1978. Cane sugar world. Palmer Publ., New York.

Sumner, M.E. 1979. Interpretation of foliar analyses for diagnostic purposes. Agron. J. 71:313–318.

Sund, K.A., and H.F. Clements. 1974. Production of sugarcane under saline desert conditions in Iran. Hawaii Agric. Exp. Stn. Res. Bull. 160.

U.S. Department of Agriculture. 1987. Sugar and sweetener situation and outlook yearbook. SSRV12N2 June. U.S. Gov. Print. Office, Washington, DC.

Chapter 18

Plant Analysis as an Aid in Fertilizing Cotton[1]

WAYNE E. SABBE, *University of Arkansas, Fayetteville*

LOWELL J. ZELINSKI, *University of California, Fresno*

The vegetative and reproductive growth patterns of cotton (*Gossypium hirsutum* L.) are readily adapted to plant analysis. The indeterminate growth pattern over a relatively long growing season allows plant analysis to be used for nutritional studies, monitoring of selected plant parts, and corrective fertilization for observed deficiencies during the growing season.

I. NUTRIENT UPTAKE PATTERNS

A. Dry Matter Accumulation

It is essential to understand the growth and mineral uptake patterns of a specific crop prior to interpreting plant analysis data. Not only do species differ, but there are also varietal and locality differences. The two main cotton-growing areas of the USA are the irrigated dryland areas of the Southwest Great Plains and the humid areas of the Southeast. These regions represent a range of climatic, varietal, and soil differences that are reflected in growth and nutrient uptake.

Research in Georgia (Olson & Bledsoe, 1942) and California (Bassett et al., 1970) showed similarities as well as contrasts in the pattern of dry matter production and nutrient uptake. In both areas, a slow rate of early growth was followed by a period of rapid increase in dry matter production and nutrient uptake. Dry matter production in Georgia was 3.1, 8.0, 37.6, and 51.2% of the total producing during the planting to seedling, seedling to early square, early square to early boll, and early boll to maturity stages, respectively. Under California conditions, cotton planted on 1 April produced 2 to 4% of the total dry matter at first square, an additional 7 to 10% by first

[1] Contribution from the Univ. of Arkansas-Fayetteville and the Univ. of California and published with the approval of the Director of the Arkansas Agric. Exp. Stn. and the Director of the California Ext. Serv.

flower on 25 June and a further 66% during the 6-wk period between 1 July and 15 August. Cotton in irrigated areas grew proportionally more in the later growth stages than did cotton in non-irrigated areas.

Figure 18–1 (Halvey, 1976) represented the dry matter accumulation by aboveground/aerial plant parts during a growing season. Under the irrigated conditions, the growth period between full flower (72 d) and boll opening (112 d) accumulated 75% of the total dry matter. This sigmoidal pattern of growth was also reported by Bassett et al. (1970).

Wanjura and Sunderman (1976) reported that dry matter accumulation on a per plant basis was greater in a 100-cm row width than a 25-cm row width. However, the accumulation on a per unit area basis was greater for the narrower row width. They also reported that an increase in N rates reduced the vegetative dry matter loss from peak bloom to maturity. Gardener and Tucker (1967) indicated that development of vegetative plant parts was reduced by early season N deficiency especially under a one-peak flowering period. However, under an elongated two-peak flowering period, an early season N deficiency can be compensated by an adequate supply of N during

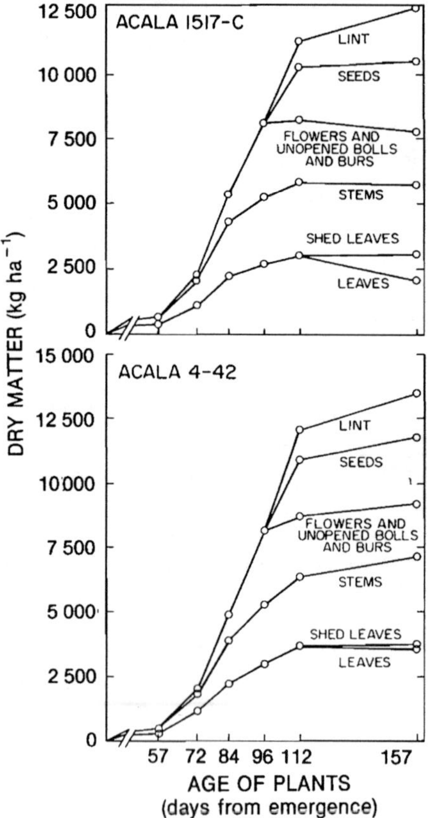

Fig. 18–1. Accumulation of dry matter in the aerial plant parts of cotton cv. Acala 1517-C and Acala 4-42 (Halvey, 1976).

the late season. Koli and Morrill (1976) observed both a prolonged vegetative growth period and fruiting period when a fertilizer rate of 45 kg of N ha^{-1} was compared to 0 kg of N ha^{-1}.

B. Nutrient Accumulation

1. Total Uptake

In general, early season nutrient uptake by cotton proceeded more rapidly than dry matter production. Cotton grown under irrigation absorbed about 15% of the total N, P, and K while dry matter accumulation was <10% of the total accumulated (Bassett et al., 1970). In humid areas, cotton absorbed 22 and 30% of the total N and K, respectively, during the accumulation of about 11% of the total dry matter (Olson & Bledsoe, 1942).

A comprehensive study on the seasonal uptake of N, P, and K among plant parts is reported by Bassett et al. (1970). Table 18-1 shows a continuous uptake of these nutrients during the growing season, the quantities/amounts of these nutrients during the growing season, and the extent of their translocation from leaves and stems to the more metabolically active region of seed formation. The greatest changes in nutrient uptake occurred between 60 and 120 d after planting. Uptake of N, P, and K increased about 10-fold during this 60-d period.

Table 18-1. Uptake and distribution of N, P, and K in plant parts (Bassett et al., 1970).

	Average no. days after planting						
	60 15 June	75 1 July	90 15 July	105 1 Aug.	120 15 Aug.	135 1 Sept.	165 1 Oct.
	kg ha^{-1}						
	Nitrogen						
Stems	1.0	3.8	7.1	9.9	12.7	12.1	18.3
Leaves	9.3	19.3	33.3	38.5	45.5	73.2	37.9
Burrs		(3.5)†	(14.5)†	(27.0)†	15.7	14.1	11.6
Seed					34.4	55.8	75.8
Total	10.3	26.6	54.9	75.4	108.3	119.2	143.6
	Phosphorus						
Stems	0.2	0.6	1.3	1.7	2.2	2.0	3.1
Leaves	1.1	1.7	3.2	3.8	3.9	3.4	3.6
Burrs		(0.5)†	(2.5)†	(4.6)†	.28	2.7	2.0
Seed					5.8	9.1	12.2
Total	1.3	2.8	7.0	10.1	14.7	17.2	20.9
	Potassium						
Stems	1.6	5.0	13.7	18.9	28.5	25.0	27.5
Leaves	5.7	12.1	22.4	29.8	31.4	25.2	20.6
Burrs		(1.3)†	(10.4)†	(22.3)†	25.4	44.4	52.5
Seed					13.0	17.0	20.6
Total	7.3	18.4	46.5	71.0	98.3	111.6	121.2

† Entire boll—includes burrs and seed.

Sabbe and Hansen (1983) reported that the total plant content of Zn, Fe, Mn, and Cu increased two- to threefold during the 7-wk period following first bloom. However, the rate of dry matter accumulation during this period was proportionally greater than the micronutrient uptake, thereby resulting in a decrease in micronutrient concentration among most plant parts.

2. Plant Parts

Correct selection of the proper plant part for the determination of the nutrition status of the whole plant requires an understanding of the partitioning within the plant as well as the seasonal distribution of specific nutrients.

A direct relationship exists between the leaf and petioles for the more mobile nutrients (i.e., N, P, and K); whereas, for the less-mobile nutrients the relationship are inversed especially for the top leaf and petiole at the later sampling dates (Table 18-2). Data in Table 18-2 indicate that the selection of a plant part must consider not only plant age but also the nutrient.

3. Nitrogen

Cotton requires N for both the vegetative and the reproductive phases of growth. Oosterhuis et al. (1983) demonstrated (Fig. 18-2) the pattern of N uptake and distribution in aboveground plant components. Their study demonstrated that the response to the initial rate of N increased N content in all plant parts. The N content increase was concomitant with a linear decrease in the N concentration of each plant part with time. This decrease in N concentration resulted from the rate of dry matter accumulation being

Table 18-2. Nutrient concentration of cotton plant parts at various sampling dates (Harris, 1960).

Date	Element	Whole plant	Top leaf	Top petiole	Bottom leaf	Bottom petiole
				$g\ kg^{-1}$		
29 June	N	43.5				
27July	N		43.4	26.6	37.9	22.3
4 Aug.	N		45.7	24.2		
29 Aug.	N		37.3	10.5	24.4	8.9
29 June	P	5.1				
27 July	P		4.6	4.2	3.0	2.2
4 Aug.	P		4.5	4.2		
29 Aug.	P		3.1	1.9	2.5	1.4
29 June	K	20.8				
27 July	K		18.2	19.5	20.3	24.5
4 Aug.	K		12.6	19.9		
29 Aug.	K		6.9	17.1	16.4	27.5
29 June	Ca	15.1				
27 July	Ca		12.3	14.3	39.0	15.7
4 Aug.	Ca		12.3	12.7		
29 Aug.	Ca		17.0	8.9	33.7	12.2
29 June	Mg	4.0				
27 July	Mg		2.0	5.4	2.8	5.4
29 Aug.	Mg		5.0	2.4	1.7	2.5

greater than the N uptake. Wanjura and Sunderman (1976) obtained similar results for the leaves, stem, and fruiting bodies. Additionally, they found that the seed N concentration increased from peak bloom to maturity. These authors reported that plant N concentration did not differ between the 25- and 100-cm row spacings.

Thompson et al. (1976) tagged newly initiated leaves and fruit during early, middle, and late season to determine the N concentration among these age classes. The youngest leaves and fruits had initially higher N concentrations than later-appearing leaves and fruits. The N concentration of early

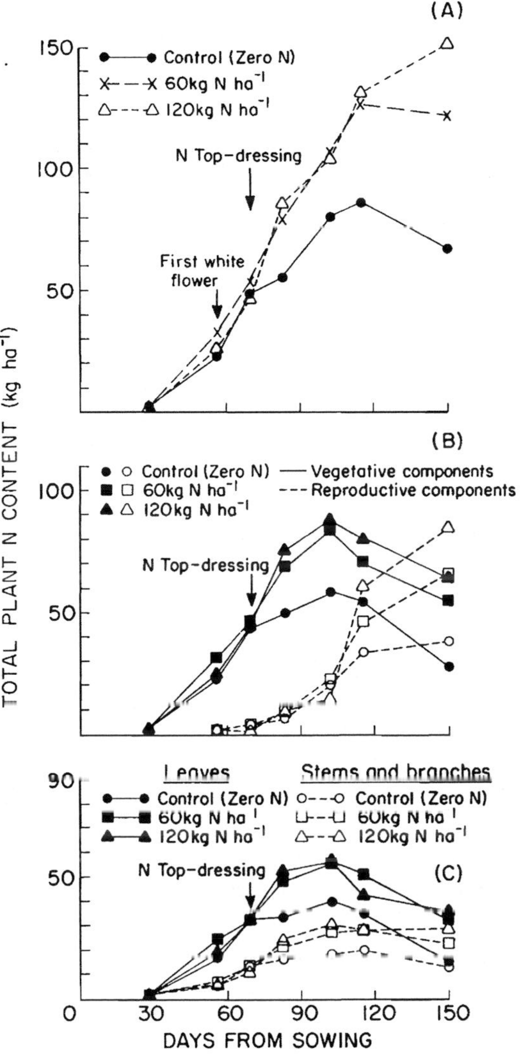

Fig. 18-2. Changes with age in aboveground N contents of total shoot, vegetative, and reproductive components of the cotton plant (Oosterhuis et al., 1983).

Table 18-3. Phosphorus concentration in cotton plant parts at various growth stages (Peacock, 1960).

Growth stage	Leaves	Petioles	Stems	Roots	Branches	Squares	Bolls
				g P kg^{-1}			
Four leaf stage	3.3	2.7	2.4	2.2			
Early bloom	3.5	2.2	2.1	1.5	2.9	4.4	
Early boll	2.7	1.7	1.6	1.5	1.5	4.9	3.7

season leaves decreased rapidly during their first 20 d; however, the leaves initiated later in the season did not decrease in N concentration during this first 20 d. Regardless of date of initiation, leaf ages between 20 and 40 d did not decrease in N concentration unless no N fertilizer had been applied. When N fertilizer had been applied the N concentration of fruits decreased for the first 40 d. In fruit older than 40 d, the N concentration remained constant. The authors concluded that the three phases of N utilization were: (i) rapid growth, (ii) photosynthetically active growth, and (iii) senescence. Therefore, a leaf would exhibit a rapid increase in N, level off, then decrease during senescence. However, if the leaf N concentration was low initially, it maintained a constant concentration until senescence.

4. Phosphorus

Patterns of P uptake and distribution among plant parts are similar to those of N (Halvey, 1976). As with N, the seed contained a high fraction of plant P at maturity (Basset et al.,1970). Phosphorus accumulation in the seed during formation was greater than total plant accumulation of P (Halvey, 1976). Data in Table 18-3 indicates the P concentration in plant organs during a growing season (Peacock, 1960). The peak translocation period for P occurred during the early boll development and may cause deficiency symptoms when other plant parts are marginally sufficient.

5. Potassium

There were two differences between the uptake pattern of K and those of N and P. The highest total plant accumulation occurred approximately at first boll opening and decreased slightly until maturity. The K in both the leaves and stems was translocated to the reproductive organs (Halvey, 1976). Hsu et al. (1978) reported that, although the rates of K decline in both leaf blades and petioles were dissimilar, each was a function of maturity and not a function of K fertilization rate. A constant decline in K occurred in leaf blades from early bloom to early boll; in petioles the early season decline was less than the decline from early bloom to early boll (Table 18-4). Apparently the bolls were greater sinks than flower buds.

6. Other Nutrients

Sabbe and Hansen (1983) observed a large increase in total Zn, Cu, Fe, and Mn during the reproductive stage (first bloom to 7 wk past first bloom).

Table 18-4. Potassium concentration of leaf-blades and petioles with stage of maturity (Hsu et al., 1978).

Plant part	Potassium concentration: Early square	Early bloom	Early half-grown boll	Percentage decrease: Early square to early bloom	Early bloom to early half-grown boll
	———	$g\ K\ kg^{-1}$	———	———	% ———
Apical immature blade	15.1	10.2	6.7	32	34
Young mature blade	12.5	10.7	7.8	14	27
Old mature blade	--	10.4	7.2	--	31
Apical immature petiole	39.6	38.0	28.1	4	26
Young mature petiole	33.1	26.6	10.9	20	59
Old mature petiole	--	25.5	10.6	--	58

However, the increases were less than the dry weight accumulation, thereby causing a concentration decrease during this growth stage. The exception was an increase in leaf Fe and Mn concentration at 7 wk past first bloom. The majority of the plant Fe and Mn were in the leaf; whereas, the majority of the Cu and Zn were in the fruiting tissue.

In four cotton cultivars and one cotton hybrid studied by Bhatt and Appukuttan (1982) the leaves, stems, and roots of 10-d-old seedlings had the maximum Mn concentration of the six growth stages sampled. Thereafter, rapid decline in Mn occurred until the square stage (from 2200 to 302, 825 to 68, and 3300 to 275 mg kg^{-1} for the leaf, stem, and root, respectively) followed by a period of rather constant Mn concentration from peak flowering until harvest (140, 16, and 30 mg of Mn kg^{-1} at harvest for leaf, stem, and root, respectively).

7. Summary

Patterns for dry matter accumulation and nutrient uptake by cotton can be summarized as (i) an initially slow growth until first flower followed by (ii) a rapid increase in both dry matter and nutrient uptake during fruit development. During the latter period, translocation of nutrients occurs to the fruiting bodies (sinks). Of the total nutrients accumulated during the growth season, about 50% of the N, 50 to 70% of the P, and 16 to 20% of the K were recoved from the field at harvest. These rather high harvest values indicate the need for maintaining a healthy plant capable of season-long nutrient uptake. A model has been described for the N balance and the carbohydrate growth requirements (Jones et al., 1974). The N requirement was based on the growth allowed by the carbohydrate supply. The model depended upon the soil N supply, daily N absorption rate, and the concentration in each plant part.

II. FACTORS INFLUENCING NUTRIENT CONCENTRATION IN COTTON

The concentration of nutrients in cotton tissues is related to environmental factors that influence their uptake. Knowledge of how these factors affect uptake and concentration is important for proper interpretation of plant analysis.

A. Nitrogen

Nitrogen is the most important nutrient for cotton production, and its effectiveness is dependent upon the environment.

1. Soil Factors

Most researchers investigating N fertility of cotton have observed a decline in N concentration in vegetative tissues over time. This decline may be due to dilution or remobilization (Thompson et al., 1976) or depletion of soil N and reduced uptake (Maples et al., 1977). Research in Arizona has indicated that the decline could be reversed by N fertilization through a drip system (Pennington & Briggs, 1983). This observation supports the hypothesis that depletion of soil N leads to a decline in plant N concentration.

The level of N in soil affects N levels in cotton plants (Amer & Abuamin, 1969; Baker et al., 1972; Cope, 1984). However, Bassett et al. (1970) found that total N uptake was not effected by rate of N application (45 and 134 kg of N/ha). Grimes et al. (1973) reported that irrigation enhanced N uptake and that coarse-textured soils were less responsive to in-season applications than were fine-textured soils. Maples et al. (1977) also found that irrigation or rainfall may increase petiole NO_3. Time of application was also investigated by MacKenzie et al. (1963), who found that petiole NO_3 responded to fertilizer applied in either July or August. Petiole NO_3 response was less to a July application (as the plants were entering cutout) than to a August application (as the plants were leaving cutout). Other research found that a mid-season application of N fertilizer had little effect on petiole NO_3 (Oosterhuis & Morris, 1979). The form of N used as a fertilizer has been investigated for its effect on petiole NO_3 concentration (Amer & Abuamin, 1969). Calcium nitrate was found to be the most effective, followed by NH_4-NO_3, then urea. Nitrogen volatilization was suggested as a reason for the relative efficiency of the fertilizers.

Soil salinity affects N uptake and growth of cotton. Work with cotton in solution culture (Pessarkli & Tucker, 1985) found a decrease in N uptake with cotton grown under high-salinity stress (-1.2 MPa) but an increase under mild-salinity stress (-0.4 MPa). These researchers also found that plant growth and dry matter production were influenced to a greater extent than was N uptake.

Spatial variability of soil and plant N levels has recently been investigated in a statistical fashion. Research in Arizona (Tabor et al., 1984) found that neither soil nor petiole NO_3 were randomly distributed in a cotton field but were related to the direction of cultural practices such as irrigation and row orientation. In a companion paper (Tabor et al., 1985), petiole NO_3 was most closely related to soil texture (i.e., clay percentage) and poorly related to soil NO_3. Thus, in order for sampling to be representative, samples should be collected as far apart from each other as possible and still within the defined statistical range yet not within the same row.

3. Plant Factors

Plant factors such as variety, plant part, and growth stage may affect nutrient concentration. The often cited work of Joham (1951) found the petiole of the most recently mature, fully expanded leaf to be the most appropriate tissue for nutrient analysis. The work of Tabor et al. (1984) has confirmed these results and has added that if the age of an individual petiole is questionable, it is best to sample the older one. Others have disagreed on the usefulness of petiole monitoring for N in cotton. Most of the disagreement has come from rainfed areas of the USA (Baker et al., 1972; Sunderman et al., 1979; Touchton et al., 1981). Work from Rhodesia (Zimbabwe) also concluded that petiole monitoring was not useful (Oosterhuis & Morris, 1979). These researchers were unable to consistently correlate yield and petiole NO_3 levels. Where a correlation existed, there was insufficient time to correct the N deficiency. Oosterhuis and Bate (1983) found that NO_3 reductase activity of the most recently mature, fully expanded sympodial leaf was a better indicator of N status than was petiole NO_3. In trials that looked at the difference between petioles and blades of the most recently mature, fully expanded leaf, Zelinski (1986) compared leaf NO_3, NH_4, and total N levels as well as petiole NO_3 and NH_4 levels to lint yield. None of the leaf N measurements were related to yield, and petiole NO_3 was closely related in only one year out of two. Unpublished work by the same author (L.J. Zelinski, Univ. of California Farm Advisor, Fresno County) found no relationship between leaf amino N and yield.

Research not directly involved in N fertilization has provided information on the N levels in other cotton plant tissues and has shown that N levels in vegetative parts decline as the season progresses, whereas N levels in seeds increase (Bassett et al., 1970; Elmore et al., 1979; Oosterhuis et al., 1983). The work by Elmore et al. (1979) found that timing of N application (either 100% preplant or 50% preplant and 50% sidedressed, or 100% sidedressed) had no effect on final seed N concentration.

Petiole NO_3 concentrations have been found to decrease up the stem (Joham, 1951; Pennington et al., 1983; Woon, 1977). The work by Tabor et al. (1984) indicates that the greatest variability of petiole nitrate occurs in immature leaves in the upper portion of the plant.

The effect of plant density on N levels is not clear. Oosterhuis and Morris (1979) and Grimes et al. (1973) showed little effect from plant population, whereas work by Sunderman et al. (1979) showed a linear decrease in petiole NO_3 as the number of cotton rows per 102-cm bed increased from one to four. Wanjura and Sunderman (1976) concluded that row spacing had little effect on plant N levels.

There are varietal differences in cotton N concentration (Oosterhuis & Morris, 1979; MacKenzie et al., 1963), but the effects of plant and soil factors are similar.

B. Phosphorus

1. Soil Factors

Phosphorus concentration in cotton plants may be increased by P fertilization (Cope, 1984; Khasawneh & Copeland, 1973; Sharpley & Reed, 1982; Stelley & Morris, 1953). In greenhouse sand culture trials, an increase in P in the solution had a direct effect on the level of P in the plants (Ergle & Eaton, 1957; Joham, 1951). However, other research has not been able to demonstrate this relationship (Clark, 1964; Janat & Stroehlein, 1986; Kapp et al., 1953).

Depth of P fertilizer placement did not have a consistent effect on P concentration (Kapp et al., 1953), and method of application (either drilled or broadcast) did not affect P concentrations (Stelley & Morris, 1953). Stelley and Morris (1953) found that liming acid soils had no effect on P concentration in cotton.

Maples et al. (1977) indicate that as petiole nitrate levels decline, petiole P concentrations increase if there is no water stress. However, in water-stressed plants, petiole P declines with petiole NO_3. Sharpley and Reed (1982) found that in water-stressed cotton, the rate of leaf P decline was slower than in nonwatered-stressed plants.

Drip irrigation could be used to apply P fertilizer throughout the season. This is not common, and research by Janat and Stroehlein (1986) found that P concentration of petioles was not increased by P fertilization through the drip system. Further, they found that the amount of P uptake by furrow and drip-irrigated cotton plants was similar.

The response of P concentration in cotton in N fertilization has been inconsistent. This response has been found to decrease (Joham, 1951), increase (Zelinski, 1986), or not affect (Clark, 1964; Zelinski, 1986) P concentration. Joham (1951) also found that K fertilization decreased petiole P concentrations. The interaction between P and Zn fertilization is well known, and Janat and Stroehlein (1986) indicated a possible reduction in petiole P by Zn fertilization.

2. Plant Factors

A decrease in P concentration of cotton tissues throughout the season has been found and is considered normal (Bassett et al., 1970; Janat & Stroehlein, 1986; Sharpley & Reed, 1982); however, exceptions have been reported (Maples et al., 1977). In contrast to the usual decline, P level in seeds increased over time (Bassett et al., 1970). From the information provided by Bassett et al. (1970) and Kapp et al. (1953), cotton tissue can be listed in order of increasing P concentration: roots, stems, mature leaves, immature leaves, fruiting parts, and finally seed.

Root growth and P uptake were found to be closely related (Khasawneh & Copeland, 1973). The amount of P taken up per unit length of root was the same regardless of P concentration in solution.

C. Potassium

1. Soil Factors

Soil K levels have been found to have a direct effect on K concentration in vegetative structures (Bennett et al., 1965; Cope, 1984). Fertilization with K has also been shown to increase K concentration in cotton (Fullmer & Stromberg, 1964; Hsu et al., 1978; Joham, 1951; Page et al., 1963). Cassmen et al. (1986) suggest that the level of NH_4OAc extractable K at the 20- to 40-cm depth is a more reliable predictor of available K than is extractable K in the 0- to 20-cm depth.

Method of fertilizer application was investigated by Fullmer and Stromberg (1964). On soils with low cation exchange capacity (CEC), broadcasting the fertilizer produced higher petiole K levels than did banding, but on soils with high CEC, drilling was more effective. Fertilizer application anytime before peak bloom was acceptable.

Substantial research has not been done on the effects of other fertilizer on K levels in cotton. Work by Zelinski (1986) showed no effect on petiole or blade K concentration as a result of N fertilization. Joham (1951) also found no effect from either N or P fertilization.

2. Plant Factors

Throughout the season, K concentration in cotton tissue declines in vegetative plant parts (Stromberg, 1960; Joham, 1951). Bassett et al. (1970) found that K concentration in cottonseed was constant, and Bennett et al. (1965) indicated that K concentration of reproductive structures (squares and bolls) did not change significantly throughout the season.

Differences in K concentration due to variety have been investigated (Weir et al., 1986; Page et al., 1963). There are few differences in either concentration or pattern throughout the season.

D. Secondary Nutrients

Research on secondary nutrients is considerably less than on N, P, and K. Therefore, our information is incomplete and many areas of potential research are available.

1. Calcium

Calcium availability in soils is most frequently related to the pH of the soil. If soil pH is low, there is a possibility of Ca deficiency. However, toxic levels of Al, H, Mn, or Fe are more likely. There has been little research involved with Ca fertilization that was not directly related to soil pH modification, but we can conclude that most soil conditions do not limit the Ca uptake by cotton.

Fertilization with other nutrients may have an effect on the concentration of Ca in cotton leaf blades or petioles. Fullmer and Stromberg (1964) and Joham (1951) showed that K fertilization reduced the Ca level in cotton petioles. Joham (1951) indicated that fertilization with P increased Ca concentration. Zelinski (1986) found that N fertilization either decreased or had no effect on Ca concentration in cotton leaf blades collected at peak bloom.

McHargue (1926) analyzed mature cotton plants in Mississippi and found the following Ca concentrations (g kg^{-1}) leaves (44), stems (14), kernels (1.9), hulls (1.4), and fiber (1.3). Research by Fullmer and Stromberg (1964) found that Ca concentration remained fairly uniform throughout the season.

2. Magnesium

Magnesium deficiencies can occur in the southeastern USA (Gheesling & Perkins, 1970). The same authors, working with solution culture, indicated that increases in substrate Mg increase the Mg level in cotton blades, petioles, and stems. They also found that Mg concentration in leaf blades that were adequately supplied with Mg remained fairly constant throughout the experiment (i.e., at least 100 d after planting). The Mg concentration in petioles and stems varied to greater extent than in blades, and the authors indicated that the blade was probably the best tissue for determining deficiency. Fullmer and Stromberg (1964) found that Mg concentration in petioles increased throughout the season.

The effects of N, P, and K fertilization on the Mg levels in blades or petioles have been investigated. Zelinski (1986) found that N fertilization increased leaf blade Mg in one year but not in the next. Joham (1951) found that N fertilization increased Mg levels, that P had no effect on Mg levels, and that K fertilization decreased leaf Mg levels. Fullmer and Stromberg (1964) found that K fertilization decreased the Mg concentration of cotton petioles.

3. Sulfur

In soils, available S comes from the decomposition of organic matter, soluble sulfate, and from release of sulfates that have been adsorbed to the mineral fraction of the soil. Fertilization with S-containing materials has been shown to increase the levels of S in leaf blades (Barton et al., 1982) and in both leaf blades and petioles (Jordan, 1964). In a solution culture experiment, Gaines and Phatak (1982) showed that as the substrate level of S increased, there was an increase in total plant S.

Research by McHargue (1926) found the following S concentration (g kg^{-1}): leaves (22), stems (4.3), kernels (3.6), and hulls (0.38).

The effect of other fertilizer nutrient additions on the S concentration has not been widely studied. Zelinski (1986) found that N fertilization slightly lowered the sulfate-S levels in leaf blades in one year but had no effect in the other year of the study.

E. Micronutrients

Field research on micronutrient concentration in cotton tissue is difficult due to soil variability, and much of the information has been developed using solution culture. Therefore, much of the interpretation relies on extrapolation of field conditions.

Increases in substrate level generally increased the levels of nutrients in cotton tissue. This increase was demonstrated for Zn by El-Gharably and Knezek (1980), Joham and Rowe (1975), Ohki (1975b) and Rehab and Wallace (1978), for Mn by Foy et al. (1969), Gheesling and Perkins (1970), and Ohki (1974), for B by Ohki (1974) and Oertli and Roth (1969), and for Mo by Amin and Joham (1960). Other research by Amin and Joham (1958) found a good correlation between water extractable Mo in soils and tissue concentration of Mo. Foy et al. (1969) also found in soils that NH_4OAc-extractable Mn was not related to plant Mn concentration.

The effect of liming on the concentration of nutrients in cotton tissues indicated that liming a soil from pH 4.4 to 7.3 decreased the Mn concentration but had no effect on Fe concentration (Hati et al., 1979). Scott et al. (1975) found that liming reduced that B concentration in cotton 30- to 65-d old.

The effect of other nutrients has received little study, but a few relationships have been found. Ohki (1974) found that high levels of B slightly increased the Mn concentration, but that B level had little effect on the concentrations of Zn, Cu, or Fe (Ohki, 1975a). Rehab and Wallace (1978) found no effect on Zn levels by increasing substrate levels of Cu and Mn, but that higher levels of Co decreased Zn concentration.

Unlike the macronutrients, the concentration of micronutrients can remain fairly constant throughout the season. Joham and Rowe (1975) found that Zn concentration in cotton leaf tissue was 75, 84, and 68 mg kg^{-1} at 54, 72, and 130 d after planting, respectively. Gheesling and Perkins (1970) found that Mn concentration in leaves remained between 70 and 90 mg kg^{-1} for the first 120 d after planting. Total plant B concentration also remained stable (i.e., 9–12 mg kg^{-1}) in cotton 30- to 65-d old (Scott et al., 1975).

Different varieties have not been shown to affect the concentration of nutrients. This was shown to be the case for Zn by Rehab and Wallace (1978), for Mn by Foy et al. (1969), and for Mo and Mn by Bhatt and Appukuttan (1982).

III. USE AND INTERPRETATION OF COTTON TISSUE ANALYSIS

A. Introduction

The interpretation of nutrient uptake and foliar analysis has been approached differently in the humid and semiarid regions of the USA. Leaf blades are taken for analysis in the humid region while petioles are taken in the semiarid area. The less-determinate varieties grown in desert areas lend themselves to diagnosis and treatment during the growing season. On the other hand, most nutritional problems diagnosed in more determinate cotton varieties of the Southeast cannot usually be corrected during the same growing season due to their short effective bloom and boll-setting period. A prognosis for seasonal correction of soil fertility difficulties is often possible with irrigated cotton in arid regions because of either a longer bloom period or, in some areas, two distinct bloom periods. Because of the predictable desert environment, determination of transitory nutrients (as found in the petiole) has proven successful. However, petiole sampling programs have been adapted to some humid areas in Arkansas and Georgia (W.E. Sabbe, 1985, personal communication). The leaf blade has been selected for analysis where the seasonal and daily climate has proved to be less predictable. Weather fluctuations affect nutrient content of leaf blades less than nutrient content of petioles. Because of the differences in culture, variety, and plant part between the cotton-growing areas, the discussion on plant analysis has been divided into petiole and leaf blade analysis.

B. Diagnosis

1. Leaf Blade Analysis

Perhaps the greatest use of leaf blade analysis is to judge the efficiency of a fertilizer practice, since leaves are accumulators of nutrients and tend to be less affected by climate and seasonal changes than petioles. The more stable nutrient concentration enables the producer to sample his crop in a more relaxed manner than would be possible with petioles because of the less rapid changes in concentration of nutrients in the leaf blades.

Concentration of nutrients in leaf blades is affected by sampling dates or stage of growth. Whole plant analyses show that early uptake of nutrients is greater in proportion to the dry weight produced (Table 18–2). The top leaf and top petiole differ in their accumulation of Ca and Mg. Whereas the level of these two nutrients decreased in the petioles with time, it increased in the leaves. A smaller reduction occurred with time in the boll compared with other plant parts. Bolls are becoming larger in the intervals between sampling, and the accumulation of K during this period resulted in lower concentration elsewhere in the plant. Potassium fertilizer increased the level of K in all plant parts, and the magnitude of the increase at the various sampling dates was dependent upon the fertilizer rate (Table 18–5).

It can be speculated that nutrient concentration ratios of top to bottom leaves may be better indicators of plant nutrient status than concentration in a single leaf from either position. Such ratios should be closer to unity in normal plants than in those with a deficiency, since translocation of nutrients to points of active growth in deficient plants is expected to result in wider ratios (Sabbe, 1974).

Analysis of cotton leaves collected by surveys in Arkansas and Georgia and from research plots in Alabama indicated fair agreement for most

Table 18-5. Effect of K fertilizer levels on K changes in plant parts (Sabbe & MacKenzie, 1973).

Plant part	Date	kg K ha^{-1}					
		0	69	138	276	414	552
		g K kg^{-1}					
New growth and leaves	24 July	16.2	19.4	22.0	24.7	28.5	28.3
	22 Aug.	8.5	11.5	16.9	21.2	21.1	24.7
Top mature leaves	24 July	10.1	14.8	19.4	25.8	27.6	31.4
	22 Aug.	6.6	9.9	14.4	16.0	20.7	29.0
Old leaves plus petioles	24 July	11.3	23.0	32.3	41.9	41.8	47.0
	22 Aug.	7.7	13.9	15.8	31.9	35.0	40.9
Stems	24 July	12.4	19.4	26.2	30.4	32.8	34.9
	22 Aug.	3.7	8.3	12.1	16.5	20.0	17.2
Bolls	24 July	18.8	21.2	23.2	24.3	25.9	25.2
	22 Aug.	15.0	17.5	19.2	22.0	22.0	22.5
Percentage decrease (avg.) between 24 July and 22 Aug.		40	38	36	27	24	19

nutrients (Table 18-6). However, Cope (1984) indicated that sampling over several years may be necessary to determine the variation among locations and years. Therefore, the paired plot technique (good and bad areas within a field) may improve the diagnosis of a plant analysis.

Sufficiency ranges for cotton nutrients in Arkansas and Georgia are given in Table 18-7. In spite of differences in time and methods of sampling, the two values are quite similar. Sabbe et al. (1972) reported that differences occurred in leaf Fe and Mn concentrations between cotton grown on alluvial soil and that grown on loessial soil. Other nutrient concentrations were similar between the two soil groups in Arkansas.

Kallinis and Kouskoleka (1967) reported that the Mo concentration in mature leaves of cotton grown in nutrient culture needs to be above 1.9 mg kg^{-1}. Leaves at this level showed incipient deficiency symptoms whereas leaves containing 2.4 mg kg^{-1} were healthy. Kouskoleka and Kallinis (1968)

Table 18-6. Nutrient concentrations of cotton leaves within the humid cotton growing region.

Date	Samples	N	P	K	Ca	Mg	Mn	Fe	B	Cu	Zn
		g kg^{-1}					mg kg^{-1}				
				Arkansas (Sabbe et al., 1972)							
23 July 1968	384	40.1	5.0	13.7	25.3	5.5	74	88			56
5 Aug. 1970	259	33.3	5.0	15.8	28.4	7.2	180	177			61
12 Aug. 1969	223	30.2	3.7	13.8	29.9	7.1	126	151			46
				Georgia (Anderson et al., 1971)							
June–July 1971	284	41.8	3.3	20.9	25.6	5.2	225	139	27	9	24
				Alabama (Cope, 1984)							
Early bloom 1971	384	44.4	3.7	14.0	21.7	8.1	160	71	40		24

Table 18-7. Selected state sufficiency ranges for nutrients in cotton leaves.

Nutrient	Arkansas†	Georgia‡
	$g\ kg^{-1}$	
N	30.0–43.0	35.0–45.0
P	3.0–6.4	3.0–5.0
K	9.0–19.5	15.0–30.0
Ca	19.0–35.0	20.0–30.0
Mg	3.0–7.5	3.0–9.0
S	3.0–7.5§	2.5–8.0
	$mg\ kg^{-1}$	
Mn	30–300	25–350
Fe	30–300	50–250
B	20+	20–60
Cu	NA¶	5–25
Zn	20–100	20–200

† Sabbe and MacKenzie (1973.)
‡ Plank (1979).
§ W.E. Sabbe (1985, unpublished data).
¶ NA = Not available.

found the critical Fe concentration to be 85 to 112 and 57 to 88 mg kg^{-1} in young and mature cotton blades, respectively. These values were obtained in solution culture and are considerably higher than the lower end of the sufficiency ranges for this nutrient in Georgia (50 mg kg^{-1}) and Arkansas (30 mg kg^{-1}). Another possible explanation of this lack of agreement is a differential variety response to substrate Fe concentration similar to that demonstrated for Mn. Foy et al. (1969) concluded that cotton varieties selected from a high-Mn soil exhibit a greater tolerance for Mn than cotton developed on a low-Mn soil. For example, a cotton variety in Arkansas showed an 8% decline in yield when 3700 mg kg^{-1} of Mn was present in leaves, whereas the yield of a western cotton was reduced 13% when the leaves contained a somewhat lower level of 2456 mg kg^{-1}. At 100% relative yield, the Mn concentrations were 243 and 162 mg kg^{-1} for the Arkansas and western varieties, respectively. In nutrient culture, Gheesling and Perkins (1970) found leaf blades to be the best indicator of the Mn status. Manganese deficiency was evident when the level was <15 mg kg^{-1}. Joham and Amin (1967) classified cotton as a highly tolerant species based on its tissue Mn level.

The critical B concentration in young cotton leaves was about 15 mg kg^{-1}, whereas it was 5 to 10 mg kg^{-1} higher in mature leaves (Murphy & Lancaster, 1971). Miley et al. (1969) found that B-deficiency symptoms occurred in later growth stages when the concentration was 18 mg kg^{-1} in petioles sampled at the early bloom and early boll stages and 13 mg kg^{-1} in leaf blades at early boll stage. Work in Zambia (Rothwell et al., 1967) indicates a leaf blade concentration of 28 mg kg^{-1} B to be sufficient and a 5 mg kg^{-1} B level to be deficient. Maximum growth was obtained in a nutrient culture containing 10 mg kg^{-1} B, whereas a 30 and 40 mg kg^{-1} B solution concentration reduced the growth to 40 and 10% of the maximum, respectively (Oertli & Roth, 1969). The B concentrations (plant tops) were approximately 160, 250, and 420 mg kg^{-1}, respectively. Ohki (1974) found that low concentration of substrate B increased the critical level of Mn in older leaf blades.

According to Elliot et al. (1968) problems with secondary and micronutrient deficiencies or toxicities are not common in cotton production. Studies in Georgia (Anderson & Harrison, 1970) show that plant tissue micronutrient (Mn, Fe,Zn, B, Cu, and Mo) concentration variations occurred among cotton varieties at each of three soil series locations as well as among locations. No correlation was found among total soil Mn, total soil Fe, or soil pH and Fe and Mn tissue levels.

Jordan (1964) increased the S concentration in cotton leaves and petioles by S fertilization. He concluded that a S concentration <2 g of S kg^{-1} in either the leaf and petioles at midseason was deficient. In Arkansas, increases in leaf blade S were obtained with S fertilizer (Maples & Keogh, 1974). However, no yield increases were obtained when leaves contained at least 4 g of S kg^{-1} in this 3-yr study. Gheesling and Perkins (1970) stated that below 2 g of S kg^{-1} in the leaf blades deficiency symptoms would occur.

2. Petiole Analysis

Plant nutrient concentrations considered sufficient for optimum production at various cotton growth stages have been reported. Determination of P and K tissue levels appears to be most helpful for predicting the status of these nutrients in the following cotton crop (Fullmer & Stromberg, 1964). However, N analysis of cotton petioles does offer the most promise by forewarning the grower of imminent N deficiencies or surpluses. Both of these conditions can be determined early in the season and prevented, usually by adjustment of the current N fertilizer program. This approach has been used in the southwestern irrigated desert area where a long seasonal growth period allows the practice to be effective. A late-season petiole analysis can also be used to evaluate the current fertilization program and aid in plans for the next year.

a. Nitrogen. The pattern of seasonal NO_3-N level in petioles in Arkansas is shown in Fig. 18–3. For optimum production, there is a characteristic level of NO_3-N in cotton petioles of each variety and at each location. Nitrogen fertilization programs can be adjusted to maintain the recommended NO_3-N concentration.

Joham (1951) recognized that petioles containing <55 mg of NO_3-N kg^{-1} were associated with lower cotton yields in Texas. MacKenzie et al. (1963) defined the needed NO_3-N concentrations in petioles as 16 000, 8000, and 2000 mg kg^{-1} for plants in early, mid-, and late-blossom stage of growth, respectively, for the lower desert areas of California. In Arizona, Tucker (1963) recommended levels of 15 000, 12 000, 6 to 8000, and 4000 mg of NO_3-N kg^{-1} for the respective growth stages of first square, first flower, first boll, and first open boll.

Under Arkansas growing conditions, the sufficiency range for petiole NO_3-N was 12 to 28, 8 to 24, 4 to 15, and 2 to 6 g kg^{-1} for 1 July, 15 July, 1 August, and 15 August, respectively (Maples et al., 1977). These authors and Amer and Abuamin (1969) suggested that the petiole level as a criterion

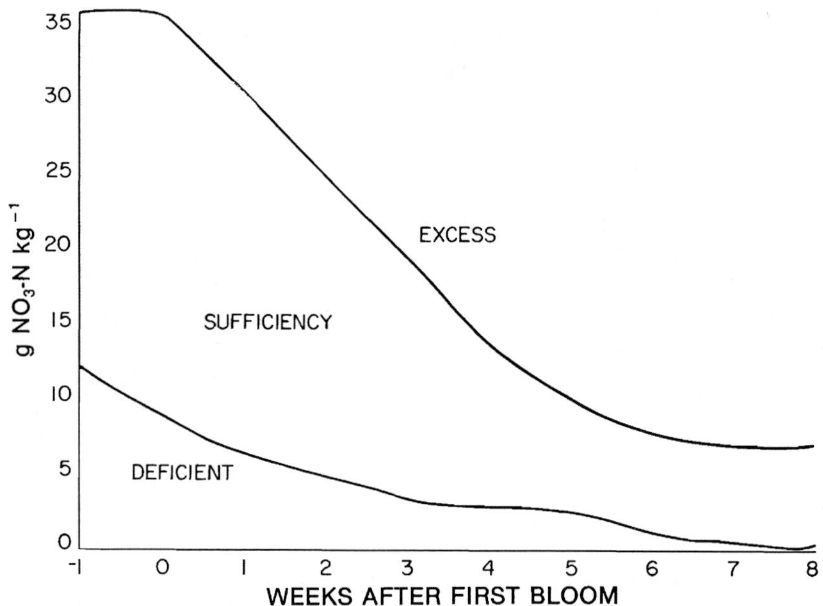

Fig. 18-3. Sufficiency range for cotton petiole NO_3 in Arkansas (Maples et al., 1977).

for plant N status was useful only if the crop was not affected by uncontrollable factors such as salinity, drought, insect, and diseases.

Regardless of whether petioles are used for diagnostic or monitoring purposes, sampling variability has to be considered (Tabor et al., 1984). In commercial, irrigated fields, petioles from the first mature leaf were best, but if the maturity of the leaf is in question, the petiole from the next older leaf should be taken.

Grimes et al. (1973) designed a model for estimating the desired NO_3-N levels in cotton petioles using N fertilization, time of season, and water management. Other authors reported that petiole NO_3 differ among cultivars, rowing spacings, and N application rate (Sunderman et al., 1979); however, no differences were found when foliar N was applied (Jenkins et al., 1982). Barton et al. (1982) reported that the concentration of plant S affected both the leaf N and petiole NO_3 levels. Therefore, knowledge of the plant S status is necessary prior to interpretation of petiole NO_3 for cotton N status.

b. Phosphorus. Concentration of P in cotton has not been studied as completely as that of N and K because many of the cotton soils contain ample available P. Consequently, fertilization with this nutrient has not always raised yields. Nevertheless, nutrient status levels have been established for P (Jones & Bardsley, 1968). In the southwestern USA desert area, petioles will contain, throughout the season, at least 0.15% P on a dry weight basis. A level of 1000 mg kg^{-1} petiole P until after peak bloom is recommended as a sufficiency level by Basset and Mikkelsen (1962). Joham (1951) suggested

0.16% P on a fresh petiole weight basis as the critical level needed for a 90-d-old, field-grown plant. Using field test methods, Wickstrom et al. (1964) suggested that, for maximum production, P levels should be high during early bloom and boll development, whereas medium levels are satisfactory during the late growth periods. They also suggested using the petiole of the basal leaf until early bloom and the petiole of the first mature leaf after bloom.

c. **Potassium.** Joham (1951) has reported the K concentration in dried petioles of high-yielding greenhouse cotton plants at 60, 90, and 145 d after emergence to be 6.0, 4.1, and 3.0%, respectively. In Brazil, de Mello et al. (1960a, b) found that, during the early bloom stage (90 d after planting), K levels of 1.8% in petioles were necessary for maximum production. In California, Fullmer and Stromberg (1964) concluded that petioles should contain at least 4.5% K at early bloom and 1% K at 130 d after planting. Using field test techniques, Wickstrom et al. (1964) recommended high K levels in petioles during the early bloom and boll development periods; however, a medium level was adequate for later growth periods.

C. Monitoring Nutrient Uptake

Enough information is available in certain cotton-producing areas for the interpretation of cotton plant analysis. Nutrient levels corresponding to deficiency and optimum growth conditions have been established for the most common limiting nutrients (N, P, and K). However, interpretation of plant analysis results is not entirely straightforward because of interactions among these and other nutrients (Belousov & Sdizhenskaja, 1954; Dastur, 1959; Hardy, 1962; Joham, 1951).

Characteristically, the N, P, and K levels in cotton petioles or blades will be high during the early stages of growth and then decrease during fruiting and maturation (Fullmer & Stromberg, 1964; Joham, 1951; MacKenzie et al., 1963). For this reason, plant analysis levels for various growth stages, such as early, mid-, and late-bloom, are important because they will indicate whether the crop has sufficient nutrients for continued optimum production. Since cotton production involves the growth of many varieties in different soil, climatic, and management environments, it is unreasonable to expect sufficiency levels for one cotton-growing area to be applicable to another region. Differences in the nutrient sufficiency levels necessary for optimum cotton yields are determined mainly by local conditions; they must, therefore, be measured locally.

Plant analysis has been used by cotton growers in the desert areas of the southwestern states of the USA to assist in controlling the N, nutrition of their crops. The soils of the area are typically low in N and cotton production, therefore, depends upon fertilizer N. Petiole analysis indicates N needs approximately 2 wk before visual deficiencies appeared (Gardener & Tucker, 1967). Depletion of N at the end of the season is desirable for easier defoliation and harvesting. Two petiole samples are recommended as a guide for N application, the first at early fruiting stage, and the second at early flower stage.

The use of petiole NO_3 to ascertain the N nutritional status of the cotton plant has been successful in irrigated fields in semiarid regions where soil N is a limiting factor, but it has not proven as successful in humid areas. Maples et al. (1977) developed a petiole monitoring program that has been used in Arkansas since 1977. This program is based not only on the petiole analysis for NO_3 but also on the petiole analysis for P, knowledge of the plants fruiting status, the soil moisture level, and previous N fertilization. Maples and Miley (1978) described the program as a N management system in which the N inputs are regulated throughout the season (10 weekly petiole samples starting 1 wk before fruit bloom) and the goal is to maximize yields in favorable growing seasons and minimize problems in unfavorable seasons.

Researchers in other states or countries have not been as enthusiastic regarding petiole nitrates as an indicator of plant N status. In research studies where no yield response was obtained with fertilizer N (Oosterhuis & Morris, 1979), where yields were also effected by rainfall (Touchton et al., 1981), or where yield did not correlate with petiole NO_3 (Koli & Morrill, 1976) a plant NO_3 monitoring program was not recommended. Additionally, Oosterhuis and Bate (1983) suggested that the NO_3 reductase activity may be a convenient, reliable, and sensitive indicator of plant N status. Researchers in Mississippi reported that while no consistent relationship existed between petiole NO_3 and maximum yield, the petiole NO_3 could be used to identify N deficiency/excess during the fruiting period (Thom & Spurgeon, 1982). The recommended rate (100–115 kg of N ha^{-1}) did produce optimum petiole NO_3 concentrations under the environmental and soil conditions of the study. For cotton grown under irrigation on the Texas high plains, Sunderman et al. (1979) concluded that within the same year and within a location that the petiole N level could be an indicator of both the N fertilizer rate and the N status of the plant.

Growers interested in their fertilizer programs are starting plant analysis operations of their own or are subscribing to the services of commercial laboratories (Tabor et al., 1984). Phosphorus analysis may be included. Some fertilizer dealers also are offering plant analysis as a part of their service to customers. Either field tests or laboratory analysis may be offered in these cases depending upon the individual dealer involved. A big factor in the success of these commercial plant analysis operations is the turn-around service. This allows the grower to immediately evaluate the fertility status or trend of his crop and to make adjustments when necessary to ensure a well but efficiently fertilized crop.

D. Fertilizer Recommendations

The response of cotton to fertilizer follows that of a nonleguminous crop (i.e., the amount of N sets the level of yield). Engelstad and Terman (1966) reported that significant cotton yield responses in Mississippi to N, P, and K fertilization occurred at an application frequency of 70, 5, and 20%, respectively. The University of Arkansas Soil Testing Program recommended N, P, and K fertilizer on 98, 45, and 67%, respectively, of the 46 000 soil samples from cotton fields submitted for fertilizer recommendations during 1980 to 1984 (W.E. Sabbe, 1985, unpublished data). Maples and Frizzell (1985) concluded that the yield potential for different cotton cultivars was a valid parameter for modifying general fertilizer N recommendations. Additionally, excessive N rates increased the management needed to avoid yield losses, and that made yield losses on non-uniform fields unavoidable because of the varying management practices needed for each soil type (Maples et al., 1987).

Since N is related directly to yield potential, the irrigated desert USA areas require about 150 kg of N ha^{-1}, the rainfed humid southeast USA requires about 100 kg of N ha^{-1}, and the upland plain areas of Texas require about 50 kg of N ha^{-1} (Tucker & Tucker, 1968). As management practices improve and as higher yields are required, the application rate will continue to increase. However, Maples and Keogh (1971) concluded that when excess N was applied in the humid areas, not only was maturity delayed but the residual N affected yields in subsequent years.

The fertilizer rate for P and K fertilizer is determined not only by crop response but also by soil characteristics. The availability of P and K during a growing season can be estimated via a soil test method adapted for that particular soil physiographic area (Jones, 1980).

Annual P fertilizer applications up to 55 kg of P ha^{-1} are required in the southeast USA where the soils have the capacity of "fix" P. The yields in the irrigated southwest area have recommendations up to 40 kg of P ha^{-1} (Jones & Bardsley, 1968). Cope (1984) reported that a soil test for P gave a good measure of soil-available P and was correlated with the fertilizer-applied P.

The recommended rate of K fertilizer varies up to 330 kg of K ha^{-1}, the actual rate depends mainly on soil characteristics such as fixation and texture (Kamprath & Welch, 1968). A residual K study indicated a significant correlation among K fertilizer rates, soil test K, and leaf K and yield with the best response closely related to the soil test K (Cope, 1984). A characteristic of the yield response to fertilizer K is a response only to the first increment of fertilizer K (Lombin & Mustafa, 1981). However, a positive correlation between fertilizer K and petiole K concentration was observed in California (L.J. Zelinski, 1986, personal communication). With the exception of the Atlantic Coastal Plains, S deficiencies usually are not present. In that soil region, an application of about 20 to 30 kg of S ha^{-1} is recommended for a two bale crop (Standford & Jordan, 1966). Barton et al. (1982) reported that the concentration of plant S affected both the leaf N and petiole NO_3 levels. Therefore, knowledge of the plant S status is necessary prior to interpretation of petiole NO_3 for cotton N status. Miley et al. (1969) and Maples and Keogh (1974) obtained very few responses to S fertilization on alluvial soils in Arkansas.

A characteristic of yield responses to fertilizer applications was reported by Cope (1984) on two fertility studies that encompassed an 18- or 24-yr period, respectively. The 67 kg of N ha^{-1} application produced more than

90% of the top yield, whereas a soil test proved to be a reliable indicator of the P and K fertilizer needs.

The deficiency symptoms of other secondary and micronutrients are known (Marcus-Wyner & Rains, 1982); however, most production cotton fields appear to have adequate levels for optimum yields (Hinkle & Brown, 1968).

In summary, plant analysis monitoring programs have been established in both cotton-growing areas of the USA. The greatest success appears to be where some of the environmental conditions are predictable/controlled and where N management is a season-long consideration.

REFERENCES

Amin, J.V., and H.E. Joham. 1958. Molybdenum in the cotton embryo. Soil Sci. 86:293–297.

Amin, J.V., and H.E. Joham. 1960. Growth of cotton as influenced by low substrate molybdenum. Soil Sci. 89:101–107.

Amer, F., and H. Abuamin. 1969. Evaluation of cotton response to rates, sources, and timing of nitrogen application by petiole analysis. Agron. J. 61:635–637.

Anderson, O.E., and R.M. Harrison. 1970. Micronutrient variation within cotton leaf tissue as related to variety and soil location. Commun. Soil Sci. Plant Anal. 1:163–162.

Anderson, O.E., H.F. Perkins, R.L. Carter, and J.B. Jones, Jr. 1971. Plant nutrient survey of selected plants and soils of Georgia. Georgia Agric. Exp. Stn. Rep. 102.

Baker, J.M., R.M. Reed, and B.B. Tucker. 1972. The relationship between applied nitrogen and the concentration of nitrate-N in cotton petioles. Commun. Soil Sci. Plant Anal. 3:345–350 (1972).

Barton, M.J., R.G. Hanson, and U.U. Alexander. 1982. Sulphur fertilizer effect on cotton: I. Nitrogen and sulphur status and petiole nitrate-nitrogen. Commun. Soil Sci. Plant Anal. 13:819–834.

Bassett, D.M., W.D. Anderson, and C.H.E. Werkhoven. 1970. Dry matter production and nutrient uptake in irrigated cotton. Agron. J. 62:299–303.

Bassett, D.M., and D.S. Mikkelsen. 1962. Use of phosphorus on cotton. p. 31. *In* Summaries of papers. Statewide conference on role of phosphorus in crop production. Div. of Agric. Sci., Univ. of California, Davis.

Belousov, M., and Z. Sdizhenskaja. 1954. Methods of diagnosing the conditions of cotton nutrition. Khlophovodstovo 4:30–35.

Bennett, O.L., R.D. Rouse, D.A. Ashley, and B.D. Doss. 1965. Yield, fiber quality, and potassium content of irrigated cotton plants as affected by rates of potassium. Agron. J. 57:296–299.

Bhatt, J.G., and E. Appukuttan. 1982. Manganese and molybdenum contents of the cotton plant at different stages of growth. Commun. Soil Sci. Plant Anal. 13:463–471.

Cassmen, K.G., B.A. Roberts, L.R. Stallings, and T.A. Kerby. 1986. Influence of available potassium distribution in the soil profile on cotton yield in the San Joaquin Valley. p. 401. *In* J. Brown (ed.) Proc. Beltwide Cotton Production Res. Conf., Las Vegas. 4–9 Jan. Natl. Cotton Counc. of Am., Memphis.

Clark, R.E. 1964. Direct and residual effects of two mixtures of nitrogen and phosphorus upon the growth of acala cotton in the Pecos Valley. Agron. J. 56:18–20.

Cope, J.T. 1984. Relationship among rates of N, P, and K, soil test values, leaf analysis, and yield of cotton at six locations. Commun. Soil Sci. Plant Anal. 15:253–376.

Dastur, R.H. 1959. The need for study of potash and phosphate requirements of the cotton plant. Indian Cotton Grow. Rev. 13:1–10.

de Mello, F.A.F. de, M.O. Brasil, H.P. Haag, and S. Arzolla. 1960a. A diagnose foliar do alfodeiro (*Gossypium hirsutum* L., Var. I.A.C. 817) no Brasil. Agros (Pelotas, Braz.) 43:233–239.

de Mello, F.A.F. de, T. Coury, M.O.C. Brasil, and E. Malavalta. 1960b. Application of foliar diagnosis to the cotton-plant. Fertilité 9:3–9.

El-Gharably, G., and B.D. Knezek. 1980. Critical levels of Zn in cotton plants. p. 166. *In* Agronomy abstracts. ASA, Madison, WI.

Elliot, J.C., M. Hoover, and W.K. Porter, Jr. (ed.). 1968. Advances in production and utilization of quality cotton: Principles and practices. Iowa State Univ. Press, Ames.

Elmore, C.D., W.I. Spurgeon, and W.O. Thom. 1979. Nitrogen fertilization increases N and alters amino acid concentration of cottonseed. Agron. J. 71:713–716.

Engelstad, O.P., and G.L. Terman. 1966. Fertilizer nitrogen: Its role in determining crop yield levels. Agron. J. 58:536–539.

Ergle, D.R., and F.M. Eaton. 1957. Aspects of phosphorus metabolism in the cotton plant. Plant Phys. 32:106–113.

Foy, C.D., A.L. Fleming, and W.H. Armiger. 1969. Differential tolerance of cotton varieties in excess manganese. Agron. J. 61:690–694.

Fullmer, F.S., and L.K. Stromberg. 1964. The use of plant and soil analysis as a guide to potassium needs for cotton in California. p. 120–129. *In* C. Bould et al. (ed.) Plant analysis and fertilizer problems IV. Am. Soc. for Hortic.-Sci., St. Joseph, MI.

Gaines, T.P., and S.C. Phatak. 1982. Sulfur fertilization effects on the constancy of the protein N:S ratio in low and high sulfur accumulating crops. Agron. J. 74:415–418.

Gardener, B.R., and T.C. Tucker. 1967. Nitrogen effects on cotton: II. Soil and petiole analyses. Soil Sci. Soc. Am. Proc. 31:785–791.

Gheesling, R.H., and H.F. Perkins. 1970. Critical levels of manganese and magnesium in cotton at different stages of growth. Agron. J. 62:29–32.

Grimes, D.W., W.L. Dickens, H. Yamada, and R.J. Miller. 1973. A model for estimating desired levels of nitrate-N concentration in cotton petioles. Agron. J. 65:37–41.

Halvey, J. 1976. Growth rate and nutrient uptake of two cotton cultivars grown under irrigation. Agron. J. 68:701–705.

Hardy, G.W. 1962. Tissue analysis of cotton. Better Crops Plant Food 46(3):1–5.

Harris, W.D. 1960. Nutrient concentration response of cotton leaf blades and petioles to changes in the substrate. M.S. thesis. Univ. of Arkansas-Fayetteville.

Hati, N., T.R. Fisher, and W.J. Upchurch. 1979. Liming of acid soil: II. Effect on plant available manganese and iron. J. Indian Soc. Soil Sci. 27:394–398.

Hinkle, D.A., and A.L. Brown. 1968. Secondary and micronutrients. p. 281–320. *In* F.C. Elliot et al. (ed.) Advances in production and utilization of quality cotton: Principles and practices. Iowa State Univ. Press, Ames.

Hsu, H.H., J.D. Lancaster, and W.F. Jones. 1978. Potassium concentration in leaf-blades and petioles as effected by potassium fertilization and stage of maturity of cotton. Commun. Soil Sci. Plant Anal. 9:265–277.

Janat, M.M., and J.L. Stroehlein. 1986. Yield, dry matter production and phosphorus uptake of drip irrigated cotton. J. Fert. Issues 3:124–128.

Jenkins, J.N., J.R. Nichols, Jr., J.C. McCarty, Jr., and W.L. Parrott. 1982. Nitrates in petioles of three cottons. Crop Sci. 22:1230–1233.

Joham, H.E. 1951. The nutritional status of the cotton plant as indicated by tissue tests. Plant Phys. 26:76–89.

Joham, H.E., and V. Rowe. 1975. Temperature and zinc interactions on cotton growth. Agron. J. 67:313–317.

Jones, U.S., and C.E. Bardsley. 1968. Phosphorus nutrition. *In* F.C. Elliot (ed.) Advances in production and utilization of quality cotton: Principles and practices. Iowa State Univ. Press, Ames.

Jones, J.B. 1980. Handbook on reference methods for soil testing. Counc. on Soil Testing and Plant Anal., Athens.

Jones, J.W., J.D. Hesketh, E.J. Kamprath, and H.D. Bowen. 1974. Development of a nitrogen balance for cotton growth models: A first approximation. Crop Sci. 14:541–546.

Jordan, H.V. 1964. Sulfur as a plant nutrient in the southern United States. USDA Tech. Bull. 1297. U.S. Gov. Print. Office, Washington, DC.

Kallinis, T.L., and H.V. Kouskoleka. 1967. Molybdenum deficiency symptoms in cotton. Soil Sci. Soc. Am. Proc. 31:507–509.

Kamprath, E.J., and C.D. Welch. 1968. Potassium. p. 256–280. *In* F.C. Elliot et al. (ed.) Advances in production and utilization of quality cotton: Practices and principles. Iowa State Univ. Press, Ames.

Kapp, L.C., R.J. Hervey, J.R. Johnston, and J.C. Smith. 1953. Response of evergreen sweet clover and cotton to phosphorus application on Houston black clay. Soil Sci. 75:109–118.

Khasawneh, F.E., and J.P. Coleland. 1973. Cotton root growth and uptake of nutrients: Relation of phosphorus uptake to quantity, intensity, and buffering capacity. Agron. J. 37:250–254.

Koli, S.E., and L.G. Morrill. 1976. Influence of nitrogen, narrow rows, and plant population on cotton yield and growth. Agron. J. 68:897–901.

Kouskoleka, H.V., and T.L. Kallinis. 1968. Iron deficiency in cotton in relation to growth and nutrient balance. Soil Sci. Soc. Am. Proc. 32:253–257.

Lombin, G., and S. Mustafa. 1981. Potassium response of cotton on some inceptisols and oxisols of northern Nigeria. Agron. J. 73:724–729.

MacKenzie, A.J., W.K. Spencer, K.R. Stockinger, and B.A. Krantz. 1963. Seasonal nitrate-nitrogen content of cotton petioles as affected by nitrogen applications and its relationship to yield. Agron. J. 55:55–59.

Maples, R., and M. Frizzell. 1985. Effects of varying rates of nitrogen on three cotton cultivars. Arkansas Agric. Exp. Stn. Bull. 882.

Maples, R., and J.L. Keogh. 1971. Cotton fertilization studies on loessial plains soils of eastern Arkansas. Arkansas Agric. Exp. Stn. Rep. Ser. 194.

Maples, R., and J.L. Keogh. 1974. Fertilization of cotton with potassium, magnesium, and sulfur on certain delta soil. Arkansas Agric. Exp. Stn. Bull. 787.

Maples, R., J.G. Keogh, and W.E. Sabbe. 1977. Nitrate monitoring for cotton production on a loring-calloway silt loam. Arkansas Agric. Exp. Stn. Bull. 825.

Maples, R.L., and W.N. Miley. 1978. Impact of nitrate monitoring on cotton production. Ark. Farm Res. 27(2):8.

Maples, R.L., W.E. Sabbe, R.F. Ford, M. Appleberry, and R.J. Mahler. 1987. Response of cotton to nitrogen in problem fields. Arkansas Agric. Exp. Stn. Bull. 901.

Marcus-Wyner, L., and D.W. Rains. 1982. Nutritional disorders of cotton plants. Commun. Soil Sci. Plant Anal. 13(9):685–736.

McHargue, J.S. 1926. Mineral constituents of the cotton plant. Agron. J. 18:1076–1083.

Miley, W.N., G.W. Hardy, M.B. Sturgis, and J.E. Sedberry, Jr. 1969. Influence of born, nitrogen, and potassium on yield, nutrient uptake and abnormalities of cotton. Agron. J. 61:9–13.

Murphy, B.C., and J.D. Lancaster. 1971. Response of cotton to boron. Agron. J. 63:539–540.

Oertli, J.J., and J.A. Roth. 1969. Boron nutrition of sugar beet, cotton and soybean. Agron. J. 61:191–195.

Ohki, K. 1974. Manganese nutrition of cotton under two boron levels II. Critical Mn level. Agron. J. 66:572–575.

Ohki, K. 1975a. Mn and B effects on micronutrients and P in cotton. Agron. J. 67:204–207.

Ohki, K. 1975b. Lower and upper critical zinc levels in relation to cotton growth and development. Physiol. Plant. 35:96–100.

Olson, L.C., and R.P. Bledsoe. 1942. The chemical composition of the cotton plant and the uptake of nutrients at different stages of growth. Georgia Agric. Exp. Stn. Bull. 222.

Oosterhuis, D.M., and G.C. Bate. 1983. Nitrogen uptake of field grown cotton. II. Nitrate reductase activity and petiole nitrate concentration as indicators of plant nitrogen status. Exp. Agric. 19:103–109.

Oosterhuis, D.M., J. Chipamaunga, and G.C. Bate. 1983. Nitrogen uptake of field grown cotton. I. Distribution in plant components in relation to fertilization and yield. Exp. Agric. 19:91–101.

Oosterhuis, D.M., and W.J. Morris. 1979. Cotton petiole analysis as an indicator of plant nitrogen status in Rhodesia. Rhodesia Agric. J. 76:37–42.

Page, A.L., F.T. Bingham, T.J. Ganje, and M.J. Garber. 1963. Availability of added potassium in two California soils when cropped to cotton. Soil Sci. Soc. Am. Proc. 27:323–326.

Peacock, R.H. 1960. The effect of water-solubility of phosphorus on phosphorus concentration in different parts of the cotton plant at various stages of growth. M.S. thesis. Univ. of Arkansas-Fayetteville.

Pennington, D.A., and R. Briggs. 1983. Nitrogen fertility and soil salinity in drip irrigated cotton. p. 49–50. *In* J.M. Brown (ed.) Proc. Beltwide Cotton Production Res. Conf. San Antonio, TX. 2–6 Jan. Natl. Cotton Counc. of Am., Memphis.

Pennington, D.A., J.A. Tabor, and A.W. Warrick. 1983. Variation in nitrate-nitrogen levels in cotton petioles. p. 50–53. *In* J. Brown (ed.) Proc. Beltwide Cotton Production Res. Conf., San Antonio, TX. 2–6 Jan. Natl. Cotton Counc. of Am., Memphis.

Pessarkli, M., and T.C. Tucker. 1985. Uptake of nitrogen-15 by cotton under salt stress. Soil Sci. Soc. Am. J. 49:149–152.

Plank, C.O. 1979. Plant analysis handbook for Georgia. Univ. Georgia Coop. Ext. Bull. 735.

Rehab, F.I., and A. Wallace. 1978. Excess trace metal effects on cotton 1. Copper, zinc, cobalt and manganese in solution culture. Commun. Soil Sci. Plant Anal. 9:507–518.

Rothwell, A., J.W. Bryden, H. Knight, and B.J. Coxe. 1967. Boron deficiency of cotton in Zambia. Cotton Grow. Rev. 44:23–28.

Sabbe, W.E. 1974. Nutrient profile as a determinant in soybean fertilizer studies. p. 393–402. *In* Proc. 7th Int. Colloquium on Plant Analysis and Fertilizer Problems. September. Hanover, West Germany.

Sabbe, W.E., and T.L. Hansen. 1983. Micronutrients in soybeans and cotton. Ark. Farm Res. 32(1):12.

Sabbe, W.E., J.L. Keogh, R. Maples, and L.H. Hileman. 1972. Nutrient analysis Arkansas cotton and soybean leaf tissue. Ark. Farm Res. 21:2.

Sabbe, W.E., and A.J. MacKenzie. 1973. Plant analyses as an aid to cotton fertilization. p. 299–313. *In* Soil testing and plant analysis. rev. ed. SSSA, Madison, WI.

Scott, H.D., S.D. Beasley, and L.F. Thompson. 1975. Effect of lime on boron transport to and uptake by cotton. Soil Sci. Soc. Am. Proc. 39:1116–1121.

Sharpley, A.N., and L.W. Reed. 1982. Effect of environmental stress on the growth and amounts and forms of phosphorus in plants. Agron. J. 74:19–22.

Stanford, G., and H.V. Jordan. 1966. Sulfur requirements of sugar, fiber, and oil crops. Soil Sci. 101:258–266.

Stelley, M., and H.D. Morris. 1953. Residual effect of phosphorus on cotton grown on cecil soil as determined with radioactive phosphorus. Soil Sci. Soc. Am. Proc. 17:267–269.

Stromberg, L.K. 1960. Need for potassium fertilizer on cotton determined by leaf and soil analysis. Calif. Agric. 14(4):4–5.

Sunderman, H.D., A.B. Onken, and L.R. Hossner. 1979. Nitrate concentration of cotton petioles as influenced by cultivar, row spacing, and N application rate. Agron. J. 71:731–737.

Tabor, J.A., A.W. Warrick, D.E. Myers, and D.A. Pennington. 1985. Spatial variability of nitrate in irrigated cotton. II. Soil nitrate and correlated variables. Soil Sci. Soc. Am. J. 49:390–394.

Tabor, J.A., A.W. Warrick, D.A. Pennington, and D.E. Myers. 1984. Spatial variability of nitrate in irrigated cotton: I. Petioles. Soil Sci. Soc. Am. J. 48:602–607.

Thompson, A.C., H.C. Lane, J.W. Jones, and J.D. Hesketh. 1976. Nitrogen concentration of cotton leaves, buds, and bolls in relation to age and nitrogen fertilization. Agron. J. 68:617–621.

Touchton, J.T., F. Adams, and C.H. Burmester. 1981. Nitogen fertilization rates and cotton petiole analysis in Alabama field experiments. Alabama Agric. Exp. Stn. Bull. 528.

Tucker, T.C. 1963. Soil and petiole analyses can pinpoint cotton's nitrogen needs. Cotton Gin Oil Press 64(2):36–37.

Tucker, T.C., and B.B. Tucker. 1968. Nitrogen nutrition. p. 183–211. *In* F.C. Elliot et al. (ed.) Advances in production and utilization of quality cotton: Principles and practices. Iowa State Univ. Press, Ames.

Wanjura, D.F., and H.D. Sunderman. 1976. Vegetative-fruiting development and N concentration of a dwarf determinate cotton cultivar. Agron. J. 68:744–748.

Weir, B.L., T.A. Kerby, B.A. Roberts, and R.H. Garber. 1986. Potassium deficiency syndrome of cotton. Calif. Agric. 40:13–14.

Wickstrom, G.A., N.D. Morgan, and A.N. Plant. 1964. Ask the plant in the field. Better Crops Plant Food 48(3):18–24.

Woon, C.K. 1977. Seasonal changes in stem carbohydrate and petiole nitrate in cotton. Ph.D. diss. Univ. of Arizona, Tuscon (Diss. Abstr. 78-05776).

Zelinski, L.J. 1986. Effects of varying nitrogen fertility levels on the concentration of nutrients in cotton leaf blades. p. 401–402. *In* J. Brown (ed.) Proc. Beltwide Cotton Production Res. Conf., Las Vegas. 4–9 Jan. Natl. Cotton Counc. of Am., Memphis.

Chapter 19

Plant Analysis as an Aid in Fertilizing Small Grains

D. G. WESTFALL, *Colorado State University, Fort Collins*

D. A. WHITNEY, *Kansas State University, Manhattan*

D. M. BRANDON, *Rice Experiment Station, Biggs, California*

Production constraints on producers and public concern over the environmental impact of fertilizer use emphasize the need for efficient fertilization programs. Proper fertilizer application ensures efficient plant use and minimum movement into non-target areas. Time and method of fertilizer application can greatly influence plant utilization and fate of fertilizers in the environment. In current agricultural production systems, we must integrate the public concerns over the environment, as affected by agriculture, and the producer's concern for achieving optimum economic return for his fertilizer investment. Soil testing is one diagnostic tool that has been used successfully for many years to improve fertilizer efficiency and economic return in small grains. More recently, plant analysis has been used as another tool to complement soil testing in developing efficient small-grain fertilization programs. Its use, in conjunction with soil testing, can greatly improve fertilizer efficiency and minimize losses into the environment when tempered with observation and experience. Small-grain production and quality are greatly influenced by fertilization. Proper fertility favors high yields of quality grain. The objective of this chapter is to provide the scientific basis for small-grain plant analysis and its use to assess the nutrient status and fertilizer requirements of these crops. The small grains discussed are wheat (*Triticum aestivum* L.), barley (*Hordeum vulgare* L.), oat (*Avena sativa* L.), and rice (*Oryza sativa* L.).

 Soil Testing and Plant Analysis, 3rd ed.—SSSA Book Series, no. 3.

I. PREVIOUS USE OF PLANT ANALYSIS

An interest in the chemical composition of small grains was reported as early as 1939 by Miller who collected plant samples from winter wheat fields. Over a 4-yr sampling period, he found large decreases in N and K concentrations as the plants approached maturity. The change in P concentration was much less. The largest changes occurred from Feekes stage 3 to 10 (tillering to boot; Large, 1954). The Feekes scale will be used throughout the remainder of this chapter and will be referred to as stages of growth from 1 to 11.4 (emerging to maturity).

Extensive work on the nutrient concentration of wheat at various locations was reported by Johnson and Schrenk (1953). Plant samples collected from various locations over a 14-yr period showed an average composition of whole plants at tillering (stage 2) as follows: 47.2 g kg^{-1} of N; 2.2 g kg^{-1} of P; 32 g kg^{-1} of K; 3.6 g kg^{-1} of Ca; 1.2 g kg^{-1} of Mg; 92 mg kg^{-1} of Mn; 208 mg kg^{-1} of Fe; and 20 mg kg^{-1} of Cu. Variation in nutrient concentrations with time was also clearly shown, with N, K, Ca, and Mg decreasing the most as plants approached maturity.

A review of the effect of soil types and fertilizers on the chemical composition of small grains (Vandecaveye, 1940) reported that the nature and quantity of available nutrients in the soil greatly influenced the concentration in the plant. Past cropping sequence, soil texture, organic matter, pH, and parent material were factors that influenced nutrient availability to the greatest extent. Nitrogen fertilizer increased N concentration in wheat plants sampled at various growth stages. At maturity, however, protein concentration of the grain was not always increased by N fertilizer. Nitrogen fertilizer was reported to decrease P and Ca concentrations and to occasionally increase N concentration in oat sampled at the hay stage (stages 10.1–10.2).

Johnson and Schrenk (1953) reported that fertilizer application had a direct influence on the nutrient concentration of wheat. Plants receiving no P fertilizer contained 1.9 g kg^{-1} of P at tillering (stage 3) compared to 2.6 g kg^{-1} P for those receiving P. Plant samples collected during the second year of study showed the residual effect of fertilization on nutrient concentration.

Greaves et al. (1940) found large variations in nutrient concentrations between varieties and years in spring and winter wheat varieties. Green manure treatments had limited influence on nutrient concentrations. Spring wheat varieties were found to contain slightly higher nutrient concentrations than winter wheat varieties. More recently, Ward et al. (1973) summarized the influence of various factors on nutrient concentration in plants and suggested uses and interpretation of plant analysis in maximizing economic return for fertilizer use.

Early studies on the use of plant analysis for rice were conducted in Asian countries. The need for proper management of N was recognized and initial work involved this problem. An asparagine test to determine mid-season needs for supplemental N was developed by Ozaki (1955). Absence of asparagine in the youngest leaf at 3 to 4 wk before heading indicated a need for additional N. The asparagine test was reported to be effective for determining the need for mid-season N, but the appearance of asparagine in the plant was influenced by light (Mitsui et al., 1959) and NH_4 (Sing et al., 1960). Early studies to correlate total N in the leaves to N fertilizer needs yielded variable results. Maume and Dulac (1954) concluded that a high level of N in rice leaves was essential for high yields, but the correlation between N concentration and yield was not very good. They also found that the highest concentrations of P and K were associated with high rice yields, but again, the relationships observed between yields and P and K concentrations in the leaf were not good.

There have been numerous other reports on plant analysis of rice in the world. Tanaka and Yoshida (1970) found that the critical level of N in the leaf blade at the tillering stage was 26 g kg^{-1} of N. Lian (1972) used plant analysis to determine the N requirement of Asian rice with some success. Phosphorus deficiency in rice was diagnosed in Taiwan by investigating P distribution changes in various plant parts at different growth stages (Houng et al., 1971). Ishizuka and Tanaka (1951) showed that K deficiency existed in rice straw when its K_2O concentration was <10 g kg^{-1}. Kiuchi and Ishizuka (1961) found that the straw content of 20 g kg^{-1} of K was required for high rice yield. The critical K concentration in the leaf blade was shown to be 10 g kg^{-1} at tillering and harvest by Tanaka and Yoshida (1970). Studies in Taiwan, by Sheng et al. (1964), indicated that a K concentration of 17 to 18 g kg^{-1} was necessary for near-maximum rice yield. Su (1976) found that the critical concentration of K in the straw depended on N and P status of the soils in the Philippines and Indonesia. Wallihan and Sharpless (1974) found that the N/S ratio and the S content were both useful indicators of the S status of the rice plant. Wang (1976) reported the critical S concentration in rice straw grown in the Amazon basin of Brazil was 0.5 g kg^{-1} at harvest. Other investigators have found plant analysis to be valuable in assessing the Zn (Mikkelsen & Brandon, 1975; Forno et al., 1975; Sedberry et al., 1980) and Si status (Lian, 1976) of rice.

Rice plant analysis studies were initiated in the 1960s in the USA. Thenabadu (1966) published the first tentative critical N concentration values for rice. Shortly after this, Mikkelsen et al. (1970) reported tentative critical concentrations of N, P, and K in California. Critical N, P, and K concentrations developed by Mikkelsen and Hunziker (1971) were used as a guide in surveying the nutritional status of 325 California rice fields. The critical N concentration in the "Y" leaf (most recently mature leaf) at panicle initiation used in their studies was 24 to 26 g kg^{-1} of total N. The critical concentrations of P and K were 0.6 to 0.8 g kg^{-1} of PO_4-P and 6 to 7 g kg^{-1} of K, respectively. However, the critical nutrient levels developed in California were not applicable to cultivars grown in the south-central USA because of cultivar, cultural system, and climatic differences (Ward et al., 1973). Normal N concentrations in the most recent fully expanded leaf of eight rice cultivars grown in Texas varied with cultivar and time of sampling.

More recent research shows that rice plant analysis is a valuable diagnostic aid in determining the nutritional status and fertilizer requirements of rice. Brandon et al. (1980) and Mikkelsen (1983) reported both the critical concentrations and adequate ranges of N, PO_4-P, and K for semidwarf rice cultivars currently grown in California. Critical concentrations and adequate ranges were reported by growth stage because concentrations of these plant nutrients change with sequential stages of plant development. The Y leaf should be collected at the mid-tillering, maximum tillering, and panicle initiation stages to monitor concentration changes over time. Multiple sampling times improved the precision of estimating N, P, and K fertilizer requirements. Plant analysis is currently used in California to refine rice fertilization practices because of beneficial effects on pest management (Flint, 1983) and fertilizer efficiency. Critical N concentrations have been developed specifically for major commercial rice cultivars widely grown in California (Brandon et al., 1981a).

The relationship between Y leaf N concentration and grain yield of rice cultivars grown in the southern USA has been reported by Brandon and Wells (1986), Brandon et al. (1981a), and Brandon et al. (1982). Their studies showed progressively greater N concentrations in the rice plant with incremental increases of N fertilizer and a close relationship between Y leaf N concentration at different plant growth stages and grain yield. The critical N concentration range was identified at 25 to 32 g kg^{-1} of total N, depending on rice cultivar and plant growth stage (Brandon & Wells, 1986). Long-grain, semidwarf Indica rices usually required higher plant N concentrations than long- and medium-grain tall rice varieties for maximum grain yield. Field studies confirmed the efficacy of plant analysis for N in both dry- and water-seeded cultural systems used in Louisiana (Helms, 1984). Considerable research has been conducted on the use of plant analysis in small grains. The vast majority of this research simply relates plant nutrient levels to soil nutrient availability (fertility level) and plant development. Few researchers have attempted to establish critical nutrient levels or to field-verify critical nutrient levels over different soils, environments, and years. Additional work in this area is needed. The next section summarizes factors that influence nutrient concentrations in the plant. When one reviews the literature in this area, an appreciation is gained of the problems and challenges encountered in establishing critical nutrient levels.

II. FACTORS INFLUENCING THE CONCENTRATION OF PLANT NUTRIENTS

Nutrient concentrations in plants are greatly influenced by soil and plant factors and their interactions with environmental conditions. An understanding of these environmental influences is necessary to improve the predictive precision of plant analysis as an aid in fertilizing small grains.

A. Soil Factors

Nutrient assimilation by plants is primarily through the root system. Root surface area, soil nutrient availability, and the soil environment for root growth have a major impact on nutrient assimilation. Among the soil factors are fertility status, temperature, and moisture.

1. Soil Fertility

Numerous investigators have found that small grains respond to fertilizer when the native soil nutrient availability is low or adverse environmental conditions occur. Where grain yields have been increased by fertilization, an increase in the plant nutrient concentration has been frequently found if samples were collected before heading (stage 10.3). Working with rice, Sims and Blackmon (1967) and Lin et al. (1973) found that the NH_4-N concentration in soils after 6 d of incubation under anaerobic conditions was highly correlated with N uptake and content. Bajaj et al. (1967) found that available soil N, based on an alkaline permanganate test method, was generally correlated with N uptake by rice.

Several studies have also shown that timing and method of N and P application influence the plant nutrient content. Phosphorus placed near the wheat seed as a starter, or in a concentrated zone, gives higher plant P concentrations than broadcast P applications (Peterson et al., 1981; Westerman & Edlund, 1985; Leikam et al., 1983; Maxwell et al., 1984, Westfall et al., 1987). Relatively late applications of SO_4-S have proven to be effective in increasing yield of soft red winter wheat and yield and S concentration in the upper two leaves at Feekes stage 7 in Arkansas (Wells et al., 1986). Brandon et al. (1982) reported that the time of N fertilizer application in relation to permanently flooding the soil under rice production greatly influenced its N content. Proper timing of N fertilizer and water management were necessary to conserve N in the flooded soil system and positively influence N uptake by the plant. The concentrations of P (Houng et al., 1971) and K (Tanaka & Yoshida, 1970) in rice are directly influenced by soil nutrient content. The relationship between soil Zn level and Zn concentration in rice has been widely studied. Tanaka and Yoshida (1970) and Mikkelsen and Brandon(1975) showed a close relationship between available Zn and Zn concentration in rice seedlings. Studies by other investigators (Sedberry et al., 1971; Mikkelsen & Kuo, 1977) confirmed that increased soil Zn availability increased the Zn concentration in rice.

The application of one nutrient has also been found to influence the plant concentration of other nutrients. Singh et al. (1986) found that P fertilization increased soil P levels and tissue P concentration of spring wheat at Feekes stages 6 and 10, but resulted in a significant decrease in tissue Zn concentration. Westfall et al. (1973) found that the concentrations of Mg, Mn, Fe, and Ca were lower in N-deficient than in N-sufficient rice plants. The concentration of Zn in the rice plant depends on the soil factors of pH, oxidation-reduction potential, organic matter, and the soil concentrations of HCO_3, P, and organic acids (Mikkelsen & Kuo, 1977; Yoshida & Tanaka, 1969). Nielson et al. (1960) reported that the application of K increased the K concentration of oat plants at the 7-leaf (Feekes 3) and heading (Feekes 10.3–10.5) stages, but depressed Mg and Ca concentrations, especially Mg, while having little effect on N and P. The critical concentration of K in rice depends on the N and P status of the soil (Su, 1976). Su (1976) showed that K in rice straw was rather constant under low soil N and P conditions and that plant analysis for K was valid only when adequate N and P were available in the soil.

Vandecaveye (1940) reported that past cropping sequences and several soil properties such as texture, organic matter, pH, and parent material also influence nutrient availability. Crooke and Knight (1971) used a granitic soil limed in increments from pH 4.5 to 7.5 to determine the variation in nutrient composition of oat, barley, and winter wheat. Liming produced an increase in N and P concentrations in the three cereals to a soil pH of 6.5 or greater. Sulfur and Fe concentrations changed little with liming, while Ca, Mg, and K increased. Manganese concentration was reduced with liming, leveling off at about 30 mg kg^{-1} of Mg. Additional nutrient interactional effect data could be cited; these examples illustrate the interactional effects that can occur.

2. Soil Temperature

Soil temperature has a large effect on plant growth through its effect on root growth and respiration. The N, P, K, and Mg concentrations in oat were lowest at a soil temperature of 5 °C and highest at 27 °C (Nielson et al., 1960). With 'Sacramento' barley, Power et al. (1970) found that the N concentration was unaffected by soil temperature in the range of 9 to 22 °C at several growth stages. The P concentration was usually highest at 9 °C. The 9 °C soil temperature did, however, slow plant growth compared to two higher soil temperatures.

Under field conditions where early season soil temperatures were lowered by stubble mulch tillage compared to clean seedbed preparation, Smika and Ellis (1971) found that the nutrient content of hard red winter wheat was not affected by soil temperature when plant samples were collected at the same stage of growth (end of tillering, stage 5 and at heading, stage 10.5).

Ambient temperature affects nutrient uptake in rice because of its influence on both plant metabolism and soil processes. Low temperatures decrease the respiratory rate and retard nutrient uptake (Yoshida, 1981). The percentage reduction in nutrient uptake caused by low temperatures is not the same for all nutrients (Takahashi et al., 1955). The inhibitory effects of low temperatures on nutrient uptake by rice suggested by Yoshida (1981) is of the order $PO_4 > H_2O > NH_4 > SO_4 > K > Mg > Ca$. High temperatures also retard absorption of NH_4, PO_4 and K (Baba, 1954).

Studies on plant growth response of rice to Zn at different temperatures have been reported. At a 16 °C temperature, which simulated normal soil temperatures during rice stand establishment in California, greatly reduced dry matter yield and Zn concentration in the shoots were observed (Mikkelsen & Brandon, 1972). Addition of Zn at 16 °C did not appreciably affect Zn concentration or total Zn uptake. However, at 24 °C, Zn concentrations were increased by Zn addition. Cool soil temperatures have frequently been associated with increased incidence and severity of Zn deficiency on soils low in available Zn (Mikkelsen & Kuo, 1977). Lindsay (1972) suggested two mechanisms for temperature sensitivity for all crops: (i) root elongation and nutrient uptake are diminished in cool soils which reduces the feeding zone of roots and, (ii) the availability of Zn from organic matter is reduced because of reduced mineralization rates in cold soils.

3. Soil Moisture

Soil moisture affects nutrient absorption in several ways. Soil moisture influences the relative concentration of nutrients in the soil solution, plus it affects root growth. Optimum soil moisture environments are dramatically different for rice than for other small grains. Therefore, the effect of soil moisture on nutrient uptake can be markedly different.

Power et al. (1961a) found that dry soil conditions increased N concentration in the aboveground portion of hard red spring wheat at later stages of growth. They (Power et al., 1961b) also observed that the amount of P assimilated by the aboveground portions of spring wheat was not consistently influenced by soil moisture or P fertilization. Boatwright et al. (1964) studied the effect of soil moisture on P uptake by wheat at various growth stages. Plants did not absorb fertilizer P from a dry soil. There was a time lag in P absorption when a dry soil was wetted, especially in older plants. They concluded that little P fertilizer will be used by wheat unless the soil around the fertilizer remains wet for 3 to 4 d.

Karlen et al. (1978) studied the effect of soil moisture on the cation composition of hard red winter wheat emerging from dormancy. Wet soil conditions tended to lower Ca and Mg concentrations in the aboveground tissue at Feekes stage 2, but had no effect on K concentration. In contrast, Letey et al. (1962) showed that Ca and Mg concentrations in barley were not greatly modified by soil O_2 supply.

The influence of soil submergence on the availability of plant nutrients to rice has been summarized by Ponnamperuma (1972). Flooding of the soil and subsequent depletion of O_2 (reduction), results in increased accumulation of NH_3 (Patrick & Mahapatra, 1968; Ponnamperuma, 1972; Shapiro, 1958; Mahapatra & Patrick, 1969; Brandon, 1977; Sah & Mikkelsen, 1986), Fe, and Mn (Sajwan & Lindsay, 1986) in the soil solution and rice plants. In addition, Sajwan and Lindsay (1986) showed that soil submergence reduces Zn availability and uptake because of increased reduction and solubilization of Fe and Mn which have an antagonistic effect on Zn uptake.

The effects of flooding and draining on nutrient availability, particularly P, of rice and crops following rice have been studied by several investigators in recent years. Martin et al. (1971) first suggested that P deficiency in safflower (*Carthamus tinctorius* L.) planted after rice was caused by soil submergence during rice culture. Phosphorus deficiency is common in crops grown in rotation with rice even though the rice crop may show no evidence of P deficiency (Brandon, 1977; Brandon & Mikkelsen, 1979; Sah & Mikkelsen, 1986). These researchers suggested that P deficiency in barley, wheat, and safflower was caused by increased solubilization of Fe during the flooded cycle (reduced) of rice production, and subsequently, by Fe adsorption of soil-solution P during the early part of the drained cycle (reoxidized). Soil and fertilizer P are rapidly adsorbed by reoxidized amorphous Fe in the drained cycle which renders soil P unavailable to crops following rice (Brandon, 1977). The findings of Willett et al. (1977) in Australia were supported by those of Brandon and Mikkelsen (1979) in California, Singhania and Goswani (1978) in India, and Griffin and Brandon (1983) in Louisiana. The extensive research on the increase of soil-solution P and P uptake by rice during soil submergence and decrease of soil-solution P and plant uptake by upland crops following rice documents the primary effect of soil moisture on P nutrition in rice-based cropping systems.

4. Other

Soil salinity can also influence plant nutrient concentrations with the effect depending on degree of salinity and the composition of soluble salts. By adding increments of salts to increase soil salinity, Hassan et al. (1970) found the following effects on nutrient concentrations: (i) decreased P, Ca (depending on nature of salinity), Cu, and Fe; (ii) increased Mg, Na, Zn, and Mn; and (iii) increased K at low salinity and reduced K at higher salinity.

The references cited in this section point out the large effects various factors have on the nutrient concentrations in small grains. It is not surprising that much diversity in nutrient levels in small grains has been reported.

B. Plant Factors

Plant species, cultivar, plant maturity, and plant part all have an effect on plant nutrient concentration. Interpretation of plant analysis data must be made taking these factors into consideration.

1. Species Differences

Small-grain species have been shown to differ in nutrient concentration when sampled at similar stages of maturity. Bishop and MacEachern (1971) compared the nutrient concentration of wheat and barley by analyzing the second, third, and fourth leaves from the bottom of the plant when heads were 90 to 100% emerged (stage 10.5). Phosphorus concentrations in wheat leaves were higher than those in barley leaves, while Ca and Mg concentrations in barley were approximately double those in wheat. Nitrogen concentrations in wheat were considerably higher than those found in barley.

Trends of higher Ca concentrations in barley compared to wheat, and higher K levels in oat plants compared to barley or wheat were observed in greenhouse studies (Crooke & Knight, 1971). Concentrations of N, P, Mg, and Mn were similar for the three crops. Five small grain species were investigated for species effect on Ca, Mg, K, and Na concentrations of seedlings under growth chamber conditions [(S.C. Brubaker, 1979) The effect of soil water and potassium on grass tetany related components of cereal forages.

(M.S. thesis, Kansas State Univ., Manhattan)]. Oat had significantly lower K concentrations than the other four species at most sampling dates while barley and triticale had the highest K concentrations. Oat consistently had the highest Ca and Mg concentrations compared to the other species. Barley was found to have a substantially higher Na concentration.

Baker and Tucker (1971) studied NO_3 accumulation in oat, wheat, barley, and rye forage and concluded that there was little difference in NO_3 accumulation. Nitrate concentration was found to be a function of N and P fertilization and date of sampling.

Brown and Stark (1986) compared the response of three classes of wheat—soft white winter, hard red winter, and hard red spring—for their response to N fertilization. They found that flag leaf N concentration at flowering was a good predictor of N-fertilizer utilization within a location, but found a variety by location interaction that hindered use of flag leaf N as an indicator of wheat N status.

Viets et al. (1954) found little difference in the response of wheat, oat, and barley to Zn fertilization, but did find slightly higher Zn concentrations in the upper leaves of barley at heading. In contrast, Gladstone and Loneragan (1967) reported little difference in the concentration of Zn in wheat and barley.

2. Cultivar Differences

Significant wheat and barley cultivar differences in the accumulation of most nutrients, with the exception of Zn, Ca, Mo, and Fe, have been reported by Kleese et al. (1968). The range of nutrient concentrations in the leaves of barley compared to the range in wheat cultivars were similar for the two species. In South Dakota, nutrient concentrations of four barley cultivars sampled at heading (stage 10.2–10.3) contained similar concentrations of N, P, Mg, S, Fe, Zn, and Mn. Potassium concentrations ranged from 15.2 to 20.2 g kg^{-1} and Ca from 4.4 to 7.9 g kg^{-1} (Diebert et al., 1968. Unpublished Pamphlet no. 1, Plant Sci. Dep., S. Dakota State Univ., Brookings).

Donohue and Brann (1984) reported only small differences between three soft red winter wheat cultivars for optimum N concentrations for entire aboveground plant samples at Feekes stage 4 and for flag leaf samples (stage 10). They concluded from these locations that a N-sufficiency range of 40 to 50 g kg^{-1} could be used for samples collected at both growth stages.

Tanaka et al. (1959) and Ward et al. (1973) reported a rather wide range of N content in the leaves of rice cultivars of different maturity. The N concentration in the most recent fully expanded leaves of eight southern USA rice cultivars at the mid tillering and panicle initiation growth stages ranged from 24 to 50 g kg^{-1} (Ward et al., 1973). Brandon et al. (1981b) reported the critical N concentrations at panicle initiation in a tall California cultivar, S6, and semidwarf cultivar, M9, were 32.5 and 36.5 g kg^{-1} of N, respectively. The critical N concentrations in the southern U.S. tall (Lebonnet) and semidwarf (Bellemont) rice cultivars in Louisiana were reported to

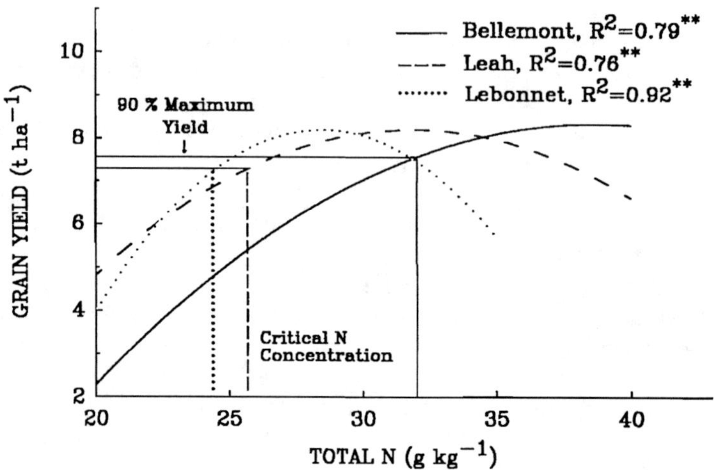

Fig. 19–1. Differences in critical total N levels of tall (Lebonnet) and semidwarf (Bellemont and Leah) rice cultivars at panicle initiation stage of growth.

be 25 and 32 g kg^{-1} of N (Brandon et al., 1981a), respectively, at the panicle initiation growth stage (Fig. 19–1). Maximum grain yields of the tall and semidwarf cultivars were observed at 28 and 38 g kg^{-1} of N, respectively, at the panicle initiation growth stage in these studies.

Definite cultivar differences in P absorption, growth, and yield of rice on P-deficient acid soil in Thailand were reported by Koyama et al. (1973). The rice cultivar, Dawk Mali 3, grew better and had greater yields than other cultivars because of higher P absorption. Ponnamperuma (1976) subsequently observed cultivar differences in P absorption and yield of a wide range of cultivars. The effect of Zn deficiency on plant Zn concentrations and yield of different tropical rice cultivars was reported by Yoshida et al. (1973). Their studies showed that cultivars most susceptible to Zn deficiency were less vigorous and had less tillering capacity. Yoshida (1981) reviewed critical, normal, and toxic concentrations of Fe in rice plant tissue. Plant chlorosis induced by Fe deficiency was evident at an Fe concentration of 70 mg kg^{-1} on a dry matter basis. Plant concentrations >70 mg kg^{-1} of Fe were adequate for normal plant growth (Yoshida, 1981). Accumulation of an excessive amount of Fe in rice leaves is characterized by Fe toxicity. Typical symptoms of Fe toxicity may be observed when the Fe content of leaf blades is >300 mg kg^{-1} (Tanaka et al., 1966). Yoshida (1981) reported differences among 12 rice cultivars in their ability to absorb Fe. Ponnamperuma (1976) showed that tropical rice cultivars also differ in their tolerance of Fe toxicity.

Cultivar and species effects on nutrient concentration in small grains exist. There are, however, several studies showing that the differences between cultivars are relatively small. For example, Ylaranta et al. (1979) found no clear differences in the micronutrient contents between spring and winter wheats over the entire growing season, while Engel and Zubriski (1982) reported significant differences in N contents of different cultivars for some years but not others. The data indicates that species and cultivar effects cannot be taken for granted and must be investigated before recommendations are made and critical values established. This seems especially true as high-yielding cultivars are developed through introduction of germplasm from sources outside the traditional genetic pool.

3. Growth Stage and Seasonal Variation

The use of plant analysis to diagnose nutrient deficiencies can be grossly misinterpreted if the user does not understand the variation in nutrient concentrations that can occur as a function of growth stage and between growing seasons. Researchers in Europe, using plant analysis for determining the second N dressing for 723 000 ha of winter cereals (Neubert et al., 1977), revealed that site and variety have a small influence on the relationship between plant N and yield of wheat, barley, and rye, as compared to seasonal weather conditions. Beringer and Hess (1979) found that the total N content could vary by 40% within a week's time in the early stages of growth, indicating the sensitivity of N concentration to environmental conditions. Ylaranta et al. (1979) sampled spring and winter wheat twice a week during the growing season to obtain information on seasonal micronutrient variations. The highest content of B, Cu, and Fe were recorded early in the stage of growth, followed by a rapid decrease towards the end of the growing season. Boron decreased from 4 to 12 mg kg^{-1} to <2, Cu from 6 to 10 mg kg^{-1} to 3 to 4, and Fe from 80 to 210 to 30 to 75 mg kg^{-1}. Manganese tended to increase in concentration towards harvest. The behavior of Zn content was different from other micronutrients, exhibiting a slight tendency to increase in the latter growth stages. A rapid drop in N content of spring wheat was reported between Feekes growth stage 3 to 10.5, with a curvilinear decrease between stages 5 to 10.5 and a linear relationship at stage 3.

The references point out the importance of evaluating variations within a season as well as between seasons in interpreting plant analysis data of small grains. Failure to consider these factors will lead to misinterpretations.

III. SAMPLING FOR PLANT ANALYSIS

The usefulness of plant analysis to assess the nutritional status of small grains largely depends on sampling procedures. Information provided by plant analysis is only as representative as the samples collected. Plant growth stage and part of the plant sampled have a tremendous influence on nutrient concentration. Several stages of growth and plant parts have been used by various researchers, but no stage of growth and plant part has been found superior for all nutrients with all small grains.

A. Stage of Growth

Small grains have been sampled from tillering (stage 3) to boot (stage 10) or early head emergence (stage 10.3). With wheat, oat, and barley,

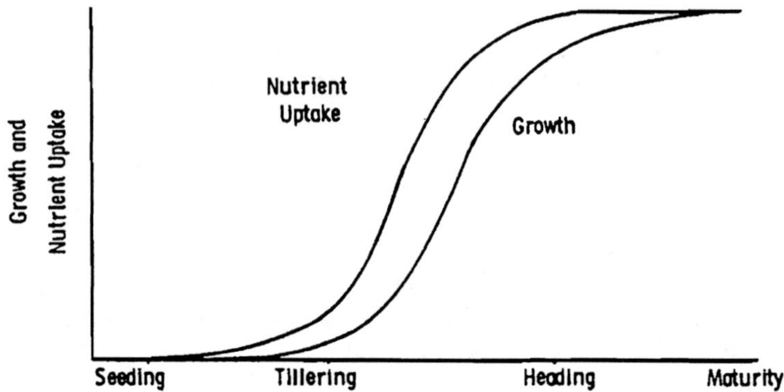

Fig. 19–2. Typical relative growth and nutrient uptake curves for small grain crops.

sampling at tillering may allow time for correction of nutrient deficiencies on the current crop. Boot and early head emergence sampling is normally too late in the season for correction of nutrient deficiencies on the existing crop.

A typical growth and development curve for small grains is shown in Fig. 19–2. Nutrient uptake precedes dry matter accumulation with a rapid increase in nutrient and dry matter accumulation occurring between tillering and head emergence. Plants sampled at heading (stage 10.3–10.4) should reflect the nutrient sufficiency at the time flowering is completed. The nutrient concentrations in plants generally decline between tillering and heading.

Critical nutrient concentrations change with stage of plant development. Therefore, it is advantageous to collect multiple samples for monitoring N-fertilizer requirements. Sampling rice between tillering and panicle differentiation provides tissue analysis results early enough for corrective N fertilization (Mikkelsen, 1983). To obtain maximum precision, it is desirable to collect a series of rice leaf samples, preferably at: (i) mid-tillering stage (40–50 d after planting for 140- and 150-d cultivars, respectively); (ii) maximum tillering stage (50–60 d after planting); (iii) panicle initiation stage (60–80 d after planting); and (iv) panicle differentiation stage (70–90 d after planting) (Brandon et al., 1981a; Mikkelsen, 1983).

Growth stage is an important factor in interpreting plant analysis data. Care should be taken to ensure that nutrient concentrations from plants are interpreted accurately with critical levels at the proper growth stage. Improper growth stage identification can result in errors in interpretation.

B. Plant Part

When sampling wheat, barley, and oat plants at tillering (stage 3), the entire aboveground portion of the plant should be sampled. At heading (stage 10.3), either whole plants, upper leaves, or flag leaves are commonly sampled.

Nutrient concentration differences have been shown between plant parts sampled at the same growth stage. Flag leaves of hard red winter wheat sampled at heading (stage 10.3) had nearly twice the N concentration as whole aboveground plant samples at the same stage (D.A. Whitney and G.A. Peterson, 1971, unpublished data), while P and K concentrations were essentially the same. Similar results were observed for oat in South Dakota (J.H. Pfeifer, 1971, unpublished data).

The influence of plant part on nutrient concentrations in rice has been reported. Large differences in nutrient concentrations occur between the most recently matured leaf and the whole plant (Westfall et al., 1973; Ward et al., 1973). Thenabadu (1966) established that the first and second most fully expanded leaves were the most sensitive indicators of the N status of rice. However, the most recently matured leaf, commonly referred to as the Y leaf, has been shown to be the best indicator of N status (Brandon et al., 1980; Hafez & Mikkelsen, 1981; Mikkelsen, 1983; Ward et al., 1973; Wallihan & Moomaw, 1967). This leaf tends to be erect, which makes sampling easier and more accurate. The Y leaf is widely used for determining the nutrient status of commercial rice crops.

C. Sample Size

Sample size for plant analyses must be large enough to provide a representative sample of the crop and sufficient dry matter for chemical analysis. Sampling procedure to minimize sample variability has been delineated, but limited research information is available on optimum sample size for wheat, barley, and oat. Chapman (1964) suggests 200 or more leaves are required for a good sample. For whole-plant samples at heading, most guidelines are for 20 to 30 plants taken randomly across the field (Jones et al., 1971; Whitney et al., 1985). Collection of 50 to 100 Y leaves from rice plants in early stages of plant development, from 30 to 60 locations within a field, provides sufficient dry matter and minimizes variability if crop growth is uniform. A sample of 30 to 60 individual Y leaves is sufficient in later stages of plant development.

Weisset et al. (1982) studied the variation in plant results as influenced by sampling pattern and sample size in 20 winter grain fields ranging in size from 34 to 109 ha. When a sample was evenly taken (one sample ha^{-1}), the coefficient of variation in total N content was 10 to 12%. They concluded that the traditional zig-zag route is not necessary. The sampling area should be no larger than 50 ha. Regardless of the sampling route used, it should be marked when multiple sampling procedures are to be collected. Separate samples should be taken from atypical areas characterized by abnormal plant growth, plant symptoms, plant nutrient status, or other specific problems. Close observation of the crop and judgment based on experience should be used in determining the size of the sampling area represented by a sample and the number of samples required in a field. The more subsamples collected, the more likely the sample will represent the field.

IV. DIAGNOSING NUTRIENT DEFICIENCIES

A. Diagnostic Approach

Information cited above indicates that large variations can exist in nutrient concentrations in plants due to soil-plant-environment relationships. Use of plant analysis to help diagnose problems that exist in a certain area of a field can be a valuable tool. This technique is termed the *diagnostic approach* of plant analysis. Samples should be collected as soon as poor growth or nutrient-deficiency symptoms become evident. Deficiency symptoms of N, P, and Zn usually become visible in small grains at early stages of growth. Plant samples should be collected from both the field area where plant symptoms are apparent and areas where plants appear normal. A sample from transitional areas between the two extremes may be useful in diagnosing the problem. Soil samples from the same area should be analyzed to increase the accuracy of plant analysis interpretation. Simple and rapid soil analysis parameters, such as pH and salinity, often reveal the soil factors that can be identified with plant growth problems. Information about fertilization, weed and insect control, tillage, and climatic conditions should be recorded. These factors often have a major influence on plant growth problems. Comparison of the nutrient concentration in plant and soil samples from the two contrasting areas in the field can be extremely helpful in pinpointing the nutrient disorder that has resulted in poor growth. However, caution should be exerted because plant symptoms that are similar to nutrient-deficiency symptoms can be caused by insects and diseases. Plant roots should always be examined to ensure that they are developing normally and insects or diseases have not adversely affected root development, thereby limiting nutrient uptake.

The diagnostic approach will often provide answers to plant nutritional problems that are difficult to analyze by other means. For example, in California, rice researchers (Mikkelsen & Brandon, 1975; Mikkelsen & Kuo, 1977) used the diagnostic approach to identify Zn deficiencies that usually occurred in only parts of the field. However, Westfall et al. (1971) reported that Zn and Fe chlorosis were corrected by Zn and Fe application to rice, but the concentration of Zn and Fe in the plant was not increased.

B. Monitoring Approach

Collecting plant samples at a given stage of development and comparing the plant analyses with established deficiency levels is called the *monitoring approach* to plant analysis. This approach requires that specific guidelines be followed in respect to species and cultivar, growth stage, plant part sampled, and environmental conditions. Over the past years, progress has been made in establishing critical levels, but it is difficult to give generalized critical levels that are applicable to broad geographic regions. The importance of cultivar variation is not well understood and has not been well documented for the small grains. Chapman (1966) demonstrated the variation in reported critical nutrient levels. More recently, Reuter and Robinson (1986) conducted an extensive literature review and summarized the various concentration ranges reported in the literature, from critical to toxic, of various nutrients for a wide range of crops. Again, their data point out the large variations in critical nutrient concentrations that had been reported by researchers throughout the world. For this reason, it is impossible to report one critical concentration level for a specific plant nutrient that is applicable to a wide geographic region of diverse environments. The authors of this chapter, therefore, are reporting critical nutrient levels that have been found to be generaly acceptable for specific regions. The transfer of critical levels from one region to another should be done with caution, and preferably only after field verification has been conducted.

Traditionally, total N concentrations have been used in the monitoring approach. However, recently there has been an interest in the use of stem NO_3 in monitoring the N status of small grains. Gardner and Jackson (1976) in Arizona found that the NO_3-N concentration of the lower 5 to 8 cm of the stem was a good indicator of N status of the wheat plant. The sufficiency ranges proposed were 7000 to 12 000 mg kg^{-1} of NO_3-N at the 3 to 4 leaf stage, 5000 to 10 000 mg kg^{-1} at the jointing stage, and 3000 to 9000 mg kg^{-1} at the boot stage of growth. Nitrogen concentrations lower than these would indicate a need for N-fertilizer applications. Levels greater than those may result in reduced yields due to excessive N availability. Visual N deficiency symptoms appeared as stem NO_3-N levels dropped to 2000 mg kg^{-1}. The determination of P in the lower portion of the wheat stem was not found to be suitable for indicating the P status of the crop. Brown and Jones (1979) reported that the NO_3-N concentrations necessary for maximum production should be at least 3500 mg kg^{-1} at the 3 to 4 leaf stage and 1000 mg kg^{-1} at jointing. Beringer and Hess (1979) found that the correlation between grain yield and NO_3 content of the stem became low after the tillering stage. However, Papastylianou (1986) found a high correlation between dry matter production and stem NO_3. Stem NO_3 was more sensitive than total leaf N as an indicator of N nutrition status of barley when samples were collected at the tillering stage of growth. Based primarily upon results of Brown and Jones (1979), Dow (1980) published critical nutrient ranges for stem NO_3 of small grains in the Pacific Northwest (Table 19–1). These levels correspond well with those reported by Gardner and Jackson (1976) in Arizona. The critical nutrient ranges of other nutrients for small grains in the Pacific Northwest are also reported in Table 19–1. Dow defined the critical nutrient range (CNR) to be "that range of concentrations above which we are reasonably sure the crop is amply supplied and below which we are reasonably sure the crop is deficient." Thus, within the CNR is a range of uncertainty. Dow points out that the accuracy of prediction is enhanced by monitoring, i.e., by taking samples periodically during a growing season. Seasonal monitoring then considers the changes in nutrient levels during the growing season. The approach of comparing across years in a particular environment can be useful in establishing critical nutrient levels for various crops that are regionally and environmentally specific.

Table 19-1. The critical nutrient range (CNR) for small grains in the Pacific Northwest (Dow, 1980).†

Nutrient	Growth stage‡	Plant part	CNR
		Wheat	
N	Joint (GS 6)	Total tops	25–30 g kg^{-1}
	Boot (GS 10)	Top two leaves	23–27 g kg^{-1}
NO_3-N	3–4 leaf (GS 2)	Underground stem	2.5–3.5 g kg^{-1}
	Jointing (GS 6)	First 2 in. aboveground	0.8–1.5 g kg^{-1}
P	12 wk (GS 4)	Total tops	3.0 g kg^{-1}
	Jointing (GS 6)	Total tops	3.2–4.0 g kg^{-1}
	Early boot (GS 9)	Total tops	1.5–2.0 g kg^{-1}
K	Jointing (GS 6)	Total tops	20–25 g kg^{-1}
	Early boot (GS 9)	Total tops	15–20 g kg^{-1}
		Barley	
NO_3-N	3–4 leaf (GS 2)	Below-ground stem	0.6–0.7 g kg^{-1}
Mn	Early tillering (GS 2.5)	Flag leaf	18–20 mg kg^{-1}
Cu	Tillering (GS 3)	Leaves	4.0 mg kg^{-1}
		Oat	
NO_3-N	3–4 leaf (GS 2)	Below-ground stem	0.6–0.7 g kg^{-1}
Mn	Early tillering (GS 2.5)	Flag leaf	22–25 mg kg^{-1}

† Prepared by the Northwest Soil and Plant Testing Work Group.
‡ Approximate Feekes scale growth stage.

The sufficiency range for wheat in the southern USA, specifically taken from data of Plant (1988) in Georgia, are presented in Table 19–2. These researchers have been able to establish sufficiency ranges for several nutrients over a wide range of growth stages for wheat.

Nutrient levels for wheat, oat, and barley for the Great Plains of the USA are shown in Table 19–3. These levels have been previously reported by Ward et al. (1973). Those values in Table 19–3 are for a late growth stage, and little yield response would be expected from nutrient application at this time. More recently, Vaughan et al. (1989) reported critical N levels for hard red winter wheat in the western Great Plains region (Table 19–4). Suggested critical N levels are given for total N and stem NO_3-N. However, the "best" N recommendation model was found to be the spring-available soil NH_4-N plus NO_3-N to 60-cm soil depth and Feekes 5 leaf N concentration. This model was accurate at low required N rates, but more conservative at higher N rates. Vaughan et al. (1989) found it difficult to use tissue analysis alone as a predictive tool because of geographic and environmental variability, thus, the reason for the necessity to include spring soil N data. The variability in stem NO_3-N values were large and the use of this plant parameter was not as accurate as total N tissue values. This resulted in large "transitional concentrations" in stem NO_3-N in Table 19–4.

A considerable amount of work has been conducted on S deficiencies of small grains in Australia. The critical S levels proposed by Spencer and Freney (1980) are presented in Table 19–5. The use of SO_4-S and total S as an indicator of nutrient sufficiency has met with varied success. However,

Table 19-2. The sufficiency range for wheat in Georgia (Plank et al., 1988).

Nutrient	Growth stage†	Plant part	Sufficiency range, g kg^{-1}
N	GS 3	Whole plant	40-50
	GS 4-6	Whole plant	35-45
	GS 7-8	Whole plant	30-40
	GS 9-10	Whole plant	25-35
	GS 10	Flag leaf	35-45
P	GS 3-5	Whole plant	4-7
	GS 6-10	Whole plant	2-4
	GS 10	Flag leaf	3-5
K	GS 3-4	Whole plant	32-40
	GS 5-8	Whole plant	20-35
	GS 9-10	Whole plant	18-30
	GS 10	Flag leaf	20-30
Ca	GS 3-10	Whole plant	2-5
	GS 10	Flag leaf	3-5
Mg	GS 3-10	Whole plant	1.5-5.0
	GS 10	Flag leaf	2-6
S	GS 3	Whole plant	2.2-5.5
	GS 4-6	Whole plant	1.9-5.5
	GS 7-8	Whole plant	1.7-5.5
	GS 9-10	Whole plant	1.5-4.0
	GS 10	Flag leaf	1.9-5.5
			—mg kg^{-1}—
Mn	Unspecified	Whole plant	35-475
Fe	Unspecified	Whole plant	25-100
B	Unspecified	Whole plant	3-20
Cu	Unspecified	Whole plant	5-25
Zn	Unspecified	Whole plant	15-70
Al	Unspecified	Whole plant	<200

† Feekes scale growth stage.

Table 19-3. Interpretation of plant analysis for oat, wheat, and barley based on aboveground samples collected as the head emerges from the boot (GS 10.1) in the Great Plains of the USA (Ward et al., 1973).

Nutrient	Deficient	Low	Sufficient	High
			g kg^{-1}	
N (Winter grain)	<12.5	12.5-17.5	17.5-30	>30
(Spring grain)	<15.0	15-20	20-30	>30
P	<1.5	1.5-2	2.0-5.0	>5
K	<12.5	12.5-15	15-30	>30
Ca (Except barley)		<2	2-5	>5
(Barley)		<3	3-12	>12
Mg		<1.5	1.5-5	>5
S		<1.5	1.5-4	>4
			mg kg^{-1}	
Mn	<5	5-24	25-100	>100
Zn		<15	15-70	>70
Cu		<5	5-25	>25

Table 19-4. Suggested critical N concentration for hard red winter wheat in the Western Great Plains (Vaughan et al., 1990).

Feekes growth stage	N form	Plant part	Units	Critical level	Transition concentration†
3	Total N	Aboveground‡	g kg^{-1}	40	
5	Total N	Aboveground	g kg^{-1}	32	25–33
5	Total N	Leaf§	g kg^{-1}	38	34–38
5	NO_3-N	Stem	mg kg^{-1}	866	639–2750
7	Total N	Aboveground	g kg^{-1}	27	19–27
7	Total N	Leaf	g kg^{-1}	35	34–41
7	NO_3-N	Stem	mg kg^{-1}	619	

† The "medium" range where sufficiency and deficiency could coexist.
‡ Aboveground plant material.
§ Two most fully developed leaves.

Table 19-5. The critical S level in wheat in Australia (Spencer & Freney, 1980).

	Growth stage		
	GS 2	GS 5	GS 7
Sulfate S (g kg^{-1})	140	120	110
Sulfate S conc. (mg kg^{-1})	400	360	190
Total S conc. (g kg^{-1})	3	2.8	1.5
N/S ratio	15	16	19

Table 19-6. Critical and adequate concentrations of N in "Y leaves† of southern U.S. grown rice cultivars at different plant growth stages, Crowley, LA.

		Total N, g kg^{-1}			
		Panicle initiation		Panicle differentiation	
Rice cultivar by maturity	Grain type	Critical‡	Adequate	Critical‡	Adequate
		Very early group			
Labelle	Long	28–30	30–35	23–25	25–30
L-201	Long	25–27	26–30	20–22	22–27
M9	Medium	25–27	26–30	20–24	24–28
M-201	Medium	30–32	32–35	25–27	27–31
		Early group			
Lebonnet	Long	26–28	28–31	22–24	24–28
Lemont	Long	30–32	32–35	26–28	28–32
Belmont	Long	32–34	34–38	28–30	30–34
Saturn	Medium	24–26	26–28	20–22	22–24
Mars	Medium	26–28	28–32	22–24	24–28
Brazos	Medium	26–28	28–32	22–24	24–28
		Mid-season group			
Starbonnet	Long	25–27	27–32	22–24	24–28
Newbonnet	Long	26–28	28–30	24–26	26–28

† Kjeldahl N on dry wt. basis of most recently matured leaves.
‡ Plants with critical concentration of N produced 90% of maximum yield.

Table 19-7. Critical and adequate concentrations of N, P, and K in medium- and short-grain rice cultivars at different plant growth stages in California.

Plant growth stage	Total N, g kg^{-1}		Extractable PO_4-P, mg kg^{-1}		Extractable K, g kg^{-1}	
	Critical	Adequate	Critical	Adequate	Critical	Adequate
Mid-tillering	38–40	40–48	1000	1000–1800	14	14–28
Maximum tillering	34–36	36–42	1000	1000–1800	12	12–24
Panicle initiation	30–32	32–36	800	800–1800	10	12–24
Flag leaf	26–28	28–32	800	800–1800	10	12–22

† Analysis on dry wt. basis of most recently matured leaves for Kjeldahl N, 20 g kg^{-1} of HAc-extractable PO_4-P and K.

the N/S ratios have been found to be less affected by environmental factors. Rassmussen et al. (1977) and Wells et al. (1986) have revealed the N/S ratio proposed by Spencer and Freney (1980) are applicable to many environments under which wheat is grown in the USA.

The critical and adequate concentrations of N in the Y leaf of rice cultivars grown in the southern USA are shown in Table 19–6 for the panicle initiation and differentiation stages of growth. Critical and adequate concentrations of N, P, and K in rice cultivars grown in California are shown in Table 19–7. A more rapid decline in nutrient concentrations over time can be expected if the soil has low levels of residual fertilizer nutrients while a gradual decrease of plant nutrient concentration, as a function of plant development, is typical in soils with high levels of residual N, P, and K.

Monitoring the nutritional status of small grains is a valuable tool in establishing the optimum amount of fertilizer required for optimum yield. A well-planned program of plant analysis permits identification of adequate levels of fertilizer required for optimum yields when used with fertilization and cropping history. Plant analysis may also prevent costly application of excessive amounts of nutrients or applications of nutrients not needed. It should be remembered that identification of nutrient deficiencies late in the growing season may be so late in the plant's stage of development that application of the deficient nutrient may not result in maximum return to the fertilizer investment. Therefore, the monitoring approach must encompass both preplant or in-season soil testing, as well as plant nutrient monitoring during the early stages of growth.

C. New Techniques

Some new techniques to aid in the interpretation of plant analysis have been suggested recently. Brooks and Reisenauer (1985) found that soluble reduced N, expressed as a fraction of the protein N, was a useful predictor of relative growth rate. Leigh and Johnston (1983a, b, 1985, 1986) have reported the usefulness of expressing concentrations of P and K on the basis of tissue water. Less variation during the growing season occurs in the concentration of these two elements if they are expressed on the basis of concentration in the tissue water. However, they concluded that N concentrations based on this method are unlikely to be useful in determining the N requirements of barley and that N% in the dry matter was more useful. When K concentrations were calculated on the basis of tissue water, they remained relatively constant during vegetative growth and then increased when water loss was rapid during senescence. Also, whereas K concentrations in the dry matter were affected by the supply of N, P, K, and water, those concentrations based upon tissue water content were influenced only by K supply. Similar relationships were reported for P.

D. Quick Test

The use of a "quick test" for assessing plant nutrient status can be a useful tool for spot-checking fields when laboratory facilities required for more accurate analysis are not readily available. Different degrees of success have been reported by various investigators using quick tests. Guettinger and Koehler (1963) reported that a quick NO_3-N test on winter wheat was related to the available soil N and could be used to determine top-dress needs of the crop. The quick test results were closely related to total N in the plant material. Hart and Pettygrove (1985) used commercially marketed "EM Quant Test Strips" as a method of field assessment of the N status of wheat. The test strips were able to separate high from low levels.They concluded the strips were useful in evaluating the amount of NO_3 in wheat plants, and that this method could be used in other situations where critical levels have been previously established for a cultivar at a particular stage of growth. However, the test strips were not effective in determining the NO_3 concentration when the wheat plant was under moisture stress or if the samples were allowed to dry slightly before testing.

A quick test for N in rice tissue is currently used in California to determine N status. Hafez and Mikkelsen (1981) and Mikkelsen (1983) evaluated an Orange-G colormetric method for protein determination in rice leaves. The amount of dye absorbed by dry, ground Y leaves at three stages of rice development was highly correlated (r^2 = 0.81–0.98) with percent N as determined by the Kjeldahl method. They found that 0.2 g of plant material and a 20-mL dye ratio was best correlated with Kjeldahl N if one regression equation was used for all stages of rice plant growth. The Orange-G dye method is a rapid, simple, inexpensive, and acceptable method for determining the N status of rice. This test may be substituted for the Kjeldahl method in monitoring N concentration changes in rice and in determining N fertilizer requirements in mid-season.

V. PRACTICAL APPLICATIONS OF PLANT ANALYSIS

Plant analysis can be useful in diagnosing plant nutrient deficiencies and toxicity problems of small grains when used in combination with soil analysis, plant symptomology, and experience. An effective plant nutrient monitoring program will identify specific plant nutrients required and their optimum rates over time for maximum economic yield. However, under most situations, plant analysis is not as effective in guiding fertilizer usage on annual crops as is soil testing. Soil testing provides an opportunity for the producer to identify potential nutrient deficiencies before planting, thereby ensuring that plants do not become deficient early in the growing season. Plant analysis allows in-season monitoring and adjustment, and thus ensures optimum nutrient relationships during the entire growing season. Wide use of both soil testing and plant analysis will be beneficial to most producers. However, care should be taken to ensure that plant analysis results are interpreted from appropriate data that are applicable to the crop and growing environment.

REFERENCES

Baba, I. 1954. Breeding of rice suitable for heavy manuring. Absorption and assimilation of nutrients and its relation to adaptability for heavy manuring and yield in rice variety. Jpn. J. Breed. 4:167–184.

Bajaj, J.C., M.L. Gulati, and R.V. Tamhane. 1967. Correlation studies for soil tests for available nitrogen with nitrogen uptake and responses of paddy and wheat. J. Indian Soc. Soil Sci. 15:29–33.

Baker, J.M., and B.B. Tucker. 1971. Effects of rates of N and P on the accumulation of NO_3-N in wheat, oats, rye and barley on different sampling dates. Agron. J. 63:204–207.

Beringer, H., and G. Hess. 1979. Feasibility of plant analysis for estimating late N requirements of winter wheat. Landwirtsch. Forsch. 32:384–394.

Bishop, R.F., and C.R. MacEachern. 1971. Response of spring wheat and barley to nitrogen, phosphorus, and potassium. Can. J. Soil Sci. 51:1–11.

Boatwright, G.O., H. Ferguson, and P.L. Brown. 1964. Availability of P from superphosphate to spring wheat as affected by growth stage and surface soil moisture. Soil Sci. Soc. Am. Proc. 28:403–406.

Brandon, D.M. 1977. Phosphorus transformations in alternately flooded soils and their effects on rice-rotation crops. Ph.D. thesis. Univ. of California, Davis (Diss. Abstr. 7809215).

Brandon, D.M., H.L. Carnahan, J.N. Rutger, S.T. Tseng, C.W. Johnson, J.F. Williams, C.M. Wick, W.M. Canevari, S.C. Scardaci, and J.E. Hill. 1981a. California rice varieties: Description, performance and management. Div. Agric. Sci., Univ. of California, Spec. Publ. 3271, p. 17–21.

Brandon, D.M., and D.S. Mikkelsen. 1979. Phosphorus transformations in alternately flooded soils: I. Cause of plant phosphorus deficiency in rice rotation crops and correctional methods. Soil Sci. Soc. Am. J. 43:989–994.

Brandon, D.M., D.S. Mikkelsen, C.M. Wick, J.F. Williams, and L.A. Post. 1980. Relationship between "Y" leaf nitrogen concentration and yield components of the short stature rice variety M9. p. 86–87. *In* 18th Rice Tech. Work Group, Davis, CA. Texas Agric. Exp. Stn., Texas A&M Univ., College Station.

Brandon, D.M., R.P. Mowers, R.H. Brupbacher, H.F. Morris, W.J. Leonards, Jr., T.R. Laing, S.M. Rawls, and N.J. Simoneaux. 1982. Nitrogen requirements of new rice varieties and relationship between "Y" leaf N and grain quality yield. p. 52–100. *In* 74th Annu. Prog. Rep. Rice Res. Stn. Louisiana State Univ., Crowley.

Brandon, D.M., and B.R. Wells. 1986. Improving nitrogen fertilization in mechanized rice culture. Fert. Res. 9:161–170.

Brandon, D.M., F.E. Wilson, Jr., P.E. Schilling, R.H. Bruphacher, H.F. Morris, Jr., and W.J. Leonards, Jr. 1981b. Nitrogen requirements of new rice varieties and relationship between "Y" leaf N and grain yield. p. 31–74. *In* 73rd Ann. Prog. Rep. Rice Res. Stn., Louisiana State Univ., Crowley.

Brown, B.D., and J.P. Jones. 1979. Tissue analysis for nitrogen fertilization of irrigated soft white winter wheat. Idaho Coop. Ext. Ser., Current Info. Ser. 461.

Brown, B.D., and J.P. Jones. 1979. Nitrogen tissue tests for barley and oat. p. 169. *In* Agronomy abstracts. ASA, Madison, WI.

Brown, B.D., and J. Stark. 1986. Hard red and soft white winter wheat response to nitrogen. *In* Proc. 37th Annu. NWPFA Fert. Conf., Boise, ID. 8–9 July.

Brooks, S.L., and H.M. Reisenauer. 1985. Plant nitrogen as indicators of growth rate. J. Plant Nutr. 8:63–71.

Chapman, H.D. 1964. Foliar sampling for determining the nutrient status of crops. World Crops 16:36–46.

Chapman, H.D. 1966. Diagnostic criteria for plants and soils. Div. Agric. Sci., Univ. of California, Berkeley.

Crooke, W.M., and A.H. Knight. 1971. Crop composition in relation to soil pH and root cation exchange capacity. J. Sci. Food Agric. 22:235–241.

Donohue, S.J., and D.E. Brann. 1984. Optimum N concentration in winter wheat grown in the Coastal Plain Region of Virginia. Commun. Soil Sci. Plant Anal. 15:651–661.

Dow, A.J. 1980. Critical nutrient ranges in northwest crops. Western Reg. Publ. 43. Washington State Univ. Irrigated Agric. Res. and Ext. Ctr., Prosser, WA.

Engel, R.E., and J.C. Zubriski. 1982. Nitrogen concentrations in spring wheat (*Triticum aestivum*) at several growth stages. Commun. Soil Sci. Plant Anal. 13:513–544.

Flint, M.L. 1983. Integrated pest management for rice. Div. Agric. Sci. Univ. of California Publ. 3280.

Forno, D.A., C.J. Asher, and S. Yoshida. 1975. Zinc deficiency in rice. I. Soil factors associated with the deficiency. Plant Soil 42:551–562.

Gardner, B.R., and E.B. Jackson. 1976. Fertilizer, nutrient composition and yield relationships in irrigated spring wheat. Agron. J. 68:75–78.

Gladstone, J.S., and J.F. Loneragan. 1967. Mineral elements in temperature crop and pasture plants. I. Zinc. Aust. J. Agric. Res. 18:427–466.

Greaves, J.E., A.F. Bracken, and C.T. Hirst. 1940. The influence of variety, season and green manure upon the composition of wheat. J. Nutr. 19:179–186.

Griffin, J.L., and D.M. Brandon. 1983. Effect of lowland rice culture on subsequent soybean response to phosphorus fertilization. Field Crops Res. 7:195–201.

Guettinger, D.L., and F.E. Koehler. 1963. Nitrogen relationships in wheat plants. Agron. J. 55:409–410.

Hafez, A.A., and D.S. Mikkelsen. 1981. Colormetric determination of nitrogen for evaluating the nutritional status of rice. Commun. Soil Sci. Plant Anal. 12:61–69.

Hart, J.M., and G.S. Pettygrove. 1985. A quick test for nitrate status of wheat. *In* Proc. 36th Annu. Northwest Fert. Conf., Salt Lake City, UT. 16–17 July.

Hassan, N.A.K., J.V. Drew, D. Knudsen, and R.A. Olson. 1970. Influence of soil salinity on production of dry matter and uptake and distribution of nutrients in barley and corn: I. Barley (*Hordeum vulgare* L.). Agron. J. 62:43–45.

Helms, R.S. 1984. A rice cultural management study. Ph.D. diss. Louisiana State Univ., Baton Rouge (Diss. Abstr. DA8511753).

Houng, K.H., T.P. Liu, C.T. Wei, and T.T. Chen. 1971. Changes in distribution of macroelements among various tissues of rice plants with growth. II. Distribution of P and K. J. Chin. Agric. Chem. Soc. 9(9):67–72.

Johnson, W.J., and W.G. Schrenk. 1953. The effect of fertilizer treatment on the composition of wheat at different stages of growth in southeast Kansas. Trans. Kans. Acad. Sci. 56:456–464.

Jones, J.B., Jr., R.L. Large, D.B. Pfleider, and H.S. Klasky. 1971. The proper way to take a plant sample for tissue analysis. Crops Soils 23(8):15–18.

Karlen, D.L., R. Ellis, Jr., D.A. Whitney, and L.L. Grunes. 1978. Influence of soil moisture and plant cultivar on cation uptake by wheat with respect to grass tetany. Agron. J. 70:918–921.

Kiuchi, T., and Y. Ishizuka. 1961. Effect of nutrients on the yield of constituting factors of rice. J. Soc. Soil Manure (Jpn) 32(5):198.

Kleese, R.A., D.C. Rasmusson, and L.H. Smith. 1968. Genetic and environmental variation in mineral element accumulation in barley, wheat and soybeans. Crop Sci. 8:591–593.

Koyama, T., C. Chammek, and P. Snitwongse. 1973. Varietal difference of Thai rice in resistance to phosphorus deficiency. Tech. Bull. 4. Trop. Agric. Res. Ctr., Ministry of Agric. and Forestry, Japan.

Large, E.C. 1954. Growth stages in cereals: Illustration of the Feekes scale. Phytopathology 3:128–129.

Leigh, R.A., and A.E. Johnston. 1983a. Concentrations of potassium in the dry matter of tissue water of field-grown spring barley and their relationships to grain yield. J. Agric. Sci. 101:675–685.

Leigh, R.A., and A.E. Johnston. 1983b. The effects of fertilizer and drought on the concentrations of potassium in the dry matter and tissue water of field-grown spring barley. J. Agric. Sci. 101:741–748.

Leigh, R.A., and A.E. Johnston. 1985. Nitrogen concentrations in field-grown spring barley: An examination of the usefulness of expressing concentrations on the basis of tissue water. J. Agric. Sci. 105:397–406.

Leigh, R.A., and A.E. Johnston. 1986. An investigation of the usefulness of phosphorus concentrations in tissue water as indicators of the phosphorus status of field-grown barley. J. Agric. Sci. 107:329–333.

Leikam, D.F., L.S. Murphy, D.E. Kissel, D.A. Whitney, and H.C. Moser. 1983. Effects of nitrogen and phosphorus application method and nitrogen source on winter wheat grain yield and leaf tissue phosphorus. Soil Sci. Soc. Am. J. 47:530–535.

Letey, J., L.H. Stolzy, H. Valoras, and T.E. Szuszkiewiez. 1962. Influence of soil oxygen on growth and mineral concentration of barley. Agron. J. 54:538–540.

Lian, S. 1972. Response of rice yield to nitrogen top-dressing during panicle initiation stage and the diagnosis. J. Taiwan Agric. Res. 21:11–17.

Lian, S. 1976. Silica fertilization of rice. p. 205–206. *In* The fertility of paddy soils and application for rice. ASPAC Food and Fert. Tech. Ctr., Taipei, Taiwan.

Lin, C.F., A.H. Chang, and C.S. Tseng. 1973. The nitrogen status and nitrogen supplying power of Taiwan soils. J. Taiwan Agric. Res. 22(3):186–203.

Lindsay, W.L. 1972. Zinc in soils and plant nutrition. Adv. Agron. 24:147–186.

Mahapatra, I.C., and W.H. Patrick, Jr. 1969. Inorganic phosphorus transformation in waterlogged soil. Soil Sci. 197:281–287.

Martin, W.E., R.L. Sailsbery, D.M. Brandon, and R.T. Petersen. 1971. Answering the riddle of poor safflower after rice. Calif. Agric. 25(9)1:4–6.

Maume, L., and J. Dulac. 1954. Controle de la nutrition du rize par le diagnostic foliare. J. Riz (Arles) 4:93–100.

Maxwell, T.M., D.E. Kissel, M.G. Wagger, D.A. Whitney, M.L. Cabrera, and H.C. Moser. 1984. Optimum spacing of preplant bands of N and P fertilizer for winter wheat. Agron. J. 76:243–247.

Mikkelsen, D.S. 1983. Diagnostic plant analysis for rice. p. 30–13. *In* H.M. Reisenauer (ed.) Tissue testing in California Bull. 1879.

Mikkelson, D.S., M.D. Miller, M. Brandon, C. Wick, and J. Lindt. 1970. Plant analysis, a technique for diagnosing the nutritional status of rice. p. 56. *In* Proc. 13th Rice Tech. Work. Group, Beaumont, TX. 24–26 Feb. Texas Agric. Exp. Stn., Texas A&M Univ., College Station.

Mikkelsen, D.S., and D.M. Brandon. 1972. Zinc deficiency in the soil-water-rice system. p. 81 *In* Proc. 14th Rice Tech. Work. Group, Davis, CA. 20–22 June. Texas Agric. Exp. Stn., Texas A&M Univ., College Station

Mikkelsen, D.S., and D.M. Brandon. 1975. Zinc deficiency in California rice. Calif. Agric. 29(9):8–10

Mikkelsen, D.S., and R.R. Hunziker. 1971. A plant analysis survey of California rice. Agrichem. Age 14(6):18–22.

Mikkelsen, D.S., and S. Kuo. 1977. Zinc fertilization and behavior in flooded soils. Spec. Bull. 5. Commonwealth Bureau of Soils, Commonwealth Agric. Bureau, Farnham Royal, England.

Mitsui, S., K. Kumazama, and M. Singh. 1959. Asparagine test in relation with nitrogen nutritional status of crop plants. 1. Rice. Soil Plant Food (Tokyo) 4:189–195.

Nielson, K.F., R.L. Holstead, A.J. MacLean, R.M. Holmes, and S.J. Bourget. 1960. The influence of soil temperature on the growth and mineral composition of oats. Can. J. Soil Sci. 40:255–263.

Neubert, P., V.P. Vielemeyer, I. Hudt, and G. Vanselow. 1977. Investigations on the parameters of the plant analysis technique for winter cereals. Tagungsber. Akad. Landwirtschaftswiss. D.D.R. 149:123–130.

Ozaki, K. 1955. Determination of top-dressing for rice in field by detection of asparagine. Soil Plant Food (Tokyo) 1:88–90.

Papastylianou, I. 1986. Diagnosis of nitrogen deficiency in barley grown in different rotation systems by plant analysis. Fert. Res. 9:291–250.

Patrick, W.H., Jr., and I.C. Mahapatra. 1968. Transformation and availability to rice of nitrogen and phosphorus in waterlogged soils. Adv. Agron. 20:323–259.

Peterson, G.A., D.H. Sanders, P.H. Grabowski, and M.L. Hooker. 1981. A new look at row and broadcast phosphate recommendations for winter wheat. Agron. J. 73:13–17.

Plank, C.O. 1988. Plant analysis handbook for Georgia. Coop. Ext. Serv. Univ. of Georgia, Athens.

Ponnamperuma, F.N. 1972. The chemistry of submerged soils. Adv. Agron. 24:29–96.

Ponnamperuma, F.N. 1976. Screening rice for tolerance to mineral stresses. p. 341–353. *In* M.J. Wright and S.A. Ferrari (ed.) Proc. of Workshop on Plant Adaption to Mineral Stress in Problem Soils, Beltsville, MD. 22–23 Nov.

Power, J.F., D.L. Grunes, and G.A. Reichman. 1961a. The influence of phosphorus fertilization and moisture on growth nutrient absorption by spring wheat: I. Plant growth, N uptake and moisture use. Soil Sci. Soc. Am. Proc. 25:207–210.

Power, J.F., G.A. Reichman, and D.L. Grunes. 1961b. The influence of phosphorus fertilization and moisture on growth and nutrient absorption by spring wheat: II. Soil and fertilizer P uptake in plants. Soil Sci. Soc. Am. Proc. 25:210–213.

Power, J.F., D.L. Grunes, G.A. Reichman, and W.D. Willis. 1970. Effect of soil temperature on rate of barley development and nutrition. Agron. J. 62:567–571.

Rassmussen, P.E., R.E. Famig, L.G. Ekin, and C.R. Rohde. 1977. Tissue analysis guidelines for diagnosing sulfur deficiency in white wheat. Plant Soil 46:153–163.

Reuter, D.J., and J.B. Robinson (ed.). 1986. Plant analysis, an interpretation manual. Inkata Press, Melbourne, Australia.

Sah, R.N., and D.S. Mikkelsen. 1986. Transformation of inorganic phosphorus during the flooding and draining cycles of soil. Soil Sci. Soc. Am. J. 50:62–67.

Sajwan, K.S., and W.L. Lindsay. 1986. Effects of redox on zinc deficiency in paddy rice. Soil Sci. Soc. Am. J. 50:1264–1269.

Sedberry, J.E., F.J. Peterson, E. Wilson, A.L. Nugent, R.M. Engler, and R.H. Brupbacher. 1971. Effect of zinc and other elements on the yield of rice and nutrient contents of rice plants. Louisiana Agric. Exp. Stn. Bull. 653.

Sedberry, J.E., F.J. Peterson, F.E. Wilson, D.B. Mengel, S.D. Maika, P.E. Schilling, and R.H. Brupbacher. 1980. Influence of soil reaction and zinc application on yield and zinc contents of rice plants. Louisiana Agric. Exp. Stn. Bull. 724.

Shapiro, R.E. 1958. Effect of flooding on availability of phosphorus and nitrogen. Soil Sci. 85:190–197.

Sheng, C.Y., N.R. Su, T.C. Lin, and M.P. Feng. 1964. Correlation among soil PK values, plant analysis and response of rice to add fertilizer in Taoyuan Latosols. J. Agric. Assoc. China. New Ser. 48:18.

Sims, J.L., and B.G. Blackmon. 1967. Predicting nitrogen availability to rice. II. Assessing available nitrogen in silt loams with different previous year crop history. Soil Sci. Soc. Am. Proc. 31:676–680.

Sing, M., K. Kumazama, and S. Mitsui. 1960. Asparagine test in relation with nitrogen nutritional status of crop plants. III. Rice Soil Plant Food (Tokyo) 5:167–173.

Singh, J.P., R.E. Karamonos, and J.W.B. Stewart. 1986. Phosphorus-induced zinc deficiency in wheat on residual phosphorus plots. Agron. J. 78:668–675.

Singhania, R.A., and N.N. Goswani. 1978. Transformation of applied phosphorus under simulated conditions of growing rice and wheat in sequence. J. Ind. Soc. Soil Sci. 26:192–197.

Smika, D.E., and R. Ellis, Jr. 1971. Soil temperature and wheat straw mulch effects on wheat plant development and nutrient concentration. Agron. J. 63:388–391.

Spencer, K., and J.R. Freney. 1980. Assessing the sulfur status of field-grown wheat by plant analysis. Agron. J. 72:469–472.

Su, N.R. 1976. Potassium fertilization of rice. p. 117–148. *In* The fertility of paddy soils and fertilizer applications for rice. ASPAC Food and Fert. Tech. Ctr., Taipei, Taiwan.

Takahashi, J., M. Yanagisawa, M. Kono, F. Yazawa, and T. Yoshida. 1955. Studies on nutrient absorption by crops. Bull. Natl., Inst. Agric. Sci. 4:1–83.

Tanaka, A. 1961. Studies on the nutri-physiology of leaves of rice plants. J. Fac. Agric. Hokkaido Imp. Univ. 51(3):449–550.

Tanaka, A., S. Patnaik, and C.T. Abichandani. 1959. Studies on the nutrition of rice plant. IV. Growth and nitrogen uptake of rice varieties of different durations. Proc. Ind. Acad. Sci. B 49(4):217–226.

Tanaka, A., R. Loe, and S.A. Navasero. 1966. Some mechanisms involved in the development of iron toxicity symptoms in the rice plant. Soil Sci. Plant Nutr. 12:158–162.

Tanaka, A., and S. Yoshida. 1970. Nutritional disorders of the rice plant in Asia. Int. Rice Res. Inst., Tech. Bull. 10.

Thenabadu, M.W. 1966. The nitrogen nutrition of *Oryza sativa* and its status as evaluated by plant analysis. Ph.D. diss. Univ. of California, Davis (Diss. Abstr. 66-13, 158).

Vandecaveye, S.C. 1940. Effects of soil type and fertilizer treatments on the chemical composition of certain forage and small grain crops. Soil Sci. Soc. Am. Proc. 11:305-308.

Vaughan, B., D.G. Westfall, K.A. Barbarick, and P.L. Chapman. 1990. Spring N fertilizer recommendation models for dryland winter wheat. Agron. J. 82:561-565.

Viets, F.G., Jr., L.C. Brown, and C.L. Crawford. 1954. Zinc contents and deficiency symptoms of 26 crops grown on a zinc-deficient soil. Soil Sci. 78:305-316.

Wallihan, E.F., and J.C. Moomaw. 1967. Selection of index leaf for studying the critical concentration of nitrogen in rice plants. Agron. J. 59:473-474.

Wallihan, E.F., and R.G. Sharpless. 1974. Effect of sulfur supply on the optimum concentration of nitrogenin leaves of the rice plant. Soil Sci. 118:304-307.

Ward, R.C., C.A. Whitney, and D.G. Westfall. 1973. Plant analysis as an aid in fertilizing small grains. p. 329-348. *In* L.M. Walsh and J.D. Beaton (ed.) Soil testing and plant analysis. SSSA, Madison, WI.

Wang, C.H. 1976. Sulfur fertilization of rice. *In* The fertility of paddy soil and fertilizer applications for rice. ASPAC Food and Fert. Tech. Ctr., Taipei, Taiwan.

Weissert, P., P. Neubert, and H-P. Vielemeyer. 1982. Results of methodological studies on plant sampling in winter grain stands. Arch. Acker-Pflanzenbau Bodenkd. 26:93-100.

Wells, B.R., R.K. Bacon, W.E. Sabbe, and R.L. Sutton. 1986. Response of sulfur deficient wheat to sulfur fertilizer. J. Fert. Issues 3:72-74.

Westfall, D.G., W.B. Anderson, and R.J. Hodges. 1971. Iron and zinc response of chlorotic rice grown on calcareous soils. Agron. J. 63:702-705.

Westfall, D.G., W.T. Flinchum, and J.W. Stansel. 1973. Distribution of nutrients in the rice plant and effect of two nitrogen levels. Agron. J. 65:236-238.

Westfall, D.G., J.M. Ward, C.W. Wood, and G.A. Peterson. 1987. Placement of phosphorus for summer fallow dryland winter wheat production. J. Fert. Issues 4:114-121.

Westerman, R.L., and M.G. Edlund. 1985. Deep placement effects of nitrogen and phosphorus on grain yield, nutrient uptake, and forage quality of winter wheat. Agron. J. 77:803-809.

Whitney, D.A., J.T. Cope, and L.F. Welch. 1985. Prescribing soil and crop nutrient needs. *In* O.P. Engelstad (ed.) Fertilizer technology and use. 3rd ed. SSSA, Madison, WI.

Willett, I.R., W.A. Muirhead, and M.L. Higgins. 1977. The effects of rice growing on soil phosphorus immobilization. Australian J. Exp. Agric. Anim. Husb. 18:270-275.

Ylaranta, T., H. Johnson, and J. Sippolo. 1979. Seasonal variations in micronutrient contents of wheat. Ann. Agric. Fenn. 18:218-224.

Yoshida, S. 1981. Mineral nutrition of rice. p. 111-176. *In* Fundamentals of rice crop science. Int. Rice Res. Inst., Los Baños, Philippines.

Yoshida, S., J.S. Ahn, and D.A. Forno. 1973. Occurrence, diagnosis, and correction of zinc deficiency in lowland rice. Soil Sci. Plant Nutr. 19:83-93.

Yoshida, S., and A. Tanaka. 1969. Zinc deficiency in the rice plant in calcareous soils. Soil Sci. Plant Nutr. 15(2):75-80.

Chapter 20

Plant Analysis as an Aid in Fertilizing Corn and Grain Sorghum[1]

J. BENTON JONES, JR., *University of Georgia, Athens*

HAROLD V. ECK, *USDA-ARS, Bushland, Texas*

REGIS VOSS, *Iowa State University, Ames*

The literature probably contains more plant analysis data for corn (*Zea mays* L.) than for any other field crop, a marked contrast to that available for grain sorghum [*Sorghum bicolor* (L.) Moench] which is quite limited. Although many researchers have used data for corn to evaluate plant analyses for grain sorghum, such evaluations have been questioned. Lockman (1972a, b) found minimal differences between the analyses of the whole tops of corn and sorghum seedlings. Differences, however, increased as growth progressed. Bennett (1971) also indicated that corn plant analysis data should not be used to evaluate grain sorghum plant analyses. As more data become available, such as that compiled by Reuter and Robinson (1986), grain sorghum should be treated separately from corn. However, because the two crops contain similar amounts of most elements and because the literature related to grain sorghum is scant, this chapter treats the two crops similarly unless specific data are available that distinguish between the two.

Plant analysis has only recently come into its own as a technique for evaluating field nutrient problems. Tyner (1946) proposed critical N, P, and K values for corn leaf analyses some years ago. Several books, bulletins, and review articles have been written on plant analysis (Goodall & Gregory, 1947; Lundegardh, 1951; Ulrich, 1952; Wallace, 1956; Reuther, 1960; Smith, 1962; Chapman, 1966; Wallace, 1961; Hardy, 1967; Greer, 1970; Bates, 1971; Davidescu & Davidescu, 1972; Jones, 1974; Reisenauer, 1978; Jones, 1984a, Reuter & Robinson, 1986; Martin-Prevel et al., 1987). Much of the current surge in interest in the plant analysis technique stems from the rather significant developments in methods for conducting the laboratory assay of plant

[1] Contribution from the Univ. of Georgia, Athens, GA, the USDA Conserv. and Production Res. Lab., Agric. Res. Serv., Bushland, TX, and Iowa State Univ., Ames, IA.

 Soil Testing and Plant Analysis, 3rd ed.—SSSA Book Series, no. 3.

tissue. Atomic absorption, flame emission, and direct reading spark and plasma emission spectroscopy have greatly simplified analytical procedures. Analyses for most of the essential elements can be made quickly and at relatively low cost. Plant analysis services to farmers in the USA are being provided by an ever-increasing number of state-supported and commercial laboratories. Surveys conducted by the USDA-Extension Service (Jones, 1985) since 1980 have found that approximately 500 000 plant tissue samples are being assayed annually for farmers.

Goodall and Gregory (1947) found that research related to plant analysis could be classed into four categories: (i) investigation of nutritional disorders manifested by definite symptoms, (ii) interpretation of the results of field trials, (iii) development of rapid testing methods for use in advisory work, and (iv) use of plant analysis as a method of nutritional survey. These categories are still applicable today. Practical application of the plant analysis technique in the field primarily relates to confirming suspected elemental deficiencies or toxicities, monitoring the crop to uncover significant changes in elemental composition which could lead to deficiencies or excesses (Jones, 1983), and predicting fertilizer needs (Baker et al., 1966). An interesting application of the plant analysis technique yet to be fully exploited is the evaluation of genetically controlled element uptake and utilization (Munson, 1969).

Although their utilization is quite different, plant analysis has frequently been compared to soil testing in field application. The plant's ability to absorb an element from the soil solution is reflected in the concentration of the element in the plant or in one of its specific parts. Therefore, a plant analysis can be a means of measuring the soil-plant nutrient element environment. Along these lines, Krantz et al. (1948) gave four principal objectives for a plant analysis: (i) to aid in determining the nutrient-supplying power of the soil, (ii) to aid in determining the effect of treatment on the nutrient supply in the plant, (iii) to study the relationship between the nutrient status of the plant and crop performance as an aid in predicting fertilizer requirements, and (iv) to help lay the foundation for approaching new problems or surveying unknown regions to determine where critical plant nutritional experimentation should be conducted.

None of these objectives fits the current use of plant analysis. Today, most farmers use it primarily to diagnose suspected essential element insufficiencies. The second most common use for the plant analysis is probably best described by Objective iii, although it is usually limited to use with fruit and nut trees. Munson and Nelson (1973) describe three uses for a plant analysis, verification of deficiency symptoms, for comparing assays of normal vs. abnormal plants, and for crop logging. Each one is an important application for a plant analysis result.

Although most soil tests give reliable information regarding the supply of plant-available nutrients in the soil, they are indirect measures of the sufficiency of nutrients in plants and, thus, have limited usefulness in diagnostic situations. Plant analysis gives direct information and, when properly interpreted, is the most reliable diagnostic test. However, when used together, the combined results can effectively evaluate the sufficiency of the soil-plant nutrient status to satisfy the crop requirement, and can form the basis for specifying needed corrective treatments (Jones, 1986).

The plant analysis technique involves a sequence of steps from sampling to sample preparation and laboratory analysis, and then to an interpretation of the laboratory result. Each step requires particular skill by the user. The desired plant part must be properly identified and collected; decontaminated, if necessary; and prepared for analysis without contaminating or altering its chemical content. Interpretation of an analysis is to some degree an acquired art, gained from experience on how best to apply known interpretative values to evaluate a particular analysis result. This chapter is a review of the current state of the art for the plant analysis technique when specifically applied to corn and grain sorghum.

I. SAMPLING

The validity and usefulness of a plant analysis hinges on the care and method used to obtain the required plant tissue sample. Unfortunately, less research has been devoted to sampling than to other aspects of the plant analysis technique; yet sampling, if not properly done, can invalidate the analysis result.

Plant part selected and time of sampling must correspond to the best relationship that exists between element concentration and physical appearance of the plant or yield (Bates, 1971). It means selecting a specific part at a specific location on the plant and at a definite growth stage. The number of parts, such as leaves, to collect per plant and the number of plants chosen for sampling deserve careful consideration. For both corn and grain sorghum, only one leaf per plant is normally collected. However, the number of plants from which a tissue sample is to be taken is dictated by the condition of the plants. Steyn (1961) observed that to obtain representative samples, plants in poor condition required more intensive sampling than those free from nutritional stress. Consequently, he recommended taking samples from as many plants as practical.

Most elements are not evenly distributed in the corn leaf blade (Sayre, 1952). Jones (1970) found that certain elements tend to accumulate in the leaf margins while some element concentrations are lower in the midrib than those found in the leaf blade. Therefore, an elemental content assay of the entire leaf could be affected by the shape of the leaf blade. An analysis of a typical corn leaf and its components are given in Table 20–1. Chapman (1964) must have recognized this problem of uneven elemental distribution in the corn leaf because he suggested that only the mid-third section of the leaf be analyzed to minimize the influence of margin and midrib contents on a whole-leaf analysis. Although Chapman's sampling procedure is worthy of adoption, all of the significant interpretative data for corn are given based on a whole-leaf analysis. In light of this heterogeneous distribution of elements in the corn leaf, it is important that a leaf sample consists of the entire leaf or a major portion of it. Sumner (1977b) found that the entire, or

Table 20-1. Distribution of 14 elements in the various parts of a corn ear leaf at the time of stilk (Jones, 1970).

Element	Whole ear leaf	Leaf part		
		Margin†	Blade	Midrib
	g kg^{-1}			
N	25.4	25.8	25.0	7.4
P	2.7	2.9	3.2	1.4
K	16.8	12.1	17.3	18.2
Ca	8.0	9.0	8.8	4.2
Mg	5.6	3.9	6.2	5.5
	mg kg^{-1}			
Al	95	88	99	52
B	14	51	8	6
Ba	8	6	8	11
Cu	11	12	12	8
Fe	148	181	171	79
Mn	146	272	124	70
Mo	2.6	2.0	3.0	2.5
Sr	30	22	28	27

† Consists of the 4-mm part along the edge of the leaf.

at least three-quarters, leaf must be taken from the middle of the corn plant if a valid interpretation is to be made using the Diagnosis and Recommendation Integrated System (DRIS) interpretative method.

The specific sampling procedures for corn and grain sorghum are quite different because of the nature of the two species. Historically, the plant part designated for sampling was probably selected more on the basis of its ease of identification than on the significance of its composition as related to either the plant's appearance or yield. Sampling procedures for many crops, including corn and grain sorghum have been described by Jones et al. (1971). Specific sampling procedures for both crops are given in Table 20-2.

In a diagnostic situation when a nutrient element insufficiency is suspected, it is advisable to collect, and keep separate, identical tissues from

Table 20-2. Sampling procedures for corn and grain sorghum.

Crop	Plant part	Time of sampling	No. of plants to sample
Corn or grain sorghum	Whole aboveground portion of plant	Seedling stage, <30 cm (12 in.) tall	20-30
Corn or grain sorghum	The entire, fully developed leaf below the whorl	Prior to tasseling or heading stage	12-25
Corn†	The entire leaf at the ear node (or immediately above or below it)	From tasseling to the silk initiation	12-25
Grain sorghum†	Second leaf from the top of the plant	At heading	15-25

† Recommended sampling procedure when determining nutrient status of plant as it relates to fertility status of the soil, fertilizer treatments, and yields.

both affected as well as nearby normal-appearing plants so that a comparison of analysis results can be made as suggested by Munson and Nelson (1973). However, there is some danger in this approach if the affected plants have been in that condition for an extended period (Jarrell & Beverly, 1981). Therefore, considerable care is required when selecting plants for sampling to ensure that both stage of growth as well as degree of development are reasonably similar, and that similarly mature leaf tissue exists for sampling.

There are as many instructions on what not to sample as on what to sample. Sampling is not recommended when the plant part is soil or dust covered, damaged by insects, mechanically injured, or diseased. Dead plant tissue should not be included in collected samples. In addition, sampling is not recommended when plants have been under prolonged moisture or temperature stress.

Seed is not normally a useful plant part to assay for determining the nutritional status of the sampled plant. Although the element concentration in the grain can be affected by fertilizer treatment and the nutrient status of the plant, grain yield is usually affected more than grain composition. Also, little interpretative data is available based on the elemental content of grain, with the exception of the work by Pierre et al. (1977a, b) and Russell and Pierre (1980) who studied the relationship between corn grain yield and the N content of the grain.

II. SAMPLE PREPARATION AND ANALYSIS

This topic is covered in more detail in chapter 15 of this book (Jones and Case). However, it may be well to discuss some of the procedures that apply specifically to the analysis of corn and grain sorghum plants and their leaf tissues.

A. Decontamination

Decontamination by washing plant and leaf tissue is not necessary unless the plants are unusually dusty or coated with spray residues. If Fe is an element of interest for determination, the tissue *must be* washed (Jones, 1963; Baker et al., 1964; Wallace et al., 1980). Surface-irrigated plants, or those growing under negligible rainfall conditions, require washing. Plants growing under normal rainfall conditions probably would not require washing. Steyn (1959) recommended washing fresh leaf material with a mild 0.1 to 0.3% detergent solution and then rinsing in pure water. Similar decontamination procedures have been described by Wallace et al. (1980), and Sonneveld and van Dijik (1982). Jacques et al. (1974) reported that washing grain sorghum plant tissue parts with deionized distilled water reduced Fe concentrations of blade, sheath, and head tissue but did not affect Ca, Cu, Mg, Mn, and Zn concentrations in the tissue.

Washing should be done quickly to avoid long periods of contact between the washing solution and plant material that can lead to leaching of

Table 20–3. Element concentration found between washed and unwashed corn leaves.

Element	Baker et al. (1964)		Jones (1963)	
	Unwashed	Wiped with detergent solution	Unwashed	Washed in distilled water
	g kg^{-1}			
Ca	8.7	8.5	4.8	4.5
K	20.4	21.7	12.2	12.6
Mg	1.6	1.6	3.9	4.1
N	--	--	29.8	31.1
P	2.7	2.9	2.2	2.2
	mg kg^{-1}			
B	17.2	17.0	11	10
Cu	29.5	28.8	9	10
Fe	136	134	96	85
Mn	--	--	73	64
Mo	--	--	1.1	1.2
Zn	23.1	30.5	22	22

some elements (Bhan et al., 1959). Examples of the effect of washing on corn leaf assays are given in Table 20–3.

B. Drying and Particle-Size Reduction

Drying should be rapid to minimize dry weight losses (Lockman, 1970), but at a temperature that will not cause the tissue to decompose. Corn and grain sorghum plant tissues can be easily oven dried at 80 °C in 48 h or less in a dust-free, forced-draft oven. The temperature should not exceed 85 °C because higher temperatures can result in significant thermal losses (Steyn, 1959). Tissue can also be dried in a microwave oven (Carlier & van Hee, 1971; Shuman & Rauzi, 1981) in about 10 min, although only small quantities of material can be dried at a time. Isaac and Johnson (1983) successfully microwave dried corn leaves prior to protein N assay by near infrared spectroscopy.

Reducing the dried tissue to a particle size suitable for laboratory analysis, and at the same time, ensuring a greater degree of homogeneity for the laboratory-prepared sample, is done either by the cutting action of a Wiley or hammer mill, by crushing in a ball mill, or by abrasion in a cyclone (UDY) mill. Most mills will contaminate the sample with particles of the contact surfaces, such as Cu and Zn from brass fittings, and even Fe from fittings or cutting and crushing surfaces made of steel. To avoid Fe contamination, tissue sample reduction must be done either by hand cutting or crushing in an agate ball mill. Grinding devices with Al, plastic (will add Na), or rubber (will add Zn) fittings are potential sources for contamination.

A Wiley mill fitted with a 20-mesh screen is commonly used for milling plant tissue. Particle-size reduction to $<$20 mesh is not necessary unless aliquots of $<$0.5 g are to be assayed in the laboratory. The finer the screen (40 or 60 mesh), the more homogeneous the milled sample will be, but finer grinding requires longer milling time. The longer contact time between the tissue and mill surfaces increases the potential for contamination from the mill surfaces (Hood et al., 1944).

With most milling procedures, segregation of the finer particles is likely. This can be partially controlled by eliminating static electricity build-up and by carefully mixing the milled sample (Smith et al., 1968; Nelson & Boodley, 1965). Adherence of the finer plant particles to mill components can be partially overcome by a vacuum system attached to the mill (Graham, 1972), or by using pulsing air (Ulrich, 1984). When using a Wiley mill for milling corn and grain sorghum plant tissues, which are often fibrous and will easily segregate according to particle size, sufficient time should be allowed for the entire sample to pass through the mill. Ball or cyclone milling corn or grain sorghum leaf tissues will usually produce a more homogeneous laboratory sample than milling in a Wiley-type cutting mill.

C. Organic Matter Destruction

A more complete discussion on how best to destroy the plant tissue's organic matter content may be found in chapter 15 (Jones and Case). Either 8-h muffling at 500 °C followed by hot acid solution (Munter & Grande, 1981), or wet digestion in various mixtures of acids (Tolg, 1974), are possible methods for the destruction of organic matter in corn and grain sorghum plant tissues. Block digestion procedures, described by Zasoski and Burau (1977) and Havlin and Soltanpour (1980), have particular appeal when large numbers of samples are to be digested by wet oxidation. Books by Gorsuch (1970) and Bock (1978) are good references on methods used for organic matter destruction.

D. Laboratory Analysis

Advances in instrumental analytical chemistry since 1970 have had a great impact on methods suitable for the assay of plant tissue digests for their elemental contents. Traditional wet chemistry procedures described by Piper (1942), Jackson (1958); Johnson and Ulrich (1959), as well as flame emission and atomic absorption spectroscopy described by Isaac and Kerber (1971) and by the Technicon AutoAnalyzer[2] (manufactured by Technicon Corp., Tarrytown, NY 10591), an instrumental procedure for the determination of Ca, Mg, P, and K in plant tissue digests (Steckel & Flannery, 1971) are being superseded by more rapid analytical methods.

Emission spectroscopy has been and continues to be the method of choice for the elemental assay of plant tissue digests, using a progression of excitation sources from AC and DC arcs (Mitchell, 1964), to AC spark (Baker et al., 1964; Jones, 1976), and more recently to either inductively coupled plasma (Dalquist & Knoll, 1978; Munter & Grande, 1981; Soltanpour et al., 1982) or DC plasma (DeBolt, 1980). In many of today's plant analysis laboratories, elements in plant tissue digests, from trace to percent concentrations, are being easily and quickly determined by inductively coupled plasma optical emission spectroscopy (ICP-OES) (Fassel & Kniseley, 1974; Scott et al.,

1974; Walsh, 1983; Montaser & Golightly, 1987). If the spectrometer is evacuated, S can also be determined. Today, analysts prefer using high-speed multielement analyzers with computer or microprocessor control, requirements that are met by the ICP excitation spectrometer (Jones, 1984b).

X-ray emission spectroscopy (Alexander, 1965; Kubota & Lazar, 1971; Mudrock & Mudrock, 1977) is another technique applicable for the elemental assay of plant tissues. Although this technique of analysis is nondestructive, matrix effects have seriously hampered its acceptance and broad use.

Kjeldahl digestion, an analytical technique with more than 100-yr history (Morries, 1983), is the usual procedure for N determination. A standardized Kjeldahl procedure is described in the *AOAC Manual* (Williams, 1985), and a semi-automated procedure is given by Isaac and Johnson (1983). There is an excellent review article by Nelson and Sommers (1980) on the Kjeldahl method as well as a more recent one by Jones (1987).

Various procedures for the determination of S have been described by Beaton et al. (1968), and a turbimetric procedure using an AutoAnalyzer[2] (manufactured by Technicon Corp., Tarrytown, NY 10591) by Wall et al. (1980). The LECO Sulfur Analyzer[2] (manufactured by LECO Corp., St. Joseph, MI 49085-2396) procedure has been described by Jones and Isaac (1972), and more recently by Hern (1984).

III. INTERPRETATION OF ANALYSIS

A. Elemental Distribution in Plant

Most of the essential elements are not evenly distributed in the corn plant, nor does their concentration remain constant in any one plant part or section of the plant during its life cycle. The normal range in concentration for 14 elements is given in Table 20-4. Hanway (1962a) followed the uptake and distribution of N, P, and K in different parts of the corn plant during the growing season. He also followed their concentration in various plant parts (Hanway, 1962b). Benne et al. (1964) determined the composition of corn plants at different stages of growth and accumulation of essential elements in corn on a land area basis. Gorsline et al. (1965) followed the change in concentration and uptake of 11 elements in the leaves, stalks, tassels, and ears of corn plants from emergence to maturity. Clark (1975a, b) determined the concentrations of various elements in corn leaves and the whole plant with age. Similarly, Jones (1983) monitored N and P contents in corn and grain sorghum at growth stages defined by Vanderlip and Reeves (1972). Ohki (1975) found that critical Mn values varied considerably among the leaf blades of grain sorghum, increasing with increasing age (top vs. bottom leaves). All of these studies clearly show that element content varies among plant parts and will change within the same plant part with age. As was mentioned earlier,

[2] Trade name and company used to provide specific information and does not constitute endorsement by the authors.

Tabel 20-4. The normal expected range in element concentration for various parts of a corn plant.†

	Average range in concentration				
			Stalk in silk		
Element	Whole plants at 3 to 4 leaf stage	Ear leaf at silk	Above ear node	Below ear node	Grain at maturity
	$g\ kg^{-1}$				
N	35-50	27-35	--	--	10.0-25.0
P	4-8	2-4	1-2	1-2	2.0-6.0
K	35-50	17-25	10-20	20-30	2.0-4.0
Ca	9-16	4-10	1-3	1-3	0.1-0.2
Mg	3-8	2-4	1-3	1-3	0.9-2.0
S	2-3	1-3	--	--	--
	$mg\ kg^{-1}$				
Al	100-200	10-200	10-25	50-100	--
B	7-25	4-15	4-12	4-9	1-10
Ba	--	0-50	5-20	2-15	--
Cu	7-20	3-15	3-15	3-10	1-5
Fe	50-300	50-200	50-75	50-100	30-50
Mn	50-160	20-250	20-70	50-100	5-15
Na	--	1-400	1-100	1-100	--
Sr	--	10-100	10-50	10-30	--

† Based on numerous analyses of corn plants collected by J.B. Jones and analyzed at the Ohio Plant Analysis Laboratory, Wooster, OH during the years 1965 to 1969.

certain elements will also be unevenly distributed within the corn leaf as is shown in Table 20-1.

These changing patterns of element concentration and distribution within the plant and among its specific parts emphasize the care required when collecting tissue for analysis. The heterogeneous nature of elemental content requires that specific sampling procedures be employed. Whole plant analyses, except for very young plants, are of limited value since the mixing of leaves, stalks, and other plant parts results in heterogeneous mixtures of questionable value.

Element uptake and dry matter production are not steady-state processes. Concentration and dilution occur due to differences between plant growth and element absorption as well as from the movement of the elements within and between plant parts (Bates, 1971). Terman and Noggle (1973) and Terman et al. (1977), to define the early growth rate-element concentration relationship, studied element concentration changes in corn as affected by dry matter accumulation with age and response to applied fertilizers. Under normal growing conditions, elemental absorption and plant growth parallel each other during most of the vegetative growth period. Exceptions occur only during the early growth period shortly after germination, and then after seed set and at the beginning of senescence. However, if the normal growth rate is interrupted, then elemental accumulation or dilution can occur (Jarrell & Beverly, 1981).

Although the concentrations of various elements in the corn plant can fluctuate considerably during the growing season, total uptake (concentration × dry matter) of the whole plant plotted against time provides fairly smooth curves (Ritchie & Hanway, 1982; Al-Ansari, 1985). Most of the variations in elemental content are due to concentration or dilution effects associated with dry matter accumulation (Jarrell & Beverly, 1981). Therefore, it is essential that the time of sampling be known and considered when interpreting a plant analysis result.

B. Other Factors Affecting Interpretation

Two other important factors to consider when interpreting a plant analysis result are: (i) the influence of applied fertilizers and (ii) possible varietal effects on elemental composition. Bennett et al. (1953) correlated the N, P, and K content of the ear leaf and grain to yield at varying rates of N fertilization. They obtained a significant multiple regression of ear leaf N and P contents vs. grain yield. When the ear leaf N content was 2.8% or greater, no further increase in grain yield occurred. Similarly, Gallo et al. (1968) studied the relationship between leaf composition and grain yield.

Baker et al. (1966), from analyses of more than 50 000 corn leaf samples, reported that the elemental content of hybrids differed greatly, indicating that the level of accumulation of elements in the ear leaf was under partial genetic control. Munson (1969), Gorsline et al. (1968), and Ali and Johnson (1979) have shown that all corn plants do not absorb every element equally. This suggests inherited characteristics for element uptake. Rivard and Bandel (1974) reported that, although varietal differences in the concentrations of N, P, K, and Ca in field corn were statistically significant, those differences were not large enough to interfere with the interpretation of a plant analysis result. Recently, Russell and Pierre (1980) found a relationship between corn single crosses and their parent inbred lines for N content of grain. Kamprath et al. (1982) found a greater N use efficiency by an improved corn population than that by an unimproved one. Therefore, corn genotype may be an important factor in the interpretation of a plant analysis, requiring determination as to what effect genotype has on interpretative data, whether these data are critical values, sufficiency ranges, or DRIS norms.

C. Critical Values

Earlier methods for interpreting a plant analysis centered around single values, known as "critical" or "standard" values (Kenworthy, 1961). A critical value has been defined in numerous ways (Macy, 1936; Tyner, 1946; Ulrich, 1952; Gallo et al., 1968; Melsted et al., 1969), but essentially, it is that concentration correlated with a 10% yield decrease or that concentration where visual deficiency symptoms appear. Tyner (1946) published critical values for corn as 2.90% for N, 0.295% for P, and 1.30% for K; while Melsted et al. (1969) suggested 3.00% for N, 0.25% for P, 1.90% for K, 0.40% for Ca, and 0.25% for Mg. All these values were based on the assay of the sixth leaf from the base of the plant at the time the plant was silking. Others have published critical values for several elements, giving values for various plant parts and times of sampling as shown in Table 20-5. As previously indicated, plant analysis data for grain sorghum is quite limited. It is also comparatively recent when compared to that available for corn. Most of the work in the USA has dealt with sufficiency ranges and DRIS norms. However, in Australia, Weir (1983) has determined critical values for macronutrients and most of the micronutrients in grain sorghum, and others in the USA and India have published critical values for some of the micronutrients (Table 20-5). Though of some use, critical values are significantly limited in that they only designate single points between elemental deficiency and sufficiency.

D. Sufficiency Ranges

If the full benefits of the plant analysis technique are to be realized, yield response curves as a function of element concentration must be established for all those elements that affect plant growth. These response curves are similar to those that have been described by Prevot and Ollangnier (1961) and Smith (1962) as shown in Fig. 20-1. Although, research has not been as extensive as that involving the determination of critical values, sufficient study has been done for most elements to describe the nature of the entire response curve which establishes the basis for setting the limits of the sufficiency range. The sufficiency range may be defined as that elemental concentration range between the critical value and when an excess or toxic concentration occurs. It may also be defined as that range in elemental concentration where no yield reduction occurs, nor nutrient element stress symptoms appear.

Element concentration sufficiency ranges for corn, published by Jones (1967), Neubert et al. (1969), Escano et al. (1981a, b) and Cornforth and Steele (1981), are given in Table 20-6. Ranges for grain sorghum published by Lockman (1972a, b) and Whitney (1970) are presented in Table 20-7. Some of the upper limits of these ranges may be based on partially known or inadequately defined response curves, leaving some question as to their reliability. However, these sufficiency ranges form the basis for the more commonly used procedure when interpreting a plant analysis result for corn and grain sorghum. It is important to remember that all of these interpretive ranges are designated for a particular plant part taken at a specified time and, therefore, cannot be applied to an assay result of a different plant part or from a different time of sampling. A compilation of interpretative ranges for corn and grain sorghum leaf analyses may be found in books by Chapman (1961), Reuther and Robinson (1986), and Martin-Prevel et al. (1987).

Researchers have emphasized the complexities that can be associated with an interpretation of corn leaf analysis. Dumenil (1961) and Hanway and Dumenil (1965) noted, for instance, that there is an interaction between N and P in corn leaves that affects the critical value of the one element depending on the concentration of the other. Jones (1963) observed changes

Table 20-5. Published critical values for elements found in various corn and grain sorghum plant tissues.

Element	Time	Plant part	Critical value	Reference
		Corn, g kg^{-1}		
N	Not given	Not given	7.0	Goodall & Gregory, 1974
	At tassel	Ear leaf	30.0	Melsted et al., 1969
	At tassel	Leaves	29.0	Gallo et al., 1968
	At silk	Sixth leaf from base	29.0	Tyner, 1946
	40-60 cm ht.	Entire plant	25.6	Rehm et al., 1983
	At silk	Second leaf below ear	22.5	Rehm et al., 1983
P	8 wk	Whole plant top	1.4	Terman et al., 1972
	Late summer	Lower stems	0.044	Goodall & Gregory, 1947
	At tassel	Leaves	23.0	Gallo et al., 1968
	At silk	Sixth leaf from base	2.95	Tyner, 1946
	At tassel	Ear leaf	2.5	Melsted et al., 1969
K	Late summer	Lower stems	0.83	Goodall & Gregory, 1947
	At tassel	Ear leaf	19.0	Melsted et al., 1969
	At tassel	Leaves	17-27	Gallo et al., 1968
	At silk	Sixth leaf from base	13.0	Tyner, 1946
	At silk	Leaf opposite and just below ear shoot	20.0	Hanway, 1962
Ca	At tassel	Ear leaf	4.0	Melsted et al., 1969
Mg	At tassel	Ear leaf	2.5	Melsted et al., 1969
			1.5	Peaslee & Moss, 1966
		Corn, mg kg^{-1}		
B	At tassel	Ear leaf	10.0	Melsted et al., 1969
	Not given	Upper leaves	12.2	Berger et al., 1957
	5+ wk	Entire plant	1.8	Makarim & Cox, 1983
Cu	At tassel	Ear leaf	5.0	Melsted et al., 1969
Fe	At tassel	Ear leaf	15.0	Melsted et al., 1969
Mn	At tassel	Ear leaf	15.0	Melsted et al., 1969
Mo	Not given	Stems	0.11	Dios & Broyer, 1965
	Not given	Entire plant	0.10	Peterson & Purvis, 1961
Zn	5 wk	Entire plant	1.8	Makarin & Cox, 1983
	At tassel	Ear leaf	15.0	Melsted et al., 1969
	At tassel	Ear leaf	17.2	de L. Beyers & Coetzer, 1969
	At tassel	Leaf at second node below ear	15.0	Veits, 1953
	At silk	Second leaf below ear	14.9	Grunes et al., 1961
		Grain sorghum, g kg^{-1}		
N	Full heading	Second blade below apex	25.0	Weir, 1983
P	Full heading	Second blade below apex	2.5	Weir, 1983
K	Full heading	Second blade below apex	18.0	Weir, 1983
S	Full heading	Second blade below apex	1.5	Weir, 1983

(continued on next page)

Table 20-5. Continued.

Element	Time	Plant part	Critical value	Reference
		Grain sorghum, g kg^{-1}		
Ca	Full heading	Second blade below apex	2.0	Weir, 1983
Mg	Full heading	Second blade below apex	1.5	Weir, 1983
Cl	After head emergence	Fourth blade below flag leaf	<7.1	Francois et al., 1984
		Grain sorghum, mg kg^{-1}		
Cu	30 d after seeding	Middle blades	3.6	Agarwala & Sharma, 1979
Zn	Stage 3†	Youngest mature blade	10.0	Ohki, 1984
	49 d after seeding	Middle blades	20.0	Agarwala & Sharma, 1979
	Full flowering	Second blade below apex	18	Weir, 1983
Mn	Stage 5† (boot)	Flag leaf	10	Ohki, 1975
	Stage 5† (boot)	Second leaf below flag leaf	15	Ohki, 1975
Fe	35 d after seeding	Whole shoot	65	de Boer & Reisenauer, 1973
B	35 d after seeding	Middle blades	10	Agarwala & Sharma, 1979
	Full heading	Second blade below apex	5	Weir, 1983

† See Vanderlip and Reeves (1972).

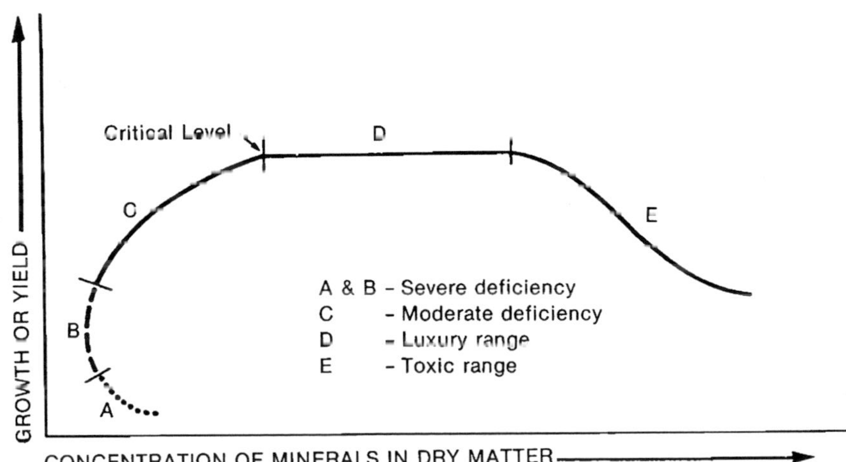

Fig. 20-1. General relationship between plant growth or yield and elemental content of the plant (Smith, 1962).

Table 20-6. Nutrient sufficiency ranges for corn ear leaf taken at silk as given by several workers.

Element	Jones (1967)	Neubert et al. (1969)	Cornforth and Steele (1981)	Escano et al. (1981a)
		$g\ kg^{-1}$		
N	27.6–35.0	26.0–40.0	22.5–33.0	21.0–36.0
P	2.5–4.0	2.5–5.0	1.8–3.2	2.6–3.5
K	17.1–25.0	17.0–30.0	17.0–30.0	17.0–23.0
Ca	2.1–10.0	2.1–10.0	4.0–8.0	3.5–4.9
Mg	2.1–6.0	3.1–5.0	1.3–2.5	1.7–3.0
S	--	2.1–5.0	1.2–3.0	1.5–2.3
		$mg\ kg^{-1}$		
Al	200	--	--	200
B	4–25	15–90	4–25	--
Cu	6–20	8–20	8–20	13–19
Fe	21–250	21–250	--	70–120
Mn	20–150	34–200	18–140	32–170
Mo	--	0.6–1.0	--	--
Zn	20–70	50–150	22–85	32–52

in the elemental content of the ear leaf with increasing rates of N, P, and K fertilization as did Al-Ansari (1985) with different soil levels of available P and K. In a nutrient culture study, Clark (1970) found wide variance in the elemental content of the corn plant as various elements became deficient or excessive. Loue (1987) evaluated the equilibrium relationship that exists among Ca, Mg, and K in corn leaves. The importance of these interactions as they relate to yield has been reported by Peck et al. (1969), Walker et al.

Table 20-7. Nutrient sufficiency ranges for grain sorghum at several stages of growth.

	Lockman (1972a, b)				Whitney (1970)
Element	Whole plant, 23–39 d after planting	Youngest fully developed leaf 37–56 d after planting	Bloom stage third leaf below head 66–70 d after planting	Dough stage third leaf below head, 82–97 d after planting	Last fully extended leaves
			$g\ kg^{-1}$		
N	35.0–40.0	32.0–42.0	33.0–40.0	30.0–40.0	30.0–50.0
P	3.0–6.0	2.0–6.0	2.0–3.5	1.5–2.5	2.5–4.0
K	30.0–45.0	20.0–30.0	14.0–17.0	10.0–15.0	27.5–40.0
Ca	9.0–13.0	1.5–9.0	3.0–6.0	2.0–6.0	--
Mg	3.5–5.0	2.0–5.0	2.0–5.0	1.0–5.0	--
			$mg\ kg^{-1}$		
B	1–13	1–10	1–10	1–6	--
Cu	8–15	2–15	2–7	1–3	5–15
Fe	160–250	55–200	65–100	40–80	--
Mn	40–150	6–100	8–190	8–80	--
Zn	30–60	20–40	15–30	7–16	15–40

(1971), and Walker and Peck (1972, 1974). Until recently when the DRIS concept was introduced, little had been done to incorporate these interactions into the interpretation of plant analysis data.

E. Diagnosis and Recommendation Integrated System

Diagnosing plant nutrient disorders using a critical value or sufficiency range requires that the elemental concentration result be compared with a standard value for that same tissue taken at that specified stage of growth. Because element concentrations vary depending on plant part, growth stage, genotype and geographic location, this approach has recognized limitations. A new concept for plant analysis interpretation has been proposed by Beaufils (1971, 1973) as a means to overcome some of these difficulties. The DRIS approach was designed to: (i) provide a valid diagnosis irrespective of plant age or tissue origin, (ii) rank nutrients in their limiting order, and (iii) stress the importance of nutrient balance.

DRIS is a unique approach that uses the principle of elemental interrelationships by determining, in descending order, the elements that are most limiting. Beaufils used the survey approach by using the world's published literature and plotting elemental leaf content vs. yield, a distribution that is normally skewed. To normalize the distribution curve, the yield component is divided into low- and high-yield groups. Walworth et al. (1986) suggested that the data bank for determining DRIS norms have at least several thousand entries, be randomly selected, and that at least 10% of the population be in the high-yield subgroup. It is also important that the cutoff value used to divide the low- from the high-yielding subgroups be such that the high-yield data subgroup remains normally distributed. Selecting the elemental content mean, the ratio and product of elemental means are determined. The ratio or product selected for DRIS calculations is the one with the largest variance which in turn maximizes the diagnostic sensitivity.

The means and coefficients of variation (CVs) for DRIS reference parameters in high-yielding subpopulations are used in a special calibration formula described by Beaufils (1973). The calibration formula calculates relative indices for nutrients that range from negative to positive values depending on whether elements are relatively insufficient or excessive with respect to each other with the sum of all the indices always being zero. The more negative the index value for an element, the more limiting it is. When all the elements are available at similar concentrations to those found in the desirable populations, the DRIS index for each element approaches zero. Sumner (1977a, b, 1979, 1981) has discussed the DRIS system in considerable detail, including its application to corn. In addition, Walworth and Sumner (1987) have published an excellent review paper on DRIS.

The first list of DRIS diagnostic norms for 11 elements in corn leaves was published by Elwali et al. (1985a, b) and they are given in Table 20-8. A major premise of the DRIS system is the universal applicability of DRIS norms regardless of genotype, age of crop at sampling and plant location. However, Escano et al. (1981a, b), and Cornforth and Steele (1981), each

Table 20-8. The Diagnosis and Recommendation Integrated System (DRIS) foliar diagnostic reference norms for corn† (Elwali et al., 1985b).

Parameter	No.	Mean	SD	Parameter	No.	Mean	SD
N/P‡	1909	9.035	2.136	10N/Zn	1526	11.797	4.459
NK	1908	1.463	0.426	Zn/10P	1527	0.883	0.420
P/K	1909	0.169	0.054	Zn/10K	1526	0.140	0.068
Ca/N	1553	0.160	0.057	10Ca/Zn	1524	1.919	1.087
Ca/P	1554	1.447	0.612	10Mg/Zn	1527	0.830	0.504
Ca/K	1553	0.237	0.122	10S/Zn	760	0.952	0.365
Mg/N	1556	0.071	0.029	Fe/Zn	1268	4.464	1.837
Mg/P	1557	0.639	0.330	Mn/Zn	1520	1.716	1.175
Mg/K	1556	0.104	0.063	Cu/10N	1401	0.031	0.013
Mg/Ca	1554	0.465	0.182	Cu/10P	1402	0.277	0.140
S/N	788	0.084	0.019	Cu/10K	1401	0.045	0.022
S/P	788	0.703	0.225	10Ca/Cu	1402	6.022	3.511
S/K	787	0.114	0.029	10Mg/Cu	1402	2.768	1.935
Ca/S	785	1.978	0.893	Cu/10S	664	0.375	0.211
S/Mg	788	1.195	0.395	Cu/Fe	1236	0.079	0.036
Fe/10N	1297	0.394	0.097	Cu/Mn	1395	0.260	0.174
Fe/10P	1298	3.588	1.177	Cu/Zn	1372	0.356	0.200
Fe/10K	1297	0.568	0.201	B/10N	402	0.024	0.012
10Ca/Fe	1298	0.410	0.189	B/10P	403	0.269	0.135
10Mg/Fe	1298	0.190	0.098	B/10K	402	0.043	0.033
Fe/10S	687	4.868	1.419	B/10Ca	403	0.153	0.076
Mn/10N	1459	0.151	0.087	B/10Mg	403	0.335	0.152
Mn/10P	1550	1.416	1.063	10S/B	112	3.185	1.039
Mn/10K	1549	0.218	0.140	B/Fe	389	0.068	0.036
Mn/10Ca	1547	1.048	0.676	B/Mn	399	0.173	0.150
Mn/10Mg	1550	2.485	1.780	B/Zn	410	0.265	0.134
10S/Mn	782	0.648	0.351	B/Cu	401	0.950	0.620
Mn/Fe	1293	0.405	0.249				

† Number of observations, means, and standard deviations (SD) of DRIS reference parameters in the subpopulation yielding >10.0 Mg of grain ha^{-1}.
‡ Nutrient concentrations were expressed in g kg^{-1} for N, P, K, Ca, Mg, and S and in mg kg^{-1} for Fe, Mn, Zn, Cu, and B.

using small databases, concluded that local calibration improves the accuracy of a DRIS diagnosis. Elwali et al. (1985a) also noted that biological estimates of this type can only be considered, at best, approximations based on the extent of available data with modification and refinement coming in time. Walworth and Sumner (1987) also published DRIS norms for corn from six different databases. They cautioned on universal use of DRIS norms derived from but one geographical area, stating that the CVs may not reflect the extent of normal variation in all geographical areas. They concluded that by pooling databases covering most geographic areas, derived DRIS norms should be applicable to a range of specific locales and conditions.

Kelling and Schulte (1986) have found the DRIS system for interpreting corn leaf analysis a supplement rather than a substitute for the sufficiency range concept when evaluating farmer received samples at the Wisconsin Soil Testing and Plant Analysis Laboratory.

Arogun (1978) derived DRIS norms for grain sorghum on the basis of a data bank comprised of 907 sets of observations of leaf analyses and yield. The data bank was developed from the literature and unpublished results

Table 20-9. Chemical analysis for the high-yielding subpopulation of sorghum crops and resulting norms selected for DRIS indices† (Arogun, 1978).

Element	Mean	CV	Parameter	Mean	CV
	g kg^{-1}	%			%
N	30.3	17	P/N	0.112	19
P	3.4	15	N/K	2.355	23
K	13.1	11	P/K	0.259	21
Ca	4.4	20	N/Ca	7.20	30
Mg	2.4	24	P/Ca	0.795	31
			K/Ca	3.080	24
			Mg/N	0.079	26
			P/Mg	1.518	45
			Mg/K	0.183	26
			Mg/Ca	0.553	30

† Means and coefficients of variation in the subpopulation (135 of 907 crops) yielding >7.1 Mg of grain ha^{-1}.

supplied by several researchers, with most of the analysis data originating from experimental plots located in Kansas. Using average analyses for high- (>7100 kg of grain ha^{-1}) and low- (<7100 kg of grain ha^{-1}) yielding populations of sorghum crops, he calculated and selected foliar diagnostic reference norms shown in Table 20-9. The formulae for calculation of DRIS indices for grain sorghum are as follows:

$$\text{N index} = \frac{-f\,(P/N) + f\,(N/K) + f\,(N/Ca) - f\,(Mg/N)}{4} \quad [1]$$

$$\text{P index} = \frac{f\,(P/N) + f\,(P/K) + f\,(P/Ca) + f\,(P/Mg)}{4} \quad [2]$$

$$\text{K index} = \frac{-f\,(N/K) - f\,(P/K) + f\,(K/Ca) - f\,(Mg/K)}{4} \quad [3]$$

$$\text{Ca index} = \frac{f\,(N/Ca) + f\,(P/Ca) + f\,(K/Ca) + f\,(Mg/Ca)}{4} \quad [4]$$

$$\text{Mg index} = \frac{f\,(Mg/N) - f\,(P/Mg) + f\,(Mg/K) + f\,(Mg/Ca)}{4} \quad [5]$$

$$\text{in which } f\,(P/N) = \left[\frac{P/N}{p/n} - 1\right]\frac{1000}{CV} \text{ when } P/N > p/n$$

$$\text{or } f\,(P/N) = 1 - \left[\frac{p/n}{P/N}\right]\frac{1000}{CV} \text{ when } P/N < p/n$$

where P/N is the ratio of P% and N% in the tissue under diagnosis and p/n is the value of the corresponding norm for the high-yielding population and CV is the corresponding CV. The other functions are calculated in a similar manner.

Tissue concentrations of N, P, K, Ca, and Mg were used to derive the norms, and Arogun (1978) tested them with sorghum plants grown in a greenhouse experiment. Sumner et al. (1983) further tested Arogun's DRIS norms on field data obtained by Reneau et al. (1983), finding them valid and usable for diagnostic purposes.

IV. PRACTICAL APPLICATION AND FUTURE POTENTIAL

In the USA, some 500 000 plant tissue samples (Jones, 1985) are analyzed yearly for farmers by both state-sponsored and commercial laboratories. To some people, this indicates a significant under-utilization since about four million soil samples are analyzed for farmers each year in these same laboratories. To most farmers, plant analysis use is still primarily viewed as a diagnostic device to determine the source of a suspected nutrient element insufficiency. And yet its most useful application is as a means of evaluating the soil/plant nutrient element status, a use first suggested by Krantz et al. (1948), and more recently put into a systems approach by Jones (1986). Without a plant analysis result, the farmer is not able to adequately evaluate whether all the essential elements, as well as several nonessential elements that can affect plant growth, are at satisfactory levels in the plant. Environmentally, a plant analysis can provide the means of assessing heavy metal contents of plants when waste products (animal manure, sewage sludge, and industrial wastes) have been or are being soil applied, warning of possible excesses or imbalances that can either reduce yields or adversely affect product quality or safety (Logan & Chaney, 1983; Adriano, 1986).

Yearly plant analysis results can also be "tracked" as has been discussed by Clements (1960) for sugarcane (*Saccharum officinarum* L.) and by Jones et al. (1980) for peanut (*Arachis hypogaea* L.). As shifts in nutrient element concentrations are observed as moving out of the sufficiency range, changes in crop production practices can be made before yield-affecting insufficiencies occur. Repeated plant analyses can also serve as a means of evaluating the effects of mineral cycling on the elemental content of the plants as crop residues are returned to the soil. Tracking requires systematic and standardized sampling and analysis procedures along with the appropriate sufficiency ranges for the elements being tracked. With experience, the farmer will be able to customize this system of plant analysis used to fit his own individual soil/plant management situation, thereby minimizing yield losses due to nutrient element insufficiencies. Regardless of how a plant analysis result is to be used and interpreted, knowledge of those factors affecting nutrient availability and uptake by plants is also required if the course of action to correct a plant nutritional problem is to be effective (Barber, 1984).

With recent significant analytical advancements now providing fast and low-cost plant analysis service, and the current sufficiency of interpretative data, farmers should be assaying their crops on a regular basis, combining the results of soil tests and plant analyses to assess the nutrient element status of their soil/crop system (Jones, 1986).

V. TISSUE TESTING

Although the terms *tissue testing* and *plant analysis* have sometimes been used interchangeably, they denote two different techniques. Plant analysis normally denotes a total analysis performed in the laboratory, using methods and procedures discussed in some detail in this chapter, while tissue testing denotes a semiquantitative quick test on plant sap extracted from fresh tissue, and normally performed in the field. Since there has been and continues to be considerable interest in tissue testing for corn, a brief description of the technique and its application deserves some attention.

Much of the early tissue testing research was done between the mid-1920s and early 1950s. The tests using extracted sap are mostly colorimetric methods with developed color intensity being compared with standard color charts serving as the means of interpretation. Although the testing procedures are relatively simple, the results obtained are frequently not easy to evaluate. With practice, however, this system of testing can be used effectively. The user must work with plants of known nutrient status to develop the proper sampling, chemical test procedures, and interpretation skill required. It is important that only fresh reagents and test papers be used at all times.

Krantz et al. (1948) were the first to describe field-testing procedures for corn. Using sap pressed from fresh tissue, they gave instructions for the semiquantitative determination of NO_3, PO_4, and K employing test papers, vials, reagents, and color charts. Wickstrom (1967) also has discussed the use of tissue tests as a means for field diagnosis of corn plant nutrient status. Syltie et al.(1972) have given procedural details for conducting tissue tests in the field for the elements N, P, K, Mg, and Mn. Instructions for the preparation of reagents and techniques for conducting the tests are also included. Scaife and Stevens (1983) have found "Merckoquant" test strips suitable for NO_3 determination in the field as a means of assessing the N status of vegetable crops. This method could be applied to corn tissue.

The relationship between the NO_3-N content of basal corn stems and grain yield has been evaluated, but research is limited and critical stem NO_3-N contents for given levels of grain yield have not been well established. Rauschkolb et al. (1974) found 0.40 to 0.60% NO_3-N (dry wt. basis) adequate for maximizing yields. Iversen et al. (1985) found the critical stem NO_3-N content for corn varied from 1.10 to 1.60%, and they found significant differences between plants with adequate and deficient levels of stem NO_3-N for maximum yields as early as the V-5 growth stage. McClenahan and Killorn (1988) used 5-cm stem segments immediately above the soil surface at the V-6 growth stage (Hanway, 1982) for NO_3-N determinations and found critical levels of 0.90 and 1.78% (dry wt. basis) for maximum yields on soils derived from loess and glacial till, respectively. Although further calibration data are needed, this procedure would permit detection of N needs early enough during the growing season to permit necessary applications of fertilizer N.

Iron can be determined by a tissue test conducted in the field, a procedure developed by Bar-Akiva et al. (1978) and later modified by Bar-Akiva (1984). Peroxidase activity is measured by floating leaf discs in a reactive solution with the development of a blue color indicating an adequate Fe status in the plant tissue. Many believe that only this type of test, or ones similar to it, can satisfactorily determine the Fe status of a plant (Chaney, 1984).

Compared to a laboratory conducted plant analysis, field-conducted tissue tests cost less and provide immediate results, therefore, some see this as a distinct advantage for tissue tests in preference to a plant analysis. It should be remembered, however, that most tissue tests are not entirely quantitative. Rather, they provide the tester with a qualitative "yes" or "no" evaluation of the crop; that is, the element being evaluated is either present or not present at a particular concentration. It takes considerable practical experience and repeated observations to feel confident when making interpretations based on a tissue test result. To some, the tissue test and its interpretation is considered an acquired "art" rather than a strict quantitative analytical procedure. Although the test procedures themselves are based on sound analytical chemistry, it is their utilization and interpretation that requires skill gained only by repeated practical experience.

Quick tests of soil and plant tissue conducted in the field combined with field observations have been coined "The Diagnostic Approach," a procedure of observation, testing, and evaluation discussed in a special issue of *Better Crops With Plant Food* published in 1984 (published by the Potash and Phosphate Institute, Atlanta, GA).

VI. SUMMARY

Corn and grain sorghum plants have been found to contain similar amounts of most elements. Since the data for grain sorghum is scant, the two crops can be treated similarly unless specific data are available that distinguish between them. Although, to farmers, the primary use for a plant analysis is still to diagnose suspected essential element deficiencies, they need to see its broader application and to use it in a systems approach to maintain their soil/crop nutrient element status within an established sufficiency range. While plant analysis is not the total answer to detecting and solving all nutritional disorders in corn and grain sorghum plantings, it is a better tool now than it was in previous years. For a plant analysis result to be meaningful, it must be analytically correct, therefore, the care required to select, prepare, and analyze the gathered tissue sample. There have been considerable advances made on how best to prepare samples for analysis and significant technological advancements in analytical equipment used to assay plant tissue digests.

The "critical level" and "sufficiency range" concepts are generally satisfactory for the interpretation of a plant analysis result, but they are limited to a particular plant part taken at a specified time. Even with plant part and age defined, concentrations can be affected by interactions among elements and may be different for various genotypes, thus necessitating consideration of factors other than simple critical levels or sufficiency ranges. Therefore, the interpreter must be knowledgeable and experienced.

DRIS is a new unique approach for the interpretation of a plant analysis result that applies the principle of elemental interrelationships as a means for determining by order elements most limiting. The DRIS concept of interpretation is based on the balance among elements rather than specific levels of elements. Essentially, elemental balances (in the form of indices) within the tissue under diagnosis are compared with norms calculated from large data banks of analysis results vs. yield. This approach reportedly lessens or negates consideration of plant part, age, and genotype when interpreting a plant analysis result. Data banks of analyses from plants in "high-yielding populations" must be available and the tissue under diagnosis must be analyzed for two or more elements for the DRIS system to apply. Calculation of results can be cumbersome, but this can be overcome by the use of computer programs. From practical experience, a combination of DRIS and sufficiency range procedures seems to provide the most useful approach for the interpretation of a plant analysis result at the present time.

Tissue tests, consisting of rapid semiquantitative determinations, can be valuable when a quick diagnosis in the field is required. However, effective utilization and interpretation does require considerable skill on the part of the user.

Improvements continue to be made in all aspects of the plant analysis and tissue testing techniques with particular research interest on the effect genotype has on nutrient element requirement and efficiency of utilization. Further advances in the DRIS approach to interpretation will no doubt be made. Future research will probably lead to nutrient diagnosis at the cellular level by in situ monitoring of element concentrations.

REFERENCES

Adriano, D.C. 1986. Trace elements in the terrestrial environment. Springer-Verlag New York, New York.

Agarwala, S.C., and C.P. Sharma. 1979. Recognizing micronutrient disorders of crop plants on the basis of visible symptoms and plant analyses. Lucknow Univ., India.

Al-Ansari, A.S. 1985. The effect of P and K availability in soil on nutrient concentrations, uptake, and distribution in corn plants throughout the growing season. Ph.D. diss. Iowa State Univ., Ames (Diss. Abstr. 85-14368).

Alexander, G.V. 1965. An x-ray fluorescence method for the determination of calcium, potassium, chlorine, sulfur, and phosphorus in biological tissues. Anal. Chem. 37:1671–1674.

Ali, H.C., and M.W. Johnson. 1979. Quantitative inheritance of Ca, K, Mg, and Zn concentration in maize. Maydica 24:67–73.

Arogun, J.O. 1978. Application of the DRIS system to sorghum and millet. M.S. thesis. Univ. of Wisconsin, Madison.

Baker, D.E., B.R. Bradford, and W.I. Thomas. 1966. Leaf analysis of corn: Tool for predicting soil fertility need. Better Crops Plant Food 50(2):36–40.

Baker, D.E., G.W. Gorsline, G.B. Smith, W.I. Thomas, W.E. Grube, and J.L. Ragland. 1964. Technique for rapid analyses of corn leaves for eleven elements. Agron. J. 56:133–136.

Bar-Akiva, A. 1984. Substitutes for benzidine as H-donors in the peroxidase assay for rapid diagnosis of iron deficiency in plants. Commun. Soil Sci. Plant Anal. 15:929–934.

Bar-Akiva, A., D.N. Maynard, and J.E. English. 1978. A rapid tissue test for diagnosing iron deficiencies in vegetable crops. HortScience 13:284–285.

Barber, S.A. 1984. Soil nutrient bioavailability. John Wiley and Sons, New York.

Bates, T.E. 1971. Factors affecting critical nutrient concentrations in plants and their evaluation: A review. Soil Sci. 112:116–130.

Beaton, J.D., G.R. Burns, and J. Platou. 1968. Determination of sulphur in soils and plant material. Tech. Bull. 14. Sulphur Inst., Washington, DC.

Beaufils, E.R. 1971. Physiological diagnosis—A guide for improving maize production based on principles developed for rubber trees. Fert. Soc. S. Afr. J. 1:1–30.

Beaufils, E.R. 1973. Diagnosis and recommendation integrated system (DRIS). A general scheme of experimentation based on principles developed from research in plant nutrition. Soil Sci. Bull. 1. Univ. of Natal, Pietermaritzburg, South Africa.

Benne, E.J., E. Linden, J.D. Grier, and K. Spike. 1964. Composition of corn plants at different stages of growth and per-acre accumulations of essential nutrients. Mich. State Agric. Exp. Stn. Q. Bull. 47:69–85.

Bennett, W.F. 1971. A comparison of the chemical composition of the corn leaf and the grain sorghum leaf. Commun. Soil Sci. Plant Anal. 2:399–405.

Bennett, W.F., G. Stanford, and L. Dumenil. 1953. Nitrogen, phosphorus, and potassium content and yield of corn leaf and grain as related to nitrogen fertilization. Soil Sci. Soc. Am. Proc. 17:252–258.

Berger, K.C., C.T. Heikkinen, and B. Vube. 1957. Boron deficiency, a cause of blank stalks and barren ears in corn. Soil Sci. Soc. Am. Proc. 21:629–632.

Bhan, K.C., A. Wallace, and D.R. Hunt. 1959. Some mineral losses from leaves by leaching. Proc. Am. Soc. Hortic. Sci. 73:289–293.

Bock, R.A. 1978. Handbook of decomposition methods in analytical chemistry. Int. Textbook Co., Glasgow, Scotland.

Carlier, L.A., and L.P. van Hee. 1971. Microwave drying of lucerne and grass samples. J. Sci. Food Agric. 22:306–307.

Chaney, R.L. 1984. Diagnostic practices to identify iron deficiency in higher plants. J. Plant Nutr. 7:47–67.

Chapman, H.D. 1964. Foliar sampling for determining the nutrient status of crops. World Crops 16:35–46.

Chapman, H.D. (ed.). 1966. Dignostic criteria for plants and soils. Univ. of Calif., Div. of Agric. Sci., Berkeley.

Clark, R.B. 1970. Effects of mineral nutrient levels on the inorganic composition and growth of corn. Ohio Agric. Res. Dev. Ctr. Res. Circ. 181.

Clark, R.B. 1975a. Mineral element concentrations in corn leaves by position on the plant and age. Commun. Soil Sci. Plant Anal. 6:439–450.

Clark, R.B. 1975b. Mineral element concentrations of corn plants with age. Commun. Soil Sci. Plant Anal. 6:451–464.

Clements, H.F. 1960. Crop logging of sugarcane in Hawaii. p. 131–137. *In* W. Reuther (ed.) Plant analysis and fertilizer problems. Publ. 8. Am. Inst. Biol. Sci., Washington, DC.

Cornforth, I.S., and K.W. Steele. 1981. Interpretation of maize leaf analyses in New Zealand. N.Z. J. Exp. Agric. 9:91–96.

Dalquist, R.L., and J.W. Knoll. 1978. Inductively coupled plasma-atomic emission spectroscopy: Analysis of biological materials and soil for major trace, and ultra-trace elements. Appl. Spectro. 32:1–30.

Davidescu, D., and V. Davidescu. 1972. Evaluation of fertility by plant and soil analysis. Abacus Press, Turubridge Wells, England.

de Boer, G.J., and H.M. Reisenauer. 1973. DTPA as an extractant of available soil iron. Commun. Soil Sci. Plant Anal. 4:121–128.

DeBolt, D.C. 1980. Multielement emission spectroscopic analysis of plant tissue using DC argon plasma source. J. Assoc. Off. Anal. Chem. 63:802–805.

de L. Beyers, C.P., and F.J. Coetzer. 1979. The influence of zinc fertilizing on yield of maize and the mineral composition of the leaves. Agric. Sci. S. Agric. Agroplantae 1:41–46.

Dios, R.V., and T.V. Broyer. 1965. Deficiency symptoms and essentiality of molybdenum in corn hybrids. Agrochinica 9:273–284.

Dumenil, L. 1961. Nitrogen and phosphorus composition of corn leaves and corn yields in relation to critical levels and nutrient balance. Soil Sci. Soc. Am. Proc. 25:295–298.

Elwali, A.M.O., G.J. Gascho, and M.E. Sumner. 1985a. DRIS norms for 11 nutrients in corn leaves. Agron. J. 77:506–508.

Elwali, A.M.O., G.J. Gascho, and M.E. Sumner. 1985b. DRIS norms for 11 nutrients in corn leaves—Errata. Agron. J. 77:654.

Escano, C.R., C.A. Jones, and G. Uehara. 1981a. Nutrient diagnosis in corn grown on hydric dystrandepts: I. Optimum tissue concentrations. Soil Sci. Soc. Am. J. 45:1135–1139.

Escano, C.R., C.A. Jones, and G. Uehara. 1981b. Nutrient diagnosis in corn grown on hydric dystrandepts: II. Comparison of two systems of tissue diagnosis. Soil Sci. Soc. Am. J. 45:1140–1144.

Fassel, V.A., and R.N. Kniseley. 1974. Inductively coupled plasmas. Anal. Chem. 46:1155A–1164A.

Francois, L.E., T. Donovan, and E.V. Mass. 1984. Salinity effects on seed yield, growth and germination of grain sorghum. Agron. J. 76:741–744.

Gallo, J.R., R. Hiroce, and L.T. DeMiranda. 1968. Leaf analysis in corn plant nutrition. Part I. Correlations of leaf analysis and yield. Bragantia 27:177–186.

Goodall, D.W., and F.G. Gregory. 1947. Chemical composition of plants as an index to their nutritional status. Tech. Commun. 17. Imp. Bur. Hortic. and Plantation Crops, East Malling, Kent, England.

Gorsline, G.W., D.E. Baker, and W.I. Thomas. 1965. Accumulation of eleven elements by field corn (*Zea mays* L.). Pennsylvania State Exp. Stn. Bull. 725.

Gorsline, G.W., W.I. Thomas, and D.E. Baker. 1968. Major gene inheritance of Sr-Ca, K, P, Zn, Cu, B, Al-Fe, and Mn concentration in corn (*Zea mays* L.). Pennsylvania State Exp. Stn. Bull. 746.

Gorsuch, T.T. 1970. Destruction of organic matter. Int. Ser. of Monogr. in Analytical Chem. Vol. 39. Pergamon Press, New York.

Graham, R.D. 1972. Vacuum attachment for grinding mills. Commun. Soil Sci. Plant Anal. 31:167–173.

Greer, R. (ed.). 1970. Proceedings symposium in plant analysis. Int. Mineral & Chem. Corp., Skokie, IL.

Grunes, D.L., L.C. Boawn, C.W. Carlson, and F.G. Viets, Jr. 1961. Zinc deficiency of corn and potatoes as related to soil and plant analyses. Agron. J. 53:68–71.

Havlin, J.L., and P.N. Soltanpour. 1980. A nitric acid plant tissue digest method for use with inductively-coupled plasma spectrometry. Commun. Soil Sci. Plant Anal. 11:969–980.

Hanway, J.J. 1962a. Corn growth and composition in relationship to soil fertility. II. Uptake of N, P, and K and their distribution in different plant parts during the growing season. Agron. J. 54:217–222.

Hanway, J.J. 1962b. Corn growth and composition in relation to soil fertility: III. Percentages of N, P, and K in different plant parts in relation to stage of growth. Agron. J. 54:222–229.

Hanway, J.J. 1982. How a corn plant develops. Spec. Rep. 48. Iowa State Univ., Ames.

Hanway, J.J., and L. Dumenil. 1965. Corn leaf analysis—The key in correct interpretation. Plant Food Rev. 11:5–8.

Hardy, G.W. (ed.). 1967. Soil testing and plant analysis: Plant analysis. Part 2. SSSA Spec. Pub. 2. SSSA, Madison, WI.

Hern, J.L. 1984. Determination of total sulfur in plant materials using an automated sulfur analyzer. Commun. Soil Sci. Plant Anal. 15:99–107.

Hood, S.L., R.Q. Parks, and C. Horwitz. 1944. Mineral contamination resulting from grinding plant samples. Ind. Eng. Chem. Anal. Ed. 16:202–205.

Isaac, R.A., and W.C. Johnson. 1983. Determination of protein nitrogen in plant tissue using near infrared spectroscopy. J. Assoc. Off. Anal. Chem. 66:506–509.

Isaac, R.A., and J.D. Kerber. 1971. Atomic absorption and flame photometry: Techniques and uses in soil plant and water analysis. p. 17–37. *In* L.M. Walsh (ed.) Instrumental methods of analysis of soils and plant tissue. SSSA, Madison, WI.

Iversen, K.V., R.H. Fox, and W.P. Piekielek. 1985. The relationships of nitrate concentrations in young corn stalks to soil nitrogen availability and grain yields. Agron. J. 77:927–932.

Jackson, M.L. 1958. Soil chemical analysis. Prentice-Hall, Englewood Cliffs, NJ.

Jacques, G.L., R.L. Vanderlip, D.A. Whitney, and R. Ellis, Jr. 1974. Nutrient contents of washed and unwashed grain sorghum plant tissues compared. Commun. Soil Sci. Plant Anal. 5:173–182.

Jarrell, W.M., and R.B. Beverly. 1981. The dilution effect in plant nutrition studies. Adv. Agron. 34:197–224.

Johnson, C.M., and A. Ulrich. 1959. Analytical methods for use in plant analysis. California Agric. Exp. Stn. Bull. 766:26–78.

Jones, C.A. 1983. A survey of the variability in tissue nitrogen and phosphorus concentrations in maize and grain sorghum. Field Crops Res. 6:133–147.

Jones, J.B., Jr. 1963. Diagnosing nutrient element need by plant analysis. p. 15–22. *In* Proc. 23rd Annu. Corn and Sorghum Res. Conf. Publ. 23, Washington, DC.

Jones, J.B., Jr. 1967. Interpretation of plant analysis for several agronomic crops. p. 49–58. *In* G.W. Hardy (ed.) Soil testing and plant analysis. Part 2. SSSA Spec. Publ. 2. SSSA, Madison, WI.

Jones, J.B., Jr. 1970. Distribution of 15 elements in corn leaves. Commun. Soil Sci. Plant Anal. 1:27–34.

Jones, J.B., Jr. 1974. Plant analysis handbook for Georgia. Georgia Coop. Ext. Bull. 735.

Jones, J.B., Jr. 1976. Elemental analysis of biological substances by direct reading spark emission spectroscopy. Am. Lab. 8:15–20.

Jones, J.B., Jr. 1983. Soil test works when used right. Part 2. Solutions 27:61–70.

Jones, J.B., Jr. 1984a. Soil testing and plant analysis: Guides to the fertilization of horticultural crops. p. 1–67. *In* J. Janick (ed.) Horticultural reviews. Vol. 7. AVI Publ. Co., Westport, CT.

Jones, J.B., Jr. 1984b. A turnkey laboratory concept for agricultural testing. Am. Lab. 16:64–72.

Jones, J.B., Jr. 1985. Recent survey of number of soil and plant tissue samples tested for growers in the United States. *In* Proc. 10th Soil Plant Analyst's Workshop. Counc. Soil Testing and Plant Analysis, Athens, GA.

Jones, J.B., Jr. 1986. Crop essential element status assessment in the field. p. 311–348. *In* W.G. Gensler (ed.) Advanced agricultural instrumentation—Design allure. NATO AS1 Series, Ser. E. Appl. Sci. 11. Martinus Nijhoff Publ., Dordrecht, Netherlands.

Jones, J.B., Jr. 1987. Kjeldahl nitrogen determination—What's in a name. J. Plant Nutr. 10:1675–1682.

Jones, J.B., Jr., and R.A. Isaac. 1972. Determination of sulfur on plant material using a LECO sulfur analyzer. J. Agric. Food Chem. 20:192–194.

Jones, J.B., Jr., R.L. Large, D.B. Pfeiderer, and H.S. Klosky. 1971. How to properly sample for a plant analysis. Crops Soils 23:114–120.

Jones, J.B., Jr., J.E. Pallas, and J.R. Stansell. 1980. Tracking the elemental content of leaves and other plant parts of the peanut under irrigated culture in the sandy soils of South Georgia. Commun. Soil Sci. Plant Anal. 11:81–92.

Kamprath, E.J., R.H. Moll, and N.E. Rodriguez. 1982. Effects of nitrogen fertilization and recurrent selection on performance of hybrid populations of *Zea mays* L. Agron. J. 74:955–958.

Kelling, K.A., and E.E. Schulte. 1986. DRIS as a part of a routine plant analysis program. J. Fert. Issues 3:107–112.

Kenworthy, A.L. 1961. Interpreting the balance of nutrient elements in leaves of fruit trees. p. 28–43. *In* W. Reuther (ed.) Plant analysis and fertilizer problems. Publ. 8. Am. Inst. of Biol. Sci., Washington, DC.

Krantz, B.A., W.L. Nelson, and L.F. Burkhart. 1948. Plant tissue tests as a tool in agronomic research. p. 137–156. *In* H.B. Kitchen (ed.) Diagnostic techniques for soils and crops. Am. Potash Inst., Washington, DC.

Kubota, J., and V.A. Lazar. 1971. X-ray emission spectrograph: Techniques and uses for plant and soil studies. p. 67–83. *In* L.M. Walsh (ed.) Instrumental methods for analysis of soils and plant tissue. SSSA, Madison, WI.

Lockman, R.B. 1970. Plant sample analysis as affected by sample decomposition prior to laboratory processing. Commun. Soil Sci. Plant Anal. 1:13–19.

Lockman, R.B. 1972a. Mineral composition of grain sorghum plant samples. Part I. Comparative analysis with corn at various stages of growth and under different environments. Commun. Soil Sci. Plant Anal. 3:271–282.

Lockman, R.B. 1972b. Mineral composition of grain sorghum plant samples. Part III: Suggested nutrient sufficiency limits at various stages of growth. Commun. Soil Sci. Plant Anal. 3:295–304.

Logan, T.J., and R.L. Chaney. 1983. Utilization of municipal wastewater and sludge on land-metals. p. 235–323. *In* A.L. Page (ed.) Utilization of municipal wastewater and sludge on land. Univ. of California, Riverside.

Loue, A. 1987. Corn. p. 598–631. *In* P. Martin-Prevel et al. (ed.) Plant analysis as a guide to the nutrient requirements of temperate and tropical crops. Lavosier Publ., New York.

Lundegardh, H. 1951. Leaf analysis (translated by R.L. Mitchell). Hilger and Watts, London.

Macy, P. 1936. The quantitative mineral nutrient requirements of plants. Plant Physiol. 11:749–764.

Martin-Prevel, P., J. Gagnard, and P. Gautier. 1987. Plant analysis as a guide to the nutrient requirements of temperate and tropical crops. Lavosier Publ., New York.

McClenahan, E.J., and R. Killorn. 1988. Relationship between basal corn stem nitrate N content at V6 growth stage and grain yield. J. Prod. Agric. 1:322–326.

Melsted, S.W., H.L. Motto, and T.R. Peck. 1969. Critical plant nutrient composition values useful in interpreting plant analysis data. Agron. J. 61:17–20.

Mitchell, R.L. 1964. The spectrographic analysis of soils, plants and related materials. Tech. Commun. 44A. Commonwealth Bureau of Soils, Harpenden, Herts, England.

Montaser, A., and D.W. Golightly. 1987. Inductively coupled plasmas in analytical atomic spectrometry. JCH Publ., New York.

Morries, P. 1983. A century of Kjeldahl (1883–1983). J. Assoc. Publ. Anal. 21:53–58.

Mudrock, A., and O. Mudrock. 1977. Analysis of plant material by x-ray fluorescence spectrometry. X-Ray Spectrosc. 6:215–217.

Munson, R.D. 1969. Plant analysis: Varietal and other considerations. p. 85–104. *In* F. Greer (ed.) Proceedings symposium in plant analysis. Int. Mineral and Chem. Corp., Skokie, IL.

Munson, R.D., and W.L. Nelson. 1973. Principles and practices in plant analysis. p. 223–248. *In* L.M. Walsh and J.D. Beaton (ed.) Soil testing and plant analysis. Rev. ed. SSSA, Madison, WI.

Munter, R.C., and R.A. Grande. 1981. Plant tissue and soil extract analysis by ICP-atomic emission spectrometry. p. 653–672. *In* R.M. Barnes (ed.) Developments in atomic plasma spectrochemical analysis. Heyden and Son, London.

Nelson, D.W., and L.E. Sommers. 1980. Total nitrogen analysis of soil and plant tissues. J. Assoc. Off. Anal. Chem. 63:770–778.

Nelson, P.B., and J.W. Boodley. 1965. An error involved in the preparation of plant tissue for analysis. Proc. Am. Soc. Hortic. Sci. 86:712–716.

Neubert, P., W. Wrazidlo, N.P. Vielemeyer, I. Hundt, F. Gullmick, and W. Bergmann. 1969. Tabellen Zur Planzenanalyze-Erste orientierende Ubersicht. Inst. fur Planzenernahrung Jena, Berlin.

Ohki, K. 1975. Manganese supply, growth, and micronutrient concentration in grain sorghum. Agron. J. 67:30–32.

Ohki, K. 1984. Zinc nutrition related to critical deficiency and toxicity levels for sorghum. Agron. J. 76:253–256.

Peaslee, D.E., and D.N. Moss. 1966. Photosynthesis on K- and Mg-deficient corn (*Zea mays* L.) leaves. Soil Sci. Soc. Am. Proc. 30:220–223.

Peck, T.R., W.H. Walker, and L.V. Boone. 1969. Relationship between corn (*Zea mays* L.) yield and leaf levels of 10 elements. Agron. J. 61:299–301.

Peterson, N.K., and E.R. Purvis. 1961. Development of molybdenum deficiency symptoms in certain crop plants. Soil Sci. Soc. Am. Proc. 25:111–117.

Pierre, W.H., L. Dumenil, and J. Henao. 1977a. Relationship between corn yield, expressed as a percentage of maximum, and the N percentage in the grain: II. Diagnostic use. Agron. J. 69:221–226.

Pierre, W.H., L. Dumenil, V.D. Jolley, J.R. Webb, and W.D. Shrader. 1977b. Relationship between corn yield, expressed as a percentage of maximum, and the N percentage in the grain: I. Various N-rate experiments. Agron. J. 69:215–220.

Piper, C.S. 1942. Soil and plant analysis. Hassell Press, Adelaide, Australia.

Prevot, P., and M. Ollangnier. 1961. Law of the minimum and balanced nutrition. p. 257–277. *In* W. Reuther (ed.) Plant analysis and fertilizer problems. Publ. 8. Am. Inst. Biol. Sci., Washington, DC.

Rauschkolb, R.S., A.L. Brown, J. Quick, J.D. Prato, R.E. Pelton, and F.R. Kegel. 1974. Rapid tissue testing for evaluating nitrogen nutritional status of (1) corn and (2) sorghum. Calif. Agric. 6:10–14.

Rehm, G.W., R.C. Sorensen, and R.A. Wiese. 1983. Application of phosphorus, potassium, and zinc to corn grown for grain or silage: Nutrient concentration and uptake. Soil Sci. Soc. Am. J. 47:697–700.

Reisenauer, H.M. (ed.). 1978. Soil and plant-tissue testing in California (rev. June 1978). California Coop. Ext. Serv. Bull. 1879.

Reneau, R.B., Jr., G.D. Jones, and J.B. Fredericks. 1983. Effect of P and K on yield and chemical composition of forage sorghum. Agron. J. 75:5–8.

Reuter, D.J., and J.B. Robinson (ed.). 1986. Plant analysis: An interpretation manual. Inkata Press Pty. Ltd., North Clayton, Australia.

Reuther, W. (ed.). 1960. Plant analysis and fertilizer problems. Publ. 8. Am. Inst. Biol. Sci., Washington, DC.

Ritchie, S.W., and J.J. Hanway. 1982. How a corn plant develops. Iowa State Spec. Rep. 48.

Rivard, C.C., and V.A. Bandel. 1974. Effect of variety on nutrient composition of field corn. Commun. Soil Sci. Plant Anal. 5:229–242.

Russell, W.A., and W.H. Pierre. 1980. Relationship between maize single crosses and their parent inbred lines for N content in the grain. Agron. J. 72:363–369.

Sayre, J.D. 1952. Accumulation of radio-isotopes in corn leaves. Ohio Agric. Exp. Stn. Res. Bull. 723.

Scaife, A., and K.L. Stevens. 1983. Monitoring sap nitrate in vegetable crops: Comparison of test strips with electrode methods, and effects of time of day and leaf position. Commun. Soil Sci. Plant Anal. 14:761–771.

Scott, R.N., V.A. Fassel, R.N. Kniseley, and D.E. Dixon. 1974. Inductively coupled plasma-optical emission analytical spectroscopy. Anal. Chem. 46:75–80.

Shuman, G.E., and F. Rauzi. 1981. Microwave drying of rangeland forage samples. J. Range Manage. 34:426–428.

Smith, J.H., D.L. Carter, M.J. Brown, and C.L. Douglas. 1968. Differences in chemical composition of plant sample fractions resulting from grinding and screening. Agron. J. 60:149–151.

Smith, P.F. 1962. Mineral analysis of plant tissue. Ann. Rev. Plant Physiol. 13:81–108.

Soltanpour, P.N., J.B. Jones, Jr., and S.M. Workman. 1982. Optical emission spectroscopy. p. 29–66. *In* A.L. Page et al. (ed.) Methods of soil analysis. Part 2. 2nd ed. Agronomy Monogr. 9. ASA, Madison, WI.

Sonneveld, C., and P.A. van Dijik. 1982. The effectiveness of some washing procedures on the removal of contaminates from plant tissue samples of glasshouse crops. Commun. Soil Sci. Plant Anal. 13:487–496.

Steckel, J.E., and R.L. Flannery. 1971. Simultaneous determination of phosphorus, potassium, calcium and magnesium in wet digestion solutions of plant tissue by AutoAnalyzer. p. 83–96. *In* L.M. Walsh (ed.) Instrumental methods for analysis of soils and plant tissues. SSSA, Madison, WI.

Steyn, W.J.A. 1959. Leaf analyses. Errors involved in the preparation phase. J. Agric. Food Chem. 7:344–348.

Steyn, W.J.A. 1961. The errors involved in the sampling of citrus and pineapple plants for leaf analysis purposes. p. 409–430. *In* W. Reuther (ed.) Plant analysis and fertilizer problems. Publ. 8. Am. Inst. Biol. Sci., Washington, DC.

Sumner, M.E. 1977a. Use of the DRIS system in foliar diagnosis of crops at high yield levels. Commun. Soil Sci. Plant Anal. 8:251–268.

Sumner, M.E. 1977b. Effect of corn leaf sampled on N, P, K, Ca, and Mg content and calculated DRIS indices. Commun. Soil Sci. Plant Anal. 8:269–280.

Sumner, M.E. 1979. Interpretation of foliar analysis for diagnostic purposes. Agron. J. 71:343–348.

Sumner, M.E. 1981. Diagnosing the sulfur requirements of corn and wheat using foliar analysis. Soil Sci. Soc. Am. J. 45:87–90.

Sumner, M.E., R.B. Reneau, Jr., E.E. Schulte, and J.O. Arogun. 1983. Foliar diagnostic norms for sorghum. Commun. Soil Sci. Plant Anal. 14:817–825.

Syltie, P.W., S.W. Melsted, and W.M. Walker. 1972. Rapid tissue tests for indication of yield plant composition, and soil fertility for corn and soybeans. Commun. Soil Sci. Plant Anal. 3:37–50.

Terman, G.L., P.M. Giordano, and S.E. Allen. 1972. Relationships between dry matter yields and concentrations of Zn and P in young corn plants. Agron. J. 64:684–687.

Terman, G.L., and J.C. Noggle. 1973. Nutrient concentration changes in corn as affected by dry matter accumulation with age and response to applied nutrients. Agron. J. 65:941–945.

Terman, G.L., J.C. Noggle, and C.M. Hunt. 1977. Growth rate-nutrient concentration relationships during early growth of corn as affected by applied N, P, and K. Soil Sci. Soc. Am. J. 41:363–368.

Tolg, G. 1974. The basis of trace analysis. p. 698–710. *In* E. Korte (ed.) Methodium chimicum. Vol. 1. Analytical methods. Part B. Micromethods, biological methods, quality control, automation. Academic Press, New York.

Tyner, E.H. 1946. The relation of corn yields to leaf nitrogen, phosphorus, and potassium content. Soil Sci. Soc. Am. Proc. 11:317–323.

Ulrich, A. 1952. Physiological basis for assessing the nutritional requirements of plants. Ann. Rev. Plant Physiol. 3:207–228.

Ulrich, J.M. 1984. Pulsing air enhances Wiley mill utility. Commun. Soil Sci. Plant Anal. 15:189–190.

Vanderlip, R.L., and H.E. Reeves. 1972. Growth stages of sorghum [*Sorghum bicolor* (L.) Moench]. Agron. J. 64:13–16.

Viets, F.G., Jr. 1953. Zinc deficiency in corn in central Washington. Agron. J. 45:559–564.

Walker, W.M., and T.R. Peck. 1972. A comparison of the relationship between corn yield and nutrient concentration in whole plants and different plant parts at early tassel at two locations. Commun. Soil Sci. Plant Anal. 3:513–523.

Walker, W.M., and T.R. Peck. 1974. Relationship between corn yield and plant nutrient content. Agron. J. 66:253–256.

Walker, W.M., R.D. Voss, and T.R. Peck. 1971. Relationships between corn leaf composition and high and low yield levels. Commun. Soil Sci. Plant Anal. 2:389–397.

Wall, L.L., C.W. Gehrke, and J. Suzuki. 1980. An automated turbidimetric methods for total sulfur in plant tissue and sulfate sulfur in soils. Commun. Soil Sci. Plant Anal. 11:1087–1103.

Wallace, A., J. Kinnear, J.W. Cha, and E.M. Rommey. 1980. Effects of washing procedures on mineral analysis and their cluster analysis for orange leaves. J. Plant Nutr. 2:1–9.

Wallace, T. (ed.). 1956. Plant analysis and fertilizer problems. Inst. de Res. Pour Les Huiles et Olengineux, Paris.

Wallace,T. 1961. The diagnosis of mineral deficiencies in plants by visual symptoms. Chem. Publ. Co., New York.

Walsh, J.N. 1983. A handbook of inductively coupled plasma spectrometry. Blackie, London.

Walworth, J.L., W.S. Letzsch, and M.E. Sumner. 1986. Use of boundary lines in establishing diagnostic norms. Soil Sci. Soc. Am. J. 50:123–128.

Walworth, J.L., and M.E. Sumner. 1987. The diagnosis and recommendation integrated system DRIS. p. 149–188. *In* B.A. Stewart (ed.) Advances in soil science. Vol. 6. Springer-Verlag New York, New York.

Weir, R.G. 1983. Tissue analysis for pastures and field crops. N.S.W. Dep. Agric. Advisory Note No. 11/83.

Whitney, D.A. 1970. Soil and plant analyses for corn and sorghum survey. Kansas State Univ. Mimeo. 3a 162 1 300.

Williams, S. (ed.). 1985. Official methods of analysis of the AOAC. 14th ed. Assoc. of Off. Anal. Chem., Arlington, VA.

Wickstrom, G.A. 1967. Use of tissue testing in field diagnosis. p. 109–112. *In* G.W. Hardy (ed.) Soil testing and plant analysis. Part 2. SSSA, Madison, WI.

Zasoski, R.J., and R.G. Burau. 1977. A rapid nitric-perchloric acid digestion method for multi-element tissue analysis. Commun. Soil Sci. Plant Anal. 3:425–436.

Chapter 21

Plant Analysis as an Aid in Fertilizing Vegetable Crops

C. M. GERALDSON, *University of Florida, Bradenton*

K. B. TYLER, *University of California, Parlier*

Vegetables are generally intensively grown, and often require large quantities of nutrients and moisture for optimum and dependable production. Many vegetables are short-season crops, completing the cycle from planting to harvest in 8 to 12 wk. Compared with agronomic crops that are harvested in an advanced stage of maturity, vegetable crops are usually harvested in an early stage of maturity. With most crops, the best-quality produce depends upon rapid and continuous growth. Production systems range from field-grown irrigated and nonirrigated crops to greenhouse and hydroponic cultures. Overall, most vegetables are grown in soil cultures whether it be in the field or in greenhouses. The nutritional status of vegetable crops can be altered by such factors as soil properties, fertilizer practices, rainfall and irrigation, crop utilization, and resulting interrelationships.

Yields and quality of vegetable crops can be altered and correlated with variations in the ionic root environment. The composition of the ionic root environment is influenced by the soil cation exchange complexes, organic matter, fertilizer application, and water. Any variation in quantities, placement, or source of fertilizer, results in associated changes in the ionic root environment. Rainfall and irrigation alter ionic concentration and movement. Actively growing crops selectively remove nutrients from this dynamic system. Maturity and development stage of the plant influence nutrient requirement. Optimal crop production can be attained by using procedures and developing systems that regulate the ionic root environment.

I. USE OF PLANT ANALYSIS FOR VEGETABLE CROPS

Plant analysis can aid in the development of predictable crop production program by:

 Soil Testing and Plant Analysis, 3rd ed.—SSSA Book Series, no. 3.

1. Monitoring and appraising the efficiency of a soil test-fertilizer program.
2. Measuring the effectiveness of fertilizer rate, placement, source, and time of application.
3. Determining the modifying effects of environment on nutrient availability and uptake of nutrients by the plants.
4. Studying the influence of fertilizer nutrient interactions on nutrient uptake.
5. Evaluating varietal fertilizer use efficiency.
6. Relating nutrient solution experiments to field experiments.

Plant analysis can also be a valuable diagnostic tool for identifying and confirming nutritional disorders that occur in vegetable crops production (Maynard, 1979).

Although fertilization and irrigation are practiced more extensively and intensively in vegetable crops production, the application of tissue analysis to these crops is not different than for agronomic crops. The variables that alter plant composition are essentially the same for both groups. It is possible the production systems exist or can be developed which minimize these variables and thus minimize the associated variations in plant composition. Such sophisticated production procedures would reduce the need for a diagnostic technique such as plant analysis. However, as the composition of most vegetables is normally influenced by the dynamics of plant environment systems, plant analysis is and is likely to remain a valuable diagnostic tool.

II. PLANT ANALYSIS GUIDELINES

The desired objective of plant analysis is to analyze the plant sufficiently early that nutrients can be applied if needed. But because the major portion of nutrients are used by the plant as the crop approaches maturity, abnormalities tend to develop mostly during the later stages of the growth. Thus, plant tissue analyses may be of more value in confirming the adequacy of present fertilizer management and in determining fertilizer practices for future years than in correcting abnormalities for the present crop.

Both total nutrient composition and soluble fractions have been used successfully to obtain data which can be correlated with crop production. The procedures most commonly used in the western USA are those described by Johnson and Ulrich (1961).

For most nutrients, an actively growing plant part is selected; for example, selection of the youngest fully mature leaf is excellent for tomato (*Lycopersicon esculentum* Mill), beet (*Beta vulgaris* L.), carrot (*Daucus carota* L.), onion (*Allum cepa* L.), potato (*Solanum tuberosum* L.), and many other crops. Leaf blade or petiole tissue can be used, but usually the petiole reflects greater variation in nutrient composition between deficient plants and those with adequate supply.

Nutrient concentrations found in "normal" plants that can be used as guidelines for vegetable crop production are presented in Table 21-1. The

Table 21-1. Common nutrient ranges found in plant tissue (dry wt. basis) of various vegetable crops.

Vegetable crop growth stage	Part sampled	N	P	K	Ca	Mg	Fe	B	Cu	Zn	Mn
		mg g^{-1}					mg kg^{-1}				
Asparagus,	Fern from	24	3.0	15	4	1.5	--	50	20	10	
mature fern†	45–90 cm up	38	3.5	24	5	2.0	--	100	--	60	160
Bean (snap)	Young mature	30	2.5	18	8	25	300	40	15	30	30
bud†	trifoliate leaf	60	5.0	25	3	7.0	450	60	30	60	300
Beet, mature†	Young mature	35	2.0	20	25	3.0	--	60	--	15	70
	leaf	50	3.0	40	35	8.0	--	80	--	30	200
Broccoli,	Young mature	32	3.0	20	12	2.3	100	30	1	45	25
heading†	leaf	55	7.0	40	25	4.0	300	100	5	95	150
Brussels	Upper leaves	22	2.6	24	3	2.3	--	30	--	--	--
sprouts†		42	4.5	34	22	4.0	--	40	--	--	--
Cabbage, heads	Young wrapper	30	3.0	30	15	2.4	30	30	--	20	--
half-grown†	leaf	40	5.0	40	35	4.5	60	60	--	30	--
Cabbage, heads	Midrib of young	--	--	40	14	3.0	--	--	--	--	--
half-grown†	upper leaf	--	--	60	20	5.0	--	--	--	--	--
Cabbage,	Quarter sections of	--	--	30	4	1.4	40	--	--	--	25
mature	mature heads	--	--	40	6	2.0	100	--	--	--	50
Cantaloupe†	Petiole-young	--	--	60	30	7.0	--	--	--	--	--
	mature leaf	--	--	80	50	10.0	--	--	--	--	--
Cantaloupe†	Blade	20	2.5	18	50	10.0	--	30	--	30	--
		30	4.0	25	70	15.0	--	80	--	50	--
Carrot,	Young mature	18	2.0	20	20	--	--	40	--	--	--
midgrowth†	leaf	24	3.5	40	35	--	--	60	--	--	--
Carrot,	Young mature	21	2.0	25	14	4.3	120	29	4.5	20	190
midgrowth	leaf	25	3.0	43	20	5.3	335	35	7.0	50	325
Cauliflower	Young mature	--	5.0	--	20	--	--	30	5	50	
at heading†	leaf	--	7.0	--	35	--	--	60	10	--	80
Cauliflower,	Leaf blade	30	5.4	30	7	2.4	--	--	--	43	--
buttoning		45	7.2	37	8	2.6	--	--	--	59	--
Celery,	Young mature	25	3	40	6	2	20	30	5	20	200
half grown†	leaf	35	5	70	30	5	40	50	8	50	300
Celery,	Young mature	30	3	35	4	2.5	--	--	--	--	--
half grown‡	leaf	40	6	55	15	4	--	--	--	--	--
Cucumbers	Young mature	--	--	--	--	--	--	50	--	20	--
	leaf	--	--	--	--	--	--	80	--	40	--
Kale and collads,	Young mature	40	3	30	30	--	--	50	--	--	--
almost mature†	leaf	50	6	40	40	--	--	80	--	--	--
Lettuce, heads	Wrapper leaf	35	4	60	14	5	--	25	--	--	--
half size†		40	6	80	20	7	--	45	--	--	--
Onion,	Young mature	15	2.5	--	--	--	--	30	--	10	--
midgrowth	leaf	25	4	--	--	--	--	45	--	15	--
Pea, bud to	Entire top	31	3	22	12	2.7	--	20	--	--	--
full bloom§	growth	36	3.5	28	15	3.5	--	60	--	--	--
Pea, mid-	Young mature	37	2.5	15	15	2.5	--	30	--	--	--
growth†	leaf	35	3.5	30	25	4	--	60	--	--	--
Pepper (bell)	Young mature	30	7	40	4	10	--	40	10	--	--
midgrowth§	leaf	45	3	54	6	17	--	100	20	--	--
Potato, tubers	Young mature	30	2	40	20	5	70	30	--	20	30
half grown†	leaf	50	4	80	40	8	150	40	--	40	50
Potato, tubers	Petioles of young	20	2	50	15	5	--	--	--	--	--
half grown	mature leaf	30	4	90	25	15	--	--	--	--	--
Potato, 30-cm	Young mature	40	2	35	6	8	--	--	--	--	--
tall plants	leaf	50	4	50	9	11	--	--	--	--	--

(continued on next page)

Table 21-1. Continued.

Vegetable crop growth stage	Part sampled	N	P	K	Ca	Mg	Fe	B	Cu	Zn	Mn
		— mg g^{-1} —					— mg kg^{-1} —				
Spinach, 30-50-d old‡	Young mature leaf	42	4.8	38	6	16	240	42	45	50	50
		52	5.8	52	12	18	245	63	65	75	85
Spinach, mature†	Young mature leaf	40	3	30	6	16	220	40	5	50	30
		60	5	40	10	18	245	60	7	75	60
Sweet corn, after silking	Ear leaf	28	1.8	18	16	4	60	40	8	20	100
		35	3	28	25	8	160	70	12	40	140
Sweet corn, 80% silk‡	Ear leaf	29	3	15	9	3	--	--	10	15	50
		33	8	26	13	8	--	--	15	45	106
Sweet corn, silking§	Ear leaf	26	2	18	1.5	2	--	20	--	--	--
		35	3	25	3	3	--	30	--	--	--
Sweet potato, midseason§	Mature leaf	32	2	29	7.3	4	--	--	--	--	40
		42	3	43	9.5	8	--	--	--	--	100
Turnip, midgrowth†	Young mature	35	4	--	30	--	--	30	--	--	60
		45	6	--	50	--	--	60	--	--	80
Tomato (MH), 1st mature fruit†	Young mature leaf	30	5	25	40	6	--	40	4	15	60
		60	8	40	60	9	--	80	8	30	100
Tomato staked, 1st mature fruit	Young mature leaf	40	8	30	14	4	--	--	--	--	--
		60	10	50	18	6	--	--	--	--	--
Tomato, trellised mature fruit§	Young mature leaf	25	3	30	5	6	100	30	5	--	50
		40	6	40	20	10	300	100	10	--	100
Watermelon, midgrowth§	Young mature leaf	20	2	25	25	6	--	--	4	--	--
		30	3	35	35	8	--	--	8	--	--

† O.A. Lorenz, 1970, unpublished data; Lorenz (1944); Lorenz et al., (1956); Tyler and Lorenz (1964).
‡ G.R. Klacan, 1965 unpublished data.
§ C.M. Geraldson, 1970 published in part.

concentration ranges shown do not indicate deficiency or excess limits, since such limits depend on all factors affecting nutrient uptake. The deficiency-sufficiency range of certain soluble nutrients in several vegetable crops is given in Table 21-2. Direct comparisons of results are difficult because of the influence that climatic and edaphic variables, cultural practices, and variety have on nutrient uptake. In addition, reported nutrient levels are influenced by differences in sampling, extraction, and analytical techniques.

Table 21-2. Plant analysis guide for vegetable crops.†

Crop growth stage	Plant part	Nutrient‡	Deficient	Intermediate	Sufficient
Asparagus, midgrowth of fern	10-cm tip of new fern	NO_3-N	100	250	500
		PO_4-P	800	1 200	1 600
		K	10	20	30
Bean, bush snap midgrowth	Petiole of 4th leaf from tip	NO_3-N	2 000	2 500	3 000
		PO_4-P	1 000	1 500	2 000
		K	30	40	50
Broccoli, 1st buds	Midrib of wrapper leaf	NO_3-N	5 000	7 000	9 000
		PO_4-P	2 500	3 500	5 000
		K	20	30	40

(continued on next page)

Table 21-2. Continued

Crop growth stage	Plant part	Nutrient‡	Deficient	Inter-mediate	Sufficient
Brussels sprouts, midgrowth	Midrib of mature leaf	NO_3-N	5 000	6 000	7 000
		PO_4-P	2 000	2 500	3 500
		K	30	40	50
Cabbage heading	Midrib of wrapper leaf	NO_3-N	5 000	7 000	9 000
		PO_4-P	2 500	3 000	3 500
		K	20	30	40
Cabbage, Chinese at heading	Midrib or wrapper lead	NO_3-N	8 00	9 000	10 000
		PO_4-P	2 000	2 500	3 000
		K	40	60	70
Cantaloupe, early fruit set	Petiole of sixth leaf from growing tip	NO_3-N	5 000	7 000	9 000
		PO_4-P	1 500	2 00	2 500
		K	30	40	50
Cantalopue, 1st mature fruit	Petiole of sixth leaf from growing tip	NO_3-N	2 000	3 000	4 000
		PO_4-P	1 000	1 500	2 000
		K	20	30	40
Carrot, mid-growth	Petiole of young mature leaf	NO_3-N	5 000	7 500	10 000
		PO_4-P	2 000	3 000	4 000
		K	40	50	60
Cauliflower, buttoning	Midrib of young mature leaf	NO_3-N	5 000	7 000	9 000
		PO_4-P	2 500	3 500	5 000
		K	20	30	40
Celery, mid-growth	Petiole of youngest fully elongated leaf	NO_3-N	5 000	7 000	9 000
		PO_4-P	2 000	3 500	5 000
		K	40	50	60
Celery, near maturity	Petiole of youngest fully elongated leaf	NO_3-N	4 000	5 000	6 000
		PO_4-P	2 000	2 500	3 000
		K	30	40	50
Cucumber, early fruit	Petiole of sixth leaf from tip	NO_3-N	5 000	7 000	9 000
		PO_4-P	1 500	2 000	2 500
		K	30	40	50
Eggplant, 1st harvest	Petiole of young mature leaf	NO_3-N	5 000	6 000	7 500
		PO_4-P	2 000	2 500	3 000
		K	40	50	70
Garlic, at bulbing	Youngest, fully elongated leaf	Total N	30	40	50
		PO_4-P	2 000	2 500	3 000
		K	20	30	40
Lettuce, heading	Midrib of wrapper leaf	NO_3-N	4 000	6 000	8 000
		PO_4-P	2 000	3 000	4 000
		K	20	30	40
Lettuce, harvest	Midrib of wrapper leaf	NO_3-N	3 00	4 500	6 000
		PO_4-P	1 500	2 000	2 500
		K	15	20	25
Onion, mid-growth	Tallest leaf	Total N	25	30	35
		PO_4-P	1 000	1 500	2 000
		K	25	30	40
Pepper, Chili 1st bloom	Petiole of young mature leaf	NO_3-N	5 000	6 000	7 000
		PO_4-P	2 000	2 500	3 000
		K	30	40	50
Pepper, sweet early growth	Petiole of young mature leaf	NO_3-N	8 000	10 000	12 000
		PO_4-P	2 000	3 000	4 000
		K	40	50	60

(continued on next page)

Table 21-2. Continued.

Crop growth stage	Plant part	Nutrient‡	Deficient	Inter-mediate	Sufficient
Pepper, sweet early fruit set	Petiole of young mature leaf	NO_3-N	4 000	6 000	8 000
		PO_4-P	2 000	3 000	4 000
		K	30	40	50
Potato, early season	Petiole of 4th leaf from growing tip	NO_3-N	8 000	10 000	12 000
		PO_4-P	1 200	1 600	2 000
		K	90	100	120
Potato, mid-season	Petiole of 4th leaf from growing tip	NO_3-N	6 000	7 500	9 000
		PO_4-P	800	1 200	1 600
		K	70	80	90
Potato, late season	Petiole of 4th leaf from growing tip	NO_3-N	3 000	4 000	5 000
		PO_4-P	500	800	1 000
		K	40	50	60
Spinach, mid-growth	Petiole of young mature leaf	NO_3-N	4 000	6 000	8 000
		PO_4-P	2 000	3 000	4 000
		K	20	30	40
Squash, summer zucchini, early bloom	Petiole of young mature leaf	NO_3-N	12 000	13 500	15 000
		PO_4-P	4 00	5 000	6 000
		K	60	80	100
Sweet corn, tasseling	Midrib of 1st leaf above primary ear	NO_3-N	1 500	1 000	1 500
		PO_4-P	500	750	1 000
		K	20	30	40
Sweet potato midgrowth	Petiole of 6th leaf from growing tip	NO_3-N	1 500	2 500	3 500
		PO_4-P	1 000	1 500	2 000
		K	30	40	60
Tomato (canning) early bloom	Petiole of 4th leaf from growing tip	NO_3-N	8 000	10 000	12 000
		PO_4-P	1 000	2 00	2 500
		K	30	40	60
Tomato (canning) fruit 2.5-cm diam.	Petiole of 4th leaf from growing tip	NO_3-N	6 000	8 000	10 000
		PO_4-P	1 000	2 000	2 500
		K	20	30	40
Tomto (canning) first color	Petiole of 4th leaf from growing tip	NO_3-N	2 000	3 000	4 000
		PO_4-P	80	1 200	1 800
		K	10	20	30
Tomato, cherry early fruit set	Petiole of 4th leaf from growing tip	NO_3-N	8 000	9 000	10 000
		PO_4-P	2 000	2 500	3 000
		K	90	60	70
Tomato (market) early bloom	Petiole of 4th leaf from growing tip	NO_3-N	10 000	12 000	14 000
		PO_4-P	2 500	3 000	3 500
		K	40	50	60
Tomato (market) fruit 2.5-cm diam.	Petiole of 4th leaf from growing tip	NO_3-N	8 000	10 000	12 000
		PO_4-P	2 500	3 00	3 500
		K	30	40	50
Tomato (market) full ripe fruit	Petiole of 4th leaf from growing tip	NO_3-N	4 000	5 000	6 000
		PO_4-P	2 000	2 500	3 000
		K	20	30	40
Watermelon early fruit	Petiole of 5th leaf from growing tip	NO_3-N	5 000	7 000	9 000
		PO_4-P	1 500	2 000	2 500
		K	30	40	50

† Information in this table was taken from Lorenz and Tyler (1983). Analytical methods for NO_3-N, PO_4-P, and total K are described by Johnson and Ulrich (1961). Total N for garlic and onion were analyzed by semi-micro Kjeldahl-N method.

‡ Nutrients NO_3-N, total N, and PO_4-: are mg kg^{-1}, K is mg g^{-1}.

III. INFLUENCES ON NUTRIENT UPTAKE

A. Growth Stage

When attempting to relate tissue analysis to plant nutrient requirement, careful consideration should be given to nutrient uptake patterns as affected by growth stage. During the vegetative stage of plant development, nutrient uptake closely parallels the plant growth rate as illustrated in Fig. 21-1 for celery (*Apium gravelolens* var. *dulce*) and lettuce (*Lactuca sativa* L.). Results presented in Fig. 21-2 for potato serve to indicate the correlation between tissue analysis, fertilizer application, and yield response. Comparing A and B (Fig. 21-2), note the variation in petiole NO_3-N and magnitude of the yield and both with similar fertilizer treatments. Data in Fig. 21-2B show that NO_3 concentrations may decrease to low levels during late season without a subsequent reduction in yield. Numerous tests have demonstrated that satisfactory yields can be expected if NO_3-N levels of about 9000 mg kg^{-1} are maintained during early growth.

Following floral initiation, tuber formation, or enlargement of storage organs, the rate of nutrient absorption often decreases on a dry weight basis. There is usually a transport of certain elements from the vegetative portions of the plant to the developing flower, fruit, seed, root, or tuber.

The magnitude of transport varies with each element. Data in Table 21-3 compare the relative mobility of certain elements in pea (*Pisum sativum* L.).

While developing seeds accumulate N and P at rates emulating the rate of dry matter gain, immobile elements like Ca and B are accumulated at very low rates.

B. Cultivar

The effects of stage of development and another important determinant of plant nutient status, crop (cultivar), are shown in Table 21-4. While the trends for concentrations of each of the five elements shown are similar for both pea cultivars, the actual concentrations differ noticeably. In both cultivars, concentrations of N, P, and K decrease as the plant matures, but Ca levels increase over the same period. Magnesium concentrations appear to plateau early in the plant's development and remain constant through maturity.

C. Fertilizer Application

Although N is considered a key element in sweet corn fertilization, evidence that total N content, particularly at early growth stages, may also be a poor indicator of N availability is given in Table 21-5. This study was conducted in cooperation with the Dep. of Soil Science at the Univ. of Wisconsin on a Waupun silt loam at Arlington, WI.

Although tissue P at both the eight-leaf stage and the 80% silk stage increased with increasing rates of P fertilizer, the effect did not parallel the

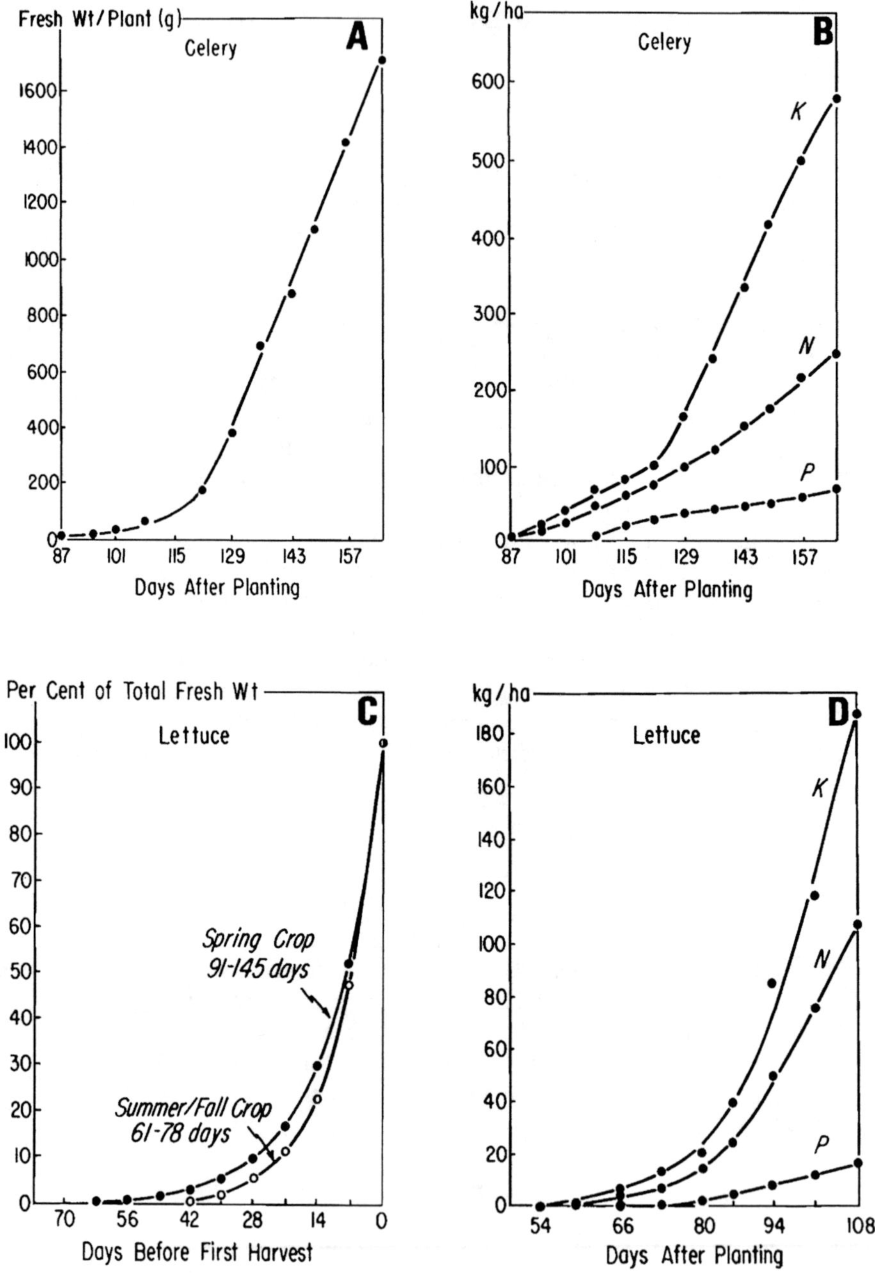

Fig. 21-1. Growth rates and nutrient absorption by celery and lettuce. (*A*) Growth of direct-seeded celery (Redrawn after Zink, 1963). (*B*) Nutrient absorption of direct-seeded celery (Redrawn after Zink, 1963.) (*C*) Growth of lettuce. (Redrawn after Zink & Yamaguchi, 1962.) (*D*) Nutrient absorption by lettuce. (Redrawn after Zink & Yamaguchi, 1962.)

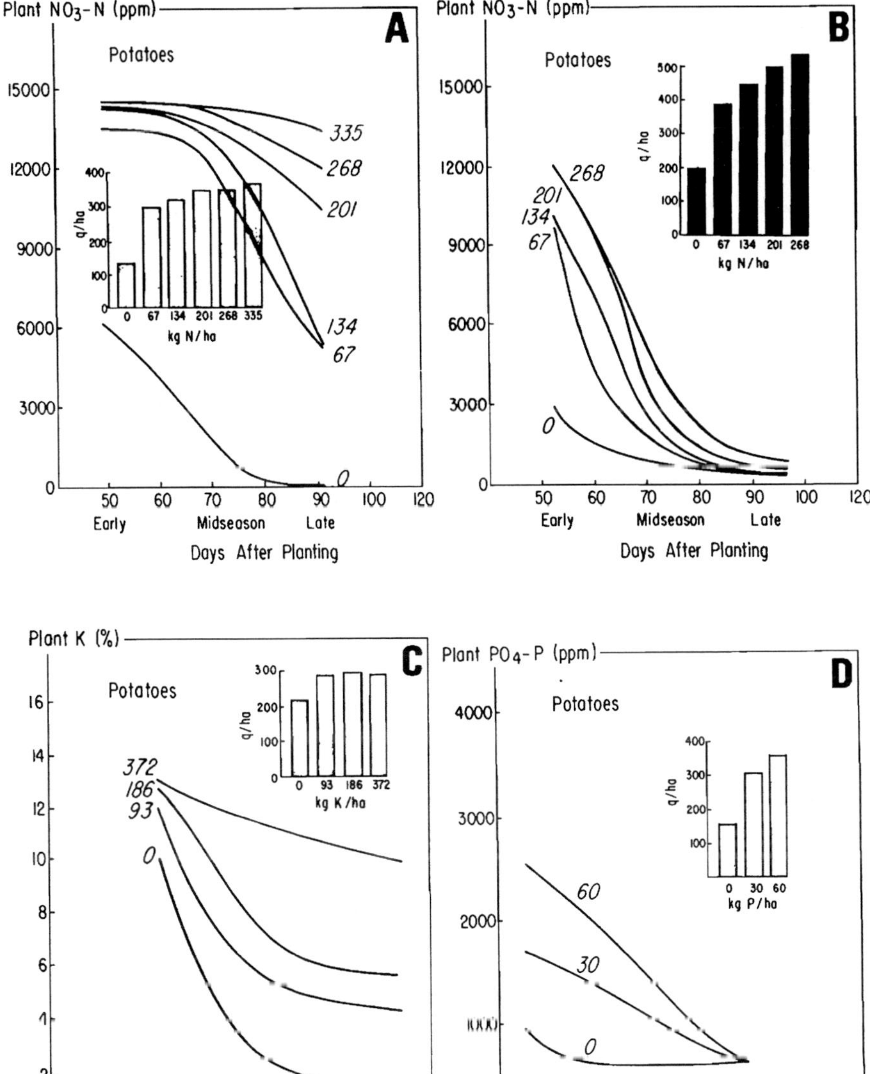

Fig. 21-2. Effect of fertilization on nutrient composition of petioles and tuber yields of potato. (Redrawn after Lorenz et al., 1964.) (*A*) Nitrate-N concentrations in petioles and tuber yields of potato fertilized with N. (*B*) Nitrate-N concentrations in petioles and tuber yields of potato fertilized with N. (*C*) Potassium concentrations in petioles and tuber yields of potato fertilized with K. (*D*) Phosphorus concentrations in petiole and tuber yields of potato fertilized with P.

Table 21-3. Nutrient concentration of N, P, Ca, and B in vines and seeds of pea (*Pisum sativum* L.) at harvest maturity (solution culture data) (Bartz, 1959).

Plant part sampled	Nutrient concentration N	P	Ca	B
	—— mg g^{-1} ——			mg kg^{-1}
Vines	32.0	3.9	8.2	19.0
Immature seeds	47.0	4.0	0.7	9.0

Table 21-4. The effect of growth stage and cultivar on nutrient concentration of whole pea plants (Bartz, 1959).

Cultivar	Growth stage at sampling	Nutrient concentration N	P	K	Ca	Mg
		—— mg g^{-1} ——				
Early June	Four node	68.0	4.4	22.0	6.0	2.5
	Eight node	41.0	4.0	30.0	12.5	4.0
	First blossom	36.0	3.5	28.0	12.0	3.0
	Full bloom	31.0	3.0	22.0	12.0	2.7
	Flat pod	24.0	2.5	17.0	16.0	3.5
	Canning maturity	15.0	1.4	15.0	19.0	3.0
Sweet type	Five node	63.0	5.0	25.0	9.0	2.5
	Seven node	49.0	3.5	24.0	18.0	3.8
	Ten node	44.0	3.0	20.0	17.5	3.5
	Fourteen node	35.0	3.1	17.0	17.0	3.9
	First blossom	33.0	3.0	16.0	18.0	3.8
	Flat pod	27.0	2.8	13.0	20.5	
	Canning maturity	23.0	2.5	13.0	21.0	4.0

Table 21-5. The effects of six rates of N, P, and K on nutrient concentration and yield of sweet corn (J.K. Hammes, 1963, unpublished data).

Element	Fertilization rate	Concentrations in whole plants sampled at: 8-Leaf	80% Silk	Yield of green corn
	kg ha^{-1}	—— N, mg g^{-1} ——		kg ha^{-1}
N	0	37	19.0	5 376
	44.8	32	17.0	11 872
	89.6	32	19.0	13 440
	134.4	38	21.0	15 008
	179.2	37	22.0	15 232
	224.0	38	22.0	15 680
		—— P, mg kg^{-1} ——		
P	0	3.3	2.8	11 648
	19.7†	3.4	2.6	14 336
	39.4	3.7	2.9	14 560
	59.1†	3.7	2.9	15 232
	78.8	3.6	2.7	15 232
	98.6†	3.7	3.3	15 456
		—— K, mg g^{-1} ——		
K	0	20	12	12 320
	37.2	46	18	15 008
	74.4	43	24	14 336
	111.6	47	32	15 232
	148.7	46	32	14 336
	185.9	48	34	15 456

Table 21-6. Influence of fertilizer placement on concentration of N,P, and K found in sweet corn seedlings (20 d from planting) (J.F. Bartz, 1959, unpublished data).

Fertilizer treatment†			Concentration of N, P, and K in sweet corn seedlings (entire top growth)		
Band	Broad-cast	Seed applied	N	P	K
	kg ha^{-1}			mg g^{-1}	
280	--	--	44.0	3.3	24.0
--	280	--	46.0	3.0	22.0
--	280	11.2	49.0	6.6	22.0
--	280	22.4	53.0	8.2	20.0
280	--	11.2	53.0	7.2	18.0
280	--	44.8	54.0	9.8	16.0
--	--	11.2	51.0	6.4	15.0

yield response curve. Somewhat surprising, the banded P did not affect the level of P found in the tissue as is often observed.

In contrast, banded K was positively reflected in plant K content and yield. At the early sampling, the tissue K showed a pronounced break between the low rates of K fertilizer before plateauing at about 4.6%. The later sampling showed a more direct relationship between fertilizer rate and tissue K.

This is not to say that N and P contents of sweet corn are unrelated to fertilizer treatment. A dramatic change in concentration of nutrients in sweet corn seedlings resulting from fertilizer placement is shown in Table 21-6. Seed-applied ammonium polyphosphate resulted in noticeable increases in N and P concentration. In this study, early growth responses (up to the eight-leaf stage) were also related to fertilizer treatment but these differences were mitigated as the season progressed and could not be measured at harvest.

As mentioned earlier, the total N for sweet corn (*Zea mays* vr. *rugosa*) grown on a loamy sand appeared to be a relevant test. Similarly, data for high-density snap bean (*Phaseolus vulgaris* L.) grown on an irrigated sand (Table 21-7) clearly indicate the tissue response of this crop to N fertilization.

Table 21-7. Influence of N and K fertilization on total N of snap bean (*Phaseolus vulgaris* L.) leaves sampled at full bloom (G.R. Klacan, 1965, unpublished data).

Fertilizer rates		
K	N	Total N in leaves
kg ha^{-1}		mg g^{-1}
55.8	89.6	48
	112.0	54
	134.4	54
	179.2	59
11.6	89.6	50
	112.0	52
	134.4	57
	179.2	57

Table 21-8. Influence of lime and K amendments on nutrient concentration of snap beans ('Bush Blue Lake') (J.F. Bartz, 1959, unpublished data).

Concentration of nutrients in leaf tissue	Limestone application, kg ha $^{-1}$	Potassium application, kg ha $^{-1}$				
		0	112	224	336	$\bar{x}$
K, mg g $^{-1}$	0	24	28	32	34	29
	4 480	20	27	32	34	28
	8 960	19	27	35	32	28
	13 440	20	28	33	32	28
	$\bar{x}$	21	27	33	33	
	CV = 11.1% "f" Lime NS; K_2O*, Line × K_2O*					
Ca, mg g $^{-1}$	0	9	9	8	8	9
	4 480	10	10	10	9	10
	8 960	11	12	12	10	11
	13 440	10	11	12	9	10
	$\bar{x}$	10	10	11	9	
Mg, mg g $^{-1}$	0	4.1	3.9	3.6	3.6	3.8
	4 480	5.6	5.3	4.4	4.2	4.9
	8 960	6.4	5.8	5.7	4.6	5.6
	13 440	6.3	5.8	5.3	4.3	5.4
	$\bar{x}$	5.6	5.2	4.8	4.2	
Mn, mg kg $^{-1}$	0	100	112	88	50	88
	13 440	50	25	20	50	30
	$\bar{x}$	75	68	54	35	

D. Nutrient Interaction

The interacting effects of dolomitic lime and KCl application on nutrient accumulation are given in Table 21-8. This trial was conducted on an irrigated loamy sand in central Wisconsin where the pre-trial pH was 5.5 and the exchangeable K level of the soil was 112 kg ha $^{-1}$. Potassium, applied as broadcast KCl, exerted a strong competitive influence on concentrations of Ca, Mg, and Mn found in the plant tissue. Lime application resulted in increased Ca and Mg concentrations but decreased the concentration of Mn in the leaf tissue. Similar responses have been observed for other vegetable crops [pea, broccoli (*Brassica oleracea* L.), and spinach (*Spinacia oleracea* L.)].

Interacting effects of lime and K applications on snap bean yields are given in Table 21-9. In this and several related studies, there was a negative yield response to K. Increasing the rate of limestone application had no effect on bean yields.

E. Problems and Limitations

The analyses presented in Tables 21-1 and 21-2 reflect an average combination of nutritional levels that have been associated with average to good crop production. Manipulation of specific variables can alter plant composition and the associated crop yield. In most production systems, however, control of single variables of even specific combinations is seldom accom-

Table 21-9. Influence of lime and K amendments on yield of snap bean ('Bush Blue Lake') (J.F. Bartz, 1959, unpublished data).

Limestone application, kg ha^{-1}	Potassium application, kg ha^{-1}				
	0	112	224	336	$\bar{x}$
0	7 538	6 966	7 347	6 608	7 112
4 480	6 944	7 090	7 190	5 835	6 765
8 960	7 314	7 829	7 874	7 482	7 627
13 440	7 291	7 892	7 213	6 877	7 314
$\bar{x}$	7 269	7 437	7 403	6 698	

plished or approached. Plant responses can be measured but the combination of variables responsible are difficult to identify and evaluate. On this basis, dependable recommendations are difficult and sometimes impossible to obtain. The rapid changes occurring in the ionic root environment further complicate identification of variables. In many cases, therefore, by the time anomalies are diagnostically confirmed, it is too late to correct the nutrient status of the root environment.

IV. CONCLUSION

The data presented indicate some of the potential complications associated with using tissue analyses for improving vegetable crop production. Immobility of Ca and B would indicate that decreasing supply or temporary deficiency in the substrate would be most difficult to detect by means of plant analysis, even when using selected plant parts. This also complicates analyses for other nutrients having to use other plant parts. The effects of cultivar, stage of growth, planting date, and seasonal fluctuations, in conjunction with fertilizer variations, indicate the potential combinations and significance of interrelated variables that affect plant composition and correlated crop responses.

It should be recognized that soil variation and moisture fluctuations can alter (and sometimes drastically) the ionic root environment. This further complicates the identity and evaluation of the combination of variables that can be correlated with a crop response.

Would it not seem logical that, to improve and approach an optimal production, systems should be developed that minimize the variables? At present, it is possible to establish and maintain an ionic root environment that is essentially nonvariable. This is accomplished by means of a nutrient gradient using a constant source of water and a constant source of nutrients with respect to crop utilization (Geraldson, 1970). It is possible with such a system to evaluate an identifiable combination of variables of a single manipulted variable.

Thus, crop response as a responding variable could be correlated precisely with identifiable combinations of constant and manipulated variables. Crop production would be the measure of crop response to identifiable variables and plant analysis could be an aid to improving the combination of variables required for optimal production. Correlations obtained in this manner attain a degree of scientific validity that is not otherwise possible.

REFERENCES

Bartz, J.F. 1959. Yield and nodule development of Alaska peas as influenced by nutrition and soil moisture. Ph.D. thesis. Univ. of Wisconsin, Madison (Diss. Abstr. AAD 59-03167).

Geraldson, C.M. 1970. Precision nutrient gradients—Component for optimal production. Soil Sci. Plant Anal. 1(6):317–331.

Johnson, C.M., and A. Ulrich. 1961. Analytical methods for use in plant analysis. California Agric. Exp. Stn. Bull. 766. p. 26–78.

Lorenz, O.A. 1944. Studies on potato nutrition: I. The effects of fertilizer treatment on the yield and composition of Kern County potatoes. Am. Potato J. 21:179–192.

Lorenz, O.A., K.B. Tyler, and F.S. Fullmer. 1964. Plant analyses for determining the nutritional status of potatoes. p. 226–240. *In* C. Bould (ed.) Plant analysis and fertilizer problems. Am. Soc. of Hortic. Sci., Alexandria, VA.

Lorenz, O.A., and K.B. Tyler. 1983. Plant tissue analysis of vegetable crop. p. 24–29. *In* H.M. Reisenour (ed.) Soil and plant tissue testing in California. Univ. of California Div. of Agric. Sci. Bull. 1879.

Maynard, D.N. 1979. Nutritional disorders of vegetable crops: A review. J. Plant Nutr. 1(1):1–23.

Tyler, K.B., and O.A. Lorenz. 1964. Diagnosing nutrient needs of melons through plant tissue analysis. Proc. Am. Soc. Hortic. Sci. 85:393–398.

Zink, F.W. 1963. Growth rate and nutrient absorption of celery. Proc. Am. Soc. Hortic. Sci. 82:351–357.

Zink, F.W., and M. Yamaguchi. 1962. Studies on the growth rate and nutrient absorption of head lettuce. Hilgardia 32:471–500.

Chapter 22

Plant Analysis as an Aid in Fertilizing Orchards

TIMOTHY L. RIGHETTI, *Oregon State University, Corvallis*

KRIS L. WILDER, *Oregon State University, Corvallis*

GEORGE A. CUMMINGS, *North Carolina State University, Raleigh*

Two major advances in technology are changing traditional approaches to nutritional research and fertility management. Sophisticated analytical equipment and microcomputer availability now allow nutritional experiments and diagnostic services that were once logistically and economically impractical. There is certainly cause for optimism. The purpose of this chapter is to present an overview of diagnostic technology. Although most examples are limited to temperate tree fruits with emphasis on our experience in the Pacific Northwest and mid-Atlantic states, the important concepts have wide application and will be useful in applying tissue analysis to orchards in general.

Optimal mineral nutrition is essential to high productivity. There are many examples where correction of nutritional problems resulted in enhanced profit. Plant analysis is a valuable tool that helps growers make better management decisions. We are strong advocates of regular sampling (at least every other year) of commercial orchards. Not all problems are nutritional, and it is no surprise that even when treatments result in a more favorable nutritional composition, tree performance may not change if other non-nutritional factors limit tree response. We view tissue analysis as a means of identifying a situation where maximum profit is not possible. When corrected, the grower has the opportunity to maximize returns, provided that a large set of other interacting factors are also optimized.

I. USES OF LEAF ANALYSIS

Perennial crops are ideally suited for application of tissue analysis. Since established plantings last many years, rapid turnaround time is not as essential and long-term management is possible. Routine monitoring of crop status followed by appropriate fertilizer adjustments is advocated by many state extension programs. Leaves are generally sampled during the summer and processed in ample time to be considered when fertilizer decisions are made for the next crop year.

Correction of acutely occurring nutritional problems is not as important as consistent long-term management. Stored nutrient pools within the tree are more important than previously believed. This discovery is especially true for N. Evidence suggests that trees have a large capacity to recycle N (Weinbaum et al., 1984) and various other nutrients (Smith, 1962). This is reflected in the occasional observation that eliminating fertilizer applications for several years may be required before any loss in tree performance occurs (Kenworthy, 1973; Smith, 1962). Although many growers would like to base management decisions involving foliar sprays on leaf analysis in the same year, recommendations based on rapidly changing mineral concentrations in spring and early summer are difficult.

The standard approach to interpreting leaf analyses is to compare observed concentrations in leaves collected at a specific time to reference values (critical concentrations or sufficiency ranges). Nutritional concentrations substantially lower or higher than reference values are associated with decreases in crop growth, yield, or quality. Extension service fertilizer guides often list critical concentrations or ranges associated with optimum performance for the nutrients most likely to cause problems for specific crops and locations. Suggested management approaches to correct these problems are also provided.

In addition to routine summer leaf samples, tissue analysis of other samples is often useful. When troubleshooting, samples need not be collected at a specific time. Similar tissues are collected from healthy and unhealthy plants. Both samples are analyzed and compared. If the problem is nutritional, obvious differences are often apparent. However, if the problem results in growth restriction there may be little or no differences in element concentration. Tissue analyses can confirm visual symptoms that suggest deficiency or toxicity. One can also determine if a fertilizer treatment has led to an increase in the specific nutrient in question. Leaf analyses are often a good indicator of the success of either fertilizer sprays or soil amendments.

Regular sampling of a specific orchard or area within an orchard is more valuable than using leaf analyses only when nutritional problems are suspected. Long-term trends are more convincing than a single sampling. A series of analyses over a period of years can indicate approaching nutritional problems. Some state programs suggest resampling the same trees. Although long-term trends are helpful, it is sometimes difficult for a grower to determine if a change in mineral status in a given year is due to altered management or seasonal shifts (Fig. 22-1). In this example, 1983 was a low-N year. Growers might interpret their results differently if they know the year as a whole is above or below average. Long-term averages, current industry averages, and where a grower's values fit relative to other local growers with the same commodity all help in interpreting data from a given year. Even though short- or long-term industry averages are helpful and should be considered, the information can sometimes be misleading if the "industry" is higher or lower than it should be.

Using leaf analysis enables growers to better manage their orchards. When specific deficiencies are identified, large responses are possible. However, the use of leaf analysis to guide fertilizer management of fruit trees is helpful but not essential. Although many people view the primary function of diagnostic services as detecting nutritional deficiencies, the major contribution to growers has been reducing the application of unneeded fertilizers.

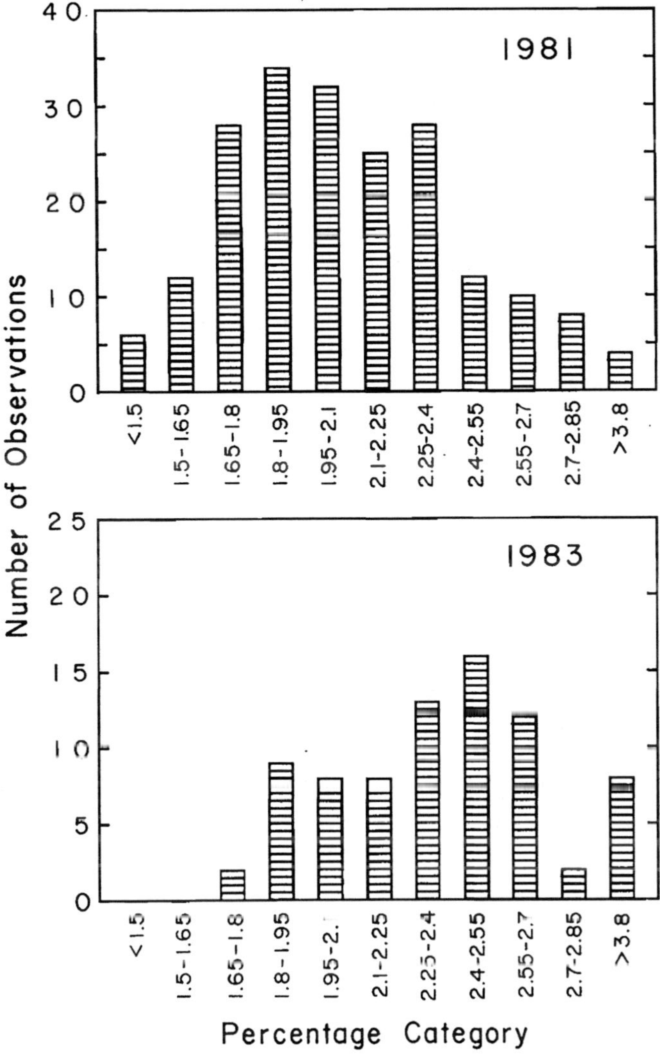

Fig. 22-1. Distribution of leaf N values in 1981 and 1983 for sweet cherry samples processed through the Oregon State University Plant Analysis Laboratory.

II. LIMITATIONS OF PLANT ANALYSIS

Plant analysis is best viewed as a useful tool rather than as a means of making a rigid diagnosis. There is no substitute for a knowledgeable evaluation that considers seasonal factors and individual characteristics of a given orchard. Although tree crops generally take longer to respond to soil-applied nutrients than other agricultural commodities, the wide use of foliar fertility programs also leads to rapid short-term changes in nutritional content. It can be important to consider the amounts, time, and method of past fertilizer applications, crop load, seasonal climatic effects, plant vigor, pruning, irrigation, method of true size control, and orchard floor fertilizer treatment.

Almost all state programs routinely generate a computer-devised recommendation for individual grower samples. There is some concern that this leads to rigid diagnostic criteria rather than flexible individual recommendations. Without this flexibility, factors that strikingly influence general physiology, leaf, and fruit mineral content are being ignored. Rigid recommendations that are based on narrow critical ranges can be misleading.

Soil testing can help reveal a cause of a nutrient problem and is most useful when it supplements a leaf analysis. Even though soil tests frequently weakly correlate with tree uptake, these tests are still valuable. Nutrient availability is pH dependent and recommendations based on leaf analyses alone assume pH is not a problem. It is foolish to design a fertilizer program without knowing soil pH. Furthermore, if leaf analysis suggests a deficiency and soil analysis suggests adequate levels, one might suspect that additional factors are important. On the other hand, if low tissue levels are accompanied with low soil levels, the grower is much more certain of the source of the problem.

III. FACTORS TO CONSIDER WHEN MAKING INTERPRETATIONS

Considerable effort has been made to define sufficiency ranges. However, it is our view that research projects to define sufficiency levels are not nearly as important as efforts to develop a more thorough understanding of the factors that may alter interpretations. Similar mineral concentrations for trees growing under different conditions do not suggest identical management approaches. Whether a particular element is within a recommended sufficiency range is only the first step in making decisions on possible management changes. Seasonal effects, crop load, vigor, pruning, orchard floor management, variety or rootstock, and marketing objectives should all be considered.

A. Seasonal Differences

Concentrations of minerals within any given tissue is a reflection of nutrient uptake, growth, transport, and remobilization of nutrients within the plant. Climatic factors affect all these processes and may explain some of the differences in mineral concentrations that occur for the same tissue in different years. Temperature and soil moisture, coupled with relative humidity, determine the amount of transpiration from plant leaves. Transpiration and uptake can have a profound effect on elemental composition. Although irrigation may lessen moisture variability, many factors cannot be controlled.

Nutrient availability for root uptake is a summation of those nutrients intercepted by plant roots, nutrients in soil solution reaching the plant root by mass flow, and those reaching the plant root by diffusion. Since diffusion appears to be the main mechanism supplying K to the plant root while mass flow amounts for a large portion of the Ca (Barber, 1962), it is apparent that leaf K and Ca levels would be differently influenced not only by soil supply, but also by soil moisture, air temperature, and transpiration. In a 4-yr pot experiment with apples comparing sheltered apple trees with trees grown under outdoor conditions, the Ca/K ratio was always higher in unsheltered trees with greater transpiration loss. Tromp (1980), using two apple (*Malus sylvestris* Mill.) rootstocks, found no difference in K foliar concentration of trees grown under high or low humidity, but Ca was higher with both rootstocks under low humidity. Data by Forshey (1969) also showed that K concentration of apple leaves was positively related to rainfall during the growing season. Boron deficiency of many crops is accelerated by dry soil conditions. Latimer (1941) noted increased occurrence of internal cork of apples, a symptom of B deficiency, in years when June and July rainfall was below normal. Probably, the increased root growth and activity in the surface soil associated with the higher rainfall and increased soil moisture could account for higher leaf B levels in wet years compared to dry years. Climatic differences also affect mineral concentration due to an effect on crop load and vigor.

B. Crop Load

The first prerequisite to understand the effect of crop load on leaf nutrient levels is to recognize the differences in elemental content of fruit compared to foliage. In both stone and pome fruits, concentration in the fruit of a mobile element, e.g., K, may approximate that in foliage while an immobile element such as Ca may be <5% of levels in foliage. For all elements listed in Table 22-1, it is apparent that a large portion of the nutrients

Table 22-1. The distribution of elements in Elberta peach trees (Cummings, 1973).

	Element								
	N	P	K	Ca	Mg	Mn	Fe	Zn	Cu
	$g\ kg^{-1}$					$mg\ kg^{-1}$			
Foliage	27.1	0.88	13.3	14.2	3.73	120	140	22	21
Fruit	17.4	1.04	9.5	0.4	0.41	36	54	26	13
Twigs	13.7	0.32	6.9	15.2	1.92	38	58	35	9
Branch	2.2	0.09	0.8	2.7	0.30	8	32	10	4
Trunk	2.1	0.10	0.7	2.6	0.24	8	35	8	5
Roots	8.2	0.97	2.7	1.6	0.63	22	140	15	4

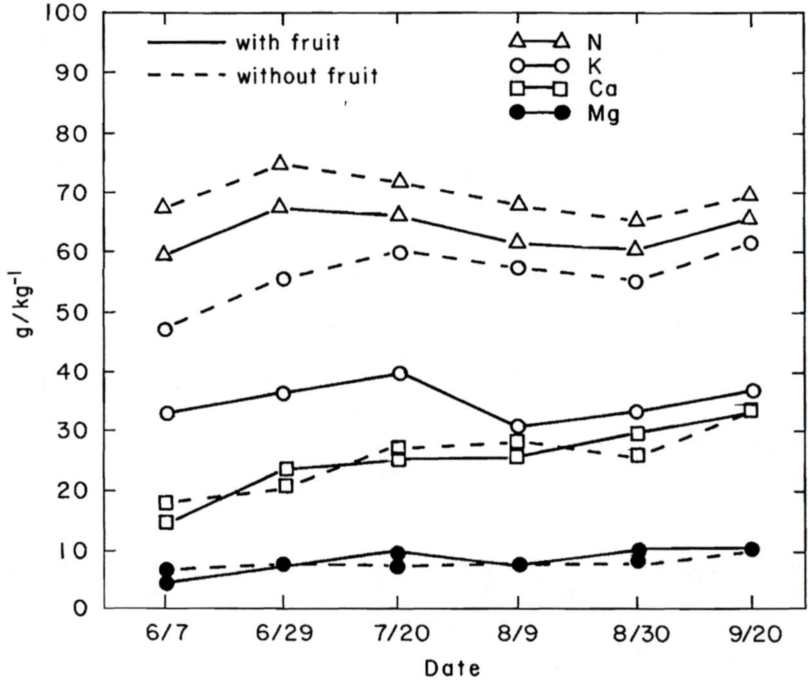

Fig. 22–2. Composition of peach leaves as related to leaf age and presence of crop.

annually taken up by peaches [*Prunus persica* (L.) Batsch] are in foliage and fruit, with much smaller amounts in the woody portions of the tree (Cummings, 1973). In peaches, concentration of K, N, P, Zn, and Cu in the fruit were similar to levels in the foliage, while Mg, Fe, and Mn were much lower in the fruit (Cummings, 1973). The greatest differences between foliage and fruit is noted with Ca.

Data in Fig. 22–2 compares seasonal leaf concentration of N, K, Ca, and Mg of bearing and nonbearing trees (McClung & Lott, 1956). Potassium was the only element where leaf levels were consistently lower in the bearing trees. The apparent higher levels of N, Ca, and Mg in bearing leaves was offset by increased weight of leaves from nonbearing trees, and the total N, Ca, and Mg per leaf was similar regardless of fruit load. Hansen et al. (1982) reported similar results with K, Ca, and Mg on 'French' prunes (*Prunus domestica*), but effects on leaf K levels were much less when alternate branches, rather than the entire tree, was nonbearing. Forshey (1969) showed the inverse relationship between K leaf levels and apple yields for a 4-yr period. It is interesting that Wallace (1931) related the severity of K-deficiency symptoms to crop load, season, and cultivar. Although increased fruit growth and size can be associated with increasing leaf K levels (Lilleland et al., 1962), it is also clear that total crop load can reduce leaf K. Certainly, leaf levels of elements other than K may be influenced by fruit load, but the drain on leaf levels with other elements is much less than with K. A similar relation

Table 22-2. Percent N, dry wt., and total amount of N in the current growth of different tissues for fruiting and nonfruiting apple trees. (Derived from Hansen, 1980.)

	Percentage N		Dry wt., kg		Amount of N, g	
Tree fruit	Fruit	No fruit	Fruit	No fruit	Fruit	No fruit
Leaves	2.85	2.18†	1.2	2.0†	34.2	42.6†
Fruits	0.50	-	5.6	--	28.0	--
Other	1.00	1.08	1.2	5.6†	12.4	60.9†
Total tree	0.93	1.35	8.0	7.6	74.6	103.5†

† Values for nonfruiting trees differ from fruiting trees.

between leaf and fruit levels holds for apples (Oberly & Kenworthy, 1961), but fruit N is commonly much less than the leaf concentration (Hansen, 1980).

In cases where the fruit concentration of an element is less than leaf concentrations, noncropping trees may have much greater total nutrient demands and differ in tissue composition (Hansen, 1980; Smith, 1962). Crop load can either increase or decrease total N demand depending on how it affects total biomass production. An example of each case will follow. Although these two examples refer to pome fruits, the principles involved are equally applicable to stone fruits.

The data in Table 22-2 are extrapolated from a published report (Hansen, 1980) where fruiting and nonfruiting (blossoms removed) apple trees were grown in pots. Concentrations of N in leaves, fruit, and remaining tissues are presented. Also listed are estimates of total annual biomass production and amounts of N per tree in this annual production. The total dry matter production of the two types of trees was similar. The major difference was that nonfruiting trees partitioned considerably more biomass into leaves and other tree parts. Increasing crop load decreased the total N requirement of the trees. Trees with no fruit required much more N than fruiting trees because more of their biomass was in high N tissue. The major message is that it takes more N to grow excess leaves, branches, and roots than fruit. Ironically, trees with the lower leaf N concentration had more total N in their biomass.

The data in Table 22-3 are from experiments conducted on mature d'Anjou pear (*Pyrus communis* L.) trees. Trees were left unthinned or had two-thirds of the fruit removed. Leaf concentrations and the total amount

Table 22-3. Effect of thinning on percent N and total amount of N in different tissues of d'Anjou pear trees. (Derived from Messaoud, 1987.)

	Percentage N		Amount of N (g/kg) branch	
Tree part	Thinned	Unthinned	Thinned	Unthinned
Shoot leaves	2.16	1.90†	2.58	3.90†
Shoot twigs	0.90	0.88	1.82	1.83
Fruit without core	0.73	0.68	1.00	1.30
Core	0.49	0.48	0.05	0.05
Total	--	--	5.45	7.08†

† Values for unthinned trees differ from fruiting trees.

of N per kilogram of branch are presented. A branch refers to the structural portion of a tree limb before bloom. Branches were otherwise similar. The expression N per kilogram of branch allowed us to compare N accumulation in different branch sizes. Using current Oregon State University (OSU) guidelines for N levels in d'Anjou pear leaves, thinned trees would have been diagnosed as normal (2.0–2.3%) and unthinned trees diagnosed as below normal (1.8–2.0%). Unlike the apple example, total dry matter production varied for different crop loads. Thinning resulted in less leaf growth and less total biomass production. The greater dry matter production in nonthinned trees created more demand that resulted in the detection of a deficiency that was not detected in trees with lower crop load. Shoot and branch growth and crop load must be evaluated in interpreting N concentrations. Crop load can either increase total N demand as it did in this study for pear, or decrease demand as illustrated in the apple example.

The issue is further complicated by changes in root and aboveground partitioning. A heavy crop often lessens the amount of carbohydrate available for root growth, which can make uptake of soil immobile elements more difficult. The relatively high K concentration in fruit and reduced root growth may both be involved in the decline of leaf K concentration with large crop loads. It is clear that reasonable recommendations require a knowledge of an orchard's cropping status.

C. Plant Vigor

General plant vigor is important. In cases where little or no growth occurs, nutrients are often concentrated and deficiencies may not be apparent. Conversely, with excessive vigor, a general dilution sometimes occurs. Leaf concentrations of clearly deficient trees can overlap healthy ones (Neilson et al., 1988; Smith, 1962). Similar nutritional composition from stunted and overly vigorous trees simply does not imply similar nutritional status. Any factor that affects overall vigor can make interpretation difficult. Nitrogen, more so than any other element under orchard conditions, is most apt to be deficient and the element most closely related to lack, or excess of, vegetative growth. The use of N is almost universal with orchard crops. Thus, it is not unusual to see leaf concentrations of other elements decrease as N levels and vegetative growth increase. The concept of nutrient balance proposed by Shear et al. (1946) must always be considered when interpreting leaf analyses. Any element deficient to the extent of limiting growth, when corrected, may result in deficiency of other elements because of the increased growth demands on a marginal supply.

D. Pruning

Pruning results in a dwarfing effect on the tree, but induces vigorous growth near the pruned area. A general concentration of all nutrients can result. Even vigorous growth is normally higher in N than unpruned trees with the same N supply. Pruning can have a greater effect on leaf N concen-

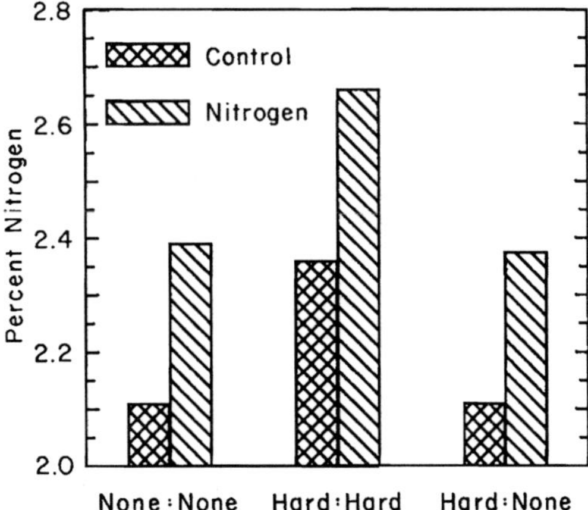

Fig. 22-3. The effects of pruning and N fertilizer on the August N content of 'Bing' sweet cherry leaves. Control treatments received no N for a 3-yr period; N treatments received 8.8 kg per tree for a 3-yr period. NONE/NONE—refers to no pruning in either of the 2 yr evaluated. HARD/HARD—refers to vigorous pruning in both of the 2 yr evaluated. HARD/NONE—refers to vigorous pruning in first year with no pruning in the second.

tration than N fertilizer. The effect of pruning on N status on cherry [*Prunus avium* (L.)] is apparent in Fig. 22-3. Using standard recommendations will suggest different management approaches. A vigorously pruned tree that has not been N fertilized and an unpruned tree that has been heavily fertilized can have similar N content. However, the pruned tree will be low in leaf N if it is not vigorously pruned in the following dormant season. There is little carryover of the pruning response beyond the summer immediately following vigorous pruning.

In essence, the objective of pruning for both stone and pome fruits is the same; to maximize yield and fruit quality for the life of the orchard. However, the pruning required to accomplish this feat varies between stone and pome fruits and among species of stone fruits. This variation makes pruning history more important in some cases than in others. However, regardless of the species or variety, pruning programs introduce considerable complexity. Not only does pruning control tree size and shape but also the amount of fruiting wood for the current year and frequently, in an older orchard with severe pruning, governs the formation of fruiting wood for future years. Recommendations that are made without regard to pruning history put the diagnostician at a clear disadvantage.

E. Orchard Floor Management

Leaf levels in fruit trees can vary with many different types of orchard floor management (Hogue & Nielsen, 1987). Cover crops change water relations, alter root distribution, change root temperatures, and compete for plant nutrients. Seasonal differences, especially in nonirrigated orchards, also alter water relations and affect root distribution. If moisture constraints restrict root activity to less-fertile subsoils, large differences in K or other surface concentrated nutrients appear. These direct and indirect effects on mineral nutrition need to be considered. Identical mineral analyses do not necessarily suggest similar management decisions. Control of orchard floor vegetation by tillage, mowing, herbicides, or a combination of these practices may influence leaf nutrient levels. The competitive effects of orchard floor vegetation is often evident when comparing mowing with clean cultivation (Bould et al., 1972). Tree growth and yield have been higher in clean cultivated plots and foliage was higher in N and K (Bould et al., 1972). The favorable influence of herbicides compared to clean cultivation or mowing on leaf levels, especially of N and K, could be related to changes in the effective rooting area as well as lack of competition. In some cases, leaves from trees with various mulch treatments were higher in K than unmulched trees (Baker,1943). Since leaf K levels increased even when the mulch materials contained little or no K, this is attributed to increased root proliferation in the surface soil where K levels would be higher than in the subsoil. In a comparison of vegetative cover, clean cultivation, and herbicides, Haynes and Goh (1980) found no effect on N leaf levels the first year. However, by the third year, N was lowest in the vegetative cover plots. Leaf K levels from trees in the clean cultivated plots were also lower than K levels in leaves from other treatments. Finally, Ries et al. (1963) observed growth and leaf N of both apples and peaches increased with chemical weed control compared to mechanical weed control or the use of black plastic mulch.

F. Variety and Rootstock

Differences in foliar concentration of elements among species, cultivars, and rootstocks were evident in some of the earliest reports concerning nutritional studies (Ballinger et al., 1966; Smith, 1962). There may be clear advantages to selecting rootstocks that exclude undesirable elements while more efficiently using needed nutrients. Much of the recent work with rootstocks in both apples and peaches has been concerned with leaf Ca levels. It is clear that Ca concentration varies with rootstock with apples (Sistrunk & Campbell, 1966; Poling & Oberly, 1979) and peaches (Knowles et al., 1984; Couvillon, 1982). However, rootstocks, such as MM 106, that produces high leaf Ca levels, may not result in high fruit Ca levels. Although rootstocks can alter Ca and other nutrient content, most variations were viewed as not important enough as to be a criterion for rootstock selection in currently popular rootstocks (Kenworthy, 1973). Regardless of rootstock, scion composition is still a realistic measure of nutrient needs.

Cultivar variations in standard values usually result from horticultural differences, market intentions, and use rather than from large differences in basic nutritional requirements. However, dwarfing rootstocks often produce radically different distributions of dry matter between fruit, leaves, and other tissues than standard trees. This can result in appreciable differences, especially with regard to N requirements. It is also conceivable that dwarfing trees could have more concentrated leaf nutrients.

G. Summary

Although much is known about factors that affect nutritional composition and leaf analysis interpretation, it is difficult to evaluate the interacting set of circumstances that occur in an orchard setting. Even if physiological or environmental factors are not a major factor, unusual mineral composition is still difficult to interpret. Low levels could be caused by insufficient nutrients in the soil, an inadequate soil pH, or nonfertility problems such as cold damage to the tree and diseased roots. Tissue analysis often indicates a condition without suggesting a real cause or the needed correction. Applying fertilizer when many other factors contribute to a deficiency can be disappointing.

IV. MAJOR NUTRITIONAL INTERACTIONS

The optimal value of one mineral element may depend on the level of another. For example, adequate Zn levels may differ, depending on P status (Mengel & Kirkby, 1982; Smith, 1962). Calcium, K, and Mg interactions also complicate simple interpretations (Mengel & Kirby, 1982; Smith, 1962). Furthermore, modifying one factor will likely alter another. Evaluating nutritional interactions before implementing management decisions is important.

A. Calcium, Potassium, and Magnesium

The most widely reported nutrient interaction, and certainly the one of most current interest, involves the principle cations (i.e., K, Ca, and Mg) found in foliage. For both pome (Vang-Petersen, 1980) and stone fruits (Cummings, 1965), increasing the application of K or Mg is reflected not only in a decrease in foliar concentration of the element not applied, but foliar levels of Ca are also often decreased. High K levels could result in lowering Ca and Mg levels by 40 and 50%, respectively (Lilleland & Brown, 1938). The sum of cations, in milliequivalents, usually is not changed. As one element is added, the other two usually decrease (Cummings, 1965). With the concerted effort to increase fruit Ca levels, all the factors that influence Ca concentration in foliage and fruit has received renewed interest. However, many factors other than K or Mg supply influence fruit levels of Ca and increasing that concentration in foliage may not result in an increased level of Ca in the fruit. Even so, this balance of foliar concentrations is so absolute it follows that prudent use of K or Mg would be mandatory. Further, in the interpretation of leaf analyses, levels of K in the mid- or upper portion of the sufficiency range, coupled with low foliar Ca, would warrant restriction in future use of K fertilizers.

B. Soil pH and Micronutrients

A substantial portion of the deficiencies of Fe, Mn, and Zn reported in the literature were from trees grown in calcareous soils (Lucas & Knezek,

1972). The most important soil test that could serve as a guide in interpretation of foliar levels of Mn, Fe, and Zn is soil pH. Normally, on mineral soils with pH below 7.0, except those low in organic matter or those with coarse texture, levels of these nutrients would be sufficient. However, Zn deficiencies in the Northwest have occurred on slightly acid soils (Neilson, 1988; Neilson et al., 1988). With low pH, excessive amounts of Mn may be taken up and lead to abnormalities in trees. One of the best examples is internal bank necrosis of apples that are characterized by localized high concentrations of Mn in the afflicted bark (Shelton & Zeiger, 1970).

C. Phosphorus/Zinc Ratios

Ballinger (1966), in his review of mineral nutrition of peaches, listed several studies where Zn deficiency was associated with high soil P levels. However, in some cases, high P/Zn ratios in leaf tissue are not always related to Zn-deficiency symptoms (Neilson et al., 1988). Olson (1972), in a detailed review concerning several crops, lists studies where evidence shows that increasing soil P levels lowers foliar levels of Zn. Evidence was presented to show that the P-induced Zn deficiency could be due to the restriction in uptake, translocation, or utilization of Zn. Though the mechanism involved in the induced deficiency remains obscure, the deficiency could be eliminated with small amounts of added Zn. In interpretation of Zn foliar concentration, it appears that soil Zn levels are of utmost importance but that the uptake, translocation, or utilization of Zn can be altered with an increase in soil P levels.

D. Nitrogen Interactions

The decrease in leaf levels of several elements by increased N supply is well-known (Smith, 1962) and is important in tree fruits where adequate N must be applied to maintain tree vigor and obtain adequate fruit size. With many elements, a simple dilution explains the results. However, the decline in leaf P is often greater than one would expect from dilution alone (Mengel & Kirkby, 1982). High concentration of N in foliage of apples is often associated with low levels of fruit Ca. Similarly, the increase in vegetative growth and increased yield associated with liberal use of N often results in lower levels of Ca in foliage.

E. Summary

In summary, not only with the cations K, Ca, and Mg is nutrient balance important, but N levels, probably primarily by dilution, and soil pH may all affect nutrient levels. As the intensity of management in orchards increases, the nutrient balance concept proposed by Shear et al. (1946) becomes of increasing importance.

Table 22-4. Sufficient nutrient ranges for sweet cherries from three different state laboratories.

	N	P	K	Ca	Mg
	dry wt., %				
Cornell Univ.	2.3–3.4	0.13–0.33	1.3–1.9	1.3–2.0	0.40–0.60
Ohio State Univ.	1.8–2.5	0.14–0.20	1.0–1.5	1.0–1.6	0.30–0.40
Oregon State Univ.	2.3–2.6	0.14–0.60	1.3–2.0	1.1–2.5	0.25–0.60
	Mn	Fe	Cu	B	Zn
	mg kg^{-1} dry wt.				
Cornell Univ.	35–135	80–500	7–12	35–50	35–50
Ohio State Univ.	30–150	30–150	3–15	30–50	10–30
Oregon State Univ.	26–160	50–400	6–20	30–75	16–80

V. DEFINING STANDARDS

Although the concept of interpreting foliar analysis is simple, the practice is more difficult. Defining sufficiency ranges involves considerable judgment. Published standard values are helpful, but differences in the sufficiency range for a given element frequently varies among geographical regions and even within a given region. For example, standard leaf values for sweet cherries that are used at various state testing laboratories (Table 22–4) differ. These differences in sufficiency ranges likely reflect philosophical differences more than regional variation in physiological requirements. Advocates of broad ranges are conservative about suggesting management changes. Advocates of narrow ranges are concerned with minimizing a potential nutrient limitation. Both strategies are likely indicative of lack of information. Probably a large part of this variation can be accounted when considering the methods used in developing standard values.

A. Historical Development

Standard values for many elements were derived from tissue analysis from high-producing orchards. Adjustments were made from results of controlled experiments where incremental amounts of nutrients were applied; growth, yield, and fruit quality measurements taken, and values correlated to foliar elemental levels. It would appear that a rather precise desired tissue concentration could be derived from this method. However, the precise value derived in one area may be quite different than the value derived from a similar study in another location or even the same study at the same location in another year. Sufficiency ranges that are developed for a specific cultivar, crop load, pruning regime, and environmental condition may not have broad applicability. Thus, standard values accepted today are recognized by researchers in fruit nutrition as a level well above the concentration where visual deficiency symptoms would occur but well below the level where damage would occur from excess. It is fortunate that the current sufficiency ranges for most elements are rather broad and slight variations from standard values are not reflected in deleterious effects on tree growth, fruit yield, or quality.

Sufficiency ranges for a given element in tables as proposed (Bould, 1984; Shear & Faust, 1980) do not vary greatly between pome fruits or even among the stone fruits listed. A comparison of values between pome and stone fruits reveals broad overlapping with values for many elements almost identical. Yet, differences do exist, not only among species but also within species and among cultivars. In addition, since trees are grafted, rootstocks as well as scion may influence concentration of elements.

B. Optimizing Yield, Growth, and Quality

In development of standard values, primary attention is directed to levels associated with optimum growth, fruit yield, and quality. Yield relationships have received the most attention. Excessive leaf N concentrations are associated with poor color development (Magness et al., 1939) and the excessive growth that results can aggravate fruit Ca problems. Even though leaf and fruit Ca levels are generally not strongly related, low leaf Ca levels cause concern for most growers. Similarly, high leaf K values are generally considered undesirable because of possible consequences relating to Ca nutrition. Sharples (1980) suggests that a minimum leaf P level (0.20%) required for satisfactory yield is less than the minimum (0.24%) for good storage quality. Most leaf levels associated with desirable quality are similar to levels associated with satisfactory yield (Sharples, 1980). Consideration of marketing objectives will likely alter an interpretation since fresh market, processing, or stored produce have different nutritional requirements. For example, lower N levels are often suggested to produce more desirable color and higher sugar content of juice.

C. Disease Incidence

Numerous studies during 1930 to 1950 noted a decrease in disease incidence related to N or K fertilization, but in most cases the trees were deficient in N or K. Probably the most frequently reported relationship was the increase in infection and severity of fire blight (*Erwinia amylovora*) with increased N fertilization and leaf N levels. In a recent review on fire blight, Zwet et al. (1979) noted that the vigorous growth resulted from N fertilization was related to fire blight infection. However, they also surmised that any cultural practice that stimulates growth was conducive to fire blight infection. Management of fertilization to maintain sufficient, but not excess vigor, is a difficult task. Season, crop load, cultivar as well as cultural practices may exert wide variation in late-season N levels. In addition to tree hardiness, N nutrition affects bud hardiness and fruit set. Zwet et al. also emphasized the importance of balanced nutrition since the deficiency of some element other than N could result in higher N levels and greater susceptibility to fire blight infection.

Numerous attempts have been made to relate sudden death of peach trees, normally called the "peach tree short life syndrome," to nutrition (Yadava & Doud, 1980). Survival has been increased by N fertilization (Cummings & Ballinger, 1972) and lime application (Cummings, 1983). However, in the N study no increase occurred in survival after the first increment of N and in the lime experiment, though foliar levels of Ca and Mg were increased, the influence of increased soil pH could affect numerous factors as well as Ca and Mg leaf levels.

D. Cold Hardiness and Winter Injury

Considerable effort has been made to evaluate relationships between mineral nutrition and winter injury or cold hardiness. Cold damage in deciduous fruits include injury from: (i) a sudden severe freeze in late fall or early winter before trees are fully dormant; (ii) low temperature during the dormant season reflected in bud kill, injury to trunk, twigs, and roots; and (iii) bud loss to freezing after rest is broken with hardiness decreasing until full bloom when buds are highly susceptible to freezing temperature.

Results of numerous studies indicate that normally, bud survival and fruit set are enhanced with increased N, but Boynton and Oberly (1966) emphasized that this favorable effect may be influenced by size of previous crop, time of N application, and season. Most concerns deal with tree longevity, since winter injury accounts for a large portion of tree losses. There is little doubt that N status of the tree is related to hardiness and a decrease in hardiness occurs with deficient as well as excess N. However, loss of low N, generally unthrifty trees, is quite different from the physical damage of trunk and branches resulting from freeze injury under high N conditions. Nitrogen levels high enough to promote winter injury may be rare. Orchardists are well aware of the risks involved when high N trees are growing vigorously at the time of the first killing frost.

E. Summary

Optimizing nutritional status to maximize yield may not necessarily mean that nutrition is optimal for other parameters, such as crop quality. Definitions of optimum ranges may depend on which yield or quality parameters are most important. Adverse effects of nutrition on disease and cold hardiness may occur with excess N, especially late in the growing season. Nutrient deficiencies or imbalance that lead to a general unthriftiness also increase susceptibility. Although some nutrients may be involved, there is not convincing evidence that variation in nutrient levels, within most accepted sufficiency ranges, is related to large differences in disease incidence, winter injury, or cold hardiness.

VI. REEVALUATION OF STANDARD VALUES

A. Changes in Orchard Management

Changes in cultivars, rootstocks, and training systems have decreased the time required for orchards to reach full production, increased, often several fold, the number of trees for a given area and also increased yield. These changes allow for more consistent annual fruit production but also require more precise management. Since the increased tree population and yield exerts a greater need for nutrients, maintenance of nutrient supply is of utmost importance. As intensity of orchard planting systems and management increases, ranges may need to be narrowed to avoid overlooking potential problems. This is especially important in view of the relatively small cost of fertility programs when compared to the major losses nutritional problems can cause. With continued improvement in analytical procedures, and increased experience in operating service laboratories, critical ranges are gradually being defined more narrowly. Fewer labs advocate broad ranges. However, even when one is comfortable with a specific set of standards, it is dangerous to rigidly interpret mineral analyses without considering many other site specific factors.

Changes in cultural practices and marketing objectives also necessitate changes in standard values. An excellent example is foliar and fruit Ca levels in apples. Prior to 1960, fruit storage conditions and Ca nutrition received less attention, and if low foliar concentrations were observed in conjunction with poor tree growth, fruit yield or quality, blame for poor performance was often attributed to low soil pH. Frequently, even though soil pH and soil Ca levels were raised by lime application only small changes in foliar Ca levels were observed and fruit Ca levels were often not affected. Even though leaf Ca status is not always related to fruit Ca levels, the prevailing attitude regarding Ca foliar levels has changed from one of little importance and concern to the attitude that it is of great importance. Calcium concentrations should be well above the critical level accepted prior to 1960.

B. Modifying Sufficiency Ranges

Although many researchers recognize the need to modify current standards, this is not an easy task. The fact that several years are often required before tree fruits respond to added fertilizer (Smith, 1962) complicates the problem. Many years of field data over a range of orchards are required to develop even tenuous critical concentrations or ranges. Three to 4 yr of field trials at several locations for each of 11 different elements on 12 tree crops quickly becomes a life-time effort. Evaluating both yield and postharvest quality further complicates field research. For all these reasons, judgment, rather than vigorous testing, is often the basis for modification of critical ranges, and tissue composition of high-producing orchards with acceptable quality is regularly used as a benchmark. Approaches that incorporate a measure of normal variation to define critical ranges (Kenworthy, 1973) rather than a judgmental decision have not been widely accepted. Although surveying desirable orchards is an easy approach, it does have severe disadvantages. Basing desirable ranges solely on survey data introduces considerable biases. Growers may expend much and receive little benefit to achieve levels that represent luxury consumption by high-producing orchards.

One would not expect strong, significant relationships between mineral content and yield, growth, or quality in survey data. The situation pictured in Fig. 22–4a clearly shows the problem. This example relates tissue P to yield in corn (Walworth et al., 1986), but the same principle applies to tree fruit. Although conventional correlation coefficients would not be significant, there is an important relationship among the 8000 data points. High-yielding plots (>12 tons/acre) are always within a given range of P concentration. Plots with poor nutritional status (outside the range) have no chance of obtaining high yields. Good nutrition is essential to obtain maximum production, but other factors often limit yield even when nutrition is optimal. Weather, bees, gophers, insects, and a host of other factors can cause poor yields in nutritionally sound orchards. This fact will destroy any chance the conventional statistician has of detecting significant relationships. It is more logical to search survey data for the characteristic triangular pattern. Although far fewer data points are available, we have used this approach to verify current critical ranges for tree fruits. An example for sweet cherries in Fig. 22–4b, clearly suggests that high yields are not possible when Ca concentrations are outside of the OSU 1.10 to 2.50% range (Table 22–4). However, Ca nutrition is not necessarily the causal factor. Calcium values outside of such a range usually indicate some substantial problem of which Ca status is a symptom rather than a cause. A boundary line approach need not be limited to yield. It may be possible to quickly derive sufficiency ranges for other parameters. In Fig. 22–4c, Cu concentrations of 12 mg kg^{-1} or greater are associated with a lower incidence of leaf spot. Lower Cu concentrations may reflect an absence of beneficial Cu-containing sprays. Although altering nutritional status may not lead to enhanced yield or quality, it is unwise to maintain nutrient levels that are outside a sufficiency range and suggest one has no chance of obtaining maximum production or profit.

C. Ratio-Based Interpretations

There is some concern that elements should not be evaluated individually. Recent studies suggest that ratio-based approaches (Diagnosis and Recommendation Integrated System, DRIS) that use nutritional balance may be appropriate. The DRIS procedures have been successfully used on many crops (Beaufils & Sumner, 1976; Sumner, 1977a, b), including some tree fruits (Alkoshab et al., 1988; Beverly et al., 1984; Davee et al., 1986). A diagnosis is based on values of a given element relative to other elements rather than a rigidly defined sufficiency range.

The DRIS indices for individual elements are based on the deviation of each important ratio (in which elements are either in the numerator or denominator) from its optimum value. Optimum values are defined as the mean

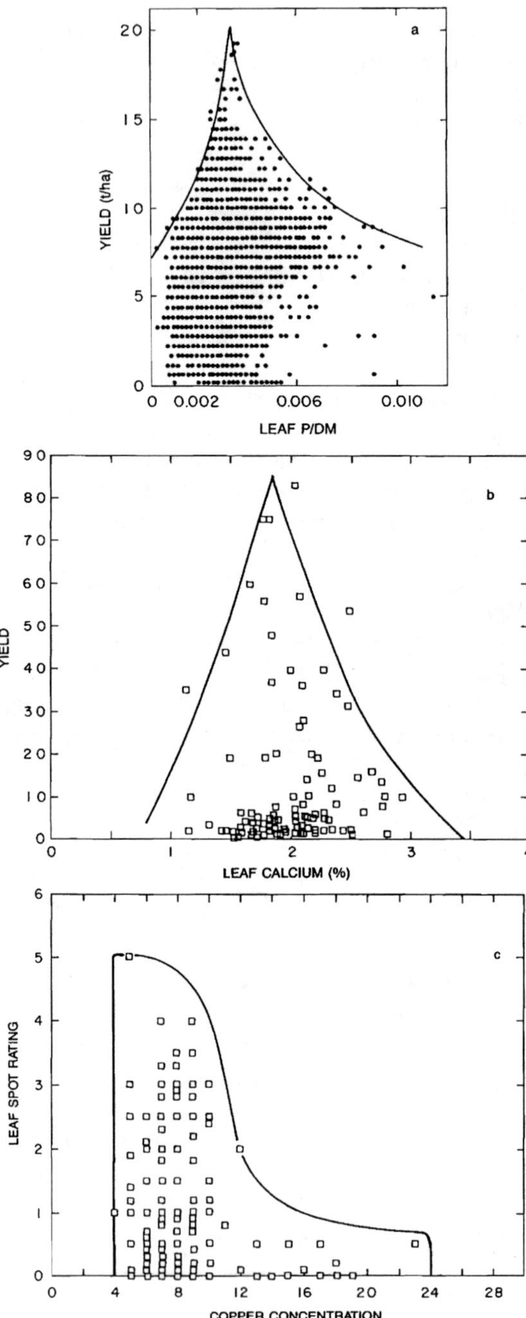

Fig. 22-4. Boundary evaluations to determine sufficiency ranges. (*A*) Relationship between corn yield and plant P for 8000 observations. (Data after Walworth et al., 1986.) (*B*) Relationship between sweet cherry yield and leaf Ca for 113 observations. (*C*) Relationship between leaf spot in sweet cherry and leaf Cu for 318 observations.

ratio for a desirable subpopulation. A DRIS index can be calculated for each element by averaging a series of intermediate functions calculated from important ratios of which the element was a component. Intermediate functions can be calculated using either negative or positive versions of the two formulas originally proposed by Beaufils and Sumner (1976). If the observed ratio value S_i is greater than the reference value M_i, the following equations apply:

$$FN_i = (S_i/M_i - 1) \quad CV_i$$

$$FD_i = -(S_i/M_i - 1)/CV_i$$

If the observed ratio value S_i is less than the reference value M_i, the following equations apply:

$$FN_i - (1 - M_i/S_i)/CV_i$$

$$FD_i = -(1 - M_i/S_i)/CV_i$$

where FN_i = intermediate function for the element in the numerator of an important ratio i, DF_i = intermediate function for the element in the denominator of an important ratio i, S_i = ratio i of elements in a tissue sample, M_i = optimum value for ratio i defined as the mean value of that parameter in a selected "desirable" population, and CV_i = coefficient of variation for ratio i in the desirable population. An individual element's index value will be more negative or positive depending on the degree of relative deficiency or excess. Thus, DRIS not only identifies deficiencies or excesses, but also places elements most likely to be deficient or excessive in a relative order. High negative values indicate deficiency, high positive values indicate an excess, and the sum of all values, irrespective of sign, gives a measure of the intensity of the imbalance (Nutritional Imbalance Index, NII). Perfectly balanced trees have an NII of zero. The larger the imbalance index, the greater the imbalance among nutrients. Although details of the approach are presented elsewhere (Beaufils & Sumner, 1976; Sumner, 1977b), Fig. 22–5 is an example of a balance evaluation on 500 sweet cherry samples. Although not all trees with balanced nutrition (Region A) are high yielding because of many other important factors, high yields do not occur with imbalanced nutrition (Region B). It is also possible to empirically verify critical values by determining which sufficiency ranges would be in best agreement with a DRIS diagnosis (Alkoshab et al., 1988; Davee et al., 1986; Righetti et al., 1988b).

A DRIS diagnosis for mulching treatments on Royal Anne sweet cherries appears in Table 22–5. Standard mineral analyses and yields for the same plots appear in Table 22–6. In this example, K and P were relatively deficient (most negative) in control trees. Both mushroom compost and black plastic resulted in more favorable K status, but yields with black plastic were still low. The ratio-based evaluation provides a possible explanation. Although

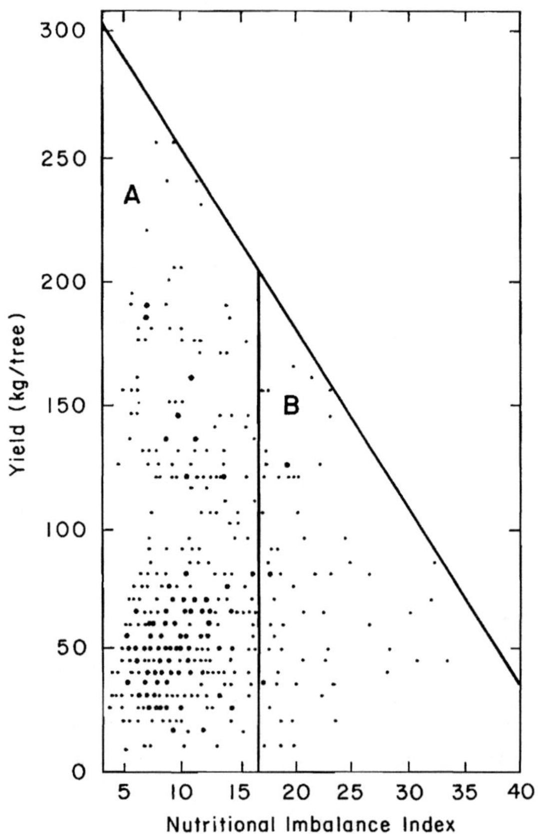

Fig. 22–5. Balance relationships where Nutritional Imbalance Indices (NII) are plotted against yield of 'Royal Anne' sweet cherries. Region A represents a zone where non-nutritional factors limit yield. In Region B, the imbalance threshold is exceeded and nutritional factors limit yield. Large points represent multiple observations.

K status improved, Ca and Fe imbalances became more severe and the NII was not improved. Changes in Ca and Fe status may be a symptom of excessive soil moisture under plastic mulch. The relative Ca deficiencies and Fe excesses were not apparent using sufficiency ranges. In 1982, N concentrations would have been diagnosed as deficient for all plots. However, all minerals were lower in 1982 and N levels relative to other minerals have not changed considerably. This is likely due to lower crop load and greater shoot growth. When a ratio-based diagnosis accompanies critical value evaluations, this is clear. Nitrogen was one of the least likely elements to become limiting. Thus, ratios may be especially helpful when seasonal or management factors result in a general concentration or dilution of elemental composition. Expressing mineral contents on a fresh weight, unit area, or total ash basis (Robinson, 1980; Rogers et al., 1953) has also been suggested to reduce concentration or dilution effects for individual nutrients.

Table 22-5. DRIS indices for mulching treatments on 'Royal Anne' sweet cherry.

Mulching treatment‡	DRIS indices†										
	N	P	K	Ca	Mg	Mn	Fe	Cu	B	Zn	NII
				1981							
Control	1.46	−2.72	−4.27	−0.94	−0.60	0.72	−0.17	2.52	2.05	1.97	17.4
Mushroom compost 1978	−0.93	−2.64	−0.30	0.28	−1.91	1.81	−0.79	0.78	1.51	2.18	13.1
Mushroom compost 1980	−0.23	−2.53	−1.32	0.06	−2.23	2.42	−0.33	1.15	1.03	1.98	13.3
Black plastic with trickle irrigation	0.03	−2.56	−1.54	−4.27	−0.94	1.04	3.88	1.37	2.02	1.08	18.9
				1982							
Control	2.87	−1.97	−5.25	−1.96	0.09	2.02	0.76	2.74	−0.13	0.83	18.6
Mushroom compost 1978	0.22	−1.88	−0.99	−1.49	−1.48	3.72	1.21	1.36	−1.20	0.52	14.1
Mushroom compost 1980	−0.30	−2.10	−1.48	−1.71	−1.66	3.41	1.28	1.88	−0.68	1.36	15.9
Black plastic with trickle irrigation	0.92	−1.93	−1.98	−5.56	−0.64	2.56	5.96	1.61	−0.26	−0.70	22.1

† If DRIS indices are placed in increasing order, nutrients most likely to be deficient (most negative) and most likely to be excessive (most positive) are readily identified. Ratios used to calculate indices are the same as those used in Fig. 22-1.

‡ All mulches applied in 1980, except mushroom compost 1978, which was applied 2 yr earlier.

Table 22-6. Mineral elements, yield, and relative yield for mulching treatments† on 'Royal Anne' sweet cherry.

Mulching treatment	Mineral concentration										Yield	Relative yield
	N	P	K	Ca	Mg	Mn	Fe	Cu	B	Zn		Yield treatments / Yield control
	dry wt., %					mg kg^{-1} dry wt.					kg/tree	
					1981							
Control	2.58	0.20	0.82	1.84	0.60	79	129	9	92	45	110	1.00
Mushroom compost 1978	2.44	0.19	1.69*	1.95	0.46*	88*	109	8	85	54	188*	1.70*
Mushroom compost 1980	2.43	0.20	1.28*	1.99	0.44*	99*	130*	8	84	44	144*	1.30*
Black plastic with trickle irrigation	2.54	0.21	1.31*	1.61	0.52*	82	263*	9	95	32	125	1.13
					1982							
Control	2.18	0.16	0.59	1.24	0.49	60	103	7	44	16	33	1.00
Mushroom compost 1978	2.00	1.16	1.24*	1.10	0.33*	67	100	8	40	13	48*	1.45*
Mushroom compost 1980	2.02	0.18	1.23*	1.27	0.36*	76	128	10	49	23*	51*	1.54*
Black plastic with trickle irrigation	1.91	1.16	0.91*	1.05	0.38*	60	200*	7	43	12	29	0.88

* Values followed by an asterick are significantly different from the control treatment ($P = 0.05$) for a given year.

Ratio-based approaches may have some advantages. In much of the literature where DRIS was superior to sufficiency range approaches, sufficiency ranges may have been inappropriate rather than any possible advantages for a ratio-based system. We have found excellent agreement between the two approaches (Alkoshab et al., 1988; Davee et al., 1986; Righetti et al., 1988a). Furthermore, DRIS may not detect all deficiencies or excesses (Righetti et al., 1988a). The range of index values obtainable varies considerably among elements (Righetti et al., 1988a). If the sum of DRIS indices, regardless of sign, is used as a criterion to identify imbalances, relative deficiencies or excesses for some elements are masked. The relationship between DRIS indices and the respective concentrations varies for different elements. Some indices so depend on the concentration of the element involved that a ratio-based diagnosis would have little advantage over a sufficiency range diagnosis (Righetti et al., 1988a). Other elements have indices that so depend on the concentrations of other elements that the index is of little use in evaluating nutritional status (Righetti et al., 1988a). Even with its shortcomings, DRIS is useful because severely imbalanced trees are never high yielding. The approach is survey-based and easy to develop. It may be a useful supplement to sufficiency ranges when severe imbalances are detected and it provides a means of independently evaluating tentative critical values for either yield or other parameters with little effort, if appropriate databases exist.

VII. INTERPRETATION COMPLEXITY

Although detailed discussion of all the difficulties one is likely to encounter is not possible, a brief description of how OSU currently uses leaf analysis to make fertilizer recommendations and the factors that are important will be useful. A summary of the results of tissue analyses on Oregon sweet cherries conducted at the Department of Horticulture's Plant Analysis Laboratory is presented in Table 22–7, and suggests the range of values one might expect. Most private tissue analysis laboratories also place elements in one of five ranges: deficient, below normal, normal, above normal, and excessive. Although a cursory view of Table 22–7 indicates that deficiencies and excesses are rare, this can be misleading. At least 25% of the samples received were deficient for some element, 25% were excessive for some element, and more than 33% were either deficient or excessive for some element.

Initiating a fertilizer program or increasing current fertilizer application rates are generally advised when values are below normal or deficient. Decreasing the amount of fertilizer currently applied or discontinuing fertilizer applications is generally suggested when values are above normal or excessive. However, this simple rule of thumb is not rigidly followed. Recommendations vary for each element and involve considerable judgment. A tissue analysis by itself is of limited value. Only when considering recent trends in a given orchard, long-term industry averages for an area (providing sample numbers are sufficient), whether this year's averages are above or below

Table 22-7. Percentage sweet cherry samples from Oregon analyzed over the past 15 yr that are in deficient, below normal, normal, above normal, and excess categories for leaf concentrations.†

	Deficient	Below normal	Normal	Above normal	Excess
		N			
Conc. range	<1.70	1.71–2.29	2.30–2.60	2.61–3.50	>3.50
Percent in category	4	54	30	12	1
		Phosphorus			
Conc. range	<0.10	0.11–0.12	0.13–0.40	0.41–0.65	>0.65
Percent in category	1	3	86	4	6
		Potassium			
Conc. range	<1.00	1.0–1.30	1.31–2.00	2.01–3.00	>3.00
Percent in category	16	11	48	24	1
		Sulfur			
Conc. range	<0.07	0.08–0.10	0.11–0.15	0.16–0.50	>0.50
Percent in category	1	28	61	10	0
		Calcium			
Conc. range	<0.50	0.51–1.09	1.10–2.50	2.51–3.00	>3.00
Percent in category	0	3	68	22	7
		Magnesium			
Conc. range	<0.18	0.19–0.24	0.25–0.60	0.61–2.00	>2.00
Percent in category	2	6	64	28	0
		Manganese			
Conc. range	<20	21–49	50–160	161–400	>400
Percent in category	0	5	63	24	8
		Iron			
Conc. range	<40	41–50	51–400	401–500	>500
Percent in category	0	0	93	4	3
		Copper			
Conc. range	<3	4–5	6–20	21–50	>50
Percent in category	1	14	83	1	1
		Boron			
Conc. range	<25	26–30	31–75	76–100	>100
Percent in category	3	2	72	16	6
		Zinc			
Conc. range	<10	11–17	18–80	81–300	>300
Percent in category	8	34	50	8	0

† Concentrations for N, P, K, S, Ca, and Mg are percent dry matter. Concentrations for Mn, Fe, Cu, B, and Zn are mg kg^{-1}.

normal, assessing possible imbalances, evaluating orchard performance especially with regard to crop load and vigor, considering climatic conditions, and knowing something about soil characteristics can the best use of tissue analysis be made. The following guidelines for sweet cherries are presented to demonstrate interpretation complexity rather than exhaustively cover all possible diagnoses.

A. Nitrogen

Understanding the dynamics of N in the tree is essential if one is going to interpret N levels in leaf tissue. Elemental concentrations can be misleading. This is especially true because leaf N values vary over a relatively narrow range. A 15% difference (2.0–2.3%) is enough to radically change one's interpretation. Unfortunately, these small changes in nutrient concentration do not necessarily correspond to altered total amounts of N within a tissue, nor can they be interpreted as changes in the uptake of N from the soil. To illustrate this, one can imagine two situations where total growth is different. A N concentration of 2.4% in leaves from a tree with <3 cm of new growth is simply not comparable to a N concentration of 2.0% in leaves from a tree with more than 60 cm of growth. The total amount of N on a per-ha or per-tree basis is much greater in the trees with greater growth despite smaller concentration. It is a mistake to diagnose these trees with above-average growth as N deficient even if concentrations are below desired levels.

From a practical perspective, N concentrations can be interpreted if crop and vigor are considered. We view sufficiency ranges as applying to trees with normal vigor and normal crop, and therefore modify interpretation accordingly. Plants with normal vigor and crop load should have N concentrations close to the normal ranges. If a sample falls in the deficient range, one can generally assume that N levels should be increased. However, few samples fall in this range for sweet cherry (Table 22–7). Values that are in below normal, normal, and above normal ranges (the majority of samples) must be evaluated relative to plant vigor. For plants with above-average vigor, below-normal values are often the result of dilution, and N fertilizer rates do not necessarily need to be increased. Similarly, plants with insufficient growth often have above-normal N concentrations. If growth is above average and N levels above normal, fertilizer N rates can likely be reduced. If growth is below average and N levels below normal, fertilizer N rates should probably be increased. Using both tissue analysis and vigor evaluations enhances recommendations; using either by itself can be misleading. Many factors cause insufficient growth and increasing N fertilization rates will not solve a lack of vigor caused by other factors. In situations where other factors are restricting growth, above-normal N values commonly occur. If a sample falls in the excessive range, one can generally assume N levels should be reduced with one major exception. When plants are severely stunted, N is concentrated in the leaves and excessive N values are likely a symptom of some other problem rather than a cause of the poor growth observed. In any case, excessive N values are rare (Table 22–7).

B. Phosphorus

Phosphorus responses are rare in tree fruits and have not been documented for field-grown sweet cherries. There have been some spectacular growth responses in Hood River, Oregon on pears when leaf concentrations are low. These responses are due to a unique situation where trees are growing on volcanic soils with very high P-fixing ability. Some responses from P-containing fertilizers may be due to the carrier effects (Ca and S) rather than directly to P. In general, P applications are not recommended. The small percentage of samples in the below-normal and deficient categories (Table 22–7) may benefit from trial applications of large amounts of P on selected trees, but low values are often associated with diseased or otherwise stressed trees and very low soil pH. Correcting other problems is probably more important than altering P status. Above-normal and excessive P levels are not a severe problem in themselves, however, the categories are useful. Marginal Zn concentrations should be taken more seriously when P is either above normal or excessive.

C. Potassium

Potassium deficiencies are relatively common for sweet cherries grown in the Willamette Valley, and trees have responded to K fertilizer additions. Unfortunately, K additions are difficult and expensive. Therefore, leaf K concentrations should be interpreted cautiously. Values vary considerably from year to year. Low values may be due to excessive crop load, a dry year (most orchards are unirrigated) that results in surface roots being less active, or other environmental variables that are poorly understood. Initiating an expensive K program when other explanations are likely may be unwise. Soil K analysis may be helpful. Trial applications, with monitoring over several seasons, are suggested before making a large investment. Above-normal K levels are generally not a major concern, unless Mg concentrations are also marginal.

D. Sulfur

Sulfur has only recently been routinely monitored in Oregon's orchards. Current sufficiency levels are tentative and difficult to interpret. Responses have not been observed, but the high rainfall and declining use of S-containing N fertilizers suggest the possibility that deficiencies may develop. Sulfur applications are only recommended when deficient values are accompanied by insufficient growth or S-deficiency symptoms. Above-normal S concentrations are not considered harmful. The utility of using total S concentrations as a diagnostic criterion needs to be confirmed.

E. Calcium

Although Ca may be important in fruit quality, Ca deficiency in vegetative tissue is unlikely. When below-normal Ca values are observed, they are almost always the result of water stressed or otherwise unhealthy trees. Insufficient soil Ca is probably very rare. Above-normal or excessive Ca concentrations can be important indicators of marginal K or Mg nutrition.

F. Magnesium

Magnesium deficiencies occur in Oregon. When levels are below normal or deficient and soil pH is low, dolomitic limestone is recommended. Values in either the deficient or below-normal category are rarely encountered when soil pH is considered adequate. Seasonal variability sometimes occurs and knowledge of long-term and current industry averages is useful. Mg deficiencies usually develop slowly and can be detected with a gradual decline of Mg concentrations over several years. Above-normal Mg levels suggest that marginal K concentrations should be evaluated more closely.

G. Manganese

Manganese concentrations are strongly correlated with soil pH. Below-normal Mn levels have only been detected on alkaline soils east of the Cascades. In these cases, foliar Mn sprays may be appropriate. The most useful aspect of Mn concentrations is in evaluating the need to lime. Soil pH analyses are sometimes difficult to interpret because it is not clear which depth or location around the tree is most appropriate to sample. Sampling at both 0 to 8 in. and 8 to 16 in. depths is often helpful. When Mn concentrations are above normal or excessive, soil sampling at several depths and locations is routinely recommended. A more detailed soil sampling almost always reveals a low pH that should be corrected. Similarly, if a low soil pH is accompanied by a normal Mn concentration, more detailed soil sampling often reveals that the pH problem is less severe than the single soil sample may indicate. Although the above guidelines are useful, Mn-containing fungicides can contaminate leaf samples.

H. Iron

Iron concentrations are also correlated with soil pH. However, the relationship is not as strong as for Mn. Therefore, the latter is more useful for diagnosing pH problems. Above-normal Fe can be an indication of poor soil drainage. High soil pH and visual symptoms may be a more reliable indicator of deficiencies than total leaf Fe concentration, especially since mild, but visible, Fe chlorosis is often not associated with declines in yield or quality. Since deficiencies are rare, little importance is placed on Fe concentration. However, some field evidence suggests that high Fe concentrations may suggest detrimental growing conditions such as excessive moisture (Davee et al., 1986).

I. Copper

Copper deficiencies have not been verified for sweet cherries in Oregon, but the lower limit of the sufficiency range is similar to levels where deficiency symptoms and responses have been reported for several tree crops in other areas. In view of the small cost of Cu sprays, applications may be appropriate for below-normal and deficient categories. Relatively few samples are above 20 ppm Cu and these cases can almost always be explained by the recent use of Cu-containing pesticides. Although toxicities in Oregon's sweet cherry has not been verified, avoiding additional Cu applications is suggested when leaf values are more than 50 ppm.

J. Boron

Boron is the most commonly applied micronutrient in the Northwest. Insufficient B leads to undesirable tree growth and poor yields. Excessive B is toxic. Applying appropriate amounts is essential because of the narrow range in application rates between solving a deficiency problem and creating toxicities. Either soil or leaf tissue analyses can suggest deficiencies, but, in general, August leaf analysis is the method of choice. Tissue analysis provides a direct measurement of the amount of B taken up by plants, and since foliar sprays are becoming the dominant method of B application, soil levels are receiving less attention. This may be a mistake, since B is not translocated (to any appreciable extent) from leaves to roots. Although considerable amounts of B from foliar sprays reach the soil either directly from the application or in subsequent leaf fall, some growers are not comfortable relying solely on foliar applications.

Although B deficiencies were more common in the past, August leaf values below 25 mg kg^{-1} are unusual. Values in the deficient range are relatively rare for cherries (Table 22-7). This is probably a reflection of the increased use of B sprays in the past 10 yr. Values under 30 mg kg^{-1} for sweet cherry clearly suggest a problem, but increasing evidence suggests B should be applied even at higher leaf levels to provide adequate B to developing buds on temperate trees (Callan et al., 1978; Hanson & Breen, 1985; Shrestha et al., 1987). There may be potential benefit to applying B when leaf values are in the normal range. Recent work on filberts (*Corylus* spp.) (Shrestha et al., 1987) and strawberries (*Fragaria* spp.) (Riggs et al., 1987) suggests leaf B and the B content of developing fruits may be unrelated. Similar phenomena likely occur with other fruits. In Fig. 22-6, leaf levels of B in hazelnut trees from five different similarly producing orchards that had not been treated with B sprays varied from 26 to 79 mg kg^{-1}, but the B concentration in May fruits were remarkably similar and late July fruits were almost identical (Shrestha et al., 1987). It appears that the B content of fruit tissues is generally similar and independent of leaf B across orchards with a wide range of B levels. A similar situation occurred in an experiment where various levels of soil-applied B were used on strawberries (Riggs et al., 1987). Even though total B content in strawberry plants could be doubled with the high B application, little change occurred in nonleaf tissues. Approximately 90% of the additional B in a plant resulting from soil application of B was found in the leaves. Total B content of nonleaf tissues did not change.

The above observations have altered our interpretative philosophy. Developing buds may require additional B to maximize fruit set regardless of the B status of the tree as a whole. High levels of B in young developing

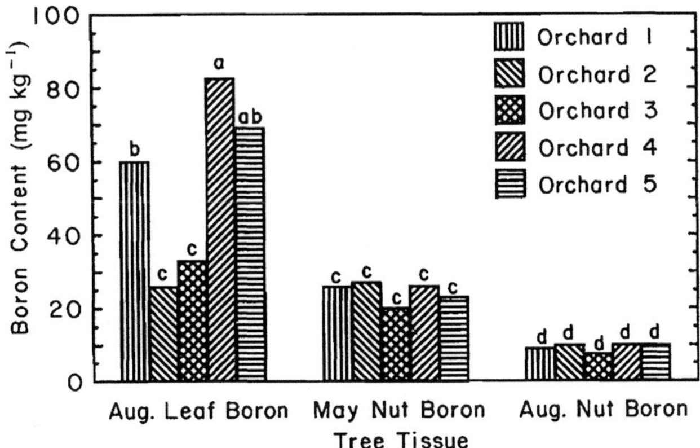

Fig. 22-6. Average B concentrations for nonsprayed filbert plots in five commercial orchards. May nut values were generally similar and August nuts varied little regardless of August leaf levels. Bars labeled with the same letter are not significantly different.

fruit may be essential and these high levels are best achieved by sprays applied directly to buds in fall or early spring. From a practical perspective, it may be valuable to apply annual B sprays to fruit crops regardless of leaf B concentration. Even if benefits do not occur every year, B applications are relatively inexpensive and a fruit set response 2 out of 4 yr would be an excellent return on a small investment. Occasional soil B applications are still recommended for the general maintenance of the tree, with annual foliar sprays providing for the special needs of reproductive tissues. Leaf analysis is still a valuable tool for B management. If annual applications are going to be made, it would be desirable to carefully monitor leaf levels and skip a year if B levels become above normal. Grower practice suggests that it is unlikely one will produce toxic levels of leaf B with recommended rates of foliar B application. However, in view of the losses excessive B can cause, prudent monitoring of leaf B is a wise decision (Neilson et al., 1985).

K. Zinc

Zinc deficiencies are common for Oregon sweet cherries. Current recommendations call for annual foliar Zn applications if the orchard has ever had visual Zn symptoms or recent concentrations have been either below normal or deficient. In view of the low cost of Zn applications and the likelihood of a response, there is no need to be conservative when applying Zn to cherries. By avoiding Zn applications when tissue concentrations are above normal or excessive, toxicity can be avoided. Since there is usually little carry-over into succeeding seasons, problems are not likely when making annual applications at recommended rates. Other tree fruits may be more sensitive to excessive Zn.

Table 22-8. Recommended U.K. leaf and fruit standards for Cox's Orange Pippin orchards. (From Sharples, 1980.)

	Percentage dry matter				
August "M" position leaf	N	P	K		Mg
For satisfactory yield	2.4–2.8	0.20–0.26	1.3–1.6		0.25–0.30
For good storage quality	2.6 max.	0.24 min.	1.6 max.		0.25 min.
Harvest fruit	mg kg^{-1}				
	N	P	K	Ca	Mg
For good storage quality	500–700	110 min.	1300–1600	50, 45	50

VIII. POTENTIAL USES OF FRUIT TESTING

Some reports suggest that leaf mineral concentrations can be used to predict postharvest quality (Sharples, 1980; Waller, 1980), but most emphasis has been placed on fruit tissue concentrations. In some fruit-growing areas, mineral fruit analyses are used to aid in postharvest management decisions (Ferguson et al., 1979; Rowe, 1980; Sharples, 1980; Waller, 1980). Fruit Ca concentrations are important. A causal effect of fruit Ca is likely in view of the success Ca sprays have in reducing postharvest disorder incidence (Sharples, 1980; Van der Boon, 1980). In some locations, an early season fruitlet analysis is used to assist in designing orchard Ca spray programs. Although fruit Ca is important, low Ca levels are also an indicator of undesirable environmental conditions and may not always be the direct cause of a disorder. Some programs establish minimum fruit mineral standards below which poor keeping quality is likely to occur. Standard values for both fruit and leaf tissues currently in use in England appear in Table 22-8. These values are specific for 'Cox's Orange Pippin' and may not universally apply, but serve as a good example. If a large percentage of potentially low- and high-quality fruit can be segregated so that each category of fruit is appropriately marketed or stored, profit may be enhanced. Fruit Ca is commonly related to postharvest disorders, but N, K, Mg, P, and B have also been implicated. Bitterpit in apples is most often predicted, but relationships may also exist with regard to senescent and low-temperature breakdown and other disorders (Bramlage et al., 1980; Sharples, 1980; Waller, 1980). Relationships between fruit mineral content and storage quality for d'Anjou pears have recently been presented (Richardson & Al-Ani, 1982; Vaz & Richardson, 1984). Cork spot in pears is Ca related. Firmness, color, and other quality parameters in pome fruits may also be related to fruit mineral composition (Fallahi et al., 1985; Rowe, 1980; Sharples, 1980). Most programs suggest sampling within 2 wk of commercial harvest. Regression equations have also been proposed as a means of identifying inferior fruit (Weis et al., 1985). Prediction equations for different locations and varieties are often dissimilar. Each specific disorder and each variety will likely require its own standard values or predictive equation for a specific locality. Computer software may be required to make storage selections based on whatever criterion a user selects as important (Fallahi et al., 1988). When fruit size and harvest maturity indices are included with mineral analysis data, the accuracy of predictions may be improved.

In some cases, it is difficult to develop critical ranges. Correlations between early season fruit minerals and quality or storage parameters are often as strong as late-season analyses, but absolute values often vary more between years than values from harvest samplings (Sharples, 1980). Even harvest samplings may not be consistent between years, suggesting that optimum fruit mineral values sometimes vary for different seasons (Perring & Holland, 1985). There is some evidence that more fruit Ca is required in high bitterpit years (Terblanche et al., 1980). In Fig. 22–7a, fruit Ca was consistently associated with bitterpit in every year, as reported by Van der Boon (1980). However, when we reorganized the existing data, no long-term relationship between mean average annual fruit Ca concentration and percent bitterpit could be obtained (Fig. 22–7b). We viewed Ca content as moderating the severity of bitterpit occurrences and replotted relative (percentile) Ca vs. relative bitterpit to create a consistent relationship (Fig. 22–7c). An evaluation of an orchard's Ca status and successful placement of fruit into high- and low-risk categories could then be made. It may be reasonable to segregate fruit for long-term storage based on relative Ca content, regardless of whether the year as a whole is above or below normal. Similar ranking approaches may also be appropriate for other nutrients. Ranked selection procedures may be useful for a variety of nutrients with respect to several quality parameters (Fallahi et al., 1988). Since a tree's mineral status with regard to one yield or quality component may be unrelated to other components (Fallahi et al., 1985), a definition of optimum nutritional status will depend on which yield or quality components are most important.

IX. TISSUE TYPES AND SAMPLE TIME

The best analysis, critical values, and interpretation are of little value without standard sampling procedures and methods of sample preparation. Leaves, petioles, stems, spurs, and fruit samples collected at different times from different locations within the tree have all been used (Smith, 1962). Sampling time and position of sample affect mineral composition. Basal, midshoot, and terminal leaf positions have different elemental concentrations (Smith, 1962). Although values obviously differ, it is difficult to choose an ideal tissue. No one tissue type or sampling time is going to be ideal for all elements important in orchard nutrition. Despite considerable research on the merits of various tissues and sampling times, the issue is far from resolved. A detailed review of the advantages and disadvantages of various approaches is beyond the scope of this presentation. Many factors are important in selecting a tissue or sampling time. The strength of correlations between mineral elements and observed responses to fertility management, the stability of mineral status over a long enough period to fit the realities

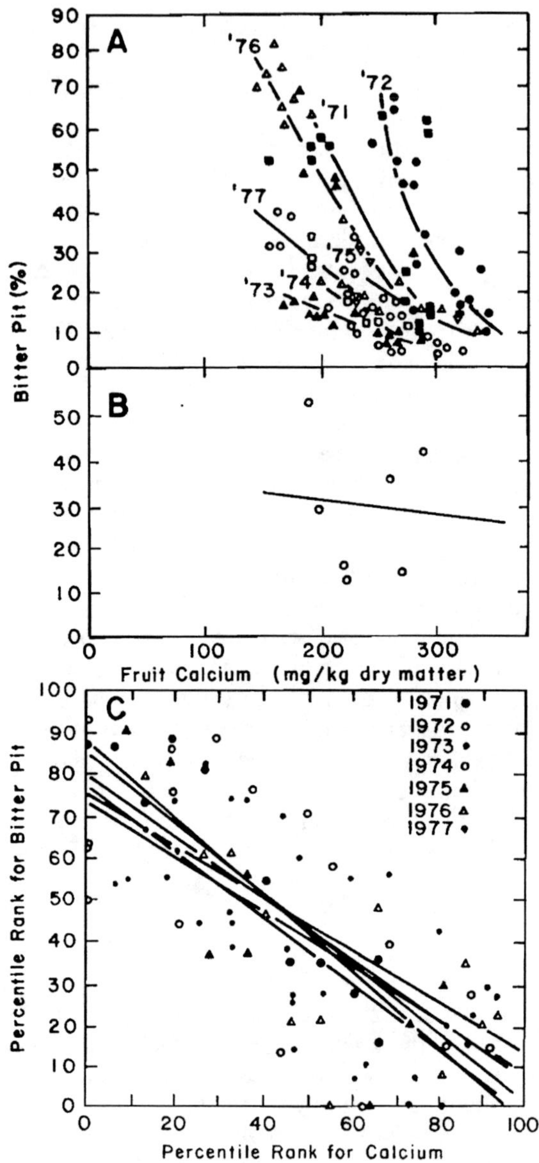

Fig. 22–7. Relationship between the percentage of bitter pit in 'James Grieve' apples and the Ca concentration in the fruit (size 70–75 mm): (*A*) Plots of Ca level vs. bitter pit in all years. (*B*) Plots of average Ca level vs. average bitter pit for each individual year. (*C*) Plots of percentile rank Ca level vs. percentile rank pitter pit in all years. (Data after Van der Boon, 1980.)

of commercial operations, and an avoidance of sampling complexity when a wide range of clients make sample collections, need to be considered. Physiologically defined sample times may have more merit than calendar-based collections (Batjer & Westwood, 1958; Kenworthy, 1969; Smith, 1962). Bloom dates, stages of fruit maturity, and other physiological events have been used to define sampling periods.

There is no doubt that other tissues or sampling times may be superior for specific nutritional disorders, but late July to early August leaf samples from the midshoot of current year's growth is a widely used compromise. Some laboratories suggest sampling 60 to 80 d after petal fall. Leaf values change little during late summer (Batjer & Westwood, 1958; Smith, 1962), adding to the attractiveness of mid-summer (July or August) sampling. In view of the expenses involved in reestablishing standard values, the move toward multielement analyses on a single sample, and orchardists' reluctance to collect a series of samples, a single midsummer analysis for all elements is likely to continue.

It is not essential to conduct elaborate long-term research to standardize a sampling method. Collection of consistently uniform samples for both standard development and tissue diagnosis is more important. Although alternative tissue may work equally well, there is value in standardizing methods to promote information interchange between locations and professionals. However, it is difficult to abandon long-term research information. State and private leaf analysis programs with a history of success are not going to change their procedures solely to facilitate industry standardization.

A similar situation exists for fruit sampling. Many different sampling schemes have been suggested. Whole fruits, skin, cortical plugs, opposite longitudinal sections, and horizontal sections, with and without skin tissue have all been proposed (Holland et al., 1975). A sampling time close to harvest is often, but not always, more informative than earlier samplings (Fallahi et al., 1985; Sharples, 1980). For winter pears, which require a minimum storage time before ripening is possible, postharvest sampling times may be appropriate. There is not as strong a requirement to make marketing decisions immediately after harvest as is the case with apples. Once again, it is unlikely that a single tissue type or sample time will be ideal for relating all elements to all disorders. Differing storage facilities (i.e., controlled atmosphere vs. standard storage) could have an effect on which sampling scheme is most predictive.

Since fruit samples are often difficult to dry, grind, and digest, schemes varying from wet digests of fresh tissue to dry ashing of freeze-dried samples have been used. Thus, the logistics involved in processing several fruit samples over a short period may dictate processing approaches. Variable means of expressing the results have also been proposed. Fresh weight, dry weight, and per fruit basis have all been suggested (Waller, 1980; Weis et al., 1985). There is general agreement that expressing mineral content on a fresh weight basis is important when samples vary in percent dry matter. However, when samples are similar, dry weight approaches work equally well. There is no easy way to define relationships between mineral compositions and fruit storability or quality. Several years of survey data where preharvest mineral analyses are related to postharvest quality are required. Since concentrations of fruit tissues change with time, analysis after storage cannot be compared to analysis at harvest unless whole fruit analysis is used (Bramlage et al., 1979).

X. PROBLEMS IN COLLECTING A REPRESENTATIVE SAMPLE

A leaf sample should represent a relatively homogeneous management unit. Some state extension services advocate a random sampling throughout the unit, while others recommend a more intensive sampling of a select group of trees (5–10) with average vigor within a sampling unit that can be annually resampled. A single sampling should not represent more than 2 ha. Only one tree age, variety or strain, and preferably only one rootstock type, should be sampled at a time. Unless leaves are unusually small, 50 to 100 leaves are enough for a sample. Leaves are generally sampled from all sides of the plant and with only one leaf from a shoot. Many different sampling schemes (leaves from the middle of the current season's terminal shoots; first fully expanded leaves; leaves or petioles opposite fruit clusters) are possible as long as a representative sample is collected for the same tissue type as used in standard development.

Fruit samples are also collected in a random manner with 25 to 50 individual fruits representing a management unit. Since mineral composition is size-dependent, sampling only fruit that are within a specific size category has been suggested (Weis et al., 1985). It is difficult to determine if specific sites within the tree are more appropriate. Since the goal of tissue sampling is to categorize on the basis of a management unit, random samplings of fruit are more often recommended than the sampling of specific locations within the tree structure. However, the large variability among fruit from a given orchard is a considerable problem. This is illustrated in Fig. 22–8

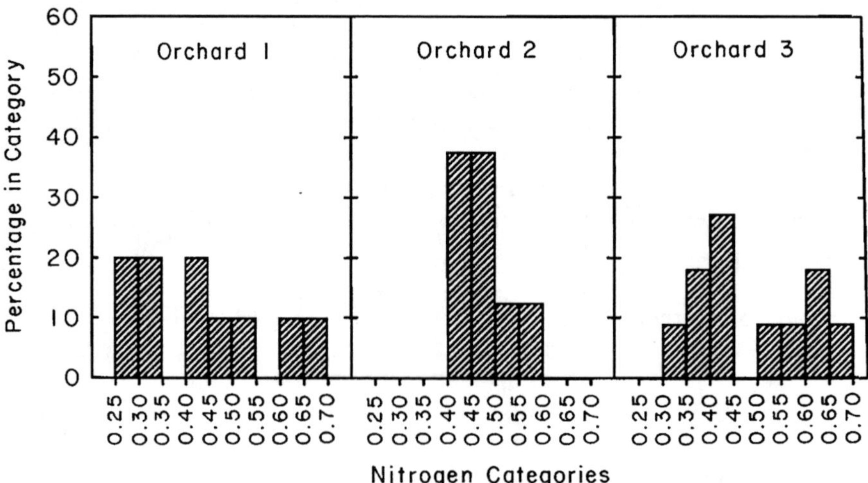

Fig. 22–8. Distribution of N values for 'Newtown' apple cortical tissue during late August for three orchards with the same mean N content.

where the N distributions of three 'Newtown' apple orchards with the same mean N concentration are shown. The actual number of high or low N fruit is clearly different despite similar means. Similar patterns exist with other mineral elements. In many cases when only a relatively small percentage of fruit suffer from a disorder, it is not possible to correlate disorder incidence to mean mineral concentrations for individual orchards. This poor correlation can occur even when the mineral concentrations of unaffected and affected fruit from the individual orchards are clearly different. Although considerable effort has been invested to determine which type of fruit tissue sample, sampling time, or means of expression will result in the strongest correlation with disorder incidence, orchard variability rather than inappropriate tissue or analytical procedures is the larger problem.

Leaves and fruit affected by insect or disease damage, phytotoxicity from pesticides, and herbicide injury are generally avoided. Cleaning of samples to remove surface contamination is not as critical (Kenworthy, 1973) as formerly believed (Taylor, 1956). Primary contaminates that should be removed are soil, dust particles, nutrient spray residues, and pesticides containing nutrient elements. Washing to remove residues of pesticides is not essential if the pesticides alter neither the determination nor the composition of the nutrients being evaluated. Calcium sprays can contaminate the skin of fruit tissues, but present little problem if internal tissue without skin is analyzed. If cleaning is to be done, washing with a mild detergent and rinsing in distilled water is usually adequate. However, if leaves have wilted or dried, thorough washing is impossible and is likely to remove soluble nutrients, especially K, and may cause more harm than good. All washing operations should be done rapidly. Washing only partially removes the contamination.

XI. CONCLUSION

Mineral analysis of leaf tissue will continue to be a valuable tool for the fruit industries. Interpretation is not simple because tree fruit nutrition is complex. However, attempting to maintain leaf elemental analyses within the appropriate ranges suggested in extension publications should enhance the probability of obtaining desirable yield and quality. Refinements to a diagnosis can be made by considering cultural practices, tree vigor, crop load, and environmental conditions. In the future, ratio-based diagnoses may become available as a useful supplement to standard critical concentration approaches. In any case, regular sampling provides more information than sporadic evaluation. A series of foliar analyses over a period of years can indicate approaching nutritional problems. Long-term averages, current industry averages, and where a grower's values fit relative to other local growers all help in interpreting data from a given year.

Mineral analysis of fruit tissue is not routinely available to U.S. growers but commercial programs are likely in the future. It may be possible to segregate fruit for storage by evaluating relative elemental contents. As increasing emphasis is placed on tree crop quality, nutritional evaluations may become more complex. Since fruit mineral status with regard to one quality component may be unrelated to other components, definitions of optimum nutrition and storage decisions will depend on which quality components are most important (Fallahi et al., 1988). Using mineral analyses to increase the probability of maximizing yield, quality, and profit is one of many management tools that should be considered.

REFERENCES

Alkoshab, O., T.L. Righetti, and A.R. Dixon. 1988. Evaluation of DRIS for judging the nutritional status of hazelnuts. J. Am. Soc. Hortic. Sci. 113:643–647.

Baker, C.E. 1943. Further results on the effect of different mulching and fertilizer treatments upon the potassium content of apple leaves. Proc. Am. Soc. Hortic. Sci. 42:7–10.

Ballinger, W.E., H.K. Bell, and N.F. Childers. 1966. Peach nutrition. p. 276–390. *In* N.F. Childers (ed.) Fruit nutrition. Horticultural Publ., New Brunswick, NJ.

Barber, S.A. 1962. A diffusion and mass-flow concept of soil nutrient availability. Soil Sci. 93:39–49.

Batjer, L.P., and M.N. Westwood. 1958. Seasonal trend of several nutrient elements in leaves and fruits of 'Elberta' peach. Proc. Am. Soc. Hortic. Sci. 71:116–126.

Beaufils, E.R., and M.E. Sumner. 1976. Application of the DRIS approach in calibrating soil, plant yield and quality factors of sugarcane. Proc. S. Afr. Sugar Technol. Assoc. 50:118–124.

Beverly, R.B., J.C. Stark, J.C. Ojala, and T.W. Embleton. 1984. Nutrient diagnosis of 'Valencia' oranges by DRIS. J. Am. Soc. Hortic. Sci. 109:649–654.

Bould, C. 1984. Methods of diagnosing nutrient disorders in plants. *In* C. Bould et al. (ed.) Diagnosis of mineral disorders in plants. Chemical Publ., New York.

Bould, C., H.M. Hughes, and E. Gunn. 1972. Effects of soil management and N, P, K, fertilizers on tree growth, yield, and leaf-nutrient composition of dessert apples. Exp. Hortic. 24:25–36.

Boynton, D., and G.H. Oberly. 1966. Apple nutrition. p. 1–50. *In* N.F. Childers (ed.) Fruit nutrition. Hortic. Publ., New Brunswick, NJ.

Bramlage, W.J., M. Drake, and J.H. Baker. 1979. Changes in calcium level in apple cortex tissue shortly before harvest and during postharvest storage. Commun. Soil Sci. Plant Anal. 10:417–426.

Bramlage, W.J., M. Drake, and W.J. Lord. 1980. The influence of mineral nutrition on the quality and storage performance of pome fruit grown in North America. p. 29–39. *In* D. Atkinson et al. (ed.) Mineral nutrition of fruit trees. Butterworths, Boston.

Callan, N.W., M.M. Thompson, M.H. Chaplin, R.L. Stebbins, and M.N. Westwood. 1978. Fruit set of 'Italian' prune following fall foliar and spring boron sprays. J. Am. Soc. Hortic. Sci. 103:253–257.

Couvillon, G.A. 1982. Leaf elemental content comparisons of own-rooted peach cultivars to the same cultivars on several peach seedling rootstocks. J. Am. Soc. Hortic. Sci. 107:555–558.

Cummings, G.A. 1965. Plant and soil effects of potassium and magnesium fertilization of 'Elberta' peach trees. Proc. Am. Soc. Hortic. Sci. 86:141–147.

Cummings, G.A. 1973. The distribution of elements in 'Elberta' peach tree tissues and the influence of potassium and magnesium fertilization. J. Am. Soc. Hortic. Sci. 98:474–477.

Cummings, G.A. 1983. Effects of soil pH, phosphorus rates, deep tillage practices, and clay additions on peach tree performance in deep sandy soils of North Carolina. J. Am. Soc. Hortic. Sci. 108:586–589.

Cummings, G.A., and W.E. Ballinger. 1972. Influence of long-time nitrogen, pruning, and irrigation treatments upon yield, growth, and longevity of 'Elberta' and 'Redhaven' peach trees. HortScience 7:133–134.

Davee, D.E., T.L. Righetti, E. Fallahi, and S. Robbins. 1986. An evaluation of the DRIS approach for identifying mineral limitations to yield in 'Napoleon' sweet cherry. J. Am. Soc. Hortic. Sci. 111(6):988–993.

Fallahi, E., T.L. Righetti, and J.T. Raese. 1988. Ranking tissue mineral analyses to identify mineral limitations on quality in fruit. J. Am. Soc. Hortic. Sci. 113(3):382–389.

Fallahi, E., T.L.Righetti, and D.G. Richardson. 1985. Predictions of quality by preharvest fruit and leaf mineral analyses in 'Starkspur Golden Delicious' Apple. J. Am. Soc. Hortic. Sci. 110:524–527.

Ferguson, I.B., M.S. Reid, and M. Prasad. 1979. Calcium analysis and the prediction of bitter pit in apple fruit. N.Z. J. Agric. Res. 22:485–490.

Forshey, C.G. 1969. Potassium nutrition of deciduous fruits. HortScience 4:39–41.

Hansen, P. 1980. Crop load and nutrient translocation. p. 201–212. *In* E. Atkinson et al. (ed.) Mineral nutrition of fruit trees. Butterworths, Boston.

Hansen, P., K. Ryugo, D.E. Ramos, and L. Fitch. 1982. Influence of cropping on calcium, potassium, magnesium, and carbohydrate status of 'French' prune trees grown on potassium limited soils. J. Am. Soc. Hortic. Sci. 107:511–515.

Hanson, E.J., and P.J. Breen. 1985. Effects of fall boron sprays and environmental factors on fruit set and boron accumulation in 'Italian' prune flowers. J. Am. Soc. Hortic. Sci. 110:389–392.

Haynes, R.J., and K.M. Goh. 1980. Some effects of orchard soil management on sward composition, level of available nutrients in the soil, and leaf nutrient content of mature 'Golden Delicious' apple trees. Sci. Hortic. 13:15–25.

Hogue, E.J., and G.H. Nielsen. 1987. Orchard floor vegetation management. p. 377–429. *In* J. Janick (ed.) Horticultural reviews. Vol. 9. Van Nostrand Reinhold Co., New York.

Holland, D.A., M.A. Perring, R. Rowe, and D.J. Fricker. 1975. Discrepancies in the chemical composition of apple fruits as analyses by different laboratories. J. Hortic. Sci. 50:301–310.

Kenworthy, A.L. 1969. Fruit, nut and plantation crops: Deciduous and evergreen. A guide for collecting foliar samples for nutrient analyses. Hortic. Rep. 11. Dep. of Horticulture, Michigan State Univ., East Lansing.

Kenworthy, A.L. 1973. Leaf analysis as an aid in fertilizing orchards. p. 381–392. *In* L.M. Walsh and J.D. Benton (ed.) Soil testing and plant analysis. Rev. ed. SSSA, Madison, WI.

Knowles, J.W., W.A. Dozier, Jr., C.E. Evans, C.C. Carlton, and J.M. McGuire. 1984. Peach rootstock influence on foliar and dormant stem nutrient content. J. Am. Soc. Hortic. Sci. 109:440–444.

Latimer, L.P. 1941. Relation of weather to prevalence of internal cork in apples. Proc. Am. Soc. Hortic. Sci. 38:63–69.

Lilleland, O., and J.G. Brown. 1938. The potassium nutrition of fruit trees. II. Leaf analysis. Proc. Am. Soc. Hortic. Sci. 36:91–98.

Lilleland, O., K. Uriu, T. Muroka, and J. Pearson. 1962. The relationship of potassium in the peach leaf to fruit growth and size at harvest. Proc. Am. Soc. Hortic. Sci. 81:162–167.

Lucas, R.E., and B.D. Knezek. 1972. Climatic and soil conditions promoting micronutrient deficiencies in plants. p. 265–283. *In* J.J. Mortvedt et al. (ed.) Micronutrients in agriculture. SSSA, Madison, WI.

Magness, J.R., L.P. Batjer, and L.O. Regeimbal. 1939. Correlation of fruit color in apples to nitrogen content of leaves. Proc. Am. Sci. Hortic. Sci. 37:39–42.

McClung, A.C., and W.L. Lott. 1956. Mineral nutrient composition of peach leaves as affected by leaf age and position and the presence of a fruit cup. Proc. Am. Soc. Hortic. Sci. 67:113–120.

Mengel, K., and E.A. Kirkby. 1982. Principles of plant nutrition. Int. Potash Inst., Worblenfen-Bern/Switzerland.

Messaoud, M. 1987. Effect of crop load on mineral uptake and partitioning in D'Anjou pears. M.S. thesis, Oregon State University.

Neilson, C.H. 1988. Seasonal variation in leaf zinc concentration of apples receiving dormant zinc. HortScience 23(1):130–132.

Neilson, D., P.B. Hoyt, and A.F. Mackenzie. 1988. Comparison of soil tests and leaf analysis as methods of diagnosing Zn deficiency in British Columbia apple orchards. Plant Soil 105:47–53.

Neilson, G.H., J. Yorston, W. Vanlierop, and P.B. Hoyt. 1985. Relationships between leaf and soil boron and boron toxicity of peaches in British Columbia. Can. J. Soil Sci. 65:213–217.

Oberly, G.H., and A.L. Kenworthy. 1961. Effect of mineral nutrition on the occurrence of bitter pit in 'Northern Spy' apples. Proc. Am. Soc. Hortic. Sci. 77:29–34.

Olson, S.R. 1972. Micronutrient interactions. p. 243–264. *In* J.J. Mortvedt (ed.) Micronutrients in agriculture. SSSA, Madison, WI.

Poling, E.B., and G.H. Oberly. 1979. Effect of rootstocks on mineral composition of apple leaves. J. Am. Soc. Hortic. Sci. 104:799–801.

Perring, M.A., and D.H. Holland. 1985. The effect of orchard factors on the chemical composition of apples. V. Year-to-year variations in the effects of NPK fertilizers and sward treatment on fruit composition. J. Hortic. Sci. 60:37–46.

Richardson, D.G., and A.M. Al-Ani. 1982. Corkspot of d'Anjou pear fruit relative to critical calcium concentration and other minerals. Acta Hortic. 124:113–118.

Ries, S.K., R.P. Larsen, and A.L. Kenworthy. 1963. The apparent influence of simazine on nitrogen nutrition of peach and apple trees. Weeds 11:270–273.

Riggs, D.I.M., T.L. Righetti, and L.W. Martin. 1987. The effect of boron application on boron partitioning in 'Tristar' and 'Benton' strawberries. Commun. Soil Sci. Plant Anal. 18(12):1453–1467.

Righetti, T.L., O. Alkoshab, and K. Wilder. 1988a. Diagnostic biases in DRIS evaluations of sweet cherry and hazelnut. Commun. Soil Sci. Plant Anal. 19(13):1429–1447.

Righetti, T.L., O. Alkoshab, and K. Wilder. 1988b. Verifying critical values from DRIS norms in sweet cherry and hazelnut. Commun. Soil Sci. Plant Anal. 19(13):1449–1466.

Robinson, J.B.D. 1980. Soil and tissue analysis in predicting nutrient needs. p. 355–364. *In* D. Atkinson et al. (ed.) Mineral nutrition of fruit trees. Butterworths, Boston.

Rogers, B.L.,L.P. Batjer, and A.H. Thompson. 1953. Seasonal trend of several nutrient elements in 'Delicious' apple leaves expressed on a percent and unit area basis. Proc. Am. Soc. Hortic. Sci. 61:1–5.

Rowe, R.W. 1980. Future analytical requirements in the fruit industry. p. 399–405. *In* D. Atkinson et al. (ed.) Mineral nutrition of fruit trees. Butterworths, Boston.

Sharples, R.O. 1980. The influence of orchard nutrition on the storage quality of apples and pears grown in the U.K. p. 17–28. *In* D. Atkinson et al. (ed.) Mineral nutrition of fruit trees. Butterworths, Boston.

Shear, C.B., and M. Faust. 1980. Nutritional ranges in deciduous tree fruits and nuts. Hortic. Rev. 2:142–163.

Shear, C.B., H.L. Crane, and A.T. Myers. 1946. Nutrient element balance: A fundamental concept in plant nutrition. Proc. Am. Soc. Hortic. Sci. 47:239–248.

Shelton, J.E., and D.C. Zieger. 1970. Distribution of manganese^{-54} in 'Delicious' apple trees in relation to the occurrence of internal bark necrosis (IBN). J. Am. Soc. Hortic. Sci. 95:758–762.

Shrestha, G.K., M.M. Thompson, and T.L. Righetti. 1987. Foliar-applied boron increases fruit set in 'Barcelona' hazelnut. J. Am. Soc. Hortic. Sci. 112:412–416.

Sistrunk, J.W., and R.W. Campbell. 1966. Calcium content differences in various apple cultivars as affected by rootstocks. Proc. Am. Soc. Hortic. Sci. 88:38–40.

Smith, P.F. 1962. Mineral analysis of plant tissues. Ann. Rev. Plant Physiol. 13:81–108.

Sumner, M.E. 1977a. Preliminary N, P, and K foliar diagnostic norms for soybeans. Agron. J. 69:226–230.

Sumner, M.E. 1977b. Use of the DRIS system in foliar diagnosis of crops at high yield levels. Commun. Soil Sci. Plant Anal. 8:251–268.

Taylor, G.A. 1956. The effectiveness of five cleaning procedures in the preparation of apple leaf samples for analysis. Proc. Am. Soc. Hortic. Sci. 67:5–9.

Terblanche, J.H., K.H. Gurgen, and I. Hesebeck. 1980. An integrated approach to orchard nutrition and bitterpit control. p. 71–82. *In* D. Atkinson et al. (ed.) Mineral nutrition of fruit trees. Butterworths, Boston.

Tromp, J. 1980. Mineral absorption and distribution in young apple trees under various environmental conditions. p. 173–182. *In* O. Atkinson et al. (ed.) Mineral nutrition of fruit trees. Butterworths, Sevenoaks, Kent, England.

Van der Boon, J. 1980. Dressing or spraying calcium for bitterpit control. p. 309–315. *In* D. Atkinson et al. (ed.) Mineral nutrition of fruit trees. Butterworths, Boston.

Vang-Peterson, O. 1980. Calcium, potassium, and magnesium nutrition and their interactions in 'Cox's Orange' apple trees. Sci. Hortic. (Canterbury, England) 12:153–161.

Vaz, R.L., and D.G. Richardson. 1984. Effect of calcium on respiration rate, ethylene production and occurrence of cork spot in d'Anjou pears (*Pyrus communis* L.). Acta Hortic. 157:227–236.

Wallace, T. 1931. Chemical investigations relating to potassium deficiency of fruit trees. J. Pomol. Hortic. Sci. 9:111–121.

Waller, W.M. 1980. Use of apple analysis. p. 383–394. *In* D. Atkinson et al. (ed.) Mineral nutrition of fruit trees. Butterworths, Boston.

Walworth, J.L., W.S. Letzsch, and M.E. Sumner. 1986. Use of boundary lines in establishing diagnostic norms. Soil Sci. Soc. Am. J. 50:123–128.

Weinbaum, S.A., I. Klein, F.E. Broadbent, W.C. Micke, and T.T. Muraoka. 1984. Effects of time of nitrogen application and soil texture on the availability of isotopically labeled fertilizer nitrogen to reproductive and vegetative tissue of mature almond trees. J. Am. Soc. of Hortic. Sci. 109:339–343.

Weis, S.A., W.J. Bramlage, and M. Drake. 1985. Comparison of four sampling methods for predicting post storage senescent breakdown of 'McIntosh' apple fruit from preharvest mineral composition. J. Am. Soc. Hortic. Sci. 110:710–714.

Yadava, U.L., and S.L. Doud. 1980. The short life and replant problems of deciduous fruit trees. Hortic. Rev. 2:1–116.

Zwet, T. Van Der, and H.L. Keil. 1979. Fire blight—A bacterial disease of rosaceous plants. USDA Handb. 510. U.S. Gov. Print. Office, Washington, DC.

Chapter 23

Plant Analysis as an Aid in Fertilizing Forage Crops

K. A. KELLING, *University of Wisconsin, Madison*

J. E. MATOCHA, *Texas Agricultural Experiment Station, Texas A&M University, Corpus Christi*

Production of forages can be influenced more by proper fertilization than many other crops since, in most cases, the hay, haylage, green chop, or grazed materials remove all of the aboveground plant parts. Consequently, especially in the nonpasturing situations, a larger portion of essential plant nutrients are removed and soils become depleted more quickly.

Plant analysis is unique from other crop diagnostic tests in that it provides a nutritional profile of the plant at the time the sample is taken. Its use is based on the principle that the nutrient levels present are the result of all the factors affecting the plant growth. However, the tissue nutrient levels are only one of the factors that influence crop growth, and in the case under question, nutrients may or may not be influencing the ultimate crop yield. For example, poor alfalfa (*Medicago sativa* L.) growth due to disease-impaired root systems may also show low K levels (Grau et al., 1989).

I. INTERPRETATIVE SYSTEMS FOR FORAGES

Most plant analysis programs currently interpret the results using the critical or sufficiency levels concept. However, since nutrient concentrations vary depending upon plant part, growth stage, variety and even geographic location, this approach has some serious limitations. Attempts have been made to reduce the influence of these problems by: (i) establishing critical values for different growth stages; (ii) correcting nutrient concentrations for increasing dry matter; (iii) using sufficiency ranges with the lower limit of the sufficient range set at about the critical level; or (iv) using nutrient ratios or simultaneous multiple regression of several elements (Munson & Nelson, 1973; Sumner, 1979).

 Soil Testing and Plant Analysis, 3rd ed.—SSSA Book Series, no. 3.

A. Critical Levels or Concentrations

Plant nutrient critical level has been defined in several ways. Jones and Eck (1973) define it as "that concentration below which yields decrease or deficiency symptoms appear." For many nutrients, yield decreases occur even before visible deficiency symptoms are observed. Because the exact concentration of a nutrient below which yields decline is difficult to determine precisely, the most common definition of critical level is the nutrient concentration in a particular plant part sampled at a particular growth stage below which a yield reduction of at least 5 to 10% is observed due to the low supply of the nutrient (Ulrich & Hills, 1967). Diagnosis using this system requires that the plant tissue being analyzed be compared with standardized critical values established for similar plant parts at the same stage of growth. Figure 23–1 illustrates the general relationship between nutrient supply, tissue concentrations, and crop yields. In most field studies, a broad transition zone in the curve between adequacy and deficiency is common, while in greenhouse-nutrient depletion experiments the curve breaks more sharply at a readily identifiable point.

The critical level is often somewhat higher in field experiments than in greenhouse studies. This is illustrated by work with subterranean clover (*Trifolium subterraneum* L.), where Bouma and Dowling (1982) reported critical values of 3.2 g of P kg^{-1} tissue for two field experiments in Australia. These were similar to results from earlier work (Bouma et al., 1969); however, the corresponding value from a greenhouse experiment was 2.2 g kg^{-1}. A comparable difference between pot and field experiments has also been shown for P levels in white clover (*T. repens* L.), where Andrew (1960) reported a critical value of 2.3 g kg^{-1} (greenhouse), while Rayment and Bruce (1979) suggested a value of 3.0 g kg^{-1} from their field experiments. Similarly, for work with crimson clover (*T. incarnatum* L.), Jordan and Bardsley (1958) suggested critical S levels of 1.0 g kg^{-1} for greenhouse work and 1.0 to 1.6 g kg^{-1} for field studies. To the extent possible, data from greenhouse experiments on nutrient ranges have been avoided in this chapter and most values given in the following tables are based on field experiments.

B. Nutrient Sufficiency Ranges

For most crops, there is an optimal range of concentrations, rather than a single point, over which yield will be maximized. If possible, one would attempt to supply nutrients at the lowest level which still provides top yields, but because of the many factors affecting yields and concentrations, this point is difficult to identify. Growers, therefore, usually strive for operating anywhere within the sufficiency range.

Jones (1967) suggested a series of nutrient zones or ranges that correspond to the general compositions illustrated in various sections of Fig. 23–1. In the deficient range, visible nutrient deficiency symptoms are likely evident, and crop yield is <75 to 80% of maximum. In the low range (80–90%

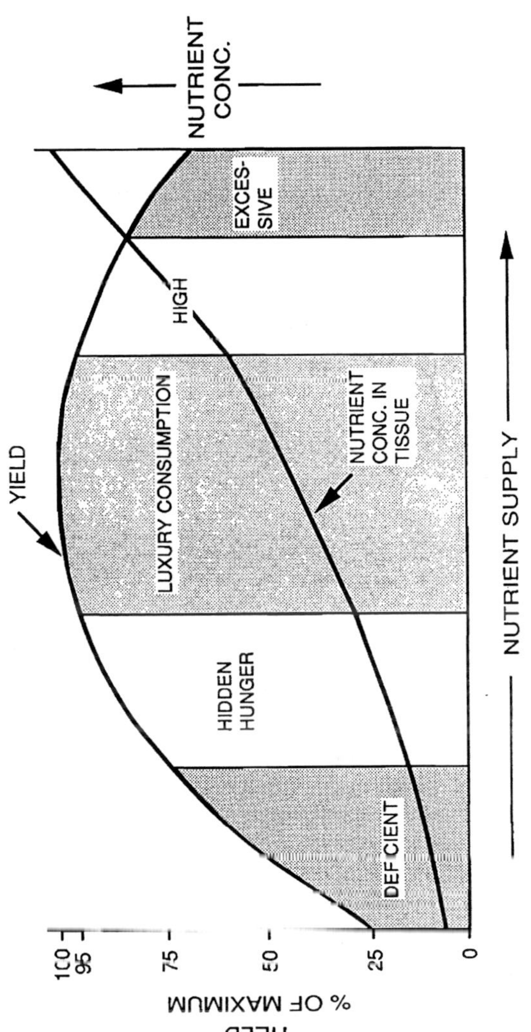

Fig. 23-1. Relationship between nutrient supply, yield, and nutrient concentration in plant tissue. (Adapted from Brown, 1970.)

of top yields), there are no clearcut deficiency symptoms, yet responses to additions of the low nutrient are likely. The sufficiency range represents the yield plateau (90–100% maximum yield). The nutrient concentration increases more rapidly as nutrient supply increases because there is little further increase in dry matter production to dilute the additional nutrients. Most nutrients have fairly broad sufficiency ranges. The lower end of the sufficiency range (or the upper end of the low range usually represents the critical concentration or value. In the high range, the plant takes in more of the nutrient than is required for maximum production. This range is sometimes referred to as the "zone of luxury consumption."

C. Nutrient Ratios—DRIS

Modern analytical multiple-element analysis gives results that lend themselves to multiple-regression analysis for interpretation. Greater information is gained since complex interactions can be evaluated with respect to yields. Although some work has been done in this way particularly for N, P, and K, this type of interpretation has not been widely accepted or incorporated into routine plant analysis programs.

A variation of the multiple regression approach recently developed is the Diagnosis and Recommendations Integrated System (DRIS) developed by Beaufils (1973) and introduced in this country by Sumner (1977a, b, 1979). As originally developed, all of the yield-determining factors that can be expressed quantitatively or qualitatively are considered simultaneously in making a diagnosis. These factors may include not only soil and plant analysis data, but also information on other crop growth-related factors such as climate, insects, disease, and varieties. However, in practice, it is the interpretations for plant analyses that have been most widely developed.

The DRIS approach to interpreting the results of plant analysis involves examination of ratios of plant nutrient concentrations corresponding to the average for the high-yielding population (norms) that are then compared to the ratios present in the sample being analyzed. A ratio of plant nutrient concentrations by itself cannot be used to diagnose plant problems, but combinations of different nutrient ratios can be combined mathematically to determine what nutrients are most likely to limit yield. The results of such calculations are the DRIS indices.

Proponents of this system point out that a major advantage of DRIS is that stage of maturity, plant part, and cultivar are less important than they are for the critical level or sufficiency range approaches to interpreting plant analyses. Thus, it should be possible to sample prebud alfalfa rather than waiting for first flower and still achieve accurate results with this method. Plant norms sufficiently established should be applicable to wherever the crop is grown, and nutrients are ranked in their order of limiting importance or yield (Sumner, 1977a), although more recent data have shown that locally developed norms may be more precise (Escano et al., 1981; Walworth et al., 1986).

A problem encountered by some users of the DRIS system is a tendency to interpret the results too dogmatically. Some regard every negative index as representing a deficiency and pay no attention to positive indices. But because all measurements are subject to some degree of uncertainty, both positive and negative indices must be evaluated, and the evaluations should not be made disregarding nutrient concentrations altogether.

Another limitation is that not all of the norms used to develop DRIS indices have been evaluated under field conditions. High-yielding alfalfa might be high in Ca, for example, because soil conditions required to obtain high yields coincidentally favored luxury consumption of Ca. The need for Ca is rarely evaluated under field conditions, this need typically regarded as being fulfilled by liming, although raising the pH of an acid soil provides many plant benefits in addition to supplying Ca.

II. SPECIAL CONSIDERATIONS WHEN USING PLANT ANALYSIS WITH FORAGES

A. Influence of Species or Cultivar

Plant analysis may give us more help in studying nutritional problems of perennial, rather than annual, species of forage crops. In the latter, plant analysis may be used in crop logging to help identify nutritional problems and determine the effectiveness of fertilizer treatments. However, this analysis gives little help for the current crop, since it is usually too late to apply remedial measures by the time forage plants are large enough to sample. Conversely, perennial forage crops, or crops subject to multiple harvests, may give valuable information as to the need of refertilization or making remedial top-dressings during the growing season. These problems are most acute where the critical value or sufficiency range method of interpretation is used.

Differences in critical levels have been noted between cultivars of the same forage species (Baligar et al., 1985). Porter and Reynolds (1975), after examination of elemental composition of 11 alfalfa cultivars at 1/10 bloom, concluded that nutrient concentrations were not good parameters for evaluating dry matter production. Similarly, Romero et al. (1981) observed that herbage K concentrations in different alfalfa cultivars differed at maximum yields and the onset of K-deficiency symptoms.

B. Influence of Nutrient Interactions

Macronutrient and secondary nutrient studies designed and executed to determine interactions among various nutrients are not as numerous as one might expect. Ideally, deficiency, sufficiency, and luxury levels of plant nutrients should be derived from the same experiment. However, this has only been possible in a few cases; therefore, cross references in plant analysis among several experiments are often necessary to obtain ranges for nutrient levels. Although a complete analysis is necessary before one can make an accurate balanced judgment of nutritional needs, it is only rarely that more than one nutrient is acutely deficient at the same time.

Plant analysis should be able to detect with reasonable success the element in lowest supply that limits growth. Other values may initially appear adequate, but must be reexamined after the first nutrient deficiency has been corrected. For example, plant analysis of grasses may correctly indicate adequate supplies of P if N is acutely deficient. Such plants will not show benefit to P applications alone and may respond only temporarily to applications of N. They will, however, respond dramatically to applications of both N and P. After correction of a single deficiency, the concentration of other nutrients nearly always decreases, often to levels below those found in the initial assay.

C. Influence of Plant Part

Ulrich (1956) suggested several attributes that a crop plant part should have for evaluating the whole plant's nutrient status. These include: (i) a wide range of composition values from deficiency to luxury uptake; (ii) ease in sampling; (ii) a relatively constant critical level on successive sampling dates; and (iv) ease in identifying and collecting field samples and providing plant material stable enough so that is will not deteriorate during shipment.

Studies have been made fractionating entire forage plants into six or more parts to determine the one that best reflects the nutrient status of the plant. Studies in California (Ulrich, 1956) and Illinois (Chamblis et al., 1970) have indicated that alfalfa midstems most effectively reflect the P status of the plant and usually dry readily and arrive in good condition. Other studies (Ulrich et al., 1967) have indicated that the leaves are often more sensitive for S assay and that the second or third fully matured leaf down from the apex of the plant should usually be taken. Workers adopting the Ohio system (Jones, 1967), use the top 15 cm (6 in.) of the alfalfa plant for all nutrients. Since plant disease or other factors may cause a leaf fall, midstems appear to offer some advantage. Additionally, it is possible to sample midstems out of commercially cut hay before or after baling and still do a reasonable evaluation of the P, K, and S status. However, on a more practical basis, most of the aboveground plant parts are used in forage crops rather than only a selected part, as is typical for other crops; therefore for many forages, the whole top is sampled for tissue analysis. The usual clipping height of approximately 5 to 8 cm is often used.

D. Influence of Maturity Stage

Stage of plant growth affects the nutritive value and elemental concentration of plant tissue. Some data describing this effect are presented where available; however, the authors emphasize that few data have been collected from well-designed fertility experiments sampled at various stages of growth that would help establish better ranges of nutrient concentrations. This chapter primarily presents concentration or plant analysis data from plants

Table 23-1. Changes in selected alfalfa composition and nutrient ratios from an early sampling (2 wk before harvest) and a sampling at harvest (first flower) for the highest yielding treatment at Hancock and Lancaster, WI, 1981.

	Hancock			Lancaster		
Parameter	Early	Harvest	Change†	Early	Harvest	Change†
			%			%
N, g kg^{-1}	41.6	31.2	−33	41.5	32.9	−26
P, g kg^{-1}	3.4	2.4	−41	3.5	2.6	−34
K, g kg^{-1}	29.9	27.4	−9	30.8	26.0	−18
S, g kg^{-1}	2.9	2.4	−20	3.0	2.5	−20
Ca, g kg^{-1}	7.7	8.8	+12	14.9	13.3	−12
Mg, g kg^{-1}	3.2	2.7	−18	3.7	2.8	−32
B, mg kg^{-1}	74	50	−48	41	49	+16
N/P	12.23	13.00	+6	11.68	12.65	+6
N/K	1.39	1.14	−21	1.35	1.27	−6
N/S	14.34	13.00	−10	13.83	13.16	−5
N/Ca	5.40	3.54	−53	2.78	2.49	−13
Mg/N	0.077	0.086	+10	0.089	0.085	−5
B/N	17.79	16.02	−11	9.88	14.89	+34
P/K	0.0114	0.088	−29	0.11	0.10	−10

† Based on the value at harvest. (From Kelling and Schulte, 1986.)

sampled at predetermined stages of growth, typically at or near the early bloom stage for alfalfa and clovers. Samples obtained from grazed pastures may well contain nutrient levels at the extreme ends of the ranges due to the variable stages of the physiological growth of the plant.

As shown in Table 23-1, samples taken at early stages of growth tend to have higher concentrations of N, P, K, S, and Mg, whereas concentrations of other nutrients such as Ca and B tend to remain stable or increase as the plant matures (Kelling & Schulte, 1986). Nutrient ratios tend to be more stable over time than are the concentrations.

Similar trends in nutrient concentrations have been observed for several leguminous forages including alfalfa (Smith, 1964; Kimbrough et al., 1971; Collins & Taylor, 1980; Collins, 1983), subclover (Ozanne et al., 1969; Jones et al., 1972; Spencer, 1978; Drlica & Jackson, 1979; Jones et al., 1980), ladino clover (Smith, 1964; Wilkinson & Gross, 1967), red clover (Smith, 1964; Collins, 1983), and birdsfoot trefoil (*Lotus corniculatus* L.) (Smith, 1964; Collins, 1983; Russelle et al., 1985; McGraw et al., 1986). This means that even though immature plants were functionally deficient in a nutrient such as P, the level present early in the season may still be much above the critical level established for the early bloom stage. For example, Jones et al. (1972) determined that critical P levels for subterranean clover decreased from 6.1 to 1.1 g kg^{-1} when plants were sampled from 41 to 151 d after clipping. Similarly, Kimbrough et al. (1971) showed that K concentrations for bud stage alfalfa were 143% of those in full bloom alfalfa and that correlations with yields were highest with the earlier sampling.

It must also be emphasized that not all nutrients behave similarly. As stated previously, Ca and B tend to accumulate as the plant matures. This was shown by Jones et al. (1976) with subterranean clover, where critical

Ca levels increased from 8.1 to 14.7 g kg^{-1} for 85- to 114-d-old plants, respectively, and defoliation tended to lower Ca critical levels. In this case, Ca/Mg ratios were no more stable than the concentrations themselves. Miller and Smith (1977) observed a continual increase in B concentration from early vegetation to bloom stage of alfalfa.

D. Influence of Environment

Environmental conditions other than soil fertility or fertilizer applied can influence nutrient concentrations of forages sampled at similar stages of growth. For example, Russelle et al. (1985) observed generally higher concentrations of most nutrients in birdsfoot trefoil grown at a northern Minnesota location compared to those present in tissue from a more southerly one. From an experiment with five legume species grown under several controlled temperature regimes ranging from 21/17 to 15/10 °C day/night temperatures, Smith (1970) noted that yields generally were highest under the cooler environments, whereas macronutrient and most micronutrient concentrations tended to be higher under the warm conditions. He made similar observations with more extensive studies with alfalfa (Smith, 1969, 1971). He concluded that higher amounts of exchangeable K are needed in soils under cool environmental conditions for alfalfa to have sufficient concentrations of K in the herbage for maximum production.

These generalizations should be viewed with some caution, however, as Gross and Jung (1981) showed that several grasses and legumes, including alfalfa, red clover, and alsike clover, had higher Ca and lower P levels in November or December than in May under similar temperature regimes. White clover was the only exception to this trend. Other environmental influences may include moisture supply, light, and growing season. These factors have been reviewed by Bates (1971), Fleming (1973), Reid and Jung (1974), and Peterschmidt et al. (1979).

With multiple-harvested crops like forages, the time of cutting cannot be separated from environment, since the growing conditions for the first cut may be substantially different from those present at third cut. Averaged over 11 alfalfa cultivars, Porter and Reynolds (1975) observed that P concentrations varied little from first to fifth cut, but K generally declined with the later harvests. Calcium and Mg were high in November, but lower in the middle of the season. Chamblis et al. (1970) noted a general decline in P concentration in the top leaves from first to third cuts, but found the midstem P levels to be much more stable.

E. Influence of Harvest Method

Plant analysis can be especially helpful in a grazing system of forage production where nutrient recycling via animal excreta has been shown to be important, and must be taken into consideration in planning a successful fertilization program (Matocha et al., 1973; Rouquette et al., 1973). In this chapter, plant analysis of forages is discussed in the context of forage removal through mechanical means rather than a biological system in which nutrient recycling is active.

F. Influence of Chemical Fraction Measured

Most elements are reported as the total amount of the nutrient present on a dry weight basis. However, as reported by Martin and Matocha (1973), it was found in California that the PO_4-P (the amount soluble in a 2% acetic acid digestion) reflected the P status as well as did total P and could be readily determined without wet or dry ashing. Although the PO_4-P value could not be traced to any distinct chemical fraction, in alfalfa it usually constitutes 85 to 90% of the total P present. A close correlation was found between total P in alfalfa hay and the levels of PO_4-P in the midstems of this crop (Fig. 23-2). Bouma and Dowling (1982) showed that P values from subterraneanclover tissue extracted with 10 *N* H_2SO_4 resulted in a curve with a more clearly defined inflection point, and correlated better with yield than did total P.

Total S measurement in plant tissue has often presented analytical problems, particularly at low S levels. With the development of the Johnson and Nishita (1952) method for readily reducible S, it is possible to directly determine small amounts of sulfate-S present in plant tissue without ashing or precipitation of barium sulfate. This fraction has been found useful in determining the S status of legume forage crops. Usually there is little sulfate present in S-deficient plants and the amount increases rapidly above the critical level. In some cases, such as with clover and alfalfa, the total S and SO_4-S values have been found to be highly correlated. However, SO_4-S levels also vary with stage of maturity (Jones et al., 1980); thus several researchers have suggested that N/S ratio may be a more definitive assay (Westermann, 1975; Spencer, 1978; Jones et al., 1980).

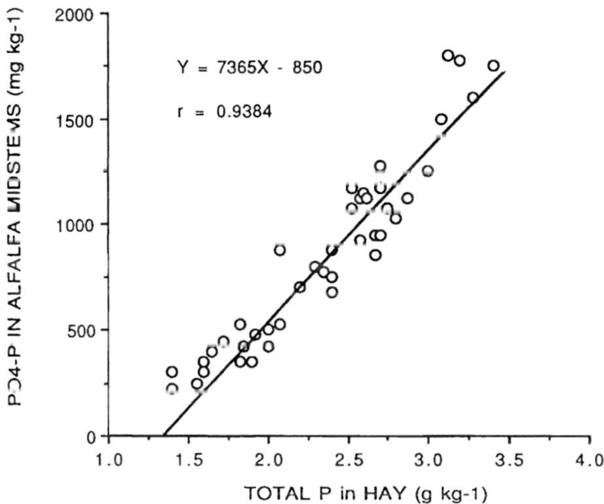

Fig. 23-2. Relationship between total P and PO_4-P in midstems (Martin & Matocha, 1973).

Total N is the usual measure of the N status of forages. However, NO_3-N may be a more sensitive indicator of nutrient status and is also a measure of the potential NO_3-toxicity hazard of forages when fed to ruminants. Nitrate-nitrogen accumulates in large amounts in many grass forages, but is usually not present in legume plants in appreciable amounts unless large amounts of manure or commercial N have been applied.

III. PLANT ANALYSIS OF LEGUMES IN PURE STANDS

A. Alfalfa Critical Values

More nutritional research has been carried out with alfalfa than any other single leguminous crop. Samples for plant analysis were usually collected at or near the early bloom stage just before usual harvest. Table 23-2 shows the critical levels of nutrients in various parts of alfalfa plants as suggested by a variety of researchers or extension workers. Ranges given in the table are similar to those suggested in the original version of this chapter (Martin & Matocha, 1973), namely:

1. *Deficient Range:* <80% yield, deficiency symptoms present
2. *Critical Range:* 80 to 90% yield, hidden hunger area
3. *Adequate Range:* 90 to 100% yield, normal yield
4. *High Range:* 100 to 70% yield, abnormally high, potentially toxic

Single-value critical levels reported by many workers will be listed as the upper limit of the "critical range."

As expected, the values for N, P, K, S, and most of the other nutrients tended to be somewhat higher where the top 15-cm portion of the plant was used for a sample rather than the whole plant top. As noted previously, critical values derived from greenhouse experiments tended to be lower than values seen with field experiments (Sallee et al., 1959), and the curves relating yield responses to tissue nutrient concentrations tended to break less sharply in field studies.

A survey of 20 midwestern, southeastern, and northeastern states and one Canadian province showed that of the 14 states that responded, 11 interpret plant analysis for alfalfa, and most of these (Indiana, Kentucky, Missouri, Nebraska, Ohio, Pennsylvania, and South Dakota) use the sufficiency ranges published by Jones (1967) as their primary criteria. In a few cases (11 out of a possible 84), a somewhat broader interpretation for N, Ca, Mn, B, or Cu were used. In four instances (Georgia, Ontario, Virginia, and Wisconsin), half or more of the diagnostic criteria being used differ somewhat from those proposed by Jones (S.M. Combs, 1989, personal communication).

The critical values for N vary as much or more with stage of maturity than for other nutrients. Although N fertilization of alfalfa is not ordinarily required in most alfalfa production areas, responses to added N occur when nodulation is poor and N_2 fixation is reduced because of acid soils or the lack of proper strains of rhizobia. Plants deficient in N are commonly low

Table 23-2. Nutrient sufficiency ranges and critical nutrient levels for alfalfa.

Chemical fraction	Plant part	Nutrient range				References
		Deficient	Critical	Adequate	High	
		g kg^{-1}				
Total N	Whole top		20			Van Lierop et al., 1980
Total N	Top 15 cm	<40	41-45	46-55	56-70	Jones, 1967
Total N	Top 15 cm	<13	13-25	26-37	>37	Kelling, 1982
Total N	Top 15 cm	<40	41-44	45-50	>50	Cornforth & Sinclair, 1982
Total N	Top 15 cm			30-50		Plank, 1988
Total P	Whole top		2.0			Van Lierop et al., 1980, OMAF, 1988
Total P	Whole top		3.0			Smith & Dobrenz, 1977
Total P	Whole top	<1.8		2.5-5.0		Chapman, 1967
Total P	Whole top	<2.0	2.1-2.2	2.3-3.0	>3.0	Sallee et al., 1959
Total P	Whole top		2.5			Bear & Wallace, 1950
Total P	Whole top		2.5			Rhykerd & Overdahl, 1972
Total P	Whole top		2.5			Gerwig & Ahlgren, 1958
Total P	Top 15 cm	<2.0	2.0-2.5	2.6-7.0	7.1-10	Jones, 1967
Total P	Top 15 cm	<2.0	2.0-2.5	2.6-7.0	>7.0	Kelling, 1982; Cornforth & Sinclair, 1982; Plank, 1988
Total P	Upper stem		3.5			Melsted et al., 1969
PO_4-P	Midstem	<500	500-800	800-2000	>2000	Sallee et al., 1959
		(above values in mg kg^{-1})				
Total K	Whole top		20			Smith & Powell, 1979
Total K	Whole top	<8	8-10	11-14	14-30	Martin et al., 1955
Total K	Whole top		20			Nelson & Barber, 1964
Total K	Whole top		17			Doll, 1962, OMAF, 1988
Total K	Whole top		12			Van Lierop et al., 1980
Total K	Whole top	<10	22			Bailey, 1983
Total K	Whole top		14			Gerwig & Ahlgren, 1958
Total K	Whole top		18			Markus & Battle, 1965
Total K	Whole top		9-11			Seay et al., 1949
Total K	Whole top		22			Walker et al., 1987
Total K	Top 15 cm	<17.5	17.6-20	20-35	35-50	Jones, 1967
Total K	Top 15 cm			20-35		Plank, 1988
Total K	Top 15 cm	<18	18-24	25-38	>38	Cornforth & Sinclair, 1982; Kelling, 1982
Total K	Upper stem		22			Melsted et al., 1969
Total K	Midstems	<6.5	6.5-8.0	8.0-15	15-30	Martin et al., 1955
Total S	Whole top		1.2			Caldwell et al., 1969
Total S	Whole top		2.2			Hoeft & Walsh, 1975, OMAF, 1988
Total S	Whole top		1.9			Hanson et al., 1984
Total S	Whole top		1.0-1.2			Martel & Zizka, 1977
Total S	Whole top	<2.0		3.0		Nelson & Barber, 1964
Total S	Whole top		2.0-2.2			Ensminger & Freney, 1966
Total S	Whole top		1.5-2.0			Westermann, 1975
Total S	Whole top	<2.0		>2.0	>3.0	Chapman, 1967
Total S	Whole top		1.2-1.5			Seim et al., 1969
Total S	Whole top	<1.7	1.7-2.1	2.1-3.0		Andrew, 1977
Total S	Whole top	<2.0	2.0-2.2	2.5-3.0	>4.0	Martin & Walker, 1965
Total S	Whole top	<2.0	2.0-2.2	>2.3		Pumphrey & Moore, 1965
Total S	Whole top		2.2			Harward et al., 1962
Total S	Whole top		2.0			Bear & Wallace, 1950
Total S	Top 15 cm	<2.0	2.0-2.5	2.6-5.0	>5.0	Cornforth & Sinclair, 1982
Total S	Top 15 cm			2.5-5.0		Plank, 1988

(continued on next page)

Table 23-2. Continued.

Chemical fraction	Plant part	Deficient	Critical	Adequate	High	References
		Nutrient range				
		——— $g\ kg^{-1}$ ———				
N/S ratio	Whole top	25–15	15–11	<11		Pumphrey & Moore, 1965
N/S ratio	Whole top	>18		<17		Westermann, 1975
SO_4-S	Whole top		2.0			Chapman, 1967
SO_4-S	Whole top		0.5			Westermann, 1975
Total Ca	Whole top		18	22		Nelson & Barber, 1964
Total Ca	Top 15 cm	<10	10–18	18–30	30–40	Jones, 1967
Total Ca	Top 15 cm	<5		5–30	>30	Kelling, 1982
Total Ca	Top 15 cm	<2.5	2.5–5.0	5.1–30	>30	Cornforth & Sinclir, 1982
Total Ca	Top 15 cm			8–30		Plank, 1988
Total Ca	Upper stem		8			Melsted et al., 1969
Total Mg	Whole top		2.0			OMAF, 1988
Total Mg	Whole top	<2	2.5			Chapman, 1967
Total Mg	Whole top		3.5			Smith & Dobrenz, 1977
Total Mg	Top 15 cm	<2	2–3	3–10	10–20	Jones, 1967
Total Mg	Top 15 cm	<2	2–3	3–10	>10	Kelling, 1982; Cornforth & Sinclair, 1982
Total Mg	Top 15 cm			2.5–10		Plank, 1988
Total Mg	Upper stem		4.0			Melsted et al., 1969
		——— $mg\ kg^{-1}$ ———				
Total B	Whole top		20			OMAF, 1988
Total B	Whole top	<15	20	>20		Chapman, 1967
Total B	Whole top	<15	15–20	20–80	200	Martin et al., 1965
Total B	Whole top	<10	10–20	20		James & Weaver, 1964
Total B	Whole top	<20	20–24	25–35	>35	Cornforth & Sinclair, 1982
Total B	Upper stem	--	30			Melsted et al., 1969
Total B	Top 10 cm	<10	20	>20		James & Weaver, 1964
Total B	Top 15 cm	<20	20–30	31–80		Kelling, 1982
Total B	Top 15 cm			20–80		Plank, 1988
Total Mn	Whole top		20			OMAF, 1988
Total Mn	Whole top	<10		>15		Chapman, 1967
Total Mn	Top 15 cm	<20	20–30	31–100	250	Jones, 1967
Total Mn	Top 15 cm	<15	15–20	21–200	200–700	Kelling, 1982
Total Mn	Top 15 cm	<20	20–30	25–35		Cornforth & Sinclair, 1982
Total Mn	Top 15 cm			25–100		Plank, 1988
Total Mn	Upper stem		25			Melsted et al., 1969
Total Fe	Whole top	<40	40–44	45–60	>60	Cornforth & Sinclair, 1982
Total Fe	Top 15 cm	<20	20–30	31–250	251–400	Jones, 1967; Plank, 1988
Total Fe	Upper stem		30			Melsted et al., 1969
Total Zn	Whole top		10			OMAF, 1988
Total Zn	Whole top	<15		20–50		Chapman, 1967
Total Zn	Top 15 cm	<11	11–20	21–70	71–100	Kelling, 1982; Cornforth & Sinclair, 1982: Plank, 1988
Total Zn	Upper stem		15			Melsted et al., 1969
Total Zn	Upper half		8	14		Boawn & Viets, 1953
Total Zn	Leaves		4–6	10–24		Lo & Reisenauer, 1968
Total Cu	Whole top		5			OMAF, 1988
Total Cu	Whole top	<1–2	5	5–14		Dye, 1962
Total Cu	Whole top	<5		5–16		Chapman, 1967

(continued on next page)

Table 23-2. Continued.

Chemical fraction	Plant part	Nutrient range				References
		Deficient	Critical	Adequate	High	
		mg kg^{-1}				
Total Cu	Whole top	<4	4-5	>5		Andrew & Thorne, 1962
Total Cu	Top 15 cm		<3	3-30	>30	Kelling, 1982
Total Cu	Top 15 cm	<5	6-10	11-30	31-50	Jones, 1967
Total Cu	Top 15 cm	<6	6-10	>10		Cornforth & Sinclair, 1982
Total Cu	Top 15 cm			5-20		Plank, 1988
Total Cu	Upper stem		7			Melsted et al., 1969
Total Mo	Whole top		0.5			OMAF, 1988
Total Mo	Whole top	<0.3		0.4-2		Chapman, 1967
Total Mo	Top 15 cm	<0.5	0.5-0.9			Cornforth & Sinclair, 1982
Total Mo	Top 15 cm			1-5		Plank, 1988
Total Mo	Upper stem		0.5			Melsted et al., 1969
Total Mo	Leaf		0.5			Reisenauer, 1956

in Mo. In addition, the N status may shift rapidly if environmental factors such as overirrigation, lack of oxygen, and high temperatures or extreme drought become unfavorable for the N_2-fixing bacteria on the plant roots.

The range of critical values for K is quite broad and tends to reflect some differences in plant part sampled, and location differences. Workers in Midwest states tended to use the top 15-cm portion of the plant, and recorded somewhat higher critical values than those reported where entire tops were used. Critical values from California experiments were often much lower than in other states. This may be influenced by the knowledge, that in cooler environments, alfalfa shows greater persistence, less winterkill, greater disease resistance, and produces better the second year where higher K levels are present. This tendency appears to be supported by the data of Smith (1969, 1970, 1971) and others who have demonstrated responses to higher soil test levels and higher plant K levels under colder temperature regimes. Although most of the K values given are for whole tops or the top 15 cm of growth, California data suggest that K content of the midstems is directly related to the K content of the whole tops at high correlation levels (Martin & Matocha, 1973).

There also appears to be some tendency for the more recent experiments to show somewhat higher K critical values. This observation was also made by Rhykerd and Overdahl (1972), and may be due to several factors including better, more responsive varieties, more frequent cutting schedules, and inclusion of forage quality considerations and stand persistence in determination of K-sufficiency level.

Sulfur deficiency of alfalfa is common in many areas of the USA and its occurrence may be readily identified either by symptoms or by plant analysis. While soil testing for SO_4 is used in several states, this method of determining S need is considered only partly effective at best, due to the various possible environmental sources of S not measured by the soil test (Kelling, 1981). Data from various states are in general agreement that 2.0 to 3.0 g kg^{-1} of total S represents the threshold of S deficiency.

Ulrich et al. (1967) has fractionated alfalfa plants grown in nutrient culture with a wide range of S nutrition from acute deficiency to high supply. They suggested use of SO_4-S rather than total S and concluded that the second to fourth leaf was the most satisfactory plant part, with a critical level of 100 to 200 mg kg^{-1} of SO_4-S. Midstems were also found to reflect nutritional status (with less sensitivity than leaves), but had a critical value of about 40 mg kg^{-1} of SO_4-S. Total S values in alfalfa are readily convertible to SO_4-S as shown in Fig. 23–2. Although several workers have suggested that N/S ratios may be more definitive in establishing S deficiencies, the critical ratios have also varied between experiments (Table 23–2).

Calcium and Mg deficiencies are not commonly encountered in areas limed to adequate levels, especially with dolomitic lime. Beneficial effects of aglime are reported in many sections of the country where lime is known to raise soil pH, reduce toxicity from Al or Mn in acid soils, and improve the growth and nodulation of alfalfa. Clearcut responses to Ca and Mg per se are relatively rare. Published criteria of nutrient ranges for these nutrients are variable and may be somewhat uncertain.

Deficiencies of some micronutrients such as B, Mn, and Mo are fairly common in alfalfa-producing areas. Plant analysis can be quite useful along with soil tests or nutrient deficiency symptoms in identifying causes of the micronutrient problems.

Reasonably close agreement on critical levels of B for plant parts sampled was found in the literature. Excess B from high-B water or excessive B fertilization results in a leaf burn. Such leaves show a white-colored marginal scorch and a B content of 150 to 200 mg kg^{-1} before leaf fall.

Manganese deficiency causes a chlorosis of alfalfa in some stages where alfalfa is grown in old lake sediments. Manganese deficiency is likely as Mn values fall below 25 to 30 mg kg^{-1}. Excess Mn is also a problem on some acid soils. Leaf values of 250 to 350 mg kg^{-1} Mn usually indicate an excess of soluble Mn, which may be alleviated by liming the soil.

Molybdenum deficiency occurs on many acidic soils and it can be corrected by either liming or by the application of small amounts of Mo fertilizers. The net result of a Mo deficiency is a deficit of plant N since the nodule rhizobia of Mo-deficient plants do not fix sufficient N. Mortvedt (1981) suggests that legumes should be analyzed for N rather than Mo since the relative increase in growth of various legumes from applied Mo is more closely related to uptake of N than of Mo, and tissue analysis of Mo is difficult. Published critical levels for Mo are in good agreement and are usually about 0.5 mg kg^{-1} of Mo in the leaves or whole alfalfa tops. Molybdenum and Cu nutriton of animals are closely related. Functional Cu deficiency is common when forage contains more than 7 to 10 mg kg^{-1} of Mo and with Cu levels below 5 mg kg^{-1}. The alfalfa plant can tolerate from 500 to 1000 mg kg^{-1} of Mo before Mo toxicity is encountered.

Copper and Zn deficiencies of alfalfa occur infrequently. Martin and Matocha (1973) suggest that published Cu values may be in error since hay samples with <5 mg kg^{-1} of Cu are common in many areas of normal production, and can be as low as 1 to 2 mg kg^{-1} with no evident plant deficiency symptoms.

Zinc deficiency can be produced in nutrient culture and it occasionally appears on alfalfa seedlings on low Zn soils where other crops fail due to Zn deficiency. Several researchers have noted that alfalfa will contain less Zn than other forage legumes when grown on low to moderate Zn soils (Millikan, 1953; Smith et al., 1974).

B. Alfalfa DRIS Norms

Table 23-3 shows the DRIS norms developed from these independently created data sets. In the case of the Wisconsin norms, the values given represent an updated revision of previously published norms (Erickson et al., 1982; Kelling et al., 1986). Note that substantial differences exist between the norms generated between all three locations, especially for those related to P, K, Mg, and B. Some of these differences may be due to differences in soil chemical properties (e.g., higher native Mg levels in Wisconsin soils), whereas other differences may be due to other factors such as the Wisconsin data being largely from specific hand sampling of pure alfalfa, compared to the Pennsylvania data being mostly from harvested forage that may contain some grass. Although one of the claimed original advantages of DRIS is that with a sufficiently broad data set the norms should be geographically universal, several researchers have since shown that when DRIS norms are established with local data the percentage of accurate diagnoses increases for samples from that area (Escano et al., 1981; Walworth et al., 1986; Erickson, 1987). Therefore, it would seem that to the extent possible, those norms which most closely match local conditions are the ones that should be used.

Independent evaluation of the Wisconsin norms by Minnesota researchers (Russelle & Sheaffer, 1986) showed them to be more accurate than critical values in diagnosing responses, and the diagnoses were not as affected by harvest management as were those performed by the critical value approach.

C. Clover Critical Values

The group of legumes commonly referred to as clovers (*Trifolium* spp.) will be discussed together. Red clover, rose clover, ladino clover, and subterranean clover are all true clovers, while bur clover is a medic closely related to alfalfa. All have similar growth habits. Ladino and red clover are perennial plants while subterranean, rose, and bur clovers are reseeding annuals. Available nutrient criteria on clovers are listed in Table 23-4.

Phosphorus values are similar for most of the species with the critical levels at approximately 1.8 to 2.5 g of P kg^{-1} tissue. Recent studies done in Wisconsin showed that critical levels for both P and K for red clover appear to be about the same as those for alfalfa (2.5 and 22.5 g kg^{-1} for P and K, respectively) (Kelling & Peters, 1989). Annual clovers usually require less K than perennial varieties. Field samples of K-deficient annual clovers

Table 23-3. DRIS norms for alfalfa generated from three independent databases.

Parameter	Georgia†		Wisconsin‡		Pennsylvania§	
	Norm	CV, %	Norm	CV, %	Norm	CV, %
(N/DM) × 100	2.95	7.2	3.26	19.7	3.13	14.7
N/P	12.45	19.1	10.68	17.5	9.49	20.3
N/K	1.50	18.2	1.35	34.3	1.16	25.6
N/S	--	--	12.21	15.0	12.42	18.3
N/Ca	2.53	11.6	2.43	23.0	2.36	27.1
Mg/N	0.055	22.6	0.104	28.9	0.075	21.3
N/Zn	0.150	24.5	0.165	16.6	0.125	20.2
B/N	15.09	19.6	11.23	26.8	8.84	28.0
N/Cu	0.458	26.1	0.508	35.3	0.469	28.2
Mn/N	--	--	9.41	35.7	16.2	41.1
(P/DM) × 100	0.243	15.5	0.301	15.9	0.330	15.7
P/K	0.124	23.8	0.127	31.8	0.123	16.3
S/P	--	--	0.883	16.9	0.677	17.4
P/Ca	0.216	23.8	0.231	22.6	0.275	30.1
Mg/P	0.672	21.3	1.17	24.5	0.639	23.5
Zn/P	90.45	55.6	65.8	21.1	70.4	19.8
B/P	185.4	26.2	119.7	32.9	77.1	31.5
P/Mn	--	--	0.0116	36.8	0.0075	38.8
P/Cu	0.035	27.8	0.048	35.6	0.054	33.0
(K/DM) × 100	2.03	19.1	2.38	33.9	2.69	18.9
S/K	--	--	0.112	36.2	0.079	18.4
K/Ca	1.94	19.3	1.97	34.5	2.29	40.1
Mg/K	0.083	31.9	0.155	56.2	0.076	33.3
Zn/K	9.75	27.0	8.25	33.0	8.41	25.1
B/K	23.13	33.7	15.26	47.3	9.08	34.8
Mn/K	--	--	12.59	48.4	15.8	42.9
Cu/K	3.43	15.5	3.01	50.2	2.25	44.2
(S/DM) × 100	--	--	0.262	17.8	0.230	18.4
S/Ca	--	--	0.201	20.8	0.201	26.2
Mg/S	--	--	1.35	25.5	0.901	20.7
Zn/S	--	--	75.06	17.7	106.0	25.0
B/S	--	--	134.2	22.1	111.0	27.6
S/Mn	--	--	0.0099	32.3	0.0049	44.3
S/Cu	--	--	0.042	33.2	0.041	31.9
(Ca/DM) × 100	1.19	13.1	1.37	17.8	1.24	26.0
Mg/Ca	0.136	23.1	0.259	14.6	0.184	26.7
Zn/Ca	18.27	24.7	15.01	25.7	19.16	29.8
B/Ca	34.06	20.7	26.74	28.6	21.6	21.3
Ca/Mn	--	--	0.052	40.5	0.030	38.3
Ca/Cu	0.141	19.3	0.213	32.9	0.188	31.9
(Mg/DM) × 100	0.161	19.2	0.351	20.4	0.218	21.4
Mg/Zn	0.0075	31.5	0.019	31.4	0.0092	24.6
Mg/B	0.0036	23.7	0.011	35.9	0.0081	29.1
Mg/Mn	--	-	0.014	46.0	0.0046	44.3
Mg/Cu	0.0227	28.0	0.055	37.1	0.034	26.1
(B/DM) × 10^6	44.18	17.9	36.3	30.1	26.4	30.4
B/Zn	2.01	47.4	1.84	26.6	1.10	28.2
B/Mn	--	--	1.31	35.0	0.653	46.6
B/Cu	7.05	37.7	5.54	35.5	3.84	41.8
(Zn/DM) × 10^6	21.44	49.8	20.1	22.1	24.4	21.3

(continued on next page)

Table 23-3. Continued.

	Georgia†		Wisconsin‡		Pennsylvania§	
Parameter	Norm	CV, %	Norm	CV, %	Norm	CV, %
Mn/Zn	--	--	1.53	34.6	1.93	41.5
Zn/Cu	2.89	29.0	3.06	32.1	3.77	29.6
(Mn/DM) × 10^6	--	--	26.32	35.8	45.6	38.6
Mn/Cu	--	--	4.67	44.4	7.77	45.4
(Cu/DM) × 10^6	6.89	26.6	6.37	31.8	6.32	29.2

† Adapted from Walworth et al. (1986).
‡ K.A. Kelling and T.W. Erickson (1987, unpublished data).
§ Adapted from Erickson (1987).

in California showed leaf values of <10 g kg^{-1} of K, while samples after correction of K deficiency usually contained 12 to 17 g kg^{-1} of K (Martin & Matocha, 1973). Although nutrient removal rates per unit of dry matter harvested tend to be similar for some of the clovers and alfalfa, the clovers are apparently less responsive to P or K additions and more efficient in uptake of K at a given soil test level (Smith & Smith, 1977; Kelling & Peters, 1989). This reduced responsiveness may be due to the more profuse surface rooting system for clover.

Sulfur and micronutrient critical values for clovers also appear to be similar or slightly lower than those suggested for alfalfa.

D. Clover DRIS Norms

Table 23-5 presents the DRIS norms developed for subterranean clover by Jones et al. (1986). These authors concluded that the DRIS system appeared to work well for diagnosing the P and S status of subclover under field conditions; however, they did not find the N or Ca indices useful and were undecided about K or Mg. Similar problems with Ca have been noted by Kelling and Schulte (1986) with alfalfa.

E. Birdsfoot Trefoil Critical Values

Several studies have been conducted where the nutrient concentrations in birdsfoot trefoil are compared with concentrations found in other leguminous forages or grasses (Beeson & MacDonald, 1951; Smith, 1964, 1970; Smith et al., 1974; Gross & Jung, 1978, 1981; Collins, 1982, 1983; Baligar et al., 1985). More recently, Russelle et al. (1985) and McGraw et al. (1986) extensively studied the patterns of nutrient accumulation and distribution in various trefoil plant parts at several stages of growth at two locations. However, all of these studies only examine the nutrient levels present and likely represent the adequate range of nutrients for trefoil. They do not establish critical levels as nutrient stress was not imposed. In only one instance (Wright et al., 1984) was a critical level found for trefoil; in this case for P at 2.4 g kg^{-1}.

Table 23-4. Nutrient sufficiency ranges and critical nutrient levels for clovers.

Kind	Species	Plant part	Nutrient	Nutrient range: Deficient	Critical	Adequate	Reference
				g kg^{-1}			
Red	*T. pratense*	Tops	P		2.5	2.5–8.0	Jackson et al., 1964
Red	*T. pratense*	Tops	P			2.0–4.0	Smith et al., 1985
Red	*T. pratense*	Tops	P		2.7		Wright et al., 1984
Ladino (white)	*T. repens*	Tops	P		2.3		Andrew, 1986
Ladino (white)	*T. repens*	Tops	P		3.0		Rayment & Bruce, 1979
Ladino (white)	*T. repens*	Tops	P		1.0–2.0	3.0	Parks et al., 1967
Ladino (white)	*T. repens*	Tops	P		2.5	2.5–3.2	Jackson et al., 1964
Ladino (white)	*T. repens*	Tops	P		1.5–2.5	3.0–3.5	Rendig et al. 1950
Ladino (white)	*T. repens*	Petioles	PO_4-P		600 (mg)	600–1200	Ulrich, 1945
Rose	*T. hirtum*	Tops	P	1.0–1.4	1.4–1.8	1.9–2.4	Martin et al., 1957
Rose	*T. hirtum*	Tops	P			2.0–2.5	Jones et al., 1970
Rose	*T. hirtum*	Leaves	P	0.7	1.9	1.9	Martin & Matocha, 1973
Sub	*T. subterraneum*	Tops	P		3.0–3.1		Spencer & Bouma, 1970; Bouma et al., 1969; Bouma & Dowling, 1982
Sub	*T. subterraneum*	Tops	P			2.0–2.8	Jones et al., 1980
Sub	*T. subterraneum*	Tops	P			2.6–3.2	Osman et al., 1977
Sub	*T. subterraneum*	Tops	P		2.5		Drlica & Jackson, 1979
Sub	*T. subterraneum*	Tops	P		1.4		Ozanne et al., 1969
Sub	*T. subterraneum*	Tops	P		0.8–1.3		Jones et al., 1972
Sub	*T. subterraneum*	Leaves	P	0.7		2.0–2.6	Martin & Matocha, 1973
Bur	*M. hispida*	Tops	P			2.5	Jones et al., 1970

Red	*T. pratense*	Tops	K		18		McNaught, 1958
Red	*T. pratense*	Tops	K		15		Smith & Smith, 1977
Red	*T. pratense*	Tops	K		22.5		Kelling & Peters, 1989
Ladino	*T. repens*	Tops	K		15		Spencer & Govaars, 1982
Ladino	*T. repens*	Petioles	K		8	8–17.5	Ulrich, 1945
Ladino	*T. repens*	Leaves	K	<17	17–19	20–24	Cornforth & Sinclair, 1982
Rose	*T. hirtum*	Leaves	K	6.5–8.7	8.7–9.5	12–18	Martin & Matocha, 1973
Sub	*T. subterraneum*	Leaves	K	5.4–7.2	7.2–9.0	9.5–15	Martin & Matocha, 1973
Sub	*T. subterraneum*	Tops	S		2.0		Drlica & Jackson, 1979
Sub	*T. subterraneum*	Tops	S			1.2–1.6	Jones et al., 1970
Sub	*T. subterraneum*	Tops	S		1.6–2.0		Spencer & Bouma, 1970; Bouma et al., 1969
Bur	*M. hispida*	Leaves	S		2.25		Jones et al., 1970
Rose	*T. hirtum*	Tops	S			1.4–1.9	Jones et al., 1970
Crimson	*T. incarnatum*	Tops	S		1.5		Jordan & Bardsley, 1958
Sub	*T. subterraneum*	Tops	SO_4-S	<100 (mg)	100–200		Jones et al.,1980
Sub	*T. subterraneum*	Tops	SO_4-S		170 mg kg^{-1}		Jones, 1962
Bur	*M. hispida*	Stems/leaves	SO_4-S	<100 (mg)	>100 mg kg^{-1}		Jones et al., 1980
Sub	*T. subterraneum*	Tops	N:S		14–16		Spencer, 1978
Sub	*T. subterraneum*	Tops	N:S		19–22		Jones et al., 1980
Sub	*T. subterraneum*	Young leaves	Ca		5.5–6.9		Jones et al., 1976
Sub	*T. subterraneum*	Old leaves	Ca		8.1–14.7		Jones et al., 1976
Red	*T. pratense*	Leaves/pet.	Cl		200 mg kg^{-1}		Whitehead, 1985
Sub	*T. subterraneum*	Tops	Zn		15 mg kg^{-1}		Riceman & Jones, 1958
Sub	*T. subterraneum*	Tops	Cu		3 mg kg^{-1}		Reuter et al., 1983
Sub	*T. subterraneum*	Tops	Mo		<0.1 mg kg^{-1}		Petrie & Jackson, 1982

Table 23-5. DRIS foliar diagnostic norms for subclover. (From Jones et al., 1986.)

Parameter	Norm	SD
P/N	0.091	0.022
N/S	14.438	3.910
K/N	0.751	0.184
Ca/N	0.317	0.130
Mg/N	0.093	0.152
P/S	1.291	0.386
K/P	8.507	2.244
Ca/P	3.647	1.416
Mg/P	1.008	0.398
K/S	10.510	2.615
Ca/S	4.695	2.516
Mg/S	1.314	0.690
K/Ca	2.641	1.036
K/Mg	9.839	4.761
Mg/Ca	0.312	0.161

IV. PLANT ANALYSIS OF GRASSES IN PURE STANDS

Since most aboveground plant parts are used in grass forages rather than a select plant part as is usually the case with agronomic crops, the whole tops are sampled for tissue analysis. Several forage grasses for which plant analysis data are available have been divided into warm-season perennial, warm-season annual, tropical grasses, and cool-season grasses/perennial and annual. Where sufficient data are available, nutrient level ranges are reported for deficient, critical, adequate, and high values. The values are presented for plants sampled as near to the usual harvest age as possible. In some cases, the effect of growth stage on nutrient levels is also presented. Certain important grasses have been omitted because adequate analysis data are not available to develop nutrient levels or descriptive conditions of the experimental studies were not given. The following information is presented in full recognition of the need for more research on correlative studies of soil and plant tissue analysis and interpretative procedures.

A. Warm-Season Grasses

1. Bermudagrasses

Although new, improved bermudagrasses have been tested and released, Coastal bermudagrass [*Cynondon dactylon* (L.) Pers.] still remains one of the principal forage grasses for the south and southeastern USA. This perennial grass with a high-production potential and drought tolerance is well-adapted to many soils and the long, hot summers in these regions. Coastal bermudagrass approaches its yield potential best on coarse-textured and well-drained soils. With its high-production capability, this grass places a high demand on the soil nutrient supply, especially since many sandy soils have low indigenous nutrient supplies. Therefore, successful production of bermudagrass may depend strongly on additions of essential nutrients.

Table 23-6. Nutrient criteria for whole tops of bermudagrass, 4 to 5 wk between clippings.

Element	Nutrient ranges				References
	Deficient	Critical	Adequate	High	
	g kg^{-1}				
	Coastal				
N	<15	18–22	25–30	>34	Adams et al., 1967a, b; Burton, 1954; Fisher & Caldwell, 1959; Jordon et al., 1966; Jordan & Bardsley, 1958; Matocha & Smith, 1980; Woodhouse, 1968, 1969; Plank, 1988; Cripps et al., 1989
P	<1.6	1.8–2.4	2.4–3.0	>4.0	
K	<13	13–15	15–18	>22	
S	<1.0	1.4–1.6	1.5–2.0	>3.0	
	Common and Midland				
N	<17	20–25	26–32	>36	Adams et al., 1967a; Hallock, 1966
P	<2.2	2.4–2.8	2.8–3.4	>4.0	
K	<15	16–18	19–22	>24	
S	<1.2	1.4–1.8	1.8–2.4	>3.2	

Relatively few data are available describing maximum-minimum yields with associated plant tissue analysis. The set of critical values shown in Table 23–6 was derived from several studies using interpolative methods in development of the nutrient ranges. The nutrient ranges established for Coastal bermudagrass tops are for 4- to 5-wk-old regrowth of established forage. Depending on soil moisture availability, the plant height would range from 22 to 45 cm at clipping. Coastal bermudagrass is highly responsive to N and moderately responsive to P, K, and S fertilization. Response to N fertilization by this grass is obviously affected by the availability of all essential plant nutrients.

a. Nitrogen, Phosphorus, and Potassium Critical Values. Early work (Fisher & Caldwell, 1959) on N response by Coastal bermudagrass where P, K, S, and water (irrigation) were supplied in adequate to excessive levels showed response up to 1200 kg of N ha^{-1}. Tissue levels of K at the maximum yield potential of the grass ranged from 18 to 20 g K kg^{-1}. Results of other work with Coastal bermudagrass (Matocha & Smith, 1980; Cripps et al., 1989) where several levels of K were studied, and the propensity for luxury consumption of K was considered, suggest lower K values ranging from 15 to 18 g kg^{-1} for the sufficiency level. The range given for the adequate level is somewhat higher and wider to accommodate seasonal fluctuations and variations due to applied nutrient ratios.

Excessive accumulation of K in the plant, or "luxury consumption," can occur due to extremely high levels of indigenous soil K or excessively high K fertilization rates. Coastal bermudagrass forage containing more than 23 g of K kg^{-1} is considered in the high or excessive category. However, bermudagrasses, unlike many grasses, are not normally accumulators of K. In many cases, the "dilution effect" caused by rapid growth under conditions of adequate N and water will reduce the instances of excessive K concentration in the plant tissue. Also, fluctuations in plant K due to season and stage of physiological maturity will usually show higher K levels during the first two clippings and lowest K levels during the drier and warmer period in August.

Concentrations of P in Coastal bermudagrass generally will show less variation among the nutrient ranges than will N or K. Typically, deficient Coastal bermudagrass will contain <1.6 g of P kg^{-1}, while the critical level will range from 2.0 to 2.5 g of P kg^{-1} (Table 23–6).

Coastal bermudagrass has shown its ability to make reasonable growth under P-stress conditions (J.E. Matocha, 1978, unpublished data). In deep, sandy soils under high N and K fertilization, Coastal bermudagrass showed little or no yield response to P for 4 yr, even though soil test values were critically low. However, these soils had a surface thickness of 180 cm. Response to applied P would have been expected to occur earlier under conditions of higher clay content and thinner surface soil.

In other situations, this warm-season grass responds markedly to P applications under conditions of adequate N, K, and S nutrition. Recent research on acid sandy soils indicated P response by Coastal bermudagrass was accentuated by lime applications (Hillard et al., 1987). In some cases, although only a slight response to P is measured by dry matter yields, a greater effect upon protein content of the forage has been measured.

Under conditions of highly intensified production on neutral to acid soils, tissue P concentrations in 4- to 5-wk-old forage can range from 2.4 to 3.0 g kg^{-1} (Fisher & Caldwell, 1959), although this may be influenced by the time of sampling. Usually clipping one will contain significantly higher P than will clipping three. With less intense production, similar age Coastal bermudagrass forage may range from 2.2 to 2.6 g of P kg^{-1} (Jordon et al., 1966; Adams et al., 1967a). Appropriate plant composition data for Coastal bermudagrass grown on calcareous soils are scarce. Generally, it appears that tissue P values will be comparable or slightly lower for forage grown under calcareous soil conditions than in an acid soil environment where adequate levels of essential plant nutrients are present according to soil test estimations.

'Common' and 'Midland' bermudagrasses are also grown for forage production in the south and southeast, but less extensively than Coastal bermudagrass. These grasses are used more for pasture than hay production. Both have a lower genetic yield potential than Coastal bermudagrass. Research evidence indicates that both Common and Midland can produce somewhat better quality forage than Coastal bermudagrass (F.M. Rouquette, Jr., 1975, personal communication).

The slower-growing varieties (Common and Midland) usually contain higher mineral concentrations than Coastal bermudagrass when grown under similar conditions (Hallock et al., 1966; Adams et al., 1967a).

b. N/S Ratios and Sulfur Nutrition. Fewer experiments have described S nutrition of grasses because S deficiencies are not as widespread as those of the macronutrients, N, P, and K. Early research (Jordan & Bardsley, 1958) suggested a critical S level of 1.4 g kg^{-1} in Coastal bermudagrass forage.

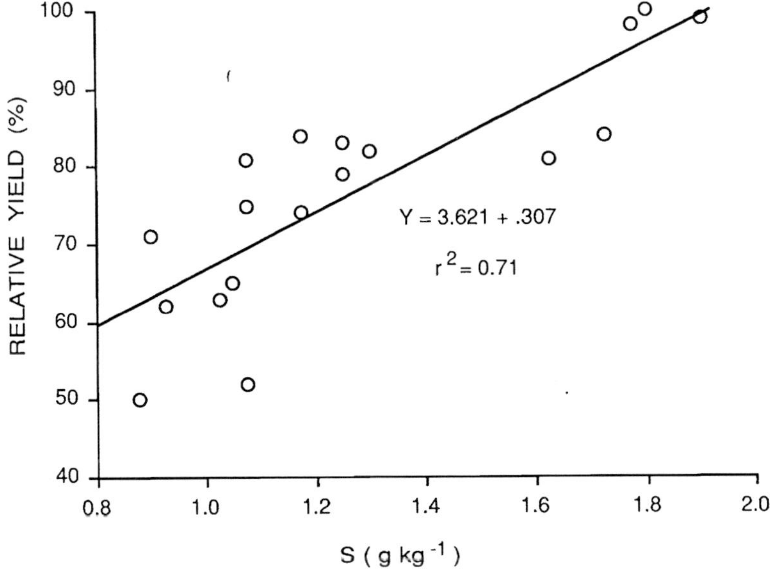

Fig. 23-3. Relative yields of Coastal bermudagrass in respect to S concentration in tops, 4 to 5 wk after clipping.

Other, more recent work in Texas from a factorial experiment with rates and sources of S, suggested a critical range of 1.4 to 1.6 g of S kg^{-1} in Coastal bermudagrass whole tops (Matocha, 1971). When subjected to high levels of soil S, Coastal bermudagrass can absorb S in luxurious quantities as is shown in the relative yield curve (Fig. 23-3). Similar to other nutrients such as N and K, S concentrations in Coastal bermudagrass can fluctuate during the season (Table 23-7). Plant age or interval between clippings will have a substantial bearing on S concentrations. Decreases in tissue S appear more rapid during the first 2 wk of regrowth than during the last 2 wk.

Sulfur is an important component of certain amino acids. Many sandy soils that are low in organic matter are low in available S (SO_4). High rates of N fertilization, a necessity for high yields of forage grasses, can cause an N/S imbalance and wide N/S ratios in the forage if low S-containing fertilizers are used. The N/S ratios in the plant has been suggested as a useful criteria in determining adequacy of S (Dijkshoorn et al., 1960). This criteri-

Table 23-7. Seasonal changes in N, P, K, and S concentrations in Coastal bermudagrass.

Element	Date of sampling				References
	June	July	August	September	
	g kg^{-1}				
N	27	27	23	27	Adams et al., 1967a; J.E. Matocha, 1975, unpublished data
P	2.3	3.0	1.9	3.0	
K	21	30	17	18	
S	2.2	1.8	1.6	2.1	

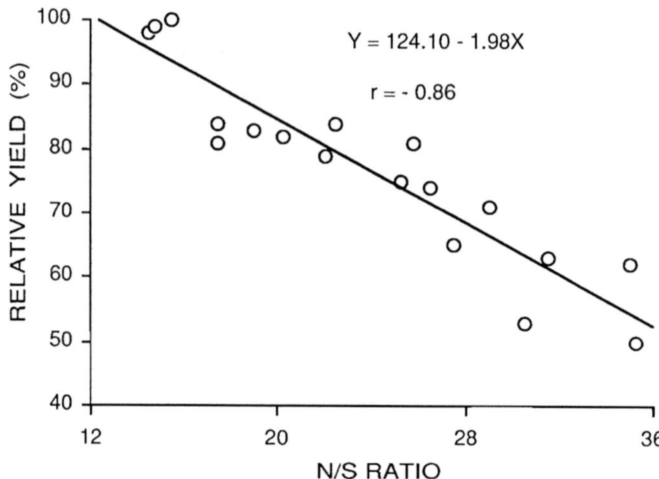

Fig. 23–4. Relative yields of Coastal bermudagrass in respect to N/S in tops, 4 to 5 wk after clipping.

on has been useful in evaluating nutrition needs of legumes (McNaught & Chrisstoffels, 1961; Pumphrey & Moore, 1965).

Coastal bermudagrass research in Texas (J.E. Matocha, 1976, unpublished data) indicates a close relationship between N/S ratios and grass forage yields (Fig. 23–4). An N/S ratio for Coastal bermudagrass ranging from12 to 16 appears ideal for medium- to high-intensity forage production. Apparently, such N/S ratios are about optimum for ruminant nutrition (Allaway & Thompson, 1966).

Under conditions of adequate P and K fertilization, response to S by Coastal bermudagrass depends upon adequate N fertilization. In many cases, an important condition for S response is adequate Mg nutrition. Research on a sandy soil receiving adequate N, P, K, and Ca, but without Mg fertilization, showed a 29% yield increase for Coastal bermudagrass due to applied S. However, with Mg fertilization, the addition of S caused a 53% increase in dry matter yields (Matocha, 1969, 1971). In this study, an application of Mg without S fertilization resulted in a slight yield depression, and the interactive effects of S and Mg accounted for approximately 89% of the yield response to S and Mg levels.

c. **Bermudagrass DRIS Norms.** Table 23–8 shows the DRIS norms for Coastal bermudagrass developed from three high-yield experiments in Louisiana. Evaluation of these norms for two independent populations showed high rates of diagnosis accuracy (75–100%) on several hundred observations (Tarpley et al., 1985). Furthermore, the DRIS analyses correctly identified S and Mg responses at another location even though the original norm data set did not include S or Mg variables.

Table 23-8. Selected DRIS norms for N, P, K, Ca, Mg, and S in oven-dried Coastal bermudagrass tissue. (Adapted from Tarpley et al., 1985).

Ratio	Norm	CV, %
N/P	10.11	11.92
N/S	11.85	16.59
N/Ca	7.61	11.60
N/Mg	14.94	14.75
K/N	0.71	17.74
P/K	0.14	21.87
S/P	0.87	17.57
P/Ca	0.76	9.65
P/Mg	1.48	11.37
S/K	0.12	21.82
K/Ca	5.37	17.54
K/Mg	10.70	25.10
Ca/Mg	1.97	11.92
S/Ca	0.65	15.02
S/Mg	1.29	18.78

2. Dallisgrass

Dallisgrass (*Paspalum dilatatum* Par), a perennial grass, is used extensively in the southeastern USA and many tropical parts of the world. It is a dependable grass for forage production over a wide range of soil and climatic conditions and is highly responsive to N fertilization.

a. Critical Values. Robinson et al. (1988) reported 86% maximum yield was obtained with 448 kg of N ha^{-1}. At this yield level, the forage contained 23.4 g of N kg^{-1}. Other work in Alabama (Ashley et al., 1965) with irrigated dallisgrass indicated response to 672 kg of N ha^{-1}, but some loss of stand was experienced with this high-intensity, maximum production.

As is generally the case with other warm-season grasses, dallisgrass shows less response to P fertilization than to N fertilization. Mays et al. (1980) reported that dallisgrass responses to P fertilization under field conditions have not been documented in the USA. Andrew and Robbins (1978) in Australia estimated critical P concentrations for dallisgrass at 2.6 g kg^{-1}. Robinson et al. (1988) showed adequately fertilized dallisgrass contained 2.9 and 22.4 g kg^{-1} of P and K, respectively. Nutrient criteria values are shown in Table 23-9.

Table 23-9. Nutrient criteria for whole tops of actively growing dallisgrass, 3 to 5 wk of age.

Element	Nutrient ranges				References
	Deficient	Critical	Adequate	High	
	g kg^{-1}				
N	<14	16–18	23–30	--	Andrew & Robbins, 1978;
P	<2.4	2.6	2.8–3.0	--	Robinson et al., 1988
K	--	--	21–24	--	

Table 23-10. Selected norms and coefficients of variation for DRIS analysis of dallisgrass with growth rate ≥110 kg ha^{-1} d^{-1}. (Adapted from H.J. Savoy and D.L. Robinson, 1989, personal communication.)

Parameter	Mean	CV, %
P/N	0.08104	16.56
P/K	0.1038	19.90
P/S	0.6586	18.36
P/Mg	0.8545	21.30
K/N	0.7913	12.76
K/Mg	8.441	23.54
S/N	0.1246	14.16
S/K	0.1585	13.44
S/Ca	0.8244	12.39
S/Mg	1.309	17.16
Ca/N	0.1526	16.89
Ca/P	1.906	14.77
Ca/K	0.1962	21.33
Ca/Mg	1.584	8.853
Mg/N	0.09756	21.75
N/10^6 DM†	29 902	9.224
P/10^6 DM	2 419	18.31
K/10^6 DM	23 513	11.10
S/10^6 DM	3 692	10.40
Ca/10^6 DM	4 526	12.84
Mg/10^6 DM	2 881	15.87

† DM = dry matter.

b. Dallisgrass DRIS Norms. Preliminary DRIS norms for dallisgrass were developed by Savoy and Robinson (1989) and are presented in Table 23-10. When these norms were used as a basis for determining N and P deficiency, accuracies of 75 to 100% were obtained. Use of the dry matter indices presented in Table 23-10 did not improve diagnostic accuracy in evaluating the N status, but did slightly improve the ability to correctly evaluate dallisgrass P status.

3. Millet, Sudangrass, Sorghum-Sudangrass Hybrids, and Johnsongrass

Sudangrass [*Sorghum sudanese* (Piper) Staph] and millet (*Pennisetum typhoides* Burn. Staph and Hubbard) are primarily used for temporary summer pastures with some used as hay and silage. Sorghum- [*Sorghum bicolor* (L.) Moench] sudangrass hybrids are also annuals used widely in the south and the Great Plains for hay, silage, and pasture production. These summer annuals are relatively high-yielding crops that can tolerate hot and droughty conditions.

Genetic improvements in sorghum-sudangrass hybrids in terms of disease tolerance and forage quality have improved the reliability of this source of temporary forage crops especially in droughty periods. The improvements in sorghum crosses have reduced the use of the perennial, Johnsongrass [*S. halepense* (L.) Pers.], as a forage crop.

a. Critical Values. Nutrient demands for millet, sudangrass, and sorghum-sudangrasses are somewhat different and this is reflected in their composition. Pearl millet has been found to contain twice the NO_3-N and plant K as sudangrass grown under identical conditions (Smith & Clark, 1968). Apparently, pearl millet has the physiological capability to obtain nutrients from low-fertility soils, and in general, may show higher NO_3 levels than the sorghums. Nitrogen and K levels in millet are usually higher than those of sorghum-sudangrass, while P levels are relatively similar. Sorghum-sudangrass values appear to agree closely with sudangrass and sorghum grasses at similar stages of growth; consequently, only one set of nutrient criteria ranges is shown.

Aging of most plants results in progressive changes in chemical composition. In the case of forage sorghum and millet, this change usually involves a decrease in nutrient concentrations and substantial increases in yields with stage of growth. Other physiological changes occurring in the sorghum-type grasses involve sugar synthesis as senescence occurs and softdough stages are attained. The change in certain mineral element content can be quite drastic in a relatively short period. Therefore, caution must be exercised in interpretation of nutrient ranges and growth stages.

Values for the perennial sorghum grass, Johnsongrass (which are usually lower than for the annuals), were available only for the boot stage (6–7 wk following clipping), and therefore, are considerably lower than those for millet, sudangrass, and sorghum-type grasses.

Data from well-designed soil fertility experiments with these grasses, complete with mineral composition analysis suitable for establishment of critical values, are limited. Many chemical composition studies have been directed toward sugar and nutrient contents of silages and are of little use in determining nutrient composition values. Data were selected from experiments with forage clippings made 4 to 5 wk following the previous clipping, or in some cases, at 50% boot stage of growth and developed into nutrient ranges presented in Table 23–11.

b. Sorghum and Millet DRIS Norms. DRIS norms for sorghum were developed by Arogun (1978), mostly from data obtained from Kansas. These norms (Table 23–12) were then tested against several varieties of sorghum and millet grown under various fertility regimes in the greenhouse. Arogun (1978) concluded that the norms developed from the sorghum data were applicable to both sorghum and millet whole plants sampled at an early vegetative growth stage (about 45 d after planting). He also saw no varietal influence in the ability of the norms to correctly identify nutrient deficiencies.

B. Tropical Grasses

Three of the principal tropical grasses used in forage production that will be discussed in this chapter are guineagrass (*Panicum maximum* Jacq.), napiergrass (*Pennisetum purpurem* Schumach.), and pangolagrass (*Digitaria decumbens* Stent.). All of these grasses are widely used in the tropics.

Table 23-11. Nutrient criteria for millet, sorghum-sudan hybrids, and sudangrass for whole tops, 4 to 5 wk after clipping, and Johnsongrass at boot stage.

	Nutrient ranges				
Element	Deficient	Critical	Adequate	High	References
	g kg^{-1}				
			Millet		
N	<19	18–20	22–30	>35	Clapp & Chambles, 1970; Clark et al., 1965; Schneider & Clark, 1970; Smith & Clark, 1968
P	<1.6	1.6–2.0	2.2–3.0	>4.0	
K	<15	16–20	23–38	>40	
S	<1.0	1.0–1.2	1.4–1.8	>2.5	
			Sorghum-sudan and Sudangrass		
N	<16	14–16	18–22	>25	Jones, 1985; Dotzenka et al., 1966; Smith & Clark, 1968
P	<1.4	1.4–1.8	2.0–3.0	>3.5	
K	<10	15–18	19–28	>25	
S	<1.0	1.0–1.2	1.4–1.8	>2.5	
			Johnsongrass		
N	<12	13–15	16–18	>22	Spooner et al., 1971
P	<1.4	1.6–2.0	2.0–2.5	>3.0	
K	10	12–15	16–18	>24	

Table 23-12. DRIS norms for high-yielding sorghum and millet sampled about 45 d after planting. (Adapted from Arogun, 1978.)

Parameter	Mean	CV, %
P/N	0.112	18.67
N/K	2.36	22.97
N/Ca	7.20	29.89
Mg/N	0.079	26.37
P/K	0.259	20.95
P/Ca	0.795	31.19
P/Mg	1.52	45.34
K/Ca	3.08	24.17
Mg/K	0.183	26.59
Mg/Ca	0.553	30.05

Table 23-13. Nutrient criteria for whole tops of pangolagrass, 4 to 5 wk between clippings.

	Nutrient ranges				
Element	Deficient	Critical	Adequate	High	References
	g kg^{-1}				
N	<10	12–15	17–20	>25	Gammon & Blue, 1952; Gomide et al., 1969; Harris et al., 1968; Kretschmer & Hayslip, 1963; Little et al., 1959; Plucknett & Fox, 1965; Rodriquez, 1949; Salette, 1970
P	<1.0	1.2–1.6	1.6–2.4	>2.8	
K	<11	12–14	16–20	>22	
S	<1.2	1.5–2.0	2.0–2.5	>3.0	

Pangolagrass is perhaps the best known and most widely grown grass throughout the Caribbean, Central America, and tropical South America. This grass adapts to a wide range of soil types and rainfall conditions where adequate soil fertility exists. It performs extremely well in both the wet and dry tropics and tolerates wide differences in soil pH. This grass regrows rapidly after overgrazing.

A strong fertilization program is required for all three grasses in the tropics due to the highly weathered nature of the soils. Careful management of fertilization is important on tropical soils that possess properties of high nutrient fixation and high leachability. Yield response and mineral composition data useful in characterizing critical values of N, P, K, and S for pangolagrass were scarce. Consequently, nutrient ranges presented in Table 23–13 are relatively wide.

When levels of P, K, and other essential nutrients are supplied at adequate levels, pangolagrass is highly responsive to N fertilizer. However, N recovery is generally low for this grass because it is characteristically low in protein. Minerals such as P are also substantially lower in nutritionally healthy pangolagrass than certain other warm-season grasses. Phosphorus readings as low as 0.6 to 0.8 g kg^{-1} in whole tops of pangolagrass were measured in greenhouse studies (W.G. Blue, 1972 personal communication). Andrew and Robbins (1978) indicated a critical value of 1.7 g kg^{-1} for P in pangolagrass. This agrees with work in Brazil (Gomide et al., 1969), which indicated similar critical values for 4-wk-old forage. It appears that nutrient levels in pangolagrass that are optimum for maximum grass production are not adequate to meet the needs of ruminants (NRC, 1958).

The N content of pangolagrass can fluctuate widely with level of plant-available K. Kretschmer and Hayslip (1963) showed a strong relationship between K fertilization and plant N concentrations. Gammon and Blue (1952) indicated that K levels of pangolagrass grown on high K-fixing soils in Florida should approximate 20 g kg^{-1} for optimum growth. Establishment of clover stands in pangola sod requires careful management of the K status of the soil due to the high K requirement of this grass.

Napiergrass and guineagrass are widely used forage grasses of the tropics (Vincente-Chandler et al., 1959). Production practices for guineagrass are somewhat similar to those for pangolagrass. However, the forage yield potential of guineagrass is substantially greater than that of pangolagrass. Napiergrass (elephant grass) is also an extremely high producer of forage when soil fertility is not limiting. Guineagrass and pangolagrass are used mostly for grazing, while napiergrass is mechanically harvested for hay and haylage in many of the tropic countries. Insufficiently documented mineral composition data on napiergrass and guineagrass prevented development of nutrient criteria.

C. Cool-Season Grasses

Cool-season annuals and many perennial grasses produce high-quality forage during the winter and early spring when warm-season grasses are normally dormant, and grazed forage availability is minimal. Production of high-quality grass forage from the standpoint of protein and soluble carbohydrates (in vitro digestible dry matter) requires good management and a sound fertilization program.

Several species of cool-season grasses are commonly grown. Principal ones, and those for which yield response and nutrient composition data are available, include annual (*Lolium multiflorum* Lam.) and perennial ryegrasses (*L perenne* L.), perennial orchardgrass (*Dactylis glomerata* L.), bromegrass (*Bromus inermis* Leyss.), tall fescue (*Festuca arundinacea* Schreb.), and bluegrass (*Poa pratensis* L.).

1. Ryegrasses

Annual and perennial ryegrasses are important grass species used as winter forages in the southeastern USA and many intensively pastured temperate grassland regions of the world. Ryegrasses produce high-quality forage over a relatively long growing season. These forage grasses produce less forage than oat in the early fall and less than wheat or rye during low winter temperatures. However, they make slow but steady growth in winter that picks up rapidly in the spring. They continue to grow later into the spring than winter cereal crops. Although ryegrasses have received considerable attention in studies describing nutrient response and yields, as well as quality parameters, only a few studies relate mineral composition to yields and provide data useful in establishment of nutrient criteria. While some differences in mineral composition may exist between annual and perennial varieties, published data are not sufficient to develop separate nutrient criteria. These grasses are highly succulent when in an active growth stage and contain relatively high levels of most mineral elements. However, mineral levels can change rapidly with stage of growth and season. Reductions of about 20 g kg^{-1} in N and K as growth of perennial ryegrass progressed from the early leafy stage to heading were measured by Fleming and Murphy (1968). The effect of relative maturity on P concentration was similar to that for N and K. Nutrient levels for ryegrasses (Table 23–14) are for whole tops sampled 4 to 5 wk following clipping.

2. Orchardgrass

A perennial cool-season grass, orchardgrass is moderately productive with yield potential somewhat higher than that of timothy (*Phleum pratense* L.) and bluegrass under similar environmental conditions. Kresge and Younts (1963) substantiated other work that indicated this grass responded fairly well to N fertilization. They found good response to N rates up to 224 kg of N ha^{-1} with no further yield increase from 448 kg of N ha^{-1}. Similar to many other grasses, orchardgrass is especially sensitive to an N and K imbalance. Optimum plant levels of P or K cannot be established with insufficient N fertilization. Likewise, if K is deficient, it is difficult to establish the adequate level of plant N. The nutrient criteria given for orchardgrass in Table

Table 23-14. Nutrient criteria for whole tops of annual and perennial ryegrasses, 4 to 5 wk between clippings.

Element	Nutrient ranges: Deficient	Critical	Adequate	High	References
	g kg^{-1}				
N	<27	30-34	35-40	<50	Robinson et al., 1987; Dijkshoorn, 1958; Fleming & Murphy, 1968; Thomas et al., 1952; Widdowson et al., 1965
P	<2.8	2.8-3.4	3.6-4.4	>5.0	
K	<21	24-28	28-32	>45	

23-15 were developed after a review of literature produced some suitable data from a limited number of experiments. These values should be regarded as estimates and may be subject to change when additional data become available.

3. Bromegrass

Several studies have been conducted that show the mineral composition of smooth bromegrass. These studies show that nutrient concentrations for bromegrass are generally lower than those found in legumes (Van Riper & Smith, 1959; Loper & Smith, 1961; Loneragan et al., 1968; Reid et al., 1970; Smith et al., 1974), that bromegrass tends to be somewhat lower in P, K, and S than some other grasses (Reid et al., 1970; Smith et al., 1974) and somewhat higher in Cu (Smith et al., 1974), and that environment, genotype, and competitive effects of mixed species can influence the mineral composition of bromegrass (Watkins, 1940; Loper & Smith, 1961; Cassler et al., 1987; Cassler & Reich, 1987). However, none of these studies were conducted as nutrient rate experiments such that critical nutrient levels could be determined; therefore, the reported values likely represent normal or sufficient levels of these elements.

In some cases, N trials have been conducted with bromegrass such that the sufficiency range could be established. Smith (1972) showed that yields were near optimum when brome contained about 31 to 34 g of N kg^{-1} tissue, whereas others optimized yields at 27 to 36 g of N kg^{-1} tissue (MacLeod, 1965; Krueger & Scholl, 1970).

Table 23-15. Nutrient criteria for whole tops of orchardgress, 3 to 4 wk between clippings.

Element	Nutrient ranges: Deficient	Critical	Adequate	High	References
	g kg^{-1}				
N	<24	27-30	32-35	>42	Gordon et al., 1962; Griffith et al., 1964; Kresge & Younts, 1963; MacLeod, 1965; Reid et al. 1967; Reynolds, 1971; Robinson et al., 1962; J.R. Brown, 1971, personal communication
P	<1.8	2.0-2.4	2.3-2.8	>3.5	
K	<20	23-25	26-30	>35	
S	<1.2	--	2.5	--	

Table 23-16. Nutrient criteria for whole tops of actively growing tall fescue, 5 to 6 wk between clippings.

Element	Nutrient ranges				References
	Deficient	Critical	Adequate	High	
	g kg^{-1}				
N	<23	24–30	32–36	>40	Hallock et al., 1966; Sabbe & Hileman, 1975; Reid & Jung, 1965
P	<2.4	2.6–3.2	3.4–4.0	>4.5	
K	<22	24–38	28–32	>38	

4. Tall Fescue Grass

Tall fescue, a perennial cool-season grass, is an important pasture and hay production grass in the upper south, southeast, and Pacific northwest portions of the USA. Research indicates that dry matter production and forage quality are highly dependent upon adequate fertilization and mineral levels in the plant tissue. Improper management has led to low palatability problems at advanced stages of maturity, and this has been identified as an undesirable characteristic of this species. A number of researchers have shown that fertilizer treatments influence palatability of forages (Boawn & Viets, 1953; Moon, 1954; Markley et al., 1959). Differences in palatability of tall fescue forage due to variable N, P, and K treatments were described by Reid and Jung (1965). Nutrient ranges for macronutrients in tall fescue are presented in Table 23-16.

5. Bluegrass

Although forage production from Kentucky bluegrass is substantially less per unit area than from other cool-season grases, when properly fertilized this grass produces forage that is palatable and of good quality. In earlier years, this grass was extensively studied for effect of applied N, P, and K fertilizers on yields and plant composition. This grass appears to respond extremely well to N and K, and to a lesser extent to P fertilization.

The nutrient values given in Table 23-17 were derived from several studies in which the effects of varying one or more nutrients on growth and chemical composition were determined. The values are estimates for actively growing Kentucky bluegrass foraged harvested some 4 to 6 wk following the previous clipping.

Table 23-17. Nutrient criteria for whole tops of Kentucky bluegrass, 4 to 6 wk between clippings.

Element	Nutrient ranges				References
	Deficient	Critical	Adequate	High	
	g kg^{-1}				
N	<21	24–38	26–32	>35	Butler & Hodges, 1967; Grava et al., 1970; Pierre & Robinson, 1937; Vincente-Chandler et al., 1959; Walker & Pesek, 1967
P	<1.8	2.4–3.0	2.8–3.6	>4.0	
K	<15	16–20	20–24	>30	
Mg	--	--	4.0–4.8	>5.0	
S	--	--	1.6–2.4	--	

V. PROBLEMS OF USING PLANT ANALYSIS ON MIXED LEGUME AND GRASSLANDS

When grasses and legumes are grown together as a plant community, problems of competition between the species for nutrients, space, light, and water can result. In addition, one must consider the reaction of each species to grazing or periodic harvest.

In the classic concept of mixed pasture, we usually include one or more grass species and one or more legumes. The legumes in this association usually provide some of the N for the needs of the grass through root leakage, decomposition of roots, nodules, and other plant parts. Fertilization of well-managed, mixed grass-legume sward of compatible species may require only the input of non-fertilizers. This system often provides good production of both grass and legumes and does so quite inexpensively. The grass fraction is often somewhat N deficient. Attempts to apply N fertilizers to a grass-legume usually result in a failure of the legume to fix N (Martin et al., 1965).

Plant analysis of a mixture of legumes and grasses will usually show a deficiency of N. Jones et al. (1970) showed that in an annual grassland range with a mixed stand, the S status of the community is best revealed by a SO_4-S measure of the legume components. The N-deficient grasses showed S values well above critical levels until enough N had been applied to overcome the N deficiency.

In irrigated pasture experiments fertilized with N and P (Martin et al., 1965), the concentration of P in the grass component was slightly greater than the levels in the legume fraction. Phosphorus values ranged from deficiency to adequacy for both species. More recently, Burmester and Adams (1983) showed that the P critical level for tall fescue was 2.0 to 3.0 g kg^{-1}, whereas the approximate critical content for a fescue-clover forage was 3.0 to 4.0 g kg^{-1}. Similarly the critical K values for fescue-clover were 20 to 25 g kg^{-1} as compared to 13 to 23 g kg^{-1} for the fescue alone. As a rule, the K concentration of grass in association with alfalfa or other legumes will be higher than that of the legume. For example, Doll (1962) found the critical K level of alfalfa to be 17 g kg^{-1}, while that of an alfalfa-orchardgrass mixture was 19 g of K kg^{-1}. Smith et al. (1974) found that legumes were higher than grasses in most macro- and micronutrients except for K and Mn. Similar observations were made by Napitupla and Smith (1979) when they proportionally mixed various amounts of orchardgrass and alfalfa.

These data emphasize the need for caution when applying individual species critical value criteria to harvested material from a mixed stand.

ACKNOWLEDGMENT

The organization and some parts of the text of this chapter are similar to the original version by W.E. Martin (deceased) and J.E. Matocha as chapter 24, "Plant Analysis as an Aid in the Fertilization of Forage Crops," for the book *Soil Testing and Plant Analysis,* rev. ed., 1973, edited by L.M. Walsh and J.D. Beaton. We wish to formally acknowledge the contribution of this earlier effort.

REFERENCES

Adams, W.E., M. Stelly, H.D. Morris, and C.B. Elkins. 1967a. A comparison of Coastal and common bermudagrass [*Cynodon dactylon* (L.) Pers.] in Piedmont region. II. Effect of fertilization and crimson clover (*Trifolium incarnatum*) on nitrogen, phosphorus, and potassium contents of forage. Agron. J. 59:281-284.

Adams, W.E., A.W. White, R.D. McCreery, and R.N. Dawson. 1967b. Coastal bermudagrass forage production and chemical composition as influenced by potassium source, rate, and frequency of application. Agron. J. 59:247-250.

Allaway, W.H., and J.F. Thompson. 1966. Sulfur in the nutrition of plants and animals. Soil Sci. 101:240-247.

Andrew, C.S. 1960. The effect of phosphorus, potassium and calcium on the growth, chemical composition, and symptoms of deficiency of white clover in a subtropical environment. Aust. J. Agric. Res. 11:149-161.

Andrew, C.S. 1977. The effect of sulphur on the growth and nitrogen concentrations, and critical sulphur concentrations of some tropical and temperate pasture legumes. Aust. J. Agric. Res. 28:807-820.

Andrew, C.S., and M.F. Robbins. 1978. The effect of phosphorus on the growth, chemical composition and critical phosphorus percentages of some tropical pasture grasses. Aust. J. Agric. Res. 22:693-706.

Andrew, C.S., and P.M. Thorne. 1962. Comparative responses to copper of some tropical and temperate pasture legumes. Aust. J. Agric. Res. 13:821-835.

Arogun, J.A. 1978. Application of the DRIS system to sorghum and millet. M.S. thesis, Univ. of Wisconsin-Madison.

Ashley, D.A., O.L. Bennett, B.D. Doss, and C.E. Scarsbrook. 1965. Effect of nitrogen rate and irrigation on yield and residual nitrogen recovery by warm-season grasses. Agron. J. 57:370-372.

Bailey, L.D. 1983. Effects of potassium fertilizer and fall harvests on alfalfa grown on the eastern Canada prairies. Can. J. Soil Sci. 63:211-219.

Baligar, V.C., R.J. Wright, O.L. Bennett, J.L. Hern, H.D. Perry, and M.D. Smedley. 1985. Lime effect on forage legume growth and mineral composition in an acid subsoil. Commun. Soil Sci. Plant Anal. 16:1079-1093.

Bates, T.E. 1971. Factors affecting critical nutrient concentrations in plants and their evaluation: A review. Soil Sci. 112:116-130.

Bear, F.E., and A. Wallace. 1950. Alfalfa—Its mineral requirements and chemical composition. New Jersey Agric. Exp. Stn. Bull. 748.

Beaufils, E.R. 1973. Diagnosis and recommendation integrated system (DRIS). A general scheme for experimentation and calibration based on principles developed from research in plant nutrition. Soil Sci. Bull. 1, Univ. of Natal, South Africa.

Beeson, K.C., and H.A. MacDonald. 1951. Absorption of mineral elements by forage plants: III. The relation of stage of growth to the micronutrient element content of timothy and some legumes. Agron. J. 43:589-593.

Boawn, L.C., and F.G. Viets, Jr. 1953. Zinc deficiency of alfalfa in Washington. Agron. J. 44:276-277.

Bouma, D., and E.J. Dowling. 1982. Phosphorus status of subterranean clover: A rapid and simple leaf test. Aust. J. Exp. Agric. Anim. Husb. 22:428-436.

Bouma, D., K. Spencer, and E.J. Dowling. 1969. Assessment of the phosphorus and sulfur status of subterranean clover pastures. 3. Plant tests. Aust. J. Exp. Agric. Anim. Husb. 9:329-340.

Brown, J.R. 1970. Plant analysis. Missouri Agric. Exp. Stn. Bull. SB881.

Burmester, C.H., and F. Adams. 1983. Fertilizer requirements of new plantings of tall fescue and white clover on two low-fertility Ultisols. Agron. J. 75:936-940.

Burton, G.W. 1954. Coastal bermudagrass. Georgia Agric. Exp. Stn. Bull. N.S. 2.

Butler, J.D., and T.K. Hodges. 1967. Mineral composition of grasses. J. Hortic. Sci. 2(2):62-63.

Caldwell, A.C., E.C. Seim, and G.W. Rehm. 1969. Sulfur effects on the elemental composition of alfalfa (*Medicago sativa* L.) and corn (*Zea mays* L.). Agron. J. 61:632-634.

Cassler, M.D., M. Collins, and J.M. Reich. 1987. Location, year, maturity, and alfalfa competition effects on mineral element concentrations in smooth bromegrass. Agron. J. 79:774-778.

Cassler, M.D., and J.M. Reich. 1987. Genetic variability for mineral element concentrations in smooth bromegrass related to dairy cattle nutritional requirements. p. 569–577. *In* W.H. Gabelman and B.C. Loughman (ed.) Genetic aspects of plant mineral nutrition. Martinus Nijhoff Publ., Boston.

Chamblis, C.G., D.A. Miller, and J.A. Jackobs. 1970. Selection of an alfalfa plant part for phosphorus analysis. Agron. J. 62:294–296.

Chapman, H.D. 1967. Plant analysis values suggestive of nutrient status of selected crops. p. 77–92. *In* Soil testing and plant analysis. Part 2. SSSA Spec. Publ. 2. ASA, CSSA, and SSSA, Madison, WI.

Clapp, J.G., Jr., and D.S. Chambles. 1970. Influence of different defoliation systems on the regrowth of pearl millet, hybrid sudangrass and two sorghum-sudangrass hybrids from terminal, axillary, and basal buds. Crop Sci. 10:345–349.

Clark, N.A., R.W. Hemken, and J.H. Vandersall. 1965. A comparison of pearl millet, sudangrass, and sorghum-sudangrass hybrids as pasture for lactating dairy cows. Agron. J. 57:266–268.

Collins, M. 1982. Yield and quality of birdsfoot trefoil stockpiled for summer utilization. Agron.J. 74:1036–1041.

Collins, M. 1983. Changes in composition of alfalfa, red clover and birdsfoot trefoil during autumn. Agron. J. 75:287–291.

Collins, M., and T.H. Taylor. 1980. Yield and quality of alfalfa harvested during autumn and winter and harvest effects on the spring crop. Agron. J. 72:839–844.

Cornforth, J.S., and A.G. Sinclair. 1982. Fertilizer and lime recommendations for pastures and crops in New Zealand. Ministry Agric./Fisheries, Wellington, NZ.

Cripps, R.W., J.L. Young, and A.T. Leonard. 1989. Effects of potassium and lime applied for Coastal bermudagrass production on a sandy soil. Soil Sci. Soc. Am. J. 53:127–132.

Dijkshoorn, W.J. 1958. Nitrogen, chlorine and potassium in perennial ryegrass and their relation to the mineral balance. Neth. J. Agric. Sci. 6:131–138.

Dijkshoorn, W.J., E.M. Lampe, and D.F.J. Van Burg. 1960. A method for diagnosing the sulfur nutrition status of herbage. Plant Soil 13:227–241.

Doll, E.C. 1962. Potassium fertilization of alfalfa in Kentucky. Kentucky Agric. Exp. Stn. Bull. 679.

Dotzenko, A.D., N.E. Humburg, G.O. Hinze, and W.H. Leonard. 1966. Effects of stage of maturity on the compositions of various sorghum silages. Colorado Agric. Exp. Stn. Tech. Bull. 87.

Drlica, D.M., and T.L. Jackson. 1979. Effects of stage of maturity on P and S critical levels in subterranean clover. Agron. J. 71:824–828.

Dye, W.B. 1962. A micronutrient survey of Nevada forage. Univ. of Nevada Tech. Bull. 227.

Ensminger, L.E., and J.R. Freney. 1966. Diagnostic techniques for determining sulfur deficiencies in crops and soils. Soil Sci. 101:283–290.

Erickson, T.W. 1987. Development of DRIS plant diagnostic norms for Wisconsin grown alfalfa. M.S. thesis, Univ. of Wisconsin-Madison.

Erickson, T., K.A. Kelling, and E.E. Schulte. 1982. Predicting alfalfa nutrient needs through DRIS. Proc. Wis. Fert., Aglime and Pest Manage. Conf. 21:233–246.

Escano, C.R., C.A. Jones, and G. Uehara. 1981. Nutrient diagnosis in corn grown on Hydric Dystrandepts. II. Comparison of two systems of tissue diagnosis. Soil Sci. Soc. Am. J. 45:1140–1144.

Fisher, F.L., and A.G. Caldwell. 1959. The effect of continued use of heavy rates of fertilizers on forage production and quality of Coastal bermudagrass. Agron. J. 51:99–102.

Fleming, G.A. 1973. Mineral composition of herbage. p. 529–566. *In* G.W. Butler and R.W. Bailey (ed.) Chemistry and biochemistry of herbage. Vol. 1. Academic Press, New York.

Fleming, G.A., and W.E. Murphy. 1968. The uptake of some major and trace elements by grasses affected by season and stage of maturity. J. Br. Grassl. Soc. 23:174–185.

Gammon, M., Jr., and W.G. Blue. 1952. Potassium requirements for pastures. Proc. Soil Sci. Soc. Fla. 12:154–156.

Gerwig, J.L., and G.H. Ahlgren. 1958. The effect of different fertility levels on yield, persistence and chemical composition of alfalfa. Agron. J. 50:291–294.

Gomide, J.A., C.H. Noller, G.O. Mott, J.H. Conrad, and D.L. Hill. 1969. Mineral composition of six tropical grasses as influenced by plant age and nitrogen fertilization. Agron. J. 61:120–123.

Gordon, C.H., A.M. Decker, and H.G. Wiseman. 1962. Some effects of nitrogen fertilizer, maturity and light on the composition of orchardgrass. Agron. J. 54:376–378.

Grau, C.R., K.A. Kelling, and R.P. Wolkowski. 1989. Observations on the effect of P, K, and S fertilization on alfalfa infected by *Phytophthora megasperma.* J. Prod. Agric. 2:136–139.

Grava, J., G.W. Randall, D.M. Larsen, and R.S. Farnham. 1970. Soil fertility investigations of grass seed production in northwestern Minnesota. Minnesota Agric. Exp. Stn. Misc. Rep. 91.

Griffith, W.K., M.R. Teel, and H.E. Parker. 1964. Influence of nitrogen and potassium on the yield and chemical composition of orchardgrass. Agron. J. 56:473–475.

Gross, C.F., and G.A. Jung. 1978. Magnesium, Ca, and K concentration in temperate-origin forage species as affected by temperature and Mg fertilization. Agron. J. 70:397–403.

Gross, C.F., and G.A. Jung. 1981. Season, temperature, soil pH, and Mg fertilizer effects on herbage Ca and P levels and ratios of grasses and legumes. Agron.J. 73:629–634.

Hallock, D.L., R.H. Brown, and R.E. Blaser. 1966. Response of Coastal and Midland bermudagrass and Kentucky 31 fescue to nitrogen in southeastern Virginia. Virginia Agric. Exp. Stn. Agron. Res. Rep. 112.

Hanson, R.G., N. Risner, and S.R. Malady. 1984. Sulfur fertilization of two aquic hapludalf soils: I. Effect on alfalfa yield and quality. Commun. Soil Sci. Plant Anal. 15:227–237.

Harris, H.D., V.N. Schroder, and R.L. Silman. 1968. Nutrient deficiency effects on yield and chemical composition of plants grown on Leon fine sand. Florida Agric. Exp. Stn. Tech. Bull. 725.

Harward, M.E., J.T. Chao, and S.C. Fang. 1962. The sulfur status and sulfur supplying power of Oregon soils. Agron. J. 54:101–106.

Hillard, J.B., V.A. Haby, F.M. Hons, J.V. Davis, and A.T. Leonard. 1987. Forage response to residual soil phosphorus and pH. II. Coastal bermudagrass. Texas Agric. Exp. Stn. Consolidated Progr. Rep. 4527:55–60.

Hoeft, R.G., and L.M. Walsh. 1975. Effect of carrier, rate and time of application of S on the yield, and S and N content of alfalfa. Agron. J. 67:427–430.

Jackson, T.L., H.H. Rampton, and J. McDermid. 1964. Effect of lime and P on the yield and P content of legumes in western Oregon. Oregon Agric. Exp. Stn. Tech. Bull. 83.

James, D.W., and W.H. Weaver. 1964. The place for boron fertilizer in central Washington field crop production. Washington State Univ. Mimeo Paper EM2510.

Johnson, C.M., and H. Nishita. 1952. Microestimation of sulfur in plant materials, soils and irrigation waters. Anal. Chem. 6:114–119.

Jones, J.B., Jr. 1967. Interpretation of plant analysis for several agronomic crops. p. 49–58. *In* Soil testing and plant analysis, Part 2. SSSA Spec. Publ. 2. ASA, CSSA, and SSSA, Madison, WI.

Jones, J.B., Jr., and H.V. Eck. 1973. Plant analysis as an aid in fertilizing corn and grain sorghum. p. 349–364. *In* L.M. Walsh and J.D. Beaton (ed.) Soil testing and plant analysis. rev. ed. ASA, CSSA, and SSSA, Madison, WI.

Jones, M.B. 1962. Total sulfur and sulfate sulfur content of subterranean clover as related to sulfur responses. Soil Sci. Soc. Am. Proc. 26:482–484.

Jones, M.B., D.M. Center, C.E. Vaughn, and F.L. Bell. 1986. Using DRIS to assay nutrients in subclover. Calif. Agric. 24(9):19–21.

Jones, M.B., P.W. Lawler, and J.E. Ruckman. 1970. Differences in annual clover responses to phosphorus and sulfur. Agron. J. 62:439–442.

Jones, M.B., J.E. Ruckman, and P.W. Lawler. 1972. Critical levels of P in subclover (*Trifolium subterraneum* L.). Agron. J. 64:695–698.

Jones, M.B., J.E. Ruckman, W.A. Williams, and R.L. Koenigs. 1980. Sulfur diagnostic criteria as affected by age and defoliation of subclover. Agron. J. 72:1043–1046.

Jones, M.B., C.E. Vaughn, and R.S. Harris. 1976. Critical Ca levels and Ca/Mg ratios in *Trifolium subterraneum* L. grown on serpentine soil. Agron. J. 68:756–759.

Jones, R.M. 1985. Forage yields and crude protein content of sudan-sorghum hybrid. p. 38–41. *In* Texas Agric. Exp. Stn. Cons. Progr. Rep. 4347.

Jordan, H.V., and C.E. Bardsley. 1958. Response of crops to sulfur on southeastern soils. Soil Sci. Soc. Am. Proc. 22:254–256.

Jordon, C.W., C.E. Evans, and R.D. Rouse. 1966. Coastal bermudagrass response to applications of P and K as related to P and K levels in the soil. Soil Sci. Soc. Am. Proc. 30:477–480.

Kelling, K.A. 1981. Sulfur—How to know where it should be used. Proc. Wis. Fert., Aglime and Pest Manage. Conf. 20:30–35.

Kelling, K.A. 1982. Alfalfa fertilization. Univ. of Wisconsin-Extension Publ. A2448.

Kelling, K.A., and J.B. Peters. 1989. Red clover fertilization. p. 108–113. *In* Proc. 13th Wis. Forage Production and Use Symp., Wisconsin Dells, WI 24–25 Jan. Wisconsin Forage Counc., Madison.

Kelling, K.A., and E.E. Schulte. 1986. DRIS as a part of a routine plant analysis program. J. Fert. Issues 3:107–112.

Kelling, K.A., E.E. Schulte, and T. Erickson. 1986. Adapting DRIS for alfalfa: What are the diagnostic norms? Better Crops Plant Food 70(Winter):18–20.

Kimbrough, E.L., R.E. Blaser, and D.D. Wolf. 1971. Potassium effects on regrowth of alfalfa (*Medicago sativa* L.). Agron. J. 63:836–839.

Kresge, C.B., and S.E. Younts. 1963. Response of orchardgrass to potassium and nitrogen fertilization on a Wickham silt loam. Agron. J. 55:161–164.

Kretschmer, A.E., Jr., and N.C. Hayslip. 1963. Evaluation of several pasture grasses on Immokalee fine sand in south Florida. Florida Agric. Exp. Stn. Bull. 658.

Krueger, C.R., and J.M. Scholl. 1970. Performance of bromegrass, orchardgrass, and reed canarygrass. Wisconsin Agric. Exp. Stn. Res. Rep. 69.

Little, S., J. Vicente, and F. Albruna. 1959. Yield and protein content of irrigated napiergrass, guineagrass and pangolagrass as affected by nitrogen fertilization. Agron. J. 51:111–113.

Lo, S.Y., and H.M. Reisenauer. 1968. Zinc nutrition of alfalfa. Agron. J. 60:464–466.

Loneragan, J.F., J.S. Gladstones, and W.J. Simmons. 1968. Mineral elements in temperate crop and pasture plants. Aust. J. Agric. Res. 19:353–364.

Loper, G.M., and D. Smith. 1961. Changes in micronutrient composition of the herbage of alfalfa, medium red clover, ladino clover, and bromegrass with advance in maturity. Wisconsin Agric. Exp. Stn. Res. Rep. 8.

MacLeod, L.B. 1965. Effect of nitrogen and potassium on the yield and chemical composition of alfalfa, bromegrass, orchardgrass and timothy grown as pure species. Agron. J. 57:261–266.

Markley, R.A., J.L. Cason, and B.R. Baumgardt. 1959. Effect of nitrogen fertilization and urea supplementation upon the digestibility of grass hays. J. Dairy Sci. 42:144.

Markus, A.K., and W.R. Battle. 1965. Soil and plant response to long-term alfalfa (*Medicago sativa* L.). Agron. J. 57:613–616.

Martel, Y.A., and J. Zizka. 1977. Yield and quality of alfalfa as influenced by additions of S to P and K fertilizations under greenhouse studies. Agron. J. 69:531–535.

Martin, W.E., and J.E. Matocha. 1973. Plant analysis as an aid in the fertilization of forage crops. p. 393–426. *In* L.M. Walsh and J.D. Beaton (ed.) Soil testing and plant analysis, rev. ed. ASA, CSSA, and SSSA, Madison, WI.

Martin, W.E., V.V. Rendig, A.D. Haig, and L.J. Berry. 1965. Fertilization of irrigated pasture and forage crops in California. California Agric. Exp. Stn. Bull. 815, p. 1–35.

Martin, W.E., A. Ulrich, M. Morse, and D.L. Mikkelsen. 1955. Potassium deficiency of alfalfa in California. Better Crops Plant Food 39(10)6–12, 46–51.

Martin, W.E., and T.W. Walker. 1965. Sulfur requirements and fertilization of pasture and forage crops. Soil Sci. 101:248–257.

Martin, W.E., W.A. Williams, and W.H. Johnson. 1957. Refertilization of rose clover. Calif. Agric. 11:7–14.

Matocha, J.E. 1969. Response of Coastal bermudagrass to various sources of sulfur on a sandy soil of East Texas. Texas Agric. Exp. Stn. Prog. Rep. 2696.

Matocha, J.E. 1971. Influence of sulfur sources and magnesium on forage yields of Coastal bermudagrass [*Cynondon dactylon* (L.) Pers.]. Agron. J. 63:493–496.

Matocha, J.E., F.M. Roquette, Jr., and R.L. Duble. 1973. Recycling and recovery of N, P, and K by Coastal bermudagrass. I. Effect of sources and rate of N under a clipping system. J. Environ. Qual. 2:125–129.

Matocha, J.E., and L. Smith. 1980. Influence of potassium nutrition on *Helminthosporium cynodontis* and dry matter yields of Coastal bermudagrass. Agron. J. 72:565–567.

Mays, D.A., S.R. Milinson, and C.V. Cole. 1980. Phosphorus nutrition of forages. p. 805–846. *In* F.E. Khasawneh et al. (ed.) The role of phosphorus in agriculture. ASA, CSSA, and SSSA, Madison, WI.

McGraw, R.L., M.P. Russelle, and J. Grava. 1986. Accumulation and distribution of dry mass and nutrients in birdsfoot trefoil. Agron. J. 78:124–131.

McNaught, K.J. 1958. Potassium deficiency in pastures. I. Potassium content of legumes and grasses. N. Z. J. Agric. Res. 1:148–181.

McNaught, K.J., and P.J.E. Chrisstoffels. 1961. Effect of sulphur deficiency on sulphur and nitrogen levels in pastures and lucerne. N.Z.J. Agric. Res. 4:177–196.

Melsted, S.W., H.L. Motto, and T.R. Peck. 1969. Critical plant nutrient composition values useful in interpreting plant analysis data. Agron. J. 61:17–20.

Miller, D.A., and R.K. Smith. 1977. Influence of boron on the other chemical elements in alfalfa. Commun. Soil Sci. Plant Anal. 8:465–478.

Millikan, C.R. 1953. Relative effects of zinc and copper deficiencies on lucerne and subterranean clover. Aust. J. Biol. Sci. 6:164–177.

Moon, F.E. 1954. The composition and nutritive value of hay grown in the east of Scotland and the influence of late applications of nitrogenous fertilizers. J. Agric. Sci. 44:140.

Mortvedt, J.J. 1981. Nitrogen and molybdenum uptake and dry matter relationships of soybeans and forage legumes in response to applied molybdenum on acid soil. J. Plant Nutr. 3:245–256.

Munson, R.D., and W.L. Nelson. 1973. Principles and practices in plant analysis. p. 223–248. *In* L.M. Walsh and J.D. Beaton (ed.) Soil testing and plant analysis. rev. ed. ASA, CSSA, and SSSA, Madison, WI.

Napitupula, J.A., and D. Smith. 1979. Changes in herbage chemical composition due to proportion of species in alfalfa-orchardgrass mixtures. Commun. Soil Sci. Plant Anal. 10:565–577.

National Research Council. 1958. Nutritive requirements of beef cattle. Natl. Acad. Sci. Publ. 579. U.S. Gov. Print. Office, Washington, DC.

Nelson, W.L., and S.A. Barber. 1964. Nutrient deficiencies in legumes for grain and forage. p. 143–180. *In* H.B. Sprague (ed.) Hunger signs in crops. Davis McKey Co., New York.

Ontario Ministry of Agriculture and Food. 1988. 1989–1990 field crop recommendations. Publ. 296. Ontario Min. of Agric. and Food, Toronto, ON, Canada.

Osman, A., C.A. Raguse, and D.C. Sumner. 1977. Growth of subterranean clover in a range soil as affected by microclimate and phosphours availability. II. Laboratory and phytotron studies. Agron. J. 69:26–29.

Ozanne, P.G., J. Keay, and E.F. Biddiscombe. 1969. The comparative applied phosphate requirement of eight annual pasture species. Aust. J. Agric. Res. 3:227–243.

Parks, C.L., H.F. Perkins, and J.T. May. 1967. A greenhouse study of P and K requirements for ladino clover establishment on kaolin strip mine spoil. Georgia Agric. Res. 9:2.

Peterschmidt, N.A., R.H. Delaney, and M.C. Greene. 1979. Effects of over-irrigation on growth and quality of alfalfa. Agron. J. 71:752–754.

Petrie, S.E., and T.L. Jackson. 1982. Effects of lime, P, and Mo application on Mo concentration in subclover. Agron. J. 74:1077–1081.

Pierre, W.H., and R.R. Robinson. 1937. The calcium and phosphate content of pasture herbage and of various pasture species as affected by fertilization and liming. J. Am. Soc. Agron. 219:477–496.

Plank, C.O. 1988. Plant analysis for Georgia. CES Univ. of Georgia, College of Agric., Athens.

Plucknett, D.L., and R.L. Fox. 1965. Effects of phosphorus fertilization on yields and composition of pangolagrass and *Desmodium intortum*. p. 1525–1529. *In* Proc. IX Int. Grassl. Congr., São Paulo, Brazil. 8–19 Jan.

Porter, T.K., and J.H. Reynolds. 1975. Relationship of alfalfa cultivar yields to specific leaf weight, plant density, and chemical composition. Agron. J. 67:625–629.

Pumphrey, R.V., and D.P. Moore. 1965. Diagnosing sulfur deficiency of alfalfa (*Medicago sativa* L.) from plant analysis. Agron. J. 57:364–366.

Rayment, G.E., and R.C. Bruce. 1979. Effect of phosphorus fertilizer on established white clover based pastures in south-east Queensland. 2. Macronutrient status and prediction of yield responses using plant chemical tests. Aust. J. Agric. Anim. Husb. 19:463–471.

Reid, R.L., and G.A. Jung. 1965. Influence of fertilizer treatments on the intake, digestibility and palatability of tall fescue hay. J. Anim. Sci. 24:615–625.

Reid, R.L., and G.A. Jung. 1974. Effects of elements other than nitrogen on the nutritive value of forages. p. 395–435. *In* D.A. Mays (ed.) Forage fertilization. ASA, CSSA, and SSSA, Madison, WI.

Reid, R.L., G.A. Jung, and C.M. Kinsey. 1967. Nutritive value of N fertilization of orchardgrass pasture at different periods of the year. Agron. J. 59:519–525.

Reid, R.L., A.J. Post, and G.A. Jung. 1970. Mineral composition of forages. West Virginia Agric. Exp. Stn. Bull. 589T.

Reisenauer, H.M. 1956. Molybdenum content of alfalfa in relation to deficiency symptoms and response to molybdenum fertilization. Soil Sci. 81:237–242.

Rendig, V.V., W.E. Martin, and F.F. Smith. 1950. Pasture forage—Effect of fertilizers on composition of forage on phosphorus-deficient soils. Calif. Agric. 4(8):6.

Reuter, D.J., J.F. Loneragan, A.D. Robson, and D.J. Tranthim-Fryer. 1983. Intraspecific variation in the external and internal copper requirements of subterranean clover. Agron. J. 75:45–49.

Reynolds, J.H., C.R. Lewis, and K.F. Lasker. 1971. Chemical composition and yield of orchardgrass forage grown under high rates of nitrogen fertilization and several cutting requirements. Tennessee Agric. Exp. Stn. Bull. 479.

Rhykerd, C.L., and C.J. Overdahl. 1972. Nutrition and fertilizer use. p. 437–468. *In* C.H. Hanson (ed.) Alfalfa science and technology. Agronomy Monogr. 15. ASA, Madison, WI.

Riceman, D.S., and G.B. Jones. 1958. Distribution of zinc and copper in subterranean clover (*Trifolium subterraneum* L.) grown in culture solutions supplied with graduated amounts of zinc. Aust. J. Agric. Res. 9:73–122.

Robinson, D.L., K.G. Wheat, and N.L. Hubbert. 1987. Nitrogen fertilization influences on gulf ryegrass yields, quality and nitrogen recovery for Olivier silt loam soil. Louisiana Agric. Exp. Stn. Bull. 784.

Robinson, D.L., K.G. Wheat, N.L. Hubbert, M.S. Henderson, and H.J. Savoy, Jr. 1988. Dallisgrass yield, quality and nitrogen recovery responses to nitrogen and phosphours fertilizers. Commun. Soil Sci. Plant Anal. 19:529–542.

Robinson, R.R., C.L. Rhykerd, and C.F. Gross. 1962. Potassium uptake by orchardgrass as affected by time, frequency and rate of potassium fertilization. Agron. J. 54:351–353.

Rodriguez, J.P. 1949. Effect of nitrogen applications on the yields and composition of forage crops. J. Agric. Univ. P. R. 33:98–117.

Romero, N.A., C.C. Sheaffer, and G.A. Malzer. 1981. Potassium response of alfalfa in solution, sand and soil culture. Agron. J. 73:25–28.

Roquette, F.M., Jr., J.E. Matocha, and R.L. Duble. 1973. Recycling and recovery of N, P, and K by Coastal bermudagrass: II. Under grazing conditions with two stocking rates. J. Environ. Qual. 2:129–132.

Russelle, M.P., R.L. McGraw, J. Grava, and P.R. Beuselinck. 1985. Elemental composition of birdsfoot trefoil. Commun. Soil Sci. Plant Anal. 16:987–1013.

Russelle, M.P., and C.C. Sheaffer. 1986. Use of the diagnosis and recommendation integrated system with alfalfa. Agron. J. 78:557–560.

Sabbe, W.E., and L.H. Hileman. 1975. Nutrient composition of fescue and bermudagrass grown in Arkansas pastures. Arkansas Farm Res. 23(6):4.

Salette, J.E. 1970. Nitrogen use and intensive management of grasses in the wet tropics. p. 404–407. *In* J.T. Norman (ed.) Proc. XI Int. Grassl. Congr., Surfers Paradise, Australia. 13–20 Apr.

Sallee, W.R., A. Ulrich, W.E. Martin, and B.A. Krantz. 1959. High phosphorus for alfalfa. Calif. Agric. 13(8):7, 8, 13.

Savoy, H.J., Jr., and D.L. Robinson. 1989. Development and evaluation of preliminary DRIS norms for dallisgrass. Commun. Soil Sci. Plant Anal. 20:655–683.

Schneider, B.A., and N.A. Clark. 1970. Effect of potassium on the mineral constituents of pearl millet and sudangrass. Agron. J. 62:474–477.

Seay, W.A., O.J. Attoe, and E. Truog. 1949. Correlation of potassium content of alfalfa with that available in soils. Soil Sci. Soc. Am. Proc. 14:245–249.

Seim, E.C., A.C. Caldwell, and G.W. Rehm. 1969. Sulfur response by alfalfa (*Medicago sativa* L.) on a sulfur-deficient soil. Agron. J. 61:368–371.

Smith, D. 1964. Chemical composition of herbage with advance in maturity of alfalfa, medium red clover, ladino clover and birdsfoot trefoil. Wisconsin Agric. Exp. Stn. Res. Rep. 16.

Smith, D. 1969. Influence of temperature on the yield and chemical composition of Vernal alfalfa at first flower. Agron. J. 61:470–473.

Smith, D. 1970. Influence of temperature on the yield and chemical composition of five forage legume species. Agron. J. 62:520–523.

Smith, D. 1971. Levels and sources of potassium for alfalfa as influenced by temperature. Agron. J. 63:497–500.

Smith, D. 1972. Influence of nitrogen fertilization on the performance of an alfalfa-bromegrass mixture and bromegrass grown alone. Wisconsin Agric. Exp. Stn. Res. Rep. R2384.

Smith, D. 1975. Effects of potassium topdressing a low fertility silt loam soil on alfalfa herbage yield and composition and on soil K values. Agron. J. 67:60–64.

Smith, D., and A.K. Dobrenz. 1977. Response of alfalfa to fertilization with P and Mg in the greenhouse. J. Plant Nutr. 5:1–13.

Smith, D., and R.D. Powell. 1979. Yield of alfalfa as influenced by levels of P and K fertilization. Commun. Soil Sci. Plant Anal. 10:532–543.

Smith, D., D.A. Rohweder, and N.A. Jorgensen. 1974. Chemical composition of three legume and four grass herbages harvested at early flower during three years. Agron. J. 66:817–819.

Smith, D. and R.R. Smith. 1977. Responses of red clover to increasing rate of topdressed potassium fertilizer. Agron. J. 69:45–48.

Smith, D.T., and N.A. Clark. 1968. Effect of soil nutrients and pH on nitrate nitrogen and growth of pearl millet [*Pennisetum typhoides* (Burm.) Staph and Hubbard] and sudangrass [*Sorghum sudanense* (Piper) Staph]. Agron. J. 60:38–40.

Smith, R.R., N.L. Taylor, and S.R. Bowley. 1985. Red clover. p. 458–470. *In* N.L. Taylor (ed.) Clover science and technology. Agronomy Monogr. 25. ASA, CSSA, and SSSA, Madison, WI.

Spencer, K. 1978. Sulphur nutrition of clover: Effects of plant age on the composition-yield relationship. Commun. Soil Sci. Plant Anal. 9:883–895.

Spencer, K., and D. Bouma. 1970. Sulphur content of clover as an indicator of pasture response. p. 341–344. *In* J.T. Norman (ed.) Proc. XI Int. Grassl. Congr., Surfers Paradise, Australia. 13–20 Apr.

Spencer, K., and A.G. Govaars. 1982. The potassium status of pastures in the Moss Vale district, New South Wales. Tech. Paper 38. CSIRO, Melbourne, Australia.

Spooner, A.E., W.R. Jeffrey, and H.J. Hunneycutt. 1971. Effect of management practices on Johnsongrass for hay production. p. 12–15. *In* Arkansas Agric. Exp. Stn. Bull. 769.

Sumner, M.E. 1977a. Use of the DRIS system in foliar diagnosis of crops at high yield levels. Commun. Soil Sci. Plant Anal. 8:251–268.

Sumner, M.E. 1977b. The DRIS approach to recommendations based on leaf and soil analysis. North-Central Soil Fert. Workshop, St. Louis. 2 Nov. Dep. of Agron., Univ. of Georgia, Athens.

Sumner, M.E. 1979. Interpretation of foliar analyses for diagnostic purposes. Agron. J. 71:343–348.

Tarpley, M.L., D.L. Robinson, B.K. Gustavson, and M.M. Eichhorn, Jr. 1985. The DRIS for interpretation of Coastal bermudagrass analysis. Commun. Soil Sci. Plant Anal. 16:1335–1348.

Thomas, B.A., A. Thompson, V.A. Oyanuga, and R.H. Armstrong. 1952. The ash constituents of some herbage plants at different stages of maturity. Exp. Agric. 22:10–22.

Ulrich, A. 1945. Critical phosphorus and potassium levels in ladino clover. Soil Sci. Soc. Am. Proc. 10:150–160.

Ulrich, A. 1956. Plant analysis as a guide to the mineral nutrition of alfalfa. Proc. VII Int. Grassl. Congr. 8:3–11.

Ulrich, A., and F.J. Hills. 1967. Principles and practices of plant analysis. p. 11–24. *In* Soil testing and plant analysis. Part 2. SSSA Spec. Publ. 2. ASA, CSSA, and SSSA, Madison, WI.

Ulrich, A., M.A. Tabatabai, K. Ohki, and C.M. Johnson. 1967. Sulfur content of alfalfa in relation to growth in filtered and unfiltered air. Plant Soil 26:235–252.

Van Lierop, W., Y.A. Martel, and M.P. Cescas. 1980. Optimal soil pH and sufficiency concentrations of N, P and K for maximum alfalfa and onion yields on acid organic soils. Can. J. Soil Sci. 60:107–117.

Van Riper, G.E., and D. Smith. 1959. Changes in the chemical composition of the herbage of alfalfa, medium red clover, ladino clover and bromegrass with advance in maturity. Wisconsin Agric. Exp. Stn. Res. Rep. 4.

Vincente-Chandler, J., S. Silva, and J. Figerella. 1959. Effects of nitrogen and fertilization and frequency of cutting on the yield of composition of guineagrass in Puerto Rico. J. Agric. Univ. P. R. 43:228–239.

Walker, W.M., D.W. Graffis, and C.D. Faulkner. 1987. Effect of potassium and boron upon yield and nutrient concentration of alfalfa. J. Plant Nutr. 10:2169–2180.

Walker, W.M., and J. Pesek. 1967. Yield of Kentucky bluegrass (*Poa pratensis*) as a function of its percentage of nitrogen, phosphorus, and potassium. Agron. J. 59:44–47.

Walworth, J.L., M.E. Sumner, R.A. Isaac, and C.O. Plank. 1986. Preliminary DRIS norms for alfalfa in the southeastern United States and a comparison with midwestern norms. Agron. J. 78:1046–1052.

Watkins, J.M. 1940. The growth habits and chemical composition of bromegrass (*Bromus inermis* Leyss.) as affected by different environmental conditions. J. Am. Soc. Agron. 32:527–538.

Westermann, D.T. 1975. Indexes of sulfur deficiency in alfalfa: II. Plant analyses. Agron. J. 67:265–268.

Whitehead, D.C. 1985. Chlorine deficiency in red clover grown in solution culture. J. Plant Nutr. 8:193-198.

Widdowson, F.V., A. Penny, and R.J.B. Williams. 1965. An experiment measuring effects of N, P, and K contents of grass. J. Agric. Sci. 64:93-99.

Wilkinson, S.R., and C.F. Gross. 1967. Macro- and micronutrient distribution within ladino clover (*Trifolium repens* L.). Agron. J. 59:372-374.

Woodhouse, W.W., Jr. 1968. Long-term fertility requirements of Coastal bermudagass. I. Potassium. Agron. J. 60:508-512.

Woodhouse, W.W., Jr. 1969. Long-term fertility requirements of Coastal bermudagrass. II. Nitrogen, phosphorus and lime. Agron. J. 61:251-256.

Wright, R.J., M.C. Carter, T.B. Kinraide, and O.L. Bennett. 1984. Phosphorus requirements for the early growth of red clover, trefoil and flat pea. Commun. Soil Sci. Plant Anal. 15:49-63.

Chapter 24

Plant Analysis as an Aid in Fertilizing Tobacco[1]

GORDON S. MINER, *North Carolina State University, Raleigh*

M. R. TUCKER, *North Carolina Department of Agriculture, Raleigh*

Plant analysis is used extensively by the tobacco (*Nicotiana tabacum* L.) industry to evaluate quality of the cured leaf. Assessment of leaf quality depends primarily on the relative concentrations of various organic constituents, such as reducing sugars and total alkaloids. Concentrations of certain inorganic constituents are also indicative of quality. High concentrations of Cl (>10 g kg^{-1}) in the cured leaf lower leaf quality by reducing burn rate and fire-holding capacity and by increasing equilibrium moisture content. High concentrations of K (>25 g kg^{-1}) in cured leaf improve leaf quality by increasing the burn rate and fire-holding capacity. Evaluation of cured leaf quality by chemical analysis is beyond the scope of this chapter and will not be included. However, it emphasizes the importance of plant analysis to the tobacco industry for purposes other than plant nutrient evaluation of the growing crop.

From the standpoint of nutrient evaluation, plant analysis has been used quite extensively on tobacco for confirming visual plant symptoms caused by deficient or toxic levels of certain nutrient elements. Practical application of plant analysis as a diagnostic tool rests essentially on the assumption that a rapid and positive relationship exists between soil nutrient supplies within the root zone and the concentration of those nutrients in the plant. Research has shown conclusively that such a relationship does indeed exist.

Tobacco has a relatively short growth duration which limits the time in which plant analysis can be used for diagnostic purposes, particularly from the standpoint of taking corrective action. The time between post-transplant recovery to the beginning of harvest is about 8 wk. It is usually within and during this period that growth abnormalities due to nutrient deficiencies or toxicities occur. Corrective soil application of fertilizer nutrients in the later

[1]Paper No. 12062 of the Journal Series of the North Carolina Agric. Res. Serv., Raleigh, NC 27695-7601.

stages of this growth period are generally impractical due to the large plant size. Some micronutrient deficiencies identified during this period can be corrected by foliar application (Elliot, 1969). Fortunately, micronutrient deficiencies or toxicities are not a common occurrence on tobacco but isolated cases of Mo (Sims & Atkinson, 1976; Sims et al., 1983) and Mn (Elliot, 1969) deficiency as well as Mn toxicity (Hiat & Ragland, 1963) have been observed.

Preventative maintenance through a systematic plant monitoring scheme would be the most effective means of identifying and dealing with potential nutritional problems. Although this applies to all nutrients, it is especially true for N in flue-cured tobacco because yield and quality are influenced more by N supply during the growing season than any other nutrient element (Breland et al., 1967; Person & Whitty, 1982; Evanylo & Sims, 1987). Excess N sometimes increases yield but lowers leaf quality due to high alkaloid and low reducing sugar concentrations, delays maturity (ripening), and produces immature and unripe grades of cured leaf. Insufficient N reduces yield and produces poor leaf quality as a result of low alkaloid and high reducing sugar concentrations coupled with a thick, close-textured cured leaf with reduced combustibility. Careful regulation of N fertilizer is required to produce cured leaves with proper chemical balance and physical properties. Nitrogen management for flue-cured tobacco is complicated by rainfall. Garner (1939) stated that rainfall distribution had a greater influence on nutrient uptake and utilization than any other variable in tobacco production. Excess rainfall leaches N, and within-season N fertilizer adjustment is sometimes required. The amount of adjustment required depends upon the amount of water percolated through the root zone, depth to the argillic horizon, and stage of growth. Drought, on the other hand, reduces dry weight accumulation and produces a crop with characteristics similar to those of excess N.

The essential factors for developing a nutrient monitoring database should include: (i) the optimum leaf nutrient concentrations of all essential nutrients, by stalk position, at each of several identifiable preharvest growth stages; (ii) the optimum nutrient concentrations in the cured leaf; and (iii) regression equations relating nutrient optima of the cured-leaf to the green leaf. Currently, the database to support a viable nutrient monitoring program that meets the above criteria has not been assembled.

Many tobacco types are grown but the two most important are burley and flue-cured. While considerable plant analysis data for these two types have been published, data related specifically to critical nutrient deficiency or toxicity concentrations are somewhat limited. Some experience has been gained in the interpretation of tobacco tissue analyses and this information will be used to supplement published data where appropriate.

I. DRY MATTER AND NUTRIENT ACCUMULATION

McCants and Woltz (1967) stated, "some knowledge and consideration of the total growth and development of the plant are essential to a complete understanding of the nutrition of the plant." Following is a description of the growth characteristics of tobacco and the corresponding nutrient relationship.

A. Flue-Cured Tobacco

The duration of a flue-cured tobacco crop ranges from 90 to 120 d. This period can be divided into three general stages of development: (i) seedling recovery (transplanting to 3 wk) during which little growth occurs; (ii) rapid vegetative growth up to flowering (from 3 to 8 wk); and (iii) leaf expansion and maturation (from 8–about 16 wk) during which leaves are harvested as they ripen, progressing from the bottom to the top of the plant.

Nutrient accumulation by a typical flue-cured tobacco crop is shown in Table 24–1. Relatively low rates of N are applied to enhance leaf quality and ensure leaf ripeness and maturity at harvest. Whereas accumulation of N is usually greater than that of other nutrients in most crops, K accumulation is greatest in flue-cured tobacco that is properly fertilized.

The accumulation of dry matter and nutrients in the aboveground portion of flue-cured tobacco over a cropping season was characterized by Grizzard et al. (1942) and by Raper and McCants (1966). Dry matter accumulation, as a percentage of the total, increased at a lower rate than that of N, P, K, Ca, and Mg prior to flowering at 8 to 9 wk after transplanting (Fig. 24–1). Discounting the small amounts of dry matter and nutrients accumulated during the first 3 wk, approximately 60% of the N, 45% of the K and Mg, and 37% of the Ca and P were absorbed during the next 3-wk period while only about 20% of the dry matter accumulated. By flowering (8–9 wk after transplanting), more than 90% of the N and K and 70 to 80% of the Ca, Mg, P, and dry matter had accumulated. Thus, by flowering, most of the N and K were absorbed whereas accumulation of Ca, Mg, and P continued until the youngest leaves reached physiological maturity.

The concentration of nutrients in the aboveground portion of flue-cured tobacco reaches a maximum at 4 to 5 wk after transplanting and declines to a stable level after 9 to 10 wk (Fig. 24–2). Rapid dry matter accumulation from the 4th or 5th wk through flowering causes the decline in nutrient concentration through a dilution effect. Since 70 to 80% of the dry matter ac-

Table 24–1. Estimates of nutrients accumulated by above- and belowground plant parts to produce a 2240 kg ha^{-1} crop of flue-cured tobacco (Hawks & Collins, 1983).

Macro and secondary nutrients		Micronutrients	
Nutrient	Content	Nutrient	Content
	kg ha^{-1}		kg ha^{-1}
N	78	B	0.08
P	13	Mn	0.8
K	90	Fe	trace
Ca	62	Zn	trace
Mg	25	Cu	0.04
S	20	Mo	trace

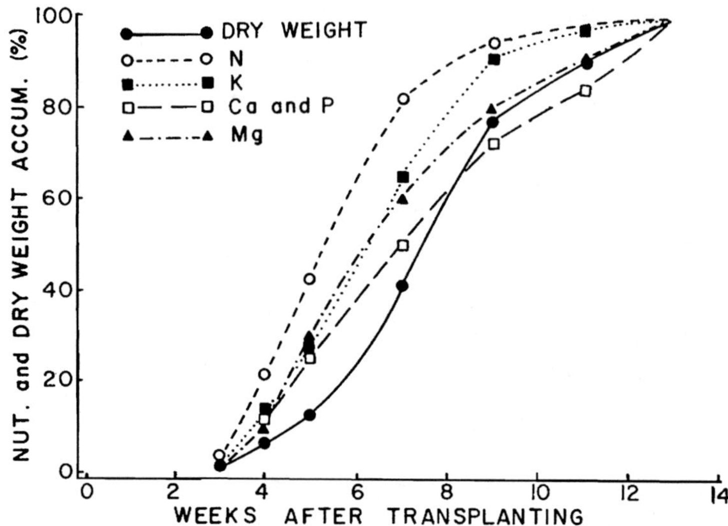

Fig. 24-1. Relative accumulation of dry matter and nutrients in the above ground portion of field-grown, flue-cured tobacco (Raper & McCants, 1966).

cumulates within the first 7 to 8 wk, a high demand for nutrients occurs within the first 5 to 6 wk. Thus, flue-cured tobacco requires high levels of readily available nutrients early in the growing season to reach its maximum growth and quality potential.

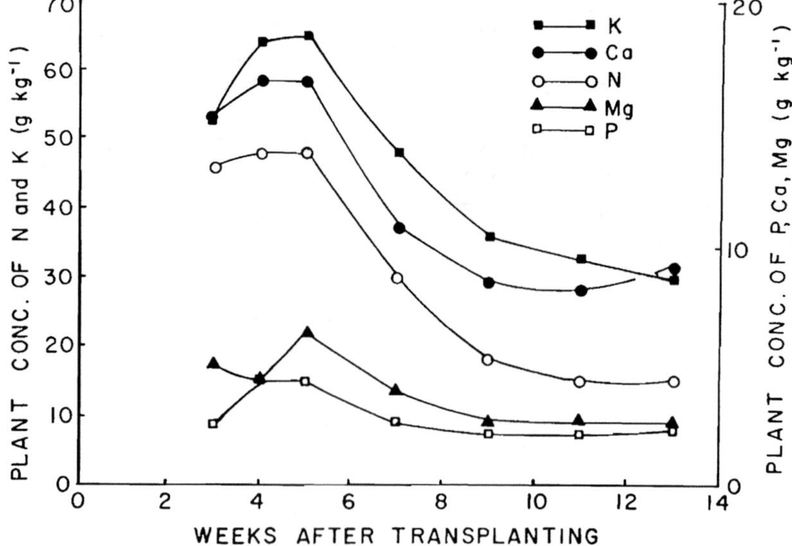

Fig. 24-2. Concentration of selected nutrients in aboveground portion of field-grown, flue-cured tobacco during the growing season (Derived from Raper & McCants, 1966).

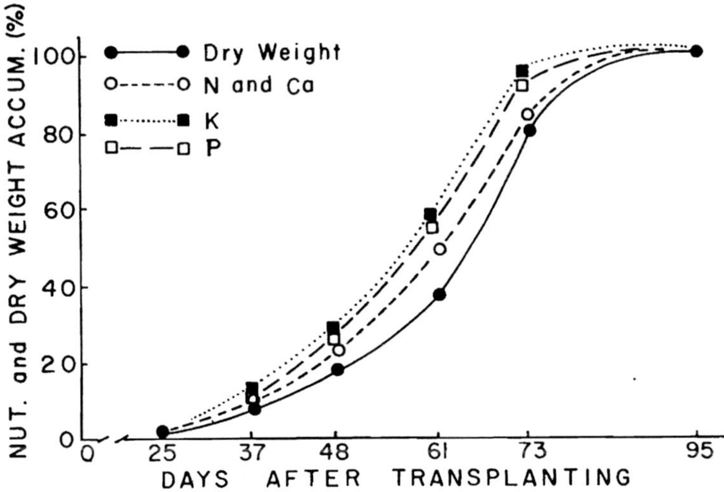

Fig. 24-3. Relative accumulation of dry matter and nutrients by burley tobacco during the growing season (Atkinson et al., 1977).

B. Burley Tobacco

Burley and flue-cured tobaccos have similar growth patterns with respect to nutrient accumulation and dry matter production (Fig. 24-1 and 24-3). The two tobacco types are also similar in that nutrient accumulation occurs at a more rapid rate than dry matter in the early growth stage. However, the data shown here suggests that flue-cured tobacco grows off more rapidly than burley. For example, 70 d after transplanting, flue-cured tobacco had accumulated more than 80% of its 91-d total dry matter, whereas burley had accumulated only 70% of its 95-d total dry matter. These differences in nutrient accumulation and dry matter production between these two types of tobacco may be attributable to differences in growing environment or germplasm. Final yields for the two tobacco types were similar at 2520 and 2700 kg ha^{-1} for flue-cured and burley, respectively (Fig. 24-1 and 24-3).

High rates of N, usually containing a relatively high proportion of NH_4^+, are applied to burley tobacco. In the studies discussed previously, 224 kg ha^{-1} of N was applied to the burley whereas the flue-cured tobacco received only 67 kg ha^{-1}. Sims and Atkinson (1973) reported that a high N rate (320 kg ha^{-1}) from NH_4NO_3 resulted in a total plant dry weight at 40 d that was less than half that where either no N or 45 kg ha^{-1} were applied for burley. After 70 d, dry weights were similar for the low and high N rates. The depressed early growth with the high N rate was attributed to reduced availability of soil Ca, Mo, and P resulting from fertilizer-induced soil acidity. In addition, nitrification was delayed at the high N rate and NH_4^+ toxicity and high soluble salts may have contributed to reduced early growth of the burley.

Nutrient monitoring by plant analysis may have even more potential for burley than for flue-cured tobacco because all leaves are harvested simultaneously at the end of the season. Excessive N is not as critical for burley as for flue-cured tobacco and a monitoring program may aid in maintaining desired nutrient levels. However, the problems associated with taking corrective action for flue-cured tobacco, early nutrient requirements and large plant size, also apply to burley.

II. SAMPLING

Obtaining a plant tissue sample that is representative of the crops nutritional status is one of the most critical aspects of plant tissue analysis. Guidelines suggested for sample selection on tobacco vary from the uppermost fully developed leaf before bloom (Jones, 1974; Plank, 1979) to the fourth leaf from the top of the plant prior to or at bloom (S.J. Donohue, 1988, personal communication). These sampling suggestions represent essentially the same stalk position and can be interpreted as the uppermost, fully developed leaf just prior to flowering. This corresponds to a leaf in which cell division is complete but cell expansion has yet to occur and whose stalk position is about four leaves below the apical meristem. The recommended number of leaves per sample varies from 6 to 12 depending on the individual laboratory.

Although there is general agreement on the plant tissue sampled for tobacco, there are no published data from definitive studies relating the selection of the most appropriate plant tissue to the nutrient status of the plant during the course of the growing season. Present guidelines reflect the generally recognized practice of choosing the most recently mature leaves for analysis. However, changing nutrient concentrations within specific plant tissue during the growing season makes interpretation difficult unless nutrient sufficiency ranges and critical nutrient values for all stages of growth are known.

Plant samples taken for problem diagnostic purposes from either small planted seedlings or juvenile plants in the field generally involve analyzing a composite of small leaflets or leaves that exhibit nutrient deficiency symptoms. Normal appearing plants from an adjacent area should also be analyzed for comparison purposes.

Critical nutrient concentrations tend to remain stable in tissue of the same physiological age (Ulrich & Hills, 1967) whereas increasing maturity of leaves tend to result in decreasing N and K concentrations, little change in P, and increasing concentrations of Ca and Mg (Munson & Nelson, 1973). The relationship between Mn concentration and physiological age of tobacco leaves is illustrated in a study by Rufty et al. (1979). Phytotron-grown plants, exposed to 120 mg L^{-1} of Mn under three temperature regimes, showed that leaf Mn increased with time at all stalk positions and was highest in the bottom leaves at each sampling date (Table 24-2). Although there was some variation, Mn concentrations were similar among leaves of similar physiological age.

Table 24-2. The effect of day/night temperature and time after transplanting on Mn concentration in tobacco leaves from various stalk positions of plants exposed to 120 mg L^{-1} Mn (Rufty et al., 1979).

Leaf group†	Days after transplanting					
	7	14	21	28	35	42
	Mn, mg kg^{-1}					
			22/18 °C			
6						2 765
5						3 353
4					4 040	5 253
3				4 423	6 323	7 590
2		5 280	5 210	7 727	8 543	9 627
1	1 743	4 830	6 173	8 437	9 430	10 693
			26/22 °C			
6						4 457
5					3 765	6 283
4				4 643	5 593	8 230
3			4 750	7 270	7 500	0 527
2		5 800	8 767	10 073	10 020	11 923
1	3 150	6 897	7 737	9 136	11 267	12 833
			30/26 °C			
6						4 800
5					5 200	6 430
4				4 710	5 860	7 865
3			6 100	7 600	8 200	9 531
2		7 000	8 600	9 300	10 560	10 920
1	4 100	6 500	8 500	8 830	10 400	11 314

† Each leaf group consists of three leaves with group one representing the oldest leaf group (bottom stalk position).

The effects of temperature on the susceptibility of tobacco to Mn toxicity were also shown in the above study (Rufty et al., 1979). Visual Mn-toxicity symptoms occurred at 3, 7, and 26 d after exposure to Mn in the low, medium, and high temperature regimes, respectively. Appearance of these symptoms corresponded to leaf Mn concentrations of 1200, 3500, and 8000 mg kg^{-1} for the respective temperatures. Thus, critical Mn-toxicity concentrations varied with temperature. The same effect may occur with critical deficiency concentrations of some nutrients. Interpretation of foliar analyses, already complicated by factors such as stage of growth and plant part sampled, is further complicated by environmental factors such as temperature and moisture stress (Grandeza and Briones, 1986).

III. NUTRIENT SUFFICIENCY VALUES

A range of nutrient sufficiency values for tobacco is used by various public agencies involved in providing plant analysis service. No distinction is made between burley and flue-cured tobacco in those states where both types are grown.

Table 23-3. Primary and secondary nutrient sufficiency ranges used by various states for the uppermost, fully developed tobacco leaf prior to flowering.

States†	N	P	K	Ca	Mg	S
	g kg^{-1}					
Alabama	35.0–42.5	2.7–5.0	25.0–32.0	15.0–35.0	2.0–6.5	2.0–6.0
Georgia	35.0–42.5	2.7–5.0	25.0–32.0	15.0–35.0	2.0–6.5	--
North Carolina	35.0–42.5	2.0–5.0	25.0–32.0	7.5–35.0	2.0–6.0	1.5–4.0
South Carolina	35.0–42.5	2.7–5.0	25.0–32.0	15.0–35.0	2.0–6.5	--
Virginia	35.0–42.5	2.5–5.0	25.0–32.0	15.0–35.0	2.0–6.5	--

† Alabama, C.E. Evans; North Carolina, M.R. Tucker; South Carolina, R.C. Lippert; Virginia, S.J. Donahue, 1989, personal communications; and Georgia, Plank, 1979.

Table 24-4. Micronutrient sufficiency ranges used by various states for the uppermost fully developed tobacco leaf prior to flowering.

States†	Mn	Fe	B	Cu	Zn	Mo
	mg kg^{-1}					
Alabama	30–250	50–200	20–50	5–60	20–80	--
Georgia	30–250	50–200	20–50	15–60	20–80	--
North Carolina	20–350	50–200	21–100	5–10	20–60	0.5–1.1
South Carolina	30–250	50–200	20–50	5–20	20–80	--
Virginia	30–250	50–200	20–50	15–60	20–80	--

† Alabama, C.E. Evans; North Carolina, M.R. Tucker; South Carolina, R.C. Lippert; Virginia, S.J. Donahue, 1989, personal communications; and Georgia, Plank, 1979.

Table 24-5. Sufficiency ranges for nutrients in uppermost fully developed tobacco leaf at bloom stage (Jones, 1974).

Macronutrients		Micronutrients	
Nutrient	Sufficiency range	Nutrient	Sufficiency range
	g kg^{-1}		mg kg^{-1}
N	35.0–42.5	Mn	30–250
P	2.7–5.0	Fe	50–200
K	15.0–32.0	B	20–50
Ca	15.0–35.0	Cu	15–60
Mg	2.0–6.5	Zn	20–80

There is some variation among tissue testing laboratories in the range of nutrient sufficiency values used for interpreting tobacco leaf analyses but values are remarkably similar (Tables 24–3 and 24–4). The similarities result from laboratories using guidelines (Table 24–5) initially published by Jones (1974) and later revised by Plank (1979). The differences result from modifications imposed by individual laboratories based on their experiences and accumulated database. Although these nutrient sufficiency ranges are useful and serve as guidelines for interpretation of the nutrient status of tobacco, they are based primarily on accumulated knowledge and experience and not on definitive published research results.

Surveys of nutrient concentrations in cured leaves are indicative of the nutrient status of field-grown tobacco. Flue-cured tobacco from 32 farms in Ontario, Canada, have been evaluated annually since 1970 and nutrient

Table 24-6. Concentrations of nutrients in flue-cured tobacco leaves grown on 32 farms in Ontario, Canada averaegd over 10 yr (Elliot et al., 1985a, b).

Macronutrients		Micronutrients	
Nutrient	Concentration	Nutrient	Concentration
	g kg^{-1}		mg kg^{-1}
N	22.1	Fe	268
P	2.9	Mn	132
K	22.2	Zn	39
Ca	20.4		g kg^{-1}
Mg	4.1	Cl	0.62

concentrations averaged over 10 yr are shown in Table 24-6. In general, these nutrient concentrations fall within the range for nutrient sufficiency values described previously.

IV. CRITICAL NUTRIENT VALUES

The critical nutrient value has been defined as the concentration of nutrient in a particular plant part sampled at a specific stage of growth in which 5 to 10% of the potential yield is lost (Sumner, 1977; Ulrich, 1943; Ulrich & Hills, 1967). From an economic perspective, Bates (1971) defined the critical value as the nutrient concentration beyond which fertilization is profitable. Realizing that the critical value is based on a specific stage of growth, interpretation of nutrient values at different growth stages becomes more complex. Proposals have been made for establishing critical nutrient values for different stages of growth (Geraldson et al., 1973; Lockman, 1972; Tserling, 1974) which would lend some flexibility to their interpretation. An example of the influence of stalk position on the Mg content in green leaves and its relationship to deficiency symptoms has been shown by Woltz (1975), Fig. 24-4. Coupled with restrictions on the interpretation due to time and

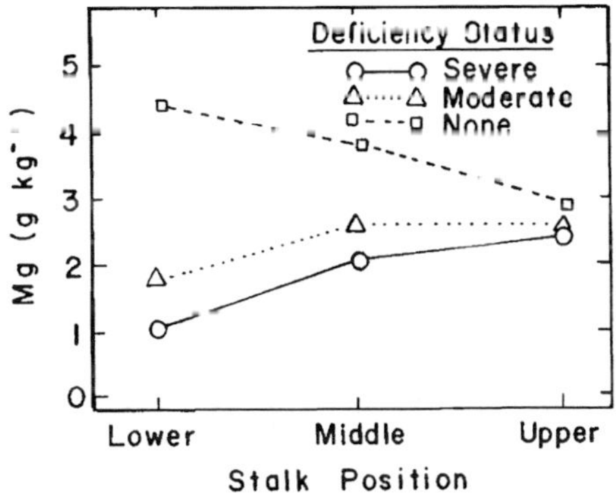

Fig. 24-4. Concentration of Mg in flue-cured tobacco leaves by stalk position at varying degrees of Mg deficiency (Woltz, 1975).

Table 24-7. Published critical concentrations of nutrients in flue-cured (FC) and burley (Bur) tobacco.

Nutri-ent	Tobacco type	Plant part	Time after transplanting	Critical concentration Toxicity	Deficiency	Reference
				mg kg^{-1}		
B	FC	Bud leaves	35–50 d		15–16	Hutcheson & Woltz, 1956
Mn	FC	Leaves		83–179	22	Bacon et al., 1950
Mn	FC				25	Tsai, 1975
Mn	FC				18	Elliot, 1969
Mn	FC	Leaves	1–6 wk	700–7000		Rufty et al., 1979
Cu	FC				<10	Bacon et al., 1950
Zn	FC				<10	Bacon et al., 1950
Mo	Bur	Cured leaves	After harvest		0.42	Sims & Atkinson, 1976
Mo	Bur	Whole plant	50 d		0.38	Sims et al., 1975
Mg		Leaves			2.0 g kg^{-1}	McCants & Woltz, 1967
Mg	FC	Leaves			2.0 g kg^{-1}	Woltz, 1975

location of the plant sample is the difficulty of evaluating and assessing the degree of yield loss.

In any event, some knowledge of critical nutrient concentration is essential for plant analyses to be an effective diagnostic tool. The critical value concept appears to be more applicable to seasonal monitoring for detecting critical nutrient concentrations prior to appearance of visual symptoms. In reality, plant analysis as a diagnostic tool for tobacco operates most frequently in the "deficient" or "sufficient" categories as opposed to a critical value concentration. This mode of operation occurs most frequently because growers generally wait until deficiency symptoms are quite pronounced before submitting samples for analyses. By this time, the nutrient concentration is well below the critical value and corrective treatment is focused on a salvage effort in which the degree of success depends upon the nutrient involved and the stage of growth.

Of equal importance from the standpoint of the effect of nutrient deficiencies on tobacco yield is their influence on cured leaf grades and market value. For example, Mn-deficient tobacco leaves are characterized by extensive leaf flecking which is visable in the cured leaves and may reduce leaf quality and market value. Consequently, for crops like tobacco, factors other than the effect of critical nutrient values on yield must be considered.

Little information is published on the critical concentration of nutrients for tobacco. Critical values published in the literature are shown in Table 24-7. Plant analysis values for a number of nutrient elements ranging from deficient to toxic concentrations are shown by Chapman (1973), some of which may be interpreted as critical concentrations. Future research efforts should focus on establishing critical nutrient concentrations for all essential nutrient elements.

V. NUTRIENT DEFICIENCY SYMPTOMS

The capability to identify nutrient deficiencies by visual observation rests on the premise that each essential nutrient element has a unique and specific function within the plant and that the plant will exhibit characteristic visual deficiency symptoms in its absence. As a general rule, the above assumptions are true, however, there are certain nutrient elements that exhibit similar deficiency symptoms and consequently could lead to a misinterpretation by visual observation. Plant analysis has the capability of making a distinction between nutrients which exhibit similar deficiency symptoms and is thus more discriminating as a diagnostic tool.

A brief description of nutrient deficiencies on tobacco are shown below. More extensive descriptions of deficiency symptoms are presented by McMurtrey (1938, 1964).

General nutrient deficiency symptoms of tobacco†

Nutrient	Deficiency symptoms
N	General yellowing and firing of lower leaves; entire plant may turn yellow in severe cases, growth is reduced, leaf size is reduced.
P	Stunted growth, abnormally green color, younger leaves stand erect.
K	Leaf margin and tip chlorosis or burning, spots of dead tissue scattered throughout leaf, leaf tips cupped inward.
Ca	Distortion of young leaves, older leaves may appear normal, in severe cases entire plant is stunted.
Mg	Older leaves chlorotic, depending on severity entire plant may be discolored. Early symptoms start at leaf tips and are similar to those of K deficiency.
S	General yellowing of young leaves very similar in color to N deficiency. On small seedling entire plant may be yellow.
Mn	Leaf flecking similar to that caused by ozone (weather fleck), lower leaves disintegrate and fall off. Younger leaves may appear normal if subsoil is acidic. Plants may be dwarfed.
Fe	Chlorosis of upper leaves.
Cu	Reduced growth and vigor, yield and quality. No specific symptom.
B	Distortion of young leaves, terminal bud dies.
Mo	Chlorosis and spotting of lower leaves.
Zn	Chlorosis of interveinal leaf area.

† The deficiency symptoms of most nutrients are generally specific in the advanced stages. Early stages of visual symptom development may be misinterpreted. Plant analysis can make the distinction in both cases.

Plant analysis has also been used to identify "hidden hunger," toxic nutrient levels, and nutrient imbalances within the plant. A more recent concept, not developed extensively for tobacco, is the Diagnostic and Recommendation Integrated Systems (DRIS) norms approach whereby the ratios between nutrient elements are calculated from the content of each nutrient element within the plant (Beaufils, 1973; Sumner, 1977). The nutritional status of the plant can then be evaluated by comparing the nutrient ratios with the DRIS norms. DRIS norms for cured burley tobacco are presented by Evanylo et al. (1988b). Evanylo et al. (1988a) have also shown the effect of the sampling period on the DRIS norms.

Although several nutrient elements exhibit characteristic deficiency symptoms, caution should be exercised in using visual symptoms alone for diagnostic purposes. Factors other than fertility can result in the appearance of symptoms similar to that of an inadequate nutrient supply.

REFERENCES

Atkinson, W.O., L.P. Bush, and J.L. Sims. 1977. Dry matter and nutrient accumulation in burley tobacco. Tob. Sci. 21:81–82.

Bacon, C.W., W.B. Leighty, and J.F. Bullock. 1950. Boron, copper, manganese and zinc requirement tests of tobacco. USDA Tech. Bull. 1009.

Bates, T.E. 1971. Factors affecting critical nutrient concentrations in plants and their evaluation: A review. Soil Sci. 112:116–130.

Breland, H.L., W.L. Pritchett, and H.W. Lundy. 1967. Effect of fertilizer treatment on composition of flue-cured tobacco. Soil Crop Sci. Soc. Fla. Proc. 27:235–242.

Beaufils, E.R. 1973. Diagnosis and recommendation integrated systems (DRIS). A general scheme of experimentation and calibration based on principles developed from research in plant nutrition. Soil Sci. Bull. 1. Univ. of Natal, Pieter-maritzburg, South Africa.

Chapman, H.D. (ed.). 1973. Diagnostic criteria for plants and soils. Dep. of Soils and Plant Nutr. Univ. Calif. Citrus Res. Ctr. and Agric. Exp. Stn., Riverside, CA.

Elliot, J.M. 1969. Effect of applied Mn, S and P on the Mn content of oats and flue-cured tobacco. Can. J. Soil Sci. 49:277–285.

Elliot, J.M., R.P. Beyaert, and R. Pocs. 1985a. A survey of flue-cured tobacco grown in Ontario in 1984. Part 2. Soil characteristics and nutrient elements. The Lighter 55(4):25–28.

Elliot, J.M., R. Pocs, and R.P. Beyaert. 1985b. A survey of flue-cured tobacco grown in Ontario in 1984. Part 1. Sugars, alkaloids, nitrogen, chlorine, and lamina weight. The Lighter 55(3):24–26.

Evanylo, G.K., J.H. Grove, and J.L. Sims. 1988a. Effect of sampling period and nutrient diagnostic indices for burley tobacco. Agron. J. 80:615–619.

Evanylo, G.K., and J.L. Sims. 1987. Nitrogen and potassium effects on yield and quality of burley tobacco. Soil Sci. Soc. Am. J. 51:1536–1540.

Evanylo, G.K., J.L. Sims, and J.H. Grove. 1988b. Nutrient norms for cured burley tobacco. Agron. J. 80:610–614.

Gandeza, A.T., and A.A. Briones. 1986. Phosphorus content of flue-cured tobacco (*Nicotiana tabacum* L.) in relation to soil moisture and adsorption conditions. J. Tob. Sci. Technol. 1:63–69.

Garner, W.W. 1939. Some aspects of the physiology and nutrition of tobacco. J. Am. Soc. Agron. 31:459–471.

Geraldson, C.M., G.R. Klacan, and D.A. Loreny. 1973. Plant analysis as an aid in fertilizing vegetable crops. p. 365–379. *In* L.M. Walsh and J.D. Beaton (ed.) Soil testing and plant analysis. Rev. ed. SSSA, Madison, WI.

Grizzard, A.L., H.R. Davies, and L.R. Kangas. 1942. The time and role of nutrient absorption by flue-cured tobacco. J. Am. Soc. Agron. 34:327–339.

Hawks, S.N., Jr., and W.K. Collins. 1983. Principles of flue-cured tobacco production. 1st ed. S.N. Hawks, Jr., and W.K. Collins, Raleigh, NC.

Hiat, A.J., and J.L. Ragland. 1963. Manganese toxicity of burley tobacco. Agron. J. 55:47–49.

Hutcheson, T.B., Jr., and W.G. Woltz. 1956. Boron in the fertilization of flue-cured tobacco. North Carolina Agric. Exp. Stn. Tech. Bull. 120.

Jones, J.B., Jr. 1974. Plant analysis handbook of Georgia. Georgia Coop. Ext. Serv. Bull. 735.

Lockman, R.B. 1972. Mineral composition of grain sorghum plant samples. III. Suggested nutrient sufficiency limits at various stages of growth. Commun. Soil Sci. Plant Anal. 3:295–306.

McCants, C.B., and W.G. Woltz. 1967. Growth and mineral nutrition of tobacco. Adv. Agron. 19:211–265.

McMurtrey, J.E., Jr. 1938. Symptoms of field-grown tobacco characteristic of the deficient supply of each of several essential chemical elements. Tech. Bull. 612. USDA, Washington, DC.

McMurtrey, J.E., Jr. 1964. Nutrient deficiencies in tobacco. p. 99–141. *In* H.E. Sprague (ed.) Hunger signs in crops. David McKay Co., New York.

Munson, R.D., and W.L. Nelson. 1973. Principles and practices in plant analysis. p. 223–248. *In* L.M. Walsh and J.D. Beaton (ed.) Soil testing and plant analysis. Rev. ed. SSSA, Madison, WI.

Person, G.D., and E.G. Whitty. 1982. Yield and quality of flue-cured tobacco as affected by nitrogen fertilization. Soil Crop Sci. Soc. Fla. 41:111–114.

Plank, C.O. 1979. Plant analysis handbook for Georgia. Georgia Coop. Ext. Serv. Bull. 735 (rev.).

Raper, C.D., Jr., and C.B. McCants. 1966. Nutrient accumulation in flue-cured tobacco. Tob. Sci. 10:109.

Rufty, T.W., G.S. Miner, and C.D. Raper, Jr. 1979. Temperature effects on growth and manganese tolerance in tobacco. Agron. J. 71:638–644.

Sims, J.L., and W.O. Atkinson. 1973. Accumulation of dry matter and nitrogen content of burley tobacco growing in fertilizer-induced acid soil. Agron. J. 6:762–765.

Sims, J.L., and W.O. Atkinson. 1976. Lime, molybdenum, and nitrogen source effects on yield and selected chemical components of burley tobacco. Tob. Sci. 20:174–177.

Sims, J.L., W.O. Atkinsons, and C. Smitobol. 1975. Mo and N effects on growth, yield and Mo composition of burley tobacco. Agron. J. 67:824–828.

Sims, J.L., M.E. Suchy, and P.L. Cornelius. 1983. Placement of molybdenum fertilizer in the transplant solution of burley tobacco. Agron. J. 75:239–242.

Sumner, M.E. 1977. Use of the DRIS system in foliar diagnosis of crops at high yield levels. Commun. Soil Sci. Plant Anal. 8:251–268.

Tsai, C.F. 1975. Studies on the Mn nutrition of flue-cured tobacco in Taiwan. Taiwan Tob. Wine Monop. Bur. Tob. Res. Inst. Res. Rep. 2:47–57.

Tserling, V.V. 1974. Principle results of investigation on nutrient diagnosis of major crops in different natural zones of the USSR. 7th Int. Coll. Plant Fert. Prof. 2:459–472.

Ulrich, A. 1943. Plant analysis as a diagnostic procedure. Soil Sci. 55:101–111.

Ulrich, A., and F.J. Hills. 1967. Principles and practices of plant analysis. p. 11–24. *In* G.W. Hardy et al. (ed.) Soil testing and plant analysis. Part 2. SSSA Spec. Publ. 2. SSSA, Madison, WI.

Woltz, W.G. 1975. Magnesium requirements of flue-cured tobacco. Proc. Soil Sci. Soc. N.C. 18:54–62.

Chapter 25

Plant Analyses as an Aid in Fertilizing Forests

G. F. WEETMAN, *The University of British Columbia, Vancouver*

C. G. WELLS, *USDA-Forest Service, Research Triangle Park, North Carolina*

This review is concerned with fertilizing natural forests and plantations; it will not cover the extensive literature on fertilization of forest nursery soils and seedlings (Duryea, 1985; Duryea & Landis, 1984). Although many of the relationships between seedling and tree foliar analysis are similar, there are two problems with foliar analyses of trees: (i) the lack of clear relationships between soil chemical analysis and foliar analysis equivalent to those that exist for forest nursery crops and (ii) the inability to apply seedling-based foliar concentration standards to older trees. Closed canopy forests have nutrient supplies from recycling of litterfall that tend to mask conventional mineral soil/foliar analysis/growth rate relationships seen in young open canopy forests and horticultural crops.

Fertilizer response prediction has been difficult. This has been due to the absence of useful soil chemical fertility/nutrient uptake relationships. The result has been: (i) the development of extensive networks of cooperative, empirical fertilizer trials to generate a pattern of stand responsiveness and (ii) increased use of foliar analysis to monitor fertilizer uptake and help predict and explain stand response patterns (Gessel et al., 1979).

Most North American and Scandinavian research has dealt with N fertilization of established stands where N tends to be limiting due to slow mineralization of soil organic N. In the southern pine stands of the USA growing on coastal marine sands low in P, this element has attracted much testing and research.

In Australia, New Zealand, South Africa, and in other countries, during the establishment of extensive areas of tropical pine plantations or establishment of plantations on old fields or moorlands, there have been a myriad of nutritional problems requiring urgent resolution (Raupach, 1967). In contrast to the N and P problems seen in closed natural forests, these new plantations have experienced a wide range of macro- and micronutrient deficiency

 Soil Testing and Plant Analysis, 3rd ed.—SSSA Book Series, no. 3.

problems. Three recent reviews are particularly relevant. Lambert (1984) which reviews foliar analysis in fertilizer research, is largely based on work in Australia. Mead (1984) provides a more general review on diagnosis of nutrient deficiencies in plantations. Van den Burg (1984) provides major literature compilation on the use of foliar analysis for determination of tree nutrient status. He includes sampling procedures and critical foliar concentrations for broadleaves and conifers.

Leaf (1973), in the introduction to an earlier version of this chapter, pointed out that the objectives of fertilization may be many: increased tree growth, biotic and abiotic disease control, insect control, increased seed production, foliar color and retention in Christmas trees, increased sugar yield from sugar maple and resin yields from southern pine, improved recreational site development, range and wildlife browse values and alteration of wood properties. Of these many objectives, the large-scale testing and aerial fertilizer applications have been largely concentrated on improved tree volume production over large areas of established forests.

While recognition of deficiency problems in fruit trees is often made by color photographs of foliar symptoms, similar color photographs of visual deficiency symptoms in forest trees are rarely published; three useful sources are Hacskaylo et al. (1969) for hardwoods in Ohio, Binns et al. (1980) for conifers of British forests and Will (1985) for *Pinus radiata* and *Eucalyptus* spp. in New Zealand, European tree species are partially covered by Baule and Fricker (1970).

I. PLANT TISSUE USED IN TREE NUTRITION DIAGNOSIS

Several criteria for sample material have been discussed for evaluation of the nutrient status of forest trees (van den Burg, 1984).

1. The tissue chosen must provide a clear separation of trees of different nutrient status. The element in question must increase tree growth when increased in supply by fertilization. That is, a response relationship must be known through experimental treatments or rigorous selection of natural trees or stands for which the growth is related to the nutrient.
2. The tissue should also have a relatively stable nutrient concentration among trees at a specific time and over a selected sampling time.
3. The ease of sample collection will increase the techniques acceptance. The difficulty of obtaining representative samples from near the top of large trees restricts the use of foliage as an assay in forests.

The lack of a clear relationship between foliar analysis, by far the most frequently used tissue sample, and growth response to fertilization has led to exploration of twigs, phloem, xylem sap, and root material as assay tissue.

Friedland and Gregory (1985) investigated the chemistry of the xylem sapstream in *Picea abies, Abies balsamea, Acer saccharum,* and *Betula papyrifera* in relation to higher than normal atmospheric deposition of N at high elevations. Carter and Larsen (1965) made an analysis of xylem sap of *Pinus taeda* and found N fertilization increased total-, amino-, and amide-N levels. In a survey of 60 tree species, glutamine was reported to be the major amino acid of the xylem sap, accounted for as much as 80% of the total amino N (Barnes, 1962, 1963).

Stark et al. (1985) extracted xylem sap from *Pseudotsuga menziesii* and evaluated the influence of pressure, crown aspect, height in the crown, tree age, time of year and day, and soil chemistry on ion concentrations. Xylem sap extraction with a pressure chamber provided a good indication of the nutrient status of a species if the trees were of similar age, and if they were sampled at the same aspect, crown height, time of day, time of years, and on the same soil. Less time was required for sampling and analysis of xylem sap than for foliage.

Phloem (inner bark) may be an alternative to foliage sampling. Will (1965) reported increased concentration of P, K, and Mg in the foliage, bark, and sapwood after P fertilization of *Pinus radiata* on a P-deficient soil. The results suggested that bark may be more sensitive to P nutrition of the tree than foliage. Timmer (1979), studying *Picea glauca*, failed to find increased P in the bark although foliar P increased. The author concluded that bark P was not increased because tree growth was not limited by N, P, and K. There was a significant linear correlation between N in the bark and foliage in the season of fertilization. Timmer (1979) cites reports for other species that show good correlations between concentrations of N in bark and needles. Alcubilla and Rehfuess (1981) indicate that N, P, K, Ca, and Mn nutrition can be evaluated by phloem analysis with similar results and accuracy as with foliar analysis.

White et al. (1970) reported that N, K, Ca, Mg, Al, Mn, and Zn in the inner bark of *Pinus taeda* showed variation with the height of sample on the main stem and correlations between concentrations of N, K, Ca, Al, and Mn but not P in bark and needles were significant. Approximately the same number of trees were required to estimate the concentration of N, P, K, and Mg within 10% of the mean in bark as in needles. More sample trees were required for bark than for needles to estimate Ca, Al, Mn, Zn, Fe, and Cu. Late summer was the most stable period in which to sample inner bark of *Pinus taeda* (White et al., 1972). Cotrufo (1983) found that xylem showed sharp increases in N in annual rings of *Pinus taeda* corresponding to N fertilization. The increased N concentration in the xylem over time was similar to that for fresh needlefall in the fall, later reported for the same plots by Cotrufo and Wells (1984).

Miller and Miller (1976) suggested that N in needlefall may serve as an indicator of the N status of forest stands. Translocation of mobile elements just prior to senescence may produce needlefall sensitive to nutrient stresses of the trees (Powers, 1984); nutrients in short supply may be more readily translocated prior to senescence. Studies were conducted in *Pinus taeda* to test the sensitivity of a number of N assay methods (Cotrufo & Wells, 1984). Assay tests were rated for sensitivity in relation to the increase in concentration of an element as a result of fertilization. In November, total N in xylem and fresh needlefall was more sensitive than that in current needles. Needles and twigs had similar sensitivity and current tissues collected in September were more sensitive than 1-yr-old tissues. Soluble- and arginine-N appeared more sensitive than total N in needles and twigs; however, these fractions were highly variable and unpublished data of Cotrufo and Wells indicate soil drainage and P may cause extreme variation in soluble-N fractions in foliage and twig tissue, rendering them unacceptable for use as a routine N assay as also suggested by Powers (1984). Imposed flooding and drought had no effect on amino-N concentration in inner park (Hodges & Lorio, 1969). Nitrogen fertilization increased N concentration of fresh needlefall in the fall over several years which corresponds to the period of increased growth rate for *Pinus taeda* following fertilization (Cotrufo & Wells, 1984).

Foliar and twig analyses were employed to evaluate N, P, K, Ca, and Mg status on rehabilitated bauxite pits. Fully expanded leaves of the current year provided the most uniform nutrient levels among the plant organs of *Eucalyptus saligna* and *E. wandoo* and showed major differences between sites (Bell & Ward, 1984).

An indirect root bioassay for P nutrition has been reported by Harrison & Helliwell (1979). The method using the rate of 32P uptake by living roots from a standard solution was tested in 10- to 18-yr-old stands of *Picea sitchensis* and *Pinus contorta* (Dighton & Harrison, 1983). The 32P uptake has been shown to be negatively related to the growth and total P content of plants. Rates of 32P uptake by roots from 10- to 18-yr-old trees were negatively related to quantities of P fertilizer previously applied and to tree height. According to the authors, the investigation suggests that the bioassay detects P status in trees not identified from needle analysis.

The review by Leaf (1973) cites important early literature on sampling material. In addition to the use of foliage, twigs, xylem, phloem, and roots, the most frequently investigated bioassays, reports are cited in which buds, flowers, and cones may be investigated for specific diagnostic purposes. In the review, Leaf cited reports concerning the analysis of leaf blades and petioles together and separately. It was concluded that for *Liriodendron tulipifera*, blades only would serve as better diagnostic tissue than either blades with petioles or petioles alone. Data were presented for comparisons of leaflets of compound leaves of *Fraxinus americana* and it was suggested that the terminal leaflet might be suitable tissue for compound leaves.

II. SAMPLING POSITION FOR FOLIAGE

Leaf (1973) enumerated sampling position factors as (i) which tree in a stand to sample, (ii) where in the tree to sample, (iii) age of foliage, (iv) growth flush of foliage of multinodal species, and (v) branch order in relation to physiological activity. In a more recent review branch aspect, foliage exposure, presence of fruit and pathogens are also discussed (Turner et al., 1978) (Table 25-1).

Although there are numerous reports of nutrient variation due to position, there is a lack of information relating nutrient concentration of foliage

Table 25-1. Foliar nutrient variation in relation to position in the tree. (From Turner et al., 1978.)

Species	Trend with tree height: Increase toward top	Decrease toward top	Stable	Max. in midcrown	Min. in midcrown	Reference
Abies alba Mill.						
healthy		Ca	N	P	K, Mg	Nemec, 1940
dying	P	Ca, N	K, Mg			
Acer saccharum		N	P, K, Ca			Wallihan, 1944
Agathis australis	P	K, Ca, Mn, Fe	N	Na	S	Peterson, 1961
Cryptomeria japonica	N, K	Ca	Mg, P			Harada et al., 1972
Picea abies	N		P, K, Ca, Mg			Swan, 1962
P. abies			P, Ca, Mg	N	K	Strebel, 1961
P. glauca	Mg	K	N, P, Ca			Swan, 1962
P. mariana	N					Gagnon, 1964
P. mariana	N, Mg	K	P	Ca		Lowry, 1970
Pinus banksiana Lamb.			P, N, K, Ca, Mg			Swan, 1962
P. elliottii			P			Humphreys & Kelly, 1962
P. nigra	N, P, K, Ca, Mg, Na					Wright & Will, 1958
P. radiata	N, P, K, & needle wt					Hall & Raupach, 1963
P. radiata		Ca, K, P	N, Mg, Na			Will, 1957
P. resinosa	N	K	P, Ca, Mg			Swan, 1962
P. resinosa						
good site	Mg	K, Ca	N		P	Madgwick, 1964
poor site	N		P, K, Ca, Mg			
P. resinosa	K					White, 1954
P. strobus	K					White, 1954
P. sylvestris	N	K	P, Ca, Mg			Swan, 1962
P. sylvestris	N, P, Mg		K, Ca, Na			Wright & Will, 1958
Populus deltoides Bart.	N, P, K	Ca, Mg, ash				Carter & White, 1971
Pseudotsuga menziesii	N, P, K, Mg	Ca				Lavender & Carmichael, 1966
P. menziesii	N	P				Brackett, 1964
Taxus baccata L.		Ca, Mg			K, P	Nemec, 1947

by position to nutrient status determined by response to fertilization. Calcium and Mg frequently decrease with height in the crown and N and P show the opposite trend; however, there are exceptions to these broad generalizations.

Variation in nutrient concentration may be greater among crown positions than between trees (Slitchter, 1984). Van den Driessche (1974) discussed extent of crown closure, rate of growth, and nutrient availability as factors influencing nutrient concentrations in the crown positions. A rationale can be made for expression of a greater sensitivity of nutrition by the lower part of the crown in contrast to the upper crown as indicated for N, P, and K in *Pinus resinosa* (Madgwick, 1964). According to van den Driessche (1974), the apical region of a plant with higher auxin levels will probably maintain a relatively high nutrient concentration at the expense of the distal regions. Other workers suggest sampling upper crown sample positions to provide samples of uniform light exposure. Detailed sampling of shade and sun leaves from lower to upper crown on *Platanus occidentalis* indicated light exposure had a greater influence on N and P concentration than crown position (Slitchter, 1984).

Auchmoddy and Hammack (1975) studied N, P, K, Ca, Mg, Mn, and Fe by crown position in four *Quercus* spp. 60 yr of age and suggested that easier-to-reach foliage from the lower crown is equally well suited for nutrient sampling. However, Lavender and Carmichael (1966) sampled foliage from four crown levels in Douglas-fir trees growing on sites of widely different productivity and found that N levels in the upper crown best reflected site productivity.

Reviews of Turner et al. (1978) and van den Driessche (1974) conclude that no differences in nutrient concentrations for conifers or hardwoods could be found in relation to the cardinal directions. A recent study of *Eucalyptus wandoo* and *E. saligna* justifies a similar conclusion for those species (Bell & Ward, 1984). Turner et al. (1978) suggest that trees or branches with fruit may have foliage that differs in nutrient concentrations from foliage not associated with fruit.

The literature suggests there may be physiological reasons for sampling a specific part of the tree crown for nutrient bioassay. A standardized sampling procedure with precisely described sample positions is essential for comparison of nutrient concentrations of trees and stands, and for calibration of tree foliage nutrition status for fertilizer recommendations.

III. SAMPLING TIME, FOLIAGE AGE, AND ANNUAL VARIATION

Forestry literature contains more information about seasonal than year-to-year change in nutrient concentrations. Nitrogen is sensitive to moisture stress. Leaf in summarizing several reports of coniferous and deciduous forest trees states that the mobile elements (N, P, and K) concentrations in first year foliage show a maximum in summer, followed by a decline as the growing season progresses; the reverse is generally the case with nonmobile elements (Ca). Magnesium concentrations follow trends intermediate to the mobile and nonmobile elements. Van den Driessche (1974) lists three categories of elements in relation to seasonal stability in foliage of *Acer pseudoplatanus, Aesculus hippocastanum,* and *Fagus sylvatica*: (i) Co, Ni, Fe, Ti, Cr, Pb, and Al, showing a fall in leaf concentration early in the season; (ii) Mn, B, Si, Ca, Sr, Ba, and Mg, showing a continual rise; and (iii) Cu, Mo, Zn, P, K, and Na, showing a gradual fall followed by a period when the concentration remained fairly constant. Sulfur seems to follow a trend similar to N, P, and K (Hoyle, 1965).

Lea et al. (1979a, b) analyzed *Acer saccharum* and *Betula alleganiensis* mid-crown foliage at eight dates from June to October and found Ca, Al, Mn, and Fe increased with age; N, P, and K were relatively stable and decreased sharply in September to October; Na and Cu decreased and then stabilized; Mg, Zn, and Co increased or were relatively stable. Aluminum with a continuous increase was an exception from the categories of van den Driessche in which Al concentration decreased early in the season. *Populus tremuloides* (Verry & Timmons, 1976) showed similar seasonal trends to the two species reported by Lea et al. (1979a); however, the sharp decreases in concentration in N, P, and K in the fall were not found, and Cu and B elements not determined by Lea et al. (1979b) were relatively stable. Day and Monk (1977) reported seasonal trends in concentration of N, P, K, Ca, Mg, and Na in foliage from five Southern Appalachian species. The N concentration of all species increased from October to November, a departure from trends in most other reports.

Variation during the year may be caused by the environment, leaf and tree growth, nutrient accumulation, redistribution within the tree and foliage leaching (Turner et al., 1978). When expressed as contents per leaf, the mobile elements increase during the first part of the growing season and decrease during the latter part; whereas, Ca tends to stabilize in the later part of the season or even increase throughout the growing season. Comparisons of N and P concentrations and content in sun and shade leaves of 5-yr-old *Platanus occidentalis* showed stable N or P concentrations at the end of the growing season for a site with low N or P foliage concentrations but declined sharply when N and P were higher (Slitchter, 1984). The difference for site was slightly greater for concentration than for content and shade leaves showed the same trend as sun leaves. The results show greater site differences in N and P concentrations at the end of the growing season than earlier but there was no stable period except for P on the site low in foliar P.

The previous discussion of seasonal nutrient element trends relates mostly to deciduous trees; however, similar trends are also reported for conifer trees. Possibly the more important consideration in sampling foliage of most coniferous and a few broad leaf species is the age of foliage in years and seasonal trends in foliage 1-yr-old and older. The reports of Mead and Will (1976) with *Pinus radiata*, Mead and Pritchett (1974) for *Pinus elliottii*, Powers (1984) for *Pinus ponderosa*, Wells and Metz (1963) for *Pinus taeda*, Lowry and Avard (1968, 1969) for *Picea mariana* and *Pinus banksiana* are examples of methodology investigations for foliage sampling of species that hold their leaves for more than 1 yr. Samples were collected seasonally from several foliar age classes on several trees from more than one site to determine when element concentrations are greatest between sites and when there is a minimal change in nutrient concentrations in a time period.

Pinus elliottii (Mead & Pritchett, 1974); *Pinus taeda* (Wells & Metz, 1963); and *Pinus radiata* (Mead & Will, 1976; Knight, 1978) foliage display concentration trends for N, P, and K that decrease during the first summer, increase in the winter and then decrease in the second summer.

Mead and Will (1976) found interactions between season and site were significant for all elements investigated, N, P, K, Ca, and Mg. Other studies also indicate interactions of season and element concentration of current and 1-yr-old foliage, which indicates seasonal changes do not follow the same pattern on all sites.

Tree species that retain their foliage more than 2 yr generally show foliage concentration peaks for N, P, and K in the first or second year followed by decreases of as much as 50% for 3- to 6-yr-old foliage (Lavender & Carmichael, 1966; Powers, 1984; Lowry, 1970).

Variation in nutrient concentrations has also been reported for different flushes within 1 yr for *Pinus taeda* (Wells, 1969). Concentration of N, P, and K increased from the first to third flush when all samples were collected at the same time during the first winter after initiation.

Annual variation in tissue of the same age sampled the same month may be expected because climatic factors are variable (Schultz, 1976). Nitrogen, P, K, Ca, and Mg of *Pinus resinosa* foliage showed significant differences between years when sampled over a 6-yr period (Leaf, 1973; Leaf et al., 1970). The highest N, P, and K levels were not in the same year and in some years the concentration of K fluctuated above and below the critical level. Most of the lower concentrations of elements were found in the years with least precipitation. Miller (1966) examined variation in N, P, and K concentration in foliage of *Pinus taeda* in relation to climatic variables and found temperature and precipitation influenced concentrations; however, Wells (1969) found variation in winter temperature had no effect on N, P, and K concentrations of the same species. In other work in 7-yr-old *Pinus taeda*, K was stable over a 5-yr period and N showed a general increase in concentration, which indicated a change in soil N availability (Wells, 1979). Leaf (1973), van den Driessche (1974), and Turner et al. (1978) (Table 25–2) cite other literature indicating annual nutrient variation in relation to climatic and environmental factors. At least some of the seasonal pattern of nutrient concentrations is a function of change in foliage mass with increasing age during at least the first 18 mo for perennial foliage coupled with differential susceptibility in leaching from foliage tissues. Nutrients such as N, P, and K which are associated with protoplasm are readily leached, whereas, Ca, a major component of the middle lamella, is not.

Investigations of several species indicate that there are within species genetic differences in tissue nutrient concentration (Knight, 1978; Turner et al., 1978). In most forest stands, genetic effects are part of the random variation; however, when rigid tissue sampling, analysis, and calibration methods are developed, influences of family, clonal, and provenance should be considered.

IV. SAMPLE PREPARATION AND ANALYSIS

Data from plant analysis are used to diagnose possible treatment needs or responses, thus contributing inputs into management decisions. To be useful, plant analysis is conducted on several samples of a population and a comparison of mean values is used to judge treatment potential or effectiveness. A comparison of means reflects sample variance. This variance may be twofold: (i) natural population variation and (ii) artifact or induced variation related to sampling errors. Plant analysis techniques need to: (i) measure natural variation that has not already been eliminated by the selection of tissues for analysis; and (ii) minimize extraneous variation that may be additive to the natural population variation, or may tend to mask this natural variation.

Induced variation may occur from several sources involving: (i) the condition of the plant tissues sampled (e.g., air-borne contamination on the tissue surfaces, insect/or disease damage to tissues) and (ii) the handling of the plant tissues from field sampling through analytical preparation, e.g., sample transporting conditions from the field to the laboratory, drying and grinding techniques, and subsampling techniques. Natural variation, on the other hand, is associated with the characteristics of the species such as genetic heterogeneity, individual plant characteristics, stage of tree development, age of tree, and the total environment.

Since overall sample variance cannot be separated into natural and induced variation, considerable care must be taken in all stages of the field and laboratory aspects of plant analysis to minimize extraneous variation. If such care is taken, it is assumed that the reported estimates of variance are associated with natural population variation. Modern laboratory analytical instrumentation (e.g., absorption and emission spectrographic instruments) has reached a level of precision of analysis of various nutrient elements that exceeds the quality of field sampling and laboratory preparation techniques. Thus, induced variation with field sampling, and laboratory preparation and subsampling outweigh that variation in the chemistry of analyses. The determination of elemental concentration levels in prepared plant tissues has quality control that is absent in the field.

The number of samples at predetermined levels of probability and precision of mean estimates that are considered acceptable limits are related to the magnitude of sample variance. Table 25–3 presents calculations of numbers of samples required at various probability levels for Douglas-fir stands (Ballard, 1985). Van den Driessche (1974) and Richter (1983) have also attempted to estimate appropriate levels of precision and confidence for various foliar nutrients in British Columbia conifers. Ballard and Carter (1983)

Table 25-2. Foliar nutrient changes during the growing season. (From Turner et al., 1978.)

Species	Form of expression	Changes during growing season: Increase	Decrease	Stable	Reference
Abies amabilis	%	Ca, Mn	N, K	P, Mg	Turner, present study
Acer platanoides L.	%	Ca, Mn, Fe	N, K	P, Mg, Cu, B	Davidson, 1960
A. pseudoplatanus L.	% & mg/leaf	Ca, Mn, B, Si,	P	Mo, K	Guha & Mitchell, 1966
	%	Sr, Sa, Mg, Fe, Pb (Co, Ni, V, Ti,Cr, Al, Zn, Na either peaked or decreased in the growing season)			
A. rubrum L.	%		K		Walker, 1955
Aesculus hippocastanum L.	% & mg/leaf	Ca, Fe, V, Ti, Cr, Pb, Al, B, Si, Co, Ba, Ma (Mn & Zn either peaked or decreased)	Cu, P, K	Sr, Ni, Mg, Mo	Guha & Mitchell, 1966
Alnus rubra Bong.	%	Ca, Mn	N, P, K	Mg	Turner, present study
Betula Britton	mg/leaf	Ca, Mg, K, N, S		P	Hoyle, 1965
	%	Ca	N, P, K, S		
B. verrucosa Ehrh.	%	Ca	N, P, K		Tamm, 1951
Crataegus phaenopyrum	%	Ca, Mn, Fe, Mg	N, P, K		Cannon et al., 1960
Euonymus alatus Maxim	%	Ca, Mn, Fe	N, K	P, B, Cu, Mg	Davidson, 1960
E. fortunei Maxim	%	Ca, Mn, Fe	N, K	P, B, Cu, Mg	Davidson, 1960
Fagus sp. L.	%	Ca	N, P, K		Henry, 1908
Fagus sylvatica L.	% & mg/leaf	Ca, Fe,V, Pb, Al, Si, Co, Sr, Na (Ti, Cr, B, Ba, Mo either peaked or decreased)		Ni, Mn, Mg, P, Zn, K	Guha & Mitchell, 1966
F. sylvatica	%	Ca, Mn, Fe, Mg	N, P, K		Olsen, 1948
	mg/leaf	Ca, Mn, Mg, Fe, N, P, K			
F. grandifolia Ehrhart	% & mg/leaf	Ca			Chandler, 1939
Gleditsia triacanthos	%	Ca, Mn, Gm, Fe	N, P, K		Cannon et al., 1960
G. triacanthos	%	Ca, Mn, Fe	N, K	P, B, Cu, Mg	Davidson, 1960
Juniperus chinensis L.	%	Ca, Mn, Fe	N, K	P, B, Cu, Mg	Davidson, 1960
J. virginiana L.	% & mg/leaf	Ca			Chandler, 1939
Larix sp. Adans.	%	Ca	N, P, K		Henry, 1908
Magnolia acuminata L.	% & mg/leaf	Ca			Chandler, 1939
Picea sp.	%	K	P	N, Ca	Touzet et al., 1969
Picea sp. A. Dietr.	%	Ca	N, P, K		Henry, 1908

P. abies	%	Ca, K	N, P		Touzet et al., 1969
P. mariana	%	Ca, Mg	N, K	P	Lowry & Avard, 1965
	mg/leaf	Ca, Mg, K	P	N	
Pinus radiata	%			B	Windsor & Kelly, 1972
P. radiata	%	Ca			Will, 1957
P. radiata	%	N, S (total), S (organic)	SO_4		Kelly & Lambert, 1972
P. radiata	%	N, P, K			Raupach, 1967
P. resinosa	%		N, K	P	White, 1954
P. strobus	% & mg/leaf	Ca			Chandler, 1939
P. strobus	%	Ca		N, P, K	Touzet et al., 1969
P. strobus	%		K		Walker, 1955
P. sylvestris	%	Ca	N, K	P	Tamm, 1955
P. taeda	%	Ca, Mg	N, K	P	Wells & Metz, 1963
	mg/leaf	Ca, Mg		N, P, K	
P. taeda	%		K	N, P	Miller, 1966
P. virginiana Mill.	%		K		Walker, 1955
Populus tremuloides Michx.	% & mg/leaf	Ca	N, P, K		Chandler, 1939
Pseudotsuga menziesii	%	Ca, Mg, N, P, K			Lavender & Carmichael, 1966
P. menziesii	%	Ca, K	P	N	Lavender, 1970
P. menziesii	%	Ca, K		P, N	Turner, present study
Quercus sp. L.	%	Ca	N, P, K		Henry, 1908
Quercus sp.	%	Ca	N, P, K	Mg	McVickar, 1949
Q. gambelii Newb.	mg/leaf	Ca, K, P			Sampson & Samisch, 1935
	mg/in.^{2}	Ca, K, P			
Q. kelloggii Newb.	mg/leaf	Ca, K, P			Sampson & Samisch, 1935
	mg/in.^{2}	Ca	K, P		
Q. palustris Moench.	%	Ca, Mn, K, Fe, Mg	N, P		Cannon et al., 1960
Syringa vulgaris L.	%	Ca, Fe, Mn	N, K	P, B, Cu, Mg	Davidson, 1960
Taxus cuspidata Sieb. et Zucc.	%	Ca, Fe, Mn	N, K	P, B, Cu, Mg	Davidson, 1960
T. media L.	%	Ca, K, Mn, Mg	N, P, B, Fe		Boonstra et al., 1957
Ulmus pumila L.	%		N		Chapman, 1941
Hardwoods	%	Ca, Si	N, P, K		McHargue & Roy, 1933
Acer, Betula	% & mg/m^2	Ca, Mg, Na	Si		Likens & Bormann, 1970
Fagus	% & mg/m^2		Ca, Mg, K		Likens & Bormann, 1970

Table 25–3. Coefficients of variation (CV) and implications for foliar sampling intensity in Douglas-fir stands. (From Ballard, 1985.)

Nutrient	No. of stands sampled	Avg. CV	95th percentile CV†	k‡ where allowable error is: 10%	20%	30%	50%
N	10	10.2	15.3	11	3	1	1
P	10	17.4	38.5	68	17	8	3
K	10	17.6	30.4	42	11	5	2
Ca	10	25.9	36.1	60	15	7	2
Mg	10	17.1	25.4	30	8	3	1
S	10	11.8	20.3	19	5	2	1
Mn	10	34.3	49.8	114	29	13	5
Zn	10	39.9	66.8	205	51	23	8
B	10	25.6	43.8	88	22	10	4
aFe§	10	27.8	46.6	99	25	11	4

† 95% of sampled stands are estimated to have a CV of this magnitude or less.
‡ k = the number of samples needed to estimate the mean within the specified allowable error, with 95% confidence (i.e., $\alpha = 0.05$), in 95% of sampled stands.
§ aFe = the physiologically active Fe fraction.

found that for routine forest stand nutrient status evaluation in British Columbia, foliage sampling of 20 trees per stand was likely to be adequate in nearly all cases.

Forest scientists have drawn more information on tissue preparation and analytical techniques in the laboratory from colleagues in agriculture than any other phase of plant analysis work. The only important difference between agricultural and forestry samples is that analytical values for many nutrient elements are considerably lower in the latter. For this reason, it is often necessary to use larger samples of tree tissues to ensure that the amounts of nutrients present are large enough for satisfactory measurement.

One aspect to note on this subject is a summary of the compilation by the working group on international comparison of methods of plant anlaysis (IUFRO working party $1.02–08 foliar analysis). In 1985, three foliage samples (one wood, one bark, and one litter) were sent to 29 laboratories for analysis. Table 25–4 presents the results of the analyses. The range of values for between-laboratory differences in analytical results were unexpectedly large.

Will (1986) has suggested the following measures would go a long way to eliminate most inaccuracies.

1. Check mathematical calculations, especially when dilutions are different from standard procedures. More than 1% of results were an order of magnitude less than, or greater than, the mean or median.

2. Develop a reliable system of handling samples, recording, and transcribing results. In at least one case it seemed that, while results for some elements had been recorded correctly, others had been interchanged between samples. One set of hand-written results was received in a barely legible form.

3. Do not develope a sense of false security because within-laboratory *repeatability* is good—this may bear no relationship to *accuracy*. A low SD within the laboratory may mean only that your errors are consistent!

Table 25–4. Summary of results of analyses from different laboratories on IUFRO 1985 standard samples (IUFRO working group S1.02-08 foliar analysis). (MAD = mean absolute deviation.) (From Will, 1986.)

	Element								
	N	P	K	Ca	Mg	Fe	Mn	Cu	Zn
	%					mg kg^{-1}			
	85/1 Liriodendron tulipifera foliage								
Range	1.50–1.86	0.108–0.140	0.94–1.67	0.80–1.78	0.26–0.45	49–364	196–4450	5–53	13–18
Mean ± SD	1.72 ± 0.09	0.128 ± 0.007	1.45 ± 0.13	1.23 ± 0.18	0.34 ± 0.03	89 ± 67	392 ± 56	46 ± 10	15 ± 2
Median, MAD	1.74, 0.04	0.130, 0.005	1.46, 0.06	1.22, 0.06	0.33, 0.01	78, 6.5	410, 18	47.5, 1.5	15, 1
n	29	27	27	27	27	20	21	18	20
	85/2 Douglas-fir foliage								
Range	0.98–1.27	0.096–0.117	0.52–0.64	0.32–0.47	0.09–0.14	15–115	203–480	3–28	16–25
Mean ± SD	1.16 ± 0.06	0.105 ± 0.006	0.58 ± 0.03	0.38 ± 0.03	0.11 ± 0.01	40 ± 20	411 ± 58	23 ± 5	20 ± 2
Median, MAD	1.18, 0.03	0.105, 0.005	0.58, 0.03	0.38, 0.03	0.11, 0.01	37, 4	425, 18	24, 1	20, 1
n	29	27	27	27	27	20	21	18	20
	85/3 *Pinus radiata* foliage								
Range	1.10–1.44	0.101–0.141	0.38–0.60	0.16–0.26	0.08–0.13	61–141	276–678	4–25	26–38
Mean ± SD	1.34 ±0.08	0.124 ± 0.008	0.53 ± 0.05	0.18 ± 0.02	0.10 ± 0.01	99 ±20	587 ± 87	20 ± 4	34 ± 3
Median, MAD	1.36, 004	0.126, 0.005	0.54, 0.03	0.18, 0.01	0.09, 0.01	100, 10	620, 25	21, 2	34, 2
n	29	27	27	27	27	20	21	18	20

(continued on next page)

Table 25-4. Continued.

	Element								
	N	P	K	Ca	Mg	Fe	Mn	Cu	Zn
	%					mg kg^{-1}			
				85/4 *Pinus radiata* wood					
Range	0.07–0.50	0.010–0.120	0.06–0.43	0.02–0.35	0.02–0.24	8–249	90–292	0.20–5.00	7–53
Mean ± SD	0.12 ± 0.08	0.026 ± 0.03	0.12 ± 0.06	0.06 ± 0.06	0.04 ± 0.04	40 ± 56	116 ± 41	1.7 ± 1	14 ± 10
Median, MAD	0.10, 0.01	0.013, 0.003	0.11, 0.01	0.05, 0	0.03, 0.01	19, 5	109, 4	1.5, 0.2	12, 1
n	29	27	27	27	27	19	21	17	20
				85/5 *Pinus radiata* bark					
Range	0.03–0.42	0.029–0.045	0.25–0.39	0.11–0.25	0.06–0.63	49–109	65–180	1.6–11	30–41
Mean ± SD	0.33 ± 0.07	0.038 ± 0.03	0.34 ± 0.03	0.20 ± 0.03	0.09 ± 0.11	74 ± 16	154 ± 25	7.9 ± 2.0	33 ± 3
Median, MAD	0.34, 0.01	0.039, 0.001	0.34, 0.02	0.20, 0.10	0.07, 0.01	73, 9.5	161, 6	8.3, 0.8	33, 1.5
n	29	27	27	27	27	20	21	18	20
				85/6 *Pinus radiata* litter					
Range	0.22–0.80	0.010–0.120	0.19–0.47	0.14–0.47	0.02–0.66	51–136	51–685	2.0–6.8	13–54
Mean ± SD	0.65 ± 0.10	0.105 ± 0.020	0.40 ± 0.05	0.37 ± 0.06	0.08 ± 0.12	77 ± 17	573 ± 130	5.0 ± 1.3	45 ± 8
Median, MAD	0.67, 0.03	0.110, 0.004	0.40, 0.02	0.37, 0.02	0.07, 0.01	78, 9	611, 42	5.2, 0.4	47, 2
n	29	27	27	27	27	20	21	18	20

4. Each laboratory should have a standard reference sample that is analyzed with every batch or every day. When results on that sample differ by more than a set of percentage from its standard value, all results for that batch/day should be disregarded and the samples analyzed again. Acceptable variation should be <5% for some elements (e.g., N and P) and <10% for others. While it is not possible to include a National Bureau of Standards (NBS) or International Union Forest Research Organization (IUFRO) sample with every batch, every laboratory should be able, at intervals, to use such a standard sample to cross calibrate an internal bulk sample which is used routinely.

5. Above all, join a routine check program, and encourage others to join so that interlaboratory checking becomes universally accepted. It would be advantageous to have the IUFRO interlaboratory comparison continued and perhaps expanded. If it is continued, should results continue to be reported anonymously?

6. Encourage editors of journals associated with IUFRO and its member organizations to question/reject papers whose authors cannot show evidence that the chemical results are from a laboratory of proven performance. Geologists have "standard" rocks which are used in publications as reference samples. Should we recognize reference foliage, wood, bark, and litter samples, and encourage the publication of results of analyses on them in each paper concerned with the chemical analysis of forest material?

A recent publication (Nicholson, 1984) provides information that may be useful in the preparation and analysis of foliage samples.

V. FIELD AND LABORATORY PROCEDURES FOR FOLIAGE SAMPLING

It is important that procedures match, as closely as possible, those used in establishing the interpretive information used in evaluation. The following procedures from Ballard and Carter (1983) are given as an example. They have been recommended for conifer foliage in British Columbia; other procedures may be required depending on the species, location, and objectives of the sampling.

A. Foliar Sampling Guidelines

Unbiased sampling that fairly represents the whole stand is desirable, but the method of approximating such sampling must be reasonably simple and efficient for routine operational use. The following procedure is suggested.

For a stand where stratified sampling is unnecessary, set up two or more well-separated transect lines that cross the stand and also cross any major topographic features. After excluding portions of the lines, by using guideline 4 (below), establish equally spaced reference points on these lines, corresponding to the number of trees to be sampled. From each reference point, select the nearest tree that meets guidelines 2 and 3 (below).

For stratified sampling, where a stratum occurs as numerous small "islands" within the stand, use the same transect method to identify islands and trees to be sampled. If a stratum occurs as one large island, use the above transect approach within the island, to identify individual trees for sampling. (This amounts to sampling each stratum as if it were an individual stand.) If a stratum occurs as two or more large islands, it is appropriate to use one transect line in each of two (or more) of the largest islands.

1. Sample conifer foliage during the dormant season, preferably between 15 September and 15 December. (Larch should be sampled before foliage color change. However, interpretive criteria for western larch data have not been found.)

2. Confine sampling to dominant and codominant trees.

3. Avoid trees likely to yield poor samples because of heavy cone production, insect, or disease damage. If such avoidance is impractical, these factors should be taken into account in laboratory analysis and data interpretation. Analysis of individual (rather than composite) samples is recommended if sample trees vary with regard to these factors.

4. Avoid sampling foliage near unpaved roads or in other situations that may lead to foliage contamination from dust. This is particularly important on well-drained calcareous soils, where Fe deficiency diagnosis is needed.

5. Estimate the distance from top to bottom of the live crown. Collect samples at about one-fourth to one-half of that distance from the top. If composite samples are to be prepared, sample two branch-ends per tree and collect a 2-g (fresh mass) subsample of the current year's foliage (terminals or laterals) from each branch. This gives 4 g per tree and, if 15 trees are sampled, 60 g per stand. If individual trees are to be analyzed, a minimum of 25 g (fresh mass) should be collected for each tree.

Color and size criteria are likely the easiest to apply when current and previous year's foliage are being separated. The color of foliage is often different for different foliage ages, and can vary from stand to stand. Current year's foliage is often a somewhat paler or yellower green. Twigs are invariably darker on previous year's than current year's shoots, as are fascicle sheaths in pines. Twig diameter is usually larger in terminal shoots that developed during the previous year; and the terminal buds of such shoots, if they have failed to flush, are often larger in diameter and somewhat longer than current-year buds. Care should also be taken to avoid sampling lammas or proleptic growth.

6. Usually it is best to put samples into labelled plastic bags, although samples collected under dry conditions can be put into labelled paper bags. Dirty samples should be discarded and replaced.

7. Drying of samples should begin as soon as possible, preferably on the day of sampling. However if samples are kept cool (1–5 °C), drying may be deferred as much as a week and perhaps longer. It may be easiest to ship samples directly to the analytical laboratory if prompt transport and handling

can be arranged. If samples must be dried before shipment, oven drying is best. If no oven-drying facilities are available, thorough air drying should be done.

B. Laboratory Procedures

If the analytical data are to be used for comparing foliar nutrient status between stands, fertilizer, or other silvicultural treatments, or if the foliage appears stunted, the mass of 100 needles (or needle bundles, in the case of pines) or the mass of needles per shoot should be determined for samples oven dried at 70 °C. If foliar data are not to be used for the above purposes or the foliage does not appear stunted, the cost of this determination may exceed the value of the information.

Dry foliage should be ground for chemical analysis. If a needle snaps cleanly in two when bent, it is dry enough for efficient grinding. Samples are most conveniently ground in a Waring blender stainless steel cup or Braun-type KSM-2 coffee grinder. The largest particle dimension should be no more than about 1 mm after grinding, and there should be no obvious separation of tissues if the tissues do separate into fibrous and powdery fractions, chopping rather than grinding of foliage should be done. After grinding, samples should be re-dried at 70 °C before storage in air-tight, screw-cap plastic bottles. The total oven-drying time for a sample ideally should be no more than about 12 h, but as much as 20 h of oven drying might not produce any detectable differences in analytical results.

The following analyses are recommended for foliage samples.

1. Total N, P, K, Ca, Mg, Fe, Mn, Cu, and Zn
2. Total S (Sulfate-S optional)
3. Total B
4. Active Fe (Zeche, 1970)
5. Total Mo

VI. DATA ANALYSIS AND INTERPRETATION

Leaf (1973) has pointed out that:

1. The complexity of interpretation of plant analysis data and subsequent extrapolation of interpretations is crucial when employing such analyses as diagnostic techniques to aid in formulating forest fertilization programs.

2. Such programs must strive for optimum effectiveness by providing the responses sought in the managerial objectives, and must be of minimum wastefulness to environmental quality and minimum monetary expenditure.

3. Plant analysis is a potentially powerful diagnostic technique when used with other appraisal criteria in the quantitative assessment of a forest site to ascertain the likelihood of a successful fertilization program.

4. Plant analysis may serve to contribute information on: (i) the determination of nutrient element limitations, if any, controlling the quantity or quality of forest vegetation, rather than some other environmental factors, physical or biotic in nature; (ii) the specific nutrient element limitation(s); and (iii) the appropriate or efficient levels of the nutrient element(s) required to correct the limitation(s). An assessment of the quality of a site on the basis of plant analysis alone is possible only to a limited extent.

5. For each nutrient element value, information is required on the tree species involved, seed source or genetic heterogeneity, physical environment of the trees, biotic environment of the trees, age and stage of development of the trees, management objectives, and any conditions, as discussed above, under which the plant's sampling and analyses were conducted.

Four approaches have been used to interpret foliar analysis data from forest stands.

A. Critical Values

This is the oldest approach based on the assumption that the values being examined, when compared with previously published critical levels for the tree species in question will result in response to fertilizer addition if values are close to critical.

Such critical values are assumed to be stable for the species and not strongly influenced by tree size or age; it is also assumed that the values are firmly based on field experimentation for foliar equivalent physiological condition. If these assumptions are met, then this method is considered to be of particular value in screening values from several untreated stands. In addition, critical values overcome some of the problems of: (i) variation in optimal ratios, (ii) lack of precision in sufficiency ranges, (iii) the requirement of field trials, (iv) the limitation to influences only on deficiency from probability functions, (v) the absence of norms for the Diagnosis and Recommendation Integrated System (DRIS) technique. However, deficiency values are often broad and lack the required field verification (Powers, 1984).

Visual symptoms of nutrient deficiencies, previously mentioned, tend to be associated with critical values (Table 25–5, 25–6, 25–7, and 25–9), but may be manifested at concentrations below critical values.

Published critical values for tree species tend to be regional or national; a complete data set is impossible in a review. Their value and reliability depends upon the amount of testing behind each value; this is often not indicated. Leaf (1973) summarized representative deficiency, low, medium, and high concentrations in various tree species. Surveys of typical foliar nutrient concentrations by species (Kimmins et al., 1985) are of little diagnostic value. Cumulative probability distributions or percentile classes for foliar nutrient concentration data such as those prepared by Zinke and Stangenberger (1979) for ponderosa pine and Douglas-fir are of more value by allowing a ranking to be done. A computerized diagnostic program has been prepared by Ballard and Carter (1983) for Pacific Northwest tree species which

Table 25–5. Interpretations of foliar macronutrient concentrations. (From Ballard and Carter, 1986.)

		Foliar concentration (%, dry mass basis)				
Element	Interpretation	Douglas-fir	Lodgepole pine	Western hemlock	White spruce	Western redcedar
N	Very severely deficient	0.00	0.00	0.00	0.00	0.00
	Severely deficient	1.05	1.05	0.95	1.05	1.15
	Slight moderate deficiency	1.30	1.20	1.20	1.30	1.50
	Adequate	1.45	1.55	1.45	1.55	1.65
P	Severely deficient	0.00	0.00	0.00	0.00	0.00
	Moderately deficient	0.08	0.09	0.11	0.10	0.10
	Slightly deficient	0.10	0.12	0.15	0.14	0.13
	Adequate	0.15	0.15	0.35	0.16	0.16
K	Very severely deficient	0.00	0.00	0.00	0.00	0.00
	Moderate-severe deficiency	0.35	0.35	0.40	0.25	0.35
	Slightly moderately deficient	0.45	0.40	0.45	0.30	0.40
	Possibly slightly deficient	0.80	0.55	0.80	0.50	0.85
Ca	Severely deficient	0.00	0.00	0.00	0.00	0.00
	Moderate-severely deficient	0.10	0.05	0.05	0.07	0.07
	Possible slight-moderate deficiency	0.15	0.06	0.06	0.10	0.10
	Little, if any deficiency	0.20	0.08	0.08	0.15	0.20
	No deficiency	0.25	0.10	0.10	0.20	0.25
Mg	Severely deficient	0.00	0.00	0.00	0.00	0.00
	Moderate-severely deficient	0.06	0.06	0.06	0.05	0.05
	Possible slight-moderate deficiency	0.08	0.07	0.07	0.06	0.06
	Little, if any deficiency	0.10	0.09	0.09	0.10	0.12
	No deficiency	0.12	0.10	0.10	0.12	0.14

compares results with published data and suggests interpretations. Table 25–5 provides examples of such interpretations for Pacific Northwest tree species.

B. Nutrient Ratios and Nutrient Flux Densities

Ingestad, working in Sweden, has examined the minimum mineral nutrient requirements for a defined maximum production in a range of plant species. He used a method of maintaining cultural tree seedling nutrient status at a constant level by adjustment of the supply of nutrients. He used supply solutions which contained nutrient substances in the proportions desired in the plant. Experiments had shown that approximately the same proportions are required by pine, spruce, and a number of other species and varied insignificantly with plant age (Ingestad, 1967; Morrison, 1974). Table 25–8 gives these proportions.

These ratios have been used to assess tree nutrient status and examine the effects of fertilization (Weetman, 1968) and appear to have diagnostic value. In more recent studies (Ingestad et al., 1981; Ingestad & Kahr, 1985), by varying the supply of balanced solutions by different relative addition rates almost total control of conifer nutrition and growth was obtained. This

Table 25-6. Interpretations of foliar micronutrient concentration.† (From Ballard and Carter, 1986.)

Element	Foliar concentrations (mg kg^{-1}, dry mass basis)	Interpretation
Mn	0	Severe deficiency
	4	Probable deficiency
	15	Possible deficiency or near-deficiency
	25	No deficiency
Fe	0	Deficiency likely
	25	Possible deficiency
	50	Low to zero probability of deficiency
Active Fe	0	Deficiency likely
	30	Deficiency unlikely
Zn	0	Probable deficiency
	10	Possible deficiency
	15	No deficiency
Cu	0	Probable deficiency
	1	Possible moderate deficiency
	2	Possible somewhat deficient
	2.6	Slight possibility of deficiency
	4	No deficiency
B	0	Deficiency likely
	10	B Possibly deficient; Possible NID‡
	15	B Probably not deficient; If N <1.5, then NID‡ possible If N >1.5, then NID‡ unlikely
	20	No deficiency
Mo	0	May be deficient
	0.1	Not deficient

† These interpretations are not yet species-specific.
‡ N = N concentration, in percent; NID = deficiency inducible by N fertilizer application.

had led to the conclusion that "the bulk of the information on mineral nutrition of plants deals with supra-optimum problems or emanates from inadequate experiments. Thus suboptimum nutrition, which is the problem of interest in natural vegetation and plant husbandry, has apparently not been produced because of the varied concentration as such but because the concentration supplied was not maintained."

It is argued that the nutrient flux during the period of fast growth in the field is the key variable for understanding and predicting forest productivity. From basic physiological nutrition-growth relationships, it is concluded that the same nutrient concentration within plants, of the same or different species, does not necessarily mean the same relative growth rates. This emphasized the importance of great care when interpreting data based on foliar diagnosis (Axelsson, 1981).

Formal test of this concept with seasonal irrigation of a closed Scots pine forest has led to the development of a nutrient flux density as the driving variable for nutrition under field conditions. Using N, maximum tree size at a stationary state occurs when the system is saturated with N and at optimum N concentration in the needle biomass.

Table 25–7. Visual symptoms of nutrient deficiencies in conifers (condensed mostly from literature reviews of Morrison, 1974; Kolari, 1979, and Mengel and Kirby, 1942). (From Ballard and Carter, 1986.)

Deficient element	Visual symptoms
N	Poor growth rate, often with shortened vegetative growth stage. Needles are often stunted with evenly distributed chlorosis generally occurring first in the older needles. Purple tipping and necrosis occur only at a rather late and severe stage of deficiency.
P	Symptoms are primarily in the older needles. Youngest needles green or yellow-green; older needles distinctly purple tinged; purple deepens with severity of deficiency; in very severe cases in seedlings, all needles may be purple.
K	Variable symptoms: often no immediate visible symptoms. Needles usually short, chlorotic, with some green near base and, in some severe cases, purpling and necrosis with top die-back; or purpling or browning or necrosis, sometimes with oldest needles most affected.
Ca	Symptoms first observed in the growing tips and youngest needles. General chlorosis followed by necrosis of needles, especially at branch tips; in severe cases deficiency may result in death of the terminal bud and top dieback; resin exudation may also occur.
Mg	Deficiencies begin in older and move to younger needles. Yellow tipping, bronzing or banding of needles, needle tip necrosis in severe cases.
S	General chlorosis of foliage.
Fe	Diffuse chlorosis (in mild cases, more evident in young foliage) with bright yellow discoloration in more severe cases. Retarded leader growth, shoot dieback, and bud disorders. Young needles are often short, twisted, and necrotic. Roots are often sparse and reddish brown in color.
Mn	Variable symptoms: needles may be slightly chlorotic; in severe cases some necrosis of needles maybe evident.
B	Shoot, tip, or bud dieback generally occurring late in growing season resulting in a proliferation of secondary shoots, retarded height growth and bushiness. Leader often turns reddish brown prior to dieback. Crooked, bent, distorted leaders, often enlarged, brown or necrotic pith. Variable foliar symptoms, if evident generally occur as chlorosis, necrosis, or bronzing at needle tips.
Zn	Extreme stunting of trees with shortening of often erect and thin branches. Chlorosis (sometimes with bronzed needle tips), premature needle fall with resultant turfting of current needles; in severe cases top-dieback and rosetting.
Cu	Crooked, distorted, downward bent, pendulous leaders. Needles twisted spirally, yellowed or bronzed; tipburn or necrosis of needle tips evident; in severe cases young shoots twisted or bent.
Mo	Needles occasionally long and blue-green in color, more often chlorotic followed by necrosis of tissue, beginning at needle tip and eventually covering whole needle.

This new approach to plant tissue analysis suggests that rather than use conventional single nutrient addition experiments to search for maximum forest fertilization response, attempts should be made to maintain optimum foliar proportions by varying the N addition rate of balanced nutrient supplies.

Table 25-8. Mineral nutrient ratios for seedling tissue (Ingestad, 1979).† (From Landis, 1984.)

Macronutrients	Douglas-fir	Sitka spruce	Western hemlock	Scots pine
N	1.00	1.00	1.00	1.00
P	0.30	0.16	0.16	0.14
K	0.50	0.55	0.70	0.45
Ca	0.04	0.04	0.08	0.06
Mg	0.05	0.04	0.05	0.06
S	0.09	0.09	0.09	0.09
Micronutrients		(same for all species)		
Fe	--	0.007	--	--
Mn	--	0.004	--	--
Zn	--	0.0003	--	--
Cu	--	0.0003	--	--
Mo	--	0.00007	--	--
B	--	0.002	--	--
Cl	--	0.0003	--	--

† To compute individual nutrient levels, multiply the N level by the decimal fraction (e.g., to determine the P level for Douglas-fir when the N level is 2.0%, multiply 2.0 × 0.30 which gives 0.6%).

C. DRIS Indexes

There has been increased interest by foresters in the use of DRIS: developed by Beaufils (1973). Sumner (1977, 1978, 1982) has used and promoted the system for agricultural crops. The 1978 article in *Fertilizer Solutions* magazine is the most illuminating. It points out that the index values that measure how far particular nutrients in the leaf and soil are from the optimum are used in the calibration to classify yield factors in order of limiting importance. The indexes used are based on nutrient ratios such as N/P which can distinguish between high and low crop yields (Sumner, 1978, 1981). Such

Table 25-9. Standard values for mineral nutrient concentrations in conifer needle tissue. (From Landis, 1985.)

Mineral	Units	Adequate range	
nutrient	(% dry wt.)	Bareroot seedlings†	Container seedlings
N	%	1.20–2.00	1.30–3.50
P	%	0.10–0,20	0.20–0.60
K	%	0.30–0.80	0.70–2.50
Ca	%	0.20–0.50	0.30–1.00
Mg	%	0.10–0.15	0.10–0.30
S	%	0.10–0.20	--
Fe	mg kg^{-1}	50–100	60–200
Mn	mg kg^{-1}	100–5000	100–250
Zn	mg kg^{-1}	10–125	30–150
Cu	mg kg^{-1}	4–12	4–20
Mo	mg kg^{-1}	0.05–0.25	0.25–5.0
B	mg kg^{-1}	10–100	20–100
Cl	mg kg^{-1}	10–3000	--

† Macronutrient values are from Youngberg (1984) and micronutrient values from Powers (1984).

ratios partially offset the problem of declining nutrient concentration with crop age; but there are still time-dependent problems with agricultural crops (Hanson, 1981; Amundson & Koehler, 1985). A computer program for calculating DRIS indices has been prepared for specified agricultural crops based on norms developed for these crops (Letzsch & Sumner, 1983). A critical review of DRIS has been prepared by Jones (1981) which points out assumptions upon which the system is based:

1. Ratios of nutrient element concentrations are often better indicators of nutrient deficiencies than are single nutrient element concentrations.
2. Some nutrient element concentration ratios are more important than others.
3. Maximum crop yields are attainable only when the values of important ratios approach an optimum value, which is approximately the mean value of the ratio in a selected high-yielding (or otherwise desirable) population.
4. Since important ratios must approach their optimum values for high yields to be attained, the variance of an important ratio is smaller in a high-yielding population than a low-yielding population. The ratio of the variance of a high-yielding population to its low-yielding counterpart can be used to select important ratios.
5. A DRIS index can be calculated for each nutrient element. This index is based on the mean deviation of each important ratio (in which that nutrient element is either the numerator or denominator) from its optimum value. Thus, the optimum DRIS index for any nutrient element is 0.0. Negative indexes indicate deficiency, and positive indexes indicate sufficiency.

Langenegger and Smith (1978) pointed out for pineapple [*Ananas comosus* (L.) Merr.] that DRIS emphasizes the importance of nutrient balance rather than concentration, but a sufficiency level approach could lead to wrong diagnoses in cases of balanced low (dilution effect) or balanced high concentrations.

The most recent presentation of DRIS, including the use of DRIS charts and a comparison of DRIS with other diagnostic systems, has been made by Walworth and Sumner (1987). The use of DRIS for forest crops has been reviewed by Mead (1984) and Powers (1984). Truman and Lambert (1980) have used DRIS to determine the balance between N, S, and P in *Pinus radiata* foliage. The indices were calculated for the following equations:

$$\text{N index} = (\text{fN/P} + \text{fN/S})/2$$

$$\text{P index} = (\text{fP/S} - \text{fN/P})/2$$

$$\text{S index} = (\text{fN/s} - \text{fP/S})/2$$

The method of calculation for one of the functions is:

$$fN/S = 100 \left(\frac{N/S}{n/s} - 1\right) \times \frac{10}{CV} \text{ when } N/S > n/s$$

or

$$fN/s = 100 \left(1 - \frac{n/s}{N/S}\right) \times \frac{10}{CV} \text{ when } N/S > n/s.$$

The N/S ratio is the ratio of the two actual elements (in this example, %N and %S) in the plant under investigation while n/s is the value of the norm derived from the mean value of the N/S ratios for a population of the plants with relatively high yields. The coefficient of variation for this population is CV. The index with the greatest negative number denotes the element in shortest supply, while the greater the total of the indices irrespective of sign, the more pronounced is the deficiency of the element in shortest supply.

Leech and Kim (1979, 1981) have used DRIS to assess the nutrient requirements of poplars and subsequently as a guide for fertilization of poplar plantations. A greenhouse standard for high- and low-yielding crops was developed using nutrient solutions. Macronutrient indices were related to extractable soil nutrient levels.

The main problem in use of DRIS for forest crops is the absence of adequate data on nutrient concentrations in high-yielding populations to derive the norms. The question arises if nutrient concentration associated with high-yielding seedling crops in mineral nutrition can be used. The standard values in Table 25–7 for adequate seedling concentration in conifer needle tissue taken from Landis (1985) do not necessarily represent the norms for high-yielding populations of closed tree stands. In addition, the ranges of concentrations are very broad. The norms can be provided by optimum nutrition trials in older stands; such trials have been run in Scots pine, Norway spruce, and jack pine (Tamm et al., 1974; Tamm, 1981; Weetman & Fournier, 1984). Agricultural norms were based on many thousands of high-yielding crop studies. Data sets on foliar analysis for high yield forest stands are rare, at best concentrations in high-yielding natural sites may be a substitute.

D. Foliar Vector Diagnosis

A graphical technique has been used in which element needle concentration, element needle content, and unit needle weight are plotted and the shifts in these parameters from an untreated status, are compared on one graph. The magnitude and direction of the shifts have been shown to be of value in assessing stand nutrient status and the probable response to fertilization (Timmer & Morrow, 1984) (Fig. 25–1). This technique is appropriate for all conifers that have determinate growth (i.e., needle number is set in the bud year prior to fertilization).

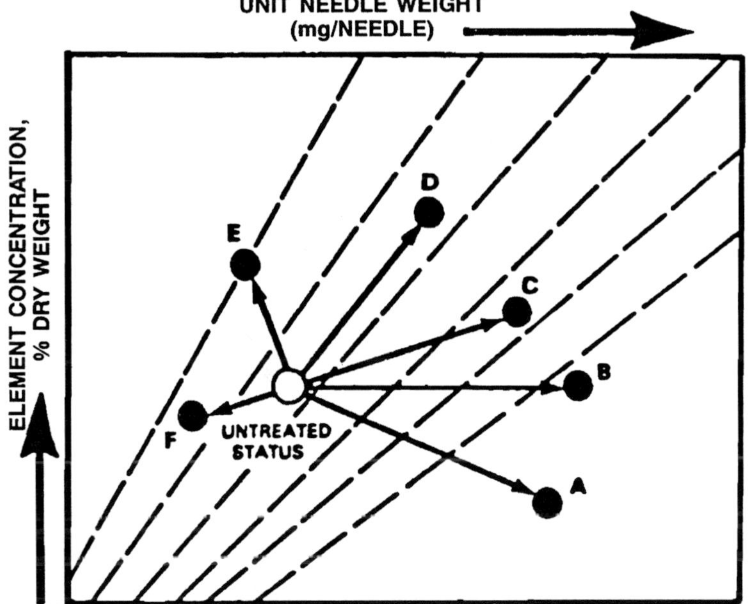

DIRECTION OF SHIFT	RESPONSE IN			CHANGE IN	
	NEEDLE WEIGHT	NUTRIENT		NUTRIENT STATUS	POSSIBLE DIAGNOSIS
		CONC.	CONTENT		
A	+	−	+	DILUTION	NON—LIMITING
B	+	0	+	UNCHANGED	NON—LIMITING
C	+	+	+	DEFICIENCY	LIMITING
D	0	+	+	LUXURY CONSUMPTION	NON—TOXIC
E	−	++	±	EXCESS	TOXIC
F	−	−	−	EXCESS	ANTAGONISTIC

Fig. 25–1. Interpretation of directional relationships between foliar concentration and absolute content of an element following treatment. (From Timmer & Morrow, 1984.)

Several studies have shown that increase in unit needle weight in the first growing season following fertilization is correlated with subsequent stand volume response. Needle biomass is determined on a unit basis making it possible to distinguish dilution effects. A conventional foliage sampling at the end of the first growing season has the additional requirement of counting and weighing of needles prior to analysis. A modification of the technique has been used in which relative values of element needle concentration, element needle content and unit needle weight, relative to the untreated status, are compared on one graph (Timmer & Morrow, 1984) (Fig. 25–2). This tech-

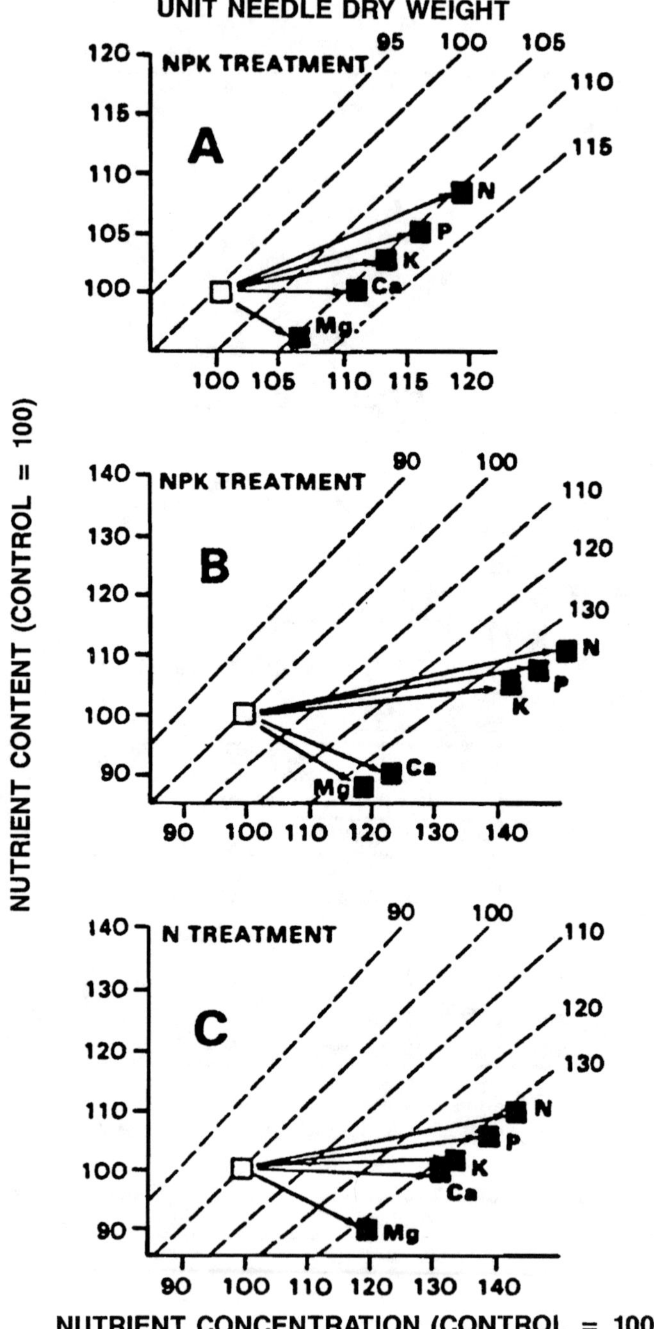

Fig. 25-2. Relative responses of nutrient composition and dry weight of current needles following fertilization with (*A*) NPK—mean of 12 screening trials, (*B*) NPK—treatments from full factorial trial, and (*C*) N—treatments from full factorial trial. (From Timmer & Morrow, 1984.)

nique permits the comparison of various elements together on one graph. The ascending order of elements along a relative unit weight line indicates the degree of deficiency for these elements.

This technique can be combined with single tree fertilizer screening trials to diagnose and predict forest stand response prior to the establishment of conventional fertilizer trials (Weetman & Fournier, 1982). Such a technique can test for fertilizer response in many candidate stands at a low cost in one growing season (Weetman, 1989).

VII. CONCLUSION

Plant tissue analyses are usually combined with soil and site analyses, in preparation of forest fertilization prescriptions. The uptake of fertilizer and the nutritional status of the tree can be verified by tissue analyses alone. This chapter has shown that there are many limitations to the use of tissue analyses. A single data set of macro- and micronutrient concentration for a forest stand is not readily interpreted unless some single element is at an extremely low concentration. Prescribed typical values for tree species, unaccompanied by site data, analytical methods used and sampling conditions, are of little value. Recent automation of analytical procedures does not necessarily result in better accuracy of analysis. International inter-lab comparisons have produced alarming variations in the analyses of the same sample.

In spite of these problems, the use and development of tissue analyses in forestry is expanding. In place of static analyses and use of critical values, increasing use is being made of concentration and leaf biomass as measures of uptake response. Use of optimum properties or ratios and the use of tissue data in dynamic growth models which predict response will probably increase.

REFERENCES

Alcubilla, M., and K.E. Rehfuess. 1981. Experiences with stem phloem analysis. Proc. 17th IUFRO World Congr., Kyoto, Jpn. Sept. Int. Union For. Res. Org., Kyoto.

Amundson, R.L., and F.E. Koehler. 1985. Utilizing the DRIS for field diagnosis of nutrient balance in winter wheat: Part II. Nutrient ratio study. p. 165. *In* Agronomy abstracts. ASA, Madison, WI.

Auchmoody, L.R., and K.P. Hammack. 1975. Foliar nutrient variation in four species of upland oaks. USDA Forest Serv. Res. Paper NE-331. Northeastern Forest Exp. Stn. U.S. Gov. Print. Office, Washington, DC.

Axelsson, B.O. 1981. Differences in yield at different sites. p. 590. *In* I.A.C. Plaizier (ed.) Proc. 17th IUFRO World Congr., Kyoto, Japan, Div. 1, Group 2. Abstract. Int. Union For. Res. Org., Kyoto.

Ballard, T. 1985. Foliar analysis for deficiency diagnosis and fertilizer prescriptions in forestry: Variability considerations. *In* Proc. 9th British Columbia Soil Science Workshop, Vancouver. 21–22 Feb. B.C. Min. of Environ. Surv. and Resource Mapping Branch, Victoria.

Ballard, T.M., and R.E. Carter. 1983. Evaluating forest stand nutrient status. British Columbia Min. of Forests Land Manage. Rep. 20.

Barnes, R.L. 1962. Glutamine synthesis and translocation in pine. Plant Physiol. 37:232–326.

Barnes, R.L. 1963. Organic nitrogen compounds in tree xylem sap. For. Sci. 9:98–102.

Baule, H., and C. Fricker. 1970. The fertilizer treatment of forest trees. BLV, Munich.

Beaufils, E.R. 1973. Diagnosis and recommendation integrated system (DRIS). Soil Sci. Bull. 1. Dep. of Soil Sci. and Agro-meteorol., Univ. of Natal, Pietermaritzburg, South Africa.

Bell, D.T., and S.C. Ward. 1984. Foliar and twig macronutrients (N, P, K, Ca and Mg) in selected species of *Eucalyptus* used in rehabilitation: Sources of variation. Plant Soil 82:363–376.

Binns, W.O., G.J. Mayhead, and J.M. Mackenzie. 1980. Nutrient deficiencies of conifers in British forests. An illustrated guide. For. Comm. Leafl. 76. UK For. Comm., Edinburgh.

Boonstra, R., A.L. Kenworthy, and D.P. Watson. 1957. Nutritional status of selected plantings of *Taxus media* Hicksi. Proc. Am. Soc. Hortic. Sci. 70:432–436.

Brackett, M.H. 1964. Patterns of elemental distribution and response of Douglas-fir foliage to nitrogen fertilizer. M.S. thesis, Univ. of Washington, Seattle.

Cannon, T.F., L.C. Chadwick, and K.W. Reisch. 1960. Nitrogen and potassium nutritional studies of *Gleditsia triacanthos inermis* "Moraine." Proc. Am. Soc. Hortic. Sci. 75:633–639.

Carter, M.C., and H.S. Larsen. 1965. Soil nutrients and loblolly pine xylem sap composition. For. Sci. 11:216–220.

Carter, M.C., and E.H. White. 1971. Dry weight and nutrient accumulation in young stands of cottonwood (*Populus deltoides* Bart.). Auburn (Alabama) Univ. Agric. Exp. Stn. Circ. 190.

Chandler, R.F., Jr. 1939. The calcium content of the foliage of forest trees. Cornell Univ. Agric. Exp. Stn. Mem. 228:1–15.

Chapman, A.G. 1941. Tolerance of shortleaf pine seedlings for some variations in soluble calcium and H-ion concentrations. Plant Physiol. 16:313–326.

Cotrufo, C. 1983. Xylem nitrogen as a possible diagnostic nitrogen test for loblolly pine. Can. J. For. Res. 13:355–357.

Cotrufo, C., and C.G. Wells. 1984. Some possible tissue assay methods for N nutrition assessment of *Pinus taeda* L. Commun. Soil Sci. Plant Anal. 14:1391–1407.

Davidson, H. 1960. Nutritional-element composition of leaves from selected species of woody ornamental plants. Proc. Am. Soc. Hortic. Sci. 76:667–672.

Day, F.P., Jr., and C.D. Monk. 1977. Seasonal nutrient dynamics in the vegetation on a Southern Appalachian watershed. Am. J. Bot. 64:1126–1139.

Dighton, J., and A.F. Harrison. 1983. Phosphorus nutrition of lodgepole pine and Sitka spruce stands as indicated by a root bioassay. Forestry 56:33–43.

Duryea, M. (ed.). 1985. Evaluating seedling quality: Principles, procedures, and predictive abilities of major tests. Proc. Workshop, Corvallis, OR. 16–18 Oct. 1984. Oregon State Univ., Corvallis.

Duryea, M., and T.D. Landis (ed.). 1984. Forest nursery manual: Production of bareroot seedling. Nijhoff and Junk Publ., The Hague, Netherlands.

Friedland, A.J., and R.A. Gregory. 1985. Elemental concentrations of xylem sap and soil water in a conifer forest. p. 218. *In* Agronomy abstracts. ASA, Madison, WI.

Gagnon, J.D. 1964. Relationships between site index and foliage nitrogen at two crown levels for mature black spruce. For. Chron. 40:169–174.

Gessel, S.P., R.M. Kenady, and W.A. Atkinson (ed.). 1979. Proc. Forest Fertilization Conf., Union, WA. 25–27 Sept. Coll. of For. Res., Inst. of For. Res., Univ. of Washington, Seattle. Contrib. no. 40.

Guha, M.M., and R.L. Mitchell. 1966. The trace and major element composition of the leaves of some deciduous trees. II. Seasonal changes. Plant Soil 24:90–112.

Hacskaylo, J., R.F. Finn, and J.P. Vimmerstedt. 1969. Deficiency symptoms of some forest trees. Ohio Agric. Res. Dev. Ctr., Wooster. Res. Bull. 1015.

Hall, M.J., and M. Raupach. 1963. Foliage analyses and growth responses in *Pinus radiata* (D. Don) showing potassium deficiencies in eastern Victoria. Appita 17:76–84.

Hanson, R.G. 1981. DRIS evaluation of N, P, K, status of determinent soybeans in Brazil. Commun. Soil Sci. Plant Anal. 12(9):933–948.

Harada, H., H. Satoo, I. Hotta, K. Hatiya, and Y. Tadaki. 1972. Study on the nutrient contents of native *Cryptomeria* forest. Bull. Gov. For. Exp. Stn. Jpn. 249:17–74.

Harrison, A.F., and D.F. Helliwell. 1979. A bioassay for comparing phosphorus availability in soils. J. Appl. Ecol. 16:497–505.

Henry, E. 1908. Les sols forestiers. Berger-Levault, Paris.

Hodges, J.D., and P.L. Lorio, Jr. 1969. Carbohydrate and nitrogen fractions of the inner bark of loblolly pines under moisture stress. Can. J. Bot. 47:1651–1657.

Hoyle, M.C. 1965. Variation in foliage composition and diameter growth of yellow birch with season, soil, and tree size. Soil Sci. Soc. Am. Proc. 29:475–480.

Humphreys, F.R., and J. Kelly. 1962. The variation of phosphorus content in *Pinus elliottii* foliage. Aust. For. 26:77–86.

Ingestad, T. 1967. Nutrient needs of seedlings and young trees. *In* Proc. Colloq. For. Fert. 1967:139–141.

Ingestad, T., A. Aronsson, and G.I. Agren. 1981. Nutrient flux density model of mineral nutrition in conifer ecosystems. Stud. For. Suec. 160:61–70.

Ingestad, T., and M. Kahr. 1985. Nutrition and growth of coniferous seedlings at varied relative nitrogen addition rate. Physiol. Plant. 65:109–116.

Jones, C.A. 1981. Proposed modifications of the diagnosis and recommendation integrated system (DRIS) for interpreting plant analyses. Commun. Soil Sci. Plant Anal. 12(8):785–794.

Kelly, J., and M.J. Lambert. 1972. The relationship between sulphur and nitrogen in the foliage of *Pinus radiata*. Plant Soil 37:395–407.

Kimmins, J.P., D. Binkley, L. Chatarpaul, and J. de Catanzaro. 1985. Biogeochemistry of temperate forest ecosystems: Literature on inventories and dynamics of biomass and nutrients. Can. For. Serv. Info. Reg. PI-X-47 E/F.

Knight, P.J. 1978. Foliar concentrations of ten mineral nutrients in nine *Pinus radiata* clones during a 15-month period. N.Z. J. For. Sci. 8:351–368.

Kolari, K.K. 1979. Micronutrient deficiency in forest trees and dieback of scots pine in Finland. A review. Folia forestale 389. Metsan tutkimuslaitos. Inst. For. Fenniae, Helsinki.

Lambert, M.J. 1984. The use of foliar analysis in fertilizer research. p. 269–292. *In* IUFRO Proc.: symposium on site and productivity of fast growing plantation, Vol. I and II, South Africa. 30 Apr.–11 May. South African For. Res. Inst., Dep. Environment Affairs, Pretoria.

Landis, T.D. 1985. Mineral nutrition as an index of seedling quality. p. 29–48. *In* M.L. Dureya (ed.) Evaluating seedling quality: Principles, procedures and predictive abilities of major tests, Proc. Workshop, Corvallis, OR. 16–18 Oct. 1984. Oregon State Univ., Corvallis.

Langenegger, W., and B.L. Smith. 1978. An evaluation of the DRIS system as applied to pineapple leaf analysis. p. 263–273. *In* A.R. Ferguson et al. (ed.) Plant nutrition. Proc. 8th Int. Colloq. on Plant Analysis and Fertilizer Problems, Auckland, NZ. 28 Aug.–1 Sept. Info. Ser. 134. N.Z. Dep. of Sci. and Ind. Res.

Lavender, D.P. 1970. Foliar analysis and how it is used—A review. Oregon State Univ. School of For. Res. Note 52.

Lavender, D.P., and R.L. Carmichael. 1966. Effect of three variables on mineral concentrations in Douglas-fir needles. For. Sci. 12:441–446.

Lea, R., W.C. Tierson, D.H. Bickelhaupt, and A.L. Leaf. 1979a. Stand treatment and sampling time of hardwood foliage. I. Macro-element analysis. Plant Soil 51:515–533.

Lea, R., W.C. Tierson, D.H. Bickelhaupt, and A.L. Leaf. 1979b. Stand treatment and sampling time of hardwood foliage. II. Micro-element analysis. Plant Soil 51:535–550.

Leaf, A.L. 1973. Plant analysis as an aid in fertilizing forests. p. 427–454. *In* L.M. Walsh and J.D. Beaton (ed.) Soil testing and plant analysis. Rev. ed. SSSA, Madison, WI.

Leaf, A.L., J.E. Berglund, and R.E. Leonard. 1970. Annual variation in foliage of fertilized and/or irrigated red pine plantations. Soil Sci. Soc. Am. Proc. 34:667–682.

Leech, R.H., and Y.T. Kim. 1979. Assessment of nutrient requirements of poplars. p. 1–15. *In* D.C.F. Fayle et al. (ed.) Poplar research, management and utilization in Canada. For. Res. Info. Paper 102. Rep. 15. Ontario Min. of Natural Resourc., Maple.

Leech, R.H., and Y.T. Kim. 1981. Foliar analysis and DRIS as a guide to fertilizer amendments in poplar plantations. For. Chron. 57(1):17–21.

Letzsch, W.S., and M.E. Sumner. 1983. Computer program for calculating DRIS indices. Commun. Soil Sci. Plant Anal. 14(9):811–815.

Likens, G.E., and F.H. Bormann. 1970. Chemical analyses of plant tissues from the Hubbard Brook ecosystems in New Hampshire. Yale Univ. School For. Bull. 79.

Lowry, G.L. 1970. Variations in nutrients in black spruce needles. p. 235–259. *In* Youngberg and C.B. Davey (ed.) Tree growth and forest soils. Oregon State Univ. Press, Corvallis.

Lowry, G.L., and P.M. Avard. 1965. Nutrient contents of black spruce needles. I. Variations due to crown position and needle age. Tech. Rep. 425, Woodl. Res. Paper 171. Pulp Paper Res. Inst., Montreal.

Lowry, G.L., and P.M. Avard. 1968. Nutrient content of black spruce needles. II. Variations with crown class and relationships to growth and yield. Woodl. Paper 3. Pulp Paper Res. Inst. Can., Pointe Claire.

Lowry, G.L., and P.M. Avard. 1969. Nutrient content of black spruce and jack pine needles. III. Seasonal variation and recommended sampling procedures. Woodl. Paper 10. Pulp Paper Res. Inst. Can., Pointe Claire.

McHargue, J.S., and W.R. Roy. 1933. Mineral and nitrogen content of the leaves of some forest trees at different times in the growing season. Bot. Gaz. 94:381–393.

McVickar, J.S. 1949. Composition of white oak (*Quercus alba*) leaves in Illinois as influenced by soil type and soil composition. Soil Sci. 68(4):317–328.

Madgwick, H.A.I. 1964. Variation in the chemical composition of red pine (*Pinus resinosa* Ait.) leaves: A comparison of well grown and poorly grown trees. Forestry 37:84–87.

Mead, D.J. 1984. Diagnosis of nutrient deficiencies in plantations. p. 259–291. *In* G.D. Bowen and E.K.S. Nambiar (ed.) Nutrition of plantation forests. Academic Press, New York.

Mead, D.J., and W.L. Pritchett. 1974. Variation of N, P, K, Ca, Mg, Mn, Zn and Al in slash pine foliage. Commun. Soil Sci. Plant Anal. 5:291–301.

Mead, D.J., and G.M. Will. 1976. Seasonal and between-tree variation in the nutrient levels in *Pinus radiata* foliage. N.Z. J. For. Sci. 6:3–13.

Mengel, K., and E.A. Kirby. 1942. A rapid method for determination of nitrogen in plant tissue. Science 96:565–566.

Miller, H.G., and J.D. Miller. 1976. Analysis of needlefall as a means of assessing nitrogen status in pine. Forestry 49:63–72.

Miller, W.F. 1966. Annual changes in foliar nitrogen, phosphorus and potassium levels of loblolly pine (*Pinus taeda* L.) with site and weather factors. Plant Soil 24:369–378.

Morrison, I.K. 1974. Mineral nutrition of conifers with special reference to nutrient status interpretation: A review of literature. Publ. 1343. Can. For. Serv. Dep. of the Environment, Sault Ste. Marie, ON, Canada.

Nemec, A. 1940. On the effect of nitrogen fertilizing of young spruce and the nutrient uptake of the needles as dependent upon the soil reaction and nutrient status of the soil. Bodenkd. Pflanzenernaehr. 17:204–235.

Nemec, A. 1947. Biochemical and pedological studies on the causes of checking Norway spruce and Scots pine in the Smrcekk circle, State Forest of Slatihary, near Chrudim. Zpr. Vyzk. Ustavu Cesk. Prum. Cukrov. Praze 1:31–52.

Nicholson, G. 1984. Methods of soil, plant, and water analysis. For. Res. Inst. Bull. 70. New Zealand For. Serv., Private Bag, Rotorua.

Olsen, C. 1948. The mineral, nitrogen and sugar content of beech leaves and beech sap at various times. C.R. Trav. Lab. Carlsberg Ser. Chim. 26:197–230.

Peterson, P.J. 1961. Variations in the mineral content of kauri (*Agathis australis* Salisb.) leaves with respect to leaf age, leaf position and tree age. N.Z. J. Sci. 4:699–678.

Powers, R.F. 1984. Estimating soil nitrogen availability through soil and foliar analysis. p. 353–379. *In* E.L. Stone (ed.) Forest soils and treatment impacts. Proc. 6th North Am. For. Soils Conf., Univ. of Tennessee, Knoxville. June 1983. Univ. of Tennessee, Knoxville.

Raupach, M. 1967. Soil and fertilizer requirements for forests of *Pinus radiata*. Adv. Agron. 19:307–353.

Richter, K. 1983. Sampling intensity for Douglas-fir foliar nutrient analysis in the Vancouver Forest Region. B.S.F. thesis. Faculty of Forestry, Univ. of British Columbia, Vancouver.

Sampson, A.W., and R. Samisch. 1935. Growth and seasonal changes in composition of oak leaves. Plant Physiol. 10:739–751.

Schultz, C.J. 1976. A review of fertilizer research on some of the more important conifers and Eucalyptus planted in a subtropical and tropical countries, with special reference to South Africa. S. Afr. Dep. For. Bull. 53.

Slitcher, T.K. 1984. Seasonal variation in nitrogen and phosphorus composition of plantation grown American sycamore foliage. Master's thesis. North Carolina Univ., Raleigh.

Stark, N., C. Spitzner, and D. Essig. 1985. Xylem sap analysis for determining nutritional status of trees: *Pseudotsuga menziesii*. Can. J. For. Res. 15:429–437.

Strebel, O. 1961. Needle analysis in old spruce stands of very good growth in the Bavarian Alpenvorland. Forstwiss. Centralbl. 80(11/12):344–352.

Sumner, M.E. 1977. Use of the DRIS system in foliar diagnosis of crops at high yield levels. Commun. Soil Sci. Plant Anal. 8(3):251–268.

Sumner, M.E. 1978. A new approach for predicting nutrient needs for increased crop yields. Fert. Solutions (Sept.-Oct.) 1978:68–78.

Sumner, M.E. 1981. Diagnosing the sulfur requirements of corn and wheat using foliar analysis. Soil Sci. Soc. Am. J. 45(1):87–90.

Sumner, M.E. 1982. The diagnosis and recommendation integrated system (DRIS). Soil/Plant Analysts Seminar. 1 Dec. Counc. on Soil Testing and Plant Analysis, Anaheim, CA.

Swan, H.S.D. 1962. The mineral nutrition of the Grand Mere Plantations. Woodl. Res. Index Contr. 131. Pulp Paper Res. Inst., Montreal.

Tamm, C.O. 1951. Seasonal variation in composition of birch leaves. Physiol. Plant. 4:461–469.

Tamm, C.O. 1955. Studies on forest nutrition. I. Seasonal variation in the nutrient content of conifer needles. II. An experiment with application of radioactive phosphate to young spruces and birches. Medd. Skogsforsk. 45(5/6).

Tamm, C.O. 1981. Above ground production of Norway spruce at optimum nutrient supply on a low potential site. p. 595. *In* Proc. 17th IUFRO Congr., Kyoto, Japan. Div. 1, Group 2. Abstract. Int. Union For. Res. Org., Kyoto.

Tamm, C.O., A. Aronsson, and H. Burgtorf. 1974. The optimum nutrition experiment Strasan. A brief description of an experiment in a young stand of Norway spruce (*Picea abies* Karst.) For. Soils Res. Note 17. Royal Coll. For., Dep. For. Ecol., Stockholm, Sweden.

Timmer, V.R. 1979. Effect of fertilization on nutrient concentration of white spruce foliage and bark. For. Sci. 25:115–119.

Timmer, V.R., and L.D. Morrow. 1984. Predicting fertilizer growth response and nutrient status of jack pine by foliar diagnosis. p. 335–351. *In* E.L. Stone (ed.) Forest soils and treatment impacts. Proc. 6th North Am. For. Soils conf., Univ. of Tennessee, Knoxville. June 1983. Univ. of Tennessee, Knoxville.

Touzet, G., J.C. Heinrich, and I. Nohn. 1969. Diagnostic foliare et vitesse de croissance. (Foliar diagnosis and growth rate.) Association Foret-Cellulose (AFOCEL), Paris. p. 18–59.

Truman, R.A., and M.J. Lambert. 1980. The use of DRIS indices to determine the balance between nitrogen, phosphorus and sulphur in *Pinus radiata* foliage. p. 369–377. *In* R.A. Rummery and F.J. Hingston (ed.) Managing nitrogen economies of natural and man-made ecosystems. Proc. of Workshop CSIRO, Mandurrah, Western Australia. 5–9 Oct. CSIRO, Canberra, Australia.

Turner, J., S.F. Dice, D.W. Cole, and S.P. Gessel. 1978. Variation of nutrients in forest tree foliage: A review. Contrib. no. 35. Coll. of For. Resourc., Inst. of For. Products, Univ. of Washington, Seattle.

van den Burg, J. 1984. Foliar analysis for determination of tree nutrient status: A compilation of literature data. Rap. Nr. 414. Rijkinstituut voor onderzoek in de bos-en landschapsbouw "De Dorschkamp", Wageningen, Netherlands.

van den Driessche, R. 1974. Prediction of mineral nutrient status of trees by foliar analysis. Bot. Rev. 40:347–394.

Verry, E.S., and D.R. Timmons. 1976. Elements in leaves of trembling aspen clone by crown position and season. Can. J. For. Res. 6:436–440.

Walker, L.C. 1955. Foliar analysis as a method of indicating potassium deficiency soils for reforestation. Soil Sci. Soc. Am. Proc. 19:233–236.

Wallihan, E.F. 1944. Chemical composition of leaves in different parts of sugar maple trees. J. For. 42:684.

Walworth, J.L., and M.E. Sumner. 1987. The diagnosis and recommendation integrated system (DRIS). Adv. Soil Sci. 6:149–188.

Weetman, G.F. 1968. The nitrogen fertilization of three black spruce stands. Woodl. Paper 6. Pulp Paper Res. Inst. Can., Pointe Claire.

Weetman, G.F., and R. Fournier. 1982. Graphical diagnosis of lodgepole pine response to fertilization. Soil Sci. Soc. Am. 46(6):1281–1289.

Weetman, G.F., and R.M. Fournier. 1984. Ten-year growth and nutrition effects of a straw treatment and of repeated fertilization on jack pine. Can. J. For. Res. 14:416–423.

Weetman, G.F.1989. Graphical vector analysis technique for testing stand nutritional status. p. 93–109. *In* W.J. Dyck and C.A. Mees (ed.) Research strategies for long-term site productivity. Proc. IEA/BE A3 Workshop, Seattle, WA. August. IEA/BE A3. Rep. No. 8. For. Res. Inst. Rotorua New Zealand Bull. 152.

Wells, C.G. 1969. Foliage sampling guides for loblolly pine. USDA For. Serv. Res. Note SE-113. Southeastern Forest Exp. Stn., Asheville, NC.

Wells, C.G. 1979. Nitrogen and potassium fertilization of loblolly pine on a South Carolina piedmont soil. For. Sci. 16:172–176.

Wells, C.G., and L.J. Metz. 1963. Variation in nutrient content of loblolly pine needles with season, age, soil, and position on the crown. Proc. Soil Sci. Soc. Am. 27:90–93.

White, D.P. 1954. Variation in the nitrogen, phosphorus and potassium contents of pine needles with season, crown position and sample treatment. Soil Sci. Soc. Am. Proc. 18:326–330.

White, J.D., L.T. Alexander, and E.W. Clark. 1972. Fluctuations in the inorganic constituents of inner bark of loblolly pine with season and soil series. Can. J. Bot. 50:1287–1293.

White, J.D., C.G. Wells, and E.W. Clark. 1970. Variations in the inorganic composition of inner bark and needles of loblolly pine tree height and soil series. Can. J. Bot. 48:1079–1084.

Will, G.M. 1957. Variation in the mineral content of radiata pine needles with age and position in tree crowns. N.Z. J. Sci. Agric. Res. 4:309–327.

Will, G.M. 1965. Increased phosphorus uptake by radiata pine in Riverhead Forest following superphosphate applications. N.Z. J. For. 10:33–42.

Will, G.M. 1985. Nutrient deficiencies and fertilizer use in New Zealand exotic forests. For. Res. Inst. Bull. 97. New Zealand For. Serv., Private Bag, Rotorua.

Will, G.M. 1986. Results from interlaboratory comparisons of chemical analysis of standard foliage samples. *In* Proc. IUFRO 18th World Congr., Ljubljana, Yugoslavia. 7–21 Sept. Int. Union For. Res. Org.

Windsor, G.J., and J. Kelly. 1972. The effects of fertilization on shoot dieback and foliar boron and sulphur concentrations in several clones of *Pinus radiata*. p. 241–257. *In* R. Boardman (ed.) Australian Forest-tree Nutrition Conf. Forestry and Timber Bur., Canberra, Australia.

Wright, T.W., and G.M. Will. 1958. The nutrient content of Scots and Corsican pines growing on sand dunes. Forestry 31:14–25.

Youngberg, C.T. 1984. Soil and tissue analysis: Tool for maintaining soil fertility. p. 75–80. *In* M.D. Duryea and T.D. Landis (ed.) Forest nursery manual: Production of bareroot seedlings. Nihjoff and Junk, The Hague, Netherlands.

Zeche, W. 1970. Nadelanalytische Unfersuchunzen über die Kalkchlorose der Waldkieler (*Pinus silvestris*). Z. Planzenernäehr. Düeng. Bodenkd. 125:1–6.

Zinke, P.J., and A.G. Stangenberger. 1979. Ponderosa pine and Douglas-fir foliage analyses arrayed in probability distributions. p. 221–225. *In* S.P. Gessel et al. (ed.) Proc. Forest Fert. Conf., Union, WA. Contrib. no. 40. Coll. of For. Res., Inst. of For. Res., Univ. of Washington, Seattle.

Chapter 26

Analytical Instruments for Soil and Plant Analysis

M. E. WATSON, *REAL, OARDC, Wooster, Ohio*

ROBERT A. ISAAC, *University of Georgia, Athens*

The development of highly sophisticated analytical instrumentation during the last 25 to 30 yr has allowed more extensive use of the basic concepts underlying soil testing and plant analysis. Major advances in electronics have allowed greater accuracy, precision, lower detection limits, and speed of analysis than ever before. These advances have allowed the readout of results to progress from simple meters, to digital readouts, to computer systems collecting the results and printing them in report form. Instruments now use electronic circuitry for self-control, self-evaluation, and self-diagnosis. Many of the modern instruments can be operated only through computer systems.

The majority of analytical instruments used for the analysis of soils and plants make use of electromagnetic radiation energy. Definite regions of the electromagnetic spectrum are used for differentiation purposes. Instruments that rely on the principles of electrochemistry also are used for soil testing and, to some extent, for plant tissue analysis. Instruments that use ion exchange are also finding their place in analysis of soils and plant tissue. Bombarding samples with high-speed neutrons and electrons has also been used to measure elements within or extracted from soils and plant tissue that may be more difficult to measure by other methods.

Coupling computer systems to instruments has allowed for automated analysis. Furthermore, since modern analytical instruments perform analysis at a rapid pace, producing vast amounts of data, a data capture and management system is required. Thus, computers, and sometimes networks of computers, are used.

Another technology that is evolving in the laboratory analysis of soil and plant tissue is robotics. Not only has instrumentation been automated and controlled by computers, but also many of the routine tasks such as pipetting, diluting, and weighing previously done by technicians can now be done by computer-controlled robots.

The purpose of this chapter is to acquaint the reader with various types of analytical instruments and related systems that are being used in laboratories, primarily for routine analysis of soil and plant tissue samples. It is not the authors' intent to go into great detail about each kind of instrument used. The advantages, disadvantages, and capabilities of each kind of instrumentation will be discussed wherever possible.

I. INSTRUMENTATION USED FOR ANALYSIS OF SOILS AND PLANT TISSUE

A. Atomic Absorption

1. Principle of Operation and Use

A summary of the various instrumentation used for the determination of element concentrations in plant tissue was presented by Jones (1972). Many of the instruments discussed then are still in use today for plant tissue analysis. However, new instruments are available that provide for a more rapid, accurate analysis than before.

Until recently, the atomic absorption spectrophotometer (AAS) was the most widely used instrument for routine determination of many elements contained in plant tissue. This AAS position of dominance is now being challenged by inductively coupled Ar plasma (ICAP) emission spectrometry.

Walsh (1955) published on the principles and usefulness of AAS. Since that time, progress in the development of AAS has occurred through incorporation of advances in electronic technology, allowing modern AAS to be faster, more sensitive, and more convenient than earlier models. Slavin (1982) has summarized the current status of AAS and has predicted changes likely to occur in AAS in the future.

Isaac and Johnson (1975) conducted a collaborative study on the determination of elements in plant tissue by AAS. Wet and dry ashing techniques were studied. In this 1975 study, seven laboratories analyzed five different tissue samples for K, Ca, Mg, Mn, Fe, Zn, and Cu. The AAS method was adopted as an official, first-action method by the Association of Official Analytical Chemists (AOAC).

Excellent publications on the use of AAS for the determination of elements in soil extracts and plant tissue digests have been authored by Isaac and Kerber (1971); Jones (1981); and Baker and Suhr (1982). There are many publications devoted to theory, principles, and use of AAS (Dean & Rains, 1969; Winefordner & Vickers, 1974; Winefordner, 1976; Price, 1979).

The AAS instrument makes use of the principle that atoms at ground-state energy levels can absorb electromagnetic radiation when radiation of appropriate wavelengths is focused upon them. The main components of an AAS instrument are: (i) hollow cathode or electrodeless discharge lamps containing the element of interest, (ii) nebulizer and burner system, and (iii) radiant-energy-detector system.

Hollow cathode lamps are usually of two types, single element or multiple elements. The element in the lamp, corresponding to the element in the sample solution to be analyzed, is excited by electrical current causing light to be emitted from the lamp. The wavelengths of emitted light are characteristic for the element contained in the lamp.

Electrodeless discharge lamps (EDL) require a microwave power supply and provide greater light output and longer life than the corresponding hollow cathode lamps. For certain elements (As and Se), EDLs provide improved sensitivity and lower detection limits.

For flame AAS, the sample solution is aspirated into a premixing chamber containing a flow spoiler where large droplets drain away, causing fine droplets to enter the flame through the burner head. The flame produces ground-state atoms of the element of interest which absorb light energy from the hollow cathode lamp that passes through the flame. Elements that can be measured provide sufficient quantity of ground-state atoms at the flame temperature. For most elements, an acetylene-air flame (2100–2400 K) is used. However, if refractory elements, such as Al, are to be measured than a higher temperature flame, such as nitrous oxide-acetylene, is used (Bowman & Willis, 1967). Refractory elements usually of concern are Si, Al, Ti, and Cr which remain as oxides in the cooler air-acetylene flame.

In the optical system, a monochromator is used to isolate the wavelength of interest by means of an adjustable grating so that wavelengths in the ultraviolet (UV) and visible regions of the electromagnetic spectrum can be isolated. In modern instruments, selection of the wavelength is done with a microcomputer. A photomultiplier tube is used to convert light energy that passes through the flame into electrical energy.

There are two basic types of AAS instruments, single beam and double beam (Fig. 26–1 and 26–2). Double-beam instruments are designed to be electronically more stable than single-beam instruments. However, double-beam instruments have more complicated optical systems than the single-beam instruments.

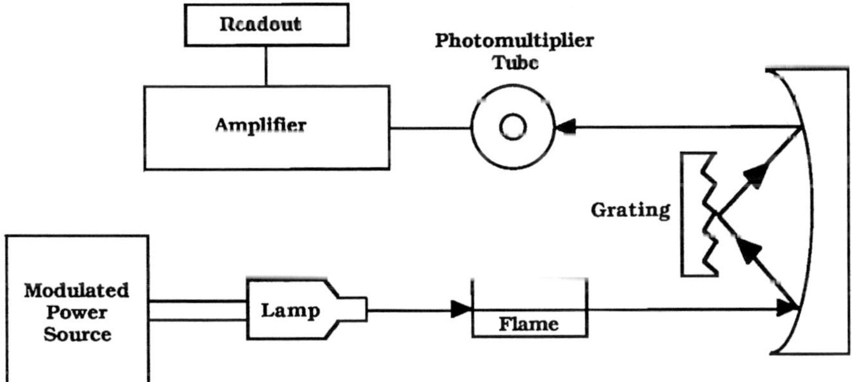

Fig. 26–1. Schematic diagram of a single-beam type AAS instrument (Skoog, 1985).

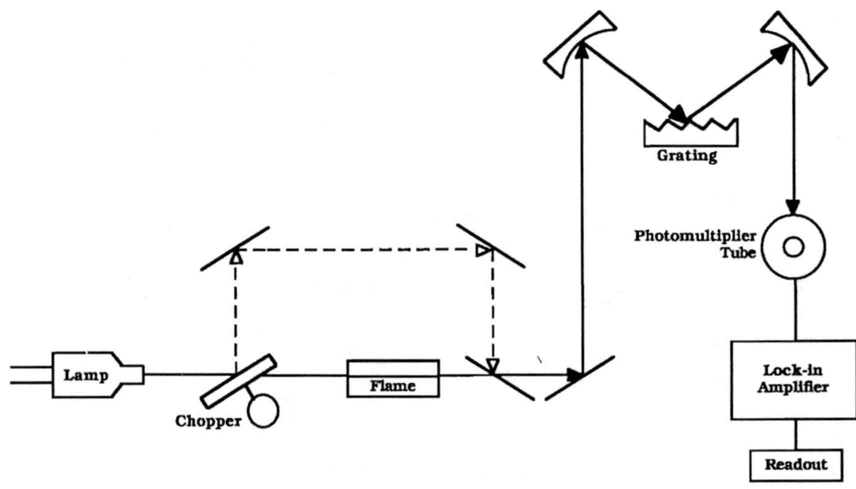

Fig. 26–2. Schematic diagram of a double-beam type AAS instrument (Skoog, 1985).

For double-beam instruments, the light energy generated by the lamp is split into a sample beam (i.e., that passing through the flame) and the reference beam (the light reflected around the flame via an optical system). Consequently, the measurement of atomic absorption by the sample atoms can be compared with that when no absorption occurs in the reference beam. Many modern AAS instruments use an electronic modulation of the lamp source to produce a pulsed direct current (dc) signal which eliminates dc flame emission interference.

The AAS may be used to measure K, Ca, Mg, Zn, Fe, Cu, Mn, Na, and Al. Often, these elements are measured individually as compared to the multielement capabilities of either plasma, arc, or spark spectrometers. Some modern AAS instruments are designed to rapidly optimize parameters for each element. Direct measurement of N and S is not possible by AAS, and sensitivity for P and B is not adequate for plant tissue. Inefficient concentrating techniques or indirect methods must be used.

Most AAS instruments are designed to operate in both absorption and emission modes. In emission mode, energy is emitted from an excited atom as the electrons of the atom return to ground-state-energy status. Thus, a hollow cathode lamp is not needed. Usually, atoms that have low ionization potentials, e.g., Na and K, are analyzed with AAS in the emission mode.

With advancements in electronic and computer systems, most AAS instruments are microcomputer controlled. The microcomputer sets the necessary parameters, such as slit width, wavelengths, and fuel/air ratios. Incorporation of a microcomputer into an AAS instrument provides reduced set-up time, rapid calibration and analysis, variable integration times, easy and fast data acqusition, and reporting. Many instruments contain turrets with multiple lamps that can be rapidly aligned by the computer permitting longer uninterrupted analysis compared to manual operation. Some modern

AAS instruments have self-contained computer programs designed for self-diagnosis of electronic problems.

Other advances in AAS instrumentation are the option of several means of background correction in addition to the continuum source for background correction. The Zeeman and Smith-Hjefie techniques offer some advantages. Background correction is used to aid in the differentiation of the analyte signal from background radiation. This feature allows accurate element analysis especially near the detection limit of the instrument.

2. Excitation Sources

Different excitation sources are used for AAS. An air-acetylene flame is the most common excitation source. However, other sources are acetylene-nitrous oxide and hydrogen flames. As mentioned previously, an acetylene-nitrous oxide flame is hotter than an air-acetylene flame and is necessary when determining refractory elements (Bowman & Willis, 1967). The cooler hydrogen flame is used for elements that are easily volatilized (e.g., As and Se).

Elements in plant tissue often determined with air-acetylene flame are K, Ca, Mg, Na, Fe, Mn, Cu, and Zn. Other elements, such as Co, can be determined with flame AAS but usually require preconcentration before measurement with AAS. Elements, often called heavy metals (Ni, Cr, Cd, and Pb), can be measured satisfactorily in many matrices provided their concentrations are sufficiently above instrument detection limits.

The graphite furnace, a flameless excitation source, is used to determine many elements, especially metals, that are present at vary low concentrations in plant tissue. Rains and Menis (1976) have compared flame and flameless atomic absorption spectrometry technology. Usually measurement of Cr, Ni, Cd, Al, Mo, Pb, and Co is more accurate with flameless AAS than with flame AAS. This increased accuracy is especially so for Mo and Pb (Wilson, 1979). An approximate range of 0.1 to 100 pg of an element can be measured with the furnace. The quantity of sample solution required is in the microliter range, compared with the milliliter range for flame AAS.

Flameless devices other than furnaces are also used, e.g., cathode sputtering, tantalum ribbons, and cold vapor. Of these, the cold vapor device may be the most useful since it is probably the best way to measure Hg (Rains & Menis, 1976).

Another device proven useful for determination of As, Se, and Sb by AAS is the hydride generator (L'Vov, 1961). The hydride form of these volatile elements is aspirated into a cool flame (hydrogen) or heated quartz tube of the AAS where atomic absorption is measured. Detection limits are low, in the nanogram range, usually allowing a measurement of 0.1 $\mu g\ L^{-1}$.

3. Advantages and Disadvantages

The primary advantages of AAS are: (i) highly specific for an element, (ii) minimum spectral interference, (iii) better detection limits for some elements than flame emission, and (iv) ease of operation.

The primary objective in using AAS is to enable a direct comparison of an unknown sample solution with a standard solution. Factors that should be considered are: (i) removal or compensation for interfering ions in the unknown sample, (ii) removal of solids to prevent clogging of the nebulizer, (iii) preparation of standard solutions to match, as nearly as possible, the matrix of the unknown sample, and (iv) addition of substances, usually organic compounds, to enhance the signal of the unknown and standard solutions.

Disadvantages of AAS are: (i) chemical interferences present for elements that form stable compounds; (ii) ionization enhancement of the signal for elements easily ionized when operating in the absorption mode, especially Na and K; (iii) matrix interferences caused by viscosity or specific gravity differences between sample and standard; (iv) much less linearity than for ICAP; (v) elements only analyzed one at a time; and (vi) not adequate for P analysis.

4. Preparation of Plant Tissue for Analysis

A small quantity of finely ground plant tissue is weighed (often 1 g for dry ashing and 0.5 g or less for wet digestion) for the determination of the elements contained in it. Dry ashing (often at temperatures of 773 K) and wet digestion are two methods used for dissolution of the sample. Various acids (HNO_3 and HCl) are used to acidify the dry ash and bring the elements of interest into solution. The solution is commonly filtered or centrifuged to remove particles that may plug the nebulizer. Acids, such as H_2SO_4, HNO_3-$HClO_4$, H_2SO_4-H_2O_2, and H_2SO_4-HNO_3-$HClO_4$, often are used to break down plant tissue by wet digestion. Because of ionic interferences, Baker and Smith (1974) recommend wet ashing over dry ashing. However, each system of sample dissolution has its advantages and disadvantages (Bock, 1979).

5. Preparation of Soil Samples for Analysis

The preparation of soil samples depends on the type of soil to be analyzed. Various soil extraction procedures have been developed in different areas that have been found to be suitable for those areas. For example, midwestern soils are generally analyzed by a Bray extractant (Bray & Kurtz, 1945) for P and an ammonium acetate extractant (Schollenberger & Simon, 1945) for mineral elements. Southeastern soils are generally analyzed by the Mehlich I or Mehlich III extractant (Mehlich, 1953, 1984). Basic western soils are analyzed by the Olsen extractant (Olsen et al., 1954) for P. The Morgan extractant (Lunt et al., 1950) is used in the northeastern USA and some western states. Some states use the DPTA extractant (Lindsay & Norvell, 1969) for micronutrient analysis. Detailed, appropriate procedures may be found in books edited by McKeague (1978) and Sabbe (1980) and bulletins edited by Dahnke (1975) and Johnson (1984). Where P extractants are mentioned above, AAS is not used for the analysis. Phosphorus is determined colorimetrically.

Soil samples are either weighed or measured by volume into appropriate vessels to which predetermined quantities of extracting solution are added. After a specified shaking period, the sample is filtered and the extract is used for mineral analysis. It is generally good analytical practice to prepare multi-element standards in the appropriate extracting solution so that the salt concentrations of the standards and extracts may be more closely matched to minimize viscosity variations.

B. Atomic Fluorescence

1. Principle of Operation and Use

Atomic fluorescence spectrophotometry (AFS) is not used extensively in laboratories for routine analysis of plant tissue or soil. However, this technique does provide important advantages over AAS (Lipari & Plankey, 1977).

The AFS relies on the principle that when atoms are excited in an energy source, either flame or nonflame, these atoms absorb radiation from an external light source causing greater atomic excitation. As these atoms undergo radiation decay back to the ground state, fluorescence can be observed at 1.58 rad to the light source. Resonance fluorescence, where excited atoms absorb and re-emit radiation at the same wavelength, is the most useful type for analytical purposes. The intensity of atomic fluorescence is directly proportional to the intensity of the light source. The light source must be of high intensity at the wavelength of interest and useful over a wide range of wavelengths. The light source must be stable, i.e., free of drift or flicker.

There are two basic types of AFS instruments, dispersive and nondispersive (Skoog, 1985). Dispersive instruments contain a modulated source, an atomizer, a monochromator or interference-filter system, a detector, a signal processor and readout. With the exception of the light source, most of these components are similar to those used for AAS. A nondispersive AFS instrument consists of a source, an atomizer, and a detector. Filters are also usually needed to overcome background radiation emitted by the flame or furnace.

2. Excitation Sources

The excitation sources used for sample solutions are generally the same as those used for AAS. Elements that can be determined with AAS also can be measured with AFS provided the light source can generate sufficient energy at the wavelength of the element of interest.

3. Advantages and Disadvantages

The advantages and disadvantages of AFS have been pointed out by Winefordner and Vickers (1964). A primary advantage is that only one light source is needed that allows rapid multi-element analyses. In AAS, different lamps are needed for each element for the most accurate analysis. Better sensitivity for far UV metals, such as Cd and Zn, can be obtained with AFS than with AAS. With AFS, wavelength scanning can be done to provide both a quantitative and rapid qualitative multi-element analysis. The AFS could provide a greater ease of operation and lower cost than AAS. However, the primary disadvantage has been lack of a sufficiently stable continuum light source. Metal vapor discharge sources have been tried for multiple wavelengths, and electrodeless discharge lamps (EDLs) have been used to some extent. However, EDLs are not available for many elements. Interferences for AFS are essentially the same as those involved in AAS. Since the primary limitation remains the lack of a very bright light source, work is underway to try to develop bright tunable lasers in the UV region of the electromagnetic spectrum. If this becomes possible on a practical scale, then the AFS will become a powerful analytical tool.

4. Preparation of Plant Tissue and Soils

The preparation of plant tissue and soils for analysis by AFS is essentially the same as that discussed for AAS. The nebulizer system can tolerate acidic solutions that are involved in the majority of the dissolution methods for plant tissue analysis.

C. Atomic Emission

1. Principle of Operation and Use

Atomic emission instruments are designed to measure the emission of radiant energy from excited atoms or ions as opposed to AAS, previously discussed, which involves measurement of the absorption of radiant energy by excited atoms. A review of optical emission spectrometry was published by Soltanpour et al. (1982).

When a neutral or partially ionized atom is excited by an energy source (i.e., flame and hot gas), valence electrons of the atom enter higher energy orbits around the nucleus. When the atom is removed from the excitation source, electrons at higher energy levels will return to their ground-state-energy orbits. Upon returning in discreet steps, increments of electromagnetic energy of specific wavelengths characteristic of the atom are emitted. Multiple wavelengths will occur because of the stepwise reduction in the energy. Light emission instruments use grating monochromators or polychromators to separate these wavelengths for measurement by a detector system. Intensity values for the emitted energy directly proportional to elemental concentration are collected and processed by a computer system.

a. Plant Tissue. Elaborations of using spark emission spectroscopy for plant tissue analysis were published by Jones and Warner (1964) and by Chaplin and Dixin (1974). Results of collaborative studies allowed the comparison of different instruments used for plant analysis. Jones and Isaac (1969) analyzed 91 plant samples that showed comparable results between spark emission spectroscopy with those obtained by AAS for several mineral elements in various kinds of plant tissue. Jones (1969) showed that for

18 laboratories participating in a plant sample exchange, agreement among laboratories using spark emission spectrographs was less than among laboratories using AAS.

A summary by Jones (1975) of a later collaborative study involving 11 laboratories showed that direct reading emission spectrographs were superior in analytical capability to the older photographic reading spectrographs. Analysis by spark emission showed coefficients of variation between laboratories of 3 to 7% for B, Cu, P, K, and Zn; while Fe, Na, and Al were >15%. These values are higher than usually found with ICAP spectrometers.

Scott and Strasheim (1975) were among the first to use an ICAP spectrometer for the determination of elements in plant tissue. Their results showed that ICAP compared favorably with AAS, spark emission, and x-ray fluorescence for plant tissue analysis. A detailed comparison of the analysis of plant tissue and other materials was made by Dahlquist and Knoll (1978). Results of the comparison showed that ICAP compared favorably with neutron activation analysis, energy dispersive x-ray fluorescence, rotating disk spark emission spectrometry, and flame AAS. They concluded that random and systematic errors of ICAP were generally smaller than or equivalent to those obtained by the other instruments. The effects of different acids and their strengths on the ICAP were evaluated as well. Both wet and dry ashing methods worked equally well provided ash particles were kept from entering the nebulizer of the ICAP.

A study of interelement and acid-content effects using 3.5% $HClO_4$ on the analysis of environmental materials was done by McQuaker et al. (1979). Results compared favorably with those obtained by Dahlquist and Knoll (1978). The nature of interference corrections required for the determination of elements in plant tissue by ICAP were studied by Spiers et al. (1983). Their studies involved single digest solutions without serial dilutions or special sample preparation techniques.

An extensive evaluation of 12 laboratories using ICAP for the analysis of plant tissue was done by Munter et al. (1984). Their study found that there were a variety of plant tissue sample preparation and instrument calibration procedures used. The best precision among laboratories was for P, Mg, Mn, Ca, and K. The worst precision was for Al, Zn, Fe, Na, Cu, and B. Much of the poor precision was caused by the diversity of sample preparation. These authors recommended that the following instrument checks be done to ensure quality of analysis: (i) spectral alignments, (ii) quantification of interferences, (iii) measurement of detection limits, (iv) calibration with high purity chemical solutions in a matrix similar to the plant-ash matrix, (v) periodic inclusion of calibration-control solutions, and (vi) adequate sampling and rinse-time intervals. It was recommended that detection limits be checked on a monthly basis to provide information on instrument performance. Spectral interferences of most concern for plant analysis and analytical wavelengths for different elements are given in the article. The plant tissue sample exchange indicated that background correction was not necessary for plant-ash matrices. The number of integrations varied from 1 to 3 with integration time generally 10 s, and wash times from 1 to 45 s. A calibration control check sample was placed at intervals of 20 to 30 among unknown samples. The acceptable variation was 2 to 6%.

Isaac and Johnson (1985) conducted a collaborative study on the determination of elements in plant tissue by ICAP spectrometry. Coefficients of variation were better than those obtained in an earlier collaborative study involving spark emission spectroscopy. In the 1985 study, 14 laboratories analyzed six different plant tissues for P, K, Ca, Mg, Mn, Fe, Al, B, Cu, Zn, and Na. All samples were dry ashed using the AOAC method 3.007a (13th ed.). The ICAP method was adopted as an official, first action method by the AOAC.

b. Soil. Jones (1977) showed the application of ICAP for the analysis of soils using the Mehlich I extractant. Soltanpour et al. (1979) demonstrated the use of ICAP for the analysis of P, K, Zn, Fe, Cu, and Mn in soil samples using the NH_4CO_3-DPTA extractant. In this comparative study between ICAP and AAS for K, Zn, Fe, Cu, Mn, and colorimetric P, no significant differences were found. However, the P results by ICAP were slightly higher than the corresponding colorimetric P results. This was due to the presence of organic P in the extracts. Munter and Grande (1980) in a comparative study showed that ICAP determination for Ca, Mg, Na, and K in an ammonium acetate extract compared favorably with AAS results. This same study showed good to excellent ICAP-AAS correlations using a DPTA extract for Cu, Zn, Fe, Mn, Cd, Ni, and Pb analysis.

2. Excitation Sources

a. Plasma. The term *plasma*, as used in conjunction with light emission instruments, refers to a hot gas in which a significant percentage of atoms have been ionized. Instruments that make use of a plasma source usually use inert Ar gas. However, N gas has also been used for specific analytical situations. The disadvantage of using Ar is its expense when compared to the cost of N. Temperatures of the Ar plasma source range from 5000 to 8000 K (Fassel, 1977). The high temperature of the plasma allows complete sample vaporization causing formation of free atoms and ions in the plasma. When Ar gas is used, the chemical environment surrounding the newly formed atoms and ions is essentially nonreactive, thus eliminating chemical interferences that are a problem with cooler energy sources.

Properties of plasma have been discussed by Babat (1947). Boumans (1978) summarized the kinds of plasma sources that are used for multielement analysis. Direct current plasma (DCP), capacitively coupled microwave plasma (CMP), radio frequency inductively coupled (ICAP), and microwave-induced plasma (MIP) have all been used. The DCP, CMP, and ICAP operated at relatively high power (0.5 to several kW) compared to MIP which operates at a few hundred watts power. The MIP instruments can be used to analyze small sample volume (1 μL to a few mL).

Instruments that have been designed to make optimum use of ICAP sources are rapidly becoming the dominant analytical instrument for measuring concentrations of many elements in plant tissue. Among the first to make practical use of the ICAP source was Greenfield et al. (1964) and Wendt and Fassel (1965). Fassel and Knisely (1974) elaborated on the principles of ICAP. Fassel (1977) discussed current and potential uses of ICAP.

The ICAP instrument passes Ar gas through a quartz torch located within an induction coil (Fig. 26–3). The induction coil is connected to a radio frequency (RF) generator. The RF generator produces power, usually at 1.5 to 3 kW and at a frequency of 27.1 MHz. The magnetic field, resulting from current flowing through the induction coil, causes electrons and ions passing through the oscillating electromagnetic field to flow at high acceleration rates within the quartz torch. As Ar gas atoms collide with the accelerated electrons and ions, ionization of the Ar gas occurs. These collisions cause the temperature of the gas to increase to 5000 to 8000 K.

Quartz torch development was critical to the routine use of ICAP instruments (Reed, 1961a). The torch can withstand extremely high temperatures because a high volume flow of Ar gas, other than that which forms the plasma, passes through part of the torch, acting as a coolant to the torch walls. Failure of the coolant Ar to pass through the torch will cause the torch to melt quickly. The consumption of Ar gas is approximately 12 L 60 s^{-1} to keep the torch walls cool and transport the sample. The torch is made of concentric channels. The outer channel is for the coolant gas, the innermost channel is for the sample, and the middle channel conducts auxiliary Ar gas which forms and sustains the plasma at about 0.5 to 1 L 60 s^{-1}.

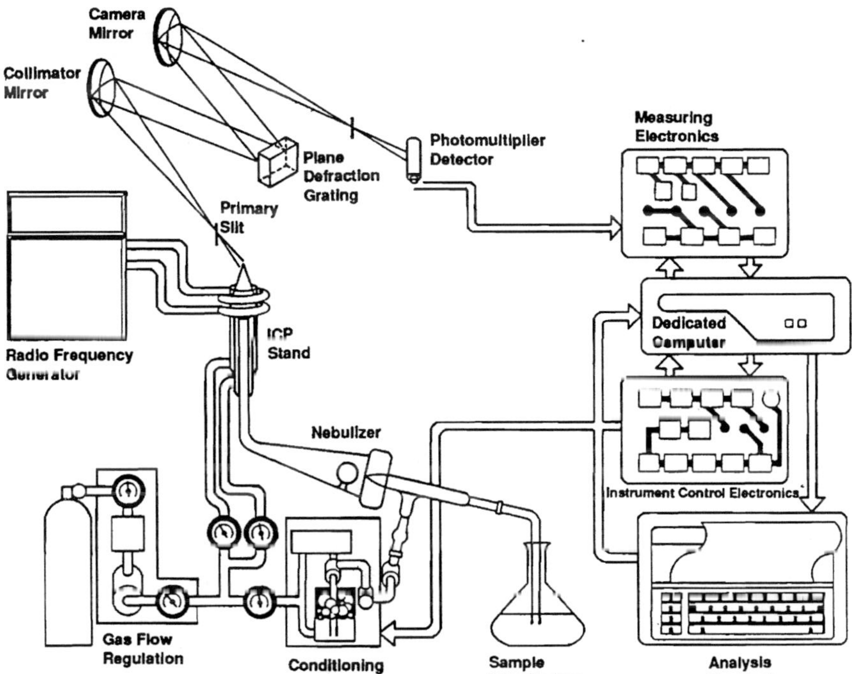

Fig. 26–3. Schematic diagram of a typical ICAP instrument (FISONS Applied Research Laboratories, Valencia, CA).

At approximately 15 to 20 mm above the coil, the emission of the analyte is observed. The optimum height for maximum sensitivity depends on the element of interest. However, in simultaneous multi-element analysis, a compromise height must be used. The residence time of the analyte in this region of the plasma is about 2 ms.

The sample is injected into the plasma through a nebulizer system. The role of the nebulizer is to separate small droplets of the sample stream from large droplets allowing the small droplets to enter the hot plasma. There are three distinct kinds of nebulizers: (i) crossflow, (ii) concentric, and (iii) Babington. Glass or tungsten carbide capillaries generally are used for cross-flow nebulizers. The V-shaped grooved Babington-type nebulizer is usually used for very high salt concentrations (Suddendorf & Boyer, 1978). Ultrasonic nebulizers also have been used to introduce sample solutions into the plasma to achieve a greater degree of atomization and, therefore, increased sensitivity. The nebulizer system is probably the most significant source of error to optimum ICAP operation. The nebulizer system must be kept very clean.

Spectrometers used in ICAP instruments are essentially the same in principle of operation as described before for the AAS instrument. Diffraction gratings are used to separate wavelengths, photomultiplier tubes are used to convert light energy to electrical energy, and readout systems under computer control are used to digitize the electrical signals.

Two basic types of spectrometers are used in ICAP instruments, those containing direct-reading polychromators for simultaneous analysis and those containing scanning monochromators for sequential analysis. ICAP instruments can be purchased that have both kinds within the same instrument. A direct-reading polychromator has a fixed grating and fixed exit slits with photomultiplier tubes located on the grating focal arc for each specific element wavelength. All wavelengths are fixed and relative to each other. Stray light in the spectrometer is reduced by using refractor plates, shields, and black internal surfaces. A scanning monochromator spectrometer has a single grating and usually a single photomultiplier tube for the measurement of intensities at all wavelengths within the range of the grating. The grating is movable and under the control of a computer, allowing the operator to use different wavelengths, one at a time.

The type of monochromator that is most useful will depend on the need. Polychromator instruments allow the measurement of many wavelength intensities of many elements simultaneously for a sample analyte. The scanning monochromator sequentially measures the wavelength intensity of each element of interest. Polychromator spectrometers are restricted to determining elements having photomultiplier tubes on the focal arc. A scanning monochromator or sequential instrument can determine any element that is amenable to ICAP analysis that is within the wavelength range of the grating of the instrument. The torches of some scanning monochromator instruments can be optimized for each wavelength, whereas with direct-reading instruments a compromise torch position is chosen, depending on the elements of interest. Wavelength selection is important for both types of ICAP, but is more critical with direct-reading instruments. Since changes are difficult, the primary objective in selecting a specific wavelength for analysis is that it be of sufficient intensity to be easily detected without interference from the intensity of other spectral lines or background radiation. Both types of ICAP instruments do this task well, depending on the resolution of the grating.

Evacuation of the spectrometer, or purging the spectrometer with Ar gas, has increased the versatility of ICAP, permitting accurate determination of elements such as N and S. Air in the spectrometer would cause reabsorption of the emission from these elements if it is not removed.

Spectral interferences can be a problem in the determination of some elements by ICAP. Emission energy consists of radiation from the analyte and interfering radiation from a variety of sources. Unwanted radiation may originate from (i) source background, (ii) stray radiation in the spectrometer, (iii) spectral line interference or band interferences, and (iv) sample matrix solution background. Source background radiation can be from (i) grating scatter due to imperfections in the grating grooves, and (ii) band interferences from overlaps of the concomitant-species spectral line with the spectral line of the element of interest. Discussions of spectral line broadening and recombination background interferences are presented by Larson and Fassel (1979). They indicate that interferences can occur even at 10 nm or more from the center of a strong emitting spectral line, such as that from Ca. Interference tables that predict spectral interferences due to the group IIIA elements have been developed by Anderson and Parsons (1984). Spectral interferences due to alkaline-earth elements are also included. Prominent analytical spectral lines were listed by Winge et al. (1979) and the selection of spectral lines to reduce interferences by Winge et al. (1982). More recently, Wohlers (1985) tabulated experimentally obtained wavelengths for ICAP instrumentation.

Detection limits for ICAP are equal to or better than AAS for many elements. The detection limit is usually defined as the analyte-concentration intensity equivalent to two times the standard deviation of the background intensity beneath the analyte spectral line. Usually a "quantitative detection limit" is used that requires the analyte concentration to be five times the standard deviation of the background intensity. Winge et al. (1979) have used three times the standard deviation of the background intensity as a practical detection limit. Detection limits are affected by the excitation source, sample matrix, nebulizer delivery efficiency, RF power, observation height above the coil, and flow rate of the sample carrier gas.

Table 26–1 shows the precision of the analysis of plant tissue over a 12-mo period in the Research-Extension Analytical Laboratory of The Ohio State University. Watters (1983) has discussed practical limits of precision for ICAP. He recommends that close examination of variance is warranted since computer software treats calibration functions as nonvariable quantities.

The ICAP instrumentation is being constantly improved. Sharicz (1983) has described advances in sequential ICAP design. Echelle gratings, discussed by Harrison (1949), minimize stray light, thus providing high resolution. The hydride generator is an important accessory system for the ICAP. Hydrides

Table 26-1. Statistics for analysis of corn leaf tissue with inductively coupled Ar plasma (ICAP).†

Element	Mean	SD	CV
		%	
P	0.44	0.019	4.32
K	2.82	0.102	3.62
Ca	1.18	0.041	2.47
Mg	0.36	0.009	2.50
	mg kg^{-1}		
Mn	73.2	2.5	3.42
Fe	96.8	4.5	4.65
B	61.6	2.0	3.25
Cu	9.5	0.3	3.16
Zn	47.9	1.6	3.34
Mo	2.2	0.5	22.7

† One hundred and forty analyses of internal check samples over a 12-mo period.

of As, Sb, Bi, Se, and Te can be introduced directly into the plasma (Hahn et al., 1982). MacDonald (1983) described a new ICAP spectrometer using concentric quartz tubes to control the flow of Ar gas into the torch. The advantages of this system are better plasma stability and fewer electronic components.

Ediger et al. (1985) describe a new ICAP that is computer controlled. The spectrometer of the ICAP is totally automated, and all multiple activities with which the analyst becomes involved during operation proceed simultaneously. The analyst can review previous data, generate reports, and define new analytical conditions, all while automated analysis proceeds. A microprocessor computer is in control of two scanning monochromators and an autosampler. Components controlled by the microprocessor are: (i) grating drive, (ii) source observation height, (iii) filter wheel, (iv) internal Hg source, (v) photomultiplier, and (vi) sensors for monochromator temperature and vacuum. The design of this ICAP instrument has allowed the determination of 50 elements per minute while scanning a 600-nm-wavelength range. Also controlled by the computer are: (i) RF power, (ii) Ar gas flow rate, (iii) ignitor, (iv) peristalic pump, and (v) safety interlocks. Five separate microprocessors are involved. A microcomputer (15 megabyte hard disk with two 360K floppy disks) communicates with all microprocessors. Spectra surrounding the analyte wavelength for each element can be retained on the computer's hard disk and mathematical smoothing may be performed on spectral data to provide a better signal-to-background noise ratio.

A new, novel approach uses a fixed grating, sequential spectrometer with multitasking software (Routh & Paul, 1985). The spectrometer determines elements sequentially by holding the grating in a fixed position while moving the photomultiplier tube to appropriate positions on the focal curve. The spectrometer uses one of two photomultiplier tube (PMT) detectors at a time. One PMT is sensitive for longer wavelengths and the other for shorter wavelengths. Multitasking software allows the analyst to set up the next task, process calibration, report data, and create a sample sequence file for the next sample set, all while the instrument is performing analysis on the present set of samples. Advantages of this new ICAP concept are high inherent light throughput, flexibility in wavelength selection (165–800 nm), intensity measurements made directly on peaks, and high rate of sample throughput.

Not only have ICAP instruments been used for the analysis of plant tissue and soil extracts, but instruments that maintain high temperature plasma with direct current (DCP) are also used. The high temperature Ar plasma is maintained between electrodes (usually three) receiving electrical current. The two electrodes of the anode are made of graphite and the cathode electrode is of tungsten. The electrode arrangement is an inverted Y, maintaining the plasma jet between them. The plasma jet is formed when the cathode electrode momentarily contacts the anode electrodes causing ionization of the Ar gas. Current then generates additional ions sustaining the plasma. A detailed discussion of this plasma source is given by Skoog (1985). Melton et al. (1978) have used DCP to analyze plant tissue for B. It was found that Li had to be added to mask emissions from alkali metals in the samples. Recovery of B from spiked samples ranged from 90.4 to 104.4%. A DCP instrument with an Echelle grating was used for trace element analysis by Johnson (1979).

b. Arc. The arc source relies on the passage of electricity across the gap between two electrodes to cause excitation of the sample. Graphite electrodes are usually used for plant tissue analysis. There are two types of arc instruments, direct current (dc) and alternating current (ac). Direct current arc instruments give poorer precision than ac-arc instruments. However, sensitivity of the dc arc is better than that of the ac arc. The spectrometer used with both arc sources is multichannel allowing many element determinations at one excitation time. Arc-source instruments have been used for plant analysis, but plasma source instruments have replaced them in most laboratories. Soltanpour et al. (1982) discuss the use of arc-source spectrographs for analysis of agricultural materials. Analysis of soils by arc-source instruments has not been widely used due to lack of sensitivity for pertinent elements and low sample throughput.

c. Spark. Alternating current-spark sources have been used for the analysis of plant tissue. Rotating disks have been used to pick up the sample solution and place it into the high voltage discharge. The spark discharge is cool and the sample solution is not consumed to any significant extent. The spark source instruments exhibit better precision but less sensitivity than arc sources. Normal excitation times are 30 to 40 s with the disk rotating at 30 rpm. Details involved in analyzing plant tissue with spark source instruments are given by Soltanpour et al. (1982). Plasma-source spectrographs have largely replaced spark source spectrographs for plant tissue analysis. The comments concerning arc-source instruments for soil analysis apply here also.

d. Flame. A flame-emission source has been used for the determination of many elements in plant tissue and soil extracts. Instruments using flame as an emission source have been used for the determination of K, Na,

Ca, and Mg for many years, and are known as flame photometers. Flame photometers were used extensively from 1950 to 1965, primarily prior to the development of AAS. A review of flame photometry and comparison with AAS, was made by Baker and Suhr (1982). The development of AAS, and more recently the multielement plasma source instruments, have replaced flame photometers in most soil and plant analysis laboratories.

3. Elements Determined

Plasma source instruments, using either polychromators or scanning monochromators, can determine many of the elements in plant tissue. The nutritionally important elements, P, K, Ca, Mg, Zn, Fe, Cu, Mn, and B are generally measured. Scanning monochromator and most polychromator instruments are able to measure other elements, such as Na, Al, Ba, Sr, Pb, Cd, Ni, Cr, and Co. Evacuated plasma spectrometers allow measurement of N and S as well. Arc and spark instruments measure P, K, Ca, Mg, Zn, Fe, Cu, Mn, and B, although not as accurately, nor as precisely, as plasma-source instruments.

Flame emission photometry may be used to measure concentrations of K, Na, Ca, and Mg, individually. This requirement is a disadvantage when compared with the multielement capabilities of emission spectrographs.

4. Advantages and Disadvantages

Acceptance of ICAP spectrometers over other instruments is due to the following advantages: (i) minimum chemical interferences, (ii) four to six orders of magnitude in linearity of intensity vs. concentration, (iii) multielement capabilities, (iv) rapid analysis, (v) accurate and precise analysis, significantly better than other emission sources, and (vi) detection limits equal to or better than AAS for many elements.

Direct current plasma instruments can essentially claim the same advantages as ICAP instruments, with the exception of reduced sensitivity of some elements by an order of a magnitude less than for ICAP. The DCP has the advantage of significantly lower Ar gas consumption compared to ICAP, is less expensive to purchase than an ICAP, and is able to accept solutions of higher salt and viscosity levels than can ICAP. A disadvantage of the DCP is the use of electrodes which must be replaced every few hours, introducing the possibility of greater drift with electrode wear. A disadvantage of the ICAP instrument is that it is expensive to purchase, especially the polychromator instrument. A disadvantage of ICAP, DCP, and other types of atomic emission instruments is the occurrence of spectral interference. However, most modern plasma spectrometer instruments use various techniques to correct or compensate for these interferences (Soltanpour et al., 1982).

With the development of plasma spectrometers, the arc and spark spectrometers are at a disadvantage since they are less accurate or precise in analyzing plant tissue digests and soil extracts than plasma-source instruments. This reduced accuracy/precision is primarily because of the many variables that affect excitation of the sample. Arc and spark spectrographs are not usually applicable to soil analysis because of this reduced sensitivity for some elements, e.g., P, K, Cu, and Zn.

Flame photometry has the advantage of simplicity and reduced cost when compared with ICAP, DCP, AC spark, or even AAS. However, the main disadvantage is that only one element can be analyzed at one time. A flame photometer, for example, would not be useful in a routine soil and plant analysis laboratory that requires the determination of many elements on thousands of samples. Furthermore, most modern AAS instruments provide emission-mode operation for the alkaline metals such as Na and K while providing absorption mode for the determination of other elements.

5. Preparation of Plant Samples

The preparation of the plant tissue samples for analysis by various emission instruments is similar to that previously given for AAS. It is important to have a uniform dissolution of the plant sample and either dry or wet ashing will suffice for most elements in the tissue. However, better recoveries of Fe with wet ash than with dry ash are usually the case. It is important to have solution standards in the same acid matrix, i.e., kind and strength of acid, in which the sample is prepared. The use of an internal standard element (often Li) in the sample is necessary when either a spark-emission spectrometer or a flame photometer is used. Since plasma spectrometers have been found to be very stable, use of an internal standard is not usually necessary.

6. Preparation of Soil Samples

Preparation of soil samples for analysis by various emission instruments is similar to that previously given for AAS. It should be noted that the filtrate should be free of any particulate matter.

D. X-Ray Fluorescence

1. Principle of Operation and Use

Fundamental principles involved in x-ray spectroscopy analysis were explained by Skoog (1985) and by Kubota and Lazar (1971). The deceleration of high energy electrons, or electronic transitions of inner orbital electrons of atoms, produces short wavelength electromagnetic radiation known as x-rays. X-rays used for analytical purposes are obtained in different ways (i) bombardment of a metal target with a beam of high-energy electrons, (ii) exposure of a substance to a primary beam of x-rays to generate a secondary beam of fluorescent x rays, and (iii) employment of a radioactive source which decays to produce x-rays. For a detailed discussion on sources and detectors refer to Skoog (1985).

The use of x-ray excitation is based on measurements similar to those used for optical spectroscopy, i.e., emission, absorption, scattering, fluorescence, and diffraction of the electromagnetic radiation. X-ray sources produce both continuous (Bremsstrahlung) and a discontinuous spectrum.

a. Emission. Electrons produced at the heated cathode of an x-ray tube are accelerated toward an anode (the target) at a potential that can be as high as 100 kV. When the electron beam collides with the target, part of the energy of the electron beam is converted to x-rays. At each collision, an electron is decelerated and a photon of x-ray energy is produced. The deceleration process usually involves a series of collisions, resulting in loss of kinetic energy after each collision. These collisions cause the energies of emitted x-ray photons to vary continuously over a wide range. Details of the x-ray photon energy are explained by Skoog (1985).

Spectral emission lines produced by resulting x-ray photons consist of two series: shorter wavelengths, the K-series; and longer wavelengths, the L-series. Elements with atomic numbers <23 produce K-series wavelengths. The x-ray spectra result from electronic transitions involving the electron orbitals nearest the atom's nucleus. When high-energy electrons from the cathode remove electrons from these orbitals in lightweight atoms, the K-series wavelengths are produced. Thus, collisions cause formation of excited ions that lose quanta of x-radiation energy as outer orbital electrons go through transitions to vacated inner orbitals. When an electron is lost from the second main quantum level, an L-series of lines result.

b. Absorption. When x-rays pass through matter, their energy is diminished when absorbed and scattered by the elements of the matter. An absorption spectrum is created and consists of a few well-defined absorption peaks indicative of the appropriate elements. Beer's law is as applicable to the absorption of x-ray radiation as it is for other types of electromagnetic radiation. In this case, then:

$$\ln(\mathrm{Po}/\mathrm{P}) = \mu_M p x$$

where x is centimeters of sample thickness, P and Po are the powers of the transmitted and incident beams, respectively; p is the density of the sample; and μ_M is the mass absorption coefficient ($cm^2\ g^{-1}$). This coefficient is characteristic of the element, but independent of the physical and chemical state of the element.

c. Fluorescence. Absorption of x-rays produces excited ions that return to ground state by transitions involving electrons from higher energy levels. As this process occurs, emission of x-ray radiation of wavelengths occurs that is identical to emission that results from excitation produced by electron bombardment. The main difference between absorption and emission is that absorption requires a complete removal of the electron, i.e., ionization, whereas emission involves transitions of an electron from a higher energy level within the atom (Skoog, 1985).

2. Types of X-Ray Fluorescence Instruments

X-ray fluorescence is used for qualitative and quantitative analysis of elements with atomic numbers greater than oxygen (>8). There are three basic types of x-ray fluorescence instruments: (i) wavelength dispersive, (ii) energy dispersive (EDXRF), and (iii) nondispersive. Of these three types, the energy dispersive instrument has recently proven most useful for routine analysis of plant tissue and soil (Williams, 1976; Knudsen et al., 1981). Wavelength dispersive instruments also have been used for determining elements in plant tissue (Allen et al., 1974a). Jones (1982) elaborated on the use of x-ray fluorescence spectrometry for soil analysis. Modern instruments are capable of detecting more than 80 elements. Birks and Gilfrich (1976) referred to three classes of x-ray fluorescence instruments. Two kinds of wavelength dispersive instruments are wavelength scanning and simultaneous wavelength, to allow measurement for many elements at one time. The third class is the energy-dispersion instrument in which a (Si [Li]) detector and multichannel analyzer record the intensity of many elements simultaneously.

A main advantage of the energy-dispersive instruments is their reduced cost compared with wavelength-dispersive instruments. The advent of high resolution (Si[Li]) detectors and minicomputers has increased their popularity as a routine analytical instrument.

Knudsen et al. (1981) used an EDXRF, multielement instrument for the analysis plant tissue. These authors provided minimum detection limits for various elements. Detection limits are adequate for all plant nutrients essential to plant growth having atomic numbers higher than 11. They found that x-ray energies were too low for determination of N and B. Approximately 100 plant tissue samples can be routinely analyzed per day for 10 or more elements.

3. Elements Determined

Knudsen et al. (1981) used EDXRF to measure Mg, Si, P, S, Cl, K, Ca, Mn, Fe, Cu, and Zn. Other elements that were determined included Cr, Ni, Pb, As, Al, Sr, Br, Cd, Co, Rb, Ba, and Mo. Minimum detection limit values of elements essential for plant growth were below concentrations found in normal or deficient tissue except for Mo, with a minimum detection limit of approximately 3 mg kg^{-1}. The elements showing the most variation were Ca, Fe, and Zn.

4. Advantages and Disadvantages

Results of x-ray fluorescence analysis of plant tissue compare favorably with AAS and ICAP analyses for many elements (Irons et al., 1976; Knudsen et al., 1981). A main advantage of EDXRF analysis is that it is a nondestructive analysis, i.e., not requiring dissolution of the sample. Consequently, a rapid analysis can be done. Additional advantages are its simplicity, the absence of collimeters, a crystal diffractor, and the closeness of the detector to the sample allowing the use of weak x-ray sources. These features decrease the cost of the EDXRF system compared to a wavelength-dispersive system. X-ray fluorescence offers an advantage in that spectral line interferences are unlikely.

A main disadvantage to using EDXRF for plant tissue analysis and soil analysis is the lack of standards made of matrix similar to the sample matrix. The matching of standard-sample matrix and the unknown-sample matrix is important due to the absorption characteristics of each sample matrix. Results of the analysis of the unknown sample can be either low or high if the matrices are not matched.

5. Preparation of Plant Tissue Samples

Dried samples should be ground to pass a 40-mesh screen, weighed, and then pelleted. It was found by Knudsen et al. (1981) that optimal pellet weights were 500 mg (32-mm diam.) and 100 mg (13-mm diam.). Weights could vary 25% without significant changes in intensities. Intensities of the light elements (Mg, Al, Si, P, and S) can be increased by using boric acid to back pellets. However, when boric acid was used to back pellets, intensities from elements of higher atomic number decreased from 60% to <20% for those of nonbacked pellets. NBS reference samples were used as calibration standards. In addition, cellulose matrix standards were used successfully as calibration standards.

6. Preparation of Soil Samples

Soil samples have been generally prepared by grinding to 100-μm size and then fusing with either borax or lithium borate.

E. Electron Microprobe

1. Principle of Operation and Use

Although electron-microprobe analysis is not used for routine analysis of plant tissue samples, this methodology is especially useful in researching the accumulations of various elements in different tissues and cellular components as well as in determining the association of heavy metals with chemical fractions or phases in soils.

The electron microprobe relies on the principle of x-ray emission, discussed previously under x-ray fluorescence. Emitted x-rays are detected and analyzed with a wavelength- or an energy-dispersive spectrometer. Modern electron microprobe instruments use three integrated beams of radiation, i.e., electron, light, and x-ray. A vacuum system providing a pressure of $<1.33 \times 10^{-3}$ Pa (Skoog, 1985) is used.

2. Elements Determined

Rasmussen and Knexek (1971) described the operation of the electron microprobe for analysis of plant tissue and soil. The elements that can be determined are essentially the same as those measured with x-ray fluorescence instruments. Scanning electron beams are useful in studying the surface chemistry of various plant tissues, measuring the elements contained therein. Lauchli and Schwander (1966) were among the first to use the microprobe to study localization of minerals in plant tissue. Miller et al. (1970) evaluated P precipitation at the root surface. Determination of K and Cl ions in chloroplasts, cytoplasm, vacuole, and nuclei of plant leaf cells was made by Pallagy (1973) using a scanning energy dispersive x-ray microprobe. Hayward and Parry (1973) used the electron microprobe to measure the concentration of Si in leaf tissue and internode regions of barley (*Hordeum vulgare* L.). Lee and Kittrick (1984) used an electron microprobe to determine elements associated with Zn and Cu in both an oxidizing and an anaerobic soil environment.

3. Advantages and Disadvantages

The advantage of this method of analysis is use of intact tissue samples. Since the electron beams are so narrow (0.1–1 μm diam.), very small areas of plant tissue or soil can be analyzed.

The main disadvantage is that the system does not lend itself to rapid analysis. Furthermore, an electron microbe is expensive. It is more suitable for research than for routine use in plant or soil analysis laboratories.

4. Sample Preparation

The preparation of the sample is unique for the kind of tissue analyzed. Usually the selected tissue is at the cellular level and often fixative methods must be used to maintain the cellular structure. Rasmussen and Knexek (1971) provided additional references for sample preparation procedures.

F. Potentiometric Methodologies

1. Principle of Operation and Use

Potentiometric methods are used to measure the concentrations of ions in plant tissues and soil extracts. These methods usually involve ion selective electrodes. Pommer (1967) discussed principles and applications of glass hydrogen-ion selective electrodes for the measurement of pH in soils. A comprehensive study of non-glass ion selective electrodes was reported by Durst (1969). Potentiometric methods for ion selective electrodes have been reviewed by Carlson and Keeney (1971), Buck (1974), Allen et al. (1974d), and Skoog (1985).

The potentiometric method of measuring ions in plant tissue and soil extracts involves the use of a differential membrane. A membrane separates the external sample solution from the internal solution, which contains the ion of interest. Electrical current is carried through the membrane by a single ion species. The ion electrode is read against a reference electrode which completes the electrical circuit.

Membrane potential is logarithmically related to the ionic activity, expressed as:

$$E = 2.3 \frac{RT}{nF} \log A$$

where E = net membrane potential, A = ionic activity, n = valence, and the remaining expression is the Nernst factor. The electrode responds to ion activity, but ionic activity is usually sufficiently related to the concentration of dilute monovalent solutions to be an acceptable measure of concentration.

There are certain properties that membranes must have to be used as an ion selective electrode. These properties are: (i) the membrane material must have minimal solubility in analyte solutions (usually aqueous), i.e., approaching zero; (ii) the membrane must exhibit small electrical conductivity; and (iii) the membrane or some ionic species contained in the membrane matrix must be able to bind selectively with the analyte ion. Usually ion-exchange, crystallization, and complexation binding are found in the membrane. There are different kinds of membranes that are used in electrodes (i) glass used for H^+, Na^+, K^+, Ag^+, NH_4^+, Li^+, Cu^{2+}, Fe^{3+}; (ii) liquid used for Cl^-, NO_3^-, Ca^{2+}, Cu^{2+}; (iii) solid ion exchange polymer membrane, used for Ca^{2+} and Mg^{2+}; (iv) crystalline solid state, used for Cl^-, F^-, I^-, S^-, Ca^{2+}, and Cu^{2+}; and (v) silicone rubber impregnated with appropriate precipitation, used for Cl^-, F^-, I^-, SO_4^{2-}, and PO_4^{3-}.

2. Elements Determined

The most commonly used electrode for soil analysis is the glass hydrogen-ion selective electrode. Virtually all soil testing laboratories today use this electrode for measuring soil pH. Methods using the hydrogen electrode have been reported by Peech et al. (1947), Schofield and Taylor (1955), and Peech (1965). Methods using the hydrogen electrode to measure pH of buffered soil suspensions to determine soil lime requirement have been developed by Shoemaker et al. (1962), Adams and Evans (1962), Yuan (1974), Mehlich (1976), and McLean et al. (1977). While the pH electrode is not generally used for plant tissue analysis, it is extremely important in soil analysis.

Probably the most common electrode used for plant tissue analysis is the nitrate electrode. Methods using the nitrate electrode are reported by Paul and Carlson (1968), Baker and Smith (1969), Raveh (1973), and Sweetsur and Wilson (1975). Methods using the nitrate electrode for soils have been reported by Meyers and Paul (1968), Bremner et al. (1968), Mahendrappa (1969), and Raveh (1973). Different extracting solutions for plant tissue have been used, but the most common is 0.025 M $Al_2(SO_4)_3$ solution (Baker & Smith, 1969). The pH of this solution should be buffered to reduce interference from other anions. This solution also contained 10 μg mL^{-1} of NO_3-N. Paul and Carlson (1968) used Al resin to reduce interference from HCO_3^- and organic anion interference. Their measurements of NO_3-N with the NO_3 electrode compared favorably with the determination of NO_3-N with the phenoldisulfonic acid method.

Chloride can interfere with the measurement of low concentrations of NO_3. Resins in the silver form have been used to reduce chloride interference. Other interfering ions, at various concentrations (Skoog, 1985) to the NO_3 determination, are ClO_4^-, I^-, ClO_3^{2-}, CN^-, Br^-, HS^-, CO_3^{2-}, $H_2PO_4^-$, HPO_4^{2-}, PO_4^{3-}, OAC^-, F^-, and SO_4^{2-}.

Chloride ions extracted from plant tissue and soils have been measured with a chloride-specific electrode (Havemann, 1984; Moody & Thomas, 1977; Krieg & Sung, 1977; LaCroix et al., 1970). Interference of chloride measurement occurred from both malic and citric acids. Other ions that can interfere, depending on the concentration, are CN^-, I^-, Br^-, $S_2O_3^{2-}$, NH_3, OH^-, ClO_4^-, OAC^-, HCO_3^-, SO_4^{2-}, and F^-. Interference from these ionic species can be substantial if their concentrations are greater than the chloride concentration.

Probably the most ideal ion selective electrode is that for fluoride. Extracts of plant tissue and soil have been analyzed for F^- with the fluoride electrode (Baker, 1972; Louw & Richards, 1972; Van Den Heede et al., 1975; Kauren, 1977; Jacobsen & Heller, 1978; Villa, 1979; Eyde, 1983). Comparative and collaborative studies were conducted by Van Den Heede et al. (1975) and by Jacobsen and Heller (1978). It was found that the F^- electrode was acceptable, only slightly more variable than the distillation method (final action) of the AOAC. The only ions that interfere directly are the OH^- at pH 8 and H^+ at pH 5. A total ionic-strength-adjustment buffer (TISAB) is used to compensate for variable ionic strength in samples as well as maintain the solution in an optimum pH range. If Al is at an elevated concentration in the plant tissue, F^- may be complexed by Al, reducing the measured F^- concentration. Citrate has been used as a complexing agent to mask out the Al (Kauren, 1977).

A gas-sensing electrode specific for NH_3 has been used to measure total N in Kjeldahl digests of plant materials (Eastin, 1976). Gas-sensing probe is a better name than electrode since it consists of a specific ion and reference electrode immersed in an internal solution that is retained by a thin membrane permeable to the gas. A description and mechanism of response is provided by Skoog (1985).

Other electrodes that have been used in plant analysis work include Br (Abdalla & Lear, 1975) and Na (McNerney, 1976) electrodes. Potassium and Ca electrodes have also been used for plant tissue and soil analysis as discussed by Carlson and Keeney (1971).

3. Advantages and Disadvantages

A primary advantage of using ion selective electrodes for determining various ion concentrations in plant tissue and soil extracts is the speed of determination. The electrodes are relatively inexpensive and simple to operate.

There are, however, some important disadvantages that should be kept in mind. Numerous interferences can occur if concentrations of the interfering ionic species are substantially greater than the ion of interest in the sample. The effects of the interfering ions can be quite large, and removing the interference can be complex and time consuming. All of the mentioned electrodes require the ion of interest to be extracted from the plant tissue or soil before measurement. Methods for extracting the ionic species are important since most methods try to reduce the effect of the interfering ion during extraction. For the hydrogen ion electrode, which is primarily used to measure the pH of soil-water suspensions, accuracy is affected by ionic strength differences between soil samples and solution pH standards, temperature, colloidal matter, suspended matter, and the presence of residual liquid-junction potentials.

Another disadvantage is that considerable care must be taken in maintaining these electrodes. Maintenance is less of a problem with the advent of the modular approach for some electrodes (NO_3^-). Most manufacturers will guarantee correct functioning of the electrode for several months.

4. Plant Sample Preparation

Preparation of plant tissue generally requires that tissue be ground to pass a 20-mesh screen, usually a 40-mesh screen is preferred. As indicated previously, specific solutions must be used to extract the ion from the plant tissue. In certain cases (e.g., F^-), the range of pH must be maintained by a buffer reagent. Interfering ions are usually removed by resin techniques or complexing with other ions. Consistent temperature conditions are also important for correct measurement. An inherent error in analysis using ion-selective electrodes is that seldom do the standard samples match the unknown samples in ionic strength. This can be difficult since little is known about the unknown sample. However, these differences can be minimized by addition of TISAB solution mentioned previously. Sensitivity of ion selective electrodes is not usually as good for elements that can be measured by other techniques.

5. Soil Preparation

Soil samples are usually ground to pass a 10-mesh sieve, and extracted with a specific solution for the ion being determined. As previously noted, soil pH measurements are usually conducted in a soil-water suspension. Various soil-water systems ranging from a paste to soil-water ratios of 1:1, 1:2, 1:5, and 1:10 have been reported. The most commonly used soil-water ratios are 1:1 and 1:2.

G. Coulometric-Amperometric Titration

1. Principle of Operation and Use

Chloride titrators have been used as an accurate and precise instrument to measure the concentration of chloride ions extracted from soil and plant tissue samples.

The principle of coulometric titrations is explained by Skoog (1985). Potentiostatic and amperostatic techniques are used for coulometric analysis. The amperostatic technique is used for the determination of Cl^- concentration in soil and plant tissue extracts. Electrical current in a coulometric titration is carefully maintained at a constant and accurately known level. The product of this current in amperes, and the time in seconds required to reach an end-point, yields the number of coulombs, which is proportional to the quantity of analyte involved. The end-point is detected amperometrically in chloride titrators.

In the chloride titrator, a constant dc is passed between a pair of Ag generator electrodes (Ag wire) in the coulometric circuit, causing release of Ag ions into the titration solution at a constant rate. Since Ag^{2+} are produced at a constant rate, they react with Cl^- in the sample to form a precipitate. The end-point is reached when all the Cl^- in the sample has been precipitated. The end-point is indicated because of increasing concentration of Ag^{2+} which causes a rising current to flow through the pair of Ag electrodes. The amperometric circuit senses the increase in current, which stops the timer. The timer runs concurrently with the generation of Ag^{2+}. Since the rate of Ag^{2+} generation is constant, the amount of Cl^- precipitated is proportional to the elapsed time.

2. Elements Determined

Some of the commercial coulometric titrators are multipurpose for the determination of a variety of inorganic species. However, most are designed for a single analysis with the chloride titrator most commonly used for soil and plant tissue analysis. Others are designed for sulfur dioxide, carbon dioxide, and water determinations.

3. Advantages and Disadvantages

The main advantage of the chloride titrator for measurement of Cl^- in soil and plant tissue extracts is the lack of interfering species, although sulfide and iron will interfere at very high concentrations. The instrument is sensitive to low chloride concentrations. The primary disadvantage is that samples high in chloride concentration may take considerable time to reach the end-point.

4. Preparation of Plant Tissue and Soil

Plant tissue is dried and ground to pass at least a 20-mesh sieve. Soil samples are dried and ground to pass a 10-mesh sieve. Two grams of plant tissue or 10 g of soil are extracted with 0.1 *M* HNO_3 acid and 10% acetic acid. The extracting solution is filtered from the sample and the chloride measured directly by the chloride titrator.

H. Neutron Activation Analysis

1. Principle of Operation and Use

Although neutron activation analysis (NAA) is not usually used for routine analysis of soil and plant tissue, it provides an analytical capability that can be extremely valuable to soil and plant analysis laboratories. It provides a means to check on the accuracy of other analytical procedures used for analyzing soil and plant tissue.

Details of the principle involved in NAA are given by DeSoete et al. (1972), Helmke (1982), and Skoog (1985). The NAA relies on measurement of radioactivity that has been induced in the sample by irradiation with neutrons. Thermal neutrons from a reactor or radioactive decay source are used to induce radiation in the sample. Properties of neutrons are a function of their kinetic energy. Slow neutrons have kinetic energies of 1 keV, while kinetic energies of fast neutrons are approximately 0.5 meV. Thermal neutrons, those in thermal equilibrium with their environment at room temperature, are used in NAA and have kinetic energies of approximately 0.025 eV (Skoog, 1985).

Nuclear reactors are generally the source for NAA. This neutron source will have high neutron fluxes allowing detection limits for many elements in the 10^{-3} to 10-μg range. Radioactive sources have also been used. The most common source is californium-252 which has a half-life of 2.6 yr. Neutrons are formed from spontaneous fission.

After neutrons are produced, they are used to bombard the target material. The two main reactions encountered at the time of bombardment are neutron capture and neutron transmutation. Neutron capture involves capture of a neutron by an atomic nucleus to give an isotope of a mass number greater than one. The emission of a gamma ray occurs simultaneously. Transmutation involves the absorption of the neutron followed by a release of a proton.

Neutron activation methods can be classified as destructive and nondestructive. For the destructive method, the sample must be dissolved chemically before the element of interest is counted. For the nondestructive method, the sample is analyzed intact without any preparatory pretreatment.

The nondestructive method requires a spectrometer that can isolate the gamma-ray signal produced by the element of interest from signals arising from other components. Adequate resolution depends on the complexity of the sample and the presence or absence of elements which produce gamma rays of about the same energy as the element of interest.

Modern NAA analysis uses computer-controlled detector/analyzer systems. In North America, there are more than 60 locations that have facilities for NAA (Katz, 1985).

2. Elements Determined

NAA is potentially applicable to 69 elements (Skoog, 1985). The elements Cu, Mn, Al, Cl, K, Na, Rb, As, Ca, Fe, Mg, Zn, Cr, Co, and Se have been determined in various plant tissues (Capar et al., 1978). Haskin and Zieg (1971) indicate possible NAA of urea in soil and plant analysis.

3. Advantages and Disadvantages

The main advantage of NAA is its extreme sensitivity for some elements. Sensitivity for some of the nutritive elements are: (i) K, 5×10^{-2} μg; (ii) Mn, 5×10^{-5} μg; (iii) Co, 5×10^{-3} μg; (iv) Cu, 1×10^{-3} μg; and (v) Zn, 1×10^{-1} μg (Skoog, 1985). Another advantage is that for many elements the nondestructive method is adequate, providing a more rapid anal-

Table 26-2. Coefficients of variation of the analysis for various elements with atomic absorption spectrophotometer (AAS), inductively coupled Ar plasma (ICAP), and neutron activation analysis (NAA) instrumentation (Capar et al., 1978).

Element	AAS	ICAP	NAA
		%	
Al	--	2.9	4.9
As	5.2	--	5.8
Ca	--	4.8	5.8
Fe	2.1	1.7	5.9
Mg	1.3	1.1	28.0
Mn	3.9	2.2	5.5
Zn	1.8	2.2	4.6

ysis than the destructive method. Furthermore, NAA is free from chemical interferences and is also a multielement technique.

A disadvantage is that a neutron source is required, which is expensive and potentially dangerous to operate. The handling of radioactive isotopes requires special knowledge and training. Destructive methods are needed for some elements to increase the sensitivity of detection. Another disadvantage is the length of time the sample is counted. For some species, hours, days, or weeks may be required for accurate analysis. Usually about 5 to 15 min is sufficient time for short half-life isotopes.

Capar et al. (1978) compared the analytical precision of NAA with that of ICAP and AAS in the determination of Al, As, Ca, Fe, Mg, Mn, and Zn in plant tissue (Table 26-2). The NAA method compared favorably with the other two methods for all elements except Mg.

4. Sample Preparation

Plant tissue samples are dried and ground to pass a 100-mesh sieve. The sample size is generally between 50 to 500 mg. Samples of plant tissue or soil are sealed in polyethylene vials for short irradiation times and in heat-sealed quartz tubing for irradiation times of several hours (Helmke, 1982). Effects of sample thickness, sample shapes, and distance to detector window have been studied by Heft (1976). Presampling contamination, sample storage containers, storage technique, irradiation procedures, and irradiation containers have been studied by Becker (1976).

I. Sulfur Analyzers

1. Principle of Operation and Use

Some plant analysis laboratories use S analyzers to measure the quantity of S in plant tissue. Before the development of S analyzers, the determination of plant tissue S often required dissolution of the sample with perchloric acid followed by a turbidimetric measurement of the sulfate in solution. Sulfur analyzers have greatly simplified S analysis. In soil analysis, sulfate is much more important since it is the form most readily available to plants. Since the S analyzers determine total S, generally they are not used for soil analysis.

Bremner and Tabatabai (1971), Jones and Isaac (1972), and Hern (1984) have described S analyzers that are used to measure S in soils and plant tissue. Early model S analyzers used induction furnaces to oxidize plant tissue or soil in a stream of oxygen, converting S to SO_2. The quantity of SO_2 was measured with iodometric titration. Modern versions of the instrument make use of resistance furnaces and use an infrared absorption cell or an amperometric titration to measure the quantity of SO_2 and thus S contained in the sample. In the IR version, the SO_2 molecule absorbs energy in the infrared region of the spectrum allowing a comparison of SO_2 produced from an unknown sample with that of a standard sample. In the titration version, SO_2 gas is scrubbed from the combustion gases by a diluent. As SO_2 is absorbed by the diluent, the current decreases. A titrant is then added to restore the current to a preset level. The amount of titrant used is proportional to the SO_2 absorbed. Modern S instruments feature computer-controlled operations and calculation of the S present in the sample.

2. Advantages and Disadvantages

The analysis is rapid and clean. In the IR version, no chemicals are involved, other than a catalyst. No accelerators are needed, as is the case when induction furnaces are used. Samples do not require any pretreatment other than drying and grinding. Modern instruments are more precise and easier to operate than those using iodometric titration. The precision of 132 plant tissue samples analyzed over 1 yr in the Research-Extension Analytical Laboratory of The Ohio State University with a S analyzer was 10.85% relative standard deviation with a mean of 0.30% S.

Disadvantages are that oxygen gas is required and the furnace operates at very high temperatures (about 3000 K). Because of the high temperature, ceramic boats are needed to hold the sample. Vanadium pentoxide is required as a catalyst to ensure complete oxidation and release of SO_2. Interferences from smoke and chlorine in the sample can occur. It is necessary to use organic matrix samples as calibration standards.

3. Sample Preparation

The plant sample should be dried and ground to pass at least a 20-mesh sieve. Soil samples should be dried and ground to pass a 10-mesh sieve. The analysis works best when very fine material is used. It is sometimes necessary to compress the sample into pellets to obtain a more uniform oxidation of the sample, especially with low bulk density samples. The quantity of sample ranges from 0.1 to 1.0 g. Usually 0.5 g is sufficient.

J. Automated Analyzers

1. Automated Macro-Kjeldahl Instrument

a. Principle of Operation. Nitrogen determination by the Kjeldahl procedure can be done by an automated instrument. The automated steps are (i) acid and catalyst mixture added to the sample, (ii) initial digestion, (iii) final digestion, (iv) cooling, dilution, and NaOH addition, (v) steam distillation and titration, and (vi) evacuation of the flask.

b. Advantages and Disadvantages. An automated instrument can analyze 150 to 200 samples per day. Once the turret is full, a N analysis is completed every 3 min. Analytical results are equal to or better than the manual Kjeldahl methods. A comparison of the automated macro-Kjeldahl instrument with other methods of analyzing plant tissue for N was made by Wall et al. (1974). It was found that N analysis with the automated macro-Kjeldahl method gave a relative standard deviation of $< 1\%$. The automated method has been adopted as a first action by the AOAC for N determination of animal feed materials (Noel, 1976).

A disadvantage of the instrument is that a mercuric oxide catalyst must be used since the digestion must be completed in 6 min. Bjarn (1980) substituted an antimony based catalyst for mercury without loss of accuracy for most plant samples.

Other macro- or semi-macro-Kjeldahl instrument systems include the same essential steps with some of the steps automated. For example, some semi-automated instruments may have only the distillation and titration steps automated. Other systems use automated continuous flow analysis to measure the NH_4^+ in the digest solution. Digestion of the sample usually is done with block digestors when these systems are used.

An automated instrument developed by the LECO Corporation (St. Joseph, MI) uses a combustion method to measure the concentration of N in plant tissue samples. A comparison between N determination by this instrument and the AOAC copper catalyst Kjeldahl-N method was done by Sweeney and Rexroad (1987). The results of their study showed that the mean N values of 14 samples was 8.61% for the LECO combustion, and 8.58% for the AOAC Kjeldahl method. Precision of analysis compared favorably for the two methods with a range in standard deviation from 0.013 to 0.052% N for the combustion and 0.006 to 0.035% N for the AOAC Kjeldahl method. The main differences occurred in the analysis of samples containing NO_3-N. The LECO instrument measured the N from the NO_3^- while the Kjeldahl did not. Once the sample is weighed (about 150 mg) into a tin capsule, the analysis proceeds automatically. Analysis time for one sample is approximately 3 min. Three gases are used: oxygen for combustion, He as a carrier, and compressed air to drive the valve system. In this system, the N_2 is evolved and measured by thermal conductivity.

c. Sample Preparation. Sample preparation is similar to that for manual macro-Kjeldahl. The sample must be dried and ground to pass at least a

20-mesh sieve. One gram of sample is usually used unless the N content is suspected to be inordinately high, then 0.5 g is used.

2. Continuous Flow Air-Segmented-Stream Instruments

a. Principle of Operation. Continuous flow systems were developed to automatically analyze solutions for selected elements. These systems use separate modules linked together to form a continuous single or multichemistry flow system. The modules generally consist of sampler, peristaltic pump, mixing manifold, colorimeter, and recorder. In addition to these basic modules, other components may include dialysis membranes, delay coils, and temperature baths. Microprocessors or microcomputers may be added to make necessary computations and manage data (Allen et al., 1974b).

Air bubbles are added to the analytical stream to segment the stream. Air bubbles aid the mixing of reagent chemicals. Rollers of the peristalic pump alternately squeeze the plastic tubing, thereby forcing the segmented stream to flow through the system. Chemical reactions that produce color occur in the mixing coils that are part of the manifold. The air bubble is removed with a debubbler just prior to the stream entering the colorimeter. Some rapid flow analyzers do not remove the air bubble, but account for the bubble by the use of an electronic optical chopper system.

b. Elements Determined. Determinations of P, K, Ca, and Mg in wet digestion solutions of plant tissue have been made with this type of automated analysis system (Steckel & Flannery, 1968). Basson et al. (1969) adapted a continuous flow system to measure B in plant tissue. The use of autoanalyzer systems for the analysis of soils and plant tissue was done by Isaac and Jones (1970). Twine and Williams (1971) determined P and N in Kjeldahl digests of plant tissue. It was also found that the flow system could handle HNO_3-$HClO_4$-H_2SO_4 digest of the plant tissue as well. Dual channel systems to measure P, K, Ca, and Mg in HNO_3-$HClO_4$ digestion solutions were discussed by Steckel and Flannery (1971). They were able to analyze 35 samples per hour. Their results compared well with those analyzed by AAS, where applicable.

Schuman et al. (1973) found the detection limit for N as NH_4^+ to be 0.5 $\mu g\ mL^{-1}$ for a continuous flow system. An automated measurement of Cl^- with the continuous flow air-segmented stream was done by Gaines et al. (1984). Automated turbidimetric methods for SO_4^{2-} using the continuous flow concept have been developed by Ferrara et al. (1965), and Sansum and Robinson (1974). Wall et al. (1980) reported an automated turbidimetric method for SO_4^{2-} in plants, soils, and fertilizer. An automated continuous flow system was used to measure S in wet digests of plant tissue (Basson & Boehmer, 1972).

c. Advantages and Disadvantages. The primary advantage of the continuous flow air-segmented analysis system is improvement in analytical precision since each sample has the same time for color development. Another advantage is that once the system is started, little attention is required for most determinations. The analysis is rapid, usually 20 to 40 samples per hour depending on the determined element. Rapid flow analyzers can analyze samples at a faster rate, e.g., NO_3-N at 120 samples/h.

A disadvantage is that relatively new pump tubes must be used. Chemical interferences can occur, but usually are minimized by using complexing agents and generally are not a problem for the concentration of constituents normally found in plant tissue. A disadvantage of the system is that the plant tissue must have had a preparatory dissolution treatment. Numerous varied chemical reagent solutions that are needed also can be considered a disadvantage.

d. Sample Preparation. Soil and plant samples must be dried and ground and given appropriate treatment to bring the element of interest into solution. Generally, plants are wet digested with H_2SO_4 or HNO_3-$HClO_4$, or are dry ashed. The quantity of plant tissue sample ranges from 100 mg to 1 g depending on the element of interest and the kind of plant tissue. After the samples are in solution, small quantities (3–10 mL) are poured into sample cups that are placed into the autosampler trays.

3. Continuous Flow, Nonsegmented Stream Instruments

Nagy et al. (1970) and Ranger (1981) have used continuous flow, nonsegmented stream instruments to measure various elements and radicals in solution, i.e., flow injection analysis (FIA).

a. Principle of Operation. The primary features of FIA are: (i) controlled sample dispersion, (ii) variable flow rates, (iii) baseline resolution between each sample, (iv) high sample throughput, and (v) absence of reaction stabilization time. The system is similar to the continuous flow, air-segmented system in that a sampler, pump, reaction manifold, and detector are needed. The main difference is that an injection valve is used to inject the sample into the carrier stream as segments. Detectors can be ion selective electrodes, AAS, ICAP, and colorimeters. Timing is the most important part of the system. Injection valves must be precise to ensure exact volumes are injected at precise intervals. When the sample is first injected, it forms a slug in the reagent carrier stream, but as the sample moves through the narrow bore tubing, the slug disperses into the carrier stream under laminar flow conditions to form a concentration gradient. The degree of dispersion depends on sample volume, tubing bore size, tubing length, flow rate, and mixing coil diameter.

Radial, rather than axial, dispersion in the stream contributes most significantly to sample dispersion. Fluid is moved both toward and away from tubing walls creating a scrubbing action on the walls, reducing sample carryover.

b. Elements Determined. The elements that can be determined are generally the same as determined by continuous flow, air-segmented systems. In addition, the FIA system has been used to supply samples to AAS and ICAP instruments (Alexander et al., 1982). Used in this way, FIA can greatly increase the speed of analysis. Also it is useful when only very small volumes of sample are available.

c. Advantages and Disadvantages. A primary advantage of FIA is speed. The absence of air bubbles has allowed analysis rates of 100 to 300 samples per hour (Skoog, 1985). Nagy et al. (1970) have indicated analyses of 500 samples per hour are possible if small bore tubing is replaced by packed-bed reactors. Elimination of air bubbles has allowed for enhanced response time, more rapid start-up and shut-down times, and simpler equipment, except for the injection system. Small sample solutions, typically 10 to 30 μL are used. Flow injection analysis is precise, accurate, and simple to operate.

A disadvantage is that precise and durable injection valves must be used to inject the sample rapidly and accurately each time. Wear of the injection valves has been a disadvantage, but, improvements to the injection system continue to be made (Skoog, 1985).

d. Sample Preparation. Preparation of soils and plant tissue for FIA is the same as for continuous flow, air-segmented stream analyzers. It is important that samples be free of particles to prevent plugging or wearing of the injection valve. The sample solution should not be so viscous that it reduces mixing with reagents.

K. Voltammetry and Polarography

1. Principle of Operation and Use

Principles of voltammetry and polarography are discussed by Allen et al. (1974c), Street and Peterson (1982), and by Skoog (1985). Voltammetry is based upon measurement of a current that develops in an electrochemical cell under conditions of complete concentration polarization. This contrasts with potentiometric methods where measurements at currents that approach zero and polarization are absent. It is different than coulometry in that means are taken to minimize or compensate for the effects of polarization. Also, in voltammetry, minimum consumption of the analyte occurs, while all the analyte is consumed in coulometry, i.e., converting it to another state.

Voltammetry constitutes a group of electroanalytical methods that measure current as a function of applied potential obtained under conditions that encourage polarization of the indicator electrode. Polarography is the most widely used voltammetric method. It uses a Hg micro-electrode, in the form of a dropping Hg electrode (DME). The resulting current-voltage curve obtained from operating the DME is called a *polarogram* and contains qualitative and quantitative information about the sample.

Various types of polarography and voltammetry have been developed. These are (i) classic polarography, 2×10^{-6} *M* detection limit; (ii) current-sampled polarography, 1×10^{-6} *M* detection limit; (iii) normal-pulse polarography, 5×10^{-7} *M* detection limit; (iv) differential-pulse polarography, 1×10^{-7} *M* detection limit; (v) fast linear-sweep polarography, 1×10^{-7} *M* detection limit; (vi) cyclic voltammetry, 1×10^{-7} *M* detection limit; (vii) alternating current, 5×10^{-7} *M* detection limit; and (viii) stripping voltammetry, 1×10^{-9} *M* detection limit.

Components of a typical general purpose cell for polarographic measurements are: (i) Hg reservoir, (ii) dropping Hg electrode, (iii) reference electrode, (iv) counter electrode, and (v) a N_2 purge port. Uniform Hg drop formation is critical for accurate measurements. There are a variety of commercially available polarographs and selection depends on the capabilities needed. Some instruments are computer controlled.

2. Elements Determined

Various elements can be determined with voltammetric methods. Adeloju et al. (1982) used stripping voltammetry to measure Se in a HNO_3-H_2SO_4 digest of plant tissue. Wolnik et al. (1985) determined Cd and Pb in various plant tissues using differential-pulse, anodic-stripping voltammetry. Other elements that can be determined are As, Zn, Ni, Co, Cu, and Fe. Voltammetry can be used for chemical species that can be made to oxidize or reduce, or that will form a stable complex or slightly soluble salt with Hg. Voltammetric methods are not generally used for soil analysis.

3. Advantages and Disadvantages

The method is extremely accurate and precise. Detection limits are low for many elements. When the DME is used, each Hg drop is an exact duplicate of the preceding drop, allowing currents to be accurately reproduced. Because the Hg surface is always renewed, there is no chance of surface fouling. Another important advantage is the high over-voltage associated with the reduction of hydrogen ions. Consequently, metal ions can be deposited from acidic solutions.

A disadvantage of the DME is the ease with which Hg is oxidized. This property limits the use of the electrode as an anode. At potentials $> +0.4$ V, Hg (I) forms, masking other oxidizable species. Another disadvantage is that the DME is cumbersome to use and tends to malfunction as a result of clogging.

4. Sample Preparation

Plant tissue samples must be brought into solution by wet digestion methods. Adeloju et al. (1982) critically evaluated some wet digestion methods for stripping voltammetric determination of Se in NBS orchard leaves. Selenium was found to be 0.078 ± 0.004 mg kg^{-1} with voltammetry compared to the certified value of 0.08 ± 0.01 mg kg^{-1}. The most effective wet digestion method was that using HNO_3 and H_2SO_4. However, Wolnik et al. (1985) used dry ashing with H_2SO_4 as an ashing aid for the determination of Cd and Pb.

L. Near Infrared Reflectance Spectrometry

1. Principles of Operation and Use

Near infrared reflectance spectrometry (NIR) uses monochromatic light directed at the plant tissue sample (Fig. 26–4). The diffuse reflected radiation is absorbed by lead sulfide detectors, converting radiation to electrical energy. Either scanning monochromators or wavelength filters are used to provide the needed wavelengths. A bandpass of 10 to 15 nm has been found satisfactory. The wavelength region that is usually used is from 1100 to 2500 nm. Reflectance data is transformed to log 1/reflectance intensity and plotted as a function of the wavelength to obtain the reflectance spectra. Peaks of spectra correspond to absorption bands of the components of the sample. Calibration of the NIR instrument is done by collecting a population of plant

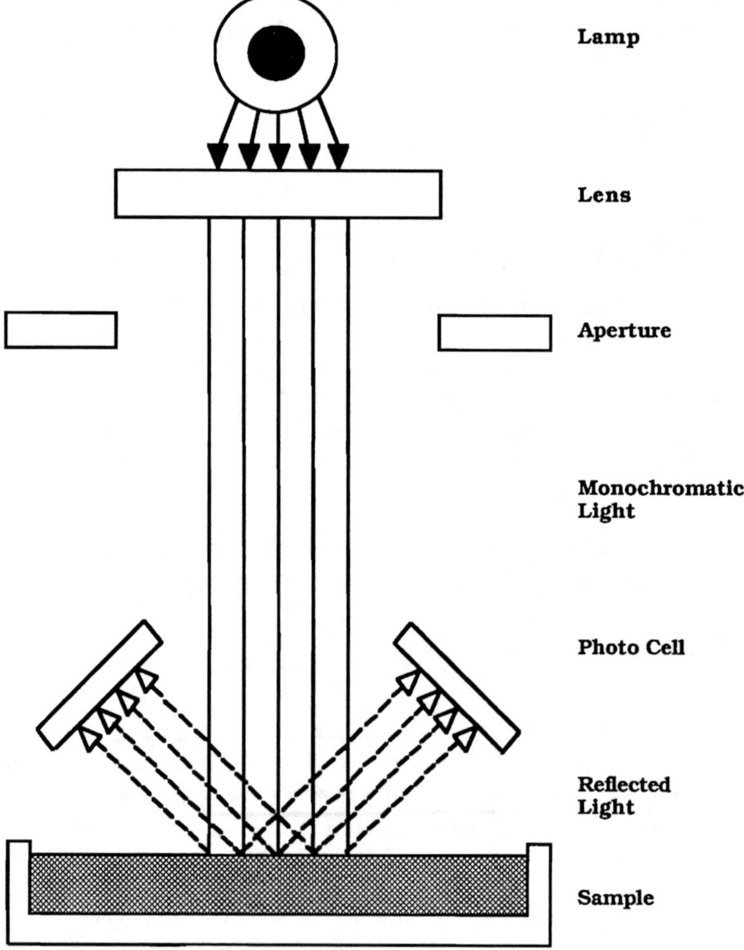

Fig. 26–4. Schematic diagram of a scanning type NIR instrument.

tissue samples, giving a range in the parameters of interest, and then performing wet chemistry to determine the values of these parameters. Multiple regression analysis is done to relate the intensity of the reflectance radiation to the values of the parameters determined by wet chemistry. The chosen regression equations are used to determine parameter values of unknown samples. A NIR is not generally used for soil analysis.

2. Elements Determined

Norris et al. (1976) and Norris and Barnes (1977) were among the first to use NIR to measure protein in plant tissue as components of animal feeds. Most NIR analyses have involved the determination of protein, acid detergent fiber, neutral detergent fiber, fats, and oils in various plant tissues (Hymowitz et al., 1974; Norris et al., 1976; Branine et al., 1983). However, determination of N contained in leaf material was made with NIR by Murray and Hall (1983) and by Isaac and Johnson (1984). From the Isaac and Johnson study involving 994 samples, 86% showed <0.2% difference in N values between NIR and Kjeldahl methods. Ninety-nine percent showed <0.3% difference. Correlation between Kjeldahl and NIR-N for tissue of various crops is given in Table 26-3.

Calcium and P have also been determined with NIR in plant tissue constituents of animal feeds. However, the degree of accuracy is usually less than for protein and N parameters.

3. Advantages and Disadvantages

A main advantage of NIR analysis of plant tissue samples is that it is a rapid, nondestructive analysis. NIR instruments are easy and safe to operate and chemicals are not involved.

The disadvantages are that the analysis can only be as accurate as the population of the samples used in the calibration. Populations used for calibration should contain at least 50 samples. Accuracy of the NIR analysis depends upon the wet chemistry methods used for the calibration samples. In some cases, high correlations are difficult to achieve between NIR and wet chemistry for plant populations which include multiple species. However, this may not always be the case. For example, work done in The Ohio State University's Research-Extension Analytical Laboratory has shown a correlation coefficient of 0.954 for N when the population used for calibration

Table 26-3. Correlation between Kjeldahl and NIR-N for various crops.

Crop	Scientific name	Multiple correlation coefficient
Peanut	*Arachis hypogaea* L.	0.955
Soybean	*Glycine max* (L.) Merr.	0.945
Bentgrass	*Agrostis palustris*	0.965
Bermudagrass	*Cynodon dactylon* (L.)	0.980
Wheat	*Triticum aestivum* L.	0.991
Pecan	*Carya illinoensis* (Wangenh.)	0.994

was a mixture of alfalfa and various grasses. In this case, a 350 sample population was used to calibrate the NIR. Also, particle size and sample moisture affect results. Moisture of the sample should be $<20\%$. Particle size of the sample should be uniform, and use of a cyclone-type grinder is recommended.

4. Sample Preparation

The plant tissue sample must be dried and ground to pass a 1-mm sieve. A cyclone-type grinder has been found to produce the most uniform particle size. For best results, a forced air oven should be used to dry the sample. It is important to process, handle, and store unknown samples in the same manner as samples used for calibration purposes.

M. Ion Exchange Chromatography

1. Principle of Operation and Use

Details of ion exchange chromatography (IC) are given by Small et al. (1975) and Skoog (1985). The IC is used largely for analysis of water. However, IC has recently been used for analysis of soil and plant tissue.

Ion exchange columns, constructed of small porous beads of divinylbenzene, are used to remove ions of interest from solution passing over the beads. The ion exchange process is based on exchange equilibria between ions in solution and ions of the same charge on the surface of the synthetic resin bead. Beads are prepared as exchange surfaces by chemically bonding acidic or basic functional groups to the bead. The most common groups are sulfonic acid and quarternary amines.

Once the ions of interest are adsorbed to the resin beads, an eluting solution is used to exchange the ions back into solution. The eluting solution is passed through another resin column (stripper column) to remove background-eluting electrolyte. The concentration of the ion of interest is measured by an electrical conductivity cell.

If a sample, for example, containing cations is injected at the head of the separator column, the cations will exit the separator column at different times in a HCl-background eluent. When this eluent solution enters the suppressor column, a resin containing a OH^- functional group would exchange with the Cl^- from the HCl resulting in H_2O and the ion of interest. Thus, the cation is measured in a conductivity cell in a background of water.

An analogous scheme for anion analysis can be made where NaOH is the eluent, an anion exchanger is the separating column, and a strong acid cation exchanger in the H^+ form is used as the stripper.

The main components of a suppressor column system are the: (i) eluent reservoir, (ii) pump, (iii) sample injection valve, (iv) precolumn and separation column, (v) eluent suppressor column, and (vi) conductivity cell and recorder (Fig. 26–5). Systems have been developed that eliminate the suppressor column. This approach depends upon the small differences in conductivity between the eluted ions and the prevailing eluent ions (Skoog, 1985).

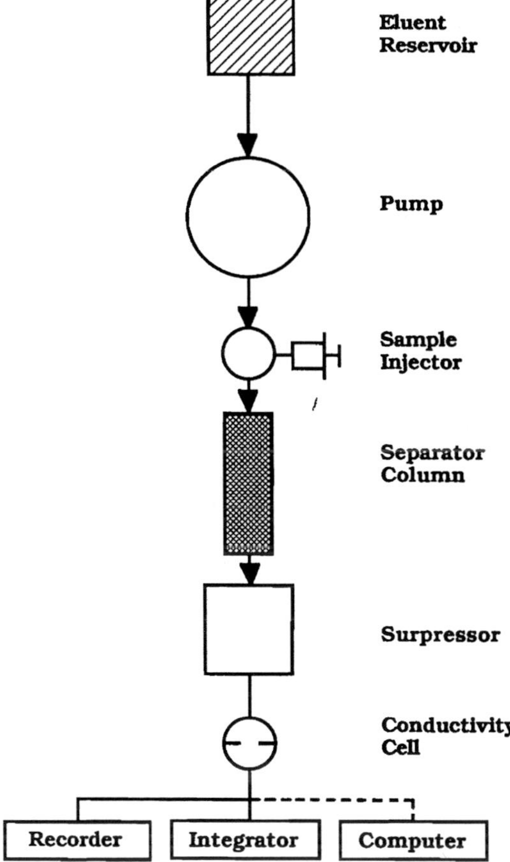

Fig. 26-5. Schematic diagram of a dual column type IC instrument.

Systems can also use optical and electrochemical detectors. Computers are used to collect and process data.

Many classes of compounds, i.e., inorganic, metals, organic, and amino acids can be determined. A major advance in IC has been the development of membrane chemical suppression devices that extend the ability of the technique to measure inorganic cations and anions (Franklin, 1985).

2. Elements Determined

Simultaneous determination of total S and chloride in plant tissue was done with IC by Busman et al. (1983). Sulfur in the plant tissue was converted to SO_4^{2-} by combusting the tissue in a Schoniger-type oxygen flask containing water. Results obtained for total S with IC compared closely with those obtained by the methylene-blue method after digestion with NaOBr. Average total S of 15 plant tissue samples by the two methods were 0.255 and 0.259%, respectively. Coefficient of variation for S analysis ranged from 2.7 to 7.1% for ryegrass and soybean stalks, respectively.

Analysis of chloride in plant tissue with IC compared well with the mercuric thiocyanate and ferric ammonium sulfate chloride method giving results of 0.380 and 0.377%, respectively. Coefficient of variation for chloride ranged from 5.2 to 10.2% for the same plant species previously mentioned.

Basta and Tabatabai (1985a) determined total K, Ca, Mg, and Na in plant tissue with IC. Results of the IC compared favorably with AAS and flame photometry. The IC is sensitive giving detection limits of 0.1, 0.05, 0.1, and 0.03 mg L^{-1} of K, Na, Ca, and Mg, respectively. Potassium and Na were determined in 6 min, and Ca and Mg in 7 min.

Dick and Tabatabai (1979) determined NO_3^- and SO_4^{2-} in soil extracts using IC. They reported detection limits of approximately 1 mg kg^{-1}. Nieto and Frankenberger (1985) reported on the simultaneous analysis of soil extracts for Cl^-, NO_2^-, NO_3^-, SO_3^-, and SO_4^{2-} in <10 min using IC: Their IC results agreed closely with those obtained by other methods with r values ranging from 0.987 to 0.998. The same authors also reported in the same journal on an IC method for the simultaneous analysis of NH_4^+, Na^+, Mg^{2+}, and Ca^{2+} in soils. Basta and Tabatabai (1985a) used IC methods to determine K, Na, Ca, and Mg in soils using an ammonium-acetate extractant. They found good agreement between the IC method and AAS methods for K, Na, Ca, and Mg and flame photometry methods for K and Na.

3. Advantages and Disadvantages

The IC is very sensitive, and is relatively free of interfering ions if the correct suppressor columns are used. Small aliquots of sample can be used (about 2 mL). The system can provide multielement determinations and can be easily automated.

A disadvantage is that conductimetric detection is temperature dependent. Most ionic solutions conduct about 2% more electricity per degree of Celsius increase in temperature. Tissue samples must be predigested and strong acids such as HNO_3 and $HClO_4$ can cause difficulties. Solutions injected should also be low in organic material and soluble salts. Sample sizes as small as 20 mg have been used.

4. Sample Preparation

Plant tissue samples should be ground to pass a 40-mesh screen. Basta and Tabatabai (1985b) dry ashed 0.1 g of plant tissue, followed by dissolution with 5 m*M* HCl. The eluent for K and Na was 5 m*M* HCl and for Ca and Mg, it was 2.5 m*M* HCl + 2.5 m*M* m-phenylenediamine dihydrochloride. Busman et al. (1983) combusted only 20 to 100 mg of plant tissue in a Schoniger-type oxygen flask for the determination of S and Cl^-.

Soil samples should be dried and ground to pass a 10-mesh screen and extracted with an appropriate extracting solution.

N. Conductivity

1. Principle of Operation and Use

Conductivity, while not generally applied to field soils, is useful when applied to heavily fertilized soils such as gardens and greenhouse mixtures. This procedure is also useful in determining the viability of salt-water flooded soils. Dunkle and Merkle (1944) and Geraldson (1957) report on the relation of conductivity to plant growth. Merkle and Dunkle (1944) discuss the use of conductivity as a diagnostic aid for greenhouse soils. Detailed discussions of conductivity are given by Bower and Wilcox (1965) and Willard et al. (1968).

Conductivity is a measure of the property of aqueous solutions to conduct an electrical current. In practice, conductivity is determined by using a Wheatstone bridge to match the resistance of an unknown solution. The measured resistance is developed between two electrodes in a conductivity cell. *Conductivity* is defined as the reciprocal of this resistance and is a function of both the cross-sectional area and distance between the electrodes. Because of this relationship, cell constants have to be determined for each pair of electrodes. For direct-reading instruments,

C = specific conductance, Ks/Measured conductance, Km where C = cell constant,
Ks = known conductivity of a pure salt solution, and
Km = measured conductivity of the pure salt solution.

The conductivity of the sample is then given by the equation $K = C \times$ measured conductance of the sample.

Other factors that influence conductivity are the total concentration of ions in the soil solution, their valence, mobility, relative concentrations, and the temperature of the soil solution.

Conductivity measurements for soil solutions are generally expressed as μ mho/cm at 25 °C or in the International System as μ siemens/cm at 25 °C. The two terms are equivalent. If the temperature of the solutions is not 25 °C, a correction factor must be applied.

Electrodes for the conductivity cells have been constructed of various materials with a platinum coated platinum-iridium alloy finding wide acceptance. Other electrode materials include stainless steel and graphite.

Commercial conductivity cells have been constructed that have cell constants varying from 0.2 to 10.0 cm. The most commonly used cell in soil testing has a cell constant of 1.0 cm.

Modern conductivity instruments feature direct reading of either conductivity or resistance with automatic temperature compensation.

2. Sample Preparation

Soil samples are dried and ground to pass a 10-mesh screen. Greenhouse mixtures do not require any pretreatment. Various water extraction systems have been used ranging from a saturated paste extract for salt-affected soils and greenhouse mixes to 1:2 soil water ratios by volume and by weight for field soils.

II. LABORATORY AUTOMATION

A. Definition

Laboratory automation refers to the performance of laboratory tasks without human intervention. There are two types of automation, discreet, and continuous. Discreet automation requires all aspects of the analysis to be carried out on an individual sample. Continuous automation refers to a continuous flowing stream, as referred to previously.

Quantitative analysis of plant tissue samples can be broken down into several steps: (i) sample preparation, (ii) weighing samples, (iii) dissolution of samples, (iv) removal of interferences, if needed, (v) measuring a property of the sample, and (vi) calculating and reporting of concentration of analyte.

B. Robotics

1. Requirements

At this time, robots are not used in routine plant or soil analysis laboratories to any great extent. However, within the next 5 to 10 yr, robots will likely play a significant role in some aspects of plant analysis. Usable laboratory robots are beginning to appear on the market at reasonable costs.

Laboratory robots usually consist of a horizontal arm with a 360 degree rotational movement. The robot contains a gripping hand that has 360 degree wrist-like movement, permitting the manipulation of vials or tubes, pouring liquids or solids, and shaking and swirling liquids in tubes. Some robots are designed to change hands relative to the task required. The robotic system is controlled by a programmable microprocessor. Some modern laboratory robots can be programmed to bring samples to a master laboratory station to be diluted, filtered, and treated with reagents (Owens & Eckstein, 1982). The robot can also be programmed to heat, shake, dispense liquids, and collect fractions.

The requirements for a computer capable of planning movements of robots are: (i) capability of considering alternative sequences of acts and to select among the most promising of these alternatives, (ii) capability of keeping track of time and to allow for the time it takes to form and execute plans, and (iii) capability of deciding which sensors to read at a particular time (McGratten & Macero, 1984).

Presently, there are two ways of directing a robot: (i) by means of a teach pendant (a hand-held device to teach the robot-required manipulation) and (ii) by directing the robot to proceed to specific coordinates.

2. Future Trends

Progress in the development of robotics is being made in (i) teaching methods, (ii) developing sensor-based interaction of the robot with its environment, and (iii) higher-positioning accuracy. Teaching robots with the aid of interactive graphics is expected in the near future. Also, robotic control languages are being developed.

C. Advantages and Disadvantages

In order for the advantage of laboratory automation to be realized, it is necessary to have a high volume of work. For laboratories in which large numbers of routine analyses are performed daily, the savings realized by automation can be great. Another advantage is that the speed of analysis is usually greater. The precision obtained by automation is better than for humans doing the same manipulations. Labor costs can be decreased as well, since automation requires fewer skilled technicians, but may require a skilled supervisor of automation. Probably the greatest disadvantage is the initial cost of the system and its set up.

III. COMPUTERIZED INSTRUMENTATION

A. Operational Modes

1. Off-Line

There are three modes in which computers can operate in conjunction with analytical instruments (Skoog, 1985). For off-line mode, the data are collected by a human operator and transferred to the computer for processing by magnetic tape, punched cards, or keyboard entry.

2. On-Line

A direct communication between the instrument and computer is made through an electronic interface. In the interface, the signal is digitized and stored by the computer. The computer remains separate but can provide data storage and instructions for processing the data. In this arrangement, the computer can also be used in the off-line mode as well. Many personal computers can be interfaced with instruments in the on-line mode.

3. In-Line

In-line operational mode is used by many modern instruments. Usually a micro- or mini-computer is contained within the instrument. The operator instructs the computer in the operation of the instrument. Sometimes the operator will be able to program the computer, but usually the manufacturer of the instrument programs the computer.

For the in-line and on-line modes, data are transferred from the instrument in real time. That is, data values are sent to the computer as they are created in the instrument. Computers with multi-tasking capability usually are not occupied fully in collecting data, but have time to perform other processing. Real-time processing involves data processing simultaneously with data acquisition. Real-time processing reduces the amount of computer storage needed and can often be used to provide adjustment of instrument parameters to improve the quality of future data signals.

Most computer-data management systems are generally designed to increase laboratory efficiency by (i) eliminating labor-intensive operations, (ii) minimizing the number of analytical errors through automation, especially the error-prone steps, (iii) providing knowledge of errors in real time to allow quick correction of the problem, and (iv) using quality-control criteria to detect gross errors.

IV. COMPUTERIZED MANAGEMENT OF DATA AND INFORMATION

A. Definition

Vast amounts of data are obtained when sophisticated modern instruments are used to analyze samples of plant tissue. As early as 1960, software was written for mainframe computers to perform automatic data collection from instruments. When computer hardware prices decreased in the 1970s, proliferation of affordable laboratory-information systems began. More recently, inexpensive personal computers have become useful in plant and soil analytical laboratories.

Computer management of data and information refers to acquisition, manipulation, statistical testing, filing, and presentation of data as well as providing operational information to the instrument. Many modern instruments allow set-up or alteration of many operational parameters of the instrument by the computer. This operational information must be retained in the computer and used at the appropriate time (Aldous et al., 1976; Anderson & Munter, 1983; Dessy, 1977). The most recently computerized management of information has been through the development of laboratory information management systems (LIMS) (Reber, 1983). Improvements on integrating the LIMS to analytical instruments are discussed by Thompson (1985).

B. Benefits

The LIMS is a database-management system that harnesses the flow of information in the laboratory, incorporating analytical instruments into the total laboratory information. The benefits of such a system have been elucidated by Reber (1983). For example, some of these benefits are automatic sample numbering, verification of entered information, test assignment for the samples, generation of work lists, data collection and reduction, data comparison with check samples, electronic filing, sample location tracking, report and billing generation, and on-line programming capability. The LIMS are of such sophistication today that information and data can be obtained from many instruments at the same time. In addition, the LIMS allow manipulation of previously accessed data while collecting new data. The more automated a laboratory becomes, the more control and monitoring capabilities the LIMS will have over sample flow and quality control.

C. Computer Networking

The interconnection of various computers or computer systems in a laboratory is referred to as a computer network. Reasons for networks of computers are the fear of total dependence on one computer and possible data-manipulation bottlenecks when many sophisticated, computer-controlled instruments are used at the same time. A computer network usually involves microcomputers, minicomputers, or mainframe computers, all exchanging information. Usually, the mainframe computer exists outside the laboratory area. However, with small, relatively inexpensive microcomputers, networking with a mainframe is not as common as in the past.

D. Future Developments

Future development of laboratory computer use will include improved computer networking to facilitate electronic transfer of information, from one laboratory to another or to the client. This will allow faster information transfer with a great reduction in time between analysis of the sample and the final report.

Progress in the development of laboratory robots will increase the use and sophistication of laboratory computers. Furthermore, more instruments will use computers to monitor electronic circuitry for failure and will allow switching to an alternate back-up circuit. This capability would greatly reduce the downtime of the instrument during peak work loads. Increased development of computer systems will provide much greater output of results since the instrument and robot systems can work 24-h days under the supervision of the computer.

V. INSTRUMENTATION IN THE NEXT DECADE

Computers will be the rule rather than the exception in the next 10 yr (Hirschfeld, 1985). Instruments will be built with computers as an integral part. These computers will be personal computers, very easy to operate. The computer will control the functions of self-diagnostics, alternate circuitry switching, adaptive instrument operation, and operation of robots that interface with automated instruments. Micro-miniaturizing sensors and excitation sources will be incorporated to make the instrument less costly to operate. More use will be made of artificial intelligence programming allowing instruments to calibrate the instrument from a set of calibration samples. An example of this is the NIR instrument. More use of different instruments of complementary capabilities interfaced through a common computer may be used as more parameters of plant samples need to be measured (Hirschfeld, 1985). Robotics will become an integral part of the total instrumentation of the laboratory.

REFERENCES

Abdalla, N.A., and B. Lear. 1975. Determination of inorganic bromide in soils and plant tissue with a bromide selective-ion electrode. Commun. Soil Sci. Plant Anal. 6:489–494.

Adams, F., and C.E. Evans. 1962. A rapid method of measuring lime requirement of red-yellow podzolic soils. Soil Sci. Soc. Am. Proc. 26(4):355–357.

Adeloju, S.B., A.M. Bond, and M.H. Briggs. 1982. Critical evaluation of some wet digestion methods for the stripping voltammetric determination of selenium in biological materials. Anal. Chem. 56:2397–2401.

Aldous, K.M., D.G. Mitchell, and J.S. Merritt. 1976. A data system for a small general analytical laboratory. Am. Lab. (Fairfield, Conn.) Sept. 8:33–40.

Alexander, P.W., R.J. Finlayson, L.E. Smythe, and A. Thalib. 1982. Rapid flow analysis with inductively coupled plasma atomic-emission spectroscopy using a micro-injection technique. Analyst 107:1335–1342.

Allen, S.E., H.M. Grimshaw, J.A. Parkinson, and C. Quarmby. 1974a. Instrumental techniques. p. 377–463. *In* S.E. Allen (ed.) Chemical analysis of ecological materials. John Wiley and Sons, New York.

Allen, S.E., H.M. Grimshaw, J.A. Parkinson, and C. Quarmby. 1974b. Instrumental techniques. p. 436–455. *In* S.E. Allen (ed.) Chemical analysis of ecological materials. John Wiley and Sons, New York.

Allen, S.E., H.M. Grimshaw, J.A. Parkinson, and C. Quarmby. 1974c. Instrumental techniques. p. 393–401. *In* S.E. Allen (ed.) Chemical analysis of ecological materials. John Wiley and Sons, New York.

Allen, S.E., H.M. Grimshaw, J.A. Parkinson, and C. Quarmby. 1974d. Instrumental techniques. p. 402–404. *In* S.E. Allen (ed.) Chemical analysis of ecological materials. John Wiley and Sons, New York.

Allen, H.E., W.R. Matson, and K.H. Mancy. 1970. Trace metal characterization in aquatic environments by anode stripping voltammetry. J. Water Pollut. Control Fed. 42:573–581.

Anderson, R.D., and R.C. Munter. 1983. Use of a microcomputer to improve sample handling on an automated ICP. Spectrochim. Acta 38B:151–155.

Anderson, T.A., and M.L. Parsons. 1984. ICP emission spectra III: The spectra of the group IIIA elements and spectral interferences due to group IIA and IIIA elements. Appl. Spectrosc. 38:625–634.

Babat, G.I. 1947. Gas reactions in an electrodeless discharge. J. Inst. Electr. Eng. Part 1. 94:27.

Baker, A.S., and R. Smith. 1969. Extracting solution for potentiometric determination of nitrate in plant tissue. J. Agric. Food Chem. 17:1284–1287.

Baker, A.S., and R.L. Smith. 1974. Preparation of solutions for atomic absorption analysis of Fe, Mn, Zn, and Cu in plant tissue. J. Agric. Food Chem. 22:103–107.

Baker, D.E., and N.H. Suhr. 1982. Atomic absorption and flame emission spectrometry. p. 13–27. *In* A.L. Page et al. (ed.) Methods of soil analysis. Part 2. 2nd ed. Agronomy Monogr. ASA and SSSA, Madison, WI.

Baker, R.L. 1972. Determination of fluoride in vegetation using the specific ion electrode. Anal. Chem. 44:1326–1327.

Basson, W.D., and R.G. Boehmer. 1972. An automated procedure for the determination of sulphur in plant tissue. Analyst 97:266–270.

Basson, W.D., R.G. Boehmer, and D.A. Stanton. 1969. An automated procedure for the determination of boron in plant tissue. Analyst 94:1135–1141.

Basta, N.T., and M.A. Tabatabai. 1985a. Determination of exchangeable bases in soils by ion chromatography. Soil Sci. Soc. Am. J. 49:84–89.

Basta, N.T., and M.A. Tabatabai. 1985b. Determination of total potassium, sodium, calcium, and magnesium in plant materials by ion chromatography. Soil Sci. Soc. Am. J. 49:76–81.

Becker, D.A. 1976. Analytical design in activation analysis: The role of accuracy and precision. p. 1143–1155. *In* P.D. LaFleur (ed.) Accuracy in trace analysis: sampling, sample handling, analysis. Vol. 2. Proc. 7th Materials Res. Symp. Spec. Publ. 422. U.S. Dep. Commerce/NBS, Gaithersburg, MD.

Birks, L.S., and J.V. Gilfrich. 1976. X-ray spectrometry. Anal. Chem. 48:273R–281R.

Bjarn, O.O.C. 1980. Kjel-Foss automated analysis using an antimony-based catalyst collaborative study. J. Assoc. Anal. Chem. 63:657–663.

Bock, R. 1979. A handbook of decomposition methods in analytical chemistry. Int. Textbook Co., London.

Boumans, P.W.J.M. 1978. "Plasma sources" for multielement analysis of solutions: Capabilities, limitation, and perspectives. Mikronchim. Acta 1:399–412.

Bower, C.A., and L.U. Wilcox. 1965. Soluble salts. p. 933–951. *In* C.A. Black (ed.) Methods of soil analysis. Part 2. 2nd ed. Agronomy Monogr. 9. ASA, Madison, WI.

Bowman, J.A., and J.B. Willis. 1967. Some applications of the nitrous oxide-acetylene flame in chemical analysis by atomic absorption spectrometry. Anal. Chem. 39:1210–1216.

Branine, M.E., B.F. Sowell, L.J. Krysl, and H.D. Radloff. 1983. Estimating nutritive value of native range forage by near infrared reflectance spectroscopy. Proc., Western Section, Am. Soc. of Anim. Sci. 34:165–167.

Bray, R.H., and L.T. Kurtz. 1945. Determination of total, organic, and available forms of phosphorus in soils. Soil Sci. 59:39–45.

Bremner, J.M., L.G. Bundy, and A.S. Agarwall. 1968. Use of selective ion electrode for determination of nitrate in soils. Anal. Lett. 1:837–844.

Bremner, J.M., and M.A. Tabatabai. 1971. Use of automated combustion techniques for total carbon, total nitrogen and total sulfur analysis of soils. p. 9–13. *In* L.M. Walsh (ed.) Instrumental methods for analysis of soils and plant tissue. SSSA, Madison, WI.

Buck, R.P. 1974. Ion selective electrode potentiometry, and potentiometric titrations. Anal. Chem. Ann. Rev. 46:28R–51R.

Busman, L.M., R.P. Dick, and M.A. Tabatabai. 1983. Determination of total sulfur and chlorine in plant materials by ion chromatography. Soil Sci. Soc. Am. J. 47:1167–1170.

Capar, S.G., J.T. Tanner, M.H. Friedman, and K.W. Boyer. 1978. Multielement analysis of animal feed, animal wastes, and sewage sludge. Environ. Sci. Technol. 12:785–790.

Carlson, R.M., and D.R. Keeney. 1971. Specific ion electrodes: Techniques and uses in soil, plant, and water analysis. p. 39–65. *In*L.M. Walsh (ed.) Instrumental methods for analysis of soils and plant tissue. SSSA, Madison, WI.

Chaplin, M.J., and A.R. Dixin. 1974. A method for analysis of plant tissue by direct reading spark emission spectroscopy. Appl. Spectrosc. 28:5–8.

Dahlquist, R.L., and J.W. Knoll. 1978. Inductively coupled plasma-atomic emission spectrometry: Analysis of biological materials and soils for major, trace and ultra-trace elements. Appl. Spectrosc. 32:1–30.

Dahnke, W.C. (ed.). 1975. Recommended chemical soil test procedures for the North Central region. North Central Regional Publ. 221. North Dakota Agric. Exp. Stn. Bull. 499.

Dean, J.A., and T.C. Rains. 1969. Flame emission and atomic absorption spectrometry. Vol. 1. Theory. Marcel Dekker, New York.

DeSoete, D., R. Gijbels, and J. Hoste. 1972. Neutron activation analysis. Wiley-Interscience, New York.

Dessy, R.E. 1977. Computer networking; A rational approach to lab automation. Anal. Chem. 49(13):1100A–1108A.

Dick, W.A., and M.A. Tabatabai. 1979. Ion chromatographic determination of sulfate and nitrate in soils. Soil Sci. Soc. Am. J. 43:899–904.

Dunkle, E.C., and F.G. Merkle. 1944. The conductivity of soil extraction in relation to germination and growth of certain plants. Soil Sci. Soc. Am. Proc. 8:185–188.

Durst, R.A. (ed.). 1969. Ion-selective electrodes. NBS Spec. Publ. 314. U.S. Dep. of Commerce, Washington, DC.

Eastin, E.F. 1976. Use of ammonia electrode for total nitrogen determination in plants. Commun. Soil Sci. Plant Anal. 7:477.

Ediger, R.D., D.A. Yates, and E. Pruszkowska. 1985. The analyst/automated ICP spectrometer interface. Instrument. Res. 1(1):51–57.

Eyde, B. 1983. Determination of acid soluble fluoride in soils by means of an ion-selective electrode. Fresenius Z. Anal. Chem. 316:299–301.

Fassel, V.A., and R.N. Knisely. 1974. Inductively coupled plasmas. Anal. Chem. 46:1155A–1164A.

Fassel, V.A. 1977. Current and potential applications of inductively coupled plasma (ICP)-atomic emission spectroscopy (AES) in the exploration, mining, and processing of materials. Pure Appl. Chem. 49:1533–1545.

Ferrara, L.W., R.S. Floyd, and R.W. Blanchair. 1965. Automation in analytical chemistry. Tech. Symp. 109–111.

Franklin, G.O. 1985. Developments and applications of ion chromatography. Am. Lab. (Fairfield, Conn.) (June) 17:65–80.

Gaines, T.P., M.B. Parker, and G.J. Gascho. 1984. Automated determination of chlorides in soil and plant tissue by sodium nitrate. Agron. J. 76:371–374.

Geraldson, C.M. 1957. Soil soluble salts—Determination of and association with plant growth. Proc. Fla. State Hortic. Soc. 71:121–127.

Greenfield, S., I.L. Jones, and C.T. Berry. 1964. High pressure plasma as spectroscopic emission sources. Analyst 89:713–720.

Hahn, M.H., K.A. Wolnik, and F.L.Fricke. 1982. Hydride generation/condensation system with an inductively coupled argon plasma polychromator for determination of arsenic, bismuth, germanium, antimony, selenium, and tin in foods. Anal. Chem. 54:1048–1056.

Harrison, G.R. 1949. The production of diffraction gratings: II. The design of Echelle gratings and spectrographs. J. Opt. Soc. Am. 39:522–528.

Haskin, L.A., and K.E. Ziege. 1971. Nuetron activation: Techniques and possible uses in soil and plant analysis. p. 185–208. *In* L.M. Walsh (ed.) Instrumental methods for analysis of soils and plant tissue. SSSA, Madison, WI.

Havemann, C.H. 1984. Corros. Coat. S. Afr. 11-29 Chem. Abstr. 1984. 101, 182942t.

Hayward, D.M., and D.W. Parry. 1973. Electron-probe microanalysis studies of silica distribution in barley (*Hordeum sativum* L.). Ann. Bot. 37:5790–5791.

Heft, R.E. 1976. Control of sample configuration as an aid to accuracy in instrumental neutron activation analysis. p. 1275–1281. *In* P.D. LaFleur (ed.) Accuracy in trace analysis sampling, sample handling, analysis. Vol. 2. Proc. 7th Materials Res. Symp. U.S. Dep. of Commerce/NBS, Gaithersburg, MD. Spec. Publ. 422.

Helmke, P.A. 1982. Neutron activation analysis. p. 67–84. *In* A.L. Page et al. (ed.) Methods of soil analysis. Part 2. 2nd ed. Agronomy Monogr. 9. ASA and SSSA, Madison, WI.

Hern, J.L. 1984. Determination of total sulfur in plant materials using an automated sulfur analyzer. Commun. Soil Sci. Plant Anal. 15:99–107.

Hirschfeld, T. 1985. Instrumentation in the next decade. Science 230:286–291.

Hymowitz, T., J.W. Dudley, F.I. Collins, and C.M. Brown. 1974. Estimations of protein and oil concentration in corn, soybean, and oat seed by near-infrared light reflectance. Crop Sci. 14:713–715.

Iron, R.D., E.A. Schenk, and R.D. Glauque. 1976. Energy-dispersive x-ray fluorescence spectroscopy and inductively coupled plasma emission spectrometry evaluated for multielement analysis in complex biological matrices. Clin. Chem. 22:2018–2024.

Isaac, R.A., and J.B. Jones. 1970. Autoanalyzer systems for the analysis of soil and plant tissue extracts. Adv. Auto. Anal. 2:57–64.

Isaac, R.A., and W.C. Johnson, Jr. 1975. Collaborative study of wet and dry ashing techniques for elemental analysis of plant tissue by atomic absorption spectrophotometry. J. Assoc. Off. Anal. Chem. 58:436–440.

Isaac, R.A., and W.C. Johnson. 1984. Methodology for the analysis of soil, plant, feed, water and fertilizer samples. Soil Testing and Plant Tissue Analysis Lab., Univ. of Georgia, p. 46–51.

Isaac, R.A., and W.C. Johnson. 1984. Near infrared reflectance spectroscopic determination of protein nitrogen in plant tissue. J. Assoc. Off. Anal. Chem. 67:506–509.

Isaac, R.A., and W.C. Johnson. 1985. Elemental analysis of plant tissue by plasma emission spectroscopy: Collaborative study. J. Assoc. Off. Anal. Chem. 68:499–505.

Isaac, R.A., and J.D. Kerber. 1971. Atomic absorption and flame photometry: Techniques and uses in soil, plant, and water analysis. p. 18–32. *In* L.M. Walsh (ed.) Instrumental methods for analysis of soils and plant tissue. SSSA, Madison, WI.

Jacobson, J.S., and L.I. Heller. 1978. Collaborative study of three methods for the determination of fluoride in vegetation. J. Assoc. Off. Anal. Chem. 61:150–153.

Johnson, G.V. (ed.). 1984. Procedures used by state soil testing laboratories in the Southern region of the United States. Agric. Exp. Stn., Oklahoma State Univ. Bull. 190.

Johnson, G.W. 1979. Trace analysis with a direct current plasma Echelle spectrometer system. Ph.D. thesis. Colorado State Univ., Fort Collins (Diss. Abstr. 40104:1673-B).

Jones, A.A. 1982. X-ray fluorescence spectrometry. p. 85–121. *In* A.L. Page et al. (ed.) Methods of soil analysis. Part 2. 2nd Ed. Agronomy Monogr. 9. ASA and SSSA, Madison, WI.

Jones, Jr., J.B. 1969. Elemental analyses of plant leaf tissue by several laboratories. J. Assoc. Off. Anal. Chem. 52:900–903.

Jones, Jr., J.B. 1972. Plant tissue analysis for micronutrients. p. 317–346. *In* J.J. Mortvedt et al. (ed.) Micronutrients in agriculture. SSSA, Madison, WI.

Jones, Jr., J.B. 1975. Collaborative study of the elemental analysis of plant tissue by direct reading emission spectroscopy. J. Assoc. Off. Anal. Chem. 58:764–769.

Jones, Jr., J.B. 1977. Elemental analysis of soil extracts and plant tissue ash by plasma emission spectroscopy. J. Assoc. Off. Anal. Chem. 58:764–769.

Jones, Jr., J.B. 1981. Analytical techniques for trace element determination in plant tissues. J. Plant Nutrition 3:77–92.

Jones, Jr., J.B., and R.A. Isaac. 1969. Comparative elemental analysis of plant tissue by spark emission and atomic absorption spectroscopy. Agron. J. 61:393–394.

Jones, Jr., J.B., and R.A. Isaac. 1972. Determination of sulfur in plant material using a Leco Sulfur Analyzer. J. Agric. Food Chem. 20(6):1292–1294.

Jones, Jr., J.B., and M.N. Warner. 1964. Analysis of plant-ash solutions by spark emission spectroscopy. p. 152–160. *In* E.L. Grove and A.J. Perkins (ed.) Developments in applied spectroscopy. Vol. 7A. Plenum Press, New York.

Katz, S.A. 1985. Neutron activation analysis. Am. Lab. (Fairfield, Conn.) (June) 17:16–28.

Kauren, P. 1977. The use of buffers in the determination of fluoride by an ion-selective electrode at low concentrations and in the presence of aluminum. Anal. Lett. 10:451–465.

Knudsen, D., R.B. Clark, J.L. Denning, and P.A. Pier. 1981. Plant analysis of trace elements by x-ray. J. Plant Nutr. 3:61–75.

Krieg, D.R., and D. Sung. 1977. Interferences in chloride determinations using the specific ion electrode. Commun. Soil Sci. Plant Anal. 8:109–114.

Kubota, J., and V.A. Lazar. 1971. X-ray emission spectrograph: Techniques and uses for plant and soil studies. p. 67–82. *In* L.M. Walsh (ed.) Instrumental methods for analysis of soils and plant tissue. SSSA, Madison, WI.

LaCroix, R.L., D.R. Keeney, and L.M. Walsh. 1970. Potentiometric titration of chloride in plant tissue extracts using the chloride ion electrode Commun. Soil Sci. Plant Anal. 1:1–6.

L'Vov, B.V. 1961. Analytical use of atomic absorption spectra. Spectrochim. Acta 17:761–770.

Larson, G.F., and V.A. Fassel. 1979. Line broadening and radiative recombination background interferences in inductively coupled plasma-atomic emission spectroscopy. Appl. Spectrosc. 33:592–599.

Lauchli, A., and H. Schwander. 1966. X-ray microanalyzer study on the localization of minerals in native plant tissue sections. Experimentia 22:503–505.

Lee, F.Y., and J.A. Kittrick. 1984. Electron microprobe analysis of elements associated with zinc and copper in an oxidizing and an anaerobic soil environment. Soil Sci. Soc. Am. J. 48:548–554.

Lindsay, W.L., and W.A. Norvell. 1969. Development of DTPA micronutrient soil test. p. 84. *In* Agronomy abstracts. ASA, Madison, WI.

Lipari, F., and F.W. Plankey. 1977. Continuum source atomic fluorescence spectrometry. Am. Lab. (Fairfield, Conn.) (Aug.) 9:11–18.

Louw, C.W., and J.F. Richards. 1972. The determination of fluoride in sugar cane by using an ion-selective electrode. Analyst 97:334–339.

Lunt, H.A., C.L.W. Swanson, and H.G. Jacobson. 1950. The Morgan soil testing system. Conn. Agric. Exp. Stn. Bull. 541:1–60.

MacDonald, J.C. 1983. A new ICP spectrometer. Am. Lab. (Fairfield, Conn.) (Sept.) 15:90–108.

Mahendrappa, M.K. 1969. Determination of nitrate nitrogen in soil extracts using a specific ion activity electrode. Soil Sci. 108:132–136.

McGratten, B.J., and D.J. Macero. 1984. Laboratory robotics. Am. Lab. (Fairfield, Conn.) (Sept.) 16:16–24.

McKeague, J.A. (ed.). 1978. Manual of soil sampling methods of analysis. 2nd ed. Canadian Soc. of Soil Sci., Ottawa, ON, Canada.

McLean, E.O., J.F. Trierweiler, and D.J. Eckert. 1977. Improved SMP buffer method for determining lime requirements of acid soils. Commun. Soil Sci. Plant Anal. 8:667–675.

McNerney, F.G. 1976. Collaborative study of the determination of sodium in dietetic foods by the sodium ion electrode method. J. Assoc. Off. Anal. Chem. 59:1131–1134.

McQuaker, N.R., P.D. Kluckner, and G.N. Chang. 1979. Calibration of an inductively coupled plasma-atomic emission spectrometer for the analysis of environmental materials. Anal. Chem. 51:888–895.

Mehlich, A. 1953. Determination of P, Ca, Mg, K, Na, and NH_4. North Carolina Dep. of Agric. Mimeo 1953.

Mehlich, A. 1976. New buffer pH method for rapid estimation of exchangeable acidity and lime requirement of soils. Commun. Soil Sci. Plant Anal. 7(7):637–652.

Mehlich, A. 1984. Mehlich 3 soil test extractant: A modification of Mehlich 2 extractant. Commun. Soil Sci. Plant Anal. 15(12):1409–1416.

Melton, J.R., W.L. Hoover, P.A. Morris, and J.A. Gerald. 1978. DC arc plasma emission spectrometric determination of boron in plants. J. Assoc. Off. Anal. Chem. 61:504–505.

Merkle, F.G., and E.C. Dunkle. 1944. The soluble salt content of greenhouse soils as a diagnostic aid. J. Am. Soc. Agron. 36:10–19.

Meyers, R.J.K., and E.A. Paul. 1968. Nitrate in electrode method for soil nitrate nitrogen determination. Can. J. Soil Sci. 48:369–371.

Miller, M.H., C.P. Mamaril, and G.J. Blair. 1970. Ammonium effects on phosphorus absorption through pH changes and phosphorus precipitation at the soil-root interface. Agron. J. 62:524–527.

Moody, G.J., and J.D.R. Thomas. 1977. The determination of chloride in vegetables, fruits and juices with ion-selective electrode. J. Food Technol. 12:193–197.

Munter, R.C., and R.A. Grande. 1980. Developments in atomic plasma spectrochemical analysis. p. 653–672. *In* R.M. Barnes (ed.) Proc. Int. Winter Conf., San Juan, Puerto Rico. Heyden, London.

Munter, R.C., T.L. Halverson, and R.D. Anderson. 1984. Quality assurance for plant tissue analysis by ICP-AES. Commun. Soil Sci. Plant Anal. 15:1285–1322.

Murray, I., and P.A. Hall. 1983. Animal feed evaluation by use of near infrared reflectance (NIR) spectro-computer. Anal. Proc. 20:75–79.

Nagy, G., Z. Zeher, and E. Pungor. 1970. Application of silicone-rubber-based graphite electrodes for continuous-flow measurements. II. Voltammetric study of active substances injected in electrolyte streams. Anal. Chim. Acta 52:47–54.

Nieto, K.F., and W.T. Frankenberger, Jr. 1985. Single column ion chromatography. I. Analysis of inorganic anions in soils. II. Analysis of ammonium, alkali metals and alkaline earth cations in soils. Soil Sci. Soc. Am. J. 49:587–596.

Noel, R.J. 1976. Collaborative study of an automated method for the determination of crude protein in animal feeds. J. Assoc. Off. Anal. Chem. 59:141–147.

Norris, K.H., and R.F Barnes. 1977. Infrared reflectance analysis of nutritive value of feedstuffs. p. 237–241. *In* Feed composition, animal nutrient requirements and computerization of diets. Proc. 1st Int. Symp. Feed Composition, Animal Nutrient Requirements, and Computerization of Diets. Utah State Univ., Logan. 11–16 July 1976.

Norris, K.H., R.F Barnes, J.E. Moore, and J.S. Shenk. 1976. Predicting forage quality by infrared reflectance spectroscopy. J. Anim. Sci. 43:889–897.

Olsen, S.R., C.V. Cole, F.S. Watanabe, and L.A. Dean. 1954. Estimation of available phosphorus in soils by extraction with sodium bicarbonate. USDA Cir. 939. U.S. Gov. Print. Office, Washington, DC.

Owens, G.D., and R.J. Eckstein. 1982. Robotic sample preparation station. Anal. Chem. 54:2347–2351.

Pallagy, C.K. 1973. Electron probe microanalysis of potassium and chloride in freeze-substituted leaf sections of *Zea mays*. Aust. J. Biol. Sci. 26:1015–1034.

Paul, J.L., and R.M. Carlson. 1968. Nitrate determination in plant extracts by the nitrate electrode. J. Agric. Food Chem. 16:766–768.

Peech, M. 1965. Hydrogen-ion activity. p. 914–926. *In* C.A. Black et al. (ed.) Methods of soil analysis. Part 2. Agronomy Monogr. 9. ASA and SSSA, Madison, WI.

Peech, M., L.T. Alexander, C.A. Dean, and J.F. Reed. 1947. Methods of soil analysis for soil-fertility investigations. USDA Cir. 757. U.S. Gov. Print. Office, Washington, DC.

Pommer, A.M. 1967. Glass electrodes for soil waters. p. 362–411. *In* G. Eisenman (ed.) Glass electrodes for hydrogen and other cations. Marcel Decker, New York.

Price, W.J. 1979. Spectrochemical analysis by atomic absorption. Heyden and Son, London.

Rains, T.C., and O. Menis. 1976. An intercomparison of flame and nonflame systems in atomic absorption spectrometry. p. 1045–1051. *In* P.D. LaFluer (ed.) Accuracy in trace analysis: Samples, sample handling, analysis. Vol. 2. U.S. Dep. of Commerce/NBS, Gaithersburg, MD. Spec. Publ. 422.

Ranger, C. 1981. Flow injection analysis: Principles, techniques, applications, design. Anal. Chem. 53:20A–32A.

Rasmussen, H.P., and B.O. Knexek. 1971. Electron microprobe: Techniques and uses in soil and plant analysis. p. 209–222. *In* L.M. Walsh (ed.) Instrumental methods for analysis of soils and plant tissue. SSSA, Madison, WI.

Raveh, A. 1973. The adaptation of the nitrate-specific electrode for soil and plant analysis. Soil Sci. 116:388–389.

Reber, S. 1983. Laboratory information management systems. Am. Lab. (Fairfield, Conn.) (Feb.) 15:78–85.

Reed, T.B. 1961a. Induction-coupled plasma torch. J. Appl. Phys. 32:821–824.

Routh, M.W., and K.J. Paul. 1985. A fixed grating sequential ICP spectrometer with multitasking software. Am. Lab. (Fairfield, Conn.) (June) 17:84–97.

Sabbe, W.E. (ed.). 1980. Handbook on reference methods for soil testing. The Counc. on Soil Test. and Plant Analysis, Univ. of Georgia.

Sansum, L., and J.B.D. Robinson. 1974. Wet digestion, manual and automated analysis of total sulphur in plant material. Commun. in Soil Sci. Plant Anal. 5:365–383.

Schofield, R.K., and A.W. Taylor. 1955. The measurement of soil pH. Soil Sci. Soc. Am. Proc. 19:164–167.

Schollenberger, C.J., and R.H. Simon. 1945. Determination of exchange capacity and exchangeable bases in soil-ammonium acetate method. Soil Sci. 59:13–24.

Schuman, G.E., M.A. Stanley, and D. Knudsen. 1973. Automated total nitrogen analysis of soil and plant samples. Soil Sci. Soc. Am. Proc. 37:480–481.

Scott, R.H., and A. Strasheim. 1975. Determination of trace elements in plant materials by inductively coupled plasma optical emission spectrometry. Anal. Chim. Acta. 76:71–78.

Sharicz, K. 1983. Advances in sequential ICP spectrometer design. Am. Lab. (Fairfield, Conn.) (Mar.) 15:62–65.

Shoemaker, H.E., E.O. McLean, and P.G. Pratt. 1962. Buffer methods for determination of lime requirement of soils with appreciable amounts of exchangeable aluminum. Soil Sci. Am. proc. 25:274–277.

Skoog, D.A. 1985. Principles of instrumental analysis. 3rd ed. Saunders College, New York.

Slavin, W. 1982. Atomic absorption spectroscopy: The present and future. Anal. Chem. 54:685A–694A.

Small, H., T.S. Stevens, and W.C. Bauman. 1975. Novel ion exchange chromatographic method using conductimetric detection. Anal. Chem. 47:1801–1809.

Soltanpour, P.N., S.M. Workman, and A.P. Schwab. 1979. Use of inductively-coupled plasma spectrometry for the simultaneous determination of macro- and micronutrients in NH_4HCO_3-DPTA extracts of soils. Soil Sci. Soc. Am. J. 43:73–78.

Soltanpour, P.N., J.B. Jones Jr., and S.M. Workman. 1982. Optical emission spectrometry. p. 29–65. *In* A.L. Page et al. (ed.) Methods of soil analysis. Part 2. 2nd ed. Agronomy Monogr. 9. ASA and SSSA, Madison, WI.

Spiers, G.A., M.J. Dudas, and L.W. Hodgins. 1983. Instrumental conditions and procedures for multielement analysis of soils and plant tissue by ICP-AES. Commun. Soil Sci. Plant Anal. 14:629–644.

Steckel, J.E., and R.L. Flannery. 1968. Automatic determination of phosphorus, potassium, calcium, and magnesium in wet digestion solutions of plant tissue. Tech. Q. 1:19–20.

Steckel, J.E., and R.L. Flannery. 1971. Simultaneous determinations of phosphorus, potassium, calcium, and magnesium in wet digestion solutions of plant tissue by autoanalyzer. p. 83–96. *In* L.M. Walsh (ed.) Instrumental methods for analysis of soils and plant tissue. SSSA, Madison, WI.

Street, J.J., and W.M. Peterson. 1982. Anodic stripping voltammetry and differential pulse polarography. p. 133–157. *In* A.L. Page et al. (ed.) Methods of soil analysis. Part 2. 2nd ed. Agronomy Monogr. 9. ASA and SSSA, Madison, WI.

Suddendorf, R.F., and K.W. Boyer. 1978. Nebulizer for analysis of high salt content samples with inductively coupled plasma emission spectrometry. Anal. Chem. 50:1769–1771.

Sweeney, R.A., and P.R. Rexroad. 1987. Comparison of LECO FP-228 "Nitrogen Determinator" with AOAC copper catalyst Kjeldahl method for crude protein. J. Assoc. Off. Anal. Chem. 70:1028–1030.

Sweetsur, A.W.M., and A.G. Wilson. 1975. An ion-selective electrode method for the determination of nitrate in grass and clover. Analyst 100:485–488.

Thompson, M.R. 1985. Integration of intelligent instruments and LIMS 2000. Instrument. Res. (March) 1(1):69–70.

Twine, J.R., and C.H. Williams. 1971. The determination of phosphorus in Kjeldahl digests of plant material by automatic analysis. Commun. Soil Sci. Plant Anal. 2:485–489.

Van Den Heede, M.A., A.M. Heyndrickx, C.H. Van Peteghem, and W.A. Van Zele. 1975. Determination of fluoride in vegetation: A comparative study of four sample preparation methods. J. Assoc. Off. Anal. Chem. 58:1135–1137.

Villa, A.E. 1979. Rapid method for determining fluoride in vegetation using an ion-selective electrode. Analyst 104:545–551.

Wall, L.L., C.W. Gehrke, T.E. Neuner, R.D. Cathey, and P.R. Rexroad. 1974. Total protein nitrogen-evaluation and comparison of four different methods. *In* 88th Annu. Meet. of Assoc. Off. Anal. Chem., Washington, DC. 15 Oct. Univ. of Missouri, Columbia.

Wall, L.L., C.W. Gehrke, and J. Suzuki. 1980. An automated turbidimetric determination of sulfate sulfur in soils and fertilizers and total sulfur in plant tissue. J. Assoc. Off. Anal. Chem. 63:845–853.

Walsh, A. 1955. Application of atomic absorption spectra to chemical analysis. Spectrochim. Acta 7:108–117.

Watters, Jr., R.L. 1983. Practical limits of precision in inductively coupled plasma spectrometry. Am. Lab. (Fairfield, Conn.) (Mar.) 15:16–25.

Wendt, R.H., and V.A. Fassel. 1965. Induction-coupled plasma spectrometric excitation source. Anal. Chem. 37:920–922.

Willard, H.H., L.L. Merritt, Jr., and J.A. Dean. 1968. Instrumental methods of analysis. 4th ed. D. Van Nostrand Co., Princeton, NJ.

Williams, C. 1976. The rapid determination of trace elements in soils and plants by x-ray fluorescence analysis. J. Sci. Food Agric. 27:561–570.

Wilson, D.O. 1979. Determination of molybdenum in wet-ashed digests of plant material using flameless atomic absorption. Commun. Soil Sci. Plant Anal. 10:1319–1330.

Winefordner, J.D. (ed.). 1976. Trace analysis, spectroscopic methods for elements. Vol. 46. Chemical analysis (a series of monographs on analytical chemistry and its applications). John Wiley and Sons, New York.

Winefordner, J.D., and T.J. Vickers. 1964. Atomic fluorescence spectrometry as a means of chemical analysis. Anal. Chem. 36:161–165.

Winefordner, J.D., and T.J. Vickers. 1974. Flame spectrometry. Anal. Chem. Annu. Rev. 46:192R–227R.

Winge, R.K., V.A. Fassel, V.J. Peterson, and M.A. Floyd. 1982. ICP emission spectrometry: On the selection of analytical lines, line coincidence tables, and wavelength tables. Appl. Spectrosc. 36:210–221.

Winge, R.K., V.J. Peterson, and V.A. Fassel. 1979. Inductively coupled plasma-atomic emission spectroscopy. Prominent lines. Appl. Spectrosc. 33:206–219.

Wohlers, C.C. 1985. Experimentally obtained wavelength tables for the ICP. ICP Info. Newsl. 10:593–688.

Wolnik, K.A., F.L. Fricke, S.G. Capar, M.W. Meyer, R.D. Satzger, E. Bonnin, and C.M. Gaston. 1985. Elements in major raw agricultural crops in the United States. 3. Cadmium, lead, and eleven other elements in carrots, field corn, onions, rice, spinach, and tomatoes. J. Agric. Food Chem. 33:807–811.

Yuan, T.L. 1974. A double buffer method for the determination of lime requirement of acid soils. Soil Sci. Soc. Am. Proc. 38:437–440.

Chapter 27

Data Processing in Soil Testing and Plant Analysis

S. J. DONOHUE AND S. W. GETTIER, *Virginia Polytechnic Institute and State University, Blacksburg, Virginia*

Computers are an important, if not essential, part of the operations of a soil testing and plant analysis program today. The diversity of samples, processing requirements, and additional environmental, forage, or water analysis components in a modern laboratory require the sophistication of computer data handling. Since 1970, tremendous advances have occurred in data processing. Computer equipment and software have not only improved, but with the expansion of computer markets to serve smaller businesses and individuals, purchase costs have decreased to affordable levels. In a 1985 survey on computer use in soil testing at Southeast U.S. land grant universities conducted by the senior author, 12 of 13 laboratories were using the computer in some capacity to process soil test data. Computer processing offers the advantages of easier manipulation of large data sets, reduced errors in calculation of recommendations and preparation of reports, more sophisticated quality control, automated invoicing and addressing, and ready access to historical data for preparation of soil test summaries. For those labs already using the computer but wishing to update their equipment or data management approach, new and better computer components and programs are available for easier, more efficient sample processing.

Computer equipment is continuing to evolve at a rapid pace. The distinctions between mainframe, mini-, and micro-computers have become blurred as speed of processing, available memory, and quantity of storage have increased. There is a tremendous diversity of hardware (the physical components of a computer system) available with which to design a system. A computer system today includes the computer, communications devices (internal or external modems, parallel ports, and serial ports), storage devices (magnetic, magnetic-optical, optical, or paper tape), cables (metal wire or fiber optics), data input devices (keyboard, optical scanners, data loggers, or voice interpreters), monitors (color or monochrome), plotters, and printers

 Soil Testing and Plant Analysis, 3rd ed.—SSSA Book Series, no. 3.

(impact, ink jet, laser, or thermal). The type of system best suited for a laboratory has to be assessed based on present needs and projected future requirements.

Because of the many different aspects of a computerized soil testing or plant analysis system, it would be helpful to evaluate the individual steps involved in developing such a system.

I. STEPS IN COMPUTERIZING A PROGRAM

A. Organization/Preparation of Recommendations for Computer Computation

An important prerequisite to computerizing a soil testing or plant analysis program is the orderly organization of fertilizer and lime recommendations. All inputs used to calculate recommendations must be listed along with specifics on how they alter the recommendation. A typical set of inputs may include crop to be grown; second crop; previous crop; soil nutrient levels; last crop's N, P, and K fertilization rates; time and amount of last lime application; and soil series or soil's yield potential. The recommendations for these inputs can be written in table or equation form. The necessary time should be taken to prepare recommendations for as many crops as one realistically expects to be grown in the region. Requests for recommendations for crops not already in the computer program will necessitate unnecessary but time-consuming hand calculation.

B. Evaluation of Computer Equipment Needs

The degree to which laboratories should computerize depends on sample volume, location, and user services to be offered. In general, large sample volume laboratories offering many types of analyses have a need for more computer sophistication and automation than low sample volume laboratories. This section discusses some of the options available in computer equipment to the user. It should be noted that the computer field is advancing rapidly and keeping abreast of the computer literature is necessary if one is to remain fluent in this area. Before final selections are made regarding system components, computer hardware and software specialists should be consulted.

1. Data Acquisition

The simplest method of transferring test results from an instrument to a computer is by hand typing the data on a keyboard. Although simple, errors in transcription may occur and speed of data entry is relatively slow. Electronic data capture offers many advantages since it is well suited for automation, eliminates transcription errors, provides immediate data access, and usually enhances productivity and accuracy. An added advantage of electronic data capture is the ability to generate quality control reports for analytical documentation. This computer documentation is especially important for laboratories working with regulatory agencies.

Many laboratory instruments are controlled with on-board or slave (dedicated) computers which are integral parts of the equipment. Some instrumentation, such as the Inductively Coupled Plasma (ICP) Emission spectrometer, cannot be operated without computers. Most other laboratory equipment, from balances to sophisticated analytical instruments, can be purchased with computer interfaces that allow direct data capture by a computer. Most analytical equipment contains a standard RS232 computer interface. For the actual linking of the instrument to the computer, a custom program usually needs to be developed (Janney et al., 1988). Standard packages are becoming available which will reduce the frustration, delays, and expense of having to develop custom programs for each instrumental interface.

2. Computer Recording of Sample History Information

Each sample arriving at a laboratory must be "logged-in" for processing. The early keypunching of paper "computer cards" has been replaced by electronic data entry where oftentimes the input screen resembles, if not duplicates, the input form. Data can either be entered directly into an interactive data management program which will prompt the operator for the correct information or the data can be entered into a storage file for later access by a computer program for data processing. These two approaches can also be used in conjunction with each other such that part of the system is interactive and part is file based.

With an interactive data entry system, operators require minimal computer training. Information is immediately processed and available for monitoring sample progress through the system as analytical tests are performed. Interactive systems also have the advantage of immediate error detection/correction and immediate access to calculated data to permit ready information retrieval. Interactive data entry systems are well suited for personal computers (PCs) and in-house mini-computers. On mainframe computers, interactive data entry is expensive and system down time ("crashes") can cause serious delays in sample processing.

A file-based system, until recently, was the approach found most desirable by large soil testing laboratories. This system, usually mainframe based, minimizes the use of expensive mainframe computer time, and facilitates data backup. However, file-based systems are giving way to interactive systems as in-house PCs and mini-computers become more powerful, making interactive systems faster to operate and data backup programs easier to execute.

3. Data Processing Equipment

The three major types of data-processing equipment are mainframe computers, mini-computers, and PCs. In general, mainframes are large computers with fast processors, often with gigabytes of memory, and have the capability of handling many users simultaneously. Mini-computers hold the middle ground between mainframes and PCs. The mini-computer has the ability to handle a number of users simultaneously and has greater processing speed and more memory than a PC. The PC is usually a single user system but can be connected to a mainframe (via a data communication device such as a modem) and emulate a terminal/workstation or be linked to other PCs/terminals as part of a local area network (LAN). Frequently, laboratories will use a combination of computers to handle their data processing needs. Most small to medium sized laboratories can manage their data processing entirely with in-house PCs or mini-computers. Some laboratories may find it effective to integrate PCs with mainframe computers.

There are certain characteristics of computers that need to be considered regardless of laboratory size. It is much easier to design and maintain a computer operation by choosing a standard operating system that offers the needed processing power, multitasking capabilities, ease of use, and the ability to run the appropriate software (O'Malley, 1988). Examples of such operating systems are MS-DOS, Macintosh, OS/2, and UNIX. Use of more than one operating system is undesirable since it can cause compatibility problems between the different system components and will force the learning of multiple systems. The computer hardware chosen should be compatible with the equipment in the laboratory and also to the host system, if this approach is used. The host system is defined as a mainframe computer or a network. A slight cost savings in computer hardware will not be a savings if the system is hard to use and does not handle the computing needs of the laboratory.

4. Computer Printers

The printing of laboratory reports, worksheets, and invoices can be handled with a variety of printers. A letter-quality impact printer will produce good results with multi-part forms. However, superior quality of printing and greater form design flexibility can be obtained with laser printers. Although laser printers can only print on single-part forms, they are fast enough that printing multiple copies is feasible. The major drawback of laser printers is their higher initial purchase price and greater operating costs compared to impact printers. However, the laser printer eliminates the need for custom-printed forms as they can rapidly print high-quality letterheads, logos, and graphics. By creating computer-generated forms, only plain paper is required and changes can be implemented immediately since there is not an inventory of printed forms with which to contend.

C. Preparation of Master and Supporting Computer Programs

Master computer programs will be needed for calculation of recommendations and preparations of reports. Secondary or support programs will also be needed for data input, error checking, record searching, invoicing, and preparation of data summaries. A computer programmer/systems analyst (P/SA) will be required to design the system and write the programs.

Although a soils specialist or agronomist may have training in computer programming, it is more desirable to hire an expert in the computer area who is well versed in the latest in computer technology, particularly one who has experience in the computer equipment and language(s) to be used. In the preparation of programs, it is imperative that the P/SA be instructed to provide clear, easy-to-understand documentation to permit future modifications to the system. Oftentimes, the P/SA hired to update a system will not be the same person who designed the system.

The P/SA has the choice of designing a data processing system with custom written programs using one or more of the several available computer languages, e.g., BASIC, C, COBOL, FORTRAN, Pascal, PL1, or purchasing an already developed program adaptable to individual laboratories. Data processing could also be handled by spreadsheets (e.g., Lotus 1-2-3, Quattro) or programmable databases (e.g., dBase, Paradox). The type of data processing selected is dictated by the needs of the system and personal choice of the programmer.

D. Designing Forms

An important aspect of a soil testing and plant analysis program is the conveying of information to the user of the laboratory. It is important that a laboratory develop forms that are easy to read and simple to understand. Preparation of such forms is no small task. Oftentimes, the designer of the forms is the scientist in charge of the program. While this person may be an expert in his field, careful review of the forms by individuals representing the users of the laboratory is essential for the program to be successful. "Success" is defined here as the completeness with which a form is filled out and the extent to which an individual uses the information that is provided by the laboratory.

1. Sample Information Sheets

Sample information sheets or questionnaires are those which accompany the samples that are sent to the laboratory and contain information that will assist in the processing of the sample, e.g., tests requested, previous treatments in the sampled area, and where to send results. Information should not be requested if it does not directly affect the recommendation to be made, e.g., the amount of fertilizer applied 3 yr ago provides little, if any, useful information that would affect the recommendation to be made. Choices should be given for answers, where possible, rather than using a "fill-in-the-blanks" approach to encourage the user to provide as much information as possible. The more information provided by the grower, the more customized the recommendations will be for a more successful program.

2. Laboratory Reports

Laboratory reports usually present three types of information: (i) a brief listing of the previous history of the sampled area, as provided on the information sheet accompanying the sample, (ii) the analytical results, with an interpretation as to the adequacy of the test level with respect to crop growth, and (iii) the fertilizer and soil amendment recommendations. The previous history section provides information that is useful for record-keeping purposes. The analytical results and "sufficiency" rating provide the actual evaluation of the ability of the soil to sustain plant growth. An explanation of the tests performed should be provided to the grower either on the report itself (normally on the back side) or in accompanying material. It should be noted that any extra material accompanying the basic report will not be given the same attention as the report itself and should, therefore, be kept to a minimum. With respect to recommendations, this information should be kept as concise as possible for greater acceptance and use by the grower. A farmer is not as interested in a discussion of the "need" to apply a certain nutrient as much as he is in knowing "how much" to apply.

E. Putting the System into Operation

One of the more important aspects of computerizing a program (or modifying an existing system) is the successful implementation of the program itself. For this to occur, the system must first be tested thoroughly for smoothness of operation. Errors in the program will cause problems that could result in delays in sample processing and a loss of credibility to the laboratory. All combinations of input data should be checked and any errors corrected. Data transfer into the computer, whether done by hand (manually) or directly from laboratory instruments should be evaluated beforehand to locate potential problems which could impede sample flow. For example, the system needs to be able to intercept or detect out-of-range values for operator review and, if necessary, correction. The more thorough the error checking, the more confidence one will have in the system.

When putting a new system into operation, consideration must be given to the education of clientele in the use of new materials for submitting samples and in interpreting new or modified recommendations or report formats. This education should be directed towards both the users of the laboratory (e.g., farmer and homeowner) and those individuals who work with the user (e.g., county extension agents and agribusiness personnel). In general, the more familiar the latter group is with the new system, the better it will be received by the user.

F. System Maintenance

It usually will be necessary to make modifications to the system after it is in operation, be it in data entry (e.g., by acquisition of new equipment or changes in analyses offered), data processing (e.g., modifications of computer components) or in the programs themselves (e.g., changes in recommendations). A system should be adaptable to changes in instrumentation, analyses offered, and upgrades in computer equipment or problems will occur. With respect to changes in the computer programs, prior documentation will help ensure that changes are made smoothly and errors are kept to a minimum. Hiring of competent individuals for installation of new equipment or computer program modifications is needed to prevent recurrent problems that may impede sample processing.

II. COMPUTER USE IN SOIL TESTING PROGRAMS

Most soil testing laboratories today are using the computer in some aspect of their program. To illustrate how the computer is used in a soil testing laboratory, the following is a discussion of the computer operation at Virginia Tech.

In the mid-1970s, a decision was made to switch to computer data processing at Virginia Tech. Up to that time, fertilizer and lime recommendations were hand calculated by extension agents, a very time-consuming process for a large soil testing program (> 100 000 samples analyzed/yr). To initiate the switchover process, fertilizer and soil amendment recommendations were first reviewed and revised where necessary. The review involved more than 25 crop/soil faculty; recommendations were expanded to cover more than 160 crops since it was felt that with more information in the computer, less hands-on work would be required. In addition to developing basic rate-type recommendations, information was prepared on time and method of fertilizer application. If this supplemental information was brief and concise, it was put into a computer "comment" format for printing on the report itself. More lengthy information was printed as a one-sheet flyer or "Soil Test Note" which was designed to accompany the basic soil test report. A total of more than 150 "comments" and 22 Soil Test Notes were developed for the program.

Also included in the review was an evaluation of previous cropping history inputs. Those inputs not directly affecting the recommendations were eliminated while additional needed inputs were added.

With respect to computer forms, soil sample information sheets and soil test report forms were reviewed from more than 25 laboratories and the basic forms were developed. An extensive review of the forms by the state's extension agents was then conducted following which the final modifications were made. The forms used for the program are presented in Fig. 27–1 and 27–2.

In regard to computer processing, three types of equipment were evaluated: data recording, processing, and printing. Since a decision had already been made to use the university mainframe computer for data processing because of the high initial cost of in-house mini-computers, compatible recording equipment was selected which consisted of two Hewlett-Packard (HP) terminals with (i) cassette tape recording devices for off-line data entry and (ii) a screen forms option. For data transfer to the mainframe, the HP terminals were connected directly to the mainframe's interactive computer system. For printing, a medium-speed in-house HP printer was used with the university's high-speed printers as an emergency backup.

With respect to computer programs, flowcharts were first prepared to diagram the various steps involved in the calculation of recommendations. A master program was then written which was made up of a number of smaller programs or subroutines, each performing a certain task. Additional support programs were also written for initial input data error checking, printing soil test reports, identification of additional tests requested, and transfer of records/data sets throughout the system. All programs were tested extensively as they were developed to prevent problems upon switchover. Several computer languages were used, including FORTRAN, SAS, and Exec, an

WRITE ONLY IN BOXES BELOW

Ext. Form 124

Cooperative Extension Service
Virginia Polytechnic Institute & State University
Soil Testing Laboratory

SOIL SAMPLE INFORMATION SHEET
FOR
COMMERCIAL CROP PRODUCTION

PLEASE PRINT
DATE___

INSTRUCTIONS: Follow sampling instructions on box. Fill out this sheet as completely as possible. Place check marks(✓) where appropriate. Use other forms for home lawns, gardens, etc. Send samples to: Extension Agronomist, Soil Testing Laboratory, Smyth Hall, VPI & SU, Blacksburg, Virginia 24061.

Grower's Name___________
Street, Route___________
City___________ Zip ___ County___________

Extra Copy For: (Dealer, etc.)___________
Street Route___________
City___________ Zip___________

For Office Use Only
Unit Code:
CAM LRO

SAMPLE IDENTIFICATION, ACREAGE

Lab No. (Leave blank)	Your SAMPLE NUMBER (Up to 5 digits)	ASC Farm No.	No. of Acres

RECOMMENDATIONS REQUESTED FOR

Crop to be grown		Next Crop	
Code (see back)	Name	Code	Name
☐☐☐		☐☐☐	

PAST HISTORY OF SAMPLED AREA

Last Crop		Last Crop's Yield (100 bu/a, etc.)	Last Crop's Fertilizer Application, lb/a			Last Lime Application	
Code	Name		N	P_2O_5	K_2O	Months Previous	Rate, T/A
☐☐☐			☐☐☐	☐☐☐	☐☐☐	☐ --- ☐ 0- 6 ☐ 7-12 ☐ 13-18 ☐ 18+	☐ 0 ☐ 0.1-1.0 ☐ 1.1-2.0 ☐ 2.1-3.0 ☐ 3.0+

SOIL INFORMATION

Soil Type Sandy (S) Loamy (L) Clayey (C) Organic (O)	Soil Name	Slope Level (L) Slight (S) Moderate (M) Steep (St)	Soil Prod. Group (see back)
☐ S ☐ L ☐ C ☐ O	___________	☐ L ☐ S ☐ M ☐ St	☐ I ☐ II ☐ III ☐ IV

OTHER INFORMATION

Will Manure Be Used?	ASC Cost-Share	Special Tests Needed? (see back)
☐ no ☐ yes	☐ no ☐ yes	______ ______ ______ ______

NOTE any unusual conditions of soil or crops___________

Fig. 27-1. Virginia Tech soil sample information sheet.

in-house interactive computer program. Languages were selected based on the amount of support provided by the university computing center. The programming itself was done by three part-time P/SA's and the professor in charge of the overall program. Figure 27–3 provides an overview of the soil test computer system which was developed.

Before initiation of the new system, one-half day training sessions were conducted with extension agents and various agribusiness personnel throughout Virginia. A new computer recommendation guidebook was distributed during the meeting along with handout materials which provided further

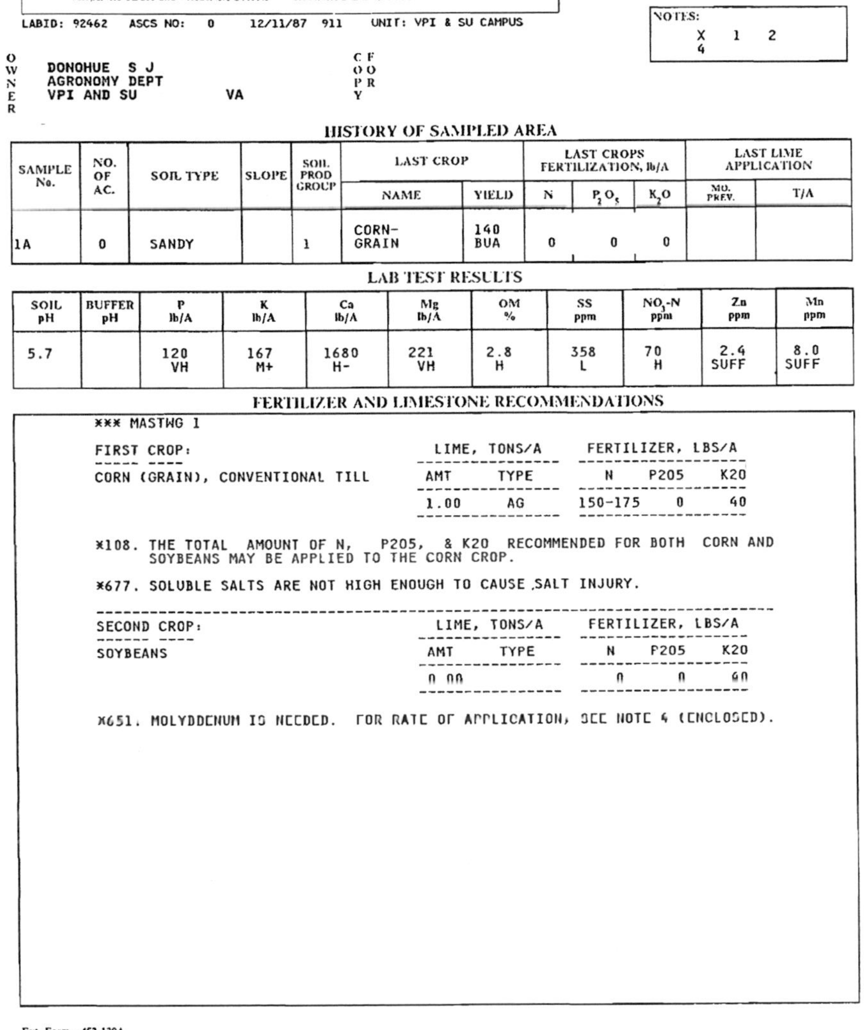

Virginia Cooperative Extension Service SOIL TEST REPORT 1

VIRGINIA TECH and VIRGINIA STATE - VIRGINIA'S LAND GRANT UNIVERSITIES

LABID: 92462 ASCS NO: 0 12/11/87 911 UNIT: VPI & SU CAMPUS

NOTES: X 1 2 4

OWNER: DONOHUE S J, AGRONOMY DEPT, VPI AND SU VA

COPY

HISTORY OF SAMPLED AREA

SAMPLE No.	NO. OF AC.	SOIL TYPE	SLOPE	SOIL PROD GROUP	LAST CROP NAME	LAST CROP YIELD	LAST CROPS FERTILIZATION, lb/A N	P_2O_5	K_2O	LAST LIME APPLICATION MO. PREV.	T/A
1A	0	SANDY		1	CORN-GRAIN	140 BUA	0	0	0		

LAB TEST RESULTS

SOIL pH	BUFFER pH	P lb/A	K lb/A	Ca lb/A	Mg lb/A	OM %	SS ppm	NO_3-N ppm	Zn ppm	Mn ppm
5.7		120 VH	167 M+	1680 H-	221 VH	2.8 H	358 L	70 H	2.4 SUFF	8.0 SUFF

FERTILIZER AND LIMESTONE RECOMMENDATIONS

*** MASTWG 1

FIRST CROP: CORN (GRAIN), CONVENTIONAL TILL

LIME, TONS/A AMT	TYPE	FERTILIZER, LBS/A N	P2O5	K2O
1.00	AG	150-175	0	40

*108. THE TOTAL AMOUNT OF N, P2O5, & K2O RECOMMENDED FOR BOTH CORN AND SOYBEANS MAY BE APPLIED TO THE CORN CROP.

*677. SOLUBLE SALTS ARE NOT HIGH ENOUGH TO CAUSE SALT INJURY.

SECOND CROP: SOYBEANS

LIME, TONS/A AMT	TYPE	FERTILIZER, LBS/A N	P2O5	K2O
0.00		0	0	60

*651. MOLYBDENUM IS NEEDED. FOR RATE OF APPLICATION, SEE NOTE 4 (ENCLOSED).

Ext. Form - 452-130A

Fig. 27–2. Virginia Tech soil test report.

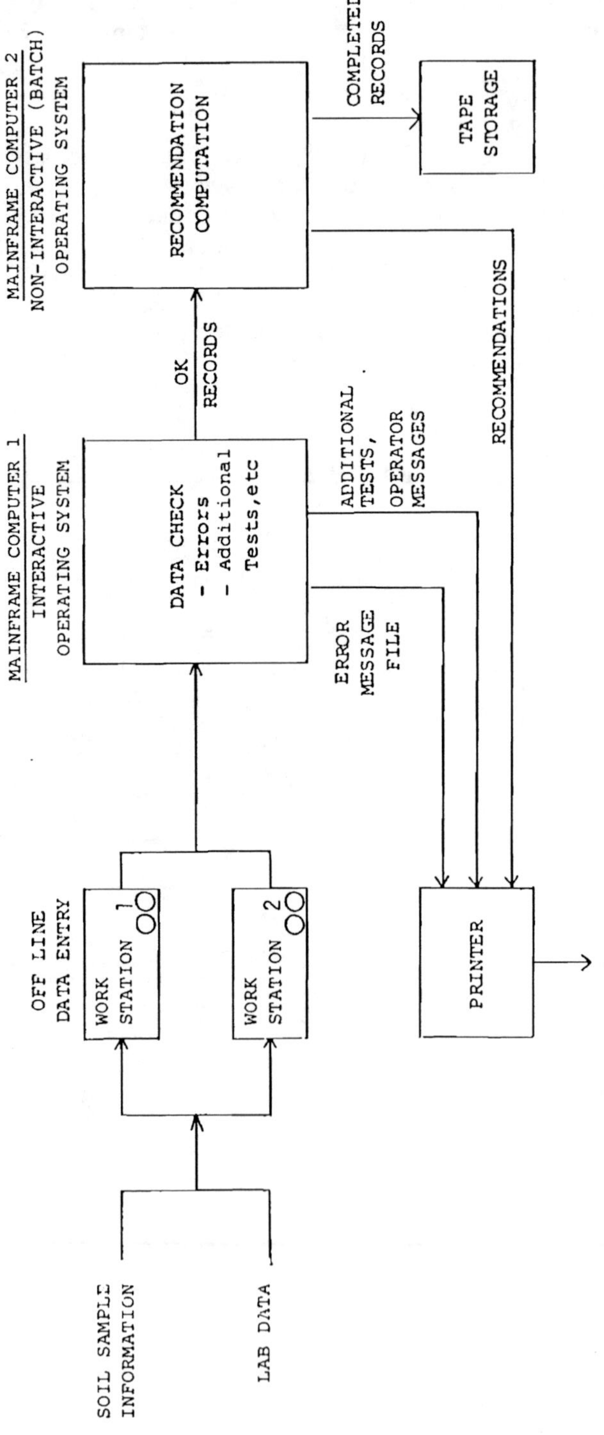

Fig. 27-3. Operation of the Virginia Tech soil test computer system.

Table 27–1. Long-term acidification of farmland in six Washington and six northern Idaho counties. (Adapted from Mahler et al., 1985.)

Soils testing below pH 6.0	Eastern Wash.			Northern Idaho		
	1960	1972	1980	1960	1972	1980
%	no. of counties					
<15	12	9	6	12	3	2
15–45	0	2	5	6	4	2
46–65	0	1	0	0	4	2
>65	0	0	1	0	1	6

details regarding the new system. In addition, farmer and homeowner slidesets with scripts were introduced at the meetings and made available for use at local meetings. Finally, information was provided to extension agents for use as news releases or for radio and television programs. When the new system was introduced in January 1980, problems were minimal and acceptance of the new computerized approach was high.

Since the early 1980s, computer recommendations have been periodically modified and changes have been made to the computer system to enhance processing. These include: (i) increasing the efficiency of data flow through the system to reduce operating costs and processing time, (ii) changeover to the university's high speed laser page printer for more flexibility in printing reports, (iii) automating data capture through instrument upgrade to an ICP spectrometer system, (iv) tie-in to the university's electronic mail system for automatic access of soil test reports by all county/city extension offices in Virginia, (v) replacement of the two data entry terminals with three hard-drive PCs to speed up data entry and improve in-lab backup data storage.

In addition to the use of computers to process soil test data, they can be used to organize these data by region, crop, nutrient level, and year and prepare summaries to evaluate the fertility status of the soil. For a survey conducted in 1985, 40 of 46 land grant university or government soil testing laboratories prepared summaries periodically, i.e., once every 1 to 10 yr or so (Donohue, 1987). Soil test summaries were used by workers in Washington and Idaho to document the acidification of the soil in the eastern Washington-northern Idaho area over the last 25 years (Table 27–1). This acidification was brought about by the change in crop rotation from a cereal crop-legume sequence to mostly continuous wheat (*Triticum aestivum* L.) rotation with acid-forming N fertilizer as the N source. The number of counties testing <pH 6.0 increased dramatically between 1960 and 1980. Because of this documentation, greater emphasis is now being placed on lime needs for this region in educational programs.

In Virginia, long-term changes in P soil test levels for corn (*Zea mays* L.) were evaluated to determine the effectiveness of the extension program (Table 27–2). Prior to 1976, P fertilizer was recommended at the "very high" test level in Virginia in a soil fertility maintenance program. In 1976, with high fertilizer prices and the more important fact that one would not expect

Table 27-2. Long-term changes in P soil test level for corn grown in Virginia.

Area	Years	No. of samples	P Soil test level			
			Low	Medium	High	Very high
			% of samples			
State	1970, 1971	13 508	7	28	44	22
	1983, 1984	29 572	3	26	59	12
Southampton County,	1970, 1971	1 295	0	6	58	36
Virginia	1983, 1984	4 781	0	8	79	12

a crop response at this level (i.e., no return would be expected on the fertilizer investment), P recommendations were discontinued at "very high". At county producer meetings in the late 1970s, response to P fertilizer at different soil test levels was emphasized. In comparing a 1970 and 1971 (2-yr) summary for Virginia (G.W. Hawkins, 1973, unpublished data), with a more recent 1983 and 1984 summary (Donohue, 1985), this change in P recommendations was documented in the decrease in the number of samples testing in the "very high" category. In Southampton County, Virginia's largest agricultural county, a similar pattern was found. The Extension educational program, considered to be responsible for at least part of this decrease, was deemed to be successful.

III. COMPUTER USE IN PLANT ANALYSIS PROGRAMS

Computers are important components of a plant analysis program. In addition to within-the-laboratory data capture and transfer, newer approaches to determining the nutrient status of the plant, such as the Diagnosis and Recommendation Integrated System (DRIS) (Beaufils, 1973), require the use of computers. The integration of numerous factors which affect a plant's nutrient status are well suited for computer techniques. The plant analysis laboratory at the Univ. of Wisconsin provides an example of a typical computerized plant analysis operation. This laboratory has had a service-oriented plant analysis program since the late 1960s. In 1971, due to an increase in the volume of samples, the laboratory was computerized. The program prints the plant analysis test results, interprets the data for sufficiency levels of nutrients, and makes recommendations for fertilizer when applicable.

In the Wisconsin program, it is recommended that a soil sample accompany the plant tissue when a sample is submitted for analysis. It was found that an analysis of the soil, along with the tissue analysis, provided a broader base for the interpretation of results and making fertilizer recommendations.

The Wisconsin plant tissue sample information sheet (Fig. 27-4) can accommodate two samples from the same field providing they both represent the same crop. It is recommended that the customer fill in the information sheet completely. This facilitates giving more accurate fertilizer and lime recommendations.

PLANT ANALYSIS INFORMATION SHEET
(Use a Separate Sheet for Each Field and Crop)

PLEASE PRINT OR TYPE

DATE July 9, 1985

FARMER'S NAME AND ADDRESS

1. Name B. A. Hybrid
2. Street, Route No. Rt. 2
3. City, State, Zip Code Cobb, WI 53526
4. Charge to account number: 800

SEE REVERSE SIDE FOR INSTRUCTIONS

Amount Enclosed $ 30.00 Cash

OFFICE USE
County Code

SAMPLE INFORMATION

5. Field designation: Corn Field (Limit to 6 Characters)
6. Crop: Corn
7. Date planted: May 9, 1985
8. Date sampled: July 9, 1985
9. Stage of growth: waist high
10. If abnormal plants are included as one of the samples, what percent of the total plants in the field are affected? 80-90
11. Plant part sampled *(check one)*: Leaves ☒ Stems ☐ Petioles ☐ Whole plant ☐ Top 6 inches ☐ Grain ☐
12. Position of plant part sampled:
    for corn *(check one)* Ear leaf ☐ Leaf below whorl ☒
    for fruit trees Current season growth ☐
    for vine fruit crops Mid-shoot leaves ☐
    for other crops *(check one)* Top ☐ Middle ☐ Bottom ☐

13. Sample location: County Iowa State WI

14.

Sample number:	Plant appearance: Normal	Abnormal
1	☒	☐
2	☐	☒

SPECIAL SOIL TESTS

15. In addition to routine soil analyses (pH, organic matter, phosphorus, potassium), check special soil tests requested:
    ☐ Calcium & Magnesium ☐ Zinc
    ☐ Boron ☐ Sulfur
    ☐ Manganese
    (extra fee charged for each special test requested)

FIELD INFORMATION

16. Soil Name: Dodgeville
17. Drainage: Poor ☐ Fair ☐ Good ☒
18. Has drainage been provided? Yes ☐ No ☒
19. If yes, *check one:* Tiled ☐ Ditched ☐ Surface drained ☐
20. Is the field irrigated? Yes ☐ No ☒

CROPPING HISTORY

21. Previous crop: Corn
22. Condition of previous crop: Poor
23. Weather: Rainfall last 30 days:
    Below normal ☒ Normal ☐ Above normal ☐
24. Air temperature last 10 days:
    Below normal ☐ Normal ☐ Above normal ☒
25. Spray or dust applied to crop being sampled:
    Weed chemicals: Yes ☐ No ☒
    Insecticides: Yes ☐ No ☒
    Fungicides: Yes ☐ No ☒

FERTILIZATION HISTORY

26. For each nutrient applied indicate the method(s) of applicaton and amount of nutrient applied.

Nutrient	Broadcast	Row	Sidedressed	Amt. lbs/a
Nitrogen (N)	☒	☒	☐	131
Phosphate (P_2O_5)	☐	☒	☐	24
Potash (K_2O)	☐	☒	☐	24

27. If any of the following secondary nutrients were applied, *check the appropriate box(es)*
    Calcium ☐ Magnesium ☐ Sulfur ☐
28. If any of the following micronutrients were applied, *check the appropriate box(es)*
    Boron ☐ Zinc ☐ Manganese ☐ Copper ☐

SOIL AND PLANT ANALYSIS LABORATORY
Soils Department, University of Wisconsin
5711 Mineral Point Road
Madison, Wisconsin 53705
(608) 262-4364

Fig. 27-4. Wisconsin plant analysis information sheet.

When the plant tissue and soil analyses are run and the information sheets are coded, the data are entered into the computer. The processed data, along with fertilizer and lime recommendations, are then printed out on the plant analysis report (Fig. 27-5). The Wisconsin plant analysis report also lists the relative levels along with the actual concentrations of the various elements in the "Laboratory Analysis for Each Plant Sample" section. These interpretations are based both on sufficiency ranges and DRIS indices for crops for

THIS REPORT FOR:

B. A. HYBRID
RT. 2
COBB, WIS. 53526

PLANT ANALYSIS REPORT
SAMPLES ANALYZED BY
SOIL AND PLANT ANALYSIS LABORATORY
5711 Mineral Point Road
Madison, WI 53705

COOPERATIVE EXTENSION PROGRAMS
UWEX University of Wisconsin—Extension University of Wisconsin—Madison
SOILS DEPARTMENT • MADISON WISCONSIN 53706

DATE JULY 14, 1985
COUNTY IOWA

CROP CORN FIELD DESIGNATION CORN Field LAB NOS. 9 - 10 ACCT. 800

IDENTIFICATION		PLANT ANALYSES FOR EACH SAMPLE												
Sample No.	Plant Appearance	N %	P %	K %	Ca %	Mg %	S %	Zn ppm	B ppm	Mn ppm	Fe ppm	Cu ppm	Al ppm	Na ppm
1	NORMAL	2.73 L	0.30 S	0.76 D	0.53 S	0.68 H	0.22 S	23 S	9 S	52 S	128 S	9.7 S	107	47
2	ABNORMAL	2.03 S	0.36 S	0.27 D	0.74 H	1.46 H	0.25 S	26 S	12 S	110 H	601 E	11.2 S	790	20

D = Deficient, L = Low, S = Sufficient, H = High, E = Excessive

INTERPRETATION OF PLANT ANALYSIS BY THE DIAGNOSIS AND RECOMMENDATIONS INTEGRATED SYSTEM (DRIS)

Sample No.		DEFICIENT	LOW	SUFFICIENT	HIGH
1	Element DRIS Index	K -48		S ZN N P CA B CU MN -5 -5 -3 -2 1 2 3 3	MG 55
2	Element DRIS Index	K -271		ZN N S P -3 -2 4 10	B CU CA MN MG 11 12 18 42 179

SOIL ANALYSES FOR EACH SAMPLE											
Sample No.	Soil pH	Est. C.E.C.	Organic Matter T/A	P Lbs/A	K Lbs/A	Ca Lbs/A	Mg Lbs/A	S Lbs/A	Zn Lbs/A	B Lbs/A	Mn Lbs/A
1	6.7		38	65 H	135 L						
2	6.2		36	52 H	120 L						

V = Very Low, L = Low, LM = Low-Medium, M = Medium, HM = High-Medium, H = High, E = Excessive

FERTILZER APPLIED				
	Bdct.	Row	Side dress	Lbs/A
N	X	X		131
P_2O_5		X		24
K_2O		X		24
Secondary and Micronutrients				

REMEDIAL FERTILIZER RECOMMENDATIONS FOR EACH SAMPLE										
Sample No.	N Lbs/A	P_2O_5 Lbs/A	K_2O Lbs/A	Mg Lbs/A	S Lbs/A	Zn Lbs/A	B Lbs/A	Mn Lbs/A	Fe Lbs/A	Cu Lbs/A
1	80									
2										

Herbicides	NO
Insecticides	NO
Fungicides	NO

WEATHER	Below Normal	Normal	Above Normal
Rainfall	X		
Temperature		X	

SUPPLEMENTAL SAMPLE INFORMATION			
Date Sampled 07/09/85	Plant Part LEAF	Soil Series DODGEVILLE	
Date Planted 05/09/85	Plant Position LEAF	Drainage GOOD	Irrigated	Previous Crop CORN
Growth Stage WAIST HI	% Abnormal 80	Type of Drainage	Condition of Previous Crop POOR

COMMENTS:

These plants are low or deficient in nitrogen, possibly as a result of inadequate N or P fertilization, excessively wet soil conditions, excessive leaching on sandy soils or excessive K fertilization.

These plants are low or deficient in potassium. Possible causes of this are low K soil test levels, inadequate K fertilization, poor drainage or soil compaction.

These plants are excessive in iron and aluminum. This most likely results from contamination of the tissue with soil particles and does not reflect the true iron and aluminum content.

DEALER'S COPY

Fig. 27-5. Wisconsin plant analysis report.

which data are available. Nutrient levels are coded as either deficient (D), low (L), sufficient (S), high (H), or excessive (E).

If the plant samples submitted for analysis are from a crop that was more mature in growth than is recommended, a statement will appear directly under the plant sample section (Fig. 27-5) which indicates the interpretation of results may be questionable. The stage of maturity is determined from the data given on the sample information sheet (Fig. 27-4).

In the "Comments" section (Fig. 27–5), specific statements are printed about each sample. For example, if some of the test results are not in the sufficient to high range for the plant samples, or medium to high for the soil samples, reference is made to printed statements which appear on the back of the report form. These statements give some of the reasons why a particular nutrient may not be in the optimum range.

IV. FUTURE NEEDS AND OPPORTUNITIES IN DATA PROCESSING

Efforts need to be continued towards the integration of equipment and computers to reduce costs, facilitate data processing, and eliminate the need for custom-design hardware packages. Error checking of data should be enhanced to permit automatic rerun of samples with suspect test results. With improved data processing and system control, sample throughput and accuracy will be improved.

Another area requiring attention is the need for rapid transfer of soil and plant reports to the user. On farm or in-business computers with modems are already in place to make this a reality. Laboratories may elect to either use commercial computer systems as a link between the user and the lab or develop an in-house information retrieval system which is accessible by phone. Expert systems are also needed for various crops which couple field data such as insect and disease evaluations, weather, and weed pressure with soil and plant tissue data to permit the development of crop management programs.

REFERENCES

Beaufils, E.R. 1973. Diagnosis and recommendation integrated system (DRIS). A general scheme for experimentation and calibration based on principles developed from research in plant nutrition. Univ. of Natal Soil Sci. Bull. 1:1–132.

Donohue, S.J. 1985. Virginia soil test summary for fiscal year 1984. Virginia Polytechnic Inst. and State Univ. Ext. Publ. 452–015.

Donohue, S.J. 1987. The value and use of soil test summaries. p. 133–144. *In* J.R. Brown (ed.) Soil testing: Sampling, correlation, and interpretation. SSSA Spec. Publ. 21. ASA, CSSA, and SSSA, Madison, WI.

Janney, R., C. Davis, E. Broncovich, and T. Gilmore. 1988. Instrument data acquisition. Am. Lab. 20:34–41.

Mahler, R.L., A.L. Halvorson, and F.E. Koehler. 1985. Long-term acidification of farmland in northern Idaho and eastern Washington. Commun. Soil Sci. Plant Anal. 16:83–95.

O'Malley, C. 1988. UNIX: Tomorrow's operating system? Personal Computing 12:100–108.

SUBJECT INDEX